TOP 100
Exotic Food Plants

TOP 100
Exotic Food Plants

ERNEST SMALL

CRC Press is an imprint of the
Taylor & Francis Group, an **informa** business

CRC Press
Taylor & Francis Group
6000 Broken Sound Parkway NW, Suite 300
Boca Raton, FL 33487-2742

First issued in paperback 2017

ISBN 13: 978-1-138-11666-5 (pbk)
ISBN 13: 978-1-4398-5686-4 (hbk)

Visit the Taylor & Francis Web site at
http://www.taylorandfrancis.com

and the CRC Press Web site at
http://www.crcpress.com

Contents

Preface

Our species, *Homo sapiens*, has existed for several hundred thousand years, and for most of that time, much of the food that people consumed was available only seasonally and from a very restricted local region. By definition, hunter-gatherers were limited to local food. Indeed, most people subsisted in the past on foods produced or acquired within a day's travel from their home. Increasingly during the last several centuries, international trade has witnessed a rising tide of importation of foreign foods. Spices were the first class of food plants to be imported in large amounts. Spices by their nature are high-value, low-weight, compact commodities with long shelf life, hence easily and relatively cheaply transported. For centuries, spices were relied on in the absence of refrigeration to preserve food and to flavor otherwise unpalatable but cheap food. However, with the exception of spices, the lack of refrigeration and other technologies, slowness of transportation, and poverty of most people prevented many foreign culinary delights from being imported, except as luxury items available only to the rich and privileged elite of society. Globalization has dramatically altered this. With progress in food science since the late nineteenth century, there has been an increasing worldwide flow of goods, including numerous plant-based food commodities. The North American food industry now offers more than 300,000 food items, and an average supermarket often stocks more than 30,000. Many foreign-origin foods, such as banana, chocolate, coconut, date, and pineapple, have become dietary staples almost everywhere. Moreover, with the growing establishment of ethnocultural populations in Western countries, there has been increasing exposure of everyone to the unique dietary offerings that are standard fare in foreign places and increasing acceptance of many items by the general population. There is still another consideration contributing to a greatly increased diversity of food: the growing recognition of the value of eating a wide variety of plant foods. It is now clear that (1) the healthiest human diet is based mostly on consumption of plant foods, (2) eating a highly varied diet is an ideal strategy to ensure adequate intake of the numerous plant constituents that contribute to health, and (3) a varied diet contributes to avoidance of the monotony that leads to obesity. Fruits in particular are now considered to be especially healthful when consumed fresh, and fresh exotic fruits are increasingly being imported.

Of course, in the modern world, transportation not only brings foreign foods to our local supermarkets; the reverse also occurs, with people migrating to other countries and necessarily being exposed to foreign foods while there. Thus, tourists, immigrants, work transferees, and business travelers are exposed to foreign cuisines for a shorter or longer period and often acquire permanent tastes for international flavors and ingredients.

Despite a huge increase in availability of foreign-origin, plant-based foods, there have been countercurrent forces contributing to a restriction of the typical Western diet. The most important problem is the increasing consumption of "fast foods" and "convenience" (prepared) foods, which suffer from "seven deadly fast food sins": (1) too salty; (2) too fatty (with saturated fats); (3) too sweet (with added sugar); (4) too calorific and too much; (5) too deficient in nutrients; (6) too contaminated with additives, preservatives, enhancers, dyes, and the like; and (7) too monotonous. People naturally crave salt (often difficult to obtain in precivilization times), the satisfying mouth feel of fats (fats are calorie rich, hence desirable back in the era when food was difficult to obtain), and sweet things (a genetic heritage of the times our simian ancestors depended on fruit trees; unfortunately, sugar is incredibly cheap and lacking in nutritional value). Most modern food products have been synthesized to sell at low cost in a mass market. Too many people have become conditioned to consumption of commercial food preparations that have been engineered with remarkably limited relevance to long-term human health, as evidenced by the obesity epidemic. Although the mass marketing of fast and convenient foods is still mainly a Western phenomenon, multinational companies are now peddling the same unhealthy foods in most areas of the world, diminishing the diversity of local indigenous cuisines. A chief goal of this book is to familiarize many with a wide range of natural plant foods that are capable of contributing to a healthier diet.

Another concerning trend has been the growing dependence on a very small number of food plants. Although at least 20,000 plant species have been used as sources of food by humans and approximately 5000 are eaten regularly, only approximately 150 food plants have entered world commerce, and only 12 species provide 75% of the world's food. Just three species—wheat, rice, and corn (maize)—account for approximately 60% of the calories and 56% of the protein that humans get directly from plants. This dependence on such a restricted food base is alarming because (1) crop failures, due for example to the emergence of new diseases, may have disastrous consequences, in terms of both economic loss and famine; (2) the dependence on major crops has become associated with subsidies and trade restrictions that have distorted efficient production and distribution of key commodities and indeed have led to serious conflicts among nations; and (3) huge monocultures have very negative effects on the world's biodiversity, usurping increasingly scarce land and often requiring wasteful inputs of water, biocides, and energy. Fortunately, many now recognize the serous economic, social, and environmental problems associated with the dependence on a restricted number of food plants, and there are attempts underway to increase the diversity of food crops. Naturally, there needs to be market demand for plant foods, especially those that are unfamiliar to most consumers and, indeed, even many food crop specialists. Toward the goals of stimulating researchers to develop new crops for their regions and encouraging the public to become interested in attractive new food plant possibilities, this book is intended to introduce as wide an audience as possible to a selection of particularly valuable and interesting exotic food plants.

The majority of the crops treated in this book grow in semitropical and tropical regions of the world, more often than not in developing countries. Sad facts of life are that food is a major consideration in the geopolitics of the world and that most crop development is in technologically advanced, rich countries of the temperate world. Accordingly, many of the crops examined in this book suffer from inadequate development, tending to keep agriculture in a relatively depressed state in many countries. Although food is abundant in most of the temperate world, the same is not true for semitropical and tropical regions—a situation that is not only morally reprehensible but threatens the political stability of the entire world. Hopefully, highlighting the exotic plants detailed in this book will contribute to efforts to improve agriculture in the more exotic parts of the planet where they are cultivated.

A welcome development of the global transfers of food and food cultures is the increased sophistication of culinary offerings that have become commonplace, not just by famous chefs in high-end restaurants but also by amateur cooks. At least in some quarters, cookery has become more refined along with culinary taste. This is not just a matter of reproducing foreign meals (often in a Westernized version): there has been a hybridization of traditional foods of different countries, often resulting in delightful culinary novelties. Although there are innumerable cookbooks available today that deal with the traditional foods of foreign countries and of course uncountable numbers of cookbooks dealing with traditional Western fare, there is a dearth of published information that encourages combining the two cuisines (so-called "multicultural cookery" or "fusion cooking"). In addition to amateur and professional cooks, the prepared food industry has become very competitive and is currently seeking new products and ingredients to produce new flavors and attributes. Toward the goal of contributing to the continuing exploration of new culinary preparations based on combining familiar and foreign traditional foods, this book provides extensive information on culinary matters related to each food plant discussed in detail.

Globalization—the result of revolutionary technological innovations powered by international capitalism—is one of the dominant and controversial issues of our time. The result is a growing homogenization of cultures, and in the case of food, this means that many regional cuisines are disappearing. Although the task of preserving the incredible variety of ethnic foods that grace the dinner tables of the world is one that belongs mainly to cooks, cookbook writers, and the cultures themselves, hopefully this book on exotic food plants will sensitize many to some of the wonderful ethnic dishes for which information is presented.

Aside from the relevance to specialists, the general public should find this book attractive because of the huge interest today in health, travel, and cooking, and hopefully this work will serve as a vehicle for public education in the realm of science and technology. Indeed, with a large audience in mind, the book has been prepared in a style and level of language that most people will find user-friendly.

As will be discussed, just what constitutes an "exotic food plant" depends on where one lives, how familiar given food plants have become in that location, and how unusual that plant or the food furnished from it is. The choice of the plants highlighted in this book was biased toward crops that are (a) economically important, (b) encountered in north-temperate countries, and (c) interesting. The numbers of tropical and subtropical fruits and vegetables are huge, and the choice of these has been guided by the likelihood that they would be encountered either at the supermarket or during travels in foreign lands.

The inclusion of a number of species as "food plants" in this book will likely be questioned by some, and this requires some comment about just what food is. First, the old idea that equates food with nutritional value is valid but is too restrictive for the real world. Food is what society at large considers to be food, and this includes a range of plant materials taken by mouth that are marginally, if at all, nutritional. High-caffeine beverages, including coffee, tea, and cola drinks are all chiefly "stimulants," certainly not consumed for their nutritional value, and the same is true for guarana, which is discussed in this book. However, these are so familiar as snack or mealtime beverages that few would challenge their inclusion as "food." Nor would most people question that the vast numbers of culinary herbs and spices, which are ingested simply for their flavor (not for their nutritional content), are also food items. Most would also not have difficulty accepting that plants used primarily to produce alcoholic beverages also qualify as food plants, although it is probably safe to say that today alcohol is rarely consumed for its nutritional significance. Some plants that are included in this book are really more consumed for their medicinal values than as food (examples in this book include the Hawaiian noni and the acai berry), but at least they have some tradition of culinary usage. Opium poppy and hemp (marijuana) are controversial subjects because they are "narcotic" plants, but their seeds consumed as food are not narcotic. The really controversial inclusions are the "masticatories" (chewed materials) coca, khat, and betel, which are regulated or illegal (at least in some jurisdictions). The reader is cautioned to reserve judgment until these chapters are read; in the context of their cultures, these are as legitimate as coffee is to most people. Probably the least defensible inclusion is tobacco, but as will be appreciated from this book, people have been eating plants with toxins for millennia, and in most cases appropriate culinary preparation prevents harm.

Acknowledgments

Brenda Brookes selected most of the illustrations and expertly electronically enhanced all of them. Her skill and patience in carrying out this very large and challenging task are greatly appreciated.

The original, copyright-free drawings on which the majority of drawings in this book are based are from publications produced from the late seventeenth century to the very early twentieth century (all sources are individually acknowledged in the captions, and original publications from which illustrations were taken are cited in Appendix 1). Beginning in the late nineteenth century, photography quickly and very substantially reduced the drawing of plants for scientific presentation. Despite the obvious accuracy of photography, biologists are well acquainted with the fact that drawings are usually vastly superior to photographs for showing details. Fortunately, during the golden age of plant illustration referred to above, masterpieces of botanical art were prepared of most food plants, and enhanced versions of many of these are presented here. The contributions of the original artists, albeit involuntary, are appreciated.

Twenty-seven of the drawings are by artists associated with the research facilities of Agriculture and Agri-Food Canada, Ottawa. These were prepared for the species for which I was unable to acquire high-quality drawings from other sources. I particularly wish to thank the following artists, who contributed the following numbers of drawings: Barry Flahey (15), Marcel Jomphe (6), Susan Rigby (3), and Brenda Brookes (3). The drawings they prepared are individually acknowledged in the figure captions. Several photographs and drawings obtained through the courtesy of various individuals and institutions are acknowledged in the corresponding captions. I particularly thank our staff photographer Eric Johnson, who prepared four photos.

Author

Dr. Ernest Small received a doctorate in plant evolution from the University of California at Los Angeles in 1969, and has since been employed with the Research Branch of Agriculture and Agri-Food Canada, where he presently holds the status of Principal Research Scientist. He is the author of 13 books and over 300 scientific publications on plants. Dr. Small's career has included dozens of appearances as an expert botanical witness in court cases, acting as an adviser to national governments, presenting numerous invited university and professional association lectures, supervising post-graduate students at various universities, participating in international societies and committees, journal editing, and media interviews. He has received several professional honors, including the G.M. Cooley Prize of the American Association of Plant Taxonomists for work on the marijuana plant; the Agcellence Award for distinguished contributions to agriculture; the George Lawson Medal, the most prestigious award of the Canadian Botanical Association, for lifetime contributions to botany; and the Lane Anderson Award, a $10,000.00 prize for science popularization, received for "Top 100 Food Plants," the companion volume to this book.

Executive Summary

This book features exotic food plants. Such plants are typically foreign, unfamiliar, strange, and exciting. However, many foreign edible plants like banana, chocolate, coconut, date, and pineapple have become so mainstream that they are no longer exotic in most of these respects and are not included. The plants examined here are, for the most part, quite unfamiliar to most people. Crops were chosen for inclusion primarily on the basis of (1) likelihood to be encountered in the English-speaking Western world or by travelers to the remainder of world; (2) importance, either globally or in particular regions; and (3) limitation of English-language information with respect to culinary aspects. Additionally, in recent times many previously obscure edible plants have become prominent, the result of sensationalistic media reports, principally because they are inherently entertaining or socially notable, and a selection of the most interesting of these is also included. This book reviews scientific and technological information about 100 of the world's most significant exotic food plants and their culinary uses. Categories of food plants covered include those that produce fruits, vegetables, spices, legumes, culinary herbs, nuts, and extracts such as starch, gums, and sweeteners. An introductory chapter reviews important historical, economic, geopolitical, health, environmental, and ethical considerations associated with exotic food plants. A user-friendly standard format is used for each of the 100 chapters that deal with a particular species or group of species. The initial section entitled "Names" provides extensive information on scientific and English common names of the plants. Next is a section called "Plant Portrait," which comprises a description of the plant, its history, and its economic and social importance. This is followed by "Culinary Portrait," which details food uses and gives practical information on storage, preparation, and potential toxicity. In the "Culinary Vocabulary" section, information is given on the names of especially important foods prepared from the plants and on a variety of related culinary words, phrases, and terms. This is followed by "Curiosities of Science and Technology," which contains notable and interesting scientific or technological observations and accomplishments that complement the main textual material. Finally, a section called "Key Information Sources" provides selected references to books and articles on the subject of each chapter. A subsection of this entitled "Specialty Cookbooks" presents references to food preparation using the particular plant in question. There are more than 2000 literature citations in this book, and the text is complemented by more than 200 drawings of very high quality.

Common Name Guide to Exotic Food Plants Discussed in Detail

The 100 common names or name combinations used as chapter headings are given in boldface. The most frequently encountered synonyms are also given here. For additional synonyms as well as names of plants mentioned incidentally in the book, see Complete Index of Common Names at the end of the book

Scientific Name Guide to Exotic Food Plants Discussed in Detail

This index presents the correct scientific names for the plants discussed in detail. For scientific synonyms of these as well as for scientific names of plants mentioned incidentally in the book, see Complete Index of Scientific Names at the end of the book.

Cautions

Some food plants contain chemicals that are potentially dangerous for some people, particularly pregnant women, infants, the elderly, and those with medical conditions (see, e.g., the chapter on sarsaparilla, which contains compounds that act like hormones, and the chapter on persimmon, which can produce dangerous bezoars—stone-like masses of undigested material—in people with certain stomach conditions). Some food plants contain chemicals that are hazardous to all people but can be made edible by expert preparation (see the chapter on cycads). Some plant products can produce dermatitis or even severe skin damage unless properly collected and prepared (e.g., cashew, ceriman, gingko). Sometimes some parts of a plant are edible, while others are not (e.g., the horseradish tree, the roots of which serve as a condiment, provided that the poisonous root bark is removed). Sometimes some varieties of a species are edible while others can be quite toxic (e.g., grass pea, neem). Occasionally, a species accumulates harmful quantities of elements, depending on growth conditions (e.g., Brazil nut can accumulate quite poisonous levels of selenium). Rau ram is an example of a number of plant species that can produce "phytophotodermatitis" in some individuals—a skin rash that is only developed in the presence of sunlight. Sometimes, an edible species is easily confused with a poisonous species (e.g., anise). Some foods should only be consumed in very limited amounts or sickness will result (e.g., saffron, nutmeg). Those familiar with the effects of drinking too much coffee will appreciate that the high-caffeine beverage plant discussed in this book—guarana—should be consumed in moderation. A few plants traditionally consumed as food are currently considered to be narcotics or so toxic that no one should consume them (see chapters on tobacco, khat, betel, and kava). Despite extensive research to date, the toxicity of most plant constituents to date has not been established (see stevia for an example of a food plant with chemicals—steviosides—of controversial safety). Virtually everyone is potentially liable to experience allergic reactions to some foods not previously consumed, and so cautious sampling of new foods over a period of time is recommended to establish personal tolerances. Many of the plants described in this book are meant to be used in small amounts at any meal and sparingly in the total diet. Information on the toxic potential of most of the plants discussed in this book is given in the chapters dealing with them, but it should not be assumed that the safety information is complete. However, the impression should not be left that the world of previously uneaten food plants is a mine field; exploring new plant foods to increase variety in the diet is one of the best ways of adding to the joy of eating as well as improving health.

People are usually ambivalent in their willingness to try new foods. Human beings are naturally omnivorous, adapted to consuming a wide variety of foods, and the ability to adapt to various diets has been a key factor that has allowed our species to populate almost all regions of the earth. However, many plants are poisonous, so eating an unfamiliar one was potentially deadly in past times. Of course in modern times it is very unlikely that plant foods obtained in the marketplace will be toxic, but our biological heritage includes a degree of distrust of unfamiliar foods as well as an interest in trying something different. Another important factor that tends to discourage the consumption of unfamiliar foods is the natural early-life habituation to foods served by our mothers, which leads to each of us acquiring a palate of acceptable tastes as well as tastes that at least at first seem objectionable. Accordingly, marketers and professional cooks are faced with particular challenges in persuading people to try new foods, so it is critical that this be done with great care. For most people, a single negative exposure to a new food results in its being rejected forever. Keys to expansion of interest in new foods are providing appropriate education and consumer-tested recipes; explanatory literature in supermarkets, demonstrations by professional cooks, and information packages for the media can also be helpful. Marketers should be careful to stock material in good condition and to discard supplies as soon as they have significantly deteriorated. Consumers need to be aware that a period of acquiring a taste for many unfamiliar foods is needed, so consumption may best be limited in initial attempts. Also, in accord with the old adage that "a spoonful of sugar makes the medicine go down," combining new and unfamiliar foods is often a good idea. An excellent tactic to expand one's range of culinary interests is to do so at an entertaining culinary event, such as an ethnic festival or food fair. Perhaps the most challenging new food to those who have not experienced it is the durian (see the chapter on this), a delicious fruit that seems repellent because of its spiny covering and nauseating smell. I personally became fond of it thanks to a colleague who held a durian party, during which the reassuring presence of others served to overcome people's inhibitions about eating "the skunk of the orchards."

1 Introduction

HOW THE CHOICE OF THE LEADING 100 EXOTIC FOOD PLANTS WAS MADE

"Exotic" food plants are typically foreign, unfamiliar, strange, and exciting. The plants examined here are indeed, for the most part, "foreign" and more or less unfamiliar to most people, at least from the viewpoint of inhabitants of the English-speaking, temperate-climate Western world, the perspective adopted in this book. Of course, those who live elsewhere are likely to consider a different set of food plants as "exotic." For example, Pandey and Rai (2005), in an article on vegetables considered to be "exotic" in India, concentrated on broccoli, cabbage, and cauliflower, which are quite mundane to Westerners. Similarly from the perspective of western Africa, Alpern (2008) documented the presence of dozens of "exotic plants" such as corn (maize), which are simply well-known crops introduced by Europeans.

Crops were chosen for inclusion in this book primarily on the basis of (1) likelihood to be encountered as imports in the English-speaking, temperate-climate Western world, or by travelers to the remainder of the world; (2) importance, either globally or in particular regions; and (3) limitation of English language information with respect to culinary aspects. Additionally, in recent times many previously obscure edible plants have become prominent, the result of sensationalistic media reports, principally because they are inherently entertaining or socially notable, and a selection of the most interesting of these is also included. Although the choice of the "top 100" exotic food plants has been somewhat personal and arbitrary, the average reader should find them to be fascinating subjects that are likely to be encountered in day-to-day life and to evoke inquiries about origin, uses, and properties.

I have excluded tropical and subtropical species discussed in detail in my book "Top 100 Food Plants" (Small, 2009) on the basis that these have become so commonplace that they can no longer be considered to be "exotic." These include not only mainstream tropical food plants like banana, chocolate, coconut, date, pineapple, and vanilla but also lesser known edible plants that are extremely important in foreign lands but have also become common in ethnic markets in Western countries (e.g., cassava, fonio, malanga, plantain, quinoa, shea nut, taro, teff). The previous book provides information in a parallel format to this volume and can be consulted by the interested reader.

Several criteria could be used to assess the relative importance of exotic food plants, such as reported market value, total produced dry or fresh weight, nutritional values, and numbers of people who depend on the crops for income or indeed for survival. Unfortunately, relevant data or analyses that would allow such an evaluation are simply not available for most exotic food plants. World production information for the 10 food crops discussed in this book, for which comprehensive data are available, is reported in Table 1.1.

In this book, a "food plant" usually is just one species, but sometimes it includes several species that are closely related or produce a very similar product that is marketed under the same name. This is so for the following chapters: asafetida, bamboo, cacti, cape gooseberries and tomatillo, cherimoya and atemoya, cumin, cycads, galangal, gum arabic, Japanese vegetables, kiwi, kumquat, loofah, melons, myrrh, passion fruit, persimmon, quinine, sapote, sarsaparilla, spinach, water chestnut, Vietnamese herbs, and wonderberry and garden huckleberry. In most cases where several species are included in a chapter, one species is of chief importance, and the others have relatively

FIGURE 1.1 *The Fruit Bearers* by British artist Aubrey Vincent Beardsley (1872–1898). (From Beardsley, 1896.) This odd ink drawing by one of the most controversial, eccentric, and talented of artists captures the fascinating strangeness of exotic food plants.

TABLE 1.1
World Production Statistics (Mean of 10 Years, 1998–2007) of the Exotic Crops Examined in This Book for Which Data Are Available (in Metric Tons)

Carob	197,413
Cashew nut[a]	2,437,044
Cashew apple[a]	1,760,775
Clove	106,459
Ginger	1,108,469
Kiwi[a]	57,298
Nutmeg, mace, and cardamom[b]	72,374
Okra	5,416,881
Opium poppy (seed)	66,807
Persimmon	2,751,284
Tobacco (unmanufactured)	6,543,214

Source of raw data used to determine the mean: Food and Agriculture Organization of the United Nations (FAOSTAT): faostat.fao.org/site/567/DesktopDefault.aspx?PageID=567.

[a] Cashew nut and cashew apple are both products of one plant (see Chapter 20).

[b] Nutmeg and mace are also both spice products of one plant (see Chapter 68). Statistics for nutmeg and mace that are separate from the spice cardamom, which is discussed by Small (2009), are not available.

limited value. However, it should be noted that frequently the plants that are of minor importance on a global scale are of great importance in particular countries.

SPICES: ESSENTIAL EXOTIC FLAVORS

Spices are the most exotic class of food plants, originating in most cases from tropical lands. Nineteen percent of the plant foods discussed in this book are spices (statistical information is given in Chapter 2). They usually contribute only minor amounts of nutritional substances to the diet and are mainly significant simply for flavoring other foods. Spices (1) have very high value on a weight basis by comparison with most other foods, (2) typically have very long shelf life, (3) usually can only be grown in very hot climates, and (4) are in high demand. These considerations guarantee that importation of spices from southern to northern nations will continue (although some flavor chemicals of spices have been synthesized and so can be produced anywhere). Most spices come from southern Asia. India is by a considerable margin the world's chief producer of spices, followed by China, Bangladesh, Pakistan, and Nepal. Spices were in great demand historically because of (1) their preservative value for food in the absence of refrigeration (a technology that started in the mid-nineteenth century; however, note that some modern scholars have concluded that the use of spices in medieval Europe to preserve meat has been greatly exaggerated); (2) their ability to mask the flavor and odor of poorly preserved, bland, and bad-tasting foods; (3) their reputed medicinal values; and (4) their trade value to nations, which became associated with national military power and the status of royalty. Before the seventh century, spices were mainly imported westward overland on the Silk Road of Asia, until they reached the eastern Mediterranean, and were subsequently shipped to Europe. With the rise of Islam in the seventh century, control of the Silk Road came under the control of Arabs, and most overland transfer of spices to Europe ceased (as did most trade between Christian Europe and the Muslim East until the end of the twelfth century). By the thirteenth century, Venetian traders had made arrangements with Arab traders, and Venice gained near-total control of the spice trade into Europe. To circumvent the monopoly that had developed, explorers, notably from Spain, Portugal, and the Netherlands, established shipping routes to the spice lands during the European Age of Discovery, beginning in the late fifteenth and early sixteenth centuries. The overland route from Asia reopened in the sixteenth century, but by then most spices came to Europe by sea. The major sea-faring nations of Europe competed viciously for the spice trade during the late Middle Ages, and triumphant countries became extremely wealthy. Major consequences included the colonization of the Americas by European powers, the widespread expansion of slavery, and the Industrial Revolution. There are numerous excellent analyses of the importance of the spice trade in determining the historical evolution of the modern world (see, e.g., Corn, 1999; Czarra, 2009; Dalby, 2000; Keay, 2005).

COMMERCIAL ASPECTS

For the most part, international trade on the basis of exotic food plant products is carried out in the same capitalistic context as trade in other goods; that is, it is governed by the World Trade Organization, the international accords, and the willingness of individual countries to accept particular materials. As discussed in the Fair Trade section, allowances are often made that amount to subsidies for "emerging" market economies but in the main commerce is governed by the self-interest of nation states, characterized by protectionism and all too often by subsidies that maintain the local production of goods that can be produced more cheaply and efficiently elsewhere. Trade in agricultural products in particular is strongly controlled by major countries and associations (notably the European Economic Union) and by a relatively restricted number of multinational interests. In the following sections, related issues of health, ethics, and environment are examined, but in the final analysis, the success of exotic food plant products that are imported into the marketplaces of the Western world is determined simply by the laws of supply and demand. There are thousands of food plant products from exotic areas that simply do not have characteristics that make them successful as merchandise in the West. There are also likely hundreds of promising plant foods that deserve to be tested for their marketability (see, e.g., National Research Council, 1989; Quah, 1996). Brack Egg (1999) recognized 782 species of edible plants in Peru alone, very few of which are exported. However, as mentioned earlier, numerous exotic plant foods such as bananas, lemons, and chocolate have become so common and indispensable that they are no longer though of as exotic. These and indeed most exotic plant foods can only be produced in tropical and semitropical climates, and at least in this respect most of the producing countries have a continuing advantage.

There are special problems associated with food production in the tropics, particularly with regard to storage of harvested material (Muller, 1988). Hot, dry weather wilts fruits and vegetables. High temperatures also increase respiration of living plant foods, resulting in weight loss. Additionally, heat can increase uptake of odors, such as from petrochemicals, and this may be particularly noticeable in fat-containing foods such as nuts. During relatively cool, wet nights moisture will be absorbed by many plant foods, making them susceptible to decay organisms, and rotting is accentuated by the hot temperatures during the day. Many animals inhabit the tropics, including insects, birds, rodents, and monkeys, all of which are highly attracted to stored food. The poor roads that must often be used for transportation can damage fruits and vegetables, leading to rotting. Sanitation conditions and hygiene practices are poor in many areas, promoting contamination of food.

As noted in the following sections, international trade in food has at times been associated with health issues, exploitation of people, and degradation of the environment. "Green politics" refers to a global political movement that generally advocates policies that proponents believe promote both societal and environmental well-being. Often the movement seems hostile to modern economic and technological developments, including global trade (see, e.g., Audley, 1997; Clapp and Dauvergne, 2005; Lappé, 1971; Torgerson, 1999). However, trade in food has frequently been beneficial to society, and the importation of exotic plants to Western countries has been and should continue to be of considerable benefit to both suppliers and consumers.

THE HEALTH DIMENSION

> Safety is relative; it is not an inherent biological characteristic of a food. A food may be safe for some people but not for others, safe at one level of intake but not another, or safe at one point in time but not later. Instead, we can define a safe food as one that does not exceed an *acceptable* level of risk. Decisions about acceptability involve perceptions, opinions, and values, as well as science.
>
> **(Nestle, 2003, p. 16)**

Evaluating food safety today involves risk assessment of several hazards, including microbial infections, pesticide residues, allergens, additives, natural toxins, and most recently terrorist contamination of the food supply. Particular attention needs to be given to vulnerable groups, including the elderly, pregnant women, young children, and the immunocompromised. Despite considerable scientific knowledge of hazards, extensive regulations, and careful surveillance of foods crossing borders, "food scares" and even occasional outbreaks of food-borne illness occur, sometimes with deadly results and often alarming the public. Western countries are usually quite stringent in their safety requirements for imported foods; sometimes regulations are too stringent, acting as *de facto* tariff barriers to importation of some exotic foods. However, (1) some exotic foods require knowledge to be consumed safely; cooking procedures that detoxify the food may be required; (2) sometimes a poorly known foreign food plant simply has not been studied sufficiently, and problems become apparent after some years of importation; and (3) although importers and vendors are generally extremely concerned about the possibility of bad publicity and legal liability resulting from importation of harmful foods, nevertheless there are occasional problems. Very rarely, a poisonous spider, frog, or snake is transported along with tropical produce, more often than not resulting in sensational media reports. However, health inspectors are vigilant to prevent importation of any kind of animal because they could transport infectious diseases or represent new, potentially damaging alien pests (note International Plant Protection Convention, 1996).

Natural Toxins

Wild plants contain natural chemicals that frequently appear to protect them against microorganisms, insects, and animal predators. In the course of thousands of years of domestication, mankind has reduced or eliminated significantly toxic chemicals from major crop plants (this reduction in natural protective chemicals is one of the reasons why many major crop plants need the application of pesticides). Many of the crops in developing nations have not undergone intensive selection and retain extraordinarily dangerous levels of toxins (see, e.g., Chapters 32 and 41). The lack of such selection is due to economic conditions that have not allowed much investment in tropical crops, and in some cases to the deliberate maintenance of high-toxin varieties, because they are resistant to pests and therefore do not require the application of pesticides, which are often not affordable in developing nations. In almost all cases, food products from these naturally poisonous plants can be detoxified by appropriate pretreatment or cooking. In some cases, the food product should be consumed only in limited quantities (e.g., saffron). Sometimes fruits are toxic unless mature; the most interesting case of this is akee, underripe fruit of which regularly poisons and indeed sometimes kills people. Information about toxicity is provided in the discussion of each food plant discussed in this book. However, in some cases knowledge is limited, and a plant may be discovered to be dangerous only after it has been consumed for years (e.g., kava, discussed later).

Microbiological Safety

Microbial contaminants generally represent the greatest risk posed by food (James, 2006; Sapers et al., 2005, 2009), and there is concern that this may be a problem when food products are imported from countries with lower health and safety standards. Contamination of foods by molds and bacteria is a particular hazard under primitive tropical conditions. Of course, contamination of all food (exotic or not) can occur at any stage between production and consumer usage, and there is no substitute for vigorous inspection and enforcement of food safety regulations. Many exotic food plant products are sold in small "health food" markets, specialty food stores, and ethnic restaurants, and sometimes sanitary conditions are less reliable than in larger establishments associated financially with "deep pockets" subject to lawsuits. "Washing" (usually just rinsing) sometimes cannot be relied on to remove contamination of fresh fruits and vegetables when their surfaces are rough (providing places where the microbes remain). At all times, consumers need to be vigilant.

Nutritional Deterioration

Shelf life is a critical consideration for all marketed foods but is especially important for imported exotic plant foods, which must travel long distances and often need to be capable of surviving rough handling. This is usually not a problem for spices, most of which will last without significant deterioration for months, sometimes for years; indeed, before refrigeration, spices were highly valued for their ability to preserve other foods. As a rule, dense, hard parts of plants last much longer than do relatively fragile soft parts; thus, edible roots, bark, stems, and seeds usually last longer than edible leaves, flowers, and fleshy fruits (although there are exceptions). Exotic spices are mostly seeds and nonfleshy fruits, sometimes stems or bark, all of which are long lasting. Imported exotic vegetables are mostly underground stems and roots, which are also long lasting. There are very few imported exotic fresh edible leaves (bay leaves are the most valuable imported leaf product, but are dried). There are hundreds of tropical species with fleshy fruits that are often consumed in tropical and subtropical areas (Prevost and Pitkanen, 1980), but very few of these fruits have good keeping qualities.

FIGURE 1.2 *Spaghetti Tree* by Barry Flahey.

In the future, it may well be that research into storage technology will result in more fruit species being imported. The relatively high cost of most imported fruits has resulted in only those with good keeping qualities being imported and considerable efforts by vendors to sell their stock as quickly as possible. Most fruits tend to lose vitamin content as they age and decay, and consumers need to be vigilant in purchasing material that is as fresh as possible. Nuts, spices, and root vegetables lose nutritional value (and taste) much more slowly than fruits, but it is often extremely difficult to know that they are too old. The best guarantee is to patronize reliable vendors.

Marketing Hyperbole

Because factual information about foreign food plants is often limited, some unprincipled marketers have greatly exaggerated the virtues of some species. Just how gullible people can be is perhaps best illustrated by what has been judged to be the greatest April Fools' Day joke of all time, "The Spaghetti Harvest" report, broadcast on April 1, 1957, by the British Broadcasting Corporation on Panorama, a television series. Allegedly in Switzerland on the border of Italy, women were shown harvesting long strands of limp spaghetti growing on "spaghetti trees" and drying the pasta in the sun. Millions of viewers accepted the report as truthful. The hoax was successful because at the time in Britain, spaghetti was an exotic delicacy available only in imported cans. (The original 3-min broadcast is available on the Internet, e.g., at http://news.bbc.co.uk/onthisday/hi/dates/stories/april/1/newsid_2819000/2819261.stm.)

The age of modern marketing has witnessed the appearance of a new class of "superfood" plants that allegedly dramatically improve health, physical appearance, and/or vigor and so seem to be better consumer choices. Several exotic food plants have been claimed to have extraordinary health benefits.

Antioxidant-Rich Fruits

Much of the "hype" concerning the health benefits of exotic foods has been concerned with the presence of antioxidants in obscure fruits from foreign lands. In fact, antioxidants are common in most fruits of the world and also occur in other edible parts of plants (they frequently serve to keep other plant chemicals from deteriorating). "Free radicals" are highly reactive "bad" chemical fragments produced as by-products of such metabolic functions of the body as breathing, digesting, and exercising. Free radicals can impair cell function and are believed to be harmful, increasing the risk of cancer, heart disease, other diseases, and premature aging. There are also "good" chemicals called antioxidants that occur naturally, and because they can disarm damaging free radicals, they seem to be useful in the body's fight against diseases. Thousands of antioxidants occur naturally in foods, and several vitamins are in fact antioxidants. It is clear that humans (and indeed probably most plant-eating animals) have become dependent on regular consumption of antioxidants, although much study is needed to know which ones are essential and in what quantities.

In the last decade, extracted antioxidants have been widely marketed as "fruitaceuticals" (a play on the word "pharmaceutical"). In the 1990s, after some studies suggested that antioxidants can prevent or slow the development of cancer and other diseases by protecting cells against damage by free radicals, marketing and consumption of supplements containing antioxidants became widespread, and today 10% to 20% of adults in North America and Europe take antioxidant supplements, particularly vitamins A, C, and E. However, some studies have suggested that there may be little benefit in antioxidant supplements (Bjelakovic et al., 2007; Hu, 2003; Lichtenstein and Russell, 2005; Shenkin, 2006). Many researchers have concluded that antioxidants work best when they are consumed naturally in food rather than pills. The reasons for this are not clear; it has been suggested that other factors in foods, perhaps in interaction with antioxidants, are responsible for beneficial effects. In any event, there is near-universal agreement that eating fresh fruits is health promoting in part because of the presence of antioxidants, and a number of common fruits (e.g., blueberry) have been recommended because of their high content of antioxidants.

Fresh fruits are not nearly as profitable as supplements or food products manufactured from fruit, so the marketplace continues to be flooded with fruitaceuticals and food products made from plants alleged to have high levels of "phytonutrients" (nutritional plant compounds that promote human health). This is particularly evident with respect to a number of "superstar fruits" originating from exotic places. Commonly, marketers claim that the fruit in question has been used by tribal groups, not just for food but also as a medicinal tonic that has had miraculous benefits. Not uncommonly, vigorous agrarian workers and remarkably healthy very old people are shown consuming the fruit. From time to time, motion picture and television stars have claimed that a given fruit has rejuvenated their health, stimulating a minor flood of interest in the fruit. Also not uncommonly, seemingly legitimate research articles are cited that are interpreted as confirming health claims. Exotic fruits discussed in this book that have been sensationalized in this manner for their alleged health benefits include acai berry, goji, kava, mangosteen, and noni.

The acai berry, the fruit of the South America assai palm, *Euterpe edulis* C. Mart., is the premier example of an exotic fruit that has rapidly become very popular in Western countries as a "superfood." The blue-black berries have been marketed as a specialty health food, consumed as a juice, smoothie, frozen purée, or supplement. Acai berries have received considerable exaggerated publicity as the "king of antioxidants."

The mangosteen (not to be confused with mango) is one of the world's most delicious fruits. Not content with its culinary virtues, many retailers are touting it as a "miracle fruit" (but see Chapter 64), useful for treating major diseases including cancer, heart disease, stroke, diabetes, and arthritis and having antiviral, antiaging, antihistamine, antibiotic, and anti-inflammatory properties, capable of reducing and preventing pain, reducing hypertension and depression, eliminating asthma, fighting bad breath, improving stomach and urinary function, lowering cholesterol—the list goes on and on. Thousands of Web sites are peddling mangosteen as a miracle cure, along with testimonials from doctors, stories of how people in exotic lands have relied on it for centuries, vague references to research—all reminiscent of snake oil salesmen of old times. Mangosteen, like many other fruits, is a rich source of healthful nutrients and is a welcome addition to one's diet, but there really is no reason to believe that it is superior to, for example, apples.

The latest egregious example of alleged fruit panaceas is goji, shrubs or small trees of China and adjacent countries (see Chapter 40). The small berries are used as condiments and medicines in Asia. In the Western world, juice from the berries (often mixed with much cheaper fruit juices) is used to make beverages that cost as much as a good bottle of red wine. Other Western products include dried fruits sometimes sold as trendy trail mixes, teas, confections, cereals, and extracts in capsule form sold as "nutraceuticals" (nutritional supplements). Several preliminary studies have been conducted in Asia, suggesting medicinal values of goji, particularly because of the presence of high levels of antioxidants. The widespread marketing claims that goji promotes youthfulness, good health, and sex drive need to be demonstrated experimentally. In the United States, the Food and Drug Administration has warned some goji purveyors that their advertisements suggesting that goji prevents, mitigates, treats, or cures disease are claims that can only be legally made for drugs subject to governmental approval. More research and less rhetoric are desirable to establish the place of goji among healthful foods of the world.

Dangers of Insufficient Risk/Benefit Analysis in the Marketing of Allegedly Healthful Exotic Plant Foods

Noni (see Chapter 67) is a tropical, evergreen shrub or tree, native to Malaysia, Australia, and Polynesia, which has been widely naturalized in southern Asia, tropical America, and the West Indies. Its fruit tastes very bad and has an unpleasant, foul odor which has been compared with vomit and putrid cheese. Nevertheless, noni has risen to star status in the fruitaceutical world. On the Pacific Islands, especially in Hawaii, there developed a tradition of using it as a medicinal plant for diabetes, high blood pressure, cancer, and numerous chronic disorders, although such applications have not been scientifically validated. Nevertheless, in recent years, noni has become a popular folk medicine, with claims that it can help cure high blood pressure and cancers. Numerous companies have taken advantage of noni's cure-all reputation and are now marketing juice from the ripened fruits or dried extracts in capsule form. Noni appears to have some beneficial effects but is being advertised in the tradition of the infamous snake oil salesman, with obviously exaggerated claims that it will cure almost everything. An Internet search for "noni" reveals thousands of Web sites, almost all of which advertise products alleging remarkable medical virtues. Noni products are now sold not only as juice, tea bags, extracts, powders, and capsules but also as facial cleansers, bath gels, and soaps. Many of the claims about scientific research conducted on noni are deceptive, and indeed there have been convictions in the United States for fraudulent advertising. It appears that noni has potential health benefits, but that at present its wide usage is a matter of concern because there may also be potential health risks.

Kava (see Chapter 51) is an example of an alleged magical cure for stress (similarly very widely marketed in the same manner that noni is being sold today) that turned out to be a liver poison. The underground stems have been used by natives of islands in the South Pacific for more than 3000 years to prepare a brownish, often bitter brew, drunk as a traditional, intoxicating, social beverage, rather like wine. Kava reduces anxiety much like the well-known Valium and is a potent muscle relaxant. Kava products have been heavily marketed in Western countries to reduce anxiety, depression, and insomnia. In 2002, the U.S. Food and Drug Administration cautioned that there is potential risk of severe liver injury from the use of dietary supplements containing kava, and a number of countries subsequently have banned the sale of kava products. This illustrates the dangers of popularizing exotic plant products before the risk/benefit profile is adequately determined.

THE ETHICAL DIMENSION

Food is fundamental to life and not surprisingly has been a frequent basis for conflicts among individuals, peoples, and nations. The history of colonialism, and to a lesser extent postcolonialism, is sadly marked by exploitation of resources and peoples—exploitation of the "South" by the "North" and of the "East" by the "West." The past several centuries of European domination of southern and eastern nations has left in many of the latter countries a legacy of continuing economic domination, and exotic food plants have been among the most important of the exploited resources.

Because food is such a fundamental resource, it is not surprising that issues concerning the control, production, and distribution of food are controversial and politicized, pitting rich against poor, capitalists against socialists, social activists and environmentalists against large, transnational corporations (or "the industrial agro-food system" or "the giants of global agribusiness"), and romanticists against Darwinian economists. The following is meant simply as a brief overview of the most important considerations. For more detailed analyses, the following may be consulted: Allen and Albala (2007), Bittman (2008), Maye et al. (2007), Nestle (2007), Pollan (2006, 2008), and Schlosser (2001).

The dominant current ethical issue concerning international trade is "fairness" to the producing source: Is the country that supplies the goods being adequately rewarded, and more particularly, are the workers who produce and process the goods in the supplying country receiving adequate care and remuneration? The concern is that rich, Western nations benefit inordinately from the trade relationships, which often cause extraordinary hardships for the workers. Although the analogy is overdone, to some it seems that the actual slavery that was common in colonial times has been replaced by a sort of economic slavery.

There are two distinct but related trends that are impoverishing farmers in many poor producing nations. First is the tendency of large corporations or rich individuals to gain control of agricultural production, reducing farmers to hired workers whose income is maintained at minimal levels. Second is the progressive control by corporations of factors essential to production, harvest, transportation, processing, and sale and accordingly of limiting the wages of workers. These trends are hardly unique to farmers in poor nations, but given the relatively limited social support systems and the simple fact that most of these nations are agrarian, a far greater proportion of the population has been negatively impacted compared with Western countries. In most poor countries, agriculture employs over half of the population and is responsible for about a third of the gross domestic product. The conflict is most frequently manifested between individual farmers and small cooperatives on the one hand, and semi-monopolistic, often international firms on the other. Once again, the problem is that a few rich and powerful controlling groups often conduct business in ways that severely penalize numerous, poor, weak, rural, small producers who are sometimes of a particular ethnicity or gender.

FAST-FOOD IMPERIALISM

The West now imports exotic foods from the far corners of the globe, and in general these foods are welcome; by contrast, much of the food that is now being exported from the West is the basis of a major problem. "Cultural imperialism" is the imposition by a culture, society, or nation, of its cultural practices and institutions upon other cultures, thereby weakening or destroying traditional elements of those cultures. The food and unique food consumption traditions (i.e., cuisine) of the non–Western nations of the world are being subjected to a particularly insidious phenomenon of Western nations: fast food. (Indeed, even among Western nations, the exportation of the fast-food industry is of concern, as witnessed by the development of American-owned fast-food franchises in France and the consequent lowering of the fabulous food standards of the French.) The spreading of fast food to other nations has derisively (and perhaps unfairly) been termed "McDonaldization" and "Coca-Colonialism." One of the more ironic manifestations of such food imperialism is the mass marketing of "ethnic foods" by chain restaurants, the concocted versions being very pale imitations of the genuine dishes.

As noted elsewhere in this book, today's fast food and junk food are usually bad food. The over-marketing of highly processed, cheap foods is the most important factor contributing to the obesity and diabetic epidemics and the deterioration of the health of people. That one may expect the same trends to develop in non–Western nations is disheartening, but there is an associated danger. There are thousands of species of exotic food plants; most of these have not yet achieved market success in the West, and many have only regional or limited significance in non–Western nations. The danger is that the unique cuisines on the basis of such foods will become extinct as local populations turn to cheap fast foods. This would be a tragedy not only for the inhabitants of the nations in which the unique foods are currently consumed but also for the remainder of the world, which might forever be denied the possibility of obtaining new culinary delights.

THE ENVIRONMENTAL DIMENSION

The production of food is responsible for about a quarter of the negative impacts human beings have on the environment, and the choice of diet has the potential of reducing environmental damage. Meat production and fishing currently are associated with major environmental issues, and some have estimated that a vegetarian diet can reduce the carbon footprint of humans by 40%. However, consumption of animal products is not a relevant issue here because this book is entirely about plant foods. Two basic recommendations have been made with regard to making consumption of plant-based foods friendlier to the environment. These are stated and examined in the following:

> By means of glasses, hotbeds and hotwalls, very good grapes can be raised in Scotland, and very good wine too can be made of them at about 30 times the expense for which at least equally good can be brought from foreign countries. Would it be a reasonable law to prohibit the importation of all foreign wines, merely to encourage the making of Claret and Burgundy in Scotland?*
>
> **Adam Smith (1723–1790)**
> *Scottish moral philosopher, political economist, "the father of modern economics"; in* An Inquiry into the Nature and Causes of the Wealth of Nations, *1776*

1) *Eat foods that are locally produced and when they are harvested; this reduces transportation and storage, which consume energy, resulting in depletion of nonrenewable energy resources and pollution of the atmosphere.*

The most famous protest against an imported plant food is the Boston Tea Party of 1773, when American patriots dumped three British shiploads of tea into Boston Harbor. For a period, several wild herbs indigenous to the United States became popular tea substitutes, but eventually another imported high-caffeine plant, coffee, became the principal beverage of the country. Clearly local food plants cannot always satisfactorily replace all imports, but attempts have been made recently to do this.

"The 100-mile diet" (Smith and MacKinnon, 2007; cf. the similar report by Kingsolver et al., 2007) showed how one could eat well on food produced in a limited geographical area (sometimes called a "foodshed") and has been used to support the philosophy of relying on local food to reduce the energy costs of importing food from more distant areas as well as to obtain superior taste and nutritional value. Epitomizing these objectives, the term "locavore" (sometimes "localvore") was coined in 2005 by Seattle-based writer Sage van Wing and accepted as the Oxford American Dictionary word of the year for 2007. Local foods probably do tend to be superior in taste and nutritional value because they are fresh, can be picked at full maturity (many fruits are picked immature as they withstand the rigors of transportation better), and need not undergo a period of storage. Other benefits of local consumption have been claimed, including economic support of local farmers and the business community, education of the public to the nature of food production, and social cohesion. The choice of Vancouver, British Columbia, as the center of the restricted area that Smith and Mackinnon (2007) used as a test of their 100-mile diet is hardly fair because the city borders the sea, is blessed with climatic conditions that support many crops, and is very heterogenous ethnically, with local production of a remarkable range of foods that cater to extremely varied tastes. The key issue, however, is whether relying on locally produced food results in savings of energy. If so, this would constitute an important reason for avoiding the importation of exotic food plants.

The first point to note with regard to the 100-mile diet is that one is literally comparing apples (a cold-temperate crop) and oranges (a semitropical crop). It makes sense to eat local foods in season

* "Glasses" are greenhouses; "hotbeds" are large boxes (often glazed) in which plants are grown, the soil heated (in old times) by a layer of rotting manure in straw beneath the roots of the plants, or alternatively warmed by pipes; hotwalls were masonry walls within which were embedded heating flues, the heated walls radiating warmth to nearby plants.

as staple components of one's diet, but it also makes sense to add imported foods for variety, especially out of season.

The next point is the astonishing difference in productivity of plants that grow in temperate regions and those in tropical regions. In the tropics, plants have a growing season that can last an entire year; in temperate regions, of course plants can only grow during the warm season. In the tropics, the sun is more or less directly overhead, providing intense sunlight, and many plants species have a special (C_4) photosynthetic biochemistry that is adapted to using full sunlight and elevated temperatures; by contrast, in temperate regions, sunlight is never as intense, and plants cannot benefit to the same extent. Crops in the tropics tend to be perennials (a life style suited to continuously hospitable growing conditions) and do not need to be planted anew each year. By contrast, most temperate region crops (not including fruit trees and fruit shrubs) are annuals (a life style suited to alternating seasons of hospitable and inhospitable growing conditions), requiring great expenditures to clear away weeds, cultivate the soil, and establish new plants each year.

The result of climate advantages in hot countries is that crop production requires far less energy and yields much more for the same effort and area of land than is the case in temperate climates. The costs to heat buildings is also much less in hot climates, adding to the much greater use of energy in temperate climates. The only energy advantage of locally produced food is that it does not have to be transported. Container ships are available today that are very efficient in transportation (using about one-thirteenth the energy per unit of weight transported per unit of distance compared with trucks, and about one-fortieth compared with cars). Spices are a special case: many commercial spices intended for industrial (i.e., large-scale) usage are now imported as "oleoresins," which are extracts containing the flavor molecules of the spice; oleoresins typically represent only 10% of the weight of a spice, and so 90% of shipping costs are eliminated. Nevertheless, moving most exotic plant foods to market is a significant cost, especially if shipping is by air freight, and this does result in adding to greenhouse gases. Desrochers and Shimizu (2008) have detailed their argument that the extra so-called "food miles" associated with transported foods is far outweighed by the production advantages of hotter climates in reducing both costs and associated environmental damages of food production. In reviewing the arguments for local food production and consumption, Standage (2009) stated, "The food-miles debate is making consumers and companies pay more attention to food's environmental impact. But localism can be taken too far" and "An exclusive focus on local foods would harm the prospects of farmers in developing countries to grow high-value crops for export to foreign markets. To argue that they should concentrate on growing staple foods for themselves, rather than more valuable crops for wealthy foreigners, is tantamount to denying them the opportunity of economic development."

Crops of the cactus family (see Chapter 14) represent a special case for arguments about consuming plant foods in ways that are supportive of environmental values. Cacti grow in hot deserts, usually in places that are considered exotic, and indeed cactus fruits are popular exotic imports. Cacti have a special photosynthetic pathway, "crassulacean acid metabolism," that permits them to carry out photosynthesis with minimum loss of water. More than 99% of the approximately 1700 cactus species possess this biochemical pathway, whereas only approximately 7% of all other plant species have this method of photosynthesis. Most plants lose water through closable pores (stomates). The pores need to be open to obtain carbon dioxide for photosynthesis, but this often results in considerable loss of water. By contrast, plants with crassulacean acid metabolism keep their pores closed during the day, opening them during the night and actually carrying out most of their photosynthesis in the dark. This is one of the methods cacti use to survive with minimal water usage; cacti also have anatomical features that minimize water loss. Because cacti are incredibly efficient in their use of water, an increasingly scarce resource in the world, and one that requires considerable energy to obtain, growing cacti for food is of great benefit to the ecology of the world. Of course food products of cacti need to be transported to temperate regions, but this points out that whether or not local or distant foods are better for the environment requires consideration of advantages as well as disadvantages.

FIGURE 1.3 Cactus plants. As noted in the text, cacti are valuable crops in hot, dry environments, producing food products with very little use of water, and therefore of great benefit to the ecology of the world. The environmental cost of transporting such foods to northern markets is insignificant by comparison. (Adapted from Rhind, W. *A history of the vegetable kingdom*. p. 720, Blackie and Son, London, 1855.)

> 2) *Buy organic foods grown without pesticides, insecticides, and herbicides that pollute the environment; buy from ethically certified sources, ensuring that good environmental practices have been followed.*

Most plant food on the planet is now produced as monocultures, that is, as single crop species grown on an area to the total exclusion of all other plant species. Almost all such crops require considerable "agricultural inputs"—one or more of water (irrigation), biocides (herbicides, pesticides, fungicides, and bactericides), and fertilizer; moreover, mechanized tilling of the soil, planting, and harvesting additionally require the input of fuel. These activities result in chemical pollution of soil, water, and atmosphere and displacement or destruction of native animals, plants, and soil organisms. The extraordinary efficiency of monocultural production (including tropical crops, such as oil palm) has enabled the planet to accommodate its huge population exceeding 6 billion people, and indeed without monocultures there would be widespread famine and death. A high output of food sufficient to feed the world is apparently only possible with a high input of chemicals, water, and energy, which inevitably results in environmental problems, although it is possible to alleviate the damage by using a variety of farming practices (see, e.g., Standage, 2009).

Although industrial monocultures are now indispensable, there are sound reasons to support small-scale farms. "Agrarianism" is a movement advocating a rural and semirural way of life, both as a means of achieving personal satisfaction and coexisting in a sustainable fashion with the environment. Small, mixed farms are almost always a relatively friendly environment for both wild plants and animals, providing different habitats among the different crops that support wild species. The continuing demise of the small family farm and the growth of huge corporate farms is therefore a growing concern for the environment. In much of the Third World, small "subsistence" farms are still the norm. Although subsistence agriculture is very much less efficient than factory farming, it is much more compatible with the environment. In tropical and subtropical countries that have many more species than found in temperate regions, subsistence and small-scale farms assist in the survival of vanishing biodiversity, a key issue facing the world. A related matter is the supply of "land races" (localized breeding selections) of many useful plants that are maintained on small farms, many of which would disappear with the elimination of the farms. Small, rural farms tend to use less, if any, biocides and make more use of natural fertilizers (especially manure) and in this respect too are environmentally beneficial. Although many of these advantages are not measurable in monetary terms, they are of great value to the health of the planet and its wild species. In view of these benefits, the support of rural production of exotic food plants by small-scale farmers in poor nations, discussed in the next section, is a very good idea.

FAIR TRADE

"Fair trade" is the phrase that has become established for a social movement assisting developing countries to export goods to developed countries (see, e.g., Decarlo, 2007; Nicholls and Opal, 2004; Raynolds et al., 2007; Stiglitz and Charlton, 2005). Implicit in the concept of fair trade is the contention that at least a proportion of normal free trade is *unfair*. With fair trade, the goods in question are typically competitively priced or sold at a slightly higher price than obtained normally, and the consumer has the satisfaction of knowing that marginalized producers and workers are assisted in establishing viable and self-sufficient, small-scale business operations in which they have vested interests and that a high-quality product has been produced in an environmentally acceptable, sustainable fashion.

Within Fair Trade, there are two types of organizations: *product certification* is carried out by the Fairtrade Labelling Organizations International (FLO), which sets standards for the supply chains of specific (primarily agricultural) products from point of origin to point of sale. Independent certification company FLO-CERT audits products and supply chains against the FLO Fairtrade Standards. Marks (as illustrated in Figure 1.5) are used to certify that products have been produced in accordance with these standards (www.info.fairtrade.net). Traders in the United States are certified by Transfair USA.

Organizational evaluation is carried out by the World Fair Trade Organization (formerly International Fair Trade Association) and the Fair Trade Federation, which evaluate organizations for their commitment to fair trade principles. Handicrafts were once the most important fair trade

FIGURE 1.4 Small-scale rooibos tea farmer and member of the Heiveld Cooperative (South Africa), Drieka Kotze, holding sheaves of cultivated rooibos tea. The Heiveld Cooperative supplies high-quality organic rooibos tea that is fairly and sustainably produced. This photo was taken by Rhoda Malgas on a farm in the Suid Bokkeveld and is reproduced with her permission (see Chapter 80).

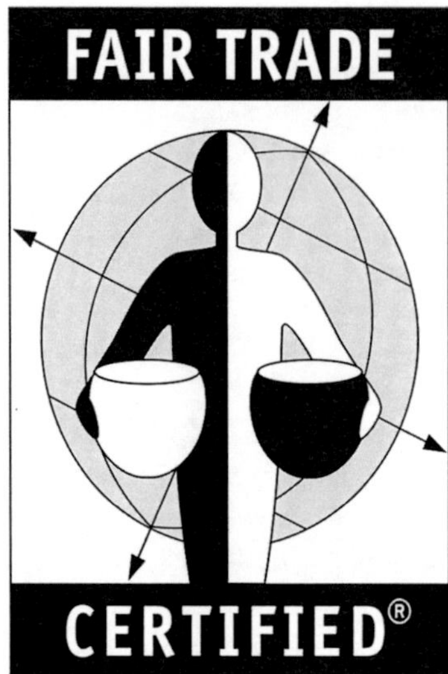

FIGURE 1.5 Fair Trade label issued by Transfair Fair Trade USA for a range of products in the United States. For information, see transfairusa.org/.

FIGURE 1.6 A peasant selling melons by the roadside in nineteenth century Chile. (From Gay, 1854.) Although labor intensive, such simple trade requires little investment and supports millions of families in developing regions.

items, but agricultural goods are now dominant, including tea, coffee, wine, cocoa, sugar, dried and fresh fruits and vegetables, juices, rice, quinoa, cotton, honey, oranges, spices, nuts, and oil seeds.

As of 2008, it has been estimated that certified fair trade sales amount of more than $4billion, benefiting approximately 8 million farmers in approximately 60 countries (there are additional benefits to a large number of artisans, particularly women). Fair trade products constitute only a minute proportion of world sales. Economists are divided on whether or not fair trade has long-term benefits. Obvious short-term benefits include alleviation of poverty in poor nations and alleviation of guilt of affluent consumers. Supporters of fair trade point out that giving preference to exports from poor nations is a way of helping innumerable poor farmers whose only source of income is their small land plots. Critics argue that when such subsidization is denied, the agricultural products would mostly be sold within the country and in neighboring countries, and production and trade would therefore be rationally guided by the rules of supply and demand of the normal marketplace. However, this ignores the undeniable facts that (1) heavy financial subsidization of Western agricultural products has unfairly disadvantaged farmers of developing nations, and (2) the rich nations of the world and more particularly the top transnational biotechnology corporations invest heavily in agricultural research and development that exacerbate the competitive disadvantage and resulting impoverishment of the poor farmers of the world.

BIOPIRACY

Biopiracy (a contraction of biological + piracy) is a somewhat imprecise term that generally refers to the commercial exploitation of unique resources and/or traditional knowledge about the resources of a given country by foreigners without permission of that country (see, e.g., Mgbeoji, 2006; Mushita and Thompson, 2006; Shiva, 1997). There has been resentment, particularly in developing countries, when the unique animals, plants, or techniques for utilization of species have been usurped by foreign companies to produce profitable pharmaceuticals, crop varieties, and industrial products, without authorization and/or without sharing the profits with the countries from which the species originated, or with the indigenous peoples who developed cultural knowledge of how to use the species. "Bioprospecting," the search for utilizable resources and indigenous knowledge, is now widely practiced by pharmaceutical companies, particularly in tropical countries because they are "megabiodiverse," that is, populated by of the order of 10 times as many species as temperate region

FIGURE 1.7 Discovery of America by Christopher Columbus, from a stamp engraved on copper by Th. De Bry, published in "Grands Voyages" in 1590, and available online from Wikipedia as a public domain document. Columbus had been granted the rights of "discovery and conquest" on April 17, 1492, by Queen Isabella and King Ferdinand of Spain. The resulting subjugation of Native American civilizations and the one-sided transfer of goods from the New World to Europe are reminiscent of biopiracy, as discussed in the text.

countries. As with fair trade, the concern is primarily with providing adequate compensation to relatively impoverished people in developing countries. Difficulties arise in interpreting what is allowable in a legal sense, and sometimes moral rights are not clear.

In 1992, the United Nations "Earth Summit" conference in Rio de Janeiro, Brazil, led to the "Convention on Biological Diversity," which was signed by 150 countries, representing more than 90% of United Nations member countries; the agreement came into effect in 1993 and has since been ratified by more than 175 countries. Article 15 of the convention deals with access to genetic resources, providing legal rights to countries to limit access to their plant and animal species and to obligate foreign parties to enter into agreements for usage.

Biopiracy is a critical concern for exotic food plants. Climate change requires that new varieties of food crops be available that can withstand changing local climates that become hotter, colder, drier, or wetter and soils that become eroded or more saline. Wild species and land races (relatively primitive cultivated forms maintained in a region by local farmers) can be essential for breeding new crops, and therefore their use is of great importance to undeveloped countries. Additionally, many food plants are potential sources of pharmaceuticals, some of which may have huge value as patented medicines.

Historical Transfers of Crop Cultivation among Nations

Biopiracy, discussed earlier, generally refers to *recent* exploitation of the unique resources and knowledge of an *underprivileged* people or region without appropriate compensation, as judged by *present-day* norms and agreements and laws. Although it may seem like historical revisionism to some, biopiracy (of a kind that was either accepted or tolerated) also occurred widely in historical times, particularly during the colonial periods of ascendant nations. For example, common tea was surreptitiously taken from China by Britain in the seventeenth century, despite very extensive efforts by the Chinese to prevent foreigners from getting hold of the plants; now, tea is grown in plantations in dozens of countries, with little thought to the debt that is owed to China. More recently, macadamia nut (see Chapter 60), Australia's only important native horticultural crop, was taken to Hawaii, where commercial plantations were initiated in the 1950s; Hawaii has become the world's second largest source of the nuts, and for many years Australians resented this "theft" of their natural resource. The pineapple is another interesting case of geographical migration: it is native to South America, but Hawaii became the world's chief source in the first half of the twentieth century, after which the Philippines became the leading source. The sunflower is still another instructive case: it probably originated in the United States and/or Mexico but did not develop into a significant crop until the late 1800s, when Russians selected varieties with very high oil content and Russia became the leading producer. In the 1940s, Americans brought the Russian varieties to the United States, and the sunflower then developed into a major American crop. Except for the recent concern that poor people and nations should be compensated for use of their indigenous crops, nations of the world have deemed it their right to cultivate all of the world's economically important plant species, without compensating other nations from which they obtained the plants. (Particular cultivars and selections, however, may be protected by international accords.) There is a curious legacy of the transfer of crops from their indigenous areas to other places: frequently, the plants are much more productive in the new areas. This has been explained as due to escape from natural predators (especially insects) and diseases that reduced yield in the native area. However, it is also true that for some plants, the natural soils and climates of foreign countries are often superior to their native conditions, and of course costs of labor, farmland, agricultural inputs, and transportation can also influence where a crop can most efficiently be cultivated. Today, almost every country depends heavily on their farmers cultivating crops that originated in other countries. Many exotic crops have not yet been tested in all countries where they could be grown, and likely the future will witness the waning of their importance in some regions and their growing importance in other places. Regardless of the uncharitable nature of much of human history, it is well to reflect on the fact that during the last 10,000 years of agricultural development, we owe a huge debt to our countless ancestors who discovered, transported, and developed the wonderful array of food plants that we now enjoy (Kiple, 2007).

REFERENCES

Allen, G., and Albala, K. (Eds.). 2007. *The business of food. Encyclopedia of the food and drink industries.* Greenwood Press, Westport CT. 439 pp.

Alpern, S.B. 2008. Exotic plants of western Africa: where they came from and when. *Hist. Afr.* 35:63–102.

Audley, J.J. 1997. *Green politics and global trade: NAFTA and the future of environmental politics.* Georgetown University Press, Washington, DC. 212 pp.

Bittman, M. 2008. *Food matters: a guide to conscious eating.* Simon & Schuster, New York. 314 pp.

Bjelakovic, G., Nikolova, D., Gluud, L.L., Sionetti, R.G., and Gluud, C. 2007. Mortality in randomized trials of antioxidant supplements for primary and secondary prevention. Systematic review and meta-analysis. *JAMA.* 297:842–857.

Brack Egg, A. 1999. *Diccionario enciclopédico de plantas útiles del Perú.* PNUD/CBC, Lima, Peru. 550 pp.

Clapp, J., and Dauvergne, P. 2005. *Paths to a green world: the political economy of the global environment.* MIT Press, Cambridge, MA. 327 pp.

Corn, C. 1999. *Scents of Eden: a history of the spice trade.* Kodansha, New York. 368 pp.

Czarra, F. 2009. *Spices: a global history.* Reaktion, London. 128 pp.

Dalby, A. 2000. *Dangerous tastes: the story of spices.* British Museum Press, London. 184 pp.

Decarlo, J. 2007. *Fair trade: a beginner's guide.* Oneworld Publications, Oxford, England. 192 pp.

Desrochers, P., and Shimizu, H. 2008. *Yes we have no bananas: a critique of the food miles perspective.* Mercatus Policy Series Policy Primer, No. 8. Mercatus Center at George Mason University, Arlington, VA. 21 pp.

Hu, F.B. 2003. Plant-based foods and prevention of cardiovascular disease: an overview. *Am. J. Clin. Nutr.* 78(8):544S–551S.

International Plant Protection Convention. 1996. *International standards for phytosanitary measures. Part 1. Import regulations.* Food and Agriculture Organization of the United Nations, Geneva, Switzerland. 2 vols.

James, J. 2006. *Microbial hazard identification in fresh fruits and vegetables.* Wiley-Interscience, Newark, NJ. 312 pp.

Keay, J. 2005. *The spice route: a history.* John Murray, London. 268 pp.

Kingsolver, B., Hopp, S.L., and Kingsolver, C. 2007. *Animal, vegetable, miracle: a year of food life.* HarperCollins, New York. 370 pp.

Kiple, K.F. 2007. *A movable feast: ten millennia of food globalization.* Cambridge University Press, Cambridge, UK. 368 pp.

Lappé, F.M. 1971. *Diet for a small planet.* Ballantine Books, New York. 479 pp.

Lichtenstein, A.H., and Russell, R.M. 2005. Essential nutrients: food or supplements? Where should the emphasis be? *JAMA.* 294:351–358.

Maye, D., Holloway, L., and Kneafsey, M. (Eds.). 2007. *Alternative food geographies. Representation and practice.* Elsevier, Amsterdam, the Netherlands. 358 pp.

Mgbeoji, I. 2006. *Global biopiracy: patents, plants and indigenous knowledge.* Cornell University Press, Ithaca, NY. 222 pp.

Muller, H.G. 1988. *An introduction to tropical food science.* Cambridge University Press, Cambridge, UK. 316 pp.

Mushita, A., and Thompson, C.B. 2006. *Biopiracy of biodiversity: global exchange as enclosure.* Africa World Press, Trenton, NY. 330 pp.

National Research Council (U.S.). Advisory Committee on Technology Innovation. 1989. *Lost crops of the Incas: little-known plants of the Andes with promise for worldwide cultivation.* National Academy Press, Washington, DC. 415 pp.

Nestle, M. 2003. *Safe food: bacteria, biotechnology and bioterrorism.* University of California Press, Berkeley, CA. 356 pp.

Nestle, M. 2007. *Food politics. How the food industry influences nutrition and health.* 2nd ed. University of California Press, Berkeley, CA. 510 pp.

Nicholls, A., and Opal, C. 2004. *Fair trade. Market-driven ethical consumption.* SAGE, London. 277 pp.

Pandey, A.K., and Rai, M. 2005. Popularizing exotic vegetable for nutrition and food diversification. *Indian J. Hortic.* 50(2):19–22, 36.

Pollan, M. 2006. *The omnivore's dilemma: a natural history of four meals*. Penguin Press, New York. 450 pp.
Pollan, M. 2008. *In defence of food: an eater's manifesto*. Penguin Press, New York. 256 pp.
Prevost, R., and Pitkanen, A.L. 1980. *Tropical fruits: travel in foreign lands & foods*. 3rd ed. Lemon Grove, CA. 368 pp.
Quah, S.C. 1996. *Underutilised tropical plant genetic resources: conservation & utilization*. Penerbit Universiti Pertanian Malaysia, Malaysia. 353 pp.
Raynolds, L.T., Murray, D.L., and Wilkinson, J. (Eds.). 2007. *Fair trade: the challenges of transforming globalization*. Routledge, New York. 240 pp.
Sapers, G.M., Gorny, J.R., and Yousef, A.E. (Eds.). 2005. *Microbiology of fruits and vegetables*. CRC Press, West Palm Beach, FL. 648 pp.
Sapers, G., Solomon, E., and Matthews, K.R. (Eds.). 2009. *The produce contamination problem: causes and solutions*. Academic Press, New York. 496 pp.
Schlosser, E. 2001. *Fast food nation: the dark side of the all-American meal*. Houghton Mifflin, New York. 383 pp.
Shenkin, A. 2006. Micronutrients in health and disease. *Postgrad. Med. J.* 82:559–567.
Shiva, V. 1997. *Biopiracy: the plunder of nature and knowledge*. South End Press, Boston, MA. 148 pp.
Small, E. 2009. *Top 100 food plants. The world's most important culinary crops*. NRC Research Press, Ottawa, ON. 636 pp.
Smith, A., and MacKinnon, J.B. 2007. *The 100-mile diet: a year of local eating*. Vintage Canada, Toronto, ON. 288 pp.
Standage, T. 2009. *An edible history of humanity*. Walker & Company, New York. 269 pp.
Stiglitz, J.E., and Charlton, A. 2005. *Fair trade for all: how trade can promote development*. Oxford University Press, New York. 315 pp.
Torgerson, D. 1999. *The promise of green politics: environmentalism and the public sector*. Duke University Press, Durham, NC. 222 pp.

2 Statistical Summary and Format of Presentation for the 100 Exotic Food Plants

ANALYSIS OF THE PRINCIPAL FOOD USES AND COMMERCIAL PRODUCTION AREAS OF THE 100 EXOTIC FOOD PLANTS

The following four tables summarize categories of food use and principal production areas for the 100 exotic food plants examined in this book. As detailed in Table 2.1 and summarized in Tables 2.2 and 2.3, 31% of the plants are used principally as sources of fruits, 28% as sources of vegetables, and 20% as sources of spices. Although they are the world's most important food plants, cereals and pseudocereals (plants not in the grass family that produce grain) are excluded from this book because they are too well known to be considered "exotic" (for information on tropical and semitropical cereals and pseudocereals, see Small, 2009, cited in Chapter 1). As shown in Table 2.4, 90% of the principal commercial production areas of the crops examined in this book are in tropical and/or subtropical areas, whereas just 10% are in temperate areas; more than half of the crops are raised on several continents, with Asia being the predominant source of exotic crops discussed in this book.

TABLE 2.1
Principal Food Uses and Commercial Production Areas of the Top 100 Exotic Food Plants

Common Names	Scientific Name	Principal Food Use	Principal Commercial Production Area
Acai berry	*Euterpe oleracea*	Fruit	Brazil
Acerola	*Malpighia emarginata*	Fruit	Subtropics
Akee	*Blighia sapida*	Fruit and vegetable	Jamaica
Allspice	*Pimenta dioica*	Spice	Jamaica
Arrowroot	*Maranta arundinacea*	Starch	Caribbean (especially St. Vincent), Brazil, Thailand, West Africa
Asafetida	*Ferula* species: *Ferula assa-foetida*, *Ferula foetida*, *Ferula narthex*	Spice	Iran, Afghanistan, India
Bamboo	Numerous genera, especially *Psyllostachys* and *Bambusa*	Vegetable	Asia, especially China
Baobab	*Adansonia digitata*	Multipurpose	Africa
Bay	*Laurus nobilis*	Spice	Turkey, Greece
Betel nut	*Areca catechu*	Masticatory	Tropical Asia
Breadfruit	*Artocarpus altilis*	Vegetable	Pacific islands
Cacti	Cactus pear (*Opuntia ficus-indica*)	Fruit and vegetable	Mexico
	Various species of the cactus family, especially *Hylocereus undatus* and *Selenicereus megalanthus*	Fruit	Southeast Asia (especially Vietnam), southern Asia, Latin America, Israel
Candlenut	*Aleurites moluccana*	Nut	Tropical Asia
Cape gooseberries and tomatillo	Cape gooseberry (*Physalis peruviana*)	Fruit	Central and South America, South Africa, Australia, New Zealand, China, India, Malaya
	Dwarf cape gooseberry (*Physalis grisea*)	Fruit	Eastern North America
	Tomatillo (*Physalis philadelphica*)	Vegetable	Latin America; tropical and subtropical areas of the Old World
Caper	*Capparis spinosa*	Spice	Mediterranean countries
Carambola	*Averrhoa carambola*	Fruit	Tropical and subtropical world
Carob	*Ceratonia siliqua*	Food extracts	Spain, Italy, Portugal, Morocco, Cyprus, Greece, Turkey
Cashew	*Anacardium occidentale*	Nut	India, Mozambique, Tanzania, Brazil
Cassabanana	*Sicana odorifera*	Fruit and vegetable	Mexico, Central America, South America
Ceriman	*Monstera deliciosa*	Fruit	Central America, Australia, Florida, California
Chayote	*Sechium edule*	Vegetable	Costa Rica, Guatemala, Mexico, Dominican Republic, and many other countries
Cherimoya and atemoya	Cherimoya (*Annona cherimola*)	Fruit	Subtropical and tropical areas, especially Spain, South America, Mexico, Central America, Hawaii, India, Australia, Israel, Italy
	Atemoya (*Annona squamosa* × *A. cherimola*)	Fruit	Subtropical and tropical (cooler areas than cherimoya)
Chinese artichoke	*Stachys affinis*	Vegetable	China, Japan, France, Belgium

TABLE 2.1 (continued)
Principal Food Uses and Commercial Production Areas of the Top 100 Exotic Food Plants

Citron	*Citrus medica*	Spice	Sicily, Corsica, Crete, Puerto Rico
Clove	*Syzgium aromaticum*	Spice	Zanzibar (Tanzania), West Indies, Madagascar (Malagasy Republic), Mauritius, Sumatra (Indonesia), the Moluccas (Indonesia), Penang (Malaysia), Guiana, Brazil
Coca	*Erythroxylum coca*, *E. novogranatense*	Masticatory	South America (mostly Columbia)
Coco de Mer	*Lodoicea maldivica*	Nut	Seychelles
Culantro	*Eryngium foetidum*	Culinary herb	Puerto Rico, Trinidad, tropical Africa and Asia
Cumin and black cumin	Cumin (*Cuminum cyminum*)	Spice	China, India, Morocco, Cyprus, Turkey, Iran, southern Russia
	Black cumin (*Nigella sativa*)	Spice	Mediterranean area, Orient
Cycads	Cycadaceae family: 11 genera with edible species	Starch	Subtropical and tropical areas
Durian	*Durio zibethinus*	Fruit	Thailand, Indonesia, Malaysia, Philippines
Epazote	*Chenopodium ambrosioides*	Culinary herb	Latin America (especially Mexico)
Feijoa	*Acca sellowiana*	Fruit	New Zealand
Fenugreek	*Trigonella foenum-graecum*	Spice	Mediterranean countries, Near East, Middle East, northeastern Africa to Ethiopia, Arabia, Turkey, India, China, Japan
Galangal	Greater galangal (*Alpinia galanga*), Kaempferia galangal (*Kaempferia galanga*), Lesser galangal (*A. officinarum*)	Spice	Southern Asia
Ginger	*Zingiber officinale*	Spice	West Indies (especially Jamaica), India, West Africa (especially Sierra Leone), Malaysia, Sri Lanka, Fiji, Hawaii, Japan, Queensland (Australia)
Ginkgo	*Ginkgo biloba*	Nut	China, Korea, Japan
Goji	*Lycium barbarum*	Fruit	China
Grass pea	*Lathyrus sativus*	Vegetable	India, Pakistan, Bangladesh, Nepal, Ethiopia
Guarana	*Paullinia cupana*	Beverage	Brazil
Guava	*Psidium guajava*	Fruit	India, Mexico
Gum arabic	*Acacia* species: Babul gum arabic (*A. nilotica*), Senegal gum arabic (*A. senegal*), Shittimwood (*A. seyal*)	Food extracts	Tropical Africa, tropical Asia
Hemp	*Cannabis sativa*	Food extracts	Temperate world
Horseradish tree	*Moringa oleifera*	Vegetable and spice	Tropical world
Jackfruit	*Artocarpus heterophyllus*	Fruit and vegetable	India, Burma, Sri Lanka (Ceylon), southern China, Malaya, East Indies, Philippines, central and eastern Africa, Surinam, Australia

continued

TABLE 2.1 (continued)
Principal Food Uses and Commercial Production Areas of the Top 100 Exotic Food Plants

Japanese vegetables	Gobo (*Arctium lappa*)	Vegetable	Japan
	Garland chrysanthemum (*Glebionis coronaria*)	Vegetable	China, Japan, Taiwan
	Honewort (*Cryptotaenia canadensis*)	Vegetable	North America
	Mitsuba (*Cryptotaenia japonica*)	Vegetable	Japan, Korea, China, Taiwan, Indonesia
	Water dropwort (*Oenanthe javanica*)	Vegetable	Asia, Hawaii
Jicama	*Pachyrhizus erosus*	Vegetable	Mexico, Philippines, China
Jujube	*Ziziphus jujube*	Fruit	China
Kava	*Piper methysticum*	Beverage	Polynesia
Khat	*Catha edulis*	Masticatory	Yemen, Ethiopia, Kenya, Somaliland, Tanzania, Madagascar, Uganda
Kiwi	*Actinidia* species: Grape kiwi (*A. arguta*), Kiwi (*A. chinensis*), Kiwi (*A. deliciosa*)	Fruit	New Zealand, China, Italy, France, Spain, United States, Chile, Australia, Japan, Korea
Kumquat	*Fortunella* species: Marumi kumquat (*F. japonica*), Meiwa kumquat (*F. crassifolia*), Nagami kumquat (*F. margarita*)	Fruit	China, Japan, Taiwan, southern Europe, United States, Puerto Rico, Guatemala, Surinam, Colombia Brazil, Australia, Israel, South Africa
Lemon grass	*Cymbopogon citratus*	Culinary herb	Southern and eastern Asia (especially India), western Africa, West Indies, Central America, Brazil, Australia
Lemon verbena	*Aloysia citriodora*	Culinary herb	Algeria, Tunisia, South Africa, parts of Asia
Loofah	*Luffa* species: Angled loofah (*L. acutangula*), Smooth loofah (*L. aegyptiaca*)	Vegetable	Subtropical and tropical Asia
Loquat	*Eriobotrya japonica*	Fruit	Spain, Algeria, Turkey, Israel, Italy, Japan, China, India, Brazil
Lychee, longan, and rambutan	Lychee (*Litchi chinensis*)	Fruit	China, India, Pakistan, Bangladesh, Burma, Taiwan, Thailand, Mauritius, Japan, Philippines, Australia, Madagascar, Brazil, South Africa
	Longan (*Dimocarpus longan*)	Fruit	Thailand, China, Taiwan
	Rambutan (*Nephelium lappaceum*)	Fruit	Thailand, Indonesia, Malaysia, Philippines
Macadamia nut	*Macadamia* species: Rough-shell(ed) macadamia nut (*M. tetraphylla*), Smooth-shell(ed) macadamia nut (*M. integrifolia*)	Nut	Hawaii, Australia

TABLE 2.1 (continued)
Principal Food Uses and Commercial Production Areas of the Top 100 Exotic Food Plants

Mangosteen	*Garcinia mangostana*	Fruit	Thailand, Vietnam, Burma, Malaya, Philippines, Indonesia, Singapore
Medlar	*Mespilus germanica*	Fruit	Temperate Europe (rarely marketed)
Melons (exotic species)	Bitter melon (*Momordica charantia*)	Vegetable	Southeast Asia, China, the Caribbean
	Horned melon (*Cucumis metuliferus*)	Fruit and vegetable	New Zealand, California, Kenya, Israel
Miracle fruit	*Synsepalum dulcificum*	Sweetener	Tropical West Africa
Myrrh	*Commiphora* species: Abyssinian myrrh (*C. habessinica*), Myrrh (*C. myrrha*)	Spice	Northeastern Africa, Arabian Peninsula
Neem	*A. indica*	Multipurpose	India, Myanmar (Burma), Southeast Asia, West Africa
Noni	*Morinda citrifolia*	Vegetable and beverage	Malay Peninsula, Pacific Islands (especially Hawaii)
Nutmeg and mace	*Myristica fragrans*	Spice	Indonesia, Grenada
Oca	*Oxalis tuberosa*	Vegetable	Venezuela to Argentina (at high altitudes), Mexico, New Zealand
Okra	*Abelmoschus esculentus*	Vegetable	Warmer parts of Africa, the Mediterranean, and the Americas
Opium poppy	*Papaver somniferum*	Spice	Netherlands, Poland, Romania, Czech Republic, former Yugoslavia, Russia, India, Iran, Turkey, Argentina, Australia, many Asian and Central and South American countries
Palmyra palm	*Borassus flabellifer*	Sweetener	India, Southeast Asia, Malaysia
Passionfruit	*Passiflora* species: Banana passion fruit (*P. tripartita* var. *mollissima*), Giant granadilla (*P. quadrangularis*), Maypop (*P. incarnata*), Passionfruit (*P. edulis*), Sweet calabash (*P. maliformis*), Sweet passionfruit (*P. ligularis*), Water lemon (*P. laurifolia*), Wingstem passion flower (*P. alata*)	Fruit	Tropics and subtropics: South America, the Caribbean, Mexico, Australia, New Zealand, Southeast Asia, Taiwan, India, Africa, the Mediterranean area, Hawaii, South Africa
Peach palm	*Bactris gasipaes*	Vegetable and fruit	Costa Rica, Brazil, Ecuador, Peru, Bolivia, Colombia, Guyana, Surinam, Venezuela, Panama, Guatemala, Dominican Republic
Pepino	*Solanum muricatum*	Fruit	Chile, New Zealand, Australia, United States (California, Hawaii)
Perilla	*Perilla frutescens*	Culinary herb	Asia

continued

TABLE 2.1 (continued)
Principal Food Uses and Commercial Production Areas of the Top 100 Exotic Food Plants

Persimmon	Oriental (Japanese) persimmon (*Diospyros khaki*)	Fruit	China, Japan, Brazil, Korea, Italy
	Date plum (*Diospyros lotus*)	Fruit	Italy, west Asia (especially Japan), China
	American persimmon (*Diospyros virginiana*)	Fruit	Southeastern United States
Pomegranate	*Punica granatum*	Fruit	Egypt, China, Afghanistan, Pakistan, Bangladesh, Iran, Iraq, India, Burma, Saudi Arabia
Quinine	*Cinchona* species: Quinine (*C. calisaya*), Redbark (*C. pubescens*)	Beverage	Africa (small amounts from Peru, Bolivia, Ecuador)
Rooibos tea	*Aspalathus linearis*	Beverage	South Africa
Rose apple	*Syzygium jambos*	Fruit	Tropical and near-tropical climates (mostly in home gardens)
Roselle	*Hibiscus sabdariffa*	Multipurpose	Tropical and subtropical areas, particularly Sudan, China, Thailand, Mexico, Jamaica
Saffron	*Crocus sativus*	Spice	Spain, Kashmir, Iran Greece
Sago palm	*Metroxylon sagu*	Starch	Southeast Asia (especially Malaysia and Indonesia), Oceania
Sapodilla	*Manilkara zapota*	Fruit	India, Philippines, Sri Lanka, Malaysia, Australia, Mexico, Venezuela, Guatemala
Sapote	Black sapote (*Diospyros digyna*), Green sapote (*Pouteria viridis*), Mamey sapote (*Pouteria sapota*), South American sapote (*Quararibea cordata*), Yellow sapote (*Pouteria campechiana*), White sapote (*Casimiroa edulis*), Woolly-leaf white sapote (*Casimiroa tetrameria*)	Fruit	Mexico to South America; some species also grown in Old World tropics
Sarsaparilla	*Smilax* species: Ecuadorian sarsaparilla (*Smilax febrifuga*), Honduran sarsaparilla (*Smilax regelii*), Mexican sarsaparilla (*Smilax aristolochiifolia*)	Beverage	Tropical Central and South America
Sea buckthorn	*Hippophae rhamnoides*	Fruit	Eurasia
Spinach (exotic species)	Malabar spinach (*Basella alba*)	Vegetable	Tropical and subtropical areas, particularly eastern Asia and India
	New Zealand spinach (*Tetragonia tetragonioides*)	Vegetable	A minor crop of most parts of the tropics and temperate regions
	Water spinach (*Ipomoea aquatica*)	Vegetable	Southeast Asia, especially Malaysia and Indonesia

TABLE 2.1 (continued)
Principal Food Uses and Commercial Production Areas of the Top 100 Exotic Food Plants

Stevia	*Stevia rebaudiana*	Sweetener	China, Taiwan, Laos, Thailand, Korea, Japan, Malaysia, Indonesia, Brazil
Sweet sop and sour sop	Sweet sop (*Annona squamosa*)	Fruit	India; other tropical and subtropical regions
	Sour sop (*Annona muricata*)	Fruit	Numerous tropical countries, including Mexico, India, Southeast Asia, Polynesia
Tamarind	*Tamarindus indica*	Spice	India; other tropical and subtropical regions
Tobacco	*Nicotiana tabacum*	Masticatory	Temperate and subtropical world
Tree tomato	*Solanum betaceum*	Vegetable	New Zealand, South and Central America, Jamaica, Africa (particularly Kenya), southern and southeastern Asia, Australia
Turmeric	*Curcuma longa*	Spice	India, China, East Indies
Vietnamese herbs	Rau Om (*Limnophila aromatica*)	Vegetable and culinary herb	Tropical and subtropical Asia
	Rau Ram (*Polygonum odoratum*)	Vegetable and culinary herb	Southeast Asia
	Vap ca (*Houttuynia cordata*)	Vegetable and culinary herb	Temperate and tropical regions of eastern Asia
	Vietnamese balm (*Elsholtzia ciliata*)	Culinary herb	Semitropical Asia
Wasabi	*Wasabia japonica*	Spice	Japan
Water chestnut	Chinese water chestnut (*Eleocharis dulcis*)	Vegetable	Mostly China; also Japan, Taiwan, Thailand, Australia
	European water chestnut (*Trapa natans*)	Vegetable	Minor crop of India, Southeast Asia
Wax gourd	*Benincasa hispida*	Vegetable	Old World tropics, particularly India and China
Wonderberry and garden huckleberry	*Solanum* species: Garden huckleberry (*S. scabrum*), Wonderberry (*S. retroflexum*)	Fruit	Africa; a minor crop of warm-temperate area of the world
Yard-long bean	*Vigna unguiculata* subsp. *sesquipedalis*	Vegetable	Old World tropics
Zedoary	*Curcuma zedoaria*	Spice	India, Southeast Asia, China

TABLE 2.2
Summary of Categories of Principal Food Uses of the 100 Food Plants

Category	Number of Food Plants
Fruit	26
Spice	18
Vegetable	15
Fruit and vegetable	7
Nut	5
Beverage	5
Culinary herb	5
Masticatory	4
Starch	3
Sweetener	3
Multipurpose	3
Food extracts	3
Vegetable and spice	1
Vegetable and culinary herb	1
Vegetable and beverage	1
Total	100

Note: Five of the categories are for more than one use.

TABLE 2.3
Leading Usages of the 100 Exotic Food Plants

Fruit	33%
Vegetable	25%
Spice	19%

Note: A few of the crops have more than one usage (cf. Table 2.1 with Table 2.2). Where a given chapter dealt with more than one species (e.g., sweet sop and sour sop, Chapter 91), it was scored collectively (e.g., sweet sop and sour sop are both fruits, so the crop was scored collectively as fruit).

TABLE 2.4
Principal Commercial Production Areas of the 100 Exotic Food Plants[a]

A. Tropical and subtropical		B. Temperate	
World (several continents)	52	World (several continents)	6
Asia	18	Asia	2
Latin America	9	Europe	1
Africa	7	Africa	1
Pacific Islands	3		
Australia, New Zealand	1		

[a] Assignments to columns A and B were made on the basis of predominant occurrence. The degree of inclusion of temperate regions in the word "subtropical" is open to interpretation, often understood as between 20° and 35° latitude in both hemispheres.

FORMAT OF PRESENTATION

Each of the chapters dedicated to a given plant or group of plants begins with a name in boldface. This is the English common name that is most frequently encountered. Occasionally, plants are known by several names with about equal frequency, in which case one or two of the names were chosen and presented in boldface. Immediately below this in regular font is the Latin name of the plant family, followed by the most frequent English name of the family. Traditional family names end in *aceae*. For eight of the first plant families recognized by botanists, there are alternative names that do not end in this suffix. In these cases, the traditional family name is given first, followed by the alternative name in parenthesis.

The first section of the chapters is entitled "Names." This begins with the correct scientific (Latin) name and is followed, where it might lead to confusion, by some commonly encountered scientific synonyms. (Some species have dozens of old scientific names.) Following the initial presentation of the scientific name, details are presented (in bullet form) about additional English common names and about etymological details of the English and Latin names. Concealed in these names is a wealth of information about the nature and utility of the plants in question.

The familiar words in English used to refer to plants are called "common," "colloquial," and "vernacular" names. Unfortunately, many plants are known by different common names as a consequence of people in different regions adopting different names for the same plant. Common names are often unreliable, used in different places or by different people to mean different plants. Moreover, many common names have different spellings. Scientific names of plants are governed by an internationally agreed upon *Botanical Code of Nomenclature* (McNeill et al., 2006), which is revised and updated periodically. A separate code governs the application of the names of cultivated varieties (Brickell et al., 2009), but this is of limited concern in this book because almost no cultivars are mentioned.

The following scientific name illustrates some conventions that botanists use in naming plants:

Planta × *alba* A. Smith ex B. Jones subsp. *alba* var. *grandifolia* (L.) L. f.

1 2 3 4 5 6 7 8 9 10 11 12

1. The first word of a scientific name indicates the genus, *Planta* in this case. A genus may be composed of one or many species. The genus name may commemorate an individual but frequently is based on an old plant name, often from classical Latin or Greek.
2. Sometimes one sees an × (the mathematical multiplication symbol, not the letter x), which indicates that the name designates a plant of hybrid origin.
3. The scientific names of species are "binomials," that is, a combination of two words, in this example *Planta alba*. Often it is said that the second word in the name is the "species name," but this is based on ignorance (the "species name" is always a combination of two words). Technically, the second word is called a "species epithet" and is generally an adjective in Latinized form. Latin adjectives are of different types but must agree with the gender of the genus name. The people who coin names try to select a specific epithet that is descriptive of some characteristic of the species. In this case, the specific epithet *alba*, which is Latin for white, may indicate that the leaves are whitish. Sometimes the specific epithet is inaccurate as a description, but for naming purposes this does not matter.

4, 5, 6. In a complete name citation, the author or authors of a name, abbreviated as much as possible without possibility of confusion with another author of a plant name, follow the name. Sometimes, as in this example, the word "ex" is present, which indicates that an author following the ex adopted the specific epithet from a name used (but not validly published) by the person preceding the ex. In this case, A. Smith first used the word *alba* in some way relating to the present name, and it was taken up by B. Jones when he coined the name *Planta* × *alba*. For technical reasons relating to the Code of nomenclature, Smith has not been given credit as the person who named the species, but at least it is indicated that his name was the inspiration that led Jones to adopt the name.

7. The abbreviation subsp. (or ssp.) means "subspecies," which is a subdivision of a species.
8. The word *alba* here is the "subspecific epithet," a word descriptive of the subspecies, just as the specific epithet is descriptive of the species. However, a special case is illustrated here, indicated by two things: the same word *alba* is used both for the specific epithet and for the subspecific epithet, and there is no citation of author(s) following the subspecific epithet. By convention, whenever a botanist splits a species into a subspecies or lower groupings, one of the groupings must be given the same epithet as the species, and no authors for this are recognized.
9. One very rarely encounters a name in which both subspecies and varieties are given, as in this example. Normally if a plant species is divided into smaller groupings, either subspecies or varieties are recognized, not both. In this case, a subspecies has been divided into varieties.
10. The word *grandifolia* is a varietal epithet, descriptive of the variety. In this example, it is Latin for "large leaved." Although the majority of descriptive terms actually do more or less accurately indicate some characteristic of the species, words that are quite inaccurate are permitted (*Planta chinensis* need not come from China).
11. Whenever one sees an author's name in brackets in a scientific name, it indicates that this author was the first to use the epithet in connection with the species, but did so in a different way, for example, assigning the species to a different genus, or using the epithet at a different rank (e.g., subspecies instead of variety). In this case, L. stands for Linnaeus, the "father of biological nomenclature," who recognized more species names than any other individual in history.
12. In this example, L. f. is the author who took the epithet first used by the author in brackets (explained in 11) and was the first to use it in the present name. "L. f." stands for "Linnaeus fils," that is, "son of Linnaeus."

The next section, called "Plant Portrait," provides a basic botanical and agricultural description of the plant as well as of its history and its economic and social importance.

The following section, "Culinary Portrait," is concerned with food uses, particularly industrial and technological aspects. The intent is not to provide details appropriate for popular cook books but rather to indicate basic characteristics of the plant that make it useful for some applications, less so for others. Under a subsection titled "Culinary Vocabulary," information is given (in bulleted format) on the names of especially important foods prepared from the plants as well as on a variety of related culinary words, phrases, and terms. The selection of such foods and phrases is limited and is intended to illustrate some of the more important, interesting, and popular applications.

The section entitled "Curiosities of Science and Technology," again in bulleted format, contains notable and interesting scientific or technological observations and accomplishments that complement the main text.

The final section called "Key Information Sources" provides selected references to books and articles on the subject of each chapter. A subsection of this entitled "Specialty Cookbooks" presents references to food preparation using the particular plant in question. These references are intended to provide the interested reader with extensive background material. Most of the references are in

English, and non-English literature is cited basically when there are few articles and books on the subject in English, or particularly important foreign-language reviews are available. For exotic food plants, key information is often available only in foreign languages.

REFERENCES

Brickell, C.D., Alexander, C., David, J.C., et al. 2009. *The international code of nomenclature for cultivated plants*. 8th ed. International Society for Horticultural Science, Leuven, Belgium. 204 pp.

McNeill, J., Barrie, F.R., Burdet, H.M., et al. 2006. International code of botanical nomenclature (Vienna Code). Regnum Vegetabile 146. A.R.G. Gantner, Verlag KG, Koenigstein, Germany. 568 pp.

3 Acai Berry

Family: Arecaceae (Palmae; palm family)

NAMES

Scientific Name: *Euterpe oleracea* C. Mart.

- "Acai" in the name acai berry is based on the Tupi word *iwasai*, meaning "[the fruit] that cries or expels water" (the Tupi were a people of Brazil with a distinct language).
- Suggested pronunciations: ah-sah-ee, ah-sigh-ee, ah-SIGH-ee, a sigh-EE.
- Before the acai berry became popular under this name, *E. oleracea* was known principally as assai or acai palm. Acai is also spelled açaí, açai, and aqai.
- *Euterpe edulis* C. Mart., a species closely related to *E. oleracea*, is also known as assai palm.
- Other names for *E. oleracea* include cabbage palm, euterpe palm, manicole palm, and palmberry (not to be confused with the mobile communication device called a Palm-Berry).
- *Oleracea* in the scientific name is Latin meaning resembling herbs or vegetables, an allusion to the soft, edible parts of the plant.

PLANT PORTRAIT

Euterpe oleracea is a palm tree, widely distributed in northern South America, especially near the coast, but is found most commonly in the Brazilian state of Pará. It grows in considerable numbers in floodplain areas of the Amazon River. The trees have slender trunks, 7 to 23 cm (3–9 in.) in diameter and 3 to 25 m (10–82 ft.) in height. Several trunks (up to 25) typically arise from the base of the plant, so that the trees grow in clumps. However, solitary trees are occasionally encountered. The leaves are situated at the top of the plant and are 2 to 5 m (6.5–13 ft.) long, with 40 to 80 leaflets arranged on each sides of the midrib. Like most palm trees, the leaves are evergreen (i.e., long-lived), but the lower (older) leaves regularly drop off. Rings around the trunk are formed from the scars of detached leaves, from the base of the plant up to the crown of leaves. Fruits are produced throughout the year in the native area, particularly in the dry season (July to December) in bunches of 500 to 900. The ripe berries are green when immature, usually ripening to dark purple, although some plants produce berries that remain green at maturity (and are often termed "white"). The fruits are round; the size of small grapes or large blueberries: 1.1 to 1.5 cm (0.4–0.6 in.) in diameter. Most of the berry (about 80%) is seed, which is surrounded by a thin layer of edible pulp.

The "palm hearts" of the species have been collected for many years. These are the edible central tissues of the stem near the top of the shoot (for a detailed discussion of palm hearts, see Chapter 74). Collecting the heart kills the trunk from which it is harvested, but because the plants produce many trunks, they are not killed. In recent times, the species has been used primarily as a source of berries.

Most trees used for harvest of either berries or hearts are wild, but there have been recent attempts to establish plantations. Selection of cultivated varieties with superior fruits is also in its infancy. *Euterpe oleracea* is grown as an ornamental tree in frost-free locations and has been cultivated as far north as Florida. The plant can also be grown in dwarfed form in containers as a tropical houseplant.

FIGURE 3.1 Acai berry palm (*Euterpe oleracea*). (From Bailey, 1900–1902.)

FIGURE 3.2 Acai berry harvest. Prepared by B. Flahey.

In Western markets, the acai berry has been marketed as a "miracle food," which allegedly bolsters the immune system, cleanses colons, assists digestion, reduces wrinkles, lowers bad cholesterol, improves skin, increases energy and sexual performance, facilitates sleep, contributes to heart health, slows ageing, and promotes weight loss. The health claims are obviously exaggerated and misleading, and in the United States there have been lawsuits against some acai purveyors for questionable business practices. Acai products include whole fruits, pulp, juice, a variety of prepared foods, and "dietary supplements" (explained below) such as capsules, tablets, and powder. In the Western World, acai can be rather expensive. In the United States, a 25-ounce bottle of acai juice concentrate sells for about $40.00. The acai berry is the most successful of the several "superfoods" that have been marked in recent time (see Chapter 1). It has been estimated that Americans spent over $100 million on acai products in 2009. The fruits do have good concentrations of health-promoting antioxidants (including vitamins C and E) and phytonutrients (especially anthocyanins), but so do many other fruits that are much cheaper and better tasting.

CULINARY PORTRAIT

The taste of acai berries has been likened to a combination of chocolate and red wine, judged by some to be bitterish or medicinal, liked by others. Generally, the taste is modified by serving acai in combination with sweet-tasting foods. The blue-black berries have been marketed as a specialty health food, consumed in juice blends, smoothies, soft drinks, liqueurs, and frozen purée. Acai berries are rather perishable and have traditionally been used in Brazil shortly after picking. Most of the inhabitants of the Amazon region consume acai juice every day, many individuals consuming about 2 L (2 American quarts) daily. In tropical America, the fruit is processed into beverages, ice cream, and pastries. Because the berries have very limited shelf life, frozen purée is commonly imported by Western nations.

Acai palm hearts are eaten in salads and are a popular exported product, but peach palm hearts have become more popular (see Chapter 74).

Culinary Vocabulary

- "Dietary supplements" (also called nutritional supplements and food supplements) are preparations that supply concentrated amounts of food components that are necessary or at least desirable for good health. These may include vitamins, minerals, "good" fatty acids, proteins, fiber, antioxidants, and other compounds. Most dietary supplements are extracted from plants. Countries differ with respect to whether dietary supplements are governed by legislation concerned with food on the one hand or drugs on the other. The dietary supplement industry is huge and extremely profitable and engages in extensive advertising. Dietary supplements may be necessary when individuals have inadequate diets or lack sufficient enzymes to process some food constituents. However, in most cases, it is preferable to eat a balanced diet, particularly high in plant foods, rather than resort to dietary supplements, which are often not as effective as the same constituents that are naturally present in plant foods and frequently represent a needless expense (see discussion and references in Chapter 1).
- *Açaílandia* are special stores in the Amazon River area, where acai juice is extremely popular, that sell the liquid, typically in plastic bags.
- *Cuias* are gourds traditionally used in northern Brazil as serving vessels for acai mixed with tapioca and other foods.
- *Açaí na tigela* (literally "acai in the bowl") is another offering, popular in southern Brazil, that combines acai and other foods, particularly granola.

CURIOSITIES OF SCIENCE AND TECHNOLOGY

- In tropical regions, the palm family is second only to the grass family in economic importance. Palm trees are used to make numerous products, and *E. oleracea* is no exception. Wood from the trunk is used in construction, and the leaves are turned into baskets, brooms, hats, mats, and root thatch.
- The acai palm is adapted to periodically flooded and waterlogged soils, by possession of pneumatophores—special structures providing air channels from above water level down to the roots, allowing them to obtain oxygen.
- The acai berry is hardly the only local wild fruit eaten by native inhabitants of South America. Indigenous peoples of the South American rainforest have been estimated to consume about 2000 different species of fruits growing in the area.
- Harvesting acai fruits requires athletic prowess: almost all of the crop comes from wild plants, which must be climbed, and the large fruit clusters cut off by hand.
- In South America, the young leaves of *E. oleracea* are mashed, and the sappy mixture is applied to stop bleeding.

KEY INFORMATION SOURCES

Andel, T.R. van. 1998. *Palm heart harvesting in Guyana's north-west district: exploitation and regeneration of* Euterpe oleracea *swamps*. Herbarium Utrecht University, Utrecht, the Netherlands. 38 pp + plates.

Anderson, A.B. 1988. Use and management of native forests dominated by acai palm (*Euterpe oleracea*) in the Amazon estuary. *Adv. Econ. Bot*. 6:144–154.

Arujo, E.F., and Silva, R.F. de. 1994. Evaluation of quality in acai seeds stored in different packages and environmental conditions. *Rev. Bras. Sementes*,16:76–79.

Brondizio, E.S. 2004. From staple to fashion food: shifting cycles and shifting opportunities in the development of the açaí palm fruit (*Euterpe oleracea* Mart.) economy in the Amazon estuary. In *Working forests in the neotropics: conservation through sustainable management?* Edited by D.J. Zarin, J.R.R. Alavalapati, F.E. Putz, and M.C. Schmink. Columbia University Press, New York. pp. 339–365.

Brondizio, E.S. 2008. *The Amazonian caboclo and the açaí palm. Forest farmers in the global market*. New York Botanical Garden Press, Bronx, NY. 403 pp.

Buss, D. 2007. The rise and rise of acai in the US. *New Nutr. Bus*. 12(6):17–20.

Cruz Pessoa, J. D., and Silva e Silva, P.V. da. 2007. Effect of temperature and storage on acai (*Euterpe oleracea*) fruit water uptake: simulation of fruit transportation and pre-processing. *Fruits (Paris)*, 62:295–302.

De Paula, J.E. 1975. Anatomy of *Euterpe oleracea* (Palmae of Amazonia). *Acta Amaz*. 5:265–278 (in Portuguese, English summary).

Gallori, S., Bilia, A.R., Bergonzi, M.C., Barbosa, W.L.R., and Vincieri, F.F. 2004. Polyphenolic constituents of fruit pulp of *Euterpe oleracea* Mart. (acai palm). *Chromatographia*, 59:739–743.

Godoy Junior, G., and Saes, L.A. 1987. Interspecific hybrids of heart of palm plants *Euterpe oleracea* × *Euterpe edulis*. *Bragantia*, 45:343–364 (in Portuguese).

Henderson, A., and Galeano, G. 1996. *Euterpe*, *Prestoea*, and *Neonicholsonia* (Palmae). In *Flora Neotropica Monograph 72*. New York Botanical Garden, New York. 90 pp.

Jardim, M.A.G. 1997. Use of the acai palm tree *Euterpe oleracea* Mart. in the Amazonian estuary. *Boletim do Museu Paraense Emilio Goeldi Serie Botanica,* 12(1):137–144 (in Portuguese).

Jardim, M.A.G., and Macambira, M.L.J. 1996. Floral biology of the acai palm tree *Euterpe oleracea* Martius. *Boletim do Museu Paraense Emilio Goeldi Serie Botanica*. 12(1):131–136 (in Portuguese).

Melo, C.F.M., de Wisniewski, A., and Alves, S. de M. 1974. Papermaking potential of the cabbage-palm [*Euterpe oleracea*]. *Boletim Tecnico do IPEAN* 63:1–34 (in Portuguese).

Mertens-Talcott, S.U., Rios, J., Jilma-Stohlawetz, P., Pacheco-Palencia, L.A., Meibohm, B., Talcott, S.T., and Derendorf, H. 2008. Pharmacokinetics of anthocyanins and antioxidant effects after the consumption of anthocyanin-rich acai juice and pulp (*Euterpe oleracea* Mart.) in human healthy volunteers. *J. Agric. Food Chem*. 56:7796–7802.

Muniz-Miret, N., Vamos, R., Hiraoka, M., Montagnini, F., and Mendelsohn, R.O. 1996. The economic value of managing the acai palm (*Euterpe oleracea* Mar.) the floodplains of the Amazon estuary, Para, Brazil. *For. Ecol. Manage*. 87:163–173.

Nobrega, A.A., Garcia, M.H., Tatto, E., Obara, M.T., Costa, E., Sobel, J., and Araujo, W.N. 2009. Oral transmission of Chagas disease by consumption of acai palm fruit, Brazil. *Emerg. Infect. Dis.* 15(4):653–655.

Pozo-Insfran, D., del Brenes, C. H., and Talcott, S.T. 2004. Phytochemical composition and pigment stability of acai (*Euterpe oleracea* Mart.). *J. Agric. Food Chem.* 52:1539–1545.

Sabbe, S., Verbeke, W., Deliza, R., Matta, V., and Damme, P. van. 2009. Effect of a health claim and personal characteristics on consumer acceptance of fruit juices with different concentrations of acai (*Euterpe oleracea* Mart.). *Appetite*, 53:84–92.

Sabbe, S., Verbeke, W., Deliza, R., Matta, V.M., and Damme, P. van. 2009. Consumer liking of fruit juices with different acai (*Euterpe oleracea* Mart.) concentrations. *J. Food Sci.* 74(5):S171–S176.

Sanabria, N., and Sangronis, E. 2007. Characterization of the acai or manaca (*Euterpe oleracea* Mart.): a fruit of the Amazon. *Arch. Latinoam. Nutr.* 57:94–98 (in Spanish).

Sousa, M.A. da C., Yuyama, L.K.O., Aguiar, J.P L., and Pantoja, L. 2006. Acai juice (*Euterpe oleracea* Mart.): microbiological evaluation thermal treatment and shelf life. *Acta Amaz.* 36:497–501 (in Portuguese).

Strudwick, J. 1990. Commercial management for palm heart from *Euterpe oleracea* Mart. (Palmae) in the Amazon estuary and tropical forest conservation. *Adv. Econ. Bot.* 8:241–248.

Strudwick, J., and Sobel, G.L. 1988. Uses of *Euterpe oleracea* Mart. in the Amazon Estuary, Brazil. *Adv. Econ. Bot.* 6:225–253.

Specialty Cookbooks

Marsh, D.C. 2009. *Raw-riffic Foods 101 super-charged juices, shakes and smoothies: more than just a raw recipe book.* CreateSpace, Scotts Valley, CA. 74 pp.

Pratt, S.G., and Matthews, K. 2006. *Superfoods health style: simple changes to get the most out of life for the rest of your life.* Harper, New York. 352 pp.

Rodnitzky, D.J.P. 2007. *The ultimate acai smoothie cookbook: more than 120 smoothie recipes made with the age-defying acai berry.* Book Publishing Company (TN), Summertown, TN. 181 pp.

4 Acerola (Barbados Cherry)

Family: Malpighiaceae (malpighia family)

NAMES

Scientific Name: *Malpighia emarginata* DC. ("*M. punicifolia*," "*M. glabra*," not *M. glabra* L.)

- The English "acerola" is based on the American Spanish word *acerola* used for the same plant, which in turn is based on the Spanish *azarole*, referring to a quite different species, the azarole (*Crataegus azarolus* L.), which has berries of similar appearance.
- Pronunciations: ah-see-ROLL-ah, as-uh-ROH-luh.
- Acerola is also known as Amazon cherry, Antilles Cherry, Barbados cherry, chercese, chereese, French cherry, garden cherry, Jamaica cherry, native cherry, Puerto Rican cherry, Surinam cherry (but see below), West Indian cherry, and wild crapemyrtle.
- Early Spanish explorers called the acerola *cereza*, Spanish for cherry, which the fruit resembles. "Barbados" in the name "Barbados cherry" is based on the region where plants were observed in early times.
- The acerola has sometimes been confused with the Surinam cherry, *Eugenia uniflora* L.
- The genus name *Malpighia* commemorates Marcello Malpighi, (1628–1694), an Italian naturalist, physician, and Professor at Bologna, Italy.
- *Emarginata* in the scientific name *M. emarginata* is Latin for "with a shallow notch" (at the tip), a description that is normally applied to leaves, but is not accurate for most leaves of this species.

PLANT PORTRAIT

The acerola is a large, bushy, evergreen shrub or small tree 2 to 6 m (6½–20 ft.) in height, with a trunk to 10 cm (4 in.) in diameter. The plant has clusters of red or rose flowers, each 2 to 2.5 cm (about an inch) wide. The fruits are borne singly or in clusters of two or three and are somewhat flattened to round. They are cherry-like but more or less three-lobed (three furrows are evident on the outside of the fruit), 1.25 to 2.5 cm (1/2–1 in.) wide, thin skinned, bright red (rarely yellow-orange), with orange, very juicy pulp. The fruits contain three small, rounded seeds with wings. The plant is adapted to tropical to subtropical climates. It has been suggested that it originated in the Yucatan Peninsula, but it seems to be naturally distributed from South Texas, through Mexico and Central America to northern South America and throughout the Caribbean (Bahamas to Trinidad). It has been introduced to subtropical areas of the world, particularly in Southeast Asia, India, and South America. Some of the largest plantations are in Brazil. Acerola is widely cultivated on a commercial scale in the tropics for fresh fruit and juice and has been noted to be an excellent natural source of vitamin C. Sweet varieties have been planted in home gardens in Florida, and there has been very minor commercial production of the fruit in the state. Acerola is more often cultivated as an ornamental shrub than a source of edible fruit in Florida and other regions of the southeastern United States. Dwarf forms are available, and the plant is also often grown as bonsai.

FIGURE 4.1 Acerola (*Malpighia emarginata*). (From Curtis, 1805, vol. 21, plate 813.) Fruit cluster at top right adapted from Bailey, L.H. (Ed.). *The standard cyclopedia of horticulture*. Macmillan, Toronto, ON. 3 vols. 1916.

CULINARY PORTRAIT

The flavor of the sweet, red, juicy pulp of acerola is reminiscent of raspberries, and when cooked the fruit tastes like a tart apple. Acerola is often eaten fresh in areas where it is cultivated, although it is too tart for many people. The fruit is also stewed with sugar as a dessert. The seeds need to be separated from the pulp in the mouth and returned by spoon to the dish, which is a nuisance, and the cooked fruits must be strained to remove the seeds (eating large quantities of whole fruits containing the seeds has produced illness). Acerola sauce or purée is used as a topping on cake, pudding, ice cream, and fruit and is used in gelatin desserts, punch, and sherbet. The fruits make excellent syrup, jelly, jam, preserves, and pies and have also been used to produce baby food, popsicles, and wine. Acerola juice is as popular in Brazil as orange juice is elsewhere, but it does not store well. Cooking causes the bright-red color to change to brownish-red. Frozen fruit falls apart when thawed. Ripe fruit bruises easily, is highly perishable, and loses flavor and nutritional content very rapidly after harvest. As a result of poor keeping and shipping qualities, acerola is not commonly cultivated for commercial fresh fruit exportation.

Culinary Vocabulary

- "Acerola powder" is a dried, powdered form of acerola fruit that is used as to boost the nutritional level of food consumed by groups with special needs (infants, the elderly, the sick, etc.) and is also marketed in health-food outlets.

CURIOSITIES OF SCIENCE AND TECHNOLOGY

- Acerola fruit has one of the highest known vitamin C content of all fruits. An acerola fruit can have up to 4.5% vitamin C compared with 0.05% in a peeled orange. Vitamin C is an acid (ascorbic acid), and the higher the content of vitamin C, the more acidic the fruit. One or two of the cherry-sized fruits can provide the recommended daily allowance of vitamin C. Vitamin C is an antioxidant and a free radical scavenger; that is, it deactivates the harmful chemicals called free radicals that are produced by the body. Acerola concentrates can now be found in many over-the-counter multivitamin supplements. Recent research in cosmetology suggests that vitamin C is also useful for the skin, and acerola extracts are now appearing in skin-care products claimed to fight cellular aging.

- In the late 1940s and 1950s, there was an explosion of enthusiasm for acerola fruit because of its extraordinarily high content of vitamin C, and breeding programs were undertaken in several countries. However, interest plummeted when it was realized that a fruit could not become a superstar because of its vitamin C content alone. Synthesized vitamin C is much cheaper and, despite the claims that "natural" is better, synthetic vitamin C has become dominant in the marketplace.
- Fresh acerola juice applied to peeled or sliced bananas keeps them from darkening. Acerola juice is also useful for preventing the oxidation of a variety of other fruits.
- Acerola trees without adequate pollination often set seedless fruit.

KEY INFORMATION SOURCES

Alves, R.E., Chitarra, A.B., and Chitarra, M.I.F. 1995. Postharvest physiology of acerola (*Malpighia emarginata* D.C.) fruits, maturation changes, respiratory activity, and refrigerated storage at ambient and modified atmospheres. *Acta Hortic.* 370:223–229.

Alves, R.E., Filgueiras, H.A.C., Mosca, J.L., and Menezes, J.B. 1999. Brazilian experience on the handling of acerola fruits for international trade: harvest and postharvest recommendations. In *International symposium effect of pre- & postharvest factors in fruit storage*. Edited by L. Michalczuk. *Acta Hortic.* 485:31–36. (ActaHort CD-rom format).

Asenjo, C.F. 1980. Acerola. In *Tropical and subtropical fruits*. Edited by S. Nagy and P.E. Shaw. AVI Publishing, Westport, CT. pp. 341–374.

Carrington, C.M.S., and King, R.A.G. 2002. Fruit development and ripening in Barbados cherry, *Malpighia marginata* DC. *Sci. Hortic.* 92:1–7.

Cerezal-Mezquita, P., and Garcia-Vigoa, Y. 2000. Acerola—a neglected American fruit with high ascorbic acid content. *Alimentaria* (Chile), 37(309):113–125 (in Spanish).

Cooper, F.L. 1971. The acerola comes to California loaded with vitamin C. In *California Rare Fruit Growers Yearbook*, vol. 3. pp. 2–8.

Duarte, O., and Paull, R.E. 2008. *Malpighia emarginata*, acerola. In *The encyclopedia of fruit & nuts*. Edited by J. Janick and R.E. Paull. CABI, Wallingford, Oxfordshire, UK. pp. 461–466.

Du-Preez, R.J. 1997. The acerola: a natural source of ascorbic acid (vitamin C). *Neltropika Bull. (South Africa)*, 298:32–33.

Gonzaga-Neto, L., and Soares, J.M. 1994. *Acerola fruit for exports: technical production prospect*, vol. 10. Publicacoes Tecnicas Fruplex (Brazil). Brasilia, DF, Brazil. 43 pp (in Portuguese).

Harjadi, S.S. 1991. *Malpigia glabra* L. In *Plant resources of South-East Asia, 2, edible fruits and nuts*. Edited by E.W.M. Verheij and R.E. Coronel. Pudoc, Leiden, the Netherlands. pp. 198–200.

Ledin, R.B. 1958. *The Barbados or West Indian cherry*. University of Florida Agric. Exp. St., Gainesville, FL. 28 pp.

Maciel, M.I.S., Melo, E. de A., de Lima, V.L.A.G., da-Silva, M.R.F., and da Silva, I.P. 1999. Processing and storage of acerola (*Malpighia* sp.) fruit and its products. *J. Food Sci. Technol.* 36:142–146.

Marino-Netto, L. 1986. *Malpighia glabra Linn. The tropical cherry*. Nobel, Sao Paulo, SP, Brazil. 94 pp (in Portuguese).

Miyashita, R. K., Nakasone, H. Y., and Lamoureux, C.H. 1964. Reproductive morphology of acerola (*Malpighia glabra* L.). Technical Bulletin No. 63. Hawaii Agricultural Experiment Station (University of Hawaii, Honolulu). 31 pp.

Morton, J. 1987. Barbados Cherry. In *Fruits of warm climates*. Creative Resource Systems, Winterville, NC. pp. 204–207.

Moscoso, C.G. 1956. West Indian cherry—richest known source of natural vitamin C. *Econ. Bot.* 10:280–294.

Pino, J.A., and Marbot, R. 2001. Volatile flavor constituents of acerola (*Malpighia emarginata* DC.). *J. Agric. Food Chem.* 49:5880–5882.

Teixeira, A.H.de C., and Azevedo, V. de. 1995. Climate limit-indexes for acerola crop growth. *Pesquisa Agropecuaria Brasileira* (Brazil), 30:1407–1410 (in Portuguese).
Yamane, TG.M., and Nakasone, H.Y. 1961. Pollination and fruit set studies of acerola, *Malpighia glabra* L. in Hawaii. *Proc. Am. Soc. Hort. Sci.* 78:141–148.

Specialty Cookbooks

Foley, R. 2008. *The rum 1000: the ultimate collection of rum cocktails, recipes, facts, and resources.* Sourcebooks, Naperville, IL. 320 pp.
Keys, J. 1986. *We the women of Hawaii cookbook: favourite recipes of prominent women of Hawaii.* Revised edition. Press Pacific, Kalui, HI. 297 pp.
Ortiz, E.L., and Caistor, N. 1998. *A taste of Latin America: recipes and stories.* Interlink, New York. 121 pp.
Ortiz, Y. 1997. *A taste of Puerto Rico: traditional and new dishes from the Puerto Rican community.* Plume, New York. 288 pp.
Valldejuli, C.A.1984. *Puerto Rican cookery.* Pelican, Gretna, LA. 408 pp.

5 Akee

Family: Sapindaceae (soapberry family)

NAMES

Scientific name: *Blighia sapida* K.D. Koenig (*Cupania sapida* J. Voigt)

- The word "akee" has been interpreted as derived from the West African *akye fufo* (possibly based on Kru, *akee* or Akan (Twi) *a-ky*, wild cashew).
- "Akee" is the usual spelling, although "ackee" is an acceptable alternative and is usual in Jamaica, the principal area of akee use. Akie and achee are occasionally encountered. Akee is also known as akee apple.
- Akee is sometimes called "vegetable brains" because when cooked the edible part of the fruit resembles brain material; it is also occasionally called "vegetable egg" because, some say, it resembles scrambled eggs.
- In parts of the eastern Caribbean, "ackee" is the name of a fruit tree widely available in tropical America, *Melicoccus bijugatus* Jacq., also known as honeyberry, mamoncillo, and Spanish lime, although the true akee is called "Jamaican ackee."
- The genus name *Blighia* commemorates Captain William Bligh (1754–1817), governor of New South Wales (a state in southeastern Australia) and captain of HMS Bounty, the subject of the memorable film *Mutiny on the Bounty* starring Marlon Brando as Fletcher Christian (for additional information, see Chapter 13). The genus name honors Bligh because he took samples to London's Kew Gardens in 1793. Some have credited Bligh with introducing akee to Jamaica in 1793, but cuttings of the species are known to have been brought there by the captain of a slave ship in 1778 and purchased by a local physician/botanist.
- *Sapida* in the scientific name *B. sapida* is Latin for "well flavored."

PLANT PORTRAIT

Akee is an evergreen tree growing to 20 m (66 ft.) in height, but usually 8 to 15 m (26–49 ft.) high. It is native to the forests of the Ivory Coast and Gold Coast of tropical West Africa. Akee attains its greatest gastronomic popularity in Jamaica, the only country where it is a major food crop. Trees were also established in most of the other Caribbean islands, such as Trinidad, Grenada, Antigua, and Barbados, as well as Florida, Central America, South America, and elsewhere, but the fruits are only occasionally harvested for food outside of Jamaica.

The akee develops hanging clusters of leathery, pear-shaped fruits 7 to 10 cm (2¾–4 in.) long. These mature from green to shades of red or yellow, or mixtures or red and yellow (red is most common). When ripe, the fruit splits open lengthwise, starting at the apex, into three sections. Each section contains a large, round, smooth, shiny, jet-black seed about 2.5 cm (1 in.) long, each surrounded by fleshy, creamy, or yellow material (technically called an aril) about 4 mm (3/16 in.) thick, which is glossy and oily, and is the only edible portion of the fruit.

Canned akees are exported from Jamaica and to a lesser extent from Haiti, Costa Rica, and Mexico, primarily to serve ethnic markets in Canada and the United Kingdom. The Jamaican exports are an important source of revenue for the nation. In 1972, the United States banned the importation of all types of akee products from Jamaica, including canned, frozen, raw, and dried, pending

FIGURE 5.1 Akee (*Blighia sapida*). Flowering branch (from Loddiges, 1817–1833, plate 1484) and fruit (bottom right) (from Rhind, 1855).

the Jamaican government's development of a reliable measure of the amount of toxin (hypoglycin) in the fruit as well as determination of safe levels. More recently, certain companies evaluated as producing safe akee products were authorized to import akee into the United States.

In addition to the culinary use of akee, the hard, durable wood is used for construction, boxes, oars, and casks; flower extracts have been used in perfume; oil from the fruits has been used to make soap; and extracts from the leaves, fruits, and seed have been used medicinally.

CULINARY PORTRAIT

In the native West African home of the akee, the fruits are occasionally eaten raw, cooked in a soup, or fried in oil, but for the most part the fruit has never attracted much African use, and the trees are mostly used for shade and ornament, and the strong, termite-resistant wood for construction. Akees are primarily consumed in Jamaica, where they are usually prepared as a cooked fruit or vegetable, but are occasionally eaten fresh. The fresh aril is crisp and nutty flavored. Cooked akee is not only said to resemble scrambled egg in appearance, but many people think the texture and taste are similar. The oily aril does not keep for long but can be preserved by placing it in boiling water for 1 minute followed by freezing. There are numerous methods of preparation in Jamaica. Parboiling in salted water or milk and then lightly frying in butter is recommended. The arils can be combined with fish, traditionally cod; "Akee and saltfish" is the national dish of Jamaica, often prepared as a scrambled dish of soaked, flaked salt cod, onions, tomatoes, scallions, and akee. The use of salted codfish in Jamaica started when slaves were brought over from West Africa in the 1700s. The North Atlantic fisheries provided a cheap source of cod, which could be preserved by salt in the absence of refrigeration, to provide protein to the slaves. Akee is also added to meat and cheese dishes and stews, or curried and eaten with rice. Fresh akee is more flavorful than the canned product, with a smoother, more buttery taste, but canned or frozen akee is almost as good.

Underripe fruits are dangerously poisonous and consumption has resulted in fatalities. "Jamaica poisoning" and "Jamaican vomiting sickness syndrome," or more technically "toxic hypoglycemic syndrome," are phrases used to describe the result of eating the fruit while it is still in its poisonous stage. Sickness has been recorded in the West African home of the akee, and also in the Caribbean where it has been established, particularly in Jamaica. Unripe fruits and the inner red membranes of the ripe arils possess the amino acid hypoglycin A; this disrupts liver metabolism, producing low blood sugar. Hypoglycin levels fall dramatically as the fruit matures. Jamaican vomiting sickness is characterized by generalized stomach discomfort starting 2 to 6 hours after eating immature akee, succeeded by sudden vomiting, prostration lasting up to 18 hours, and possibly a second bout of vomiting, usually followed by convulsions, coma, and death (average time to death is 12.5 hours). Before the development of medical treatment, 80% of people suffering from Jamaica sickness died; the vast majority of patients who receive treatment recover completely. The toxic chemical is largely neutralized by light as the fruit opens. The pink or orange-red tissue joining the aril to the seed has often been said to be highly poisonous, but research has indicated this is not accurate. The claim that overripe fruit is dangerous also does not seem to be accurate. The seeds, however, are always poisonous, and the leathery skin of the fruit is also considered poisonous. The water in which the fruit is cooked also may be poisonous, and cases of poisoning have occurred from reusing such water.

Culinary Vocabulary

- Jamaicans traditionally distinguish a "hard" akee variety called "cheese akee," preferred because it does not disintegrate on cooking, and a "soft" variety called "butter akee."
- Quiche is an open pie (i.e., no crust on top) made with a custard of egg and cream (or milk) and other ingredients. The dish is centuries old in Europe but has become well known only since the 1950s in North America, where its snobby image has made it the ultimate food for Yuppies (Young Upwardly Mobile Professionals). Ironically, "akee quiche" has become popular with many Yuppies, despite the akee's long history of use as a food for slaves and the poorest of people.

CURIOSITIES OF SCIENCE AND TECHNOLOGY

- In Jamaica, it is widely said that before the akee is harvested, it must "smile" or "laugh," meaning it must split open naturally. This rule is a way of ensuring that the fruit is not harvested underripe, when it is poisonous.
- Akee is one of many foods brought from Africa to the Caribbean during the slave trade. There are stories told of cruelly treated slaves who offered the fruits while still in their poisonous stage to their masters.
- To avoid U.S. prohibitions on possibly poisonous akees, there have been frequent illegal attempts to import canned akees into the United States by declaring them to be something else. False labels have included "Montego Brand Jamaican Callaloo in Brine" (callaloo refers to various edible greens and soup made from them), "Tropic Ginger Beer, Product of Jamaica," and "Tropic Banana Fruit."
- In West Africa, crushed akee fruits are used as poison to stun and catch fish.
- As with many other member of the soapberry family, akee fruits produce lather in water, and are used as laundry soap.
- In Cuba, an extract of akee flowers has been used to make cologne.
- In Brazil, a water extract of the toxic seeds has been drunk to get rid of bowel parasites, followed in short order by a purgative to keep the patient from being poisoned. The procedure is cheap but dangerous.
- The risk of death makes the akee one of the most extraordinary of all the plant foods of the world. In the animal kingdom, potentially deadly foods are also known, particularly in the fish order Tetraodontoidea (an order is a group of related families). This includes ocean sunfishes, porcupine fishes, and fugu, which are among the most poisonous of all marine

life. The liver, gonads, intestines, and skin of these fish contain tetrodotoxin, a powerful neurotoxin that can cause death in about 60% of people who ingest it. Other animals, such as the California newt and the eastern salamander, also possess tetrodotoxin in lethal quantities. Fugu flesh or musculature is edible and is considered a delicacy in Japan, often commanding the equivalent of $400.00 for one meal. Despite careful preparation, about 50 people die annually in Japan from eating this fish.

KEY INFORMATION SOURCES

Akintayo, E.T., Adebayo, E.A., and Arogundade, L.A. 2002. Chemical composition, physico-chemical and functional properties of akee (*Blighia sapida*) pulp and seed flowers. *Food Chem.* 77:333–336.

Blake, O.A., Jackson, J.C., Jackson, M.A., and Gordon, C.L.A. 2004. Assessment of dietary exposure to the natural toxin hypoglycin in ackee (*Blighia sapida*) by Jamaican consumers. *Food Res. Int.* 37:833–838.

Bressler, R. 1976. The unripe akee—forbidden fruit. *N. Engl. J. Med.* 295:500–501.

Brown, M., Bates, R.P., McGowan, C., and Cornell, J.A. 1992. Influence of fruit maturity on the hypoglycin A level in ackee (*Blighia sapida*). *J. Food Saf.* 12:167–177.

Chase, G.W., Jr., Landen, W.O. Jr., and Soliman, A.G.M. 1990. Hypoglycin A content in the aril, seeds, and husks of ackee fruit at various stages of ripeness. *J. Assoc. Off. Anal. Chem.* 73:318–319.

Chase, G.W., Jr., Landen, W.O. Jr., Gelbaum, L.T., and Soliman, A.G.M. 1989. Ion-exchange chromatographic determination of hypoglycin A in canned ackee fruit. *J. Assoc. Off. Anal. Chem.* 72:374–377.

Cohen, J.E., and Paull, R.E. 2008. *Blighia sapida*, ackee. In *The encyclopedia of fruit & nuts*. Edited by J. Janick and R.E. Paull. CABI, Wallingford, Oxfordshire, UK. pp. 792–795.

Esuoso, K.O., and Odetokun, S.M. 1995. Proximate chemical composition and possible industrial utilization of *Blighia sapida* seed and seed oils. *Rivista Italiana Sostanze Grasse* (Italy), 72:311–313.

Foungbe, S., Naho, Y., and Declume, C. 1986. Experimental study of the toxicity of arils from *Blighia sapida* (Sapindaceae) in relation to intoxication of children in Katiola (Côte d'Ivoire). *Ann. Pharm. Fr.* 44:509–515 (in French).

Guevart, E., Lawson-Body-Houkportie, A., Tchedre, G., and Les Soeurs Augustines. 1994. Toxicity of akee fruit: a danger to be known. *Afr. Med.* 33(311):116–118 (in French).

Henry, S.H., Page, S.W., and Bolger, P.M. 1998. Hazard assessment of ackee fruit (*Blighia sapida*). *Hum. Ecol. Risk Assess.* 4:1175–1187.

Isaacs, S. 2000. Hazard analysis and critical control point (HACCP) development for the processing of ackee. *JAGRIST (The Jamaican Agriculturist)*, 12:30–37.

Jordan, E.O., and Burrows, W. 1937. The vomiting sickness of Jamaica, B.W.I., and its relation to akee poisoning. *Am. J. Hyg.* 25:520–545.

Larson, E., Wynn, M.F., Lynch, S.J., and Doughty, D.D. 1953. Some further studies on the akee. *Quart. J. Fla. Acad. Sci.* 16(3):151–156.

Larson, J., Vender, R., and Camuto, P. 1994. Cholestatic jaundice due to ackee fruit poisoning. *Am. J. Gastroenterol.* 89:1577–1578.

Lindsay, J. 2000. Agronomic requirements for sustainable ackee production. *JAGRIST (The Jamaican Agriculturist)*, 12:12–22.

Lynch, S.J., Larson, E., and Doughty, D.D. 1951. A study on the edibility of akee (*Blighia sapida*) fruits of Florida. *Proc. Fla. State Hort. Soc.* 64:281–284.

McTague, J.A., and Forney, R., Jr. 1994. Jamaican vomiting sickness in Toledo, Ohio. *Ann. Emerg. Med.* 23:1116–1118.

Meda, H.A., Diallo, B., Buchet, J.P., Lison, D. Barennes, H., Ouangre, A., Sanou, M., Cousens, S., Tall, F., and Van De Perre, P. 1999. Epidemic of fatal encephalopathy in preschool children in Burkina Faso and consumption of unripe ackee (*Blighia sapida*) fruit. *Lancet*, 353:536–540.

Morton, J. 1987. Akee. In *Fruits of warm climates*. Creative Resource Systems, Winterville, NC. pp. 269–271.

Omobuwajo, T.O., Sanni, L.A., and Olajide, J.O. 2000. Physical properties of ackee apple (*Blighia sapida*) seeds. *J. Food Eng.* 45:43–48.

Persaud, T.V.N. 1974. Mechanism of teratogenic action of hypoglycin-A. *Experientia*, 27:414–415.

Rashford, J. 2001. Those that do not smile will kill me: the ethnobotany of the ackee in Jamaica. *Econ. Bot.* 55:190–211.

Samuels, A., and Arias, L.F. 1979. Agronomic observations on ackee (*Blighia sapida* L.) and preliminary tests on industrial processing. *Agronomia Costarricense (Costa Rica)*, 3:79–87 (in Spanish, English summary).

Sherratt, H.S.A. 1986. Hypoglycin, the famous toxin of the unripe Jamaican ackee fruit. *Trends Pharmacol. Sci.* 7:186–191.

Tanaka, K., Kean, E.A., and Johnson, B. 1976. Jamaican vomiting sickness: biochemical investigation of two cases. *N. Engl. J. Med.* 295:461–467.

Specialty Cookbooks

Black, D. 2007. *Back to roots: a Jamaican cookbook: cooking in paradise*. AuthorHouse, Bloomington, IN. 63 pp.

Goldman, V., and Bell, P. 1992. *Pearl's delicious Jamaican dishes: recipes from Pearl Bell's repertoire*. Island Trading, New York. 132 pp.

Hamilton, J., and Shears, K. 1990. *A little Jamaican cookbook*. Heinemann, Kingston, Jamaica. 60 pp.

Scala Quinn, L. 1997. *Jamaican cooking: 140 roadside and homestyle recipes*. MacMillan, New York. 162 pp.

6 Allspice (Pimento)

Family: Myrtaceae (myrtle family)

NAMES

Scientific Name: *Pimenta dioica* (L.) Merr. (*P. officinalis* Lindl.)

- "Allspice" acquired its name because its berries have an aroma and flavor that resembles many spices, most particularly cloves, cinnamon, and nutmeg.
- Allspice is also known as clove pepper, Jamaica pepper, myrtle pepper, pimento, and pimienta (pimenta).
- The word "allspice" occurs in the names or alternate names of several aromatic shrubs, including the Carolina allspice or sweet bush (*Calycanthus floridus*, L. of the southeastern United States, which includes "*Calycanthus fertilis* Walter" of the eastern United States); the wild allspice or American spicebush (*Lindera benzoin* (L.) Blume), a shrub of eastern North America, with aromatic berries, reputed to have been used as a substitute for true allspice; and the Japanese allspice or wintersweet (*Chimonanthus praecox* (L.) Link), native to eastern Asia and planted as an ornamental.
- "Pimiento" (often rendered pimento) is derived from the Castilian Spanish word for black pepper. Allspice was initially called pimienta by the Spaniards, the only Europeans to import it during the sixteenth century. The use of the terms pimiento and pimento should not be confused with pimiento (or pimento) chile peppers (*Capsicum annuum* L.), a type of chile pepper with thick walls, used fresh in salads for color and flavor, in cooking, in canning, in stuffing olives, and for flavoring cheeses.
- Columbus called allspice "pepper," a name which still persists as "Jamaica pepper" and "clove pepper." The fruits of allspice are very similar to peppercorns (fruits of *Piper nigrum* L.; see below for how to distinguish them).
- The genus name *Pimenta* comes from Spanish *pimienta* for black pepper and traces to the Latin *pigmentum*, pigment, color, or the juice of plants. In medieval times, the Latin word pigmentum acquired the meanings "spice" and "condiment."
- *Dioica* in the scientific name *P. dioica* is from the Greek: *di* from *dýo*, two + *oica* from *oîkos*, house, a poetic way of saying that there are separate houses for male and female flowers; that is, they grow on different trees.

PLANT PORTRAIT

Allspice trees are tropical evergreens, averaging 7.6 m (25 ft.) in height, sometimes growing higher than 12 m (40 ft.). The tree is native to the West Indies and Central America and also occurs wild in Cuba and southern Mexico, and possibly Haiti and Costa Rica. The species tends to have separate male and female trees, although intermediate types of plants are also found. The white flowers develop into pea-sized fruits, about 8 mm (1/3 in.) in diameter, nearly globose, produced in clusters of a dozen or more at or near the ends of branches. The berries are fleshy, sweet, and purplish-black when ripe and have the appearance of large peppercorns. The whole dried fruit is ground to produce the allspice powder of commerce. The fruit is harvested when full size but still immature, while it is brownish green, as it is then most strongly flavored. The dried berries become reddish

FIGURE 6.1 Allspice (*Pimenta dioica*). (From Köhler, 1883–1914.)

brown. Allspice trees begin to bear fruit at 7 or 8 years of age, reach maximum production at about 15 years, and may bear fruit for up to 100 years. The average tree in Jamaica produces about 1 kg (2½ lb.) of dried product, but some trees will yield up to 25 kg (about 55 lb.) in a year.

Allspice was discovered by Columbus in 1494 but was not recognized or used as a distinctive spice until the early seventeenth century. It was introduced into Puerto Rico and Barbados as well as other locations in the tropics. From the seventeenth to the nineteenth centuries, sailors used allspice berries to help preserve meat on long voyages. In 1824, it was taken to Ceylon and later Singapore, but it did not do well in either location. By the beginning of the twentieth century, Europe alone was using four times more allspice than is harvested today because it was a popular spice for processing meats and fish and was also used in baking. There has been a decline of popularity of allspice since the Second World War. Allspice has been an important but never a critical major spice of the world. It is most popular in North America and England. Jamaica grows, produces, and exports most of the world's allspice products. Smaller amounts are produced in Guatemala, Honduras, Brazil, Mexico, and the Leeward Islands. Jamaican allspice has long been considered to be superior to other sources. Attempts were made historically to cultivate allspice trees in many other locations throughout the world, but while the trees grow and flower, there has never been a successful harvest. Allspice remains the only spice commercially produced exclusively in the Western Hemisphere.

CULINARY PORTRAIT

The taste of allspice is like a combination of cloves (the dominating flavor), cinnamon, juniper berries, nutmeg, and pepper. The oil of allspice is dominated by eugenol, the same compound found in cloves, cinnamon, and nutmeg. Allspice can replace cloves in most recipes and generally should be used sparingly to avoid overpowering recipes. It is widely used to flavor pickles and pickling mixes, relishes, sauces (including applesauce), ketchup, sausages, cold cuts, Swedish meatballs, salt beef and other cured meats, game, poultry, cured fish, gravies, stuffings, cooked vegetables, rice, fruit cakes, spice cakes, cookies, pies, plum puddings, and preserves. Allspice blends well with other spices and is present in many spice mixtures; for example, it is sometimes part of the French spice mixture *quatre épices*. Allspice is the most important spice in Caribbean cuisine. It is used in the well-known mole sauces of Mexico. ("Mole" is based on the Nahuatl Indian word *molli*, a mixture or concoction, which in Mexico is always made by cooking and always includes chile pepper.) Allspice is also used in the jerk pastes of Jamaica for pork, beef, and chicken. ("Jerk" (not to be confused with soda jerking) is derived via Spanish from the native Peruvian *charqui*, dried meat. Jerking meat is done by cutting it into long strips and drying by sun or fire. In Jamaica, meats and poultry are marinated in herbs and spices and cooked over an allspice wood fire. Commercial jerk spice preparations are available, typically made with onions, chile peppers, and allspice.) Allspice is traditionally used in German sauerbraten and in many American fruit pies (especially pumpkin pie), cookies, and cakes. The curries of India and rice dishes of Turkey also rely on allspice. Allspice goes particularly well with ketchups and other tomato-based sauces. This spice is an important ingredient in Chartreuse and Benedictine liqueurs, made in European monasteries for centuries.

Few people today grind their own spices, but allspice is about the size of peppercorns and is sometimes ground in pepper grinders, often in combination with pepper. If using a grinder with plastic parts, be aware that the oil in allspice can cloud plastic. Because the volatile oils in allspice dissipate readily, allspice should be bought in quantities small enough to use promptly and stored properly (in cool areas, in tight, light-proof containers). Whole berries will keep for up to 2 years in a cool, dark place, whereas ground allspice should be used within 6 months.

Culinary Vocabulary

- "Pimento Dram" is a spicy Jamaican liqueur flavored with allspice.
- *Poivre de la Jamaïque* is the French culinary phrase for allspice.
- "West African spice mixture," a combination of spices used throughout West Africa, generally is made up of ginger, grains of paradise, chile pepper, cubeb, black and white pepper, and allspice.

CURIOSITIES OF SCIENCE AND TECHNOLOGY

- The Mayans used allspice as an embalming agent to preserve their leaders.
- The Arawak Indians of the Caribbean used allspice to cure and preserve meats. During colonial times, Caribbean pirates used allspice to marinate meat, probably to disguise bad flavors; this meat was called buccon or *boucan*, the term the Indians used, and the pirates came to be-called "boucaniers," and eventually "buccaneers."
- Nineteenth century Russian soldiers put allspice in their boots to keep their feet warm.
- In Jamaica, allspice berries have been smoked as a substitute for tobacco, in long pipes.
- About 14,000 berries are required to make a kilogram of allspice (about 5700 berries for a pound).
- At the end of the nineteenth century, umbrella poles made out of allspice saplings became extremely popular in both Britain and the United States. This led to wanton cutting of young trees in Jamaica, and endangered the Jamaican allspice industry, until legislation was passed to prevent further harvesting of the wood.
- In Jamaica, allspice trees are commonly established by seeds dropped by birds. By thinning out trees along roadsides at intervals of about 8 m (25 ft.), picturesque "pimento walks" were commonly established.
- How to distinguish a dried allspice berry from a peppercorn (only necessary if you have a cold and cannot smell the difference, or if the allspice is very old and has lost its aroma): cut

the berry in half—allspice has two small seeds, whereas a peppercorn has one large central seed. If the allspice berry has been properly dried, just shake it near your ear and you can hear the two seeds rattling inside.

KEY INFORMATION SOURCES

Al-Rehaily, A.J., Al-Said, M.S., Al-Yahya, M.A., Mossa, J.S., and Rafatullah, S. 2002. Ethnopharmacological studies on allspice (*Pimenta dioica*) in laboratory animals. *Pharm. Biol.* 40:200–205.

Badura, M. 2003. *Pimenta officinalis* Lindl. (pimento, myrtle pepper) from early modern latrines in Gdansk (northern Poland). *Veget. Hist. Archaeobot.* 12:249–252.

Centro Interamericano de Documentacion e Informacion Agricola, 1977. *Bibliography on allspice (Pimenta dioica) (1957–1973).* Turrialba, Costa Rica. 3 pp (in Spanish).

Gayle, J.R. 1976. Pimento, *Pimenta dioica* (L) Merrill syn. *P. officinalis* Lindley. *Bull. New Ser. Minist. Agric. Fish. (Jamaica)*, 64:244–252.

Green, C.L., and Espinosa, F. 1988. Jamaican and Central American pimento (allspice; *Pimenta dioica*): characterization of flavour differences and other distinguishing features. *Dev. Food Sci.* 18:3–20.

Hamilton, M. 1925. *Allspice.* Alex Nicoll Printing Co., San Francisco, CA. 57 pp.

Henry, D.D., and Gayle, J.R. 1981. *The culture of grafted pimento. A spice crop for hilly lands of Jamaica.* Publicacion Miscelanea No. 273, IICA, Kingston, Jamaica. 31 pp.

International Organization for Standardization, Geneva (Switzerland). 1980. *Spices and condiments; pimento (allspice), whole or ground; specification.* Developed by Technical Committee ISO/TC 34: Agricultural Food Products. ISO International Standard No. 973. 4 pp.

Macia, M.J. 1998. The allspice (*Pimenta dioica* (L.) Merrill, Myrtaceae) in the Sierra Norte de Puebla (Mexico). *Anal. Jard. Bot. (Madrid)*, 56:337–349 (in Spanish).

Nabney, J., and Robinson, F.V. 1972. Constituents of pimento berry oil (*Pimenta dioica*). *Flavour Ind.* 3(1):50–51.

Peter, K.V., and Kandiannan, K.K. 1999. Allspice. In *Tropical horticulture*, vol. 1. Edited by T.K. Bose, S.K. Mitra, A.A. Farooqui, and M.K. Sadhu. Naya Prokash, Calcutta, India. pp. 702–705.

Pino, J., Rosado, A., and Gonzalez, A. 1989. Analysis of the essential oil of pimento berry *Pimenta dioica. Nahrung*. 33:717–720.

Rodriguez, M., Garcia, D., Garcia, M., Pino, J., and Hernandez, L. 1996. Antimicrobial activity of *Pimenta dioica. Alimentaria*, 34:107–110 (in Spanish).

Sulistiarini, D. 1999. *Pimenta dioica* (L.) Merrill. In *Plant resources of South-East Asia, 13, spices.* Edited by C.C. de Guzman and J.S. Siemonsma. Backhuys, Leiden, the Netherlands. pp. 176–180.

Specialty Cookbooks

Also see Akee for Jamaican cookbooks (akee and allspice are Jamaican culinary specialties).

MacLauchlan, A. 1997. *Tropical desserts: recipes for exotic fruits, nuts, and spices.* MacMillan, New York. 164 pp.

7 Arrowroot

Family: Marantaceae (arrowroot family)

NAMES

Scientific Name: *Maranta arundinacea* L.

- The English word "arrowroot" was first recorded in 1696. It was based on the Arawak word *aru-aru*, meaning "meal of meals." The Arawak people (Aruac Indians), who lived on the Caribbean islands, gave it this name because they thought very highly of the starchy, nutritious meal made from the underground part of the plant. (The Arawak still live in remote areas of Guiana, a region of mainland South America due north of Brazil.) Coincidentally, the tubers were (allegedly) used to draw poison from wounds inflicted by poison arrows. The English word "arrowroot" was based on this medicinal application, combined with the similarity of *aru-aru* to "arrow." A much less common interpretation of the origin of the word arrowroot is that it comes from a native South American word, *araruta*, meaning "root flour."
- Arrowroot is also known as Bermuda arrowroot, St. Vincent arrowroot, true arrowroot, West Indian arrowroot, and West Indian reed.
- As an ornamental, arrowroot is sometimes called obedience plant, easily confused with another garden plant, *Physostegia virginiana* (L.) Benth., known as obedience and obedient plant.
- The word "arrowroot" can mean either the starch obtained from the rhizomes of the arrowroot plant (*M. arundinacea*) or the plant itself. Although *M. arundinacea* is the main species used to obtain arrowroot starch, certain other quite unrelated species have "arrowroot" in their names and also yield edible starch called "arrowroot." "East India arrowroot" or "East Indian arrowroot" is from *Curcuma angustifolia* Roxb. (the name "East Indian arrowroot" is also applied to *Tacca pinnatifida* J.R. Forst. & G. Forst. and other species of *Tacca*, as noted below). "African arrowroot," "purple arrowroot," "Queensland arrowroot," "Sierra Leone arrowroot," and "tulema (or toleman) arrowroot" are from *Canna indica* L. (*C. edulis* Ker Gawl.). "Brazilian arrowroot" is from *Manihot esculenta* Crantz. "East Indian arrowroot," "Fiji arrowroot," "Hawaii arrowroot," "Indian arrowroot," "Polynesian arrowroot," "South Sea arrowroot," and "Tahiti arrowroot" are from *Tacca leontopetaloides* (L.) Kuntze (*T. involucrata* Schumach. & Thonn.). "Indian arrowroot" has also been used for zedoary. "Florida arrowroot" is *Zamia pumila* L. (*Z. floridana* A. DC.; see Chapter 32). "Oswego arrowroot" is an old name for corn flour (corresponding to the British "corn starch"). "Japanese arrowroot" is *Pueraria montana* (Lour.) Merr. var. *lobata* (better known as kudzu). There are also other species that have "arrowroot" in their names that are not used to produce arrowroot starch.
- "Portland arrowroot" (named after the Isle of Portland in England, which is not actually an "isle" because it is joined to the mainland by a narrow strip of land) is the Eurasian species *Arum maculatum* L., better known as cuckoopint, lords-and-ladies, and Adam-and-Eve. A starch was once obtained from its root and sold as edible Portland sago and Portland arrowroot. This starch was also used for stiffening ruffs in Elizabethan times (hence the old name starchwort, i.e., starch plant). The plant is quite poisonous and it is difficult to prepare edible starch from it.

- The banded arrowroot (*Maranta leuconera* E. Morren) is more familiarly known as the prayer plant because in darkness its leaves fold up as if in prayer (an adaptation for conserving moisture). Because of this religious association, this common house plant is also sometimes called "ten commandments." Although the plant is not toxic, it is not edible like its relative, arrowroot.
- The genus name *Maranta* commemorates Bartolo(m)meo Maranto (about 1500–1571), an Italian (Venetian) physician and botanist.
- *Arundinacea* in the scientific name *M. arundinacea* is Latin for reed like, descriptive of the flowering stem.

PLANT PORTRAIT

The true arrowroot is a tropical American perennial herb, growing to 2 m (6½ ft.) in height. Flat, long, large, pointed leaves are attached to the upright stems, giving the plants the appearance of an ornamental canna. Arrowroot starch is extracted from the rhizomes (underground stems). The species is indigenous to the West Indies and is grown in tropical countries, especially on the island of St. Vincent, as well as other areas of the Caribbean, Brazil, Thailand, and West Africa. Occasionally arrowroot is grown as an ornamental, and variegated garden forms are available.

CULINARY PORTRAIT

Arrowroot is one of the three classical thickeners (along with corn starch and flour). Arrowroot starch is a white, bland-tasting powder that looks and feels like cornstarch but is preferable because of its very neutral flavor (some say it has no flavor) that does not result in a "starchy taste." It is typically used as a thickening agent for soups, stews, sauces, and puddings. Arrowroot is considered to be the most easily digested of all starches, and its chief value is as a nourishing diet for convalescents

FIGURE 7.1 Arrowroot (*Maranta arundinacea*). (From Köhler, 1883–1914.)

and invalids, especially those with intestinal problems. It is also often recommended for people with allergies to cereals and for infants. Arrowroot cookies are commonly the first cookies or "biscuits" (the British equivalent of the American "cookies") consumed by young children. Sauces made from arrowroot starch produce a beautiful glossy appearance and have long been used to make clear glazes for fruit pies.

Arrowroot mixtures thicken at a lower temperature than mixes made with flour or cornstarch. Unlike cornstarch, arrowroot does not produce a chalky taste when undercooked. Arrowroot is preferably used at the very end of cooking because unlike other thickeners (such as corn starch or tapioca), it can break down and it does not thicken up again if reheated. It should be mixed with cool liquids before adding hot liquids, cooked until the mixture thickens, and removed immediately from the heat to prevent the mixture from thinning. Arrowroot becomes clear when cooked. Two parts of arrowroot can be substituted for three parts of cornstarch. One part of arrowroot can be substituted for three parts of flour. About 2.5 teaspoons of arrowroot powder per cup of liquid is often recommended. Arrowroot is useful for preparing very delicate sauces, especially dairy or egg-based sauces that might curdle if cooked at too high a temperature. Arrowroot can also be used to prevent ice crystals from forming in homemade ice cream.

Arrowroot and tapioca starch are very similar, and so they are often confused. In fact, much of the commercial "arrowroot starch" available is actually tapioca starch, obtained from cassava (*Manihot esculenta* Crantz). Cassava starch is sometimes called "Brazilian arrowroot."

The rhizome of the arrowroot plant can be cooked and eaten as a starchy vegetable, and this is done sometimes where it is grown.

Culinary Vocabulary

- "Arrowroot vermicelli" refers to slender white Asian noodles made from arrowroot starch.
- "Haupia" is a Hawaiian pudding prepared with arrowroot, coconut milk, and sugar. The preparation is chilled until firm, sliced into squares, and frequently served at a luau.
- "Windsor soup" is a British beef consommé made with sherry, basil, thyme, marjoram, and strips of calf's foot, and slightly thickened with arrowroot.

CURIOSITIES OF SCIENCE AND TECHNOLOGY

- In South America and the Caribbean, arrowroot has been used externally as a poultice for smallpox sores and as an antiseptic tea for urinary infections.
- Arrowroot is not used for flavor, but nevertheless it is typically found in the spice, condiment, and ground herbs section of supermarkets, and brands of arrowroot are often sold by spice companies. The reason for associating arrowroot with spices is that, like spices, a little is generally all that is required, and small packages tend to generate large profits, which is the sort of product that traditionally has interested spice companies.
- Celiac disease is a condition where the proteins from the closely related cereals wheat, rye, barley, and triticale damage the small intestines. A "gluten-free" diet is one that is free of any proteins from these cereals. Not all cereals are harmful to people with celiac disease—for example rice and corn are apparently harmless. Arrowroot is considered to be gluten-free and is often part of the diet of celiac patients.

KEY INFORMATION SOURCES

Andam, C.J. 2001. Culture and processing of arrowroot [*Maranta arundinacea* L.]. *Agriculture* (*Philippines*), 5(5):28–29.

Andersson, L. 1986. Revision of *Maranta* subgen. *Maranta* (Marantaceae). *Nordic J. Bot.* 6:729–756.

Bolt, A. 1962. Monopoly island arrowroot. *World Crops*, 14:386–388.

Carandang, D.A. 1986. *Prospects, problems and fertilizer requirement of arrowroot* (Maranta arundinacea *L.*) *as coconut intercrop*. University of the Philippines, Los Banos, Laguna, Philippines. 16 pp.

Cooke, C., Carr, I., Abrams, K., and Mayberry, J. 2000. Arrowroot as a treatment for diarrhoea in irritable bowel syndrome patients: a pilot study. *Arq. Gastroenterol*. 37:20–24.

Erdman, M.D. 1986. Starch from arrowroot (*Maranta arundinacea*) grown at Tifton, Georgia. *Cereal Chem*. 63:277–279.

Erdman, M.D., and Erdman, B.A. 1984. Arrowroot (*Maranta arundinacea*), food, feed, fuel, and fiber resource. *Econ. Bot*. 38:332–341.

Erdman, M.D., Phatak, S.C., and Hall, H.S. 1985. Potential for production of arrowroot *Maranta arundinacea* in the southern USA. *J. Am. Soc. Hort. Sci*. 110:403–406.

Forio, A.F., and Villamayor, F.G., Jr. 1988. Exploiting the yield potential of arrowroot. *Radix*, 10(1):14–16.

Kay, D.E. 1973. Arrowroot (*Maranta arundinacea*). In *Root crops*. Tropical Products Institute, London. pp. 16–23.

Laguna College of Agriculture. 1987. Culture and processing [*Maranta arundinacea*]. Laguna College of Agriculture, Laguna, Philippines. 11 pp.

Maheswarappa, H.P., and Nanjappa, H.V. 2000. Dry matter production and accumulation in arrowroot (*Maranta arundinacea* L.) when intercropped with coconut. *Trop. Agric*. 77:76–82.

Maheswarappa, H.P., Nanjappa, H.V., Hegde, M.R., and Biddappa, C.C. 2000. Nutrient content and uptake by arrowroot (*Maranta arundinacea*) as influenced by agronomic practices when grown as intercrop in coconut (*Cocos nucifera*) garden. *Indian J. Agron*. 45:86–91.

Radley, J.A. (Ed.). 1976. *Starch production technology*. Applied Science Pub. Ltd., London. 1976. 587 pp.

Raja, M.K.C., and Sindhu, P. 2000. Properties of steam-treated arrowroot (*Maranta arundinacea*) starch. *Starke*, 52:471–476.

Raymond, W.D., and Squires, J. 1959. Sources of starch in colonial territories. II. Arrowroot (*Maranta arundinacea* Linn.). *Trop. Sci*. 1:182–192.

Royal Botanic Gardens Kew. 1893. St. Vincent arrowroot. *Kew Bull*. 1893:191–204.

Villamayor, F.G., Jr., and Jukema, J. 1996. *Maranta arundinacea* L. In *Plant resources of South-East Asia 9. Plants yielding non-seed carbohydrates*. Edited by M. Flach and F. Rumawas. Backhuys Publishers, Leiden, the Netherlands. pp. 113–116.

Specialty Cookbooks

Note: Special-diet cookbooks for invalids, children, and those with allergies (especially to gluten) are excellent sources of arrowroot recipes.

Benson, R.A. 1909. *A nursery manual: the care and feeding of children in health and disease*. Boericke & Tafel, Philadelphia, PA. 190 pp.

Boland, M.A. 1893. *A handbook of invalid cooking*. The Century Company, New York. 338 pp.

Harris, D.A. 1988. *Island cooking: recipes from the Caribbean*. Revised edition. Crossing Press, Freedom, CA.162 pp.

8 Asafetida

Family: Apiaceae (Umbelliferae; carrot family)

NAMES

Scientific Names: *Ferula* species

- Asafetida—*F. assa-foetida* L.
- Asafetida—*F. foetida* (Bunge) Regel
- Asafetida—*F. narthex* Boiss.

- The name asafetida is based on the Latin *asa*, gum or resin (derived from the Farsi or Persian *aza*) + the Latin *fetida*, stinking.
- Pronunciation: ah-sah-FEH-teh-dah.
- Asafetida is also spelled asafoetida and assa-foetida and has also been called devil's dung, food of the gods, and stinking gum.
- *F. narthex* is sometimes called Chinese hing. *Narthex* is Greek for umbelliferous plant, that is, a plant of the carrot family.
- The genus name *Ferula* is the classical Latin name for the giant fennel, *Ferula communis*. A "ferula" is a whip or rod that was used to punish schoolboys and slaves, a usage not connected with the genus *Ferula*.

PLANT PORTRAIT

Ferula species, of which there are over 100, are perennial, thick-rooted herbs, native from the Mediterranean region to central Asia. Several of the Asian species are known as asafetida, and this name is also applied to a gum used as a spice and medicine that is obtained from the plants. To produce asafetida, the massive fleshy taproots of mature but nonflowering plants (4 or 5 years of age) are partly uncovered and then scored, which eventually results in the release of a milky resinous juice that dries slowly on the root. This is scraped off and collected. The brown, resin-like material is asafetida, infamous for its strong, repugnant smell which has been compared with old garlic. Asafetida was known in Babylonia at least by 750 BC, and in classical Greek and Roman times. It was a popular spice in Europe from Roman times to the Middle Ages, but not after that. Asafetida is not well known in Western countries. It is still used in Asia, particularly in India. Most asafetida is produced in Iran and Afghanistan, and it is also cultivated in India. Asafetida is available commercially in several forms: "tears," "mass" (or "lump"), "paste," and "powdered." Tears are the purest form and vary in size from 5 to 30 mm (0.2–1.2 in.). Lump asafetida is the most common commercial form. Powdered asafetida often has had gum arabic, turmeric, or flour added, to diminish the odor, prevent lumping, and/or add color.

CULINARY PORTRAIT

Asafetida is either loved or loathed, as reflected by the names food of the gods and devil's dung. Fresh asafetida has a horrible smell, although, more charitably, tolerance to its odor is said to be acquired. Nevertheless, in small amounts (less than pea size), and fried in hot oil (which is strongly recommended), the resin becomes rather pleasant and oniony in taste. The use of "powdered asafetida" dilutes the taste, makes it easier to avoid adding more than is desirable, and eliminates the need for

FIGURE 8.1 Asafetida (*Ferula assa-foetida*). (From Köhler, 1883–1914.)

FIGURE 8.2 Asafetida (*Ferula foetida*). (From Aitchison, 1888.)

FIGURE 8.3 Root of Asafetida (*Ferula narthex*). (From Engler and Prantl, 1889–1915.)

frying. One-sixteenth of a teaspoon is sufficient for a meal for one person. The aroma of powdered asafetida lasts for years, while that of the resin seems to last for decades. Asafetida is generally used as an alternative to onion and garlic. Carefully used in moderation, asafetida enhances vegetable and mushroom dishes and fried and barbecued meats and fish. In the Middle Ages, it was traditionally used to spice barbecued mutton. Today, asafetida is still a popular spice in India and Persian dishes. In India, it is most commonly used with vegetables and in curries and pickles. In Asia, the whole plant, particularly the inner portion of the full-grown stem, is sometimes used as a fresh vegetable.

Culinary Vocabulary

- *Sambaar podi* is a Tamil (South Indian) spice mixture which frequently contains asafetida.

CURIOSITIES OF SCIENCE AND TECHNOLOGY

- According to ancient Greek mythology, Prometheus stole fire from Heaven and brought it down to Earth hidden in a stalk of "fennel," actually *F. communis* (the name fennel is applied to *Foeniculum vulgare* Mill., used both as a vegetable and spice). This myth is partly based on technology that is thousands of years old. *Ferula communis* has a hollow stem occupied by pith which, when dry, can be set on fire, so that the stem burns internally very slowly, and this allowed people to bring fire from place to place.
- In ancient Rome, asafetida was stored in jars together with pine nuts. The nuts absorbed the flavor of the asafetida and were powdered and used as a spice.
- The strong smells of both garlic and asafetida have a common basis in that sulfur compounds are responsible.
- Asafetida has been used as a repellent against dogs, cats, and wildlife. However, when placed in water, it is said to attract fish.
- In India, housewives often include a lump of asafetida with their stored spices to repel insects.
- Asafetida has been widely used in medicine (e.g., it was once used to treat epilepsy) and has long been important in world trade. It has acquired the reputation as the world's most

adulterated material (although this dubious honor probably belongs to the much more expensive saffron). Gypsum, clay, sand, and stones have generally been added to increase the weight and cost.

- In India, asafetida has a considerable reputation for reducing the flatulence associated with beans and cauliflower. Asafetida has in fact been used medicinally to reduce intestinal gas.
- Remarkably, asafetida is used in perfumes, the fetid odor manipulated to appeal to rather than repel the sense of smell.

KEY INFORMATION SOURCES

Abd-El-Razek, M.H., Ohta, S., Ahmed, A.A., and Hirata, T. 2001. Sesquiterpene coumarins from the roots of *Ferula assa-foetida*. *Phytochemistry*, 58:1289–1295.

Abraham, K.O., Shankaranarayana, M.L., Raghavan, B., and Natarajan, C.P. 1979. Asafetida. IV. Studies on volatile oil [of] *Ferula*. *Indian Food Packer*, 33(1):29–32.

Abraham, K.O., Shankaranarayana, M.L., Raghavan, B., and Natarajan, C.P. 1982. Odorous compounds of asafetida. VII. Isolation and identification. *Indian Food Packer*, 36(5):65–76.

Ashraf, M., and Bhatty, M.K. 1979. Studies on the essential oils of the Pakistani species of the family Umbelliferae. XXII. *Ferula foetida* Regel (ushi) seed oil. *Pak. J. Sci. Ind. Res.* 22:84–86.

Ashraf, M., Ahmad, R., Mahmood, S., and Bhatty, M.K. 1979. Studies on the essential oils of the Pakistani species of the family Umbelliferae. XXXV. *Ferula asafoetida* Linn. (Hing) seed oil. *Pak. J. Sci. Ind. Res.* 22:308–310.

Ashraf, M., Ahmad, R., Mahmood, S., and Bhatty, M.K. 1980. Studies on the essential oils of the Pakistani species of the family Umbelliferae. XLV. *Ferula asafoetida* Linn. (Herra Hing) gum oil. *Pak. J. Sci. Ind. Res.* 23:68–69.

Balasubrahmanyam, N., Abraham, K.O., Raghavan, B., Shankaranarayana, M.L., and Natarajan, C.P. 1979. Asafetida. VI. Packaging and storage studies on asafetida products. *Indian Food Packer*, 33(3):15–22.

Bhalla, K., and Punekar, B.D. 1975. Incidence and state of adulteration of commonly consumed spices in Bombay city. II. Mustard, black pepper, and asafoetida. *Indian J. Nutr. Diet.*12:216–222.

Carrubba, R.W. 1979. The first report of the harvesting of asafetida in Iran. *Agric. Hist.* 53:451–461.

Chamberlain, D.F. 1977. The identity of *Ferula assa-foetida* L. *Edinb. J. Bot.* 35:229–233.

Eigner, D., and Scholz, D. 1999. *Ferula assa-foetida* and *Curcuma longa* in traditional medical treatment and diet in Nepal. *J. Ethnopharm.* 67:1–6.

Fatehi, M., Farifteh, F., and Fatehi-Hassanabad, Z. 2004. Antispasmodic and hypotensive effects of *Ferula asa-foetida* gum extract. *J. Ethnopharm.* 91:321–324.

Kajimoto, T., Yahiro, K., and Nohara, T. 1989. Sesquiterpenoid and disulphide derivatives from *Ferula assa-foetida*. *Phytochemistry*, 28:1761–1763.

Lehrnbecher, P. 1995. *Angelica and asafetida*. P. Lang, Frankfurt am Main, Germany. 246 pp (in German).

Raghavan, B., Abraham, K.O., Shankaranarayana, M.L., and Natarajan, C.P. 1979. Asafetida. V. Asafetida products and flavour losses during cooking and frying. *Indian Food Packer*, 33(3):11–14.

Raghavan, B., Abraham, K.O., Shankaranarayana, M.L., Sastry, L.V.L., and Natarajan, C.P. 1974. Asafoetida. II. Chemical composition and physicochemical properties. *Flavour Ind.* 5:179–181.

Rajanikanth, B., Ravindranath, B., and Shankaranarayana, M.L. 1984. Volatile polysulphides of asafoetida. *Phytochemistry*, 23:899–900.

Samimi, M.N., and Unger, W. 1979. The gum resins of asafetida producing *Ferula* species from Afghanistan provenance and quality of Afghan asafetida. *Planta Med.* 36(2):128–133 (in German, English summary).

Sefidkon, F., Askari, F., and Mirza, M. 1998. Essential oil composition of *Ferula assa-foetida* L. from Iran. *J. Essent. Oil Res.* 10:687–689.

Shivashankar, S., Shankaranarayana, M.L., and Natarajan, C.P. 1972. Asafoetida—varieties, chemical composition, standards and uses. *Indian Food Packer*, 26(2):36–44.

Syed, M., Hanif, M., Chaudhary, F.M., and Bhatty, M.K. 1987. Antimicrobial activity of the essential oils of Umbelliferae family part IV. *Ferula narthex*, *Ferula ovina* and *Ferula oopoda*. *Pak. J. Sci. Ind. Res.* 30:19–23.

Specialty Cookbooks

Note: Cookbooks specialized on foods of India are the best sources of recipes using asafetida. Some examples follow.

Batra, N. 2002. *1,000 Indian recipes*. Wiley, New York. 704 pp.

Betty Crocker Editors. 2001. *Betty Crocker's Indian home cooking*. Hungry Minds Inc., New York. 336 pp.

Bhide, M. 2009. *Modern spice: inspired Indian flavours for the contemporary kitchen*. Simon & Schuster, New York. 288 pp.

Devi, Y., and Baird, D. (illustrator). 1987. *The art of Indian vegetarian cooking: Lord Krishna's cuisine*. Bala Books, New York. 799 pp.

Jaffrey, M. 2002. *Madhur Jaffrey's world vegetarian*. Clarkson Potter, New York. 758 pp.

Jaffrey, M. 2007. *Madhur Jaffrey's quick and easy Indian cooking*. Chronicle Books, San Francisco, CA. 155 pp.

Iyer, R. 2008. *660 curries*. Workman Publishing Company, New York. 809 pp.

9 Bamboo

Family: Poaceae (Gramineae; grass family)

NAMES

Scientific Names: *Psyllostachys* species, *Bambusa* species, + others.

- "Bamboo," the genus name *Bambusa*, and the grass tribe name Bambuseae are based on *bambu*, the Malay name for bamboo. The word in Malay means "explosion." Natives used to clean foliage out of bamboo groves with fire before harvesting the stems, with the result that the hollow chambers in the bamboo stems exploded from the heat.
- The genus name *Psyllostachys* is based on the name of psyllium (*Plantago psyllium* L.) + *stachys*, spike, referring to the similarity of the flowering head. The plant psyllium provides a bulk-forming laxative with the same name, which is high in fiber and mucilage.

PLANT PORTRAIT

Bamboos are mostly big, long-lived, woody, evergreen grasses. There are more than 1200 bamboo species distributed around the world, classified into 70 genera of the grass tribe Bambuseae. Some are dwarves growing only to 15 cm (6 in.) in height, whereas others are giants rising to 37 m (120 ft.) and developing stalks 20 cm (8 in.) in diameter. Bamboos are native to Africa, Australia, and the Americas, but the largest number of species is found in Asia, with close to 300 originating in China. The flowers of most plants are quite prominent and useful in identification, but most bamboos do not flower often. The most noticeable characteristic of bamboo is the division of the stem into distinct nodes or joints with intermediate smooth sectors called internodes. The internodes are mostly hollow, whereas the nodal junctions are solid. In many parts of the world, bamboos are not only used for food but also for fodder, for construction material, and for making a great variety of useful objects from kitchen tools to paper to dinnerware. In China, 7 million hectares (17 million acres) are devoted to growing bamboo, in both natural forests and plantations.

Bamboo shoots, like asparagus, are thick, pointed shoots that emerge out of the ground and if left unharvested, develop into stems. As with asparagus and several other vegetables, earth is sometimes hilled up around the shoots to blanch them, that is, to produce whiter, more succulent material. Shoots of most bamboo species are edible, but bamboo shoots are collected primarily from several species, particularly of the genera *Psyllostachys* and *Bambusa*. Young stems are harvested when about 15 cm (6 in.) in height.

CULINARY PORTRAIT

Bamboo shoots have long been popular in Asian cooking and are often found as a sliced, crunchy, creamy white vegetable in Asian food. Spring shoots are larger and tougher than winter shoots. The winter type is lighter in color, softer, tastier, and more expensive. Fresh bamboo shoots are generally not available in North America, but canned shoots are easy to obtain. Whole shoots in cans are preferred but are sometimes also available sliced. Canned bamboo in water rather than brine is also preferable. Some cooks simmer canned shoots in broth or water for 1 to 2 minutes to reduce the canned taste. Sun-dried bamboo shoots are also sometimes available. In China, bamboo shoots are typically added to dishes such as soups or stews. In Japan, bamboo shoots, known as *takenoko*

FIGURE 9.1 Bamboo scene. (From Baillon, 1876–1892.)

(literally "child of bamboo"), are commonly consumed as featured dishes or parts of dishes—grilled, boiled in soups and stews, steamed with rice, and deep fried in tempura.

Bamboo shoots are usually cooked in salted water before eating to remove bitter, poisonous chemicals (cyanogenic glycosides). The taste of bamboo shoots has often been compared with asparagus, and in fact most of the methods appropriate for preparing asparagus are also suitable for bamboo shoots. The majority of bamboos native to temperate regions can be eaten without cooking if they are not too bitter. *Phyllostachys edulis* (Carrière) J. Houz. (*P. pubescens* Mazel ex J. Houz., known as "*moso* in China and Japan) has potentially toxic concentrations of cyanogens. It is the most important temperate region bamboo harvested for shoots, is usually somewhat bitter, and is always cooked before eating. Eating large amounts of raw shoots at one time is inadvisable unless one is sure of their safety. Generally, travelers in tropical areas should not experiment with eating unidentified bamboo shoots that they have personally collected because they may be poisonous or contain a hard outer coat or irritant hairs that must be removed.

In addition to the recently emerged asparagus-like stems (i.e., shoots), the more mature stems and leaves of some bamboos are eaten locally where the species grow. The heart or pith (internal soft tissue) of the bamboo stem can be sweet, and bamboo sugar may accumulate as a residue in the joints of some species. In the East Indies, this bamboo sugar is extracted as *tabasheer* and is considered to be medicinal. In China, beer and liquor are manufactured from sweet bamboo juice. Bamboo fruit is often pear shaped, reminiscent of a corn cob, and is considered to be a delicacy in Japan.

Bamboo leaves are used to wrap food for steaming as well as to line steamer baskets. In Chinese cuisine, they are often used to enfold a filling of glutinous rice to make dim sum (Chinese snack food).

FIGURE 9.2 Bamboos in Java. (From Marilaun, 1895.)

Bamboo seeds are sometimes crushed to produce "bamboo flour," used as a thickener in Chinese cuisine.

Culinary Vocabulary

- In Southeast Asia and India, pickled bamboo sprouts are frequently used to prepare *achar* (*achard*), a popular hot condiment or pickled salad that is variously concocted.
- Dried sticks made of soybean curd are generally simply called "bean curd" or "dried (bean curd) sticks" but are sometimes called "bamboo" (and "second bamboo") because of the resemblance to bamboo in texture. These are found in Chinese and Japanese markets and are soaked before becoming pliable enough to cook.
- A "bamboo cocktail" is a beverage made with orange bitters, sherry, and dry vermouth, stirred with ice, strained, and served in a wine glass with a lemon peel. Bartender Charlie Mahoney of the Hoffman House in New York is said to have invented it about 1910. "Bamboo juice" is slang for alcoholic beverages, a phrase that originated with the U.S. Air Force while stationed in the South Pacific during the Korean War.
- "Bamboo salt" is a salty substance extracted from bamboo stems, used mostly in China for medicinal purposes, but occasionally used as a condiment.
- The so-called "bamboo tea" is tea (typically a strong, bitter Chinese black tea) packaged in serving portions (often two portions for a pot) in bamboo leaves.

CURIOSITIES OF SCIENCE AND TECHNOLOGY

- In old China, a "bamboo wife" was a cylinder of woven basketwork upon which a sleeping person sprawled. Cool air circulating through it made sleep more comfortable.
- The world's tallest grass, sometimes growing to a height exceeding 40 m (130 ft.), is *Dendrocalamus brandisii* (Munro) Kurz, which produces huge, delicious bamboo shoots.
- Bamboo grows quite differently from trees. A tree has a layer of living tissue around the outside of its trunk just beneath the bark, which adds an ever increasing circle of wood around the central mass (the concentric annual rings in transversely cut timber). In contrast, bamboo stems emerge from the ground as buds with the same diameter as the final stem. All they do is grow taller. The thickest bamboo stems developed are only about 30 cm (1 ft.) in diameter.
- Some bamboos can grow in excess of a meter (about a yard) a day.
- The giant panda in the wild survives by eating bamboo species, particularly *Gelidocalamus fangianus* (A. Camus) P.C. Keng & Wen. Pandas spend about 12 hours a day eating bamboo, consuming 30 kg (66 lb.) daily, equal to about one-third of their body weight.
- Giant pandas lack an efficient digestive tract, absorbing less than 20% of the food energy from their diet of bamboo. A single adult male can produce 20 kg (44 lb.) of dung in a day. The 40 or so pandas that live in the Chengdu Giant Panda Breeding Base in Sichuan province, southwest China, produce about 200 tonnes of excrement a year. To lessen the annual maintenance cost of millions of dollars, an effort was initiated in 2007 to turn the fiber-rich waste into a range of products, including greeting cards, bookmarks, notebooks, and even refrigerator magnets.
- Some bamboo species flower simultaneously after several decades of growth, with all of the plants in a region blooming at the same time and then dying. This can be disastrous, both for animals that depend on the bamboos for food and habitat and for humans who rely on bamboo for wood. There is therefore some validity to the common belief throughout Southeast Asia and the East Indies that when bamboo does flower, famine approaches.

FIGURE 9.3 Reedlike bamboo (*Bambusa arundinacea*). (From Nicholson, 1885–1889.)

- Kendo is a Japanese form of fencing using bamboo foils or wooden swords.
- The first light bulb filament, made in 1882 by Thomas Edison, (1847–1931), was constructed of carbonized bamboo fiber and is still functional in the Smithsonian Institution in Washington, DC.

KEY INFORMATION SOURCES

Austin, R., and Ueda, K. 1970. *Bamboo.* Walker/Weatherhill, New York. 215 pp.

Bell, M. 2000. *The gardener's guide to growing temperate bamboos.* Timber Press, Portland, OR. 160 pp.

Chapman, G.P. (Ed.). 1997. *The bamboos.* Academic Press, New York. 312 pp.

Crouzet, Y., and Starosta, P. 1998. *Bamboos.* Evergreen, Köln, Germany. 121 pp.

Dransfield, S. and Widjaja, E.A. (Eds.). 1995. *Plant resources of South-East Asia 7. Bamboos.* Backhuys Publishers, Leiden, the Netherlands. 189 pp.

Food and Agriculture Organization and World Health Organization. 1991. *Bamboo shoots and the need for standardization.* Food and Agriculture Organization, Rome, Italy. 10 pp.

Judziewicz, E.J., Clark, L.G., Londono, X., and Stern, M.J. (Eds.). 1999. *American bamboos.* Smithsonian Institution Press, Washington, DC. 400 pp.

Kleinhenz, V., Gosbee, M., Elsmore, S., Lyall, T.W., Blackburn, K., Harrower, K., and Midmore, D.J. 2000. Storage methods for extending shelf life of fresh, edible bamboo shoots. *Postharvest Biol. Technol.* 19:253–264.

Kumar, A., Ramanuja Rao, I.V., and Sastry, C. (Eds.). 2002. *Proceedings of the 5th International Bamboo Congress and the 6th International Bamboo Workshop, Bamboo for Sustainable Development, San José, Costa Rica, 2–6 Nov. 1998.* VSP, Boston, MA. 930 pp.

Lawson, A.H. 1968. *Bamboos: a gardener's guide to their cultivation in temperature climates.* Faber, London. 192 pp.

Liese, W. 1985. *Bamboos: biology, silvics, properties, utilization.* Deutsche Gesellschaft für Technische Zusammenarbeit (GTZ) GmbH, Eschborn, Germany. 132 pp.

Liese, W. 1998. *The anatomy of bamboo culms.* International Network for Bamboo and Rattan, Beijing. 208 pp.

Maruo, K. 1975. *Culture of bamboo shoots. Agric. Hortic.* 50(1):233–237 (in Japanese).

McClure, F.A. 1993. *The bamboos.* Smithsonian Institution Press, Washington, DC. 345 pp.

Meredith, T.J. 2001. *Bamboo for gardens.* Timber Press, Portland, OR. 406 pp.

Numata, M. (Ed.) 1979. *Ecology of grasslands and bamboolands in the world.* Junk BV, The Hague, the Netherlands. 299 pp.

Ohrnberger, D. 1999. *The bamboos of the world: annotated nomenclature and literature of the species and the higher and lower taxa.* Elsevier, New York. 585 pp.

Rao, A.N., and Ramanatha Rao, V. (Eds.). 1998. *Bamboo conservation, diversity, ecogeography, germplasm, resource utilization and taxonomy. Proceedings of the Training Course Cum Workshop 10–17 May 1998, Kunming and Xishuangbanna, Yunnan, China.* International Plant Genetic Resources Institute, Rome, Italy. 275 pp.

Recht, C., and Wetterwald, M.F. 1992. *Bamboos.* Timber Press, Portland, OR. 128 pp.

Roots, C. 1989. *The bamboo bears: the life and troubled times of the giant panda.* Hyperion Press, Winnipeg, MB. 100 pp.

Shou-Liang, C., Liang-Chi, C., Shou-Liang Chen, S.L., and Chia, Liang-Chi. 1988. Chinese bamboos. In *Biosystematics, floristic and phylogeny series*, vol. 4. Timber Press, Portland, OR. 120 pp.

Takisaya, O. 1976. Production and canning of bamboo shoots. *Canners J.* 53(4):41–46.

Tewari, D.N. 1992. *A monograph on bamboo.* International Book Distributors, Dehra Dun, India. 498 pp.

Villegas, M. 2001. *Tropical bamboo.* Villegas Editores, Bogota, Colombia. 176 pp.

Young, R.A., Haun, J.T., and McClure, F.A. 1961. *Bamboo in the United States: description, culture, and utilization.* U.S. Dept. of Agriculture, Crops Research Division, Agricultural Research Service, Washington, DC. 74 pp.

SPECIALTY COOKBOOKS

Dransfield, S. 1986. *Some bamboo shoot recipes.* Royal Botanic Gardens, Kew, UK. 4 pp.

Wang, H., Varma, R.V., and Xu, T. 2003. *Inspirations—recipes featuring bamboo shoots.* International Network for Bamboo and Rattan, Beijing. 83 pp.

10 Baobab

Family: Malvaceae (mallow family; until recently, placed in Bombacaceae, the bombax or kapok family)

NAMES

Scientific Name: *Adansonia digitata* L.

- The name "baobab" is possibly from North African Arabic *bibab* (*bu hibab*), many-seeded fruit.
- The African baobab is also known as dead rat tree, Ethiopian sour gourd, Judas's bag, lemonade tree, monkey-bread tree, monkey tamarind, Senegal calabash, and upside-down tree.
- The name "upside-down tree" points out the appearance of the tree in its leafless state, when the trunk has the form of a gigantic swollen tap root, and the branches look like roots directed into the sky rather than into the ground. David Livingstone (1813–1873), Scottish missionary and explorer of Africa, called the baobab a "carrot planted upside down."
- The name "monkey-bread tree" is based on the attractiveness of the fruits to monkeys and the use of the fruit pulp to make bread. Baobab is also sometimes called Ethiopian sour bread.
- The name "dead rat tree" is based on the baobab's long fuzzy fruits hanging on long stalks, giving the appearance of dead rats hanging by their tails.
- The name "lemonade tree" reflects the lemonade taste of beverages made from the pulp.
- The genus name *Adansonia* commemorates French naturalist, philosopher, and explorer Michel Adanson (1727–1806).
- *Digitata* in the scientific name *A. digitata* is Latin for digitate, a reference to the fingerlike arrangement of the leaflets on the leaves.

PLANT PORTRAIT

There are eight species of baobab trees (of the genus *Adansonia*), six native to Madagascar, one in Australia, and one in Africa, *A. digitata*, the largest and most spectacular, and the subject of this chapter. The African baobab is native to most of the countries of Africa south of the Sahara. Humans have extended its range throughout the dry areas of Africa. The baobab was apparently known in ancient Egypt, although the tree is not native to the country. Baobab trees are bizarre in appearance, with grotesquely swollen trunks. The tree is usually massive, with a barrel-like trunk that may reach a diameter of 9 m (30 ft.), or rarely 12 m (39 ft.). Height rarely exceeds 18 m (59 ft.), but a few trees are as tall as 25 m (82 ft.). The irregular crown is made up of short, twisted branches. The root system is shallow, an adaptation to quickly collecting occasional rainfall, and roots have been observed to extend as far as 46 m (50 yd.) from the tree. Baobabs generally produce leaves during the rainy season and shed their foliage during the dry season to reduce moisture loss. If continuously watered, they will hold on to their leaves. At the end of the dry season, the tree produces large white flowers 10 to 12.5 cm (4–5 in.) across, and these hang down on long stalks. The flowers open at night and are pollinated by nectar-feeding bats. The fruit of the baobab is large, gourd-like, hard, about 15 to 30 cm (6–12 in.) long, with a velvety skin, and like the flower, hangs down by a long stalk.

FIGURE 10.1 Baoab (*Adansonia digitata*). (From Figuier, 1867.)

In addition to the use of the fruit as food, the inner, fibrous bark of the tree is stripped from the trunk and traditionally used to make string, rope, fabric, netting, baskets, and brooms and more recently filters. The wood of the trees has been used by fishermen to construct rafts, canoes, and floats. The hard fruit rinds have been used as containers for liquids. The roots furnish a soluble red dye. The leaves and fruits as well as being eaten have been used for numerous medicinal purposes. The bark is also commonly used in medicines. The seeds are used in necklaces, sometimes considered to be protective against malevolent forces. The fruit pulp is rich in vitamin C and is widely used in Africa to combat scurvy-related diseases. Baobabs are interesting ornamentals and are grown as far north as Florida.

CULINARY PORTRAIT

The fruits contain a bread-like white, mucilaginous, edible, slightly sweet but somewhat bitter pulp and seeds. The fruit pulp is acidic and has been compared with that of grapefruit. It can be ground into flour, made into a refreshing drink, and used as a seasoning. The wood pulp of the tree is also sometimes eaten, and the ash has been used as a salt-like condiment. The seeds are roasted and eaten like peanuts, pounded into a sort of peanut butter, used as the basis of a relish, and used as a source of cooking oil. The young leaves can be consumed like spinach, made into soup, or dried and crushed for flavoring and for use as a coffee substitute. The leaves are a staple of populations in many parts of Africa, especially in the central region of the continent, and are a significant source of protein and minerals. The leaves are eaten as a laxative condiment in parts of West Africa, the laxative effect related to the high content of mucilage (approximately 12%). However, the mucilage content is considered nutritionally undesirable because it flushes essential food nutrients out of the digestive system before they can be absorbed, and there has been some research into methods of preparation of baobab leaves that minimize mucilage content. The seeds and seed oil similarly contain antinutritional substances, but procedures (such as fermentation of the seeds) are being examined to eliminate the problem. The food value of baobab to the indigenous inhabitants of Africa has been critical for purposes of survival, and today attempts are being made to develop more sophisticated food products from the tree.

Culinary Vocabulary

- Independently of the names "monkey-bread tree" and "monkey-bread" for the fruit of the baobab, there is a sweet yeast bread called monkey bread, which is formed from balls of dough laid next to one another, which combine during baking. It has been suggested that this bread's name is somehow related to the monkey puzzle tree (*Auraucaria araucana* (Molina) K. Koch), but more likely the name comes from the appearance of the baked bread, which resembles a pack of monkeys jumbled together. Nancy Reagan, wife of Ronald Reagan (40th U.S. president), often baked monkey bread in the White House at Christmas and claimed the name indicated that one had to monkey around with it to prepare it properly.
- Occasionally, the African baobab has been called the cream-of-tartar tree (a name usually applied to the Australian baobab). Cream of tartar (potassium bitartrate) is a brownish-red acidic powder used in baking, especially in baking powder (made by combining baking soda with cream of tartar). It is also used to give a creamier texture to sweet preparations such as candy and frosting and to stabilize and increase the volume of beaten egg whites. Cream of tartar is commercially obtained from the residue remaining when grape juice has fermented to wine. The white matrix of baobab fruits has traditionally also been used as a source of cream of tartar, hence the name cream-of-tartar tree.
- *Miyar kuka* is a soup composed mainly of baobab leaves. This is a staple of the Hausa ethnic group of central Africa.

CURIOSITIES OF SCIENCE AND TECHNOLOGY

- In Africa, the trunks of baobab trees are often excavated to store water reserves. Older trees tend to develop hollow trunks, and such trees are usually the ones that are chosen to be more thoroughly hollowed out. The hollow of a large tree can hold 9000 L (2400 American gallons). A bung at the bottom allowed water to be taken out. A chain of such trees across the Kalahari Desert functioned like oases, allowing travelers to move through extremely dry areas. Death was often the punishment for those who left the bung out of a tree so that the water was wasted.
- Toward the end of the nineteenth century, the British waged war in Sudan against the Dervish Empire of 1896–1899 (the Dervishes belonged to a Moslem sect, centered in Turkey). In 1896, the Dervish army of Emir Mahmud was forced to retreat from the Nile River. To prevent the British from having access to the water stored in the area's baobab trees, the retreating Dervish soldiers pierced the bases of the trees to allow the water to run out.
- The hollowed-out trunks of dead baobab trees have also been used as prisons, toilets, and even as pubs. One large baobab was used as a bus shelter capable of accommodating 40 people. In some parts of Africa, people believe that poets and musicians are possessed by the devil, and if given a normal burial, their bodies will pollute the earth. To prevent this, they are disposed of inside a baobab.
- In 1636, the Irish archbishop James Ussher (often misspelled "Usher," 1581–1656) calculated, by means of the Bible, that the world had been created in 4004 BC. In 1644, John Lightfoot (ca. 1602–1675), a prominent scholar and Vice-chancellor of Cambridge University, published his view that the creation of Adam had occurred in 3928 BC. Archbishop Ussher incorporated information from Lightfoot and in 1850 published his conclusion that the world had been created during the week beginning on Sunday, October 23, 4004 BC. (Ussher also calculated the date of the Great Flood as 2501 BC and judged that Adam and Eve were expelled from the Garden on November 10, 4004 BC.) Ussher's 4004 BC date was widely accepted for many years, until eighteenth-century

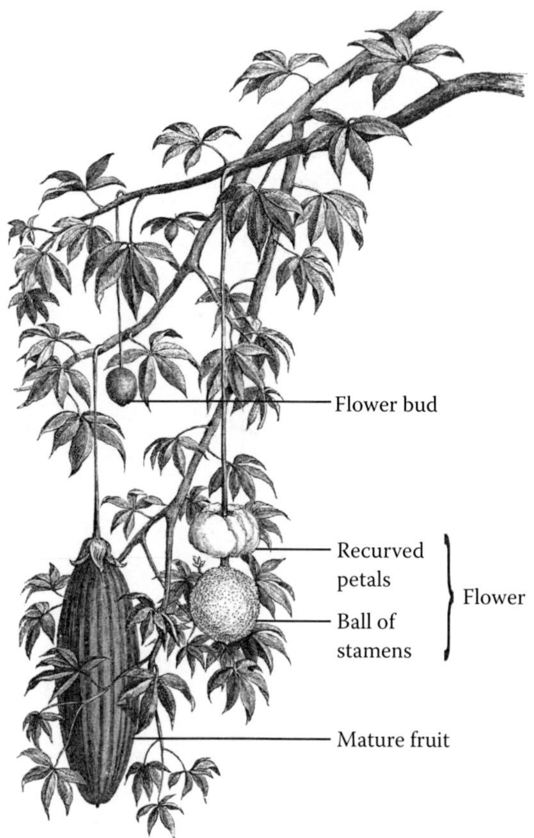

FIGURE 10.2 Baoab (*Adansonia digitata*). (From Engler and Prantl, 1889–1915.)

FIGURE 10.3 Baoab (*Adansonia digitata*). (From Schimper, 1903.)

geologists used the layers of rocks and sediment of the earth to examine scientifically the antiquity of the earth (which is about 4.5 billion years). Missionary David Livingstone (referred to above) was a follower of Ussher. French naturalist Adanson (also referred to above) had calculated that some baobabs were older than 4000 years, which would mean that Ussher was wrong. Livingstone was quite upset at this contradiction, but he himself counted growth rings, which seemed to confirm that some trees were indeed older than 4000 years. Recent studies have indicated that most baobabs are less than 400 years old, and trees rarely live more than 1000 years. The conclusion by some on the basis of tree rings that baobabs can live as long as 5000 years has been questioned; sometimes tree rings can be misleading, especially in dry regions, where extra rings can be present simply because of cycles of wet and dry months in a given year.

- In baobab trees, there are equal numbers of "right-handed flowers" (with petals that coil toward the right while still in the bud) and "left-handed flowers" (with petals that coil toward the left while still in the bud). However, right-handed flowers tend to have more male organs (stamens, which may number up to 1600 in a flower) than left-handed flowers.
- Paper, especially for monetary bills, has been made out of baobab fiber.
- In Transvaal (an old province in the northeast of South Africa), buffalo-weaver birds are reputed to always rest on the western side of baobabs, and therefore they can be used like a compass.
- In Malawi (south of the Equator and slightly inland from the east coast of Africa), when a poison arrow is withdrawn from a killed animal, the juice of baobab is poured into the wound in the belief that it neutralizes the toxin before the meat is eaten.
- Baobabs are very hardy, as evidence by their continuing to live while people have excavated their interiors to provide housing. Some trees hold in their trunks cannonballs fired during battles centuries ago. Some baobabs have continued to grow while lying flat on the ground.
- There are "authentic reports" of dead baobab trees bursting into flame from spontaneous combustion.
- The baobab has been said to be the most drought-resistant of trees. During rains the trees swell greatly, absorbing water. Large trees can store thousands of liters (over a thousand gallons) of water in their trunks for later use.
- The flowers of the baobab are white and emit a musky odor (which has been described as a sour, carrion smell) to attract bats at night (flies and moths are also attracted). The bats drink the nectar and pass pollen from flower to flower. The petals wilt within 24 hours and fall to the ground.
- Beekeepers sometimes use the hollow trunks as sites for beehives and also hang beehives on the branches to take advantage of the nectar produced by the trees.
- Baboons, monkeys, and elephants relish the pulp of the baobab fruit and distribute the seeds in their droppings.
- Elephants can be a significant problem for baobabs, some of the beasts, especially the bulls, seemingly pushing over trees for entertainment or to demonstrate their masculinity.
- After uprooting the trees and stripping off the leaves and bark for food, elephants often use the fibrous wood as a chewing gum, spitting out the woody wad after their saliva has extracted the nutrients.
- The only tree that develops a trunk larger in diameter than the African baobab is the giant sequoia (*Sequoiadendron giganteum* (Lindl.) J. Buchholz, formerly known as *Sequoia gigantea*) of California. One of these, the "General Sherman," which is the third tallest tree in the world, at breast height has a diameter of 8.8 m (28.9 ft.), greater than any other known tree in the world.

FIGURE 10.4 California sequoia (*Sequoia gigantea*). (From Flores des serres, 1853–1854, vol. 9.)

KEY INFORMATION SOURCES

Arum, G. 1989. *Baobab: Adansonia digitata*. Kengo, Nairobi, Kenya. 24 pp.

Barnes, R.F.W., Barnes, K.L., and Kapela, E.B. 1994. The long-term impact of elephant browsing on baobab trees at Msembe, Ruaha National Park, Tanzania. *Afr. J. Ecol.* 32(3):177–184.

Baum, D.A. 1995. The comparative pollination and floral biology of baobabs (*Adansonia*, Bombacaceae). *Ann. Mo. Bot. Gard.* 82:322–348.

Baum, D.A. 1995. A systematic revision of *Adansonia* (Bombacaceae). *Ann. Mo. Bot. Gard.* 82:440–470.

Baum, D.A., Small, R.L., and Wendel, J.F. 1998. Biogeography and floral evolution of baobabs (*Adansonia*, Bombacaceae) as inferred from multiple data sets. *Syst. Biol.* 47:181–207.

Caluwé, E. De, Halamová, K., and Van Damme, P. 2009. Baobab (*Adansonia digitata* L.): a review of traditional uses. In *African natural plant products: new discoveries and challenges in chemistry and quality*. Edited by H.R. Juliani, J.E. Simon, and C.-T. Ho. American Chemical Society, Washington, DC. pp. 51–84.

Davis, T.A., and Ghosh, S.S. 1976. Morphology of *Adansonia digitata*. *Adansonia*, 15:471–479.

Esenowo, G.J. 1991. Studies on the germination of *Adansonia digitata* seeds. *J. Agric. Sci.* 117:81–84.

Fenner, M. 1980. Some measurements on the water relations of baobab trees. *Biotropica*, 12:205–209.

Guy, G.L. 1971. The baobabs: *Adansonia* spp. (Bombacaceae). *J. Bot. Soc. S. Afr.* 57:31–37.

Hardy, D., and LaFon, R. 1982. The baobab, silent colossus of the African bush. *Cactus Succulent J.* 54(2):51–53.

Kerharo, J. 1969. The baobab (*Adansonia digitata* L.) as an African panacea. *Q. J. Crude Drug Res.* 9:1401–1408.

Lindsay, G. 1972. Baobab. *Pacific Discovery*, 25(4):14–19.

Maheshwari, J.K. 1971. The baobab tree: disjunctive distribution and conservation. *Biol. Conserv.* 4:57–60.

Njunga, M. 2005. The Malawian baobab. *Ulendo*, 6:64–66.

Nour, A.A., Magboul, B.I., and Kheiri, N.H. 1980. Chemical composition of baobab fruit (*Adansonia digitata* L.). *Trop. Sci.* 22:383–388.

Obizoba, I.C., and Amaechi, N.A. 1993. The effect of processing methods on the chemical composition of baobab (*Adansonia digita* L.) pulp and seed. *Ecol. Food Nutr.* 29:199–205.

Rashford, J. 1994. Africa's baobab tree: why monkey names? *J. Ethnobiol.* 14:173–183.
Reddy, A.S., Anjaria, K.B., and Rao, V.R. 2002. Baobab: an exotic tree with a promise? *Asian Agri-History*, 6:343–350.
Shukla, Y.N., Dubey, S., Jain, S.P., and Kumar, S. 2001. Chemistry, biology and uses of *Adansonia digitata*: a review. *J. Med. Aromat. Plant Sci.* 23:429–434.
Small, E., and Catling, P.M. 2003. Blossoming treasures of biodiversity. 9. African baobab: the world's fattest tree. *Biodiversity*, 4(3):27–29.
Wickens, G.E. 1982. The baobab—Africa's upside-down tree. *Kew Bull.* 47:173–209.
Wilson, T. 1988. Vital statistics of the baobab (*Adansonia digitata*). *Afr. J. Ecol.* 26:197–206.

Specialty Cookbooks

Drake, E. 2006. *A book of baobabs*. Aardvaark Press, Cape Town, South Africa. 90 pp. [Recipes in chapter 8.]

11 Bay

Family: Lauraceae (laurel family)

NAMES

Scientific Name: *Laurus* nobilis L.

- The English word "bay" is derived from the Latin *baca*, berry, originally applied to the berries of the bay tree, not the leaves.
- Bay is also called bay laurel, bay tree, Grecian laurel, laurel, laurel tree, Mediterranean laurel, noble laurel, poet's laurel, Roman laurel, royal bay, sweet bay, sweet bay tree, sweet laurel, tree laurel, true laurel, and victor's laurel.
- The plant is called "bay laurel" or similar names, whereas the commercial product is called bay leaf (plural: bay leaves), sweet bay leaf, true laurel, roman laurel, and Turkish laurel. "Bay" can be used either for the plant or the condiment.
- Bay laurel should not be confused with plants with similar sounding names, such as bayberry (*Myrica pensylvanica* Loisel.), a native of eastern North America that provides bayberry oil used in perfumes and candle making; California bayberry (*Myrica californica* Cham.), a native of the U.S. Pacific states; California bay (*Umbellularia californica* (Hook. & Arn.) Nutt.), a native of western North America, also in the Lauraceae, which produces more potent-tasting aromatic leaves than bay laurel, used for the same culinary purposes (the odor released can cause headaches, dizziness, and nausea in susceptible individuals, and as a result, the plant has acquired the name "headache tree"); *Kalmia latifolia* L., mountain laurel, a poisonous, eastern North American bog-loving plant sometimes grown as an ornamental (several other species of *Kalmia* are also called laurel; the constituent andromedotoxin in *Kalmia* is so potent it can poison honey made by bees visiting its flowers); West Indian bay tree (*Pimenta racemosa* (Mill.) J.W. Moore, a relative of allspice), a native of the Caribbean region, whose oil is used in perfumery and to make bay rum; and cherry (or English) laurel (*Prunus laurocerasus* L.), a native tree of southeastern Europe and Asia Minor. Cherry laurel has sometimes been mistaken for bay laurel with disastrous results. It is hydrocyanic; that is, it produces prussic acid, a virulent poison. Despite its toxicity, cherry laurel leaves were once occasionally used to flavor milk in England. Portuguese (or Portugal) laurel (*Prunus lusitanica* L.) is another hydrocyanic plant that has caused poisoning by virtue of being mistaken for bay laurel. Confusingly, cherry laurel has occasionally been used as a wreath of honor in place of bay laurel wreaths in some English ceremonies.
- As noted earlier, allspice (*Pimenta dioica*) is used to make bay rum. In the early days of the Pure Food Law in the United States, a regulation was adopted ordering that the leaves of the bay laurel be described as laurel leaves, not simply as bay leaves, to avoid confusion with the bay rum species.
- In Greek mythology, the multifaceted god Apollo (god of music and the arts, god of archery, and patron of medicine) was said to have pursued the reluctant nymph Daphne until the other gods turned her into a laurel tree. When Apollo learned what had become of her, he declared the tree to be sacred and wore a wreath of bay laurel leaves as a sign of his devotion. The tree was originally known in Greek as *dáphnee* and is still known in Greece as the Daphne tree. The Hebrew *aley daphna* means "leaves of Daphne."

FIGURE 11.1 Bay (*Laurus nobilis*). (From Hallier, 1880–1888, vol. 10, plate 983.)

- *Nobilis* in the scientific name *L. nobilis* is Latin for renowned or famous, reflecting the admiration with which bay was held in the past.
- The genus name *Laurus* is the classical Latin name for the bay tree.

PLANT PORTRAIT

Bay laurel is a pyramidal, evergreen, broadleaf shrub or tree that occasionally exceeds 20 m (66 ft.) in height. Its leaves are stiff. Yellowish or greenish-white flowers, about 3 mm (1/8 in.) across, are produced in the spring. The berries are about 15 mm (5/8 in.) in diameter, green when young, turning dark purple or black as they ripen toward the autumn. Bay laurel is native to western Asia Minor and perhaps parts of the European Mediterranean. It was distributed throughout the Mediterranean by people over 2000 years ago and was familiar to ancient Greek and Roman civilizations. It is now cultivated in many subtropical regions of the world, including those around the Mediterranean, the Caribbean, Central America, and the warmer parts of North America.

Bay leaves are among the most important herbs in the international spice trade. In addition to culinary uses noted below, essential oil from bay is used as a scent for soaps, perfumery, and candles. Most of the bay leaves entering international trade are from wild rather than cultivated plants. Turkey is the largest exporter of bay leaves followed by Greece. Bay laurel is cultivated commercially in these countries as well as Portugal, Spain, France, Italy, the former Yugoslavia, Morocco, China, and Central America. There is limited production in California. Although U.S. bay is the most visually attractive, its flavor is regarded as poor.

In frost-free areas, bay laurel is grown as an ornamental and as a screen and hedge, for which its tolerance to clipping makes it quite suitable. In Europe, large, well-developed potted specimens serve as "gate guards" in establishments such as restaurants and banks, with several large nurseries, mainly in Belgium, specializing in rather elaborately trained forms.

CULINARY PORTRAIT

The leaves of bay laurel are fragrant and sweet scented with a lemon and clove overtone. Fresh bay leaves are rather bitter, but the bitterness is lost after a few days of drying. The very old, dried bay leaves that most people use are regarded by purists as a poor substitute for fresh or recently dried leaves. Recently picked and dried leaves will keep in a sealed container in the dark for up to a year but should then be discarded. As the flavor of fresh leaves is stronger than that of dried leaves, half a fresh leaf should be used where a whole dried leaf is required. The taste intensifies through cooking, and bay is one of the few herbs that should be added early in cooking. Commercial bay leaves are sold whole or ground.

The leaves are important in the cooking of many Mediterranean countries, and indeed bay is used in almost every known cuisine and sometimes seems to be combined with almost every food. One leaf is sufficient to flavor family-sized portions of stews or soups; dried bay produces a concentrated flavor and should be added cautiously. The leaves are added (whole, flaked, or ground) to stuffings, soups, stews, chowders, tomato sauces, gravies, pickling spices, vinegars, puddings, salads, and vegetables as well as meat, game, fish, and poultry dishes. Ground bay is an important ingredient of liver pâté and is found in commercial products such as delicatessen-style prepared meats like corned beef, chicken loaf, mortadella (Italian bologna), pressed sausage, and luncheon meats. Ground leaves are also added to barbecue sauces, preserves, and pastries. Cracked bay leaves are a common ingredient in commercial whole pickling spice.

The essential oil of the leaves is used commercially in preparation of pickling spice, in the flavoring of vinegar, and in condiments. The oil is also used in nonalcoholic beverages, baked goods, confectionery, meats, sausages, and canned soups.

Caution should be exercised when using bay leaves in cooking. The fibrous quality remains despite long cooking. It is best to remove them before serving foods because of the possibility of throat or intestine injury from swallowing the sharp-edged leaves. Because they can block a child's throat, they are capable of cutting off the air supply and causing suffocation. Because they are so tough, they may remain intact in the digestive system, obstruct the intestines, and require surgical removal. Bay leaves should be stored away from children and pets.

Culinary Vocabulary

- *Bouquet garni* (French for "garnished bouquet") is a bundle of herbs, usually tied together with string, and used in flavoring soups, stews, and other cooked dishes, including such specialties as beef bourguigon and bouillabaisse. The bundle is removed before the dish is eaten. Although many herbs may be used, bay is the primary taste of traditional bouquet garni French dishes.

CURIOSITIES OF SCIENCE AND TECHNOLOGY

- King David, (approximately 1000–962 BC), of Israel prized bay laurel. He had his personal rooms paneled with aromatic bay wood and had bay leaf motifs used to decorate the Temple at Jerusalem.
- Dating back to medieval times, distinguished scholars and young physicians were given berries ("baca lauri," i.e., laurel berries) and wreaths of bay laurel and were called "baccalaureates," a term now denoting a bachelor's (i.e., undergraduate university) degree. The link of bay laurel with academic honors is also found in such phrases as "poet laureate" and "resting on one's laurels."
- Albertus Magnus or Saint Albert the Great (born 1193 or 1206, died 1280) was a highly accomplished scientist and philosopher. However, in past times, many of the best minds entertained beliefs that seem silly today. Magnus wrote that if someone carried a wolf's tooth wrapped in a bay leaf gathered in August, it would prevent others from speaking angry words to him.
- Bay trees were generally present in monastic gardens because it was believed that they were antiseptic.

FIGURE 11.2 Bay (*Laurus nobilis*). (From Gartenflora, 1880, vol. 29.) Such carefully manicured specimens grown in tubs are popular in Europe.

- It was fashionable at dinners during the Middle Ages to boil bay leaves with orange peels for use in finger bowls.
- Bay laurel withstands a few degrees of frost for a few days and then succumbs. This seems to have promoted the legend that in a hard winter, when the bay trees die, death will follow. Shakespeare wrote in Richard II:

  'Tis thought the king is dead.

  He will not stay

  The bay trees in our country all withered.

  (In 1399, the year of the downfall of Richard II, it was recorded that bay trees throughout England withered, although many recovered subsequently.)

- The odor of bay leaves repels many insects, especially weevils. As protection from insects, dried bay leaves are used in Turkey to wrap figs for export, in Italy to wrap licorice for shipping, and in China for packaging rice and other cereals. To keep bugs out of flour, a couple of bay leaves can simply be placed in the container.

KEY INFORMATION SOURCES

Chandoha, W. 1984. Bay: the perfect houseplant. *Organic Gardening*, 31(1):74–75.

Cheminat, A., Stampf, J. L., and Benezra, C. 1984. Allergic contact dermatitis to laurel *Laurus nobilis* isolation and identification of haptens. *Arch. Dermatol. Res.* 276:178–181.

D'Albore, G.R., and D'Ambrosio, M.T. 1982. Observations on pollinator insects of *Laurus nobilis* L. *Apicolt. Mod.* 73(3):81–87 (in Italian, English summary).

Diaz-Maroto, M.C., Perez-Coello, M.S., and Cabezudo, M.D. 2002. Effect of drying method on the volatiles in bay leaf (*Laurus nobilis* L.). *J. Agric. Food Chem.* 50:4520–4524.

Ferguson, D.K. 1974. On the taxonomy of recent and fossil species of *Laurus* (Lauraceae). *Bot. J. Linn. Soc.* 68:51–72.

Fiorini, C., Fouraste, I., David, B., and Bessiere, J.M. 1997. Composition of the flower, leaf and stem essential oils from *Laurus nobilis* L. *Flavour Fragrance J.* 12:91–93.

Ipor, I.B., and Oyen, L.P.A. 1999. *Laurus nobilis* L. In *Plant resources of South-East Asia: 13. Spices*. Edited by C.C. de Guzman and J.S. Siemonsma. Backhuys, Leiden, the Netherlands. pp. 134–137.

Kilic, A., Hafizoglu, H., Kollmannsberger, H., and Nitz, S. 2004. Volatile constituents and key odorants in leaves, buds, flowers, and fruits of *Laurus nobilis* L. *J. Agric. Food Chem.* 52:1601–1606.

Kumar, S., Singh, J., and Sharma, A. 2001. Bay leaves. In *Handbook of herbs and spices*. Edited by K.V. Peter. CRC Press, Boca Raton, FL. pp. 52–61.

Maron, R., and Fahn, A. 1979. Ultrastructure and development of oil cells in *Laurus nobilis* L. leaves. *Bot. J. Linn. Soc.* 78:31–40.

Martin, E.B. 1986. *Laurus nobilis*—bay tree. *Herbarist*, 52:25–27.

McClintock, E. 1993. Trees of Golden Gate Park, San Francisco: 48. California bay and Mediterranean laurel. *Pacific Hort.* 54(1):10–12.

Nigam, M.C., Ahamad, A., and Misrha, L.N. 1992. *Laurus nobilis*: an essential oil of potential value. *Parfumerie und Kosmetik*, 73:854–859.

Ozden, M.G., Oztas, P., Oztas, M.O., and Onder, M. 2001. Allergic contact dermatitis from *Laurus nobilis* (laurel) oil. *Contact Dermatitis*, 45:178.

Putievsky, E., Ravid, U., Snir, N., and Sanderovich, D. 1984. The essential oils from cultivated bay laurel. *Isr. J. Bot.* 33:47–52.

Raviv, M., Putievsky, E., Senderovitch, D., Snir, N., and Roni, R. 1983. Bay laurel as an ornamental plant. *Acta Hortic.* 132:35–42.

Skrubis, B.G. 1982. The drying of laurel leaves. *Perfumer Flavorist*, 7(5):37–40.

Souayah, N., Khouja, M.L., Khaldi, A., Rejeb, M.N., and Souzid, S. 2002. Breeding improvement of *Laurus nobilis* by conventional and in vitro propagation techniques. *J. Herbs Spices Med. Plants*, 9(2/3):101–105.

Takos, I.A. 2001. Seed dormancy in bay laurel (*Laurus nobilis* L.). *New Forest*, 21(2):105–114.

Tilki, F. 2004. Influence of pretreatment and desiccation on the germination of *Laurus nobilis* L. seeds. *J. Environ. Biol.* 25:157–161.

Specialty Cookbooks

Evans, H. 1996. *Basil, bay & borage*. JG Press, North Dighton, MA. 61 pp.

12 Betelnut

Family: Arecaceae (Palmae; palm family)

NAMES

Scientific Name: *Areca catechu* L.

- The word "betel" is from the Portuguese word *betel*. Portuguese explorers in the fifteenth century based their word on the Malayalam *vettila* or *verrila*, from Tamil *verrilai*. Actually, this was not the Malayalam word for the betel nut but rather for the betel leaf, which as explained below is a different species.
- Betelnut (betel nut) is also known as areca nut, areca palm, areca nut palm, betel palm, betelnut palm, catechu, catechu palm, Indian nut, penang palm, and pinang.
- "Wild betel leaf" is *Piper sarmentosum* Roxb., used as a spice in Southeast Asian cooking, and sometimes confused with betel.
- Betel (based on the Spanish word *Bethel*, meaning "House of God") is an international Christian organization that works with drugs addicts, alcoholics, and street people in various countries around the world.
- The genus name *Areca* is based on the Malabar (or Tamil) word *areec*, the common name of *A. catechu* used by the natives of Malabar (on the southwest coast of India). The word was adopted in Portuguese as *areca*.
- *Catechu* in the scientific name *A. catechu* is the common name of various Asiatic plants, including betelnut. In particular, catechu referred to a strongly astringent drug, extracted from the wood of black catechu, *Acacia catechu* (L. f.) Willd., often added to betel palm seeds when preparing chewing betel. Catechu is apparently derived from the Malay *caccu* or *kachu*, which is in turn probably from Dravidian *karaiyal* or *karaiccal*, that which is dissolved (from *karai*, to melt).

PLANT PORTRAIT

The betel palm is a tropical, Asian tree with a slender trunk typically 12 to 15 m (39–49 ft.) tall, sometimes up to 30 m (98 ft.) in height, and about 50 cm (20 in.) wide, green at first, then grayish and ringed by the remains of scars where the old leaves were attached. The fruits are generally the size and shape of small (hen's) eggs, up to 5 cm (2 in.) in diameter, hard, red-orange, yellow or scarlet, with a fibrous layer under the shell. A betel palm may produce up to 250 fruits annually. Each fruit contains one acorn-shaped (nearly round) seed about 2.5 cm (1 in.) in diameter, which is brown with fawn marbling. The seed is called "betel nut," and contains arecoline, an alkaloid chemical, which is responsible for a narcotic/stimulant effect. Also present are tannins, which produce astringency (a mouth-puckering effect), and a red dye that results in red spittle, red lips, and red feces. Laborers chew betel to provide them with energy, and betel has played much the same role where it is common as coffee and cigarettes do in the West. The species is cultivated in the Old World tropics. Its native range is obscure, possibly the East Indies and Malaysia. This palm is widely cultivated, and there are huge plantations in Asia from Pakistan and India to Malaysia, extending as far as the Pacific Southern Islands, and also in Africa. Betel palm is sometimes grown as an ornamental in subtropical southern Florida, and can also be cultivated in some of the warmest parts of California.

There is considerable commerce in betel nuts, which are mostly used in India, Southeast Asia, Malaysia, and Polynesia, but are also imported into North America.

Betel nuts are traditionally consumed with "betel leaf," which is obtained from the betel pepper, *Piper betle* L. Betel pepper is also known as betel and betelvine. It probably originated in Malesia and is cultivated along with betel palm in the Old World tropics. It is a herbaceous, evergreen, red-berried plant, the stems extending as far as 5 m (16 ft.). The betel pepper is widely cultivated in southern Asia, where betel has been chewed since ancient times. Betel pepper leaves cannot be imported into the United States.

"Betel" is a traditional, mild, legal, narcotic stimulant of Asia which, as explained below, is made by combining the seeds of the betelnut palm, the leaves of the betel pepper, and other ingredients. Chewing betel is widespread in Southeast Asia and the South Pacific islands and among those of Indian origin elsewhere in the world. It is estimated that at least 500 million people chew betel; some have contended that as much as 25% of the world's population chew betel nuts with some frequency. Betel chewing is an ancient habit that may have originated in India. The thirteenth-century traveler Marco Polo mentioned in his diaries that Indians were in the habit of consuming betel. Spitting out the red juice that is generated as well as the remaining pulpy mass is generally considered disgusting in Western countries. Naive Westerners have sometimes interpreted the spectacle as spitting blood, as was once common in people with tuberculosis. Nevertheless, this behavior has been accepted where betel consumption is practiced. However, in some areas of Asia, the staining of roads and walls of public buildings as well as health concerns has led to a growing social disapproval of betel. Even where betel chewing is permitted, it is often banned in certain public places, such as airports. Although it is difficult for those not familiar with betel chewing to understand its attraction, it is in principle no different from coffee consumption.

Arecoline from the betel nut is considered to be chiefly responsible for the stimulant effect of betel (arecoline is converted by the central nervous system to the stimulant arecaidine). Arecoline is released from the nut by saliva and lime. This chemical is in the same alkaloid group as muscarine, found in fly agaric (preparations of the fly agaric mushroom used to be left open in dishes to kill flies and are still used as a psychedelic). Arecoline is the fourth most popular natural stimulant drug, after caffeine, alcohol, and nicotine, in that order. On the basis that they contain "a poisonous or deleterious substance," primarily arecoline, betel nuts were outlawed in the United States in 1992, although the ban was partially lifted in 2000.

Lime is calcium oxide. "Burnt" or "slaked" lime (hydrated calcium oxide, calcium hydroxide) is preferable and is often prepared by burning seashells or coral. The addition of lime makes the betel mixture alkaline, and this in turn causes the arecoline (and additional alkaloids that are present) to be absorbed in the mouth more readily. Catechu, a wood preparation that is also sometimes added, presumably has the same effect. If lime is omitted from the chew, the characteristic effect is almost absent. The practice of adding lime may seem peculiar, but this is very commonly done throughout the world with other alkaloid-bearing plants to make the alkaloid more available.

Just why betel leaf (from *P. betle*) is traditionally combined with betel nut (*A. catechu*) is not entirely clear. Certainly the aromatic leaf improves the taste and is responsible for the commonly reported sweet taste and pleasant breath that results. The leaf also counters the burning, astringent taste of the betel nut, and may promote the nut's physiological effect.

There is evidence that the habit of chewing betel containing tobacco, which is frequently added to the quid, is carcinogenic (causes cancer) in humans. There is suspicion that chewing betel without tobacco is also carcinogenic in humans, and it has been suggested that betel consumption is linked to heart disease, diabetes, and asthma. An overdose of betel can cause dizziness, vomiting, and convulsions. The use of betel has decreased dramatically over the last half century and has been replaced by tobacco smoking, hardly an improvement since lung cancer has increased.

A very minor oral use of betel in Asia is as a dentifrice. The nut is burned to make a charcoal, which is pulverized and added to toothpaste.

FIGURE 12.1 Betelnut (*Areca catechu*). (From Köhler, 1883–1914.)

FIGURE 12.2 Betel pepper (*Piper betle*) shade house. (From Duthie, 1893.)

CULINARY PORTRAIT

Betel, made mostly from the seeds of the betel palm and the leaves of the betel pepper, is a masticatory (a preparation that is chewed)—indeed the world's most-used masticatory. Betel is made from slices of betel palm nuts (betel nuts) smeared on a betel pepper leaf (called "pan" or "paan," a word also sometimes used for prepared betelnut chews) together with other aromatic flavorings, lime paste, sometimes catechu (derived from the wood of black catechu, as noted above), and sometimes tobacco, all rolled up in the leaf. Condiments such as cinnamon, cardamom, cumin seeds, fennel seeds, camphor, cloves, licorice, tamarind, and dried coconut are often included. The betel wad is pressed against the inside of the cheek by the tongue and sucked or chewed (it is rarely eaten, i.e., swallowed), and saliva secretion is strongly increased. The saliva mixes with the material to generate a red juice that is spit out. The odor of betel is acrid. The taste of betel nut has been described as "a combination of cloves, citronella, and carbolic acid," and "astringent." The taste is acquired and devotees find betel nut refreshing and mildly stimulating. Frequent chewers have vivid red lips, gums, and tongues and black teeth. Habitual chewers may become toothless by the age of 25.

As with many palms that are cultivated in large plantations, the "heart" (succulent growing tip of the stem) is sometimes eaten (see Chapter 74), although this destroys the tree.

Culinary Vocabulary

- As noted above, "pan" or "paan" (pronounced phan) is sometimes used for prepared betelnut chews. This is derived from the Hindi *pān*, which originally meant "betel leaf," and is based on the Sanskrit *parná*, meaning feather or leaf.

CURIOSITIES OF SCIENCE AND TECHNOLOGY

- At the Spirit Cave archeological site in Thailand, paleobotanical remains of betel nut and betel pepper were found at the same location, providing circumstantial evidence for the practice of betel chewing in prehistoric times. These remains are between 9000 and 7500 years of age, suggesting that betel is one of the earliest known psychoactive substances used in the world.
- The black teeth that result from years of chewing betel, along with red lips, were once thought to be a sign of beauty, especially for young women. An old Southeast Asian expression held that "Only dogs, ghosts, and Europeans have white teeth!" Today, the Western ideals that white teeth are attractive (promoted by toothpaste advertising) and spitting is not are increasingly confining betel chewing among women to the elderly.
- In Papua, New Guinea, witch doctors treat "mental disorder" by providing betel and magical words. Patients are warned that the treatment will not work unless payment for the service is made.
- Generally, a knife is used to slice betel nuts, but in parts of India a special tool is used for the purpose. Known as a betel cutter, this is a hinged one-bladed instrument designed solely to cut areca nuts.
- Betelnut and some other intoxicants are far more popular in Asia than in Western nations, and by contrast high-alcohol, distilled beverages are far less popular. This difference in popularity of inebriants has been credited in part to the low levels of aldehyde dehydrogenase isozyme in Asian peoples, an enzyme in the liver that helps to metabolize alcohol.
- In 2004, a British patent application was announced for a betelnut and nicotine chewing gum. The gum is intended to be used in assisting tobacco users to escape their addiction, although the combination of two addictive substances seems contradictory for the purpose. However, betelnut has been found to be anticarcinogenic against tobacco (Padma et al., 1989).

KEY INFORMATION SOURCES

Bavappa, K.V.A. 1980. Breeding and genetics of arecanut, *Areca catechu* L.—a review. *J. Plantation Crops*. 8(1):13–23.

Bavappa, K.V.A., and Mathew, J. 1982. Genetic diversity of *Areca catechu* L. and *Areca triandra* Roxb. palms. *J. Plantation Crops*. 10(2):92–101.

Bavappa, K.V.A., Nair, M. K., and Kumar, T.P. 1982. *The Arecanut palm (Areca catechu Linn.).* Central Plantation Crops Research Institute, Kasaragod, Kerala, India. 340 pp.

Bhat, K.S. 1978. Agronomic research in arecanut—a review. *J. Plantation Crops*, 6(2):67–80.

Brotonegoro, S., Wessel, M., and Brink, M. 2000. *Areca catechu* L. In *Plant resources of South-East Asia: 16. Stimulants*. Edited by H.A.M. van der Vossen and M. Wessel. Backhuys, Leiden, the Netherlands. pp. 51–55.

Chempakam, B., Annamalai, S.J.K., and Murphy, K.N. 1982. Other uses of arecanut. *Indian Farming*, 32(9):40–43.

Dasgupta, B., Jha, S., and Sen, C. 1999. Betel vine [*Piper betle*]. In *Tropical horticulture*, vol. 1. Edited by T.K. Bose, S.K. Mitra, A.A. Farooqui, and M.K. Sadhu. Naya Prokash, Calcutta, India. pp. 592–611.

Farnsworth, E.R. 1976. Betel nut—its composition, chemistry and uses. *Sci. (New Guinea)*, 4(2):85–90.

Farooqi, A.A., and Sreeramu, B.S. 1999. Arecanut. In *Tropical horticulture*, vol. 1. Edited by T.K. Bose, S.K. Mitra, A.A. Farooqui, and M.K. Sadhu. Naya Prokash, Calcutta, India. pp. 612–630.

International Agency for Research on Cancer. 2004. *Betel-quid and areca-nut chewing; and some areca-nut derived nitrosamines*, vol. 85. IARC monographs on the evaluation on the carcinogenic risk to humans. World Health Organization/ IARC, Lyon, France. 349 pp.

Ishwar Bhat, P.S., and Rao, K.S.N. 1963. On the antiquity of the arecanut. *Arecanut J.* 13:13–21.

Joshi, Y. 1982. *Arecanut palm (Arecu catechu Linn.): an annotated bibliography up to 1981.* Central Plantation Crops Research Institute, Kasaragod, India, 116 pp.

Krochmal, C., and Krochmal, A. 1990. The betelnut, *Areca catechu*. *Bull. Natl. Trop. Bot. Gard.* 20(1):5–7.

McLeish, M.J., and Huang, J.L.A. 1990. Comparison of alkaloid levels in the nuts of *Areca catechu* Linn. *Sci. (New Guinea)*, 16(2):55–60.

Mori, H. 1987. Betel nut. In *Naturally occurring carcinogens of plant origin*. Edited by I. Hirono. Elsevier, New York. pp. 167–180.

Mujumdar, A.M., Kapadi, A.H., and Pendse, G.S.SO. 1979. Chemistry and pharmacology of betel-nut. *J. Plantation Crops*, 7(2):69–92.

Murphy, K.N. 1977. Floral and pollination biology of the betel nut palm *Areca catechu* L. *J. Plantation Crops*, 5(1):35–38.

Nelson, B.S., and Heischober, B. 1999. Betel nut: a common drug used by naturalized citizens from India, Far East Asia, and the South Pacific Islands. *Ann. Emerg. Med.* 34:238–243.

Norton, S.A. 1998. Betel: consumption and consequences. *J. Am. Acad. Dermatol.* 38:81–88.

Padma, P.R., Lalitha, V.S., and Amonkar, A.J. 1989. Anticarcinogenic effect of betel leaf extract against tobacco. *Cancer Lett.* 45(3):195–202.

Raghavan, V., and Baruah, H.K. 1957. Arecanut: India's popular masticatory—history, chemistry and utilization. *Econ. Bot.* 12:315–345.

Ranade, S.A., Verma, A., Gupta, M., and Kumar, N. 2002. RAPD profile analysis of betel vine cultivars. *Biol. Plant.* 45:523–527.

Reichart, P.A., and Philipsen, H.P. 1996. *Betel and miang, vanishing Thai habits.* Cheney/White Lotus, Bangkok. 136 pp.

Rooney, D. 1993. *Betel chewing traditions in South-East Asia.* Oxford University Press, New York. 76 pp.

Teo, S.P., and Banka, R.A 2000. *Piper betle* L. In *Plant resources of South-East Asia: 16. Stimulants*. Edited by H.A.M. van der Vossen and M. Wessel. Backhuys, Leiden, the Netherlands. pp. 102–106.

Thomas, S., and Kearsley, J. 1993. Betel quid and oral cancer: a review. *Europ. J. Cancer B. Oral Oncol.* 29(4):251–255.

Wang, C.K., Su, H.Y., and Lii, C.K. 1999. Chemical composition and toxicity of Taiwanese betel quid extract. *Food Chem. Toxicol.* 37(2/3):135–144.

Yoganathan, P. 2002. Betel chewing creeps into the New World. *N. Z. Dent. J.* 98(432):40–45.

Specialty Cookbooks

Cookbooks devoted to betelnut have not been located. Cookbooks on Asian foods frequently have recipes for preparation of paan; recipes in English are most likely to be found in cookbooks of India. Searching for betelnut recipes on the Internet almost invariably results in locating sites presenting information on the (often illegal) use for recreational inebriation. Searching for "paan" will produce genuine betelnut recipes, but the reader is cautioned that, as detailed above, betelnut consumption is harmful and indeed illegal in many regions.

13 Breadfruit

Family: Moraceae (mulberry family)

NAMES

Scientific Name: *Artocarpus altilis* (Parkinson) Fosberg
(*A. communis* J.R. Forst. & G. Forst., *A. incisus* L. f.)

- The name "breadfruit" was coined because the unripe fruit tastes and even feels like fresh bread.
- "Mexican breadfruit" and "false breadfruit" are ceriman.
- "Breadmit" refers to a breadfruit with seeds.
- The "African breadfruit," *Treculia africana* Decne, is a minor food crop in Africa, its seeds roasted and eaten as a snack or used to produce flour and cooking oil.
- The genus name *Artocarpus* is from the Greek *artos*, bread + *karpos*, fruit.
- *Altilis* in the scientific name *A. altilis* is derived from Greek meaning fat (referring to a hen or bird), descriptive of the succulent fruit.

PLANT PORTRAIT

The tropical breadfruit tree grows as tall as 26 m (85 ft.). The leaves are evergreen or deciduous, depending on climate. The species is thought to be native in the area extending from New Guinea through the Indo-Malayan Archipelago to Western Micronesia. Breadfruit is believed to have been widely spread in the Pacific area by migrating Polynesians and is a staple or subsistence crop in the Pacific islands, where cultivation has been carried out for thousands of years. In tropical areas of the world, breadfruit takes the place of such temperate region starchy staples as potatoes and cereals. Indeed, a traditional saying is that a good-sized breadfruit tree can produce enough food to sustain a family for a year. Europeans may not have seen the breadfruit until 1595. More recently, breadfruit became popular in the Caribbean region. Fruits are borne singly or in clusters of two or three at the branch tips. Good trees produce about 100 fruits annually. The fruit is oblong, cylindrical, rounded, or pear shaped, 9 to 45 cm (3½–18 in.) long, and 5 to 30 cm (2–12 in.) in diameter. The thin rind is patterned with irregular, four- to six-sided segments, which are either smooth or conical and sometimes bear a short point or spine. The rind in some varieties is rough like sandpaper. Usually, the rind is green when immature, turning yellowish-green, yellow, or yellow-brown when ripe, although one variety is lavender. In ripe fruit, the interior is cream colored or yellow and sweetly fragrant. There are varieties with seeds as well as many seedless varieties. The seeds, when present, are about 2 cm (3/4 in.) long. All parts of the tree and the unripe fruit are rich in milky, gummy latex. This sap has been used as glue. Aside from the food uses noted below, in regions where breadfruit grows, the bark is made into cloth and paper, the leaves are used for roofing and for wrapping food, and the wood is made into canoes and furniture.

The breadfruit was a central character of the most famous of all mutinies, that of the British HMS Bounty, which has generated five films, perhaps most memorably starring Marlon Brando as Fletcher Christian. Captain James Cook (1728–1779), and his crew visited Tahiti in 1769 and became acquainted with the breadfruit. Naturalist Sir Joseph Banks (1743–1820), and his botanist on that voyage, Daniel Carlsson Solander (1733–1782), extolled the virtues of the plant after returning to England, and some of the nobility came to believe that the breadfruit was a plant with great potential in the British West Indies, where it could be used to feed the slaves who worked the sugar cane fields. Banks urged King George III of England to introduce breadfruit to the West Indies, and in 1787 the King dispatched Captain William Bligh (1754–1817), on the HMS Bounty to accomplish this. The Bounty reached Tahiti in 1788, and the crew spent 6 months there collecting plants for shipment. So tireless was Bligh's pursuit of the plant that he earned the nickname "Breadfruit Bligh." The crew members, who had married Tahitians, were upset when forced to sail for the West Indies. After leaving Tahiti in 1789 with 1015 breadfruit plants, Captain Bligh was overpowered in a mutiny, and as a result he and 18 others were cast adrift in a longboat, while the breadfruit plants were tossed into the sea. Bligh's brusque personality and the crew's dissatisfaction at seeing much of the ship's supply of fresh water used for irrigating the cargo of breadfruit seedlings contributed to the mutiny. Unexpectedly, all of the outcasts except one (who was stoned to death by angry natives on the island of Tonga where the men tried to land for provisions) survived a 6700-km (3618 nautical miles), 41-day trip to the island of Timor in the East Indies. Christian and his sailors returned to Tahiti, where 16 of the 25 men decided to remain for good. Christian, along with eight others, their women, and a handful of Tahitian men then scoured the South Pacific for a safe haven, eventually settling on a desolate small island named Pitcairn. Although a British ship spent 3 months searching for them, the mutineers eluded detection. Many of their descendants still live on the island. Those who had remained on Tahiti were not so lucky. They were captured and brought to trial in England, where seven were exonerated and three were hanged. In a second voyage, Bligh was successful in bringing breadfruit to St. Vincent and Jamaica in 1792–1793. However, the slaves refused to adopt the new food as a staple part of their food supply until years after abolition. A tree that Captain Bligh planted in the St. Vincent Botanic Gardens is said to be still alive.

FIGURE 13.1 "Transplanting breadfruit from Tahiti," a 1796 painting by British artist Thomas Gosse. Captain William Bligh is shown standing in a boat.

FIGURE 13.2 Large breadfruit tree (*Artocarpus altilis*). (From Baillon, 1876–1892.)

FIGURE 13.3 Breadfruit (*Artocarpus altilis*). (From Rhind, 1855.)

CULINARY PORTRAIT

Unripe breadfruit is hard and the interior is white and somewhat fibrous. The ripe fruit is relatively soft. The breadfruit is usually picked when underripe and starchy and consumed as a cooked or boiled vegetable. This food is very versatile in its modes of consumption and preparation, whether green or ripe, raw, cooked, or juiced. The flavor is similar to a mixture of banana, melon, and papaya.

Like squash, breadfruit can be baked, grilled, fried, or boiled and served as a sweet or savory dish. When underripe fruit is cooked, the flavor tends to be bland, although some have reported that it is reminiscent of artichoke or olive. Breadfruit is also sometimes used in bread making or fermented. In some regions, such as the Marquesas Islands, unripe pulp is sun dried, reduced to powder, and after fermenting for a year in watertight pits a product resembling soft cheese is produced. Sweet, ripe breadfruit is eaten as a dessert. Although the fruit can be eaten raw, most kinds of breadfruit are laxative if eaten uncooked. Some varieties are boiled twice and the water thrown away to avoid digestive difficulties. Baked ripe fruit is said to taste like sweet potato. Cooked seeds are also consumed. In North America, breadfruit is often available fresh in some Latin and specialty produce markets and may also be purchased canned.

Tips: Select hard, evenly colored, fruit that is heavy for its size. Soak raw cut flesh in cold water before use. Avoid fruit that is soft or has black or moldy soft spots. Ripe breadfruit will normally be sold for fresh consumption, but should green breadfruit be available for cooking, beware of the milk-like, extremely sticky sap that will be released. Smear vegetable oil on hands and utensils to keep the sap from sticking. Breadfruit ripeness can be confirmed by nicking the fruit near the stalk end; the exposed flesh should be creamy to yellow rather than green.

Culinary Vocabulary

- In Hawaii, where breadfruit is especially popular, the tree and the fruit are known as *ulu*, and the use of the word in many Hawaiian culinary dishes indicates that the preparation includes breadfruit.
- In Polynesia, fermented breadfruit mash has many names, including *mahr*, *ma*, *masi*, *furo*, and *bwiru*.

CURIOSITIES OF SCIENCE AND TECHNOLOGY

- The Eocene era lasted from 34 to 54 million years ago. Fossils of breadfruit from the Eocene period have been found near Leipzig, Germany, illustrating the much larger extent of the tropical region at the time.
- Because breadfruit is a seasonal crop that produces much more than can be consumed fresh, Pacific islanders invented storage techniques. The most widespread method was to ferment the fruits by burying them in the ground. Fermented breadfruit can last for a year and can be prepared in various ways.
- In Polynesia, where breadfruit was one of the principal foods for the Islanders, it was customary to plant a tree for each baby that was born to ensure that the child would always have food to sustain its life.
- Early Hawaiians used sticky breadfruit latex to trap birds on the tips of posts. The feathers were plucked for ceremonial cloaks, after which the gummy substance was removed from the birds' feet, and they were released.
- Breadfruit latex was also used to caulk boats.
- In the Philippines, fiber from the bark it is made into harnesses for water buffalo.
- In Trinidad and the Bahamas, toasted breadfruit flowers are rubbed on gums to relieve toothache.
- Traditional Hawaiian drums are made from sections of breadfruit tree trunks 60 cm (2 ft.) long and 30 cm (1 ft.) wide and are played with the palms of the hands during Hula dances.
- Breadfruit maggots are said to be preferred by some Pacific Islander to the breadfruit itself.

KEY INFORMATION SOURCES

Atchley, J., and Cox, P.A. 1984. Breadfruit fermentation in Micronesia. *Econ. Bot.* 39:326–335.

Barrau, J. 1976. Breadfruit and its relatives: *Artocarpus* spp. In *Evolution of crop plants*. Edited by N.W. Simmonds. Longman, London. pp. 201–202.

Bennett, F.D., and Nozzolillo, C. 1987. How many seeds in a seeded breadfruit, *Artocarpus altilis* (Moraceae)? *Econ. Bot.* 41:370–374.

Coenen, J., and Barrau, J. 1961. The breadfruit tree in Micronesia. *South Pac. Bull.* 11:31–39, 65–67.

Fosberg, F.R. 1960. Introgression in *Artocarpus* (Moraceae) in Micronesia. *Brittonia*, 12:101–113.

Graham, H.D., and Negron de Bravo, E. 1981. Composition of the breadfruit *Artocarpus communis*. *J. Food Sci.* 46:535–539.

Jarrett, F.M. 1959. Studies in *Artocarpus* and allied genera. III. A revision of *Artocarpus* subgenus *Artocarpus*. *J. Arnold Arb.* 40:113–368.

Marriot, J., Perkins, C., and Been, B.D. 1979. Some factors affecting the storage of fresh breadfruit. *Sci. Hortic.* 10:177–181.

Morton, J. 1987. Breadfruit. In *Fruits of warm climates*. Creative Resource Systems, Winterville, NC. pp. 50–58.

Narashimhan, P. 1990. Breadfruit and jackfruit. In *Fruits of tropical and subtropical origin: composition, properties and uses*. Edited by S. Nagy, P.E. Shaw, and W.F. Wardowski. Florida Science Source Inc., Lake Alfred, FL. pp. 193–259.

Negron de Bravo, E., Graham, H.D., and Padovani, M. 1983. Composition of the breadmit (seeded breadfruit). *Caribb. J. Sci.* 19 (3–4):27–32.

Nwokolo, E. 1996. African breadfruit (*Treculia africana* Decne) and Polynesian breadfruit (*Artocarpus altilis* Fosberg). In *Food and feed from legumes and oilseeds*. Edited by E. Nwokolo and J. Smartt. Chapman and Hall, London. pp. 345–354.

Ragone, D. 1988. *Breadfruit varieties in the Pacific atolls*. United Nations Development Programme, Office of Project Services, Suava, Fiji. 45 pp.

Ragone, D. 1991. Ethnobotany of breadfruit in Polynesia. In *Islands, plants, and Polynesians—an introduction to Polynesian ethnobotany*. Edited by P.A. Cox and S.A. Banack. Dioscorides Press, Portland, OR. pp. 203–220.

Ragone, D. 1997. *Breadfruit, Artocarpus altilis (Parkinson) Fosberg*. International Plant Genetic Resources Institute, Rome, Italy. 77 pp.

Ragone, D. 2003. Breadfruit. In *Encyclopedia of food sciences and nutrition*. Edited by B. Caballero, L. Trugo, and P. Finglas. Academic Press, San Diego, CA. pp. 655–661.

Ragone, D., and Paull, R.E. 2008. *Artocarpus altilis*, breadfruit. In *The encyclopedia of fruit & nuts*. Edited by J. Janick and R.E. Paull. CABI, Wallingford, Oxfordshire, UK. pp. 476–479.

Rajendran, R. 1991. *Artocarpus altilis* (Parkinson) Fosberg. In *Plant resources of South-East Asia: 2. Edible fruits and nuts*. Edited by E.W.M. Verheij and R.E. Coronel. Pudoc, Leiden, the Netherlands. pp. 83–86.

Reeve, R.M. 1974. Histological structure and commercial dehydration potential of breadfruit. *Econ. Bot.* 28:82–96.

Rowe-Dutton, P. 1976. *Artocarpus altilis*—breadfruit. In *The propagation of tropical trees*. Edited by R.J. Garner and S.A. Chaudhri. Horticultural Review No. 4, Commonwealth Agriculture Bureau. pp. 248–268.

Sedgley, M. 1984. Moraceae—breadfruit. In *Tropical tree fruits for Australia*. Edited by P.E. Page. Queensland Department of Primary Industries, Brisbane, Queensland. pp. 100–103.

Spary, E., and White, P. 2004. Food of paradise: Tahitian breadfruit and the autocritique of European consumption. *Endeavour*, 28:75–80.

Stone, B. 1974. The correct botanical name for breadfruit. *J. Polyn. Soc.* 83:92–93.

Thompson, A.K., Been, B.O., and Perkins, C. 1974. Storage of fresh breadfruit. *Trop. Agric.* 51:407–415.

Wilder, G.P. 1928. *The breadfruit of Tahiti*. Bernice P. Bishop Museum, Honolulu, HI. 83 pp.

Worrel, D.B., Carrington, C.M.S., and Huber, D.J. 1998. Growth, maturation and ripening of breadfruit *Artocarpus altilis* (Park.) Fosb. *Sci. Hortic.* 76:17–28.

Zerega, N.G.C. 2003. The breadfruit trail. The wild ancestors of a staple food illuminate human migrations in the Pacific Islands. *Nat. Hist.* 112(10):46–51.

Zerega, N.J.C., Ragone, D., and Motley, T.J. 2004. Complex origins of breadfruit: implications for human migrations in Oceania. *Am. J. Bot.* 91:760–766.

Zerega, N.J.C., Ragone, D., and Motley, T.J. 2005. Breadfruit origins, diversity, and human-facilitated distribution. In *Darwin's harvest: new approaches to the origins, evolution, and conservation of crops*. Edited by T.J. Motley, N.J.C. Zerega, and H.B. Cross. Columbia University Press, New York. pp. 213–238.

Zerega, N.J.C., Ragone, D., and Motley, T.J. 2005. Species limits and a taxonomic treatment of breadfruit (*Artocarpus*, Moraceae). *Syst. Bot.* 30:603–615.

Specialty Cookbooks

Baker, E. 2000. *Taro and breadfruit (or potato!)*. E. Baker, Kapaa, HI. 70 pp.

Barnes, S. 1993. *The breadfruit in the Caribbean, 1783–1993: recipes from Caribbean cookbooks*. Agriculture and Life Sciences Division, Main Library, University of the West Indies, St. Augustine, Trinidad and Tobago. 26 pp.

Hirayama, F. 2002. *The breadfruit cookbook: the ulu cookbook*. Handworks, Kapaa, HI. 115 pp.

Pacific Tropical Botanical Garden. 1984. *Breadfruit (ulu): uses and recipes*. Revised edition. Pacific Tropical Botanical Garden, Lawai, Kauai, HI. 16 pp.

Swaby, E.M. 1979. *Breadfruit for economy*. Agency for Public Information and Scientific Research Council, Kingston, Jamaica. 32 pp.

14 Cacti

This chapter features

Cactus pear (*Opuntia ficus-indica*)
Dragon fruit (*Hylocereus undatus*, *Selenicereus megalanthus*)

The fruits of many cacti are tasty, and the stems of some species can be used as vegetables. The fruits of dozens of species can be purchased in Latin America, although only a few of these are grown on a major scale. The plants highlighted here are the most important in world commerce.

Cacti are native to North and South America and the West Indies (some species of the primitive genus *Rhipsalis* are apparently native to Madagascar and a few other parts of the Old World) and characteristically grow in hot, dry, and hostile desert areas. The family has 122 genera and approximately 1700 species, nearly all of which have succulent, spiny stems. The literature shows that Native Americans tasted virtually every species! It has been estimated that the average Mexican citizen eats as much cactus as the average American eats cauliflower.

Plants acquire carbon dioxide through pores (stomates), simultaneously losing water through these openings. Most plants keep their stomates open during the day, when sunlight is available, because the usual photosynthetic machinery requires a continuous influx of carbon dioxide while the sun is shining. Plants regulate the opening of the pores, and most plants close them at night to prevent unnecessary loss of water (keeping the pores open during the day is the price paid to acquire carbon dioxide). In arid environments, so much water can be lost when stomates are open during the day that plants may die. Six to seven percent of the world's plant species, including more than 99% of all cacti, have special photosynthetic machinery (crassulacean acid metabolism), which permits the plants to keep their stomates closed during daylight. During the much cooler nights, when water loss is much lower, the pores open, taking in carbon dioxide and storing it in a fixed chemical form (as "organic acids") until light becomes available during the day to process the organic acids further by photosynthesis. Although there are other adaptations to drought, crassulacean acid metabolism is the most significant mechanism adapting plants to dry environments. The average plant loses fives times as much water as many cacti on a surface area basis. Moreover, because the surface area of cacti is far less than that of plants that have true leaves, cacti are able to survive under conditions that would kill most plant species. Cactus pear, described in the following, produces dry matter on a surface area basis at the same rate as the average plant species, despite using only 20% as much water, and this remarkable water use efficiency is why cacti are so useful with respect to water conservation.

Cacti are favorite ornamentals, and many people specialize in collecting and cultivating these unusual plants. It has been claimed that there are more species of cacti growing in homes (especially in cold regions) than species of any other plant family (approximately 300 species of cacti are available as ornamentals). Unfortunately, many wild species have been greatly overcollected from nature. There is legislation that now protects cacti. The Convention on International Trade in Endangered Species of Wild Fauna and Flora requires appropriate permits and documentation for moving plants across international boundaries, and this applies to all cacti except certain species used in commerce. Arizona law requires that native cacti must be tagged and a fee paid before they can be moved. California also has strict requirements that deal with the movement of cacti. Candy made from pieces of wild barrel cactus was outlawed in the United States in 1952 to protect the species from being overharvested.

Before cheap synthetic aniline dyes were developed from coal tar in 1856, cactus plantations (including the cactus pear, discussed in the next section) were established for the production of cochineal (carminic acid), a dye. This is obtained by extracting the dye that the cochineal insects produce (from the females of the genus *Dactylopius*, which are much larger than the males). Approximately 32,000 female insects weigh 1 kg (70,000 weigh 1 lb.). The scale insect, which is native to Central America and Peru, feeds on the plants and develops a large quantity of the stain within its body. The dye was once used to color the robes of Aztec emperors, including those of Montezuma (1466–1520), a deep royal red. Taxes were sometimes paid in cochineal insects, and indeed the value of the dye once exceeded its weight in gold. In the sixteenth century, the export of cochineal from Mexico was second in importance and monetary value only to silver. The Spanish traders who introduced the dye to Europe kept the source secret, but Anton van Leeuwenhoek (1632–1723), the Dutch scientist who refined microscopes and first described cells, noted insect parts in the cochineal stain he was using and deduced that the dye came from insects. The dye was used for the robes of European royalty, the red jackets of British soldiers (made famous by American patriot Paul Revere when he warned in 1775 that "the redcoats are coming"), and the crimson jackets of the Northwest Mounted Police or "Mounties" (now the Royal Canadian Mounted Police). Cochineal is used to make a modern dye called carmine. Cochineal or carmine is still used in botanical stains and as a cloth dye and has some use as an edible (but tasteless and odorless) dye for foods like maraschino cherries, icings, creams, jellies, cakes, candies, wines, and liqueurs as well as in lipsticks. The dye has reacquired importance because coal-tar (aniline) dyes used for these purposes have been linked to cancer in laboratory animals.

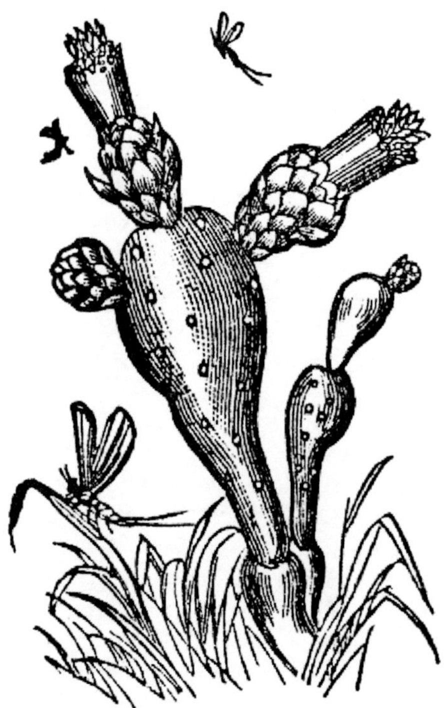

FIGURE 14.1 Cochineal insects on nopal cactus (*Opuntia cochenillifera* (L.) Salm-Dyck). (From Chambers and Chambers, 1875.)

CACTUS PEAR

Family: Cactaceae (cactus family)

Names

Scientific Name: *Opuntia ficus-indica* (L.) Mill.

- The word "cactus" is the root of the family name Cactaceae and also applies to the genus name *Cactus*. This name was used by the ancient Greeks for some spiny plant quite unrelated to cactus plants. Today, a "cactus" may be any of the approximately 1700 species in the Cactaceae.
- The cactus pear has also been called pear apple and tuna or tuna cactus (in Latin America) and Barbary fig and Indian fig (in many European countries). The names with "Indian" (for India) and "Barbary" (for the Barbary Coast, extending from the Egyptian border to the Atlantic, and including Morocco, Algiers, Tunis, and Tripoli) reflect the past tendency of Europeans to give exotic names to unfamiliar fruit (inappropriate names in this case, as the fruit comes from the New, not the Old World).
- The cactus pear has been called the "prickly pear" for many years. Unfortunately, some 20 different species of *Opuntia* are called prickly pear, so the name can be misleading.
- The phrase "prickly pear," the traditional name for the cactus pear and its fruit, evokes a negative image. To overcome this, in 1990 several people, notably Frieda Caplan, a California marketer of exotic fruits and vegetables, argued that the name "cactus pear" be used, and this recommendation was adopted by representatives of 10 countries at the Second International Conference on Tuna and Cochineal in Santiago, Chile, in 1992 (see previous section for explanation of "cochineal").
- "Tuna," a much-used Mexican term for cactus fruit, traces to the Caribbean word *tun*, fruit or seed. In Aztec Mexico the word *nochtli* was used for the fruits of cactus pears, but the Spanish conquistadors took up and established the word tuna.
- In Israel, the cactus pear fruit is called *sabra* (from the Hebrew *tzabar*), a term that was coined after the cactus was brought to the Middle East for use as a living fence. Sabra is also used to denote the cactus pear in Arabic and in northern Africa and southwestern Asia. The word has also come to be applied to native-born Israelis (sometimes to Arab women as well), by analogy meaning tough on the outside, sweet on the inside.
- Cactus pear plants are sometimes called "mission pears" because they were often grown about missions in Mexico and the southwestern United States. However, "mission pears" usually refers to very old pear varieties brought over from Europe by missionaries, especially from France, and often by Jesuits.
- In Europe, especially in Sicily, there are very tasty varieties of cactus pears with whitish or yellowish fruits. In Sicily, these are called *bastardi* or *bastadoni*.
- In Australia, where introduced *Opuntia* species are pernicious weeds, they are called "pest pears."
- The genus name *Opuntia* is based on the Greek name of a different plant that grew around the ancient town of Opus or Opuntia in Greece, the home site of a tribe called the Locri Opuntii.
- *Ficus-indica* in the scientific name *O. ficus-indica* means "fig of India," a rather imaginative description for a New World cactus. As noted below, the taste of some varieties is reminiscent of figs. However, the name has been explained as the result of Europeans considering the fruit, grown in Europe after Columbus brought back material to establish plantings, to be as tasty as figs.

Plant Portrait

The cactus pear or prickly pear has the appearance of a fleshy bush or small tree 3 to 5 m (10–16 ft.) in height. It is native to the desert zones of northwestern Mexico and southwestern United States. The plant was brought to Europe by the first Spanish colonists from Mexico and has been cultivated along the Mediterranean coast since the late seventeenth century. Numerous species of *Opuntia* are commonly called prickly pear, and in Mexico they are cultivated and harvested for food. *Opuntia tuna* (L.) Mill., commonly called "tuna" and also known as "elephant ear prickly pear," is particularly popular in Mexico because although it produces rather spiny, small fruit, the plants make an ideal hedge. ("Tuna" is also a frequent term in Mexico for cactus fruit, as discussed in Names.) However, *O. ficus-indica* is the chief species of interest, the main cactus grown outside of Mexico, and the most important edible species of cactus in the world. A mature plant can yield 100 to 200 fruits in a season. There are dozens of cultivated varieties. The cactus pear is an ideal fruit plant to raise in arid regions because it requires far less water than conventional crops. Native Americans dried cactus fruit in the sun, and these could be stored for at least a year. Cactus pears are grown for two edible commodities, as noted in the following. The plants are cultivated for fruit production on all continents except Antarctica, with at least 100,000 ha (247,000 acres) of orchards (in addition, considerable fruit is harvested from wild plants and home gardens). The species is grown in approximately 30 countries, mostly in Mexico, but with notable crops also in Chile, Bolivia, South Africa, Italy, Argentina, the United States, and Israel.

The cactus pear fruit is generally pear shaped and has a number of small spines (except in spineless varieties). The skin of commercial cactus pear fruits can be red, pink, orange, purple, green, or yellow. The fruits vary in weight from 100 to 200 g (3.5–7 oz.) and are generally 5 to 10 cm (2–4 in.) in length. The skin or rind is thick and fleshy (and represents 30%–40% of the weight of the fruit). The center of the fruit is filled with a soft juicy pulp (making up 60%–70% of the total fruit weight; approximately 85% of the pulp is water). The flesh may be green, light yellow-green, orange-yellow, deep golden, or dark red depending on variety. The sugar content of the fruits is approximately 15%. There are many hard coated, small, black seeds in the pulp (composing 5%–10% of the pulp weight). These seeds are considered very objectionable to new consumers, and there are efforts underway to breed seedless varieties.

In *Opuntia* species, the flattened branches or stem sections are technically called cladodes and popularly called pads (or sometime paddles) and joints (although cladodes are often joined together by what appear to be joints). These have the function of leaves (although some anatomists have equated the spines on the stems with the leaves of most plants). Although *Opuntia* species reproduce sexually by seeds, they also reproduce vegetatively when cladodes fall off the plant to the ground and produce roots and new daughter cladodes (indeed, commercial plantations are established entirely with cladodes or portions of cladodes). Tender young prickly pear pads, called nopales (and usually referred to as nopalitos when cut up for culinary use), have been consumed as a vegetable in central Mexico since pre-Hispanic times. Nopalitos are mostly water (more than 90%). In Mexico, most nopalitos come from *O. ficus-indica*, but in the United States, mostly in southern California and Texas, considerable amounts of nopalitos are also obtained from *Opuntia cochenillifera* (L.) Salm-Dyck (*Nopalea cochenillifera* (L.) Salm-Dyck), called the cochineal cactus and nopal cactus.

Culinary Portrait

The mild, pleasant flavor of a ripe cactus pear fruit, depending on variety, may resemble strawberries, watermelons, honeydew melons, figs, bananas, or citrus. The texture of the flesh is somewhat more granular than that of watermelon. Cactus pears may be eaten raw, at room temperature or chilled, and alone or with lemon or lime juice (desirable since the bland fruit lacks acidity). The rind is not edible. There are numerous hard seeds present in the edible flesh, and these make eating the flesh uncomfortable. The seeds can be chewed and eaten, swallowed, whole, or spit out. (Eating

FIGURE 14.2 Cactus pear (*Opuntia ficus-indica*). (From Jumelle, 1901.)

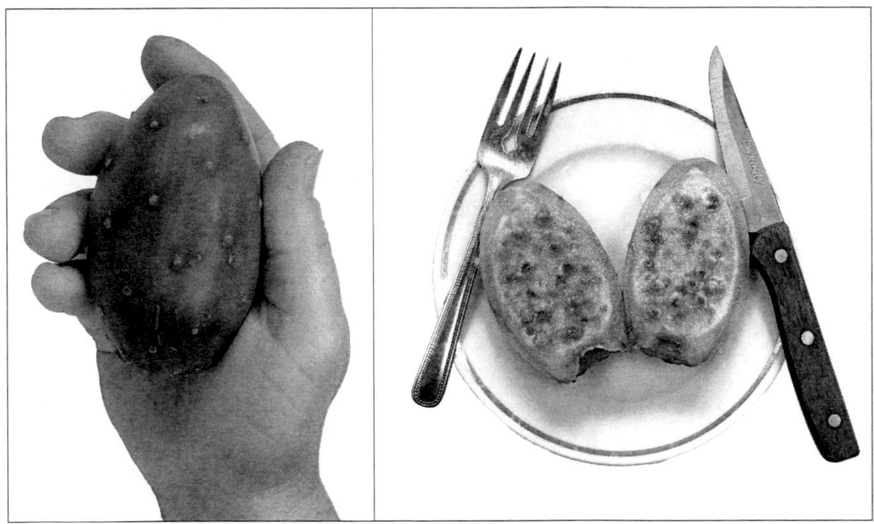

FIGURE 14.3 Cactus pear (*Opuntia ficus-indica*) fruit. Notice the large seeds in the sliced fruit. Photos by E. Small.

the seeds does not cause harm; indeed, Indians once collected the seeds, dried them in the sun, and ground them into a flourlike meal used for cooking.) The seeds can be sieved out if the fruit is to be cooked. Fresh pulp may be added to salads and various desserts. The flesh can be cooked into jams and preserved or cooked down into syrup as a base for jelly and candy—the "cactus candy" in some Mexican food stores. This syrup can be reduced even further into a dark red or black paste that is fermented into a potent alcoholic drink called *coloncha* (see Culinary vocabulary section). The fruit pulp can be dried and ground into flour for baking into small sweet cakes or stored for future use.

FIGURE 14.4 Prickly pear cactus on the plateau of Anahuac, Mexico (species not identified). (From von Marilaun, 1895.)

Cactus pears bought in supermarkets should have had the glochids (hair-like spines) removed (by scraping, singeing, or washing with a high-pressure nozzle). These may be almost invisible but can be very painful if they pierce the hand, fingers, or tongue, and once in the skin the irritation may last for several days.

The best fruits have deep, even color. Fruit should be ripened at room temperature and are ripe when they give slightly to palm pressure. Ripe fruit may be stored in the refrigerator for up to a week, but exposure to temperature below 5°C (41°F) for several days may result in injury and decay.

Some Mexican companies candy nopalitos or process them for export as pickles, sauce, or jam. The pads are said to taste something like green beans, or between that of green pepper and asparagus, although nopales in jars have a tart, pickle-like taste. Fresh nopales are sometimes available in supermarkets in North America as "cactus leaves" (although botanically they are flattened stems). These are generally harvested when 15 to 20 cm (6–8 in.) long. Nopales are highly perishable. The smaller young pads in the early spring are the most succulent, tender, and delicate in flavor, crunchy when fresh, and have the fewest spines. Immature pads also have less oxalic acid, which can be toxic in large amounts. Nopales are slippery because they exude a mucilaginous substance similar to that in okra (see Chapter 70). Because of the mucilage, diced nopalitos can help thicken soups and stew stock. (To avoid contact with the sticky fluid that oozes from the nopales, they can be steamed whole, just long enough for their color to change from bright green to olive drab. Once the color changes, they should immediately be plunged into a bowl of cool water, then sliced on a cutting board. The fluid can be mixed into dishes to enhance them.) Fresh pads are full of water and should be bright green and firm. To prepare the pad, simply hold its base and scrape the skin on both sides with a blunt knife until all the spines are removed. (Use tongs to avoid getting spines or glochids (tiny, fuzzy spines) in your fingers, and take care to remove the areoles (the places where spines develop on the pad.) Then peel the pads with a sharp knife or vegetable peeler and cut them into shoestring strips or dice them according to the needs of the recipe. They can be eaten raw in salads, boiled and fried like eggplant, pickled with spices, or cooked with shellfish, pork, chilies, tomatoes, eggs, coriander, garlic, and onions. Omelets are commonly prepared with young cactus pear pads throughout the southwestern United States. Cactus pie can be made from cactus nopalitos. The pie may taste apple-like because both the cactus and apples contain high levels of malic acid.

In Mexico, the petals from the flowers of cactus pear are added to meat dishes. Native Americans traditionally celebrated a "cactus moon" in the early spring, by collecting flower buds of wild *Opuntia* species and cooking them as a special treat.

Culinary Vocabulary

- *Huevos rancheros con nopalitos* is ranch-style eggs with cactus pads.
- *Calonche*, *colonsha*, or *clcolonshe* (pronounced kahl-ohn-chay) is an alcoholic beverage made from the fermented juice of certain cacti. Indians once used cactus pear fruits to prepare this as well as other alcoholic drinks.
- *Queso de tuna* (pronounced KEH-soh day too-nah) is a sweet paste prepared from fermented prickly pear juice, used in Mexican confections.

Curiosities of Science and Technology

- Cactus pear seeds have been found in human coprolites (mummified feces) dated at 9000 years of age from caves in Mexico, providing proof that indigenous peoples were using cacti as food in ancient times.
- The first published illustration of cactus pear appeared in 1535 in the Spanish book *La Historia General* by Oviedo y Valdés. His erroneous report that eating the fruits turned one's urine red was widely circulated (but see "pseudohematuria" in discussion of dragon fruit).
- Certain opuntias and barrel cacti can live for up to 3 years on water stored in their stems, in the absence of water uptake from the soil.
- So-called "rain roots" are produced by *Opuntia* species within a few hours of rain falling on the plants. These quickly absorb available water and die off as soon as the soil dries.
- The mucilaginous sap from the pads of the cactus pear can be used in first aid similar to the *Aloe vera* plant. Simply cut off a portion of a pad, crush it, and squeeze the juice onto a cut, burn, or bruise. In Mexico, the cactus pear is commonly used on burns and swellings.
- In Argentina, the slimy juice from cactus pear stems was traditionally used to make whitewash more adhesive. In the construction of Spanish mission buildings in California during the 1700s and 1800s, the mucilage was used to bind adobe bricks.
- Some forms of wild cactus pear form massive thickets. These have been observed to survive fires in southern California by regenerating from live stems in the center of the thickets where fire was unable to penetrate.
- Prickly pear cacti are frequently grown into hedges and fences by planting them approximately 30 cm (1 ft.) or so apart. Within several years, the plants grow together to form a spiny barrier that will repel any intruder larger than a rabbit.
- The American plant breeder and horticulturist Luther Burbank (1849–1926), was a very controversial genius, who introduced many new plants of great value, but also made some whose virtues were considered very dubious (see Chapter 100). Burbank helped develop a spineless form of cactus pear, and in 1911 he claimed that it was so good that it "promises to be of as great or even greater value to the human race than the discovery of steam." This led to numerous large plantings, especially around Los Angeles, and great disappointment when the food value of the plants proved to be much lower than expected.
- In Texas and Mexico, ranchers commonly use propane torches to singe the spines of wild *Opuntia* species so that cattle can forage on them. Cattle readily eat the singed plants, consuming cactus up to 10% of their body weight daily. Because the cacti typically contain

approximately 90% water, the cattle can be sustained for many months of drought by relying entirely on the water in the plants.

- The introduction of *Opuntia* cacti to Australia in 1832 as hedging plants caused an ecological disaster. By 1925, several species were spreading through Australia like wildfire. A variety of control measures proved ineffective until 1925, when the Argentinean moth *Cactoblastis cactorum* was introduced, the larvae of which feed on the cacti. By 1933, 90% of the cacti had been eradicated. A similar situation occurred in South Africa with *O. ficus-indica* covering a huge territory before suitable biological control agents (a weevil and a beetle as well as the moth used in Australia) were found in the 1930s. (Ironically, cactus pear is now considered a valuable crop in South Africa, and farmers growing it there have great difficulties controlling the biological control agents.)
- The cactus fruit is naturally protected by tufts of glochids (small, barbed spines). However, the biological purpose of fleshy fruits such as that of the cactus pear is to attract animals to eat the flesh, so that the seeds will be distributed (the seeds of the cactus pear are distributed principally by being consumed by birds and mammals, which spread the seeds in their droppings). As the fruit ripens, the glochids drop off, so that animals will not be discouraged from eating the fruits.
- As noted earlier, the peculiar photosynthetic mechanism of cacti involves storage of acidic compounds during the night, which are metabolized into sugars when sunlight is available. As a result, cactus tissue harvested during the night will taste more acidic than the same tissue harvested during the day. Depending on the time of day (which affects the amount of light available), the taste of nopales may differ considerably.

Key Information Sources

Annecke, D.P., and Moran, V.C. 1978. Critical reviews of biological pest control in South Africa. 2. The prickly pear, *Opuntia ficus-indica* (L.) Miller. *J. Entomol. Soc. South. Afr.* 41: 161–188.

Ayoub, T.M. 2009. *Cactus pear fruit: a promising fruit.* VDM Verlag Dr. Müller, Saarbrücken, Germany. 124 pp.

Barbera, G., Inglese, P., and Pimienta-Barrios, E. (Eds.). 1995. *Agro-ecology, cultivation and uses of cactus pear.* FAO Plant Production and Protection Paper 132. Food and Agriculture Organization of the United Nations, Rome. 216 pp.

Curtis, J.R. 1977. Prickly pear farming in the Santa Clara Valley, California. *Econ. Bot.* 31: 175–179.

Hanselka, C.W., and Paschal, J.C. 1991. Prickly pear cactus: a Texas rangeland enigma. *Rangelands*, 13: 109–111.

Inglese, P. (*Editor*), and Brutsch, M.O. (*Conference convenor*). 1997. *Third international congress on cactus pear and cochenille.* International Society for Horticultural Science, Wageningen, The Netherlands. 180 pp.

Inglese, P., Barbera, G, and La Mantia, T. 1995. Research strategies for the improvement of cactus pear (*Opuntia ficus-indica*) fruit quality and production. *J. Arid Environ.* 29: 455–468.

Jacobo, C. M. 2001. Cactus pear domestication and breeding. *Plant Breed. Rev.* 20: 135–166.

Meyer, N.B., and McLaughlin, J.L. 1981. Economic uses of *Opuntia. Cactus Succulent J.* 53: 107–112.

Mondragon-Jacobo, C., and Perez-Gonzalez, S. 1996. Native cultivars of cactus pear in Mexico. In *Progress in new crops.* Edited by J. Janick. ASHS Press, Arlingon, VA. pp. 446–450.

Nefzaoui, A., and Inglese, P. 2002. *Fourth international congress on cactus pear and cochineal.* International Society for Horticultural Science, Wageningen, The Netherlands. 356 pp.

Nobel, P.S. 1988. *Environmental biology of agaves and cacti.* Cambridge University Press, Cambridge, UK. 270 pp.

Nobel, P.S. (Ed.). 2002. *Cacti. Biology and uses.* University of California Press, Los Angeles. 280 pp.

Nobel, P.S. 2008. *Opuntia ficus-indica*, cactus pear. In *The encyclopedia of fruit & nuts.* Edited by J. Janick and R.E. Paull. CABI, Wallingford, Oxfordshire, UK. pp. 216–221.

Ortiz-Hernández, Y.D. 1999. *Pitahayas. A new crop for Mexico.* Limusa/Noriega Editores, Balderas, Mexico. 111 pp (in Spanish).

Pareek, O.P., Singh, R.S., Nath, V., and Vashishtha, B.B. 2001. *The prickly pear (Opuntia ficus-indica L. Mill.).* Agrobios, Jodhpur, India. 76 pp.

Pimienta-Barrios, E. 1993. Vegetable cactus (*Opuntia*). In *Pulses and vegetables*. Edited by J.T. Williams. Chapman and Hall, London, UK. pp. 177–191.

Russell, E.C., and Felker, P. 1987. The prickly pears (*Opuntia* spp. Cactaceae): a source of human and animal food in semiarid regions. *Econ. Bot.* 41: 433–445.

Russell, E.C., and Felker, P. 1987. Comparative cold-hardiness of *Opuntia* spp. and cvs. grown for fruit, vegetable and fodder production. *J. Hortic. Sci.* 62: 545–550.

Saenz, C. 2000. Processing technologies: an alternative for cactus pear (*Opuntia* spp.) fruits and cladodes. *J. Arid Environ.* 46: 209–225.

Savio, Y. 1987. Prickly pear cactus: the pads are "nopales," and the fruits are "tunas"—they are easy to grow and wonderful to eat. *Cactus Succulent J.* 59(3): 113–117.

Savo, Y. 1989. *Prickly pear cactus*. University of California, Davis, Family Farm Series. 6 pp.

Small, E., and Catling, P.M. 2004. Blossoming treasures of biodiversity, 11. Cactus pear (*Opuntia ficus-indica*)—miracle of water conservation. *Biodiversity*, 5(1): 27–31.

Stintzing, F.C., Schieber, A., and Carle, R. 2001. Phytochemical and nutritional significance of cactus pear. *Eur. Food Res. Technol.* 212: 396–407.

Specialty Cookbooks

Niethammer, C.J. 2004. *The prickly pear cookbook*. Rio Nuevo, Tucson, AZ. 84 pp.

Tate, J.L. 1972. *Cactus cook book; succulent cookery international*. 2nd ed. Cactus & Succulent Society of America, Reseda, CA. 126 pp.

DRAGON FRUIT

Family: Cactaceae (cactus family)

Names

Scientific Names: The main commercial species of dragon fruit belong to the genera *Hylocereus* and *Selenicereus* (featured in this chapter are *Hylocereus undatus* (Haw.) Britton & Rose and *Selenicereus megalanthus* (K. Schum. ex Vaupel) Moran)

- Species representing several genera of the cactus family are known as dragon fruit (occasionally spelled dragonfruit and rarely dragon-fruit). The curious name is based on the skin having scalelike bracts, said to resemble those of the legendary reptile.
- The most popular alternate English name for dragon fruit species is pitaya (also spelled pitahaya and pitajaya), a word derived from the native American Indian (Ta'no) name for the scaly fruit of the plants. However, in addition to the vine-like forms of *Hylocereus* and *Selenicereus* featured in this chapter, in Latin American countries, pitaya also refers to several columnar (free-standing) cacti with edible fruit. For example, one of these is *Stenocereus queretaroensis* (F.A.C. Weber) Buxb., a particularly important commercial species in Mexico. Use of the word pitaya in Latin America can be confusing. Outside of Latin America, the word pitaya is generally synonymous with dragon fruit.
- Both the plants and their fruit are also sometimes called strawberry pear and dragon pearl fruit.
- In Israel, the name Eden fruit is sometimes used.
- In Hawaii, the name Honolulu queen is sometimes encountered.
- The occasional name "fruit of the shipwrecked man of the desert" reflects the refreshing properties of the fruit.
- Because the plants often bloom at night they are often called moonflower, lady of the night, belle of the night, queen of the night, and night-blooming cereus. (The flowers are usually open for only one night. Typical of night-blooming cacti, the flowers are large, white, and fragrant to attract night-foraging animals.)

- The commercially important species *S. megalanthus* is commonly known as yellow dragon fruit, a reference to the color of the flesh of the fruit. The commercially important *H. undatus* is usually simply called dragon fruit but is sometimes known as red dragon fruit for the color of the fruit skin.
- The genus name *Hylocereus* is based on the Greek *hyle*, a wood, forest, or woodland + *cereus*; *Cereus* is a large genus of the cactus family. The name reflects occupation of wooded lands. The genus name *Selenicereus* is based on the Greek *selene*, the moon + *cereus*; the name reflects the nocturnal blooming of the flowers.

Plant Portrait

The genera *Hylocereus* and *Selenicereus* contain the main commercial species of dragon fruit. The two genera are closely related, and some specialists have recommended that *Selenicereus* should be placed in *Hylocereus*. Dragon fruit species are native to Mexico, Central America, and South America. As with most cacti, the plants are adapted to dry tropical climates with limited rainfall.

The approximately 16 species of *Hylocereus* are native from Central America to northern South America. Species of this genus have elongated, normally three-angled stems, branches with aerial roots, and large, usually white flowers. The fruits are spineless but have several or many leaf-like scales. Commercial fruits of *Hylocereus* species range in weight from 200 to 800 g (7–28 oz.). The most widely grown dragon fruit species is *H. undatus*, which generally has fruit with white pulp and pink or red skin. It is common throughout the tropics and subtropics of the New World, where it has been widely cultivated, and it is also the dominant species raised outside of Latin America.

Hylocereus costaricensis (F.A.C. Weber) Britton & Rose is another commercially important species, but mainly in Latin America. It is frequently confused with the similar *Hylocereus polyrhizus* (F.A.C. Weber) Britton & Rose (delimitation of these two species is problematical). *H. costaricensis* is native to Central America and northern South America and produces fruit with pink skin and reddish pulp. Several other species of *Hylocereus* are also cultivated for fruit.

FIGURE 14.5 Plantation of dragon fruit (*Hylocereus undatus*) in Vietnam. The treelike appearance is misleading: the vinelike stems have grown over a thick post. (Courtesy of Dr. Christian Puff, Faculty Center of Biodiversity, University of Vienna.)

The 20 species of *Selenicereus* occur in tropical America and the Caribbean region. They have ribbed or angled stems, from which aerial roots arise irregularly. The flowers are often large, and the fruits are usually reddish and covered with clusters of deciduous spines, bristles, and hairs. The most important species is *S. megalanthus*, a native of Columbia, Ecuador, Nicaragua, and Peru. Its fruit has white flesh and yellow skin, giving rise to the name "yellow dragon fruit." Commercial fruits of *S. megalanthus* range in weight from 80 to 300 g (3–11 oz.) and have skin that is yellow-orange and rough, bumpy, and prickly (the spines are easily removed at fruit maturity). In addition to Latin America, Israel is an important grower of this species.

Dragon fruit species, like most cacti, are constructed of succulent stems and branches, often with spines. The plants are vinelike, climbing over trees or trailing along the ground or over rocks. In addition to the main basal root that anchors the plants in the ground, dragon fruit species have aerial roots, assisting them to live on trees and other objects. The aerial roots often penetrate into cracks and crevices of rocks, attaching the plants and sometimes serving to obtain extra nutrients and water where they encounter pockets of soil (the aerial roots are not parasitic on other plants). The stems of some species can grow 5 m (16 ft.) in a year. The flowers can be huge—up to 30 cm (1 ft.) across. The fruits are leathery, and commercial fruits are often approximately 10 cm (4 in.) long. *H. undatus*, the principal commercial species, has three-winged, marginally wavy, fleshy, much-branched segmented stems; white flowers; and nonspiny, oblong fruit with bright red fleshy scales. There are many named varieties of dragon fruit, although breeding of new varieties is still in its infancy. At present, most commercial fruit is grown from clones (i.e., as cuttings from standard plants) rather than from cultivars (named varieties).

Wild dragon fruit species have been harvested in their native New World area by indigenous inhabitants for centuries. Dragon fruit is grown not only in its indigenous American area (especially in Nicaragua and Columbia), but in recent years cultivation has increased greatly in Southeast Asia, including Vietnam, Thailand, Philippines, and Malaysia. The fruit is also produced in Cambodia, Indonesia, Japan, Taiwan, Okinawa, Israel, China, and Reunion. There has been small-scale production in California, Hawaii, New Zealand, Spain, Australia, Ecuador, Guatemala, Mexico, and Peru. In Vietnam, the country that leads in dragon fruit production, the plants can produce six crops in a year, and 30 tonnes of fruit per hectare (13.4 tons/acre) annually. The main growing region of Vietnam is Binh Thuan province in the south of the country. Here, in 2006, more than 120,000 tonnes of fruit were harvested, and 22,000 tonnes worth 13.5 million dollars were exported. In recent years, Israel has surpassed Vietnam in exports to Europe. In Western markets, the fruit is sold as a specialty item at a premium price, although in areas of cultivation the fruit may be much less expensive.

Aside from use for fruits, some of the species are grown as ornamentals and as hedges in tropical areas. The stems and flowers of several species, particularly *Selenicereus grandiflorus*, are used to prepare drugs with a spasmolytic (spasm-relieving) effect on coronary vessels to promote blood circulation.

Culinary Portrait

Dragon fruits are eaten raw. The skin is inedible, and so are the black seeds, but the latter are very small and crunchy, producing a texture reminiscent of kiwifruit. The fruit is also used, on a relatively minor scale, to prepare juice, wine, liqueurs, and purée, which are often incorporated into ice cream. The fruit can be used in marmalades, jellies, and beverages. Red fruit is popular in expensive restaurants as centerpieces in exotic green or fruit salads. The flesh has a delicate sweet flavor reminiscent of melon, although some fruits in the marketplace are relatively bland. Some clones have produced fruit that has flesh with an objectionable odor, but others are odorless or have a pleasant smell. As with many other fruits, dragon fruit is considered to be a good source of antioxidants and vitamins.

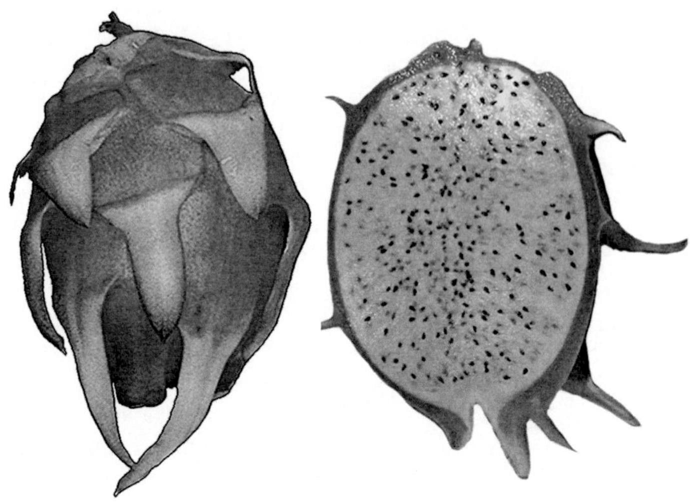

FIGURE 14.6 Dragon fruit (*Hylocereus undatus*). Left, whole fruit; right, sectioned fruit showing tiny black seeds. (Courtesy of Dr. Christian Puff, Faculty Center of Biodiversity, University of Vienna.)

FIGURE 14.7 Dragon fruit (*Hylocereus undatus*) in Vietnam: market scene. (Courtesy of Dr. Christian Puff, Faculty Center of Biodiversity, University of Vienna.)

In buying dragon fruit, it is advisable to choose those that yield slightly when lightly squeezed. Ripe fruit may be stored in the refrigerator but should be consumed within a week. The fruit should not be washed before refrigeration but should be before cutting or serving. The inedible skin may be peeled away, and the remaining flesh held with a fork. Alternatively, the fruit can be sliced in half and scooped out with a spoon. Best flavor is usually obtained by chilling the fruit or juice. Lime or lemon juice also enhances the taste.

Culinary Vocabulary

- "Dragon's Blood Punch" has traditionally been made from reddish fruit juice, fortified with alcohol or not. The use of red-fleshed dragon fruit to flavor and color the beverage adds an exotic touch.
- A "Dragotini" (not to be confused with an Ivan Drago-tini) is a Martini flavored with dragon fruit instead of an olive.

Curiosities of Science and Technology

- To raise interest in dragon fruit, Asian marketers concocted a story that it was expelled from the jaws of fire-breathing dragons thousands of years ago. However, the fruit could only be collected if soldiers killed the beast, and the fruit had to be given to the emperor. The message of this tale was that the fruit was so wonderful it could only be consumed by royalty. Indeed, a century ago, the French brought dragon fruit to Vietnam for exclusive use of the king. In later years, only the wealthy consumed the fruit. Today, production is on a large scale, and everyone can eat it.
- Dragon fruit species have been found to have a high tolerance to sulfurous gases, and accordingly in Nicaragua, they are grown commercially on the slopes of the Mount Santiago volcano.
- Pseudohematuria, a harmless reddish coloration of urine and feces, has been observed in some individuals who consume considerable amounts of red-fleshed dragon fruit (*H. polyrhizus*), reportedly because of the red pigments hylocerenin and isohylocerenin.
- The dragon fruit should not be confused with the dragon tree, *Dracaena* species, especially *Dracaena draco* L., a native of the Canary Islands, Cape Verde, Madeira, Azores, and western Morocco. Its bark and leaves, when cut, secrete a reddish resin known as dragon's blood, which was used in medieval magic and alchemy.

FIGURE 14.8 Dragon tree (*Dracaena draco*). (From Paxton's Magazine of Botany (London), 16: 45, 1849.)

Key Information Sources

Note: Dragon fruit is cultivated commercially in at least 20 countries, mostly non-English-speaking, and so agricultural guides are predominantly in foreign languages. A few guides in English have been prepared for cultivation in the subtropical parts of mainland USA, where dragon fruit is expanding in cultivation and where the demand by high-end restaurants and hotels far exceeds local supply. There are at least a half dozen extensive books in Spanish as well as monographs in other languages.

Anderson, E. F. 2001. *The cactus family*. Timber Press, Portland, OR. 776 pp.

Bauer, R. 2003. A synopsis of the Hylocereeae F. Buxb. *Cactaceae Systematics Initiatives*, 17:1–63.

Bellec, F. le, Vaillant, F., and Imbert, E. 2006. Pitahaya (*Hylocereus* spp.): a new fruit crop, a market with a future. *Fruits*, 61:237–250.

Cálix de Dios, H. 2005. A new subspecies of *Hylocereus undatus* (Cactaceae) from southeastern Mexico. *Haseltonia*, 11:11–17.

Dag, A., and Mizrahi, Y. 2005. Effect of pollination method on fruit set and fruit characteristics in the vine cactus *Selenicereus megalanthus* ("yellow pitaya"). *J. Hortic. Sci. Biotech.* 80:618–622.

Gibson, A.G., and Nobel, P.S. 1986. *The cactus primer*. Harvard University Press, Cambridge, MA. 286 pp.

Hart, G. 2005. From prickly pear to dragon fruit: the changing face of cactus-fruit growing. *Cactus Succulent J.* 77:293–299, 319.

Lichtenzveig, J., Abbo, S., Nerd, A., Tel-Zur, N., and Mizrahi, Y. 2000. Cytology and mating systems in the climbing cacti *Hylocereus* and *Selenicereus*. *Am. J. Bot.* 87:1058–1065.

Luders, L. 2001. The pitaya or dragon fruit. *Yearbook West Australian Nut and Tree Crops Assoc.* 25:78–80.

Merten, S. 2004. A review of *Hylocereus* production in the United States. *Yearbook West Australian Nut and Tree Crops Assoc.* 27:20–29.

Mizrahi, Y., Nerd, A., and Nobel, P.S. 1996. Cacti as crops. *Hortic. Rev.* 18:291–320.

Nerd, A., and Mizrahi, Y. 1997. Reproductive biology of cactus fruit crops. *Hortic. Rev.* 18:312–346.

Nobel, P.S. 2008. *Hylocereus undatus*, pitahaya. In *The encyclopedia of fruit & nuts*. Edited by J. Janick and R.E. Paull. CABI, Wallingford, Oxfordshire, UK. pp. 215–216.

Raveh, E., Weiss, J., Nerd, A., and Mizrahi, Y. 1993. Pitayas (genus *Hylocereus*). A new fruit crop for the Negev Desert of Israel. In *New crops*. Edited by J. Janick and J.E. Simon. Wiley, New York. pp. 491–495.

Small, E., and Catling, P.M. 2007. Blossoming treasures of biodiversity:26. Dragon fruit—a delicacy with exotic appeal and a hot name. *Biodiversity*, 8(4):31–36.

Tel-Zur, N., Abbo, S., Bar-Zvi, D., and Mizrahi, Y. 2004. Genetic relationships among *Hylocereus* and *Selenicereus* vine cacti (Cactaceae): evidence from hybridization and cytological studies. *Ann. Bot.* 94:527–534.

Thomson, P.H. 2002. *Pitahaya: a promising new fruit crop for Southern California*. 2nd ed. Bonsall Publications, Bonsall, CA. 46 pp.

Valiente-Banuet, A., Santos-Gally, R., Arizmendi, M.C., and Casas, A. 2007. Pollination biology of the hemiepiphytic cactus *Hylocereus undatus* in the Tehuacan Valley, Mexico. *J. Arid Environ.* 68:1–8.

Van To, L., Ngu, N., Duc, N.D., and Huong, H.T.T. 2002. Dragon fruit quality and storage life: effect of harvesting time, use of plant growth regulators and modified atmospheric packaging. *Acta Hortic.* 575:611–621.

Specialty Cookbooks

D'Leong, A. [Liang Yili]. 2008. *Savoury pitaya sensation* [*Feng wei bai bian huo long guo liao li*]. Lan tian chu ban she, Selangor, Malaysia (in Chinese and English; recipes for 36 dishes using red dragon fruit).

15 Candlenut

Family: Euphorbiaceae (spurge family)

NAMES

Scientific Name: *Aleurites moluccana* (L.) Willd. (*A. javanica* Gand., *A. triloba* J.R. Forst. & G. Forst., *Jatropha moluccana* L.)

- The candlenut (sometimes "candle nut") acquired its name based on its old use as a candle. The nuts were strung on ribs from palm leaves and used as candles in Malaysia and Indonesia as well as in Hawaii. By lighting the top nut of a string of a dozen or more nuts, it would burn for 2 to 3 minutes because of its high oil content (up to 70%), and as it was burning it would drop oil on the next nut in the chain and light that nut. Slowly, the chain of nuts would burn down like a candle.
- The candlenut is also called the candlenut tree, candleberry (a name that also refers to other species), candleberry tree, lumbang, and lumbang tree.
- The name varnish tree points out the use of the oil in varnishes.
- Because of the resemblance to walnuts, the tree is also called Belgian walnut, country walnut, Indian walnut, Tahiti walnut, and Otaheite walnut (Tahiti was formerly called Otaheite). Candlenuts are smaller than walnuts.
- In Hawaii, candlenut is known as kukui (which means candle in Hawaiian).
- The genus name *Aleurites* is based on the Greek *aleuron*, wheaten flour + *ites*, like, referring to fine meal or flour. Some of the species appear to be dusted with flour. The leaves and flowering parts of the candlenut are covered with what appears to be a mealy material.
- *Moluccana* in the scientific name *A. moluccana* means Indonesian.

PLANT PORTRAIT

The candlenut is a tropical Southeast Asian tree, usually 9 to 12 m (30–39 ft.) in height, growing as tall as 20 m (66 ft.), occasionally to 27 m (90 ft.). It bears clusters of small, whitish flowers, maple-like leaves, and nut-like, round, hard fruits 2.5 to 8 cm (about 1–3 in.) in diameter, with a thick, fleshy husk. The shell of the fruit is whitish when young, becoming black at maturity. The fruits contain one or two sometimes three waxy white kernels. This native of Malaysia, Polynesia, Malay Peninsula, Philippines, and the South Seas Islands is now widely distributed in the Tropics. The ancient Polynesians brought candlenut to the Hawaiian Islands, and it is the state's official tree. The species is cultivated in China and the Philippine Islands for a drying oil used in paints, varnishes, lacquer, and soft soap. The candlenut as well as several unrelated species is used to produce "tung oil," a superior quick-drying oil with many applications. The tung-oil tree (*Vernica fordii* (Hemsl.) Airy Shaw, formerly known as *Aleurites fordii* Hemsl.) is the principal source of high-quality tung oil.

CULINARY PORTRAIT

Fresh candlenuts contain a toxin, making them slightly poisonous, and so are not eaten raw. (One author stated that no more than two or three fresh nuts can be consumed, as more causes nausea and vomiting.) The nuts are usually roasted before being cracked open. The kernels adhere to the sides of the shell and are difficult to separate, so that a mixture of whole and cracked kernels is usually available. Candlenuts have a similar taste and texture to macadamia nuts. Candlenuts are used especially in the cuisines of Hawaii, Indonesia, and Malaysia. In Southeast Asia, the nuts are widely used as a flavoring agent, typically sautéed and or fried with other ingredients in savory dishes. Crushed nuts are also added to soups. Candlenuts are additionally used as a thickening and stabilizing agent in curries. In North America, the nuts may be available from stores specializing in Asian foods. The oil is sometimes used for cosmetic or aromatherapy purposes. It is a powerful laxative and should not be consumed. Candlenuts have a limited shelf life and may be stored in the refrigerator or freezer to slow the development of rancidity.

Culinary Vocabulary

- In Hawaii, a small amount of the pounded, roasted nuts, along with salt and sometimes chilli peppers, are used as a relish called *inamona*.
- In Indonesian cuisine, the seed is used as an indispensable spice called *kemiri*, which has little taste of its own, but enhances other flavors.
- In Indonesia, the residual oil cake is sometimes made into a snack food called *dage kemiri*.
- Chicken kpitan is a popular Indian-influenced curry dish in Malaysia, which incorporates considerable candlenut to produce a nutty taste.
- *Sambal bajak* is a spicy, pungent, Indonesian relish, made with chilli and candlenuts.

CURIOSITIES OF SCIENCE AND TECHNOLOGY

- As well as using candlenuts for candles, Hawaiians used them for torches. The end of a bamboo pole was hollowing out, filled with candlenut kernels, and set alight.
- Hawaiians covered their fishing nets with the resin from candlenut trees every 3 to 6 months to stop their nets from rotting.
- In Hawaii, the oil from green candlenuts has been used as a sealant over wounds. The mature hard shells were made into traditional Hawaiian decorative jewelry.
- Hawaiians have long made leis and bracelets by stringing candlenut shells together. The candlenut blossom is the symbol of the Hawaiian island of Molokai, and in 2000, the tree was declared to be the "official lei material" of the island. Giving a candlenut lei is only done with the understanding that it is an intimate expression of love.
- The soot from burning candlenuts was used as an ink for traditional Polynesian tattoos.
- According to belief in Southeast Asia, candlenut oil is an excellent hair treatment, leaving hair shiny and resistant to turning gray.
- In old Hawaii, candlenut oil was used to grease sled runners to increase speed going down the slopes of hills, a game played mostly by the chiefs for sport. In modern times, candlenut oil is used in high performance race cars.
- As noted earlier, the candlenut is one of the species used to produce tung oil, and so can be called "tung nut." "Tung nut" is a "palindrome"; that is, it is spelled the same backward as forward.
- Candlenuts are used in Indonesia in a gambling game in which a player tries to break an opponent's nut by hitting it with his own nut. A special cultivar with hard nuts (known as *kemiri pidak* and *muncang kelenteng*) is used for the purpose.
- New Caledonian crows (*Corvus moneduloides*) have been observed to drop candlenuts from heights to open them. Although many birds and nonhuman primates do this, these crows almost always first placed the nuts in the fork of a high tree branch located directly over a hard object, then moved the branch to release the nut. Once a suitable nut-cracking site had been located, the crows communally used the same forked branch to open nuts (see Hunt et al., 2002).

FIGURE 15.1 Candlenut (*Aleurites moluccana*). Left: flowering branch, from Lamarck and Poiret (1744–1829). Right: fruits (sectioned above). (From Engler and Prantl, 1889–1915.)

KEY INFORMATION SOURCES

Belin-Depoux, M., and Clair-Maczulajtys, D. 1974. Introduction to the study of foliar glands of *Aleurites moluccana* Willd. (Euphorbiaceae). I. the gland and its ontogenesis. *Rev. Gen. Bot.* 81:335–351 (in French, English summary).

Belin-Depoux, M., and Clair-Maczulajtys, D. 1975. Introduction to the study of foliar glands of *Aleurites moluccana* Willd. (Euphorbiaceae). II. histological and cytological aspects of the functional petiolar gland. *Rev. Gen. Bot.* 82:119–155 (in French, English summary).

Bory, G., and Clair-Maczulajtys, D. 1977. Contribution to the study of the phylloplane of *Aleurites moluccana* Willd. (Euphorbiaceae) [Petiolar glands]. *Rev. Cytol. Biol. Veg.* 40:1– 13 (in French, English summary).

Dali, J., and Gintings, A.N. 1981. *Guide to candlenut (Aleurites moluccana) planting*. Lembaga Penelitian Hutan, Bogor, Indonesia. 20 pp (in Indonesian).

Dayan, M.P., and Constantino, C.H. 1990. Lumbang (*Aleurites moluccana* (L.) Willd.). *Research Information Series on Ecosystems*, 2(1):13–25.

Eakle, T.W., and Garcia, A.S. 1977. Hastening the germination of lumbang (*Aleurites moluccana* (L.) Willd.) seeds. *Sylvatrop*. 2:291–295.

Hunt, G.R., Sakuma, F., and Shibata, Y. 2002. New Caledonian crows drop candle-nuts onto rock from communally used forks on branches. *Emu*, 102:283–290.

Hunter, M. 1990. Candlenut tree: a Polynesian legacy. *Pacific Hortic*. 51(3):54–55.

Hutabarat, B.S.M. 1977. Indonesian medicinal plants against tooth-ache: *Aleurites moluccana*, *Piper betel*, and *Cucurbita lagenaria*. *Trubus*, 8:217–218 (in Indonesian).

Legros, J. 1938. *The tung oil trees (Aleurites) and the tung oil industry throughout the world*. International Institute of Agriculture, Bureau of Agricultural Science and Practice, Section Tropical Agriculture, Rome, Italy. 237 pp.

Lopes, J.N.C., Nasi, A.M.T.T., and Lopes, J.L.C. 1976. Determination of the fatty acids in triglycerides fraction from the seeds of *Aleurites moluccana* (L.) Willd and *Pachystroma ilicifolium* Muell. *Argent Cientifica*, 4:181–184 (in Portuguese).

Malamassam, D., and Seran, D. 1993. A study on site quality indices of *Aleurites moluccana*. *Jurnal Penelitian Kehutanan*, 7(1):22–26 (in Indonesian, English summary).

Morton, J.F. 1993. The candlenut tree, handsome and wind-resistant, is a neglected ornamental in Florida. *Annu. Meet. Fla. State Hort. Soc.* 105:251–256.

Murniati, E. 1996. Internal and external influencing factors toward candle nuts (*Aleurites moluccana* Willd.) seed viability. *Keluarga Benih*. 7(1):59–65 (in Indonesian).

Siemonsma, J.S. 1999. *Aleurites molucccana* (L.) Willd. In *Plant resources of South-East Asia. 13. Spices*. Edited by C.C. de Guzman and J.S. Siemonsma. Backhuys, Leiden, the Netherlands. pp. 63–65.

Steinmetz, E.F. 1970. *Aleurites moluccana. Acta Phytotherapeutica*, 17(8):141–145.

Strauss, D. 1969. The microscopy of East Asian spices. III. a) Kemirie nuts (*Aleurites moluccana* Willd), b) Peteh beans (*Parkia speciosa* Hassk.). *Dtsch. Lebensmitt. Rundsch.* 65:210–11 (in German).

Suhardi. 1989. *Aleurites moluccana* (L) Willd as source of oil and for reforestation. In *Underutilized bioresources in the Tropics*. Edited by C.B. Lamug and R.G. Gabatin. JSPS-DOST, College, Laguna, Philippines. pp. 142–149.

Suhardi. 1989. *Aleurites moluccana* (L) Willd. as source of oil and for reforestation. In *International seminar on underutilized bioresources in the Tropics (Manilla, Philippines, 15–18 Nov. 1988)*. Edited by C.B. Lamug and R.G. Gabatin. University of the Philippines, Los Baños, Laguna, Philippines. pp. 142–149.

Suhartati. 1993. Technical [aspects] of seed treatment, nursery and plantation of kemiri (*Aleurites moluccana* Wild.). *Jurnal Penelitian Kehutanan*, 7(1):14–21 (in Indonesian, English summary).

Weismann, G. 1976. The seed oils of *Aleurites* species. *Seifen Ole Fette Wachse*, 102(3):77–78 (in German, English summary).

Young, R.G., Janolino, V.G., and Barril, C.R. 1969. The essential oil of lumbang bato *Aleurites moluccana. Philippine Agriculturist*, 53(2):95–99.

Zepernick, B. 1967. Plants used for dyeing in Polynesia: *Bischofia javanica, Casuarina equisetifolia, Ficus tinctoria, Cordia subcordata, Curcuma longa, Aleurites moluccana, Morinda citrifolia, Eugenia maire, Dianella lavarum, Dianella sandwicensis, Abutilon incanum. Willdenowia*, 5:3–97.

Specialty Cookbooks

Hyman, G.L. *Cuisines of Southeast Asia*. Thomas Woll, New York. 197 pp.

Law, R. 190. *Southeast Asia cookbook*. D.I. Fine, New York. 452 pp.

Oseland, J. 2006. *Cradle of flavor: home cooking from the spice islands of Indonesia, Malaysia, and Singapore*. Norton, New York. 384 pp.

Owen, S. 1999. *Indonesian regional food & cookery*. Frances Lincoln, London. 289 pp.

Sjahir-Hwang, C. 2004. *Singaporean, Malaysian & Indonesian cuisine*. Weico, Inc., Monterey Park, CA. 96 pp.

16 Cape Gooseberries and Tomatillo: *Physalis* Species

Family: Solanaceae (Potato Family)

This chapter features

Cape gooseberry (*P. peruviana*)
Dwarf Cape gooseberry (*P. grisea*)
Tomatillo (*P. philadelphica*)

There are almost 100 species of the genus *Physalis*, most of which are native to the New World. Some are grown for their small edible fruits, whereas others are cultivated as ornamentals, particularly for their large, brightly colored, inflated papery husks (formed from the sepals of the flowers) that surround the fruits. Most are native to warm climates and are sensitive to cold temperatures. Chinese (or Japanese) lantern, *Physalis alkekengi* L., is the best known ornamental. It produces small, white flowers followed by large, balloon-like orange-red husks (resembling miniature Chinese lanterns), inside of which are small, edible, but tasteless scarlet fruit. Some species of *Physalis* are significant weeds, and some are poisonous. Many of the species are called "ground cherries." Lists of species poisonous to house pets (dogs, cats, and birds) commonly mention that ground cherries are toxic, but rarely state which species. Although fruits of wild species are sometimes gathered by people for food, this should only be done when the plants are identified as edible with certainty. In addition to the three most important food species of *Physalis*, discussed in detail in the next section, the following two species are also consumed (Table 16.1).

PHYSALIS NAMES

- The genus name *Physalis* is based on the Greek *physa*, a bladder, so-named for the husk.
- "Ground cherry" and "husk tomato" are rather descriptive terms for all *Physalis* species used for edible fruits. Unfortunately, common names are rather confused for these species. Cape gooseberry, dwarf Cape gooseberry, and tomatillo are treated separately in this chapter because they are the most important fruit species of *Physalis*, and these names are unambiguous; however, these species are also called husk tomato and ground cherry.
- Other names found in commerce for edible *Physalis* berries include Andean cherry, cap berry, and cap gooseberry.
- The "purple ground cherry" (also known as plains Chinese-lantern and Chinese lantern of the plains) is *Quincula lobata* (Torr.) Raf. (*Physalis lobata* Torr.), a weed of the western United States. The Kiowa Indians gathered the berries to make a jelly.
- *Pubescens* in the scientific name *Physalis pubescens* is Latin for hairy.
- *Angulata* in the scientific name *Physalis angulata* is Latin for angled, referring to the stems.

TABLE 16.1
Less Important Food Species of *Physalis*

Species	English Names	Native Area	Distribution	Comments
Physalis pubescens L.	Downy ground cherry, ground cherry, hairy ground cherry, husk tomato, pops, strawberry tomato	Mexico, United States, South America	Widely naturalized in the world	Occasionally cultivated
Physalis angulata L.	Cow pops, cutleaf ground cherry, Mexican husk tomato, tomatillo ground cherry	Mexico, Antilles, Brazil	Throughout the tropics, subtropics and warm temperate regions	Primarily a weed

FIGURE 16.1 Cape gooseberry (*Physalis peruviana*). (From Curtis, vol. 27, 1808, plate 1068; vol. 53, 1826, plate 2625.)

CAPE GOOSEBERRY

Names

Scientific Name: *Physalis peruviana* L. (*Physalis edulis* Sims)

- Cape gooseberries are neither gooseberries (species of *Ribes*) nor native to the Cape. Although indigenous to South America, the Cape gooseberry has never achieved great popularity there. The name Cape gooseberry was coined by Australians who imported the fruit from South Africa. The Cape gooseberry was grown by early settlers at the Cape of

Good Hope at least by the early nineteenth century. The word "gooseberry" in its name does not reflect any relationship with the true gooseberry. "Gooseberry" has also been used to label other fruits quite unrelated to the true gooseberry; for example, the "Chinese gooseberry" was the original name of the kiwi (see Chapter 53).

- Cape gooseberry is also known as goldenberry (golden berry), golden husk, gooseberry tomato, ground cherry, husk tomato, Peruvian cherry, Peruvian ground cherry, poha, poha berry, strawberry tomato, and winter cherry.
- *Peruviana* in the scientific name *P. peruviana* is Latin for Peruvian (from where the species was first described).

Plant Portrait

The Cape gooseberry is native to Brazil but long ago became naturalized in the highlands of Peru and Chile. It has been a minor fruit in South American markets for centuries. The plant is a herbaceous or semiwoody perennial bush, usually growing 30 to 60 cm (2–3 ft.) in height, occasionally to 1.8 m (6 ft.). The fruit is globose, 1.3 to 2.5 cm (1/2–1 in.) wide, with smooth, glossy, orange-yellow skin and juicy pulp containing numerous very small yellowish seeds. The ripe fruit is sweet, with a pleasing tang. The papery, tan husk is bitter and inedible. Cape gooseberries are generally sold with the husks left on as many chefs use the husks for decorative purposes. Fruit that is picked partially green and allowed to ripen never becomes as sweet as vine-ripened fruit. The Cape gooseberry is grown as an annual in temperate regions and a perennial in the tropics (where it is considered to be an invasive weed as well as a cultivated crop). It has been widely introduced into cultivation in tropical, subtropical, and even temperate areas and is said to be growable wherever tomatoes can be raised. Good crops are produced in several Central American and South American nations, Australia, New Zealand, China, India, Malaya, and other countries. Several cultivated varieties have been selected; one of these, "Golden Berry," is said to produce juice that looks and tastes like orange juice.

Culinary Portrait

The fruit of Cape gooseberry has a tangy pineapple-like or grapelike flavor, often described as sweet and tart, with a hint of citrus. Cape gooseberries are made into sauces, used in puddings, pies, chutneys, and ice cream, and eaten fresh in fruit salads and fruit cocktails. They are also canned whole. The fruit makes excellent jellies and jams because of the very high content of pectin. The piquant aftertaste seems to go well with meats and savory foods, and the complex flavor has been claimed to complement wines and chocolates. The fruit can be substituted for raisin in cookies and cakes. Cape gooseberries are long lasting; the fresh fruits can be stored in a sealed container and kept in a dry atmosphere for several months. If purchased in a store, the fruits should have their husks, which protect them during shipping. Cape gooseberries have been dried into tasty "raisins." Unripe fruits are poisonous, at least to some people. The husk of the fruit is bitter and inedible.

Culinary Vocabulary

- Although Cape gooseberry is the established name, the attempt has been made to adopt the name goldenberry, which is relatively attractive compared with most of the other names, for commercial purposes. This is reminiscent of the (successful) replacement of the name Chinese gooseberry with the more attractive name kiwi. As noted earlier, one cultivar of Cape gooseberry has been named 'Golden Berry', Culinary preparations frequently refer to "goldenberry" (goldenberry sauce, goldenberry preserves, goldenbery jam, etc.) rather than "Cape gooseberry."

Curiosities of Science and Technology

- In the eighteenth century, native women in Peru were observed wearing perfumed Cape gooseberries as jewelry.
- In South Africa, Zulus have used a boiled infusion of Cape gooseberry leaves as an enema to relieve abdominal ailments.
- A single Cape gooseberry plant may yield 300 fruits.
- The Cape gooseberry is a recommended fruit for parrots.
- In Europe, the husk attached to the Cape gooseberry fruit has been used as a handle to dip the fruit in icing and chocolate. The husk has also been used in place of a fork when the fresh fruit is served for fondue.

Key Information Sources

Ayala, C. 1992. Evaluation of three planting distances and three systems of pruning in Cape gooseberry under greenhouse conditions. *Acta Hortic.* 310:206.

Baumann, T.W., and Meier, C.M. 1993. Chemical defence by withanolides during fruit development in *Physalis peruviana. Phytochemistry*, 33:317–321.

Carman, E. 1980. Poha jam. *Pacific Hortic.* 41(4):9–10.

Ch de Valencia, M.L. 1986. Fruit anatomy of the Cape gooseberry (*Physalis peruviana* L.). *Acta Biologica Colombiana*, 1(2):63–89 (in Spanish, English summary).

Criollo E.H., and Ibarra C.V. 1992. Germination of Cape gooseberry (*Physalis peruviana* L.) under different degrees of maturity and storage times. *Acta Hortic.* 310:183–187 (in Spanish, English summary).

Fischer, G., and Martinez, O. 1999. Quality and maturity of Cape gooseberry (*Physalis peruviana* L.) in relation to fruit coloring. *Agronomia Colombiana*, 16:35–39 (in Spanish).

Gupta, S.K., and Roy, S.K. 1981. The floral biology of Cape-gooseberry. *Indian J. Agric. Sci.* 51:353–355.

Hassanien, M.F.R. 2008. *Goldenberry: golden fruit of golden future*. VDM Verlag Dr. Müller, Saarbrücken, Germany. 100 pp.

Heinze, W., and Midasch, M. 1991. Photoperiodic reaction of *Physalis peruviana. Gartenbauwissenschaft*, 56:262–264 (in German, English summary).

Johnston, B. 1962. Cape gooseberries, a delicacy for many. *N. Z. J. Agric.* 104:372.

Klinac, D.J. 1986. Cape gooseberry (*Physalis peruviana*) production systems. *N. Z. J. Exp. Agric.* 14:425–430.

Legge, A.P. 1974. Notes on the history, cultivation and uses of *Physalis peruviana* L. *J. R. Hortic. Soc.* 99:310–314.

Leiva-Brondo, M., Prohens, J., and Nuez, F. 2001. Genetic analyses indicate superiority of performance of Cape gooseberry (*Physalis peruviana* L.) hybrids. *J. New Seeds*. 3(3):71–84.

Mayorga, H., Knapp, H., Winterhalter, P., and Duque, C. 2001. Glycosidically-bound flavor compounds of Cape gooseberry (*Physalis peruviana* L.). *J. Agric. Food Chem.* 49:1904–1908.

Mazumdar, B.C. 1979. Cape-gooseberry ground cherries—the jam fruit of India. *World Crops*, 31(1):19, 23.

Mazumder, K., and Mazumdar, B.C. 2002. Changes of pectic substances in developing fruits of Cape-gooseberry (*Physalis peruviana* L.) in relation to the enzyme activity and evolution of ethylene. *Sci. Hortic.* 96(1/4):91–104.

Micklem, T. 1949. Cape gooseberry culture in the Western Cape Province. Reprint 52. *Farming in South Africa*, (Aug):1–4.

Morton, J. 1987. Cape gooseberry. In *Fruits of warm climates*. Creative Resource Systems, Winterville, NC. pp. 430–434.

Morton, J.F., and Russell, O.S. 1954. The Cape gooseberry and the Mexican husk tomato. *Annu. Meet. Fla. State Hort. Soc.* 67:261–266.

Ramadan, M.F., and Morsel, J.T. 2003. Oil goldenberry (*Physalis peruviana* L.). *J. Agric. Food Chem.* 51:969–974.

Singh, U.R., Pandey, I.C., and Prasad, R.S. 1977. Effect of N, P, and K on growth, yield and quality of Cape-gooseberry. *Punjab Hortic. J.* 17:148–151.

Skipworth, R.G. 1944. The commercial culture of Cape gooseberries. *Rhodesia Agric. J.* 61:20–22.

Sweet, C. 1986. Cape gooseberry: it can grow here, but does anyone care? *California Grower*, 10(8):26–28.

Trinchero, G.D., Sozzi, G.O., Cerri, A.M., Vilella, F., and Fraschina, A.A. 1999. Ripening-related changes in ethylene production, respiration rate and cell-wall enzyme activity in goldenberry (*Physalis peruviana* L.), a solanaceous species. *Postharvest Biol. Technol.* 16:139–145.
Verhoeven, G. 1991. *Physalis peruviana* L. In *Plant resources of South-East Asia. 2. Edible fruits and nuts.* Edited by E.W.M. Verheij and R.E. Coronel. Pudoc, Leiden, The Netherlands. pp. 254–256.
Whitely, K.T. 1962. The Cape gooseberry. *J. Agric. West. Aust.* 3(1):59, 61.

Specialty Cookbooks

Note: Cape gooseberries are widely used in the cuisines of the world, but particularly so in Mexican cooking.

Curtis, S., Hoyer, D., and Smith, R.A. 2000. *Tacos.* Gibbs Smith, Salt Lake City, UT. 44 pp.
Macpherson, N. 1954. Cape gooseberry recipes. *N. Z. J. Agric.* 88(4):401–404.
Padmanabhan, C. 1999. *Dakshin: vegetarian cuisine from South India.* Periplus, Singapore. 176 pp.
Reader's Digest Association. 2007. *The ultimate soup cookbook.* Reader's Digest, Pleasantville, NY. 544 pp.
Schmidt, A., and Nam, I. 1996. *The book of hors d'oeuvres and canapes.* Wiley, Somerset, NJ. 320 pp.
Wolf-Cohen, E. 2000. *The book of dips and salsas.* HP Trade, New York. 96 pp.

DWARF CAPE GOOSEBERRY

Names

Scientific Name: *Physalis grisea* (Waterf.) M. Martinez (*P. pruinosa* of some authors)

- The name "dwarf Cape gooseberry" is based on the species having smaller stature (and fruits) than the Cape gooseberry, discussed earlier.
- *Physalis grisea* is also called hairy ground cherry, a name also used for *P. pubescens*, mentioned earlier.
- *Grisea* in the scientific name *P. grisea* is Latin for gray, for the grayish-green leaves.

Plant Portrait

Aside from the tomatillo and the Cape gooseberry, treated separately in this chapter, the best known species of "husk tomato" is the dwarf Cape gooseberry. This species is native to eastern North America. It is a low growing, annual plant reaching 45 to 75 cm (18–30 in.) in height when in flower. It produces cherry-sized green to yellow-gold fruit, which drop to the ground when ripe. The husks turn brown when the fruits are mature.

Culinary Portrait

Fruits of the dwarf Cape gooseberry have a pleasing and distinctive sweet, acidic flavor. They are most commonly used to make jam but are also eaten raw or used in pies and other cooked desserts, stews, sauces, and preserves. They make interesting subjects when served raw in cocktails and produce good cooked sauce for cakes and puddings. As with the other edible species of *Physalis*, the husks are inedible and should be removed before eating or cooking. For additional culinary information, see the Tomatillo and Cape gooseberry sections.

Curiosities of Science and Technology

- An ancient Indian camp site, Shawnee-Minisink located on a terrace of the Delaware River near Stroudsburg, Pennsylvania, has been found by the radiocarbon method to have been

FIGURE 16.2 Dwarf Cape gooseberry (*Physalis grisea*). (From Dillenius, 1774.)

occupied some time between 9700 and 9500 years ago. Husk tomatoes were recovered from several hearths, indicating that the fruits have been used by people in North America for most of the time that North America has been populated.

- The subject of whether tomatoes are a fruit (which they are botanically) or a vegetable (which they are in supermarkets) has been much debated. This issue bears on husk tomatoes. For trade purposes, the European Union has ruled that all *Physalis* berries are classified as "fruit," although some species are consumed as "vegetables."
- Girls in Japan once fashioned ground cherries into a kind of whistle.

Key Information Sources

Doan, A.T., Ervin, G., and Felton, G. 2004. Temporal effects on jasmonate induction of anti-herbivore defense in *Physalis angulata*: seasonal and ontogenetic gradients. *Biochem. Syst. Ecol*. 32:117–126.

Dremann, C.C. 1985. *Ground cherries, husk tomatoes, and tomatillos*. Redwood City Seed Co., Redwood City, CA. 22 pp.

Fisher, B. 1977. "Instant" fruit—it's hard to ask more [including *Physalis pruinosa*]. *Org. Garden. Farm*. 24(6):74–75.

Henry, R.D. 1990. Some embryological observations of *Physalis pruinosa* L. Solanaceae. *Phytomorphology*, 40:309–318.

Januario, A.H., Rodrigues-Filho, E., Pietro, R.C.L.R., Kashima, S., Sato, D.N., and Franca, S.C. 2002. Antimycobacterial physalins from *Physalis angulata* L. (Solanaceae). *Phytother. Res*. 16:445–448.

Mahmoud, M.H. 1995. Seed germination of *Physalis pruinosa* (L.) as affected by salinity. *Egypt. J. Hortic*. 22(2):127–135.

Martinez, M. 1993. The correct application of *Physalis pruinosa* L. (Solanaceae). *Taxon*, 42:103–104.

Menzel, M.Y. 1951. The cytotaxonomy and genetics of *Physalis*. *Proc. Am. Phil. Soc*. 95:132–183.

Menzel, M.Y. 1957. Cytotaxonomic studies of Florida coastal species of *Physalis*. *Yearb. Am. Phil. Soc*. 1957:262–266.

Nagafuji, S., Okabe, H., Akahane, H., and Abe, F. 2004. Trypanocidal constituents in plants 4. Withanolides from the aerial parts of *Physalis angulata*. *Biol. Pharmaceut. Bull*. 27:193–197.

Quiros, C.F. 1984. Overview of the genetics and breeding of husk tomato. *HortScience*, 19:872–874.

Raghava, R.P., and Raghava, N. 1994. Enhancement of fruit yield of husk tomato by seed-applied plant growth regulators and gamma rays. *Proc. Natl. Acad. Sci. India B*. 64:305–309.

Raghava, R.P., and Raghava, N. 1994. Responses of wild husk tomato, *Physalis angulata* L. to growth regulators and gamma rays on chlorophyll content and fruit yield. *Proc. Natl. Acad. Sci. India B*, 64:415–418.

Sisterson, M.S., and Gould, F.L. 1999. The inflated calyx of *Physalis angulata*: a refuge from parasitism for *Heliothis subflexa*. *Ecology*, 80:1071–1075.

Thomson, C.E., and Witt, W.W. 1987. Germination of cutleaf groundcherry (*Physalis angulata*), smooth groundcherry (*Physalis virginiana*), and eastern black nightshade (*Solanum ptycanthum*). *Weed Sci*. 35:58–62.

Waterfall, U.T. 1958. A taxonomic study of the genus *Physalis* in North America north of Mexico. *Rhodora*, 60:107–114.

Waterfall, U.T. 1967. *Physalis* in Mexico, Central America, and the West Indies. *Rhodora*, 69:82–120.

Specialty Cookbooks

Dwarf Cape gooseberries are rarely specified in recipes but can be substituted for Cape gooseberries. See the cookbooks suggested above.

TOMATILLO

Names

Scientific Name: *Physalis philadelphica* Lam. (*P. ixocarpa* of some authors)

- The word "tomatillo" is based on American Spanish, the diminutive of *tomate*, tomato (i.e., tomatillo translates as "small tomato"). Pronunciation: tohm-ah-TEE-oh or tohm-ah-TEE-yoh. The names for both tomatillo and tomato trace ultimately to the Nahuatl (language of the Aztecs) *tomatl*, a generic word for globose fruits or berries that have many seeds and watery flesh.
- Most but not all dictionaries presenting the plural of tomatillo state that it is "tomatillos." However, the usage "tomatilloes," in parallel to "tomatoes" and "potatoes," is often encountered.
- The tomatillo is also known as husk tomato (or Mexican husk tomato or Mayan husk tomato) and ground cherry. However, as noted earlier, other species of *Physalis* with edible fruits are also known as husk tomato and ground cherry, so the name tomatillo is much to be preferred.
- The tomatillo is also called Mexican green tomato and miltomate.
- In 1945, the tomatillo was publicized as the "jamberry," a new fruit introduced by scientists at Iowa State College. The name jamberry is still sometimes used. One strain was given the name "Mayan husk tomato."
- *Philadelphica* in the scientific name *P. philadelphica* is Latin for Philadelphian (although the native range is not near modern Philadelphia; geographic notions were often hazy in the eighteenth century when the species was described).

Plant Portrait

The tomatillo is a semiwoody annual, 60 to 150 cm (2–5 ft.) high, mostly growing as a sprawling plant. The species is native from southern Baja California to Guatemala. Its fruit is globose, two celled, 2.5 to 7.5 cm (1–3 in.) in diameter, smooth, and sticky. The skin color matures to yellow, yellowish-green, purple, or, more rarely reddish, or simply remains green. The flesh is pale-yellow, crisp or soft (typically firmer than a tomato), varying from acid to sweet, or insipid, and contains many tiny seeds. The fruit is reminiscent of a miniature tomato but is entirely enclosed in a thin

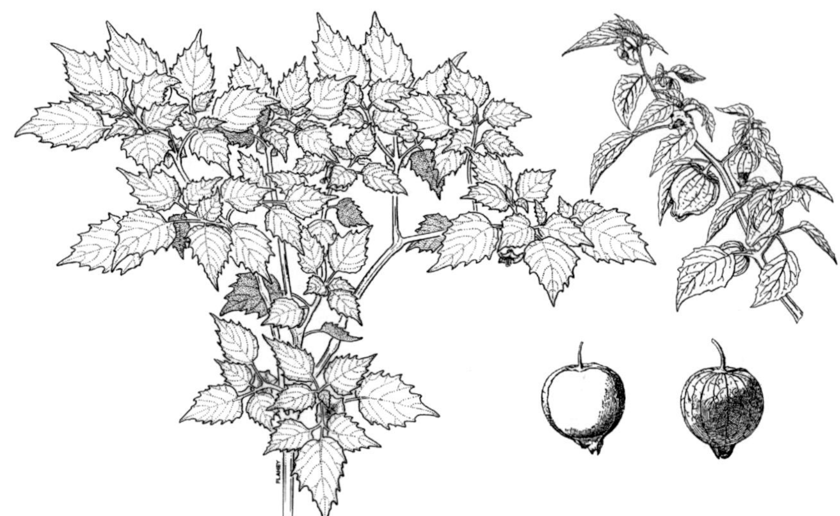

FIGURE 16.3 Tomatillo (*Physalis philadelphica*). Left: young, vegetative shoot, by B. Flahey. Right: fruiting branch and fruits. (From Bailey, 1900–1902.)

husk that is usually split by the expanding fruit and becomes straw-colored and parchment-like at maturity. The tomatillo probably was domesticated in pre-Columbian times in Mexico. It was important in the Aztec and Mayan cultures before the Spanish conquest of the New World. Indeed, in pre-Hispanic times in Mexico tomatillos were much preferred over the tomato. The tomatillo is popular in Latin America, but little is grown in the United States and Canada. However, it is cultivated to a minor extent in California, and interest has been developing in growing it in other areas of the southern United States. It was exported from the New World and is now cultivated in tropical and subtropical areas of the Old World. Tomatillos are occasionally sold, both fresh and canned, in grocery stores in the United States and Canada. Because they are an essential ingredient in Mexican cooking, they are generally available in Latino markets. Tomatillos are also available sporadically year-round in specialty produce stores and some supermarkets, especially in areas close to Mexico.

Culinary Portrait

Tomatillos are used as a vegetable, not a fruit. They are typically cooked before consumption, which enhances the zesty, tart flavor and softens the thick skin, but they are also eaten raw. The fruit is a little sweeter than tomato, and the flavor has been compared with apple with a hint of lemon and herbs, or a combination of tomato and lemon. The flavor blends well with onion, garlic, cilantro, chili, beef, pork, chicken, and cheese. Tomatillos are used in pureées as a base for chili sauces known generically as *salsa verde*, green sauce. They are also used in *salsa cruda*, a fresh salsa dish. Tomatillos contain a pectin-like substance that thickens sauces and salsa when refrigerated. In Mexico and in the Southwestern United States, salsa, *mole verde*, and other tomatillo-based sauces and dressings are used to season popular dishes such as tacos, enchiladas, tostadas, gazpacho, moles, and guacamole. Tomatillos are rather bland, and pungent chili peppers are usually present in salsa preparations to satisfy the culinary tastes associated with Mexican foods. Tomatillos are suitable for stewing, frying, baking, cooking with chopped meat, and making into soup, marmalade, and dessert sauce. The fruit is an excellent addition to salads and curries. An infusion of the husks is added to tamale dough to improve its spongy consistency. Husks are also sometimes mixed with fritters and white rice. Tomatillos are rich in vitamins A and C.

When purchasing tomatillos, it is preferable to choose those that are firm rather than soft. Husks tend to have partially opened by the time they are in a grocery store, so those with good green color can be selected (yellow indicates overmaturity; but as noted earlier some varieties are purplish). The husk should be light brown and fresh looking and generally should still be clasping the fruit. Most recipes for tomatillos require slightly unripe fruit. Fresh tomatillos can be stored in their husks in a paper bag for a month or more in the refrigerator, but like tomatoes are best used soon. Longer storage in the refrigerator requires removal of husks and placing in an air-tight (plastic) bag. Tomatillos also freeze well. Soaking fresh tomatillos for several minutes in hot tap water allows the papery husk to be removed easily. The husks should be removed one at a time and washed thoroughly. The slightly sticky surface is normal. The stem attachment point can be removed with a paring knife, and the core of the fruit then removed. Finally, the tomatillos can be chopped and added to salads or cooked as suggested earlier.

Culinary Vocabulary

- In Mexico, where tomatillo is a staple, it is called *tomate verdes* (also *tomate de milpa* and *fresadilla*).
- In the United States, salsas are often called "green taco sauce."

Curiosities Of Science And Technology

- In Mexico, a boiled solution of tomatillo husks has a reputation (not validated) for curing diabetes.
- Tomatillo, like tomato plants, produces its largest fruits from the first flowers on the main branches.
- Adventitious roots can develop from the internodes (sections of stems or branches between the nodes where the leaves arise) of many branches of tomatillo plants. When these roots contact soil, they grow into it, producing new plants that are independent of the main root system. Heavy rains have been observed to cause cultivated plants to bend down to the ground and the tips that touch the soil take root and produce new plants. Such natural vegetative reproduction is often observed in the plant kingdom.
- Unhusked tomatillo fruits stored in a cool, dry atmosphere will last for several months. In Mexico and Central America, people sometimes simply pull up entire plant with fruits attached and hang them upside-down in a dry place until the fruits are needed.

Key Information Sources

Bock, M.A., Sanchez-Pilcher, J., McKee, L.J., and Ortiz, M. 1995. Selected nutritional and quality analyses of tomatillos (*Physalis ixocarpa*). *Plant Foods Hum. Nutr.* 48(2):127–33.

Can, F., Rush, M.C., Valverde, R.A., Griffin, J.L., Story, R.N., Young, W.A., Blackmon, W.J., and Wilson, P. W. 1991–1992. Tomatillo: a potential new vegetable crop for Louisiana. *La. Agric.* 35(2):21–24.

Cantwell, M. 2000. Produce facts: tomatillo (husk tomato). *Perishables Handling Quart.* 103(Aug):21–22.

Cantwell, M., Flores-Minutti, J., and Trejo-Gonzales, A. 1992. Developmental changes and postharvest physiology of tomatillo fruits (*Physalis ixocarpa* Brot.). *Sci. Hortic.* 50:59–70.

Cartujano-Escobar, F., Jankiewicz, L., Fernandez-Orduna, V.M., and Mulato-Brito, J. 1985. The development of the husk tomato plant (*Physalis ixocarpa* Brot) cultivar Rendidora. I. Aerial vegetative parts. *Acta Soc. Bot. Pol.* 54:327–338.

Cartujano-Escobar, F., Jankiewicz, L., Fernandez-Orduna, V.M., and Mulato-Brito, J. 1985. The development of the husk tomato plant (*Physalis ixocarpa* Brot) cultivar Rendidora. II. Reproductive parts. *Acta Soc. Bot. Pol.* 54:339–349.

Cartujano-Escobar, F., Jankiewicz, L., Fernandez-Orduna, V.M., Mulato-Brito, J., and Pietkiewicz, S. 1987. The development of the husk tomato plant (*Physalis ixocarpa* Brot) cultivar Rendidora. III. Growth analysis. *Acta Soc. Bot. Pol.* 56:421–436.

Dyki, B., Jankiewicz, L.S., Staniaszek, M. 1997. Anatomy and surface micromorphology of tomatillo fruit (*Physalis ixocarpa* Brot.). *Acta Soc. Bot. Pol.* 66:21–27.

Freyre, R., and Loy, J.B. 2000. Evaluation and yield trials of tomatillo in New Hampshire. *HortTechnology.* 10:374–377.

Hernández, S.M., and Rivera, J.R.A. 1994. Tomatillo, husk-tomato (*Physalis philadelphica*). In *Neglected crops. 1492 from a different perspective*. Edited by J.E. Hernández-Bermejo and J. Léon. Food and Agriculture Organization of the United Nations, Rome, Italy. pp. 117–122.

Hudson, W.D. 1986. Relationships of domesticated and wild *Physalis philadelphica*. In *Solanaceae. Biology and systematics*. Edited by W. G. D'Arcy. Columbia University Press, New York. pp. 416–432.

Jankiewicz, L.S., and Borkowski, J. 1990. The development and cultivation of tomatillo (*Physalis ixocarpa* Brot.) under the climatic conditions of Poland: II. Flowering and fruiting. *Acta Agrobotanica*, 43:11–23.

Jankiewicz, L.S., Horodecka, E., and Borkowski, J. 1989. The development and cultivation of tomatillo (*Physalis ixocarpa* Brot.) under the climatic conditions of Poland: I. Local cultivars, growth of above-ground vegetative parts, probable resistance to *Phytophthora infestans*. *Acta Agrobotanica*, 42:5–21.

Maynard, D.N. 1994. Potential for commercial production of tomatillo in Florida. *Annu. Meet. Fla. State Hort. Soc.* 106:223–224.

Moriconi, D.N., Rush, M.C. and Flores, H. 1990. Tomatillo: a potential vegetable crop for Louisiana. In *Advances in new crops*. Edited by J. Janick and J.E. Simon. Timber Press, Portland, OR. pp. 407–413.

Morton, J. 1987. Mexican husk tomato. In *Fruits of warm climates*. Creative Resource Systems, Winterville, NC. pp. 434–437.

Mulato-Brito, J., Jankiewicz, L.S., Fernandez-Orduna, V.M., and Cartujano-Escobar, F. 1986. The root system of the husk tomato *Physalis ixocarpa* Brot. *Acta Agrobotanica*, 39:367–384.

Mulato-Brito, J., Jankiewicz, L., Fernandez-Orduna, V.M., Cartujano-Escobar, F., and Serrano-Covarrubias, L.M. 1985. Growth, fructification and plastochron index of different branches in the crown of the husk tomato (*Physalis ixocarpa* Brot.). *Acta Soc. Bot. Pol.* 54:195–206.

Myers, C. 1991. Tomatillo. Crop Sheet SMC-034. In *Specialty and minor crops handbook*. Edited by C. Myers. The Small Farm Center, Division of Agriculture and Natural Resources, University of California, Oakland, CA. 2 pp. (unpaginated).

Ostrzycka, J., Horbowicz, M., Dobrzanski, W., Janiewicz, L.S., and Borkowski, J. 1988. Nutritive value of the tomatillo fruit *Physalis ixocarpa* Brot. *Acta Soc. Bot. Pol.* 57:507–522.

Specialty Cookbooks

Arts Institutes. 2006. *American regional cuisine*. Wiley, New York. 542 pp.

Bayless, R., Brownson, J.-M., and Bayless, D.G. 2000. *Mexico one plate at a time*. Scribner, New York. 374 pp.

Curtis, S.D. 2006. *Salsas and tacos: Santa Fe school of cooking*. Gibbs Smith, Salt Lake City, UT. 96 pp.

Jamison, C.A. 1995. *The border cookbook: authentic home cooking of the American Southwest and Northern Mexico*. Harvard Common Press, Boston, MA. 512 pp.

Miller, M. 2009. *Tacos*. Ten Speed Press, Berkeley, CA. 176 pp.

Topp, E., and Howard, M. 2007. *The complete book of small-batch preserving: over 300 recipes to use year-round*. 2nd ed. Firefly Books, Richmond Hill, ON. 376 pp.

17 Caper

Family: Capparidaceae (Capparaceae; caper family)

NAMES

Scientific Name: *Capparis spinosa* L.

- The English word "caper" and the genus name *Capparis* are derived from the Greek word for the plant, *kápparis.*
- Caper is also known as caper berry, caper bud, caper bush, caper fruit, smooth caper (referring to bushes that lack spines), and spiny caper (bushes with spines).
- The word caper meaning the plant and the condiment is not related to the word caper meaning "to leap or jump about in a sprightly manner; to cut capers; to skip; to spring; to prance; to dance."
- "False capers" are nasturtium seeds (of species of *Tropaeolum*), which have on occasion been unscrupulously substituted for capers.
- *Spinosa* in the scientific name *C. spinosa* is Latin for spiny, although some spineless varieties have been selected.

PLANT PORTRAIT

Depending on authority, *C. spinosa* has been interpreted as a shrub confined to the Mediterranean region, or as also having varieties native to Africa, Asia, Australia, and islands of the Pacific. Although several of the approximately 250 species of *Capparis* produce edible flower buds ("capers") that are harvested and traded locally, *C. spinosa* is by far the most important commercial caper species. (Some *Capparis* species are poisonous and may be available as ornamentals, so one should not pick the flowers or fruits unless certain that the species is edible.) The caper is a straggling, spreading, vine-like, deep-rooted deciduous shrub, often only approximately 60 cm (2 ft.) high (sometimes up to 1.5 m or 5 ft. in height), but extending 3 m (approximately 10 ft.) or so along the ground. The leaves are leathery, thick, and shiny green. A pair of hooked spines is present at the base of each leaf stalk, although spineless forms have been selected to make gathering the capers less likely to injure hands. The mature flowers are usually white-petaled with pink or violet stamens and are 4 to 6 cm (1.6–2.4 in.) across. The species is too tender to be grown in northern regions. Capers have been eaten for millennia, and the ancient Greeks and Romans were familiar with them. They are produced commercially in Mediterranean countries, including Spain, France, Italy, Greece, Algeria, Egypt, Morocco, Tunisia, and Turkey. There is also some production in Iran and in California. The United States imports more than $10 million worth of capers annually.

The capers of commerce are the unopened flower buds. These are handpicked daily, typically in the morning, and often from wild plants, although cultivated varieties have been selected. Daily picking is necessary because the youngest flower buds (about the size of peas) have the highest quality; capers are valued in proportion to the smallness of their size. The buds are pickled in strong vinegar, brine, oil, or wine, or simply preserved in granular salt. Semimature berries ("caper berries") about the same size and color of small green olives are pickled to some extent, and sometimes eaten like small gherkins. Even young shoots with small leaves have been pickled for use as condiments but are very rarely available in the commercial trade. Capers in vinegar are traditionally packaged in tall narrow glass bottles. A glass vial of approximately 200 g (7 oz.) of good quality capers sell for approximately $5.00. Fresh capers are sold to some extent in markets near production areas.

CULINARY PORTRAIT

Pickled capers have a pungent, sharply piquant, peppery taste and a peculiar aroma. The flavor has been compared with mustard and black pepper, and in fact the caper's strong flavor comes from mustard oil (methyl isothiocyanate). Capers are used in a variety of tartar and other sauces to flavor salads, pizza, fish, meats, and other dishes. North Americans are exposed to the flavor of capers most often in tartar sauce; capers are an essential ingredient of steak tartare. French sauces, including Gribiche, Ravigote, and Rémoulade, also often are flavored with capers. Capers are used directly as a garnish for cold meats, fish, seafood, and vegetable dishes as well as sandwiches, pizza, rice, pasta, and poultry. The condiment is commonly used to flavor classical Mediterranean specialties, such as olives, arugula, and artichokes. The spice combines well with anchovies and lemon. Mayonnaise, mustard, and cold sauces are said to be particularly complemented by the sourish, somewhat bitter taste. The flavor of caper is impaired by high temperatures, and so caper-containing sauces should be added late in cooking, or after cooking. The best capers are olive green and firm to the touch, and the smaller the caper the better the flavor. Flower buds of nasturtium, marigold, and other species are sometimes fraudulently sold as capers. When picked in vinegar, capers keep for months in a refrigerator, preferably in their original nonmetal container, but tend to become softer and mushier with age. Capers in brine are best rinsed before use to flush away as much salt as possible.

Culinary Vocabulary

- "Caper sauce" is an English sauce prepared from capers, butter, flour, and the juices of the roasted meat with which the sauce is served.
- "Nonpareils" (French for "without equal") refers to small capers from France's Provence region. (The word is also used for tiny sugar pellets used to decorate cakes and confections, and for small chocolate disks coated with these pellets.)
- The Italian dish *pasta alla puttanesca*, also known as "whore's pasta," is made with a heavy sauce of tomatoes, garlic, capers, anchovies, and olives. It is said to have originated in the slums of Naples, the name reflecting usage by busy prostitutes who could make and consume the dish rapidly, between clients. Despite its questionable origin, it is popular in Italy.

CURIOSITIES OF SCIENCE AND TECHNOLOGY

- Apicius is the title of an early Roman cookbook probably compiled in the fourth or fifth century, with recipes dating to earlier centuries. It is generally considered to be the most complete early European cookbook, presenting hundreds of recipes. Caper was one of the seasonings mentioned in the book.
- Ash from burned caper roots has been used as a source of salt.
- Extracts of capers have been shown to be medicinally effective in treating enlarged capillaries and for improving dry skin. Caper extracts and pulps have been used in cosmetics, such as the cleverly named Greek preparation "Cream of Caper™." However, cases of contact dermatitis and sensitivity have been reported from some caper-based cosmetics.
- Capers contain considerable amounts of rutin. Rutin is an antioxidant, that is, a substance that counteracts "free radicals," which are damaging chemicals produced by metabolism in the body.
- Unfortunately, cases of substituting and misrepresenting condiments are not uncommon. In Colombia, buds of a species of *Cassia*, pickled in sour vinegar with cloves, have been falsely sold as capers.

- The caper is an exceptionally drought-resistant, Mediterranean shrub. Mediterranean climates have alternating wet (winter) and dry (summer) seasons. The caper is unusual in that it loses its leaves during the rainy season and retains them during the dry season; it is able to keep its leaves in the face of drought because it can obtain water from low levels in the soil. Its extensive, deep-penetrating root system makes the caper bush useful for controlling soil erosion, especially on slopes where irrigation is difficult and soil loss is more pronounced.

FIGURE 17.1 Caper (*Capparis spinosa*). (From Engler and Prantl, 1889–1915.)

FIGURE 17.2 Caper (*Capparis spinosa*). (From Turner, 1893b.)

KEY INFORMATION SOURCES

Barbera, G., and Di Lorenzo, R. 1984. The caper culture in Italy. *Acta Hortic.* 144:167–171.

Bond, R.E. 1990. The caper bush. *Herbarist*, 56:77–85.

Fici, S. 2001. Intraspecific variation and evolutionary trends in *Capparis spinosa* L. (Capparaceae). *Plant Syst. Evol.* 228:123–141.

Fici, S. 2003. The *Capparis spinosa* L. group (Capparaceae) in Australia. *Webbia*. 58:113–120.

Germano, M.P., De Pasquale, R., D'Angelo, V., Catania, S., Silvari, V., and Costa, C. 2002. Evaluation of extracts and isolated fraction from *Capparis spinosa* L. buds as an antioxidant source. *J. Agric. Food Chem.* 50:1168–1171.

Giuseppe, B. (Ed.). 1991. *Le Câpier (*Capparis *spp.)*. Commission of the European Communities, Luxembourg. 62 pp (in French).

Higton, R.N., and Akeroyd, J.R. 1991. Variation in *Capparis spinosa* L. in Europe. *Bot. J. Linn. Soc.* 106:104–112.

Innocencio, C., Alcaraz, F., Calderón, F., Obón, C., and Rivera, D. 2002. The use of flower characters in *Capparis* sect. *Capparis* to determine the botanical and geographical origin of caper. *Eur. Food Res. Technol.* 214:335–339.

Innocencio, C., Cowan, R.S., Alcaraz, F., Rivera, D., and Fay, M.F. 2005. AFLP fingerprinting in *Capparis* subgenus *Capparis* related to the commercial sources of capers. *Genet. Resour. Crop Evol.* 52:137–144.

Inocencio, C., Rivera, D., Alcaraz, F., Tomas B., and Francisco, A. 2000. Flavonoid content of commercial capers (*Capparis spinosa*, *C. sicula* and *C. orientalis*) produced in Mediterranean countries. *Eur. Food Res. Technol.* 212:70–74.

Jacobs, M. 1965. The genus *Capparis* (Capparaceae) from the Indus to the Pacific. *Blumea*, 12:385–541.

Ong, H.C., and Siemonsma, J.S. 1999. *Capparis spinosa* L. var. *mariana* (Jacq.) K. Schumann. In *Plant resources of South-East Asia. 13. Spices*. Edited by C.C. de Guzman and J.S. Siemonsma. Backhuys, Leiden, the Netherlands. pp. 88–91.

Orphanos, P.I. 1983. Germination of caper *Capparis spinosa* seeds. *J. Hortic. Sci.* 58:267–270.

Ozcan, M. 2001. Pickling caper flower buds. *J. Food Qual.* 24:261–269.

Petanidou, T., Van Laere, A.J., and Smets, E. 1996. Change in floral nectar components from fresh to senescent flowers of *Capparis spinosa* (Capparidaceae), a nocturnally flowering Mediterranean shrub. *Plant Syst. Evol.* 199:79–92.

Psaras, G.K., and Sofroniou, I. 1999. Wood anatomy of *Capparis spinosa* from an ecological perspective. *IAWA J.* 20:419–429.

Rhizopoulou, S. 1990. Physiological responses of *Capparis spinosa* L. to drought. *J. Plant Physiol.* 136:341–348.

Rhizopoulou, S., and Psaras, G.K. 2003. Development and structure of drought-tolerant leaves of the Mediterranean shrub *Capparis spinosa* L. *Ann. Bot.* 92:377–383.

Rhizopoulou, S., Heberlein, K., and Kassianou, A. 1997. Field water relations of *Capparis spinosa* L. *J. Arid Environ.* 36:237–248.

Rivera, D., Alcaraz, F., Inocencio, C., Obón, C., and Carreño, E. 1999. Taxonomic study of cultivated *Capparis* sect. *Capparis* in the western Mediterranean. In *Taxonomy of cultivated plants*. Edited by S. Andrews, A.C. Leslie, and C. Alexander. Royal Botanic Gardens, Kew, UK. pp. 451–455.

Rodrigo, M., Lazaro, M.J., Alvarruiz, A., and Giner, V. 1992. Composition of capers (*Capparis spinosa*): influence of cultivar, size and harvest data. *J. Food Sci.* 57:1152–1154.

Sozzi, G.O. 2001. Caper bush: botany and horticulture. *Hortic. Rev.* 27:125–188.

Sozzi, G.O. 2008. *Capparis spinosa*, caper bush. In *The encyclopedia of fruit & nuts*. Edited by J. Janick and R.E. Paull. CABI, Wallingford, Oxfordshire, UK. pp. 227–232.

Sozzi, C.O., and Chiesa, A. 1995. Improvement of caper (*Capparis spinosa* L.) seed germination by breaking seed coat-induced dormancy. *Sci Hortic.* 62:255–261.

St. John, H. 1965. Revision of *Capparis spinosa* and its African, Asiatic, and Pacific relatives. *Micronesia.* 2(1):25–44.

Stromme, E. 1988. The caper caper. *Pacific Hortic.* 49(4):42–44.

Zohary, M. 1969. The species of *Capparis* in the Mediterranean and the near eastern countries. *Bull. Res. Counc. Isr*, 8D:49–64.

Specialty Cookbooks

Brennan, G., and Beisch, L. 2001. *Olives, anchovies, and capers: the secret ingredients of the Mediterranean table*. Chronicle Books, San Francisco, CA.132 pp.

Hoffman, S., and Wise, V. 2004. *The olive and the caper: adventures in Greek cooking*. Workman, New York. 544 pp.

18 Carambola, Star Fruit

Family: Oxalidaceae (wood sorrel family)

NAMES

Scientific Name: *Averrhoa carambola* L.

- The English name "carambola" and the word *carambola* in the scientific name *A. carambola* are derived from *karambal*, the name in Marathi, the chief Indian language of southern and eastern Bombay state, India. "Carambola" came into English via Portuguese. The name has been traced ultimately to the Sanskrit *karmara*, "food appetizer."
- The name "star fruit" (and occasionally "five corner fruit") is based on the five-ribbed (star-like in cross section) appearance of the fruit. In Guyana and some other countries, the fruit is called "five fingers."
- Other names include coolie tamarind and Chinese jimbelin.
- "Star fruit" should not be confused with "star apple," *Chrysophyllum cainito* L., which is popular in the West Indies and Central America, where this tropical evergreen tree fruit occurs naturally. Star apple is occasionally grown commercially in parts of south Florida. The fruit is the size of an apple, showing in cross section a central star-like core of seeds imbedded in a delicious sweet pulp. Star apples are consumed fresh, and the pulp usually spooned out to avoid the bitter tasting rind. The fresh fruit is also often added to salads, beverages, and other dishes.
- Another plant known as starfruit is *Damasonium alisma* Mill., also named for its star-shaped fruit. This is an aquatic plant, widely distributed in Eurasia. In recent times, it has become a cause célèbre in southwestern England because it is rapidly disappearing due to of habitat destruction.
- Bilimbi (one of several plants also known as cucumber tree), *Averrhoa bilimbi* L., is closely related to carambola. This tree is Asian in origin but is now widely cultivated in tropical areas for its acidic fruit, used in pickles, relishes, and preserves. Immature, green-fruited carambolas are similarly astringent to bilimbi fruits and are sometimes used in the same way.
- The genus *Averrhoa* is named for the philosopher and physician Abu al-Walid Muhammad Ibn Ahmad Ibn Muhammad Ibn Rushd, more simply known as Ibn Rushd and Averrhoës (Averroës; 1126–1198), who lived in Spain during the Moorish occupation.

PLANT PORTRAIT

The carambola is a tropical to subtropical evergreen tree, 6 to 10 m (20–33 ft.) in height. The species is thought to be indigenous to Southeast Asia, probably Sri Lanka and the Moluccas. However, genuinely wild plants are not known to exist. The carambola has been cultivated in Southeast Asia and Malaysia for several hundred years. The tree is also grown throughout the Caribbean, Central and South America, Hawaii, Florida, China, Taiwan, Israel, India, Malaysia, Australia, and elsewhere. The plant produces clusters of small, sweet-smelling pink or purplish flowers. The bright yellow-green, orange-yellow, or orange fruits are elongated, 5 to 15 cm (2–6 in.) in length, 3 to 9 cm (1¼–3½ in.) in width, with usually five, sometimes four or six, longitudinal ribs. The ripe fruit is very fragrant with a thin, edible, waxy skin and crisp, juicy, yellow flesh when fully ripe. The taste is often reminiscent of a combination of apple and grape. Up to 12 flat, thin, brown seeds 6 to 12.5 mm (1/4–1/2 in.) long may be present, or the fruit may be seedless. There are more than a dozen varieties, with two classes of fruit: a smaller, very sour type (the sourness is due to oxalic acid) and a larger, so-called "sweet" type (which only has approximately 4% sugar) that is milder in flavor. The carambola, generally under the name star fruit, is becoming increasingly popular in Western markets. Aside from its attractive taste, the star-shaped, cross-sectioned slices are attractive for adorning salads and other dishes.

CULINARY PORTRAIT

The carambola should be washed well and deseeded before consumption. The fruit is delicious eaten out of hand (peeling is not necessary) and is also added to salads, curries, tarts, and other desserts, beverages, relishes, chutneys, puddings, stews, and candy, and is made into jams, jellies, preserves, and liquor. Green fruits are sometimes consumed as a vegetable or (as noted earlier) made into pickles. If cooked, this should be done for a short time only to preserve the flavor. The taste ranges from pleasantly tart and sour to slightly sweet. The juice has been described as sweet and apple like. Sweet types are used in salads, jellies, and other preparations where a sweet taste is desirable, whereas sour types are typically used in condiments. The fruit bruises easily and is often packaged by workers wearing rubber gloves. Carambola stores for up to a month in the refrigerator, or 2 weeks at room temperature. The sour types can contain 15 times as much oxalic acid as the sweet types and indeed have amounts comparable with those of high-oxalic plant foods such as spinach and rhubarb. Because of the high acidity and the danger of leaching ions into food, high-oxalic foods should not be cooked in aluminum, copper, or iron containers (stainless steel is recommended). Unripe fruit is green, half-ripe fruit is typically lemon-green, and very ripe fruit is a golden yellow. Good-quality carambola in a market is fairly firm, and the skin will be yellow with no green tinges, although slight browning along the edges is normal. (Slight browning on the fruit angles can be pared away; however, fruit with brown patches should not be purchased.)

FIGURE 18.1 Carambola (*Averrhoa carambola*), flowering branch, with a fruit at upper left. (From Lamarck and Poiret, 1744–1829, plate 385.)

Culinary Vocabulary

- *Camaranga* is a jam or other food preparation made with carambola (*kamaranga* is an Asian name for the carambola).

CURIOSITIES OF SCIENCE AND TECHNOLOGY

- The leaflets of the leaves of carambola tend to fold together at night, and when the tree is shaken. Such "nastic movements" in plants are generally adaptive, for example, protecting against damage from inclement weather.
- Sour (acidic)-fruited types of carambola have been used to clean and polish metal, especially brass, as the juice dissolve tarnish and rust. The juice can also be used to bleach rust stains from cloth and to remove stains from the hands.
- Carambola has a reputation for relieving hangovers. The validity of this has not been demonstrated.
- "Patients with renal failure, even those not yet undergoing dialysis, can develop severe and potentially fatal neurologic complications after eating star fruit ... Dr. Yung-Hsiung Lai and colleagues at Kaohsiung Medical University, in Kaohsiung, Taiwan, explain in the February issue of the American Journal of Kidney Diseases that 'substances safely ingested by healthy persons can be deleterious for uremic patients.' Because of its high potassium content, star fruit may be considered to be one such substance that uremic patients should exclude from their diets. In a review of hospital records, Dr. Lai and colleagues identified 20 uremic patients with star fruit-related toxicity in the past 10 years, of whom eight died within 5 days of ingestion despite intensive medical care" (*American Journal of Kidney Disease*, 2000, 35: 189–193; also note Neto et al., 2003).

KEY INFORMATION SOURCES

Berry, S.K. 1978. The composition of the oil of starfruit (*Averrhoa carambola* L.) seeds. *J. Am. Oil. Chem. Soc.* 55:340–341.

Campbell, C.A., Huber, D.J., and Koch, K.E. 1987. Postharvest response of carambolas to storage at low temperatures. *Proc. Fla. State Hort. Soc.* 100:272–275.

Campbell, C.W., and Marte, R.J. 1990. *Pre-production, production and post-harvest handling of carambola.* IICA Misc. Publ. Series. Inter-American Institute for Cooperation on Agriculture. Bridgetown, Barbados. 20 pp.

Campbell, C.W., Knight, R.J., Jr., and Olszack, R. 1985. Carambola production in Florida. *Proc. Fla. State Hort. Soc.* 98:145–149.

Ferguson, J.J., Crane, J.H., and Olszack, R. 1988. Growth of young carambola trees using standard and controlled-release fertilizers. *Proc. Inter-Amer. Soc. Trop. Hort.* 32:20–24.

Galán Saúco, V., Menini, U.G., and Tindall, H.D. 1993. *Carambola cultivation (*FAO Plant Production and Protection Paper*).* Food and Agriculture Organization of the United Nations, Rome, Italy. 74 pp.

Inter-American ["Interamerican"] Society for Tropical Horticulture. 1989. *Proceedings of the Inter-American Society for Tropical Horticulture, International Carambola Workshop, Georgetown, Guyana, September 4–6, 1989.* Inter-American Society for Tropical Horticulture, Ministry of Agriculture, Guyana, and Inter-American Institute for Cooperation of Agriculture, Miami, FL. 132 pp. [Also published by Print Shoppe, Miami, FL.]

Kenney, P., and Hull, L. 1986. Effects of storage condition on carambola quality. *Proc. Fla. State Hort. Soc.* 99:222–224.

Knight, R.J., Jr. 1966. Heterostyly and pollination in carambola. *Annu. Meet. Fla. State Hort. Soc.* 76:375–378.

Matthews, R.F. 1989. Processing of carambola. *Proc. Inter-Amer. Soc. Trop. Hort.* 33:83–90.

Morton, J. 1987. Carambola. In *Fruits of warm climates.* Creative Resource Systems, Winterville, NC. pp. 125–128.

Nakasone H.Y., and Paull, R.E. 1998. Carambola. In *Tropical fruits*. Edited by H.Y. Nakasone and R.E. Paull. CABI, Wallingford, UK. pp. 132–148.

Neto, M.M., Cardeal da Costa, J.A., Garcia-Cairasco, N., Coutinho, N.J., Nakagawa, B., and Dantas, M. 2003. Intoxication by star fruit (*Averrhoa carambola*) in 32 uraemic patients: treatment and outcome. *Nephrology Dialysis Transplantation*, 18(1):120–125.

O'Hare, T.J. 1993. Postharvest physiology and storage of carambola (starfruit): a review. *Postharvest Biol. Technol*. 2:257–267.

O'Hare, T.J. 1997. Carambola. In *Postharvest physiology and storage of tropical and subtropical fruits*. Edited by S.K. Mitra. CAB International, Wallingford, UK. pp. 295–307.

Oslund, C.R., and Davenport, T.L. 1983. Ethylene and carbon dioxide in ripening fruit of *Averrhoa carambola*. *HortScience*, 18:229–230.

Salakpetch, S., Turner, D.W., and Dell, B. 1990. Flowering in carambola (*Averrhoa carambola*). *Acta. Hortic*. 275:123–129.

Samson, J.A. 1991. *Averrhoa* L. In *Plant resources of South-East Asia. 2. Edible fruits and nuts*. Edited by E.W.M. Verheij and R.E. Coronel. Pudoc, Leiden, the Netherlands. pp. 96–98.

Sauco, V.G., and Paull, R.E. 2008. *Averrhoa carambola*, carambola. In *The encyclopedia of fruit & nuts*. Edited by J. Janick and R.E. Paull. CABI, Wallingford, Oxfordshire, UK. pp. 576–581.

Shaw, P.E., and Wilson, C.W. III. 1998. Carambola and bilimbi. In *Tropical and subtropical fruits*. Edited by P.E. Shaw, H.T. Chan, Jr., and S. Nagy. Agscience, Auburndale, FL. pp. 521–556.

Vines, H.M., and Grierson, W. 1966. Handling and physiological studies with the carambola. *Proc. Fla. State Hort. Soc*. 79:350–355.

Wagner, C.J., Jr., Bryan, W.L., Berry, R.E., and Knight, R.J., Jr. 1975. Carambola selection for commercial production. *Proc. Fla. State Hort. Soc*. 88:466–469.

Wilson, C.W. 1990. Carambola and bilimbi. In *Fruits of tropical and subtropical origin: composition, properties and uses*. Edited by S. Nagy. Florida Science Source Inc., Lake Alfred, FL. pp. 276–301.

Wong, K.C., Watanabe, M., and Hinata, K. 1994. Protein profiles in pin and thrum floral organs of distylous *Averrhoa carambola* L. *Sex. Plant Reprod*. 7:107–115.

Specialty Cookbooks

J.R. Brooks & Son. 1988. *Recipes that mean business: featuring carambola, Florida star fruit & other tropical favorites*. J.R. Brooks & Son, Homestead, FL. 24 pp.

Norman, J. 1992. *Exotic fruits: unusual fruits from all over the world*. Bantam, New York. 41 pp.

Strauss, S.C. 1991. *Fancy fruits and extraordinary vegetables: a guide to selecting, storing, and preparing*. Hasting House, Mamaroneck, NY. 284 pp.

19 Carob

Family: Fabaceae (Leguminosae; pea family)

NAMES

Scientific Name: *Ceratonia siliqua* L.

- The name carob traces to the Hebrew name of the plant, *kharauv*, and the Arabic *kharrūbah*, which means bean pod or carob.
- Carob is also known as algarrobo, carob bean, John's bread, locust, locust bean, and St. John's bread.
- The name St. John's Bread is probably a biblical allusion to the "locusts" that, according to Matthew 3, sustained St. John the Baptist in the desert ("John had his raiment of camel's hair, and a leathern girdle about his loins; and his meat was locusts and wild honey ... "). The word "locust" was originally applied to the carob tree, later to migratory and other grasshoppers, and now the name also refers to several other trees of the pea family with leaves and pods reminiscent of those of the carob. Probably millions of Sunday school attendees have been grossed out by the interpretation that St. John the Baptist survived by eating grasshoppers. However, the alternative interpretation that the locusts that the prophet consumed were carob pods is not entirely certain. According to the dietary laws of Leviticus 11, four kinds of insect locusts and grasshoppers are considered as "clean" for eating, and these would have been perfectly "kosher" for John to eat. In biblical times, locusts were sold as food by Judean shopkeepers, and some modern Middle Eastern and North African Jews, as well as Bedouin nomads, have eaten them. Carob was also the "husks" fed to swine which tempted the hungry Prodigal Son who was tending the livestock (Matthew 15:16). He had spent his inheritance so quickly and carelessly that he had to hire himself out as a swine tender, and lacking money for food, was tempted to eat the carob pods fed to the pigs.
- The genus name *Ceratonia* is derived from the Greek *keration*, horn or pod, a reference to the horn-shaped pods.
- *Siliqua* in the scientific name *C. siliqua* is Latin for pod and emphasizes that the pods are prominent in this species.

PLANT PORTRAIT

The carob or St. John's bread is an evergreen tree, under favorable conditions reaching 12 to 17 m (40–55 ft.) in height with a trunk up to 3.6 m (11 ft. 8 in.) in diameter. Carob trees are native to the Middle East and have been planted for thousands of years in the Mediterranean region. Numerous varieties have been selected. Most commercial cultivation is in Spain, Italy, Portugal, Morocco, Cyprus, Greece, and Turkey. The first carobs in the United States were established in 1854. Carob trees are widely grown in the southern United States, mostly as ornamentals, and there is no appreciable North American commercial production. Male trees are often preferred as shade, roadside, and park specimens because they do not litter the area with pods. On the other hand, male flowers have an unpleasant smell, and some people prefer to plant female trees for ornamental purposes. Because carob trees are male, female, or hermaphrodite, to ensure adequate pollination in commercial orchards, about one male or hermaphrodite tree is planted for every eight female trees. Sometimes the orchardist's trick of grafting a male branch to a female tree is used to provide pollen for production of the pods. Occasional trees produce as much as a ton of carob pods, and cultivated trees that are well cared for can have a productive life of 100 years. The pods and their seeds, both of which are edible, are the important product furnished by the carob. The pods are light- to dark-brown, glossy, flattened, straight or slightly curved, 10 to 30 cm (4–12 in.) long, and 11 to 2.5 cm (3/4–1 in.) wide. The pods are filled with soft, semitranslucent, pale-brown pulp, and 10 to 18 flattened, very hard seeds which are loose in their cells and rattle when the pod is fully ripe and dry. The unripe pod is green, moist, and very astringent. The ripe pod is sweet when chewed, but the odor of the broken pod is faintly reminiscent of Limburger cheese because of its content of more than 1% isobutyric acid (this compound as well as isovaleric acid contribute to the "sweaty sock" aroma of Swiss-type cheeses).

CULINARY PORTRAIT

Carob is considered to have healthful attributes as a food, to have medicinal virtues, and also to be a good substitute for chocolate, and these properties have generated a commercial market for it, especially since the 1980s. The pods and the seeds are used to make different products.

After the seeds are removed, carob pods are dried, roasted, and ground to a powder or broken into chips. Another product, carob syrup, is also commonly available. The more carob is roasted the blander its flavor and the darker it becomes. Carob flour, which is sweet and high in fiber, has been combined with wheat flour in making bread, pancakes, and candy bars. Carob flour is widely used in health foods as a chocolate substitute, although it has a slightly different taste. Carob can replace chocolate in recipes (one part chocolate is equivalent to 1¾ parts carob) but is generally not considered as flavorful. Carob dissolves less readily than chocolate (hot water is therefore recommended), melts at a lower temperature, and liquefies more readily, and these properties can present problems when preparing some dishes. Carob chocolates are widely sold with the claim that they have fewer calories and are virtually fat-free, but in fact on a weight basis, fat and calorie content of carob and chocolate are comparable. Moreover, carob's fat is mostly saturated, and many carob chocolates have had considerable fat and sugar added to enhance taste. Carob has been extensively used to flavor uncured tobacco. Carob should be stored in a very tightly sealed container to keep from becoming lumpy.

The seeds are used for several edible purposes. Tragasol gum (locust bean gum) from the seeds is an important commercial stabilizer and thickener in food products, especially used as an emulsion stabilizer and thickener in ice cream, cheese, and salad dressings. The gum is added to many soups, where its property of fully dissolving and thickening only at high temperatures is particularly useful. In sausage products such as salami and bologna it acts as a binder and lubricant. Other food uses include the manufacture of bakery products, pie fillings, powdered desserts, sauces, and dairy products. The seed residue remaining after gum extraction can be made into a starch- and sugar-free flour of 60% protein content for diabetics. The roasted seed has been used as a caffeine-free coffee substitute in Germany and elsewhere.

CULINARY VOCABULARY

- The edible, commercial gum from carob is known as carob gum, carob seed gum, locust bean gum, locust kernel gum, tragasol, gum hevo, and gum gatto.

CURIOSITIES OF SCIENCE AND TECHNOLOGY

- The uniform seeds of the carob were used as standards to balance the scales in weighing gems and gold in Oriental bazaars. The word *karat* (or *carat*), meaning the modern jewellers' karat weight (200 mg), is derived from the Greek *keration*, carob seed.
- The tap root of the carob has been reported to reach 38 m (125 ft.) underground, allowing the trees to survive where the main supply of water is very deep underground.

- Singers once chewed the pod husks of the carob in the belief that this clears the throat and voice. Carob seeds have actually been found to have the capacity to reduce inflammations in air passages in the throat and head and promote healing so that singers might actually have benefited by reduction of coughs and respiratory tract disorders.
- In Europe, carob seeds are sometimes used to make rosaries.
- Carob was the feed that sustained the cavalry of the Duke of Wellington (Arthur Wellesley, 1769–1852) during his campaign against the forces of Napoleon Bonaparte (1769–1821) in Portugal and Spain because no other feedstuff for the horses grew in the dry, rocky battle region.
- British field marshal Edmund Allenby (1861–1936), won great victories in the Middle East during the First World War, using methods that some have claimed resembled the blitzkrieg tactics of Nazi Germany during the Second World War. Although Allenby made efficient use of mechanized forces, as was the case for Wellington, the horses of his cavalry depended heavily on local carob.
- Since ancient times, carob pods have been used primarily to feed livestock but in periods of famine have also been used as human food. In southern Greece, during the Second World War, the Germany army stripped the country of livestock and most of its food, and the rural inhabitants were forced to subsist largely on carob pods.
- In North America, carob is often added to dog biscuits. Chocolate is toxic, indeed potentially deadly for dogs, but carob is quite edible, and so "carob chip cookies" are sold as a safe alternative for chocolate chip cookies. Similarly, one company sells "bark bars" for dogs as the equivalent of human chocolate bars. A Web document reported that one dog lover even went to the extreme of preparing a canine birthday cake with carob icing.
- The carob is the only Mediterranean tree with the main flowering season in autumn (September to November), like many tropical plants.
- Carob pods contain more sugar than sugar cane and sugar beet. About half of the weight of the pulp is sugar.

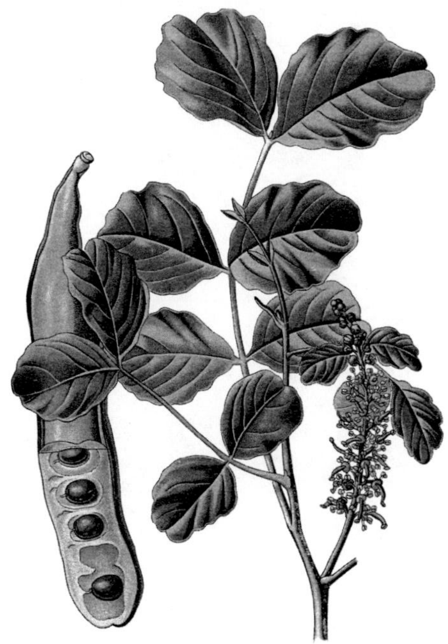

FIGURE 19.1 Carob (*Ceratonia siliqua*). (From Thomé, 1903–1905, vol. 3, plate 383.)

FIGURE 19.2 Carob (*Ceratonia siliqua*). (From Baillon, 1876–1892.)

FIGURE 19.3 Carob (*Ceratonia siliqua*). (From Turner, 1892.)

KEY INFORMATION SOURCES

Albright, F.R., Schumacher, D.V., and Major, R.A. 1978. How to detect carob in cocoa. *Food Eng.* 50(5):110–111.

Batlle, I, and Tous, J. 1997. *Carob tree,* Ceratonia siliqua *L.* International Plant Genetic Resources Institute, Rome, Italy. 92 pp.

Bosch, J., Garcia del Pino, F., Ramoneda, J., and Retana, J. 1996. Fruiting phenology and fruit set of carob, *Caratonia siliqua* L. (Caesalpiniaceae). *Isr. J. Plant Sci.* 44:359–368.

Carlson, W.A. 1980. Carob as a food—a historic review. *Lebensmitt. Wissensch. Technol.* 13(1):51–52.

Coit, J.E. 1949. *Carob culture in the semi-arid Southwest.* W. Rittenhouse, San Diego, CA. 15 pp.

Coit, J.E. 1951. Carob or St. John's bread. *Econ. Bot.* 5:82–96.

Davies, W.N.L. 1970. The carob tree and its importance in the agricultural economy of Cyprus. *Econ. Bot.* 4:460–470.

Esbenshade, H.W. 1980. Carob: a multipurpose tree crop for marginal lands intensification of food production and reduction of soil erosion. *Agrofor. Rev.* 2(2):2–5.

Esbenshade, H.W., and Wilson, G. 1986. *Growing carobs in Australia.* Goddard and Dobson, Box Hill, Victoria, Australia. 136 pp.

Federation of American Societies for Experimental Biology, and U.S. Food and Drug Administration. 1972. *Evaluation of the health aspects of carob bean gum as a food ingredient.* National Technical Information Service, Springfield, VA. 10 pp.

Greenstock, L. 1977. The protein gap—the carob-bean tree. *Int. J. Environ. Stud.* 11:163–167.

Haber, B. 2002. Carob fiber benefits and applications. *Cereal Foods World*, 47:365–369.

Haq, N. 2008. *Ceratonia siliqua*, carob. In *The encyclopedia of fruit & nuts.* Edited by J. Janick and R.E. Paull. CABI, Wallingford, Oxfordshire, UK. pp. 387–391.

Hills, L.D. 1980. The cultivation of the carob tree (*Ceratonia siliqua*). *Int. Tree Crops J.* 1:27–36.

Logan, M.D. 1960. The carob crusade. *Am. Forests*, 66(6):18–19, 63–65.

Marakis, S. 1996. Carob bean in food and feed: current status and future potentials—a critical appraisal. *J. Food Sci. Technol.* 33:365–383.

Merwin, M.L. 1981. *The culture of carob (*Ceratonia siliqua *L.) for food, fodder and fuel in semi-arid environments.* International Tree Crops Institute, Winters, CA. 18 pp.

Morton, J. 1987. Carob. In *Fruits of warm climates.* Creative Resource Systems, Winterville, NC. pp. 65–69.

Nyerges, C. 1978. The chocolate that's good for you: carob is delicious, chocolaty—and not fattening. *Org. Garden*, 25(12):122–126.

Orphanos, P.I., and Papaconstantinou, J. 1969. *The carob varieties of Cyprus.* Tech. Bull. 5. Agricultural Research Institute, Ministry of Agriculture and Natural Resources, Nicosia, Cyprus. 27 pp.

Ramon-Laca, L., and Mabberly, D.J. 2004. The ecological status of the carob-tree (*Ceratonia siliqua* L.) in the Mediterranean. *Bot. J. Linn. Soc.* 144:431–436.

Rol, F. 1959. Locust bean gum. In *Industrial gums.* Edited by R.L. Whistler. Academic Press, New York. pp. 361–375.

Spivack, E.S. 1978. Carob: nature's chocolate alternative. *Vegetarian Times*, 23(Jan–Feb):60–61.

Stylianos, M. 1996. Carob bean in food and feed: current status and future potentials—a critical appraisal. *J. Food Sci. Technol.* 33:365–383.

Thompson, P.H. 1971. The carob in California. *California Rare Fruit Growers Yearbook*, 3:61–102.

University of Adelaide. 1996. *National symposium on olives and carobs for landcare and for profit. Proceedings of a symposium held at the University of Adelaide, Roseworthy Campus, 17–18 April, 1994.* University of Adelaide, Subiaco, WA, Australia, 57 pp.

Winer, N. 1980. The potential of the carob (*Ceratonia siliqua*). *Int. Tree Crops J.* 1:15–26.

Zohary, D. 2002. The place of origin and the nature of dioecy in the carob (*Ceratonia siliqua* L.). *Nucis Newsl.* 11:38–40.

Zohary, D. 2002. Domestication of the carob (*Ceratonia siliqua* L.). *Isr. J. Plant Sci.* 50(Suppl.):S141–S145.

Specialty Cookbooks

Andersen, J., Clute, R., and Andersen, S. 1983. *Juel Andersen's carob primer: a beginner's book of carob cookery.* Creative Arts Communications, Berkeley, CA. 56 pp.

Collins, K. 1981. *The complete carob cookbook: a cookbook of delicious carob recipes.* K. Collins, Miltown, IN. 61 pp.

East, S. 1988. *The carob tree: a collection of deliciously easy vegetarian recipes.* S. East, Melbourne. 127 pp.

Goulart, F.S. 1982. *The carob way to health: all-natural recipes for cooking with nature's healthful chocolate alternative.* Warner Books, New York. 147 pp.

Hamilton, T. 1990. *Carob cookbook.* Sunstone Press, Santa Fe, NM. 112 pp.

Johnson, S. 1987. *The carob kitchen.* Hakes Pub., Salt Lake City, UT. 136 pp.

Sanders, E. 1980. *The delights of carob cooking*. Roast Pigeon Press, Porter, IN. 47 pp.
Whiteside, L. 1984. *The carob cookbook*. Revised edition. Thorsons Publ. Ltd., Wellingborough, Northamptonshire, UK. 192 pp.
Wilkin, E.E., and Kirkby, R.A. 1975. *St. John's bread: the carob cookbook*. Jurupa Mountains Cultural Center, Riverside, CA. 56 pp.

20 Cashew

Family: Anacardiaceae (cashew family)

NAMES

Scientific Name: *Anacardium occidentale* L.

- The English word "cashew" is derived from the Portuguese *cajú*, which originates from the Tupi-Indian *acajú* or the Arawak *acaeju*, meaning "to pucker the mouth," descriptive of the effect of some varieties of the fruit.
- The cashew is also called maranon, which is the Latin American Spanish word for cashew, *marañon*, presumably originating from Maranhão, Brazil, one of the first regions where cashew was observed by the Spanish.
- In West Tropical Africa, the cashew nut is known as "elephant's tick," a joke name because the nut is huge by comparison with the insect.
- The genus name *Anacardium* is usually interpreted as based on the Greek *ana*, up + *kardia*, heart, for "shaped like a heart." When the unshelled cashew nut is held with its side indentation upward, it appears to be heart-shaped, although on the whole it resembles a miniature boxing glove. The ancient Greek *anakardion* was the twig of mulberry, so this interpretation may not be correct. Another explanation is that the genus name is based on the Sanskrit *vranakart*, from *vrana*, wound + *kart* or *kar*, to do, an allusion to the fact that eating the raw fruit was considered dangerous for the mouth. "Anacardic" means pertaining to a cashew nut.
- *Occidentale* in the scientific name *A. occidentale* is Latin for western, indicating that the plant is in the Western Hemisphere.

PLANT PORTRAIT

The cashew or cashew nut is a tropical evergreen tree, sometimes growing higher than 12 m (40 ft.), but usually shorter. The plant is notable for its bushy, spreading branches and its leathery leaves. The cashew is native to tropical America and probably originated in northeastern Brazil. It was domesticated long before the arrival of Europeans at the end of the fifteenth century. The tree and its nuts and fruit had been used for centuries by indigenous tribes as food and medicine. Although the cashew has a long history as a useful plant, it did not become an important tropical crop until the twentieth century. About 400 years ago, the Portuguese brought it from Brazil to Goa in India and subsequently to their colonies in Africa. India is now the world's chief source of cashews. Considerable amounts are also produced in Mozambique, Tanzania, and Brazil. Many other Asian, African, Caribbean, and South American countries produce and export smaller amounts of cashews. The United States is the largest importer of cashew nuts. Cashew nut has been called "a poor man's crop but a rich man's food" because it is often too expensive for the workers producing it.

The cashew develops a fruit-like stalk called "cashew apple" (often also known as "cashew pear" because it looks more like a pear than an apple). The cashew apple is not a true fruit botanically (because it did not develop from the ovary, the part of the flower that contains the egg(s)), but most people call it a fruit and this practice will be followed. Most fleshy fruits contain their seeds within pulp, but this is not so with the cashew. The nut is on the *outside* of the fruit, located at the end of the fruit that is farthest away from its attachment to the stem. The cashew apple varies in size from 5 to 10 cm (2–4 in.) in length and from 3.8 to 5 cm (1½–2 in.) width. It is brilliant red, scarlet, yellowish-red, or yellow in color and has a thin, waxy skin.

The cashew nut is generally just more than 2.5 cm (1 in.) in diameter, smooth, initially olive green, turning brownish red as the fruit matures. The nut has a hard shell about 3 mm (1/8 in.) thick, which consists of two layers with an oily liquid between them. Cashew liquid is extremely caustic; it will blister the skin and is the reason that no one should ever attempt to eat a fresh (unroasted) cashew nut. Unroasted nuts will never be available in the marketplace, but where cashew trees are grown, such as in southern Florida, one could encounter fresh nuts. Touching an uncooked nut can cause skin eruptions and even the smoke given off by roasting is a poisonous irritant. The blisters that are produced are similar to those caused by poison ivy, and it will therefore not come as a surprise that poison ivy and the cashew are in the same plant family. Workers in contact with raw cashew nuts frequently develop skin problems, and so the cashew has sometimes been called "blister nut." Cashew shell liquid is made into a variety of products, used especially for brake lining compounds, waterproofing agents, preservatives, and in the manufacturing of paints and plastics. The liquid also has medicinal applications. To obtain the edible kernels known as cashews, an elaborate processing technology is necessary to rid the kernels of the caustic liquid.

CULINARY PORTRAIT

Cashew is one of the most popular nuts in the confectionery industry, indeed ranking second only to the almond in value of world trade of tree nuts. The taste is much superior to that of the common peanut. Roasted, salted nuts are the most common product, consumed primarily as a treat. Confections, cakes, and cookies also use up a large part of the cashews that are produced. The crushed nuts are popular in Asian cooking and are widely used in Chinese stir-fried dishes. The nuts are also used in Indian curries. Cashews are commonly combined with lamb, curry, and rice dishes and stews and can also be added to salads and pasta. When used in cooked dishes, it is usually preferable to add the nuts at the end of the cooking cycle to preserve their texture and flavor. Like most other nuts, cashews are high in saturated fat and so are often avoided by weight-conscious people. Cashews become rancid more rapidly than most other nuts. The freshest-tasting nuts are usually available only in vacuum-packed jars or cans, and once opened these are best stored in the refrigerator or freezer.

Cashew apples are succulent in taste, with a sweet acid flavor. Wherever cashews are grown, the fruits are eaten raw or cooked, made into jams and preserves, and fermented into a wine (which is often distilled into a potent liquor). A vinegar is also prepared. In the eighteenth century, cashew apple jam became popular for a short period in Europe. However, today cashew apples are little known outside the Tropics because they spoil easily, keeping only for 24 hours. The astringent flavor of the cashew apple is not favored outside of its native country, Brazil. Most cashew apples are left to spoil; the trees grown exclusively for the valued nuts.

Culinary Vocabulary

- Cashew nut butter is a spread made from ground, homogenized cashew nuts, which resembles peanut butter but is creamier and more delicate in taste.

CURIOSITIES OF SCIENCE AND TECHNOLOGY

- A milky sap is produced within the bark of the cashew tree, and this oxidizes on exposure to air, turning black. Wherever the tree has been grown in large numbers, people have used this as an indelible ink.
- The caustic oil from fresh cashew nuts has been used to tattoo skin in the West Indies. It has also been used in a drastic, dangerous beauty treatment, applied to the face of women,

causing extreme blistering that results in peeling away of the facial skin and hopefully producing an improved complexion.

- In Haiti, the Dominican Republic, and nearby areas, roasted cashew nuts are used as heads of dolls and other novelties. The nuts are provided with a carved mouth and eye sockets (as is done with pumpkins on Halloween) and are sometimes called "monkey nuts." Tourists once commonly imported these into the southeastern United States. Such creations were banned when cashew-headed toys caused a rash among Georgia children because of improper roasting and seepage of the caustic oil through the openings of the shell.

FIGURE 20.1 Cashew (*Anacardium occidentale*). (From Köhler, 1883–1914.)

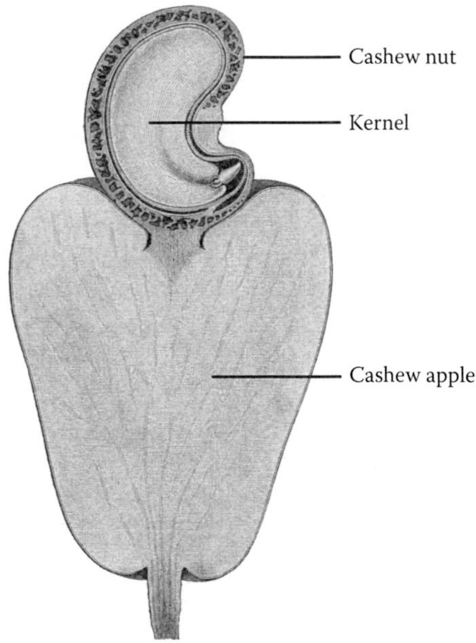

FIGURE 20.2 Cashew (*Anacardium occidentale*), long section of fruit. (From Engler and Prantl, 1889–1915.)

KEY INFORMATION SOURCES

Abdul Salam, M., and Mohanakumaran, N. 1995. *High yielding varieties of cashew*. Kerala Agricultural University, Thrissur, India. 45 pp.

Ballesteros, C.C. 1999. *Cashew nuts*. De La Salle University Press, Manila, Philippines. 45 pp.

Behrens, R. 1996. *Cashew as an agroforestry crop: prospects and potentials*. Margraf Verlag, Weikersheim, Germany. 83 pp.

Bhaskara Rao, E.V.V., and Khan, H.H. (Eds.). 1984. Cashew research and development. Proceedings of the International Cashew Symposium, Cochin, Kerala, India (12–15 March, 1979). Indian Society for Plantation Crops, Kerala, India. 314 pp.

Bopaiah, B.M. 1985. Microbial spoilage of cashew apples and its prevention. *Indian Cashew J.* 16(2):15–17.

Cantrell, G.E. 1945. *Cashew nuts*. War Food Administration, Office of Marketing Services, Washington, D.C. 53 pp.

Cundall, E.P. 1995. Cashew. In *Evolution of crop plants*. 2nd ed. Edited by J. Smartt and N.W. Simmonds. Longman Scientific & Technical, Burnt Mill, Harlow, Essex, UK. pp. 11–13.

Divakaran Pillai, M., and Haveri, R.R. 1979. *Bibliography on cashew (Anacardium occidentale L.)*. Central Plantation Crops Research Institute, Kasaragod, Kerala, India. 92 pp.

Eijnatten, C.L.M. van. 1991. *Anacardium occidentale* L. In *Plant resources of South-East Asia. 2. Edible fruits and nuts*. Edited by E.W.M. Verheij and R.E. Coronel. Pudoc, Leiden, the Netherlands. pp. 60–64.

Farooqi, A.A., and Sreeramu, B.S. 1999. Cashew. In *Tropical horticulture*, vol. 1. Edited by T.K. Bose, S.K. Mitra, A.A. Farooqui, and M.K. Sadhu. Naya Prokash, Calcutta, India. pp. 563–577.

Johnson, D. 1973. The botany, origin, and spread of the cashew, *Anacardium occidentale* L. *J. Plantation Crops*, 1:1–7.

Kannan, K.P. 1983. Cashew development in India: potentialities and constraints. Agricole Publishing Academy, New Delhi, India. 152 pp.

Kesavan, V. (Ed.). 1992. *Cashew research and development: fifth annual workshop proceedings, May 18–19, 1992, Kununurra, Western Australia*. Western Australia Dept. of Agriculture, Canberra, Australia. 138 pp.

Mandal, R.C. 1992. Cashew production and processing technology. Agro Botanical Publishers, Bikaner, India. 195 pp.

Mitchell, J.D., and Mori, S.A. 1987. The cashew and its relatives. New York Botanical Garden, Bronx, NY. 76 pp. [Memoirs N.Y. Bot. Gard. 42:1–76.]

Moncur, M.W. 1986. *Tabular descriptions of crops grown in the tropics: 15. Cashew (Anacardium occidentale L.)*. CSIRO, Institute of Biological Resources, Division of Water and Land Resources, Canberra, Australia. 54 pp.

Montealegre, J.C., Childers, N.F., Sargent, S.A., Barros, L.de M., and Alves, R.E. 1999. Cashew (*Anacardium occidentale*, L.) nut and apple: a review of current production and handling recommendations. *Fruit Var. J.* 53:2–9.

Morton, J. 1987. Cashew apple. In *Fruits of warm climates*. Creative Resource Systems, Winterville, NC. pp. 239–240.

Nair, M.K. 1979. *Cashew* (Anacardium occidentale *L.*). ICAR, Central Plantation Crops Research Institute, Kasaragod, Kerala, India. 169 pp.

Nambiar, M.C., and Pillai, P.K.T. 1985. Cashew. In *Fruits of India, tropical and subtropical*. Edited by T.K. Bose. Naya Prokash, Calcutta, India. pp. 409–438.

Nomisma. 1994. *The world cashew economy*. Nomisma, Bologna, Italy. 218 pp.

Ohler, J.G. 1966. Cashew nut processing. *Trop. Abstr.* 21:549–554.

Ohler, J.G. 1979. *Cashew*. Koninklijk Instituut voor de Tropen, Amsterdam, the Netherlands. 260 pp.

Ramos, A.D. 1992. *Cashew tree culture in the Northeast of Brazil*. Rural Industries Research and Development Corporation, Barton, A.C.T., Brazil. 306 pp.

Russell, D.C. 1969. *Cashew nut processing*. Food and Agiculture Organization, Rome, Italy. 86 pp.

Wait, A.J., and Jamieson, G.I. 1986. The cashew: its botany and cultivation. *Qld. Agric. J.* 112:253–257.

Wardowski, W.F., and Ahrens, M.J. 1990. Cashew apple and nut. In *Fruits of tropical and subtropical origin: composition, properties and uses*. Edited by S. Nagy, P.E. Shaw, and W.F. Wardowski. Florida Science Source Inc., Lake Alfred, FL. pp. 66–87.

SPECIALTY COOKBOOKS

Allen, Z. 2006. *The nut gourmet*. Book Publishing Company (TN), Summertown, TN. 256 pp.

Carder, S. 1984. *The nut lover's cookbook*. Celestial Arts, Berkeley, CA. 160 pp.

Dalass, D. 1981. *Cashews and lentils, apples and oats: from the basics to the fine points of natural foods cooking with 233 superlative recipes*. Contemporary Books, Chicago, IL. 301 pp.

Griffith, L., and Griffith, F. 2003. *Nuts: recipes from around the world that feature nature's perfect ingredient*. St. Martin's Press, New York. 332 pp.

Kaufman, W.I. 1964. *The nut cookbook*. Doubleday, Garden City, NY. 194 pp.

Planters. 1980. *The nut cookbook*. Standard Brands Inc., New York. 23 pp.

Price, S. 1979. *The nut cookbook*. S.G. Price. 160 pp.

Sampson, S. 2002. *Party nuts: 50 recipes for spicy, sweet, savory, and simply sensational nuts that will be the hit of any gathering*. Harvard Common Press, Boston, MA. 80 pp.

21 Cassabanana

Family: Cucurbitaceae (cucumber family)

NAMES

Scientific Name: *Sicana odorifera* (Vell.) Naudin (*S. atropurpurea* André, *Cucurbita odorifera* Vell.)

- The name cassabanana (less frequently spelled casabanana, and rarely cassa banana or casa banana) is of unknown origin, according to Webster's Third International Dictionary.
- Suggested pronunciations: cah-sah-bah-nah-nah, kas'uh-buh-nan'uh
- The cassabanana is not related to the banana (*Musa* species and hybrids).
- The cassabanana is also known as cassabanan, casabanana melon, melocoton, musk cucumber, and sikana. The puzzling name "zucchini melon" has also sometimes been applied. Latin American names include *coroa*, *crua*, *curuba*, and *melon de olor*. In Brazil, the charming name is *xu xu*.
- The name pepino (or pepino melon) is sometimes used in South America for the cassabanana, but this results in confusion with the fruit usually called pepino (see Chapter 75).
- 'Cassabanana Butternut' is a cultivar of the butternut squash species (*Cucurbita moschata* (Duchesne ex Lam.) Duchesne ex Poir.).
- The name cassabanana is sometimes mistakenly applied to the wax gourd (see Chapter 99).
- The genus name *Sicana* is the Peruvian name for the plant and fruit. Sicana is Ecuadorian Quechu language meaning climb or climbing, a reference to the climbing nature of the vine.
- *Odorifera* in the scientific name *S. odorifera* is Latin for fragrant.
- Suggested pronunciation of the scientific name: sik-AY-nuh OH-dor-IF-er-uh.

PLANT PORTRAIT

The cassabanana is a perennial, herbaceous, fast-growing, vine, which can climb to 15 m (49 ft.) or more and in cultivation requires a strong trellis. The plant has four-parted tendrils with adhesive discs which stick to the smoothest surface. The heart-shaped leaves have three to seven lobes and are large, sometimes more than 30 cm (1 ft.) across. The flowers are white or yellow, male flowers about 2 cm (3/4 in.) long and female flowers about 5 cm (2 in.). The fruit often has the appearance of a long, cylindrical, or ellipsoid squash or a very fat cucumber. It is 30 to 60 cm (12–24 in.) long, 7 to 11.5 cm (2¾–4½ in.) thick, and has a strong, sweet, melon-like odor. The hard shell of the fruit is smooth and glossy when ripe, orange-red, maroon, jet-black, or dark-purple with tinges of violet. The flesh is firm, orange-yellow or yellow, juicy, and reminiscent of that of a cantaloupe. The outer 2 cm (3/4 in.) of the fruit is firm fleshed, while the central cavity is filled with softer pulp, and in the center of this there is a soft, fleshy core with numerous flat, oval seeds about 16 mm (5/8 in.) long in rows the entire length of the fruit.

The cassabanana is known only in cultivation and as an escape, and its place of origin is undetermined. Two wild species of *Sicana* have been described from South America, but their status is uncertain. Although first described from Peru, *S. odorifera* probably is not of Andean origin. It may be native to Brazil or Paraguay but has been spread throughout tropical America. It was cultivated in Ecuador before the arrival of Columbus in the New World. Cassabanana is well known in Mexico and Central America and has been introduced as a curiosity in Europe, especially in France. The fruit is used for culinary purposes most frequently in Mexico, Puerto Rico, and Cuba and is also widely grown as an ornamental in tropical America. For the most part, it is grown on a small scale in South America. The cassabanana is occasionally cultivated and sold in North America.

CULINARY PORTRAIT

Cassabanana fruits are eaten when young, both raw and cooked. The cooked fruit is consumed as a vegetable or used in soups and stews. In France, the young fruits are eaten cooked, especially in soups. The ripe flesh of the young fruit is also sliced and eaten raw, much like watermelon. In Latin America, particularly in Nicaragua, the fruit is used to flavor beverages. The fruit is particularly used to make jam and preserves. The cassabanana stores well for several months (sometimes up to a year) if kept dry.

CULINARY VOCABULARY

- *Cojombro* is a well-known, cassabanana-flavored drink in Nicaragua. It is an example of what is known in Latin America as an *agua fresca* (pronounced a-woss freh-scas; sometimes shortened to just agua), a beverage made with fruits, sugar, water, and perhaps other constituents such as nuts, vegetables, and/or flowers. *Agua fresca* is Spanish for "fresh waters," reflecting the fresh taste of the liquid preparation.

CURIOSITIES OF SCIENCE AND TECHNOLOGY

- The mature cassabanana fruit exudes a strong, pleasant fragrance, reminiscent of a blend of ripe melon and peach. A single fruit can perfume a whole house. In Central and South America, the mature fruit is kept in the house as an air freshener, especially in linen and clothes closets (the fruit is thought to be moth repellent), and this practice has been taken up to a minor extent in Europe. In Central and South America, the fruit is also placed on church altars and in Christmas crèches.

FIGURE 21.1 Cassabanana (*Sicana odorifera*). (From André, 1890.)

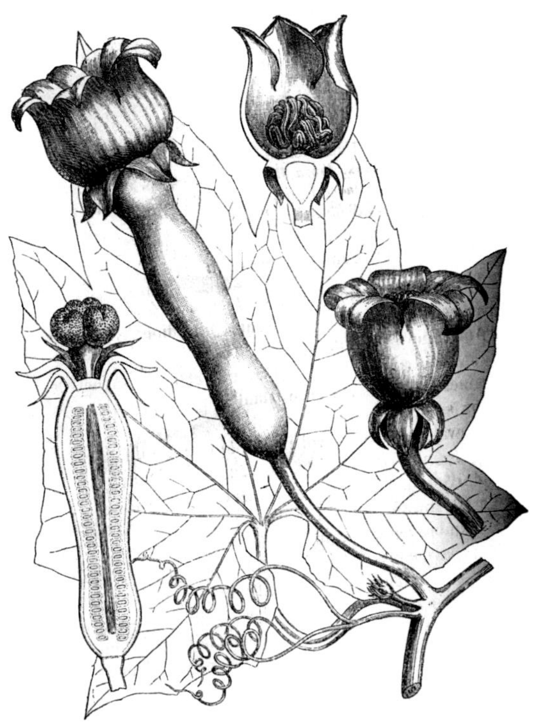

FIGURE 21.2 Cassabanana (*Sicana odorifera*) in flower and early fruit. (From Paillieux and Bois, 1892.)

FIGURE 21.3 Cassabanana (*Sicana odorifera*), by B. Flahey.

- In Latin America, a necklace of cassabanana seeds is worn around the neck in the belief that it will promote health. The health-promoting reputation of the plant is also evident in the use of an infusion of the seeds to treat fever, especially in Brazil.
- A single cassabanana fruit can produce as many as 900 seeds.
- Phytoliths are particles of hydrated silica formed in living plants, and when the plants decay, the phytoliths often remain intact for long periods. Phytoliths frequently can be used to identify species, and because they may be the only remains left, they are of great value to archeologists in assessing the species of plants that were used for various purposes by ancient cultures. There are comparatively large, distinctive, scalloped phytoliths in the fruit rinds of cassabanana that serve to identify it, and because the species is known to have been widely grown and consumed in Latin America in pre-Columbian times, the phytoliths are being used by archeologists to examine the culinary behavior of indigenous peoples in ancient times (see Piperno et al., 2000).
- All above-ground parts of the cassabanana plant are covered with minute (0.2 mm or 0.008 in. long), gland-tipped hairs. When the tiny glands on the ends of the hairs are touched by insects, they secrete a viscous liquid that quickly solidifies, adhering to the insects. The mechanism has been shown to protect the plant against certain small insects, discouraging them from feeding (see Kellogg et al., 2002).

KEY INFORMATION SOURCES

André, E. 1890. Le genre *Sicana*. *Rev Hort*. 62:515–517.

Decker-Walters, D.S., and Andres, T.C. 2002. Casabanana—an air freshener from the lowland tropics. *Cucurbit Netw. Newsl*. 9(2):1–3.

Hardy, I. 1976. General information on the fruits and seeds of the Cucurbitaceae of Venezuela. *Acta Bot. Venez*. 11(1-4):205–282 (in Spanish).

Jeffrey, C. 2001. Cucurbitaceae. In *Mansfeld's encyclopedia of agricultural plants*. Edited by P. Hanelt and Institute of Plant Genetics and Crop Plant Research. Springer, Berlin, Germany. pp. 1510–1557.

Kellogg, D.W., Taylor, T.N., and Krings, M. 2002. Effectiveness in defense against phytophagous arthropods of the cassabanana (*Sicana odorifera*) glandular trichomes. *Entomol. Exp. Appl.* (Dordrecht), 103:187–189.

Lira, R. 1991. Observaciones en el género *Sicana* (Cucurbitaceae) [Observations on the genus *Sicana*]. *Brenesia*, 35:19–59 (in Spanish).

Morton, J. 1987. Cassabanana. In *Fruits of warm climates*. Creative Resource Systems, Winterville, NC. pp. 444–445.

Parada, F., Duque, C., and Fujimoto, Y. 2000. Free and bound volatile composition and characterization of some glucoconjugates as aroma precursors in melon de olor fruit pulp (*Sicana odorifera*). *J. Agric. Food Chem*. 48:6200–6204.

Paris, H.S., and Maynard, D.N. 2008. *Sicana odorifera*., casabanana. In *The encyclopedia of fruits and nuts*. Edited by J. Janick and R.E. Paul. Oxford University Press, Oxford, UK. pp. 308–310.

Piperno, D.R., Andres, T.C., and Stothert, K.E. 2000. Phytoliths in *Cucurbita* and other neotropical Cucurbitaceae and their occurrence in early archaeological sites from the lowland American tropics. *J. Archaeol. Sci*. 27:193–208.

Robinson, R.W., and Decker-Walters, D.S. 1997. *Cucurbits*. CAB International, Wallingford, UK. 226 pp.

Rochelle, L.A. 1986. Contribution to the knowledge of crua (*Sicana odorifera* Naud.). *Sci. Agric*. 43:379–388 (in Portuguese).

Specialty Cookbooks

No English-language cookbooks with cassabanana recipes were located. When young and tender, the fruit can be treated like melon desserts. Many recipes for cooking squash are likely suitable for both young and more mature fruit. For preparing preserves, recipes for melons and citron watermelon may be adapted. For preparing jam, many fruit recipes can be adapted.

22 Ceriman

Family: Araceae (arum family)

NAMES

Scientific Name: *Monstera deliciosa* Liebm. (*Philodendron pertusum* Kunth & C.D. Bouché)

- The English name "ceriman" is based on the American Spanish name of the plant, cerimán.
- Pronunciation: SEHR-uh-muhn.
- Ceriman is also known as cut-leaf philodendron, split-leaf philodendron, false breadfruit, fruit salad plant, hurricane plant, Mexican breadfruit, monstera, Swiss cheese plant, and window leaf. Rarely, the names Japanese pineapple and locusts and honey will be encountered.
- The names cut-leaf philodendron, split-leaf philodendron, Swiss cheese plant, and window leaf are based on the prominent perforations and slits in the leaves. The name hurricane plant suggests that the openings permit the wind to pass through without damaging the foliage (or alternatively that the gaps represent damage from a hurricane).
- The genus name *Monstera* is derived from the Latin *monstrum*, monster or marvel, possibly referring to the extraordinary appearance of the leaves.
- *Deliciosa* in the scientific name *M. deliciosa* is Latin for delicious and alluring.

PLANT PORTRAIT

Ceriman is a fast-growing, stout, evergreen, herbaceous vine, which spreads over the ground and forms extensive mats if unsupported or climbs trees to a height sometimes exceeding 12 m (39 ft.), the vine occasionally known to grow to a length of more than 20 m (66 ft.). The stems are 6.3 to 7.5 cm (2½–3 in.) in diameter, roughened with the scars from the bases of fallen leaves, and they produce numerous, long, aerial roots. The leaf blades are leathery, up to 90 cm (3 ft.) in length, on flattened stalks as long as 1 m (more than 3 ft.), with deep cuts from the edges inwards and holes of various sizes on each side of the midrib. The leaves of young plants are heart-shaped, lack perforations, often overlap like shingles, and cling tightly to a support; at this stage, the plants are sometimes called "shingle plants." Young plants with unperforated leaves are often sold in the horticultural trade under the name *Philodendron pertusum*. As the plants mature, the leaves grow larger and more perforated. Several stout flowering stalks are produced, each with an enveloping white leaflike structure surrounding a green, cone-shaped fruiting head 20 to 30 cm (8–12 in.) or more in length and 5 to 9 cm (2–3½ in.) thick, reminiscent of an ear of corn or an extremely thin, elongated pineapple, or a very large, long, pine cone. Thick, hard hexagonal plates or scales cover individual ivory-colored, juicy, fragrant pulp segments. Occasionally pale-green, hard seeds the size of large peas occur, but generally the fruit is seedless. The fruit can take more than a year to mature. A fruit on the vine can be induced to ripen evenly by picking it at the point when the base of the rind is beginning to wrinkle, wrapping the whole fruit in paper, and keeping it for a few days. When unwrapped, all the little plates should fall away.

The ceriman is native to wet forests of southern Mexico, Guatemala, and parts of Costa Rica and Panama. It was introduced into cultivation in England in 1752 and in Victorian times was cultivated in English hothouses for its edible fruit. Subsequently, ceriman become familiar as an ornamental in most countries, grown outdoors in tropical areas, and cultivated as a potted plant indoors, although it tends not to bloom or produce fruit indoors. The plants are tropical, cannot tolerate frost, and grow best in semishade and high humidity. As a household ornamental plant, the ceriman is familiar to most people under the names cut-leaf philodendron, monstera, or Swiss cheese plant. The fruit is marketed to some extent in Australia and Central America; it has been grown commercially in Florida and shipped to gourmet grocers in New York and Philadelphia. There is still some cultivation for fruit in Florida, and also in California.

For those who want to raise this charming plant indoors, the following advice is offered. Before investing in a ceriman, keep in mind that this is a plant that requires considerable space because of its large leaves and climbing nature. Moreover, the leaves if chewed can poison infants and family pets, producing such symptoms as irritation of the mouth, blistering, hoarseness, and loss of voice. To encourage good leaves with plenty of dramatic holes, use a pot or other container larger than necessary (contrary to normal practice with house plants). The plants grow best in rich compost. Provide a minimum temperature of 10°C (50°F), preferably a warmer and steady temperature. Good light is best, but not direct sunlight (if light levels are too low the leaves will develop without the interesting holes). Keep just moist but allow the plant to dry out between waterings and decrease water in winter. Mist the plant occasionally and wipe the leaves with a damp sponge to remove dust. Do not cut the aerial roots (they are often ugly, but cutting them delays development), but rather direct them downward into the soil, if possible. Wrap and tie sphagnum moss around a support pole (or simply buy this setup from a garden center). When dampened, this improves the atmospheric humidity and provides a moist medium for the aerial roots (which can be directed into the sphagnum). When your plant reaches the ceiling or just outgrows the available space, propagate it by cutting just below an aerial root and potting the branch. Alternatively, air layering can be carried out, as follows. Trim the leaves from a straight length of stem where you want roots to form. With a sharp knife, make an upward slanted cut into the stem about 5 mm (1/4 in.) deep and 2. 5 cm (1 in.) long, taking care not to cut too far through the stem. Dust hormone rooting powder into the cut. Push a small piece of damp sphagnum moss into the cut area and then wrap more damp moss around the stem to make a ball that covers the wound completely. Wrap clear plastic around the moss and slightly overlap onto the stem. This will create a warm, damp atmosphere for roots to develop, which may take 2 or 3 months. When roots have formed, remove the plastic and moss and cut the stem free just below the new root formation, and you have a new plant with roots attached. Cut the stem at an angle just above a leaf node, discard the upper portion, and pot the plant in compost. Water well until established.

CULINARY PORTRAIT

The taste of ripe, raw ceriman pulp has been compared with a blend of pineapple and banana, and sometimes mango is also mentioned. Although many consider the ceriman delicious, others find that the fruit leaves an unpleasant aftertaste. It can be served as a dessert with cream, added to fruit salads and ice cream, and made into beverages, jellies, jams, and preserves. Ceriman has been used to flavor champagne in Europe.

The rind of ceriman fruit remains green but becomes lighter with maturation. When the sections of rind separate slightly at the base and bulge slightly, the fruits have been shipped to markets. If kept at room temperature, the ceriman will ripen toward the apex over 5 or 6 days, and ripe portions can be consumed as they become available. The flesh should be eaten from rind segments that have loosened and can be easily separated. The whole fruit can be ripened by wrapping in paper or plastic and kept at room temperature until the rind has loosened along the entire length of the fruit. The core is inedible. Once ripened, the fruit can be refrigerated for 7 to 10 days. Rinsing off the dark floral remnants along the fruit improves the appearance of the flesh but causes some loss of juice. Cross sections of the fruit can be held while the edible sections are nibbled from the rim.

As with other members of the arum family, needle-like crystals of oxalic acid and other potentially harmful chemicals are present in the foliage and unripe fruit, and in the floral remnants of the ripe fruit, and can cause oral and skin irritation. Some people have reported throat irritation,

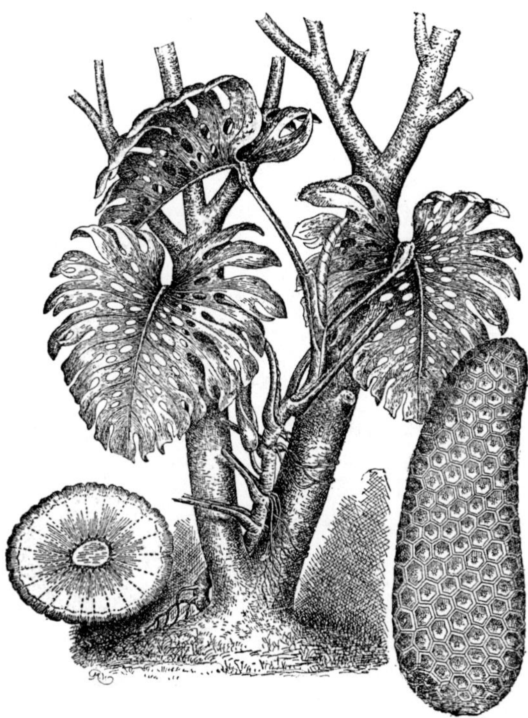

FIGURE 22.1 Ceriman (*Monstera deliciosa*). (From Bailey, 1900–1902.)

hives, diarrhea, intestinal gas, and even anaphylaxis after eating ceriman, and it is advisable not to eat much of this fruit or products made from it until confident that one will not have an undesirable reaction. Do not under any circumstances eat parts of the fruit that are immature, and if unsure, try a less hazardous fruit.

CURIOSITIES OF SCIENCE AND TECHNOLOGY

- The South American Aztecs toasted ceriman seeds to make a strong purgative (to induce vomiting).
- The aerial roots of the ceriman have been used as ropes in Peru, whereas in Mexico they are fashioned into coarse, strong baskets.
- Ceriman roots are used in Martinique as a remedy for snakebite. Snake-oil salesmen market ceriman as a cure for snakebite and arthritis.
- Young ceriman seedlings grow in the direction of the darkest area until they encounter the base of a tree on which to grow. They will then begin to climb toward the light, which is generally up into the canopy of the tree on which they are growing.
- The long cordlike dangling aerial roots of ceriman absorb water from the moist air in which the plant normally grows.

KEY INFORMATION SOURCES

Barabé, D., and Chrétien, L. 1985. Floral anatomy of *Monstera deliciosa* (Araceae). *Can. J. Bot.* 63:1423–1428 (in French).

Berry, M.H. 1970. The delicious monster. *Flower Gard.* (North. edition), 14(11):30.

Broadbent, A. 1977. Houseplants: *Monstera deliciosa*. *Pacific Hort.* 38(2):45.

Coon, N. 1975. Jungle vine [*Monstera deliciosa*]. *Plants Alive*, 3(3):17.

Fonnesbech, A., and Fonnesbech, M. 1980. In vitro propagation of *Monstera deliciosa. HortScience*, 15:740–741.

Hinchee, M.A.W. 1981. Morphogenesis of aerial and subterranean roots of *Monstera deliciosa. Bot. Gaz.* 142:347–359.

Hinchee, M.A.W. 1983. The quantitative distribution of trichosclereids and raphide crystal cells in *Monstera deliciosa. Bot. Gaz.* 144:513–518.

Khattab, M., Kamel, H., and Yacout, M. 1987. Nitrogen and potassium nutrition of *Monstera deliciosa*, Liebm. *Alex. J. Agric. Res.* 32:277–288.

Labroy, M.O. 1980. Le ceriman de Mexique (*Monstera delicosa* Liebm.) espèce frutière. *J. Trop. Agric.* 8:169–170.

Lin, D., and Duan, J. 1990. Rapid propagation of *Monstera deliciosa* in vitro. *Acta Bot. Yunnan*, 12:349–351 (in Chinese).

Morton, J. 1987. Ceriman. In *Fruits of warm climates*. Creative Resource Systems, Winterville, NC. pp. 15–17.

Mott, R.C., and Bailey, L.H. 1972. *Split-leaf philodendron*. N.Y. State Flower Ind. Inc. Bull. 20:5.

Peppard, T.L. 1992. Volatile flavor constituents of *Monstera deliciosa. J. Agric. Food Chem.* 40:257–262.

Peters, R.E., and Lee, T.H. 1977. Composition and physiology of *Monstera deliciosa* fruit and juice. *J. Food Sci.* 42:1132–1133.

Rakovan, J.N., Kovacs, A., and Szujko-Lacza, J. 1973. Development of idioblasts and raphides in the aerial root of *Monstera deliciosa* Liebm. *Acta Biol.* 24(1/2):103–118.

Ramirez, B.W., and Gomez, P.L.D. 1978. Production of nectar and gums by flowers of *Monstera deliciosa* (Araceae) and of some species of *Clusia* (Guttiferae) collected by New World *Trigona* bees, pollination. *Brenesia*, 14/15:407–412.

Sanabria, C.M.E., Camino, A.J.M., Garcia, G., and Renaud, J. 1999. Foliar morphology: *Monstera deliciosa* (Araceae) during its ontogenic development. *Ernstia*, 9:103–114 (in Spanish).

Seidemann, J. 2004. Little-known exotic fruits. 17. *Monstera deliciosa* Liebm.—an unusual fruit species. *Dtsch. Lebensm. Rundsch.* 100:184–188 (in German, English summary).

Serobian, P.A. 1974. On the anatomy and function of aerial roots of *Philodendron pertusum. Biol. Zh. Arm.* 27(6):77–81 (in Armenian).

Sharga, A.N., and Gupta, J. 1983. Ceriman: an elegant house plant. *Indian Hort.* 28(3):14–15.

Zubair, M., Inayatullah, H., and Rahman, N. 1995. Effect of different growing media and different doses of nitrogen on the growth of *Monstera deliciosa. Sarhad J. Agric.* 11:715–720.

Specialty Cookbooks

Fichter, G.S. 2002. *The sunshine state cookbook*. Pineapple Press, Sarasota, FL. 224 pp.

Raichlen, M.S. 1993. *Miami spice: the new Florida cuisine*. Workman Publishing, New York. 352 pp.

Shearer, V. 2005. *The Florida Keys cookbook: recipes and foodways from Paradise*. Globe Pequot, Guilford, CT. 296 pp.

23 Chayote

Family: Cucurbitaceae (gourd family)

NAMES

Scientific Name: *Sechium edule* (Jacq.) Sw.

- The English word "chayote" is based on the Nahuatl word for the plant and vegetable, *chayotli*. (Nahuatl is the language of the Aztecs, still spoken today in a modified form by millions in Mexico and Central America.)
- Pronunciations: chi-OH-tay, shy-OH-tay.
- The chayote is also known as brionne, chayote squash, cho-cho (chow-chow, choko), christophine (christophene; of French origin, particularly known in the French of the West Indies), climbing squash, custard marrow, Madeira marrow, mango squash, mirliton (a popular name in French-speaking districts of Louisiana), pepinello (pepinella), and vegetable pear.
- The name "vegetable pear" reflects the appearance of the fruit, which is like a very large pear.
- "Alligator pear" is sometimes applied, but this name is also used for the bitter melon (see Chapter 63) and was once used for the avocado.
- The Chinese name of chayote means "Buddha's hands" (or Buddha's hand) because the pear shape is said to look like hands clasped in prayer (this should not be confused with the "Buddha's hand" citron; see Chapter 26). "Knuckles" often appear to be present at the base of the fruit, giving the appearance of a clenched fist. Because the chayote was imported into Asia, it has become a symbol of piety among Buddhists, specifically because of the perception that its shape was perceived as the praying hands of Buddha.
- The New Orleans (Louisiana French) name *mirliton* means "toy reed flute," the shape of the fruit interpreted as suggesting children's hands cupped to blow into the instrument.
- The genus name *Sechium* is said to be based on the West Indian name *chacha*.
- *Edule* in the scientific name *S. edule* is Latin for edible.

PLANT PORTRAIT

The chayote is a perennial-rooted (above-ground parts are annual), subtropical to tropical vine, climbing by tendrils, with slender, branching stems up to 15 m (49 ft.) long, and cucumber-like leaves. A vigorous vine can produce several dozen fruit. The fruit is reminiscent of summer squash in appearance and use. It is dark green, pale green, yellowish white, or white; smooth or prickly; wrinkled with several deep longitudinal ribs; varying in length from 7.6 to 20 cm (3–8 in.) and in weight from about 100 g (4 oz.) up to 1.4 kg (3 lb.). The flesh is firm and white or whitish. There is a single, soft, flat seed, 2.5 to 5 cm (1–2 in.) long, which is edible after cooking. Although there is great variation of the fruit in native markets, the commercially grown fruit is quite uniform, generally pear-shaped, light green, smooth, often with deep folds in the skin, about 15 cm (6 in.) long and 450 g (1 pound) in weight, with a pleasing texture and sweet flavor.

The chayote was domesticated in southern Mexico and Guatemala and was a common vegetable among the Aztecs and Mayas before the Spanish conquest of Mexico. It is now an important vegetable in tropical regions, where it is often trained on a trellis or other support and produces fruit almost continuously. The leading commercial producer and exporter of chayote is Costa Rica, followed by Guatemala, Mexico, and the Dominican Republic. The plant is also grown in Algeria, Madagascar, Polynesia, China, Indonesia, Italy Spain, New Zealand, Australia and, to a limited extent, in the southern United States (notably in California, Florida, and Louisiana).

CULINARY PORTRAIT

The immature fruits have a crispy texture and a delicate mild flavor, which may require complementary seasoning. The delicate flavor is more evident when the fruit is left slightly crisp after cooking. The taste has been described as follows: a blend of zucchini, cucumber, and apple; like zucchini with a slightly citrus tang; a cross between a lima bean and an almond; and like water chestnuts. Chayote is an extremely versatile food, which can be boiled, steamed, sautéed, deep-fried, stir-fried, grilled, and frittered, and it can be creamed, buttered, or mashed. It can also be eaten raw or slightly parboiled or blanched in salads. This vegetable can be prepared in any way suitable for summer squash (denser texture requires more cooking time than zucchini) or split, stuffed, and baked like small winter squash. The fruit is often pickled, used in chutneys and marinades, or served in soups, stews, and stir-fries. In parts of Latin America, chayotes are baked in sweet pies, much like apples. Chayotes are widely available in supermarkets, especially during the winter months. For best flavor and texture, select the smallest, firm, unblemished fruits with darkest green color. Fruit with very hard skin tend to have fibrous flesh. Chayote can be refrigerated in a perforated plastic bag up to a month. Fruit quality deteriorates in storage. Maintained at room temperature, the fruit shrivels and the seed that is present sprouts (as described below). Chayote is a low-calorie food that has been recommended in weight reduction diets.

The peel is edible but, if tough, the thin skin may be removed (the skin remains firm during cooking and is usually removed, either before or after cooking). When peeling a chayote, to avoid sap staining the hands brown, some cookbooks suggest that gloves be worn, or the hands oiled, or the fruit be peeled under running water. The peculiar slick astringency that is felt when the sticky substance does get on hands is harmless and washes off in water.

Chayote microwaves well: cut into 1.3 cm (1/2 in.) cubes, place in a microwave dish, add 1/4 cup water, cover, and cook on high for 7–8 minutes or until tender.

Young leaves, young tendrils, and the large, starchy, tuberous, yam-like roots (which can weigh as much as 9 kg or 20 lb.) are also eaten traditionally by peasants, as boiled vegetables or in stews. The roots are reminiscent of Jerusalem artichoke in taste. The young shoots can be consumed like asparagus. The seeds are often considered a delicacy, especially when sautéed in butter.

Culinary Vocabulary

- *Sopa de chayote* is a Central and South American soup prepared with chayote, flavored with garlic and pepper.

CURIOSITIES OF SCIENCE AND TECHNOLOGY

- The chayote has been used medicinally for centuries. Infusions of the leaves are still used in Mexico and Central America to dissolve kidney stones and to assist in the treatment of kidney diseases.
- The vine stems of chayote are strong and flexible and have been used to make craft items, particularly baskets and hats.
- The seeds of chayote germinate inside the fruits (which are one-seeded) when they mature, even while the fruit is still on the vine, a phenomenon called vivipary. (Mangrove trees of tropical swamps are another of the few kinds of plants that do this. Seeds also sometimes germinate within old grapefruits.) Native Central American farmers have traditionally

simply planted whole fruits, knowing that this would result in the establishment of new plants. The chayote may be the only species in the gourd family (in which there are almost a thousand species), which is viviparous, and *Sechium* is the only genus that characteristically has just one seed in the fruit. It has been suggested that the reason for this is that the vines grow naturally in forests, and so the young seedling needs a good supply of

FIGURE 23.1 Chayote (*Sechium edule*). (From Paillieux and Bois, 1892.)

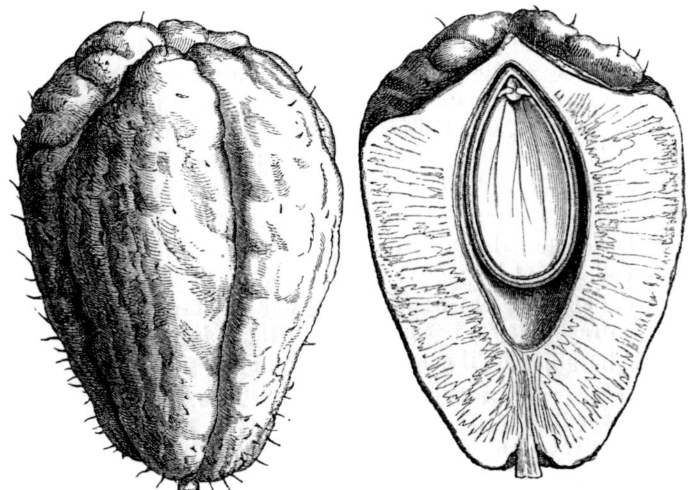

FIGURE 23.2 Fruit of chayote (*Sechium edule*). (From Baillon, 1876–1892.)

nutrition (without competition from other seedlings), which it gets from the fruit, so the vine can grow tall enough to reach sunlight and manufacture its own food. Chayote seeds do not survive in a dry state outside of the fleshy fruit, so seed is not normally available for those interested in planting this vine in their home garden. However, small plants established in pots, and whole fruits intended for planting, are often available in areas where chayote can be grown. In a supermarket, if you observe root sprouts showing at the bottom (heavy end) of a chayote, it should be avoided for food purposes, but could be used to establish a plant.

- According to a government Web site, "Chayotes can be either male or female. The female fruit has a smoother skin with a corrugated, slightly lumpy surface. The male fruit is less meaty and its surface may be covered with spines." This information is absurd—fruits are never male or female. What is true is that the flowers can be male or female.

KEY INFORMATION SOURCES

Anonymous. 1976. Growing chokos [chayote] in the home garden. *Qld. Agric. J.* 102:309–310.

Aung, L.H. 1998. Chayote. In *Tropical and subtropical fruits*. Edited by P.E. Shaw, H.T. Chan, Jr., and S. Nagy. Agscience, Auburndale, FL. pp. 486–505.

Aung, L.H., Ball, A., and Kushad, M. 1990. Developmental and nutritional aspects of chayote (*Sechium edule*, Cucurbitaceae). *Econ. Bot.* 44:157–164.

Cannon, J.M., and Koske, T.J. 1980. *Growing vegetable pears*. Coop. Ext. Pub. 2032, Louisiana State University, Baton Rouge, LA. 8 pp.

Colbert, T. 1977. The cheated chayote. *Pac. Hortic.* 38(1):6–8.

Cook, O.F. 1901. *The chayote: a tropical vegetable*. U.S. Department of Agriculture, Division of Botany, Washington, D.C. 31 pp.

Engels, J.M.M. 1983. Variation in *Sechium edule* Sw. in Central America. *J. Am. Soc. Hort. Sci.* 108:706–710.

Engels, J.M.M. 1984. Chayote: a little known Central American crop. *Plant Genet. Resour. Newsl.* 63:2–5.

Engels, J.M.M., and Jeffrey, C. 1993. *Sechium edule* (Jacq.) Swartz. In *Plant resources of South-East Asia*. 8. Vegetables. Edited by J.E. Siemonsma and K. Piluek. Pudoc Scientific Publishers, Wageningen, the Netherlands. pp. 246–249.

Hoover, L.G. 1923. *The chayote: its culture and uses*. U.S. Department of Agriculture, Washington, DC. 11 pp.

Lira Saade, R. 1996. Chayote, *Sechium edule* (Jacq.) Sw. International Plant Genetic Resources Institute, Rome, Italy. 58 pp.

Macleod, G. 1990. Volatile components of chayote. *Phytochemistry*, 29:1197–1200.

Magda, R.R. 2003. Catsup and candy from chayote. *Int. Food Market. Technol.* 17(3):8, 10.

Martindale, W.L. 1974. The choko—the perennial cucurbit. *J. Agric. Melb.* 72(12):409–410.

Maynard, D.H., and Paris, H.S. 2008. *Sechium edule*, chayote. In *The encyclopedia of fruit & nuts*. Edited by J. Janick and R.E. Paull. CABI, Wallingford, Oxfordshire, U.K. pp. 307–308.

Morton, J.F. 1981. The chayote, a perennial, climbing, subtropical vegetable. *Proc. Ann. Meet. Fla. State Hort. Soc.* 94:240–245.

Newstrom, L.E. 1985. Collection of chayote and its wild relatives. *Plant Genet. Resour. Newsl.* 64:14–20.

Newstrom, L.E. 1989. Reproductive biology and evolution of the cultivated chayote (*Sechium edule*, Cucurbitaceae). In *The evolutionary ecology of plants*. Edited by G.H. Bock and Y.B. Linhart. Westview Press, Boulder, CO. pp. 491–509.

Newstrom, L.E. 1991. Evidence for the origin of chayote, *Sechium edule* (Cucurbitaceae). *Econ. Bot.* 45:410–428.

Ortiz, C.E., and Gonzalez, A. 1997. External fruit characteristics of chayote landraces collected in Puerto Rico. *J. Agric. Univ. Puerto Rico*, 81(3/4):227–229.

Saade, R.L. 1994. Chayote. (*Sechium edule*). In *Neglected crops. 1492 from a different perspective*. Edited by J.E. Hernández-Bermejo and J. Léon. Food and Agriculture Organization of the United Nations, Rome, Italy. pp. 79–84.

Specialty Cookbooks

Marcus, G., and Marcus, N. 1982. *Forbidden fruits & forgotten vegetables. A guide to cooking with ethnic, exotic, and neglected produce*. St. Martin's Press, New York. 146 pp. (Includes a chapter on chayote with 6 recipes).

Muñoz Balladeres, C. 1990s. *Espinas y pulpa—: el chayote, planta mesoamericana: recetario*. Universidad Autónoma Chapingo, Mexico. 109 pp (in Spanish).

Willsey, E.M. 1927. *Tropical foods: chayote, yautia, plantain, banana*. Department of Education of Porto Rico, San Juan, Puerto Rico. 30 pp.

Willsey, E.M., and Ruiz, C.A. 1931. *Vegetales tropicales: chayote*. Bull. 9, College of Education, University of Porto Rico (Río Piedras), Río Pidras, Puerto Rico. 28 pp (in Spanish).

24 Cherimoya and Atemoya

Family: Annonaceae (custard apple family)

NAMES

Scientific Names: *Annona* Species

- Cherimoya—*A. cherimola* Mill.
- Atemoya—*A. squamosa* L. × *A. cherimola* Mill.

- The English name "cherimoya" is derived from the Spanish *chirimoya*, probably from the Quechua (Inca language, still spoken by Indian peoples of Peru and Bolivia) *chirimuya* or *chirimoya*, meaning "cold seeds." *Cherimola* in the scientific name *A. cherimola* has the same basis. Aside from being the name for the fruit, the Inca word cherimoya has the meaning "cold breast."
- Pronunciation: chehr-uh-MOY-ah.
- Cherimoya is also called chirimoya, chirimalla, cherimolia, custard apple, and sherbet fruit.
- The name custard apple is applied to other species of *Annona*, most notably the sweet sop (see Chapter 91). The name custard apple is often used for the cherimoya in the United Kingdom and Commonwealth countries. In Australia, the name cherimoya has been applied to the atemoya
- The atemoya (pronounced a-teh-MOH-ee-yah) is a hybrid of the cherimoya and the sweet sop. The name "atemoya" is a combination of "ate," an old Mexican name for sweet sop, and "moya" from cherimoya.
- The genus name *Annona* has been interpreted as derived either from the Latin *annona* (based on *annus*, year), the year's produce (grain or food); or on a native Brazilian or American Indian (Taino) name, *annona* or *anona*.

PLANT PORTRAIT

Cherimoya

The cherimoya is a small tree, 5 to 9 m (16–30 ft.) in height. The leaves are either evergreen or more or less deciduous, depending on the climate where the plant is grown. The fruit, resembling a giant green pine cone, is conical, ovoid, spherical, or heart shaped, 10 to 20 cm (4–8 in.), up to 10 cm (4 in.) in width, and weighing 150–600 g (5½–18 oz.), occasionally more than 2.7 kg (6 lb.), or rarely as much as 6.8 kg (15 lb.). The skin is thin or thick, smooth with fingerprint-like markings or covered with conical or rounded protuberances. Skin color varies from bronze to green, becoming yellow and almost black as the fruit ripens. The flesh is snow white (a pink-fleshed variety is available), the texture of firm custard, juicy, with numerous hard, brown or black, beanlike, glossy seeds 1.3 to 2 cm (1/2–3/4 in.) long. The cherimoya is native to Peru, Ecuador, Colombia, and Bolivia and has been cultivated for local use for hundreds of years. It is now grown in much of South America, in tropical highlands and in many subtropical areas, including Mexico, Central America, Hawaii, India, Australia, Israel, Italy, and Spain. Spain is the largest producer. Chileans consider the cherimoya to be their "national fruit," with considerable amounts produced in the Acongagua Basin. The cherimoya is cultivated commercially in California (it is said to be the third most important subtropical fruit crop in the state) but has not been commercially significant in Florida. Cherimoyas are difficult to obtain in most of North America, outside of places where they are grown, because they do not travel well.

Atemoya

The atemoya is a hybrid of the cherimoya and the sweet sop, which was created for commercial purposes in the United States in the early twentieth century, although it has also been generated in nature where the trees grown naturally near each other and in orchards where the two trees have been planted close together. Some of the best hybrids have originated from Israel. The atemoya is slightly hardier than one of its parents, the sweet sop (see Chapter 91), but not as tolerant of cold as the other, the cherimoya. The atemoya tree closely resembles the cherimoya tree, and the fruit is similar in appearance and taste, although some selections have proven much better than others. Atemoya trees, rather than cherimoya trees, are planted only in locations (typically at lower elevations) where cherimoyas cannot be grown because cherimoyas produce better-tasting fruit.

CULINARY PORTRAIT

The cherimoya has been called "the queen of subtropical fruits." It is primarily a dessert fruit, eaten fresh and fully ripe, the flesh consumed out of hand or spooned out. The flesh is pleasant smelling, juicy, slightly acidic, and somewhat granular (but less so then the pear), with a delicious flavor. The flavor of the ripe fruit is improved by chilling just before eating. The sweet taste is said to be like a combination of mango, papaya, coconut, and banana. The sugar content is high, 14% to 22%. The ripe flesh is generally eaten out of hand or scooped with a spoon from the cut-open fruit. Once sliced, the flesh discolors quickly, and to prevent oxidation, it is often sprinkled with orange juice. An interesting serving suggestion: float sections on white wine or champagne. Cherimoya can be added to fruit salads and beverages or used to flavor milkshakes, sherbet, yogurt, and ice cream. In fact, cherimoyas can be frozen and eaten like ice cream. The fruit has been fermented to produce alcoholic beverages, and also processed into juice. Slices can be dipped into batter and deep fried. The fruit can be cooked to make jelly, jam, and compôte, but cooking greatly alters the flavor. Cherimoyas bruise easily and are sensitive to extremes of heat and cold. They are usually picked and transported while unripe and can be checked for ripeness as for avocados by softness. Like avocados, they should not be chilled before they ripen. Once ripe, they can be refrigerated for up to 5 days. When overripe, the pleasing fragrance can turn into an unpleasant odor.

The fragile skin of the cherimoya is bitter and inedible. If the center of the fruit has not ripened to become soft, it too should be discarded. The seeds are inedible and toxic; they should be spit out when eating fresh fruit and need to be removed from the pulp in any process that would mechanically crush them, such as blending.

The best varieties of the atemoya are almost comparable to the cherimoya, but most are inferior. The culinary uses of the atemoya are the same as for the cherimoya.

Culinary Vocabulary

- Because of their similarity, the cherimoya, atemoya, sour sop, and sweet sop (Chapter 91) are often labeled "moya" for marketing convenience and may be encountered in supermarkets under this name.

CURIOSITIES OF SCIENCE AND TECHNOLOGY

- When Europeans arrived in western South America in the sixteenth century, the Inca Empire (1438–1525), was the largest in the world, stretching about 5000 km (3100 miles) along the Pacific shore and over the ridge of the Andes, comparable in extent with the

FIGURE 24.1 Cherimoya (*Annona cherimola*). (From Curtis, vol. 72, 1846, plate 4226.)

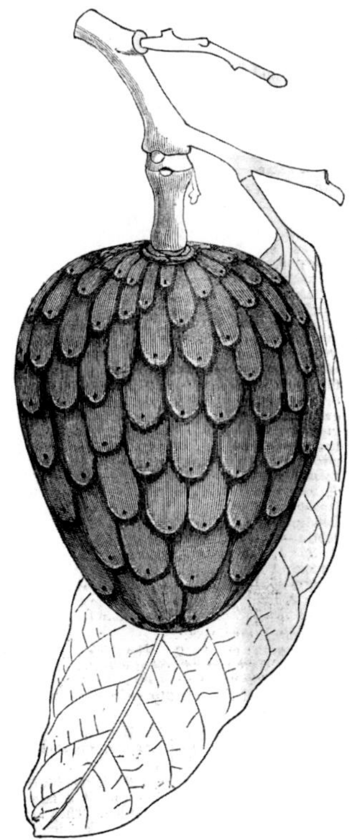

FIGURE 24.2 Cherimoya (*Annona cherimola*). (From *Gardeners' Chronicle*, April 2, 1904, p. 223.)

Roman Empire at its height. The emperor ruled over millions of people. Dozens of food plants had been domesticated, and the Spanish conquerors adopted some, including corn (maize), potatoes, and chocolate, and took them back to Europe. Cherimoya, which has been called "the jewel of the Incas," was one of the crops that the Spanish judged worthy of cultivation in Europe, and in fact modern Spain is now the world's largest producer of the fruit. However, most of the crops were ignored, and some were actively suppressed, especially those that had been prized by the Inca nobility. Farmers who dared to grow these "noble" crops were executed. These have come to be known as "the lost crops of the Incas," but in fact many have been maintained in the highlands of the Andes and the jungles of the Amazon, and are now receiving study with the prospect of commercial development.

- In rural Mexico, pulverized cherimoya seeds (which are poisonous) are toasted, ground up, and used as a potent laxative. The powder is also mixed with grease and applied to the skin to kill lice. Cherimoya seeds are sometimes crushed and used as an insecticide.

KEY INFORMATION SOURCES

Anderson, P., and Richardson, A. 1990. Which cherimoya cultivar is best? *Orchardist N. Z.* 63(11):17–19.

California Rare Fruit Growers. 1983. Cherimoya. *California Rare Fruit Growers Yearbook*, 15:5–43.

Campbell, C.W., and Phillips, R.L. 1980. *The atemoya. Fruit crops facts sheet*. Agricultural Extension Service, University of Florida, Gainesville, FL. 3 pp.

Cann, H.J. 1967. The custard apple [atemoya]. *Agric. Gaz. N. S. W.* 18:85–90.

Crane, J.H. 1993. Commercialization of carambola, atemoya, and other tropical fruits in south Florida. In *New crops*. Edited by J. Janick and J.E. Simon. Wiley, New York. pp. 448–480.

Damme, V. van, Damme, P. van, and Scheldeman, X. 1999. Promoting cultivation of cherimoya in Latin America. *Unasylva*, 50:43–47.

Damme, V. van, Damme, P. van, and Scheldeman, X. (Eds.). 1999. *First international symposium on cherimoya, Loja, Ecuador, March 16–19, 1999*. International Society for Horticultural Science, Leuven, Belgium. 383 pp (each article in English and Spanish).

Ellstrand, N.C., and Lee, J.M. 1987. Cultivar identification of chemimoya (*Annona cherimola* Mill.) using isozyme markers. *Sci. Hortic.* 32:25–31.

George, A.P., and Nissen, R.J. 1991. *Annona cherimola* Miller, *Annona squamosa* L., *A. cherimola* × *A. squamosa*. In *Plant resources of South-East Asia. 2. Edible fruits and nuts*. Edited by E.W.M. Verheij and R.E. Coronel. Pudoc, Leiden, the Netherlands. pp. 71–75.

George, A.P., Broadley, R.H., Nissen, R.J., and Hamill, S. 2002. Breeding new varieties of atemoya (*Annona* spp. hybrids). *Acta Hortic.* 575:323–328.

George, A.P., Broadley, R.H., Nissen, R.J., Hamill, S.D., and Topp, B.L. 1999. Breeding strategies for atemoya and cherimoya. *Acta Hortic.* 497:255–267.

Gomez, M.C. 1983. The cherimoya. *California Rare Fruit Growers Yearbook*, 15:5–29.

Grossberger, D. 1999. The California cherimoya industry. *Acta Hortic.* 497:119–142.

Kader, A.A., and Arpaia, M.L. 1999. Cherimoya, atemoya, and sweetsop: recommendations for maintaining postharvest quality. *Perishables Handling Q.* 97:17–18.

Lizana, L.A., and Reginao, G. Cherimoya. 1990. In *Fruits of tropical and subtropical origin: composition, properties and uses*. Edited by S. Nagy, P.E. Shaw, and W.F. Wardowski. Florida Science Source Inc., Lake Alfred, FL. pp. 131–148.

Mahbubur, R.S.M., Shimada, T., Yamamoto, T., Yonemoto, Y.J., and Yoshida, M. 1998. Genetical diversity of cherimoya cultivars revealed by amplified fragment length polymorphism (AFLP) analysis. *Jpn. J. Breed.* 48:5–10.

Merodio, C., and Plazza, J.L. de la. 1997. Cherimoya. In *Postharvest physiology and storage of tropical and subtropical fruits*. Edited by S.K. Mitra. CABI, Wallingford, Oxon, U.K. pp. 269–293.

Morton, J. 1987. Cherimoya. In *Fruits of warm climates*. Creative Resource Systems, Winterville, NC. pp. 65–69.

Morton, J. 1987. Atemoya. In *Fruits of warm climates*. Creative Resource Systems, Winterville, NC. pp. 72–75.

Palma, T., Aguilera, J.M., and Stanley, D.W. 1993. A review of postharvest events in cherimoya. *Postharvest Biol. Technol.* 2:187–208.

Paull, R.E. 1996. Postharvest atemoya fruit splitting during ripening. *Postharvest Biol. Technol.* 8:329–334.
Perectti, F., and Pascual, L. 1998. Characterization of cherimoya germplasm by isozyme markets. *Fruit Var. J.* 52:53–62.
Pino, J.A., and Rosado, A. 1999. Volatile constituents of custard apple (*Annona atemoya*). *J. Essent. Oil Res.* 11:300–305.
Schroeder, C.A. 1981. Fruit morphology and anatomy of the cherimoya. *Bot. Gaz.* 112:436–446.
Seelig, R.A., and Bing, M.C. 1990. Cherimoya. In *Encyclopedia of produce.* [Loose-leaf collection of individually paginated chapters.] Edited by R.A. Seelig and M.C. Bing. United Fresh Fruit and Vegetable Association, Alexandria, VA. 8 pp.
Smet, S. de, Damme, P. van, Scheldeman, X., and Romero, J. 1999. Seed structure and germination of cherimoya (*Annona cherimola* Mill.). *Acta Hortic.* 497:269–288.
Sweet, C. 1990. Cherimoya: its history; its future. *California Grower*, 14(3):20–23.
Sweet, C. 1990. Establishing a cherimoya grove. *California Grower*, 14(4):26–29.
Thomson, P.H. 1970. The cherimoya in California. *California Rare Fruit Growers Yearbook*, 2:20–34.

Specialty Cookbooks

Chase, D. 2007. *More smoothies for life: satisfy, energize, and heal your body.* Three Rivers Press, New York. 320 pp.
Haasarud, K., and Grablewski, A. 2006. *101 margaritas.* Wiley, Hoboken, NJ. 128 pp.
Harris, M.R. 1993. *The tropical fruit cookbook.* University of Hawaii Press, Honolulu, HI. 196 pp.
Trutter, M. 2008. *Culinaria Spain.* Könemann, Cologne, German. 487 pp.
Zaslavsky, N. 1997. *A cook's tour of Mexico: authentic recipes from the country's best open-air markets, city fondas, and home kitchens.* St. Martin's Griffin, New York. 384 pp.

25 Chinese Artichoke

Family: Lamiaceae (Labiatae; mint family)

NAMES

Scientific Name: *Stachys affinis* Bunge (*Stachys sieboldii* Miq., *Stachys tuberifera* Naud.)

- The name "Chinese artichoke" reflects the Jerusalem artichoke (*Helianthus tuberosus* L.) flavor and Chinese origin of this vegetable.
- Chinese artichoke is also known as Japanese artichoke, reflecting an association with Japan. Chinese artichokes appeared in Europe about three decades after Commodore Perry's visit to Japan, and this reinforced the association of the vegetable and Japan. (Commodore Matthew C. Perry, (1794–1858), led a U.S. expedition to Japan and the China seas during 1852–1854, and this resulted in a number of hitherto unfamiliar foods being introduced to Western culture.)
- The name "knotroot" (or knot root) has been applied because of the curious shape of the tuber, resembling a short string of fat beads.
- The French name crosne (pronounced "crone") has also been adopted as an English name, and because it sounds esoteric, it is the principal name under which Chinese artichoke is marketed in North America. Crosne was so named by Nicolas-Auguste Paillieux, (1812–1898), after his home village of Crosne near Corbeil in France. Starting in 1882, he first grew and marketed the vegetable in Europe. In 1885, Paillieux and Professor D. Bois published *Le Potager d'un Curieux* ("A Garden of Curiosities"), describing the history, culture, and uses of little known or unknown edible plants. This extremely popular book first popularized in Europe crops such as soybean (*Glycine max* (L.) Merr.), Chinese artichoke, chufa (*Cyperus esculentus* L.), and kudzu (*Pueraria montana* (Lour.) Merr. var. *lobata* (Willd.) Maesen & S.M. Almeida).
- Popular Chinese names include *gan lu zi* ("sweet dew") and *cao shi can* ("grass stone silkworm"); the Japanese name is *chorogi.*
- Chinese artichoke is unrelated to globe artichoke (*Cynara scolymus* L.) and Jerusalem artichoke (*H. tuberosus* L.).
- Chinese artichoke is occasionally called "stachys," a name based on the genus name *Stachys*. The genus name *Stachys* is derived from the Greek *stachys*, an ear of wheat or spike, for the spike-like flowering stalk of the species.
- Another occasional name is "spirals," a shape the tubers sometimes seem to take.
- *Affinis* in the scientific name *S. affinis* is Latin for allied to, a rather uninformative word because placing the species in the genus *Stachys* is already a way of stating that it is allied to the genus.

PLANT PORTRAIT

Chinese artichoke is a mint-like perennial herb, which grows wild in Northern China and Japan. The plant is 30 to 50 cm (12–18 in.) tall, with hairy (felt-like), crinkly leaves and small white or pink flowers produced in the summer. By late fall, on the ends of underground stalks up to 30 cm (1 ft.) long, there develop numerous, small, slender, edible tubers, somewhat resembling a closely threaded string of large whitish beads, just under the soil surface. These highly distinctive tubers

FIGURE 25.1 Chinese artichoke (*Stachys affinis*). (From Gartenflora, vol. 39, 1890.)

are 2.5 to 5.5 cm (1–2 in.) or more in length and 1.3 to 2.5 cm (1/2–1 in.) in thickness, with up to a dozen bulging segments (the "beads"). The flesh is white and crisp. Chinese artichoke was used as a vegetable in China at least by the fourteenth century and was cultivated there by the seventeenth century. The species was first brought to Europe in 1882, having been sent to France from China. Chinese artichokes were popular in Europe from about 1890 to 1920, but subsequently became obscure, and only in recent times has the vegetable begun to be noticed again. It is grown commercially primarily in China, Japan, France, and Belgium and may be found occasionally in home gardens or as offerings of specialty produce growers in North America (tubers and sometimes seeds are available from mail-order garden catalogues). Chinese artichoke is a rare vegetable everywhere, including China, but is expensive and considered to be a gourmet treat. Some varieties are propagated vegetatively by tubers, not by seeds, and several of these are incapable of producing seeds. For those contemplating cultivating the plant, it has been said that it is virtually impossible to find all the tubers produced, so there are always some left behind to grow the following season (whether one likes it or not). As well as being cultivated as a garden plant, Chinese artichoke is grown as a ground cover.

CULINARY PORTRAIT

The tubers are sometimes eaten raw or grated into salads, or more often boiled, baked, roasted, or steamed, much like potatoes, with which they are often compared. They have a delicate, sweet, nutty flavor that is reminiscent of salsify (*Tragopogon porrifolius* L.) and globe artichoke and a crisp texture similar to water chestnuts (see Chapter 98) and Jerusalem artichoke. The taste disappears if the vegetable is overcooked. In France, the tubers are featured in creamed soups and are steamed and eaten with butter. In Japan, the tubers are eaten salted or made into a pickle in plum vinegar. To prepare this vegetable, scrub clean (peeling is unnecessary and is very difficult in view of the odd shape), steam, boil or stir-fry lightly, and serve with butter or cream. Another serving suggestion: add French dressing, quartered tomatoes, and watercress. Chinese artichokes are very infrequently sold in North American supermarkets. If available, purchase when firm and white. The tubers do not store well, discoloring and deteriorating quickly after harvest, but can be refrigerated in a plastic bag up to a week. The tubers should not be stored in light, which darkens them and contributes to a loss of flavor.

Culinary Vocabulary

- In modern times, famous people are often trend setters, and this was also the case in the past. The well-known French playwright Alexandre Dumas (1824–1895, not to be confused with his illustrious father of the same name) in one of his plays in 1887 referred to *salade japonaise* (Japanese salad) made with Chinese artichoke, and for a period thereafter in France both the salad and the new vegetable became very popular. "Japanese salad" is still a menu item in restaurants in France, and a menu item listed as "à la japonaise" may be garnished with Chinese artichokes (see information regarding the name "Japanese artichoke").

CURIOSITIES OF SCIENCE AND TECHNOLOGY

- The dried and powdered tubers of Chinese artichoke are used medicinally in traditional herbal medicine in China, as an analgesic (pain reliever). During the days of Imperial China (i.e., when kings ruled), Chinese artichoke was highly valued. As well in China, the tubers are soaked in wine and used to treat colds and influenza. Recent research has indicated that the tubers have medicinal properties, with potential for treating kidney problems.
- Dangshan, an old historical county of China, is famous for the golden Dangshan pear, a very large, thin-skinned, juicy, crisp, sweet pear, which was thought to reduce phlegm, relieve coughs, and soothe asthma. Because of its presumed medicinal qualities, the pear was praised by Chinese doctors of old as "the Chinese artichoke of fruits, the finest cream of medicine." As noted earlier, the Chinese artichoke itself was considered to be an excellent medicine in China.
- Botanists have a very large number of technical terms to describe the parts of plants. Nevertheless, the tubers of the Chinese artichoke, which appear to be a sequence of pearly white knoblike segments, have proven difficult to describe. Among the more colorful comparisons: the Michelin tire man, the white larva of an insect, a caterpillar, a dried worm.
- Another species of *Stachys*, *Stachys floridana* Shuttlew. ex Benth., Florida betony, also produces edible tubers. This is often called "rattlesnake weed" because the tubers look like a rattlesnake's tail. In the Southeastern United States, where it is native, the tubers were consumed by Indigenous Americans and early settlers, and today they are still sometimes boiled like peanuts and eaten.
- Chinese artichoke is one of three similar, herbaceous, vegetatively propagated plants that South American Indians domesticated in the high Andes region and continue to grow for their starchy underground tubers. The other two species are mashua (añu; *Tropaeolum tuberosum* Ruiz & Pav.; Tropaeolaceae or nasturtium family) and ullucu (lisas; *Ullucus tuberosus* Caldas; Bassellaceae or basella family). The latter produces smooth spherical tubers 2 to 10 cm (1–4 in.) across, or tubers that are curved and as long as 25 cm (10 in.).

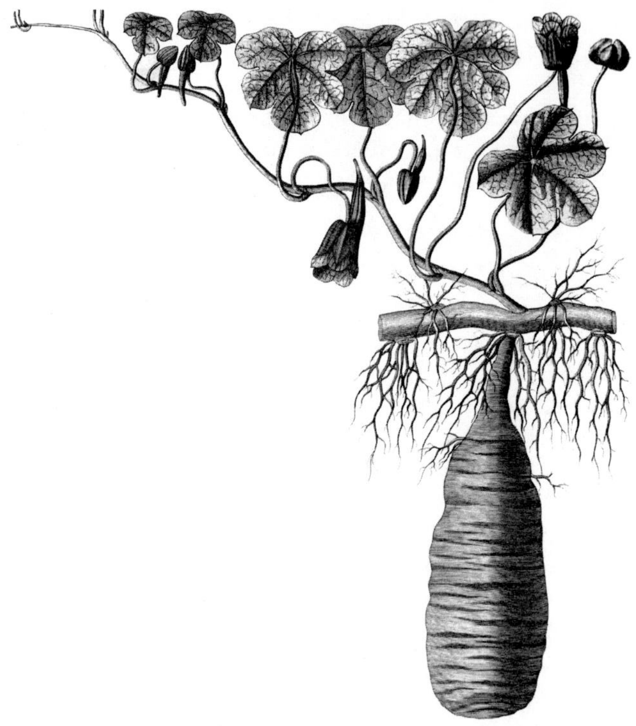

FIGURE 25.2 Mashua (*Tropaeolum tuberosum*). (From Ruiz and Pavon, 1802.)

FIGURE 25.3 Ullucu (*Ullucus tuberosus*). (From Morren, 1851–1885.)

KEY INFORMATION SOURCES

Anisomov, A.I., Davydov, V.D., and Badanov, G.P. 1981. Chinese artichoke *Stachys sieboldi*, cultivation, recommendations, food value. *Kartofel' Ovoshchi Moskva*, 7:31–32 (in Russian).

Bergh, M.H. van den. 1996. *Stachys sieboldii* Miquel. In *Plant resources of South-East Asia. 9. Plants yielding non-seed carbohydrates*. Edited by M. Flach and F. Rumawas. Backhuys Publishers, Leiden, the Netherlands. pp. 155–156.

Hosoki, T., and Yasufuku, T. 1992. In vitro mass-propagation of Chinese artichoke (*Stachys sieboldii* Miq.). *Acta Hortic*. 319:149–152.

Lagarde, J. 1970. Effect of different thermophotoperiods on dormancy in the Japanese artichoke. *Acad. Sci. Compt. Rend. Ser*. D 270:2950–2953 (in French).

Lagarde, J. 1971. Temperature effect on the tuberization of non-dormant tubers of Japanese artichoke (*Stachys sieboldi* Miq.). *Physiol. Veg*. 9:401–422 (in French, English summary).

Lagarde, J., Loiseau, J.E., Mollet, A.M., and Tort, M. 1974. Multiple correlations intervening in the growth and morphogenesis of the Japanese artichoke (*Stachys sieboldi* Miq.). *Rev. Cytol. Biol. Veg*. 37:339–351 (in French, English summary).

Li, W., Gao, H.H., Lu, R., Guo, G.Q., and Zheng, G.C. 202. Direct plantlet regeneration from the tuber of *Stachys sieboldii. Plant Cell Tissue Organ Cult*. 71:259–262.

Morand, J.C., Poutier, J.C., and Pestourie, C. 1982. Virus diseases of the Chinese artichoke. *Pepin. Hortic. Maraich*. 224:19–21 (in French).

Perko, J. 1990. Japanese artichoke: features and yield recordings. *Rev. Suisse Vitic. Arboric. Hortic*. 22:295–297 (in French, English summary).

Peron, J.Y. 1980. The Chinese artichoke. *Pepinier Hortic. Maraich*. 210:11–17 (in French).

Peron, J.Y. 1981. *The Chinese artichoke*. École Nationale d'Ingenieurs des Travaux Agricoles, Angers, France. 7 pp (in French).

Peron, J.Y. 1985. Contribution to the re-emergence of forgotten vegetables (Chinese artichoke, *Stachys sieboldii*), bulbous-rooted chervil (*Chaerophyllum bulbosum*), seakale (*Crambe maritima*)]. Bureau des Ressources Genetiques, Paris, France. *Proceedings of the Symposium: The Diversity of Vegetable Plants: Yesterday, Today and Tomorrow, Angers, France, 17–19 October 1985*. Technique et Documentation, Paris, France. pp. 135–152 (in French).

Peron, J.Y., and Briard, M. 1998. Flowering of Chinese artichoke (*Stachys sieboldii* Miq.). *Acta Hortic*. 467:143–154.

Tort, M. 1972. Rate of foliar organogenesis and leaf appearance at the shoot tip as affected by temperature and physiological state in Japanese artichoke (*Stachys sieboldi* Miq.). *Soc. Bot. France Bull*. 119:13–24 (in French).

Tort, M., and Lagarde, J. 1981. Ultrastructural study on zones in course of development of tubers and long shoots in *Stachys sieboldi* Miq. cultivated in controlled conditions. *Bull. Soc. Bot. Fr. Act. Bot*. 128:99–109 (in French).

Ueno, Y., Ikami, T., Yamauchi, R., and Kato, K. 1980. Purification and some properties of alpha-galactosidase from tubers of *Stachys affinis*. *Agric. Biol. Chem*. 44:2623–2629.

Vogel, G. 1993. Biographies of vegetables. 17. *Stachys sieboldii*. *Gartenbau-Magazine*, 2(12):59.

Yamamoto, H., Fuji, S., and Asari, Y. 2004. Production of virus-free Chinese artichoke through shoot tip culture. *J. Jpn. Soc. Hortic. Sci*. 73:82–84 (in Japanese).

Yazawa, S., Kanno, E., and Takashima, S. 1979. Studies on the growth habit of chorogi (*Stachys sieboldii* Miq.). *Bull. Exp. Farm Fac. Agric. Kyoto Prefect. Univ*. 9:24–28 (in Japanese, English summary).

Yazawa, S., Kanno, E., and Takashima, S. 1983. Dormancy of the tuber of chorogi (*Stachys sieboldii* Miq.). *Bull. Exp. Farm Fac. Agric. Kyoto Prefect. Univ*. 10:21–25 (in Japanese, English summary).

Specialty Cookbooks

Grigson, J. 1980. *Jane Grigson's vegetable book*. Penguin Books, London. 624 pp.

Larkon, J. 2008. *Oriental vegetables: the complete guide for the gardening cook*. Revised edition. Kodansha America, New York. 232 pp.

26 Citron

Family: Rutaceae (rue family)

NAMES

Scientific Name: *Citrus medica* L.

- The citron is the plant responsible for the word citrus, which includes oranges, grapefruits, lemons, limes, and many other lesser known fruits. In biblical times, Moses had specified for religious purposes the cone of the cedar, *hadar* (*kedros* in Greek) and when it fell into disfavor it was replaced by the citron, and the Palestine Greeks called the latter *kedromelon* (cedar apple). Kedros was Latinized as *cedrus*, and this evolved into citrus and subsequently into citron.
- Although the word "citron" refers to the citron in English, it refers to lemon in Czech, Dutch, French, German, Yiddish, and Scandinavian languages.
- In addition to labeling the fruit and tree of a citrus species as "citron," the word citron also is the name of a type of watermelon with inedible flesh, the rind of which is used much like that of the citron.
- A variety of citron is known as the "Leghorn citron," the name tracing to a time when the islands of Corsica and Sardinia shipped their crops to Leghorn, Italy, for processing.
- *Medica* in the scientific name is Latin for medicinal, reflecting old uses of the species to treat seasickness, lung and intestinal problems, poisoning, and other conditions.

PLANT PORTRAIT

The citron is a large, thorny, subtropical shrub or small tree, 2.4 to 4.5 m (8–15 ft.) tall. Citron fruits are elongated, with a lumpy, warty, or knobby (occasionally smooth) surface, attractive fragrance, thick greenish-yellow peel, acidic flesh divided into segments, and many seeds in the pulp. The citron is closely related to the lemon, but the fruit can be much larger—up to 9 kg (20 lb.). The peel or rind makes up most of the fruit. Two main classes of citrons are the sweet or Corsican and the acid citrons. Both have fruit that is large and elliptical in shape, with a diameter of 7.5 to 12.5 cm (3–5 in.), length of 10 to 18 cm (4–7 in.), and rind about 2.5 cm (1 in.) thick.

The Etrog is a small citron, a little larger than a lemon. It has long been used in ceremonies associated with the Jewish Feast of Tabernacles (*sukkot*), which began as a harvest festival held after tithing the first harvest to the Temple. A Jewish coin struck in 136 BC bore a representation of the citron on one side. The citron was also a popular Jewish symbol on graves and synagogues. After the Jewish rebellion against Rome in AD 66, Jews were dispersed throughout the Roman Empire and took their citron with them, planting it across the Mediterranean in Spain, Italy, Sicily, Tunisia, Algeria, and Turkey. Scarcity of the fruit in Northern Europe during the Middle Ages caused much anguish for the many Jews who had migrated there.

The "fingered citron" has a fruit divided into several finger-like sections. It is also known as "Buddha's Hand" and "Buddha's Fingers" and is used ritually in Buddhism. The highly fragrant fruit is placed as an offering on temple altars in China and Japan. This citron is cultivated in China and other eastern countries and is used as a medicine and source of perfume extracts.

Native to India and Burma, the citron has been consumed by humans for perhaps 6000 (some say 8000) years. Citrons originated in the region from south China to India. This fruit was known

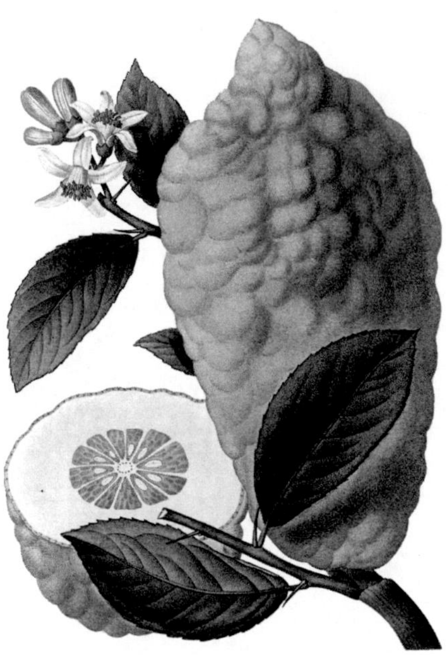

FIGURE 26.1 Citron (*Citrus medica*). (From Duhamel du Monceau, vol. 7, no. 22, 1800–1819.)

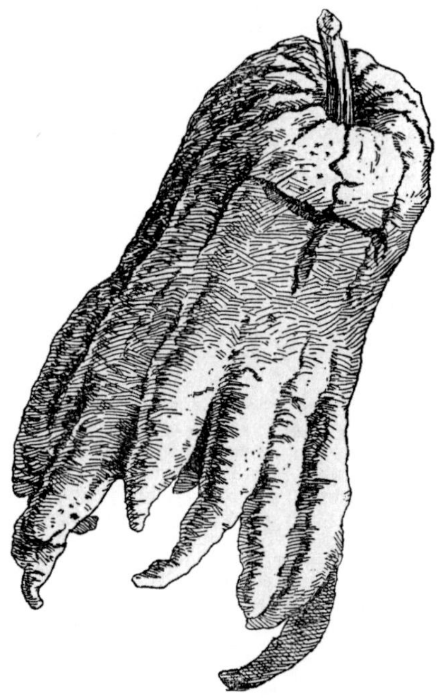

FIGURE 26.2 Fingered citron (*Citrus medica*). (From Bailey, 1900–1902.)

to the classical Arab, Greek, Roman, and Hebrew civilizations. The citron was the first citrus fruit brought to Europe—into Italy in the third century. It became a staple, commercial food in Rome in AD 301. The main producing areas of citron for food use today are Sicily, Corsica, and Crete and other islands off the coasts of Italy, Greece, and France, and the neighboring mainland. Citron is also grown commercially in Puerto Rico, and to a small extent in several other islands of the

Caribbean and in Central and South America. The United States consumes about half of the world production of the fruit. Some citron peel used in the United States is produced in California and Florida.

CULINARY PORTRAIT

The flesh of some varieties of citron is palatable, but in most it is too bitter to eat fresh. In any, case, the fruit is rarely consumed fresh because the skin and the pulp are most useful for making food products. Most citron peel is candied. It is also used to make confections, bakery products such as fruit cake, cookies, pudding, marmalade, and fruit syrups and is occasionally also employed as a base for liqueurs, such as the Corsican Cedratine. Citron is used as a flavoring for olives in Morocco and as a filling for dates in Arab countries. Candied citron purchased for use in cooking and baking should be moist and sticky, not hard and crystallized. Citron is finely diced or shaved before being mixed into a cake. Many identify citron as the most hated ingredient in traditional fruit cake, and some manufacturers have pointedly stated that their product is made without it.

The white inner part of citrus skins is the principal source of commercial pectin. Pectin is essential for making jam and jelly, and if fruits do not have enough naturally, pectin is added. Raspberries and blueberries are low in pectin content whereas apples, citrus fruits, cranberries, and currants are high. The more pectin, the less cooking is necessary to produce jelly. By adding pectin, it is even possible to make jam at room temperature, although the heat of cooking speeds up the process of setting, and also sterilizes the jam.

CULINARY VOCABULARY

- *Narthangai* is the Tamil word for a preserve made of unripe, dried citron.
- *Narthellai podi* is another Tamil specialty, made of powdered young citron leaves, chili powder, and other spices.
- *Yacha* is a Korean kind of syrup made by cooking thinly sliced citron fruit with honey or sugar and used to prepare a tea.

CURIOSITIES OF SCIENCE AND TECHNOLOGY

- The ancient Egyptians used citron in embalming mummies.
- In India, fruiting citron branches are bent down and the immature fruit put into a jar shaped like a human head (or other form) so that the mature fruit will be of the same shape. The resulting fruits are sold as curiosities.
- Branches of the citron tree are used as walking sticks in India.
- In India, citron peel is eaten to mask bad breath.
- The "Victorian language of flowers" was a secret coded language in Victorian times, with flowers and plants symbolic of certain messages, so when the flower or plant was mentioned in a letter, those who knew the code could understand the hidden information. "Citron" meant "ill-natured beauty."
- Chinese and Japanese people prize the citron for its fragrance. In central and northern China, the ripe fruits may be carried by hand or placed in a dish on a table to perfume the air of a room. Dried fruits are stored with clothing to repel moths, and in southern China, the juice is even used to wash fine linen.
- Citron is one of the well-known plants that has contributed its name to a color: a rather greenish yellow.

KEY INFORMATION SOURCES

Avtavi, A. 1976. A net canopy for citrons. *Hassadeh*, 56:786–787 (in Hebrew).

Cancel, L.E., and Hernandez, E.R. de. 1977. Chemical changes in citron fruit bars due to pH changes during processing. *J. Agric. Univ. Puerto Rico*, 61:279–289.

Cancel, L.E., and Hernandez, E.R. de. 1979. Effect of blanching and freezing on the texture and color of candied citron. *J. Agric. Univ. Puerto Rico*, 63:309–314.

Cancel, L.E., Hernandez, E-R de, and Rivera-Ortiz, J.M. 1976. Stability of vitamin C in citron slices in syrup, and in citron bars boxed and canned. *J. Agric. Univ. Puerto Rico*, 60:479–484.

Dung, N.X., Pha, N.M., Lo, V.N., Thien, N.T., and Leclercq, P.A. 1996. Chemical investigation of the fruit peel oil of *Citrus medica* L. var. *sarcodactylis* (Noot.) Swingle from Vietnam. *J. Essent. Oil Res*. 8:15–18.

Goldschmidt, E.E. 1976. Factors determining the shape of citrons. *Isr. J. Bot*. 25:34–40.

Huet, R., Dalnic, R., Cassin, J., and Jacquemond, C. 1986. The Mediterranean citron. The Corsican citron. *Fruits*, 41:113–119 (in French, English summary).

Jones, D.T. 1991. *Citrus medica* L. In *Plant resources of South-East Asia. 2. Edible fruits and nuts*. Edited by E.W.M. Verheij and R.E. Coronel. Pudoc, Leiden, the Netherlands. pp. 131–133.

Kim, I.C. 1999. Manufacture of citron jelly using the citron-extract. *J. Korean Soc. Food Sci. Nutr*. 28:396–402 (in Korean, English summary).

Lima, H., Borges, T., and Febles, C. 1977. *Estudio citotaxonomico de* Citrus medica *L*. Centro de Informacion Cientifica y Tecnica, Universidad de La Habana, La Habana, Cuba. 25 pp.

Lota, M.L., de Rocca, S.D., Tomi, F., Bessiere, J.M., and Casanova. J. 1999. Chemical composition of peel and leaf essential oils of *Citrus medica* L. and *C. limonimedica* Lush. *Flavour Fragrance J*. 14:161–166.

McCulloch, L. 1927. *Curing and preserving citron*. U.S. Department of Agriculture, Washington, DC. 8 pp.

Monselise, S.P. 1986. Citrus. In *CRC handbook of fruit set and development*. Edited by S.P. Monselise. CRC Press, Boca Raton, FL. pp. 87–108.

Morton, J. 1987. Citron. In *Fruits of warm climates*. Creative Resource Systems, Winterville, NC. pp. 179–182.

Otoi, K. 1973. Harvesting and storage techniques for citron fruit. *Kajitsu Nippon*, 28(3):52–54 (in Japanese).

Padmanabhan, D., and Radhakrishnan, R. 1985. Arrangement of cotyledons in *Citrus medica* Linn. *Current Sci*. 54:473–474.

Protopapadakis, E.E. 1987. Identification by isoenzymes of five cultivars of *Citrus medica* grafted on four rootstocks. *J. Hortic. Sci*. 62:413–419.

Saunt, J. 2000. The citron. In *Citrus varieties of the world: an illustrated guide*. Sinclair International Ltd., Norwich, U.K. pp. 130–133.

Shiota, H. 1990. Volatile components in the peel oil from fingered citron *Citrus medica* var. *sarcodactylis*. *Flavour Fragrance J*. 5:33–38.

Vekiari, S.A., Protopapadakis, E.E., Papadopoulou, P., and Argyriadou-Giannovits, N. 2002. Variation in the essential oils of citron leaves and peel infected by exocortis disease. *J. Hortic. Sci. Biotech*. 77:428–431.

Specialty Cookbooks

Susser, A. 2004. *The great citrus book: a guide with recipes*. Ten Speed Press, Berkeley, CA. 160 pp.

Vaughan, B. 1972. *Citrus cooking*. S. Greene Press, Brattleboro, VT. 32 pp.

27 Clove

Family: Myrtaceae (myrtle family)

NAMES

Scientific Name: *Syzgium aromaticum* (L.) Merr. & L.M. Perry (*Caryophyllus aromaticus* L., *Eugenia aromatica* (L.) Baill., *E. caryophyllata* Thunb., *E. caryophyllus* (Spreng.) Bullock & S.G. Harrison)

- The word "clove" ultimately derives from the Latin *clavus*, nail, but made its way into English via Old French *clou*, meaning nail. Dried cloves have a bulbous head and thinner body and look very much like the crudely formed nails of olden times.
- Similarly in German, cloves are called *Gewürznagelein*, which means spice nails, because of their resemblance to nails.
- In China, cloves were called "tongue spice," because during the second century BC of the Han Dynasty (206 BC–AD 220), courtiers were required to hold the spice in their mouths when addressing the emperor. The point of this was to deodorize bad breath so as not to displease the emperor. It has been said that this was the result of widespread use of garlic by the Chinese.
- Ripe fruits of cloves are known as "mother of clove" and are used locally as a spice in regions where cloves are produced.
- The genus *Syzgium* is based on the Greek *syzgos*, joined. The word was originally applied to an unrelated species (*Calyptranthes syzgium* Sw., a Jamaican plant) for which the name was first used and which has branches and leaves joined in pairs.
- *Aromaticum* in the scientific name *S. aromaticum* is Latin for aromatic.

PLANT PORTRAIT

The clove is a small, tropical, evergreen tree 8 to 12 m (25–40 ft.), occasionally 15 m (49 ft.) tall. Clove trees may attain an age of 70 years, but the average life of a plantation tree is not more than 20 years. Cloves are the flower buds, picked by hand while still unopened, dried, and marketed either whole or ground. Individual cloves need to be harvested when the bud is at the right size, and this means repeated picking of the same plants. Experts can judge from the size and appearance of commercial cloves whether all the buds were picked at the right time. The buds are 13 to 19 mm (1/2–3/4 in.) long. The trees begin to yield 5 to 7 years after planting, and may annually produce up to 34 kg (75 lb.) of dried buds. Cloves are native in the Moluccas (part of Indonesia, also known as the Spice Islands or Clove Islands) and the southern Philippines. Clove trees are said to flourish only near the sea. The spicy fragrance of cloves was admired by the ancient Egyptians, Chinese of the third century BC, and the classical Romans. From the eighth century on, cloves became one of the major spices in Europe. In 1524, the Portuguese took possession of the Spice Islands and established a monopoly to control the clove market. About 1600, the Dutch drove out the Portuguese and until the late eighteenth century willfully eliminated the tree from all of the islands except Amboina and Ternate. However, a Frenchman, Pierre Poivre (literally "Peter Pepper"), managed to get some plants out, and one of these is said to be the ancestor of almost all modern clove trees. Today cloves are produced in several tropical areas, including West Indies, Madagascar (Malagasy Republic), Mauritius, Sumatra (Indonesia), Moluccas (Indonesia), Penang (Malaysia), Guiana, and Brazil.

The island of Zanzibar, which is part of Tanzania, is the world's leading grower. Cloves are one of the most important of spices.

Cloves contain 14% to 20% essential oil, 70% to 90% of which is the chemical eugenol. Clove oil is distilled from the dried buds (as well as from leaves and fruits) and is brown to dark brown in color. It is used by the food, pharmacy, and perfumery industries. Dentists have traditionally used clove oil as an analgesic (pain reliever) or local anesthetic to ease toothache as well as a bactericidal, antiseptic mouthwash. Eugenol from cloves is used in germicides, perfumes, and mouthwashes, in the synthesis of vanillin, and as a sweetener or taste intensifier.

CULINARY PORTRAIT

Cloves are strongly aromatic and hot and pungent in taste (some would say fiery and burning), and when eaten alone they numb the mouth. This spice should be used sparingly. It is used in both savory and sweet dishes. Cloves are used to flavor many foods, particularly meats and bakery products, and are common constituents in sauces and pickles. They are a characteristic flavoring in Christmas holiday fare, such as wassail and mincemeat, and are used to stud ham and pork and to flavor pot roasts, delicatessen meats, meat loaf, and blood sausage. The spice is often found in stews, chutneys, marinades, gingerbread, spice cakes, fruit cakes, pies, plum puddings, candy, mulled wine drinks, stewed fruits, and all apple dishes. Cloves are sometimes used to flavor coffee. Cloves can be combined with some other spices, such as pepper, and go well with onions and garlic but are not normally mixed with herbs. Cloves are often combined with cinnamon and nutmeg and are found in such spice mixtures as garam masala and Chinese five spices. Although they are difficult to grind, it is preferable to purchase whole cloves because they retain flavor much better than ground cloves. Whole cloves can be stored indefinitely in an airtight container kept in a cool, dark place. Much of the volatile oil in old cloves that have not been properly stored will have been evaporated, and such cloves are much less effective as a flavoring agent. To test whether cloves have gone stale, place one in water: it should float vertically, and if it sinks or floats horizontally, it has become stale.

CULINARY VOCABULARY

- As well as referring to the plant and spice referred to in this chapter, the word "clove" has other culinary meanings. The segments of a garlic bulb are called cloves, the word in this case based on the same Germanic source as the English verb cleave. Another independent meaning of clove is a British unit of weight for goods such as cheese, equal to 3.5 kg (8 lb.).
- *Garam masala* (literally "hot spice") is a blend of ground spices used in Indian cookery, standardly with cinnamon, cardamom, cloves, cumin, coriander, and pepper.
- "Five-spice powder" is a mixture of spices, but the phrase has different meanings. It is most used for "Chinese five-spice powder" (five fragrance powder), which usually consists of ground cloves, fennel seeds, star anise, "cinnamon" (Chinese cinnamon, i.e., cassia), and Szechwan pepper (or ginger and/or cardamom) (six spices may be present). The blend is mostly used in China and in parts of Southeast Asia. "Tunisian five-spice powder" consists of cinnamon, cloves, grains of paradise, nutmeg, and pepper. In Bengali cuisine, "five-spice" is a mixture of cumin, black mustard, fennel, black fennel, and fenugreek.
- "Four spices" is a direct translation of the French term *quatre-épices*, referring to a mixture of four powdered spices, typically pepper, nutmeg, cloves, and cinnamon.
- The famous Worcestershire (Worcester) sauce is dominated by cloves (garlic, tamarind, paprika, or chilies are the spices most often also added). Worcestershire sauce of course comes from Worcestershire, England (where it was brought from India by Lord Marcus Sandys, ex-Governor of Bengal). Most people cannot pronounce Worcestershire sauce. Worcestershire is not pronounced wor-chester-shire or wor-sest-er-shire or woo-ster-shire.

In the United Kingdom, Worcestershire is pronounced woost-ur-shire, and Worcestershire sauce is referred to as Worcester sauce, pronounced woos-tah. In many other parts of the world, however, it is referred to as war-sest-uh-shire sauce.

CURIOSITIES OF SCIENCE AND TECHNOLOGY

- For centuries, the Chinese and all of the Arab, Phoenician, and Syrian intermediaries had a vested interested in keeping the origin of cloves a secret from Europe to maintain high prices. The Chinese convinced Italian traveler Marco Polo (1254–1324?) that cloves came from Java. Not until 200 years after the time of Polo were Europeans to discover the true source of the spice.
- In fourteenth-century England, 0.45 kg (1 lb.) of cloves was worth 7½ sheep.
- The island of Ternate, Indonesia, which was once at the center of the world's clove production, is only 9 km (5.6 miles) in diameter. Nevertheless, at least 10 fortresses of Portuguese, Spanish, English, and Dutch origin can still be visited, testimony to how spices once motivated international warfare.
- When Portugal had a monopoly on clove production in the 1500s, by control of the Molucca archipelago of the Spice Islands, the Portuguese published false nautical charts so that foreigners trying to land on the islands would be shipwrecked.
- In the sixteenth century, a German doctor recommended covering the head with cloves to cure poor circulation and cold feet.
- The "Victorian language of flowers" was a secret coded language in Victorian times, with flowers and plants symbolic of certain messages, so when the flower or plant was mentioned in a letter those who knew the code could understand the hidden information. "Cloves" meant "dignity." The Victorian language of flowers was rooted in a much older tradition in Eastern nations of using flowers to communicate messages. This tradition was

FIGURE 27.1 Clove (*Syzgium aromaticum*). (From Köhler, 1883–1914.)

introduced to England by Lady Mary Wortley Montagu, (1689–1762), who in 1716 accompanied her husband while he was ambassador to the court of the Turkish Sultan in Istanbul. She sent a Turkish love letter to England, which included the first interpretation of Eastern flower gifts. Her message:

> *Clove:* I have long loved you and you have not known it.
> *Jonquil:* Have pity on my passion.
> *Pear blossom:* Give me some hope.
> *A rose:* May you be pleased, and your sorrows mine.
> *A straw:* Suffer me to be our slave.
> *Cinnamon:* My fortune is yours.
> *Pepper:* Send me an answer.

(Lady Montagu was an eccentric genius. She introduced inoculation for smallpox into England, which she had observed in Turkey. Inoculation with smallpox has been practiced in Asia for many centuries, but vaccination became much safer when English doctor Edward Jenner discovered in 1796 that immunity to smallpox could be induced by inoculating with the much safer cowpox virus.)

- As noted earlier, Zanzibar is the world's largest producer of cloves. In 1972, to preserve its monopoly, Zanzibar made smuggling cloves out of the country a capital offense, and that year 15 people were sentenced to death for this crime.
- Indonesians are the main consumers of cloves, using up nearly 50% of the world's supply. Unfortunately, most Indonesian use is not for cooking but for smoking. Cigarettes flavored with cloves are extremely popular and are smoked by almost every male Indonesian.
- "Pomander balls" are mixtures of aromatic substances enclosed in a perforated bag or box, once used in the belief that they protected against infection, and also simply placed in handkerchiefs so ladies could sniff their sweet smell instead of bad street odors. To this day, they are still sometimes made (e.g., for keeping a closet smelling fresh) by using a toothpick to prick holes in the skin of a fruit such as an apple or orange and then placing a clove in each hole. The fruit can be dried in a cool, dark place for 2 to 3 weeks until hard. Famous American cook James Beard (1903–1985), amusingly termed the dermatitis some people develop from making pomander balls as "clove pusher's thumb."
- Synesthesia is a remarkable abnormal, inherited condition in which one sensory experience produces a simultaneous second sensory experience. For instance, people with synesthesia may "hear" colors, or "smell" numbers, or "see" music, or "taste" shapes. It can involve the joining of any of the senses. One in 25,000 people is thought to experience it. Most are females and/or lefties. Noted synesthetes include novelist Vladimir Nabokov (1899–1977), painter Wassily Kandinsky (1866–1944), and composer Franz Liszt (1811–1886). Because so much human experience includes food, those with the condition often report associations of the taste or smell of food with other senses. For example, one person with synesthesia reported that each time she saw the color green she smelled cloves.

KEY INFORMATION SOURCES

Adamson, A.D., and Robbins, S.R.J. 1975. *The market for cloves and clove products in the United Kingdom.* Tropical Products Institute, London. 37 pp.

Aiyer, A.K.Y.N., and Abraham, P. 1960. *Cultivation of cloves in India.* Indian Council of Agricultural Research, New Delhi, India. 85 pp.

Bermawie, N., and Pool, P.A. 1995. Clove. *Syzgium aromaticum* (Myrtaceae). In *Evolution of crop plants.* 2nd ed. Edited by J. Smartt and N.W. Simmonds. Longman Scientific & Technical, Burnt Mill, Harlow, Essex, U.K. pp. 375–379.

Bermawie, N., and Pool, P.A. 1997. Isozyme variation in cultivated and wild cloves (*Syzygium* sp.). *Proceedings of the Indonesian Biotechnology Conference (Challenges of Biotechnology in the 21st Century, Jakarta, Indonesia, June 17–19, 1997)*, vol. 2. Institut Pertanian, Bogor, Indonesia. pp. 709–716.

Bulbeck, D. 1998. *Southeast Asian exports since the 14th century: cloves, pepper, coffee, and sugar*. Institute of Southeast Asian Studies, Singapore. 195 pp.

Clove Growers Association. 1979. *Cloves from Zanzibar*. Clove Growers Association, Zanzibar government, Zanzibar. 20 pp.

Crofts, R.A. 1959. *Zanzibar clove industry: statement of government policy and report*. Zanzibar. 70 pp.

Donkin, R.A. 2003. *Between east and west: the Moluccas and the traffic in spices up to the arrival of the Europeans*. American Philosophical Society, Philadelphia, PA. 274 pp.

Federation of American Societies for Experimental Biology. 1973. *Evaluation of the health aspects of oil of cloves as a food ingredient*. Federation of American Societies for Experimental Biology, Bethesda, MD. 16 pp.

Hanusz, M. 2000. *Kretek: the culture and heritage of Indonesia's clove cigarettes*. Curzon, Richmond, U.K. 203 pp.

Malson, J.L., Lee, E.M., Murty, R., Moolchan, E.T., and Pickworth, W.B. 2003. Clove cigarette smoking: biochemical, physiological, and subjective effects. *Pharmacol. Biochem. Behav*. 74:739–745.

Martin, P.J. 1991. The Zanzibar clove industry. *Econ. Bot*. 45:450–459.

Martin, P.J. 1993. *Zanzibar clove cultivation manual*. Ministry of Agriculture, Livestock and Natural Resources, Unguja, Zanzibar, Tanzania. 70 pp.

Martin, P.J., Riley, J., and Dabek, A.J. 1987. Clove tree yields in the islands of Zanzibar and Pemba. *Exp. Agric*. 23:293–303.

Menon, P.S. 2000. ISSR scientists develop new clove varieties. *J. Herbs Spices Med. Plants*, 7(1):103–106.

Nurdjannah, N., and Bermawie, N. 2001. Clove. In *Handbook of herbs and spices*. Edited by K.V. Peter. Woodhead Publishing, Cambridge, U.K. pp. 154–163.

Nutman, F.J., and Roberts, F.M. 1971. The clove industry and the diseases of the clove tree. *Pest Arts News Summ*. 17:147–165.

Peter, K.V., and Kandiannan, K. 1999. Clove. In *Tropical horticulture*, vol. 1. Edited by T.K. Bose, S.K. Mitra, A.A. Farooqui, and M.K. Sadhu. Naya Prokash, Calcutta, India. pp. 712–716.

Pool, P.A., and Bermawie, N. 1986. Floral biology in the Zanzibar type clove (*Syzygium aromaticum*) in Indonesia. *Euphytica*, 35:217–223.

Pool, P.A., Eden-Green, S.J., and Muhammad, M.T. 1986. Variation in clove (*Syzygium aromaticum*) germplasm in the Moluccan Islands. *Euphytica*, 35:149–159.

Schmid, R. 1972. A resolution of the *Eugenia–Syzygium* controversy. *Am. J. Bot*. 59:423–436.

Sri Lanka Department of Minor Export Crops. 1986. *Clove—cultivation and processing*. Technical Bulletin, Department of Minor Export Crops, Peradeniya, Sri Lanka. 11 pp.

Sritharan, R., and Bavappa, K.V.A. 1981. Floral biology in clove. *J. Plantation Crops*, 9(2):88–94.

Tidbury, G.E. 1949. *The clove tree*. C. Lockwood, London. 212 pp.

Veerheij, E.W.M., and Snidjers, C.H.A. 1999. *Syzgium aromaticum* (L.) Merrill & Perry. In *Plant resources of South-East Asia. 13. Spices*. Edited by C.C. De Guzman and J.S. Siemonsma. Backhuys Publishers, Leiden, the Netherlands. pp. 211–218.

Waard, P.W.F. de. 1974. The development of clove buds and causes of irregular bearing of cloves (*Eugenia caryophyllus* (Sprengel) Bullock et Horrison). *J. Plantation Crops*, 2(2):23–31.

Wit, F. 1969. The clove tree *Eugenia caryophyllus* (Sprengel) Bullock & Harrison. In *Outlines of perennial crop breeding in the tropics*. Edited by F.P. Ferwerda and F. Wit. Misc. Paper 4, Landbouwhogeschool, Wageningen, the Netherlands. pp. 163–174.

Specialty Cookbooks

Day, A.S., and Stuckey, L. 1964. *The spice cookbook*. D. White Co., New York. 623 pp.

Humphrey, S.W. 165. *Spices, seasonings and herbs: the definitive cookbook*. Collier Books, New York. 370 pp.

Morris, S. 2001. *The essential Indonesian cookbook: aromatic dishes from tropical spice islands*. Lorenz, London. 96 pp.

28 Coca

Family: Erythroxylaceae (coca family)

NAMES

Scientific Name: *Erythroxylum* species (*Erythroxylum* is widely but incorrectly spelled *Erythroxylon*)

- *E. coca* Lam.
- *E. novogranatense* (D. Morris) Hieron.

- The word "coca" comes from a South American Indian word, either the Quechu word kúka (the Quechu were a people of central Peru believed to have descended from Aymara Indians principally from Bolivia and Peru) or the Aymara word *q'oka*, meaning "food or a meal for travelers and workers."
- "Coca" should not be confused with cacao (*Theobroma cacao* L.), the source of chocolate, or cocoa made from chocolate, or "coco butter," a misspelling of cocoa butter, obtained from the chocolate bean. Coca should also not be confused with coconut (*Cocos nucifera* L.).
- *Erythroxylum coca* is most often known simply as coca. It is also called cocaine plant and spadic. Two varieties have been recognized. *Erythroxylum coca* var. *coca* is called Bolivian coca and Huanuco coca and is widely cultivated in South America. *Erythroxylum coca* var. *ipadu* Plowman is called Amazonian coca and is cultivated in Amazonian Colombia, Peru, and Brazil.
- *Erythroxylum novogranatense* is also generally called coca and also has two varieties. *Erythroxylum novogranatense* var. *novogranatense* is called Colombian coca and is cultivated in Colombia and Venezuela. *Erythroxylum novogranatense* var. *truxillense* (Rusby) Plowman is called Truxillo coca and Trujillo coca. It is cultivated in southwestern Colombia, Ecuador, and northern Peru.
- The genus name *Erythroxylum* is based on the Greek *erythro*, red + *xylon*, wood. Some species of the genus have red wood.
- *Novogranatense* in the scientific name *E. novogranatense* is Latin for "New Granada," which from 1717 to 1819 was the name of a Spanish viceroyalty in northwestern South America, including modern Panama, Colombia, Venezuela, and Ecuador.

PLANT PORTRAIT

Coca plants are small evergreen shrubs with reddish brown bark. *Erythroxylum coca* is the major source of commercially produced coca leaves and cocaine. It is 2 to 4 m (6–13 ft.) in height but usually kept no higher than 2 m when cultivated. The leaves are 3 to 5 cm (1–2 in.) long and resemble bay and tea leaves. The foliage contains 0.1% to 1% cocaine, with higher amounts tending to occur at higher altitudes. The species is native to fertile warm valleys under 2000 m (6560 ft.) in the tropical region of the eastern Andes Mountains and is widely cultivated in South America. *Erythroxylum novogranatense* is closely related to *E. coca* but is considerably less important. This species is thought to exists only in cultivation. It is grown mostly in Colombia and Venezuela and also in northern Peru and Brazil.

South American Indians have cultivated coca plants for thousands of years for use as a masticatory (a substance that is chewed). Ceramic figurines from Ecuador of 3000 BC show men with the bulging cheeks characteristic of the coca chewer. Coca was used from Nicaragua in the north down to Chile in the south, generally only by the elite, including royalty and the ruling classes. Coca was especially employed in religious rituals, and after the Spanish conquest of South America the Catholic Church condemned the use of coca, viewing its use as a heathen exercise in spiritual life. The Spanish, however, realized that coca remarkably increased the stamina of workers and encouraged the cultivation of coca to supply their slaves in the rich silver and tin mines of Bolivia. By the late 1800s, Europeans and North Americans were aware of the stimulant reputation of coca and were using it in elixirs and patent medicines. The most infamous use of the coca plant was in the popular soft drink Coca-Cola. Today, millions of Indian in the Andes and in the western Amazon continue to chew coca leaf. Although Peru, Bolivia, and Colombia are the major producing countries, coca is now cultivated outside of South America, particularly in Africa, Ceylon, Taiwan, and Indonesia. Colombia alone is responsible for two-thirds of global coca leaf production.

Coca leaves are dried and may be powdered for use. When chewed or (more often) sucked with a pinch of lime, coca leaf releases alkaloids, principally cocaine, which exerts a stimulant action. The effect from the small amount of cocaine in the leaves is very much milder than exposure to refined cocaine. There is a numbing of sensory nerves and a dulling of hunger and pain. The Indians used coca leaves to acquire strength, endurance, and increased stress tolerance to hunger and cold. The practice of using lime to make alkaloids more available is widely practiced by different cultures using different plants (see, e.g., Chapter 12). Lime decreases the acidity in the mouth, and this makes the alkaloid more available.

Cocaine was well regarded in the late nineteenth century. Its use was advocated by such prominent public figures as Pope Leo XII, Sigmund Freud, Jules Verne, and Thomas Edison. Arthur Conan Doyle's detective character Sherlock Holmes became exceptionally insightful when he injected cocaine in *The Seven Percent Solution* (the supersleuth's preferred dosage). Coca leaf has a long history of use as a medicinal plant in South America, and cocaine has been used in Western medicine, especially as a local anesthetic. Today, the illegal marketing of refined cocaine is an extremely serious social, health, and law enforcement problem. Importation and production of cocaine have been

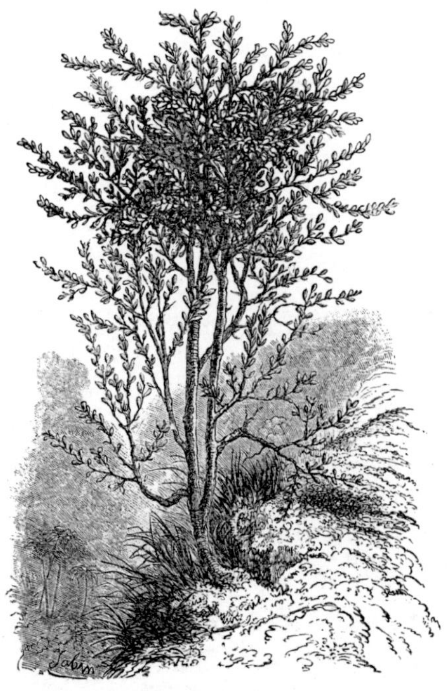

FIGURE 28.1 Coca (*Erythroxylum coca*). (From Jumelle, 1901.)

controlled by enormously powerful armed cartels such as the Medellín and Cali cartels in Colombia, which have infiltrated governments, corrupted officials, and assassinated public officials. The yearly U.S. retail cocaine market has been estimated to be worth between $30 billion and $150 billion.

The United Nations Convention Against Illicit Traffic in Narcotic Drugs and Psychotropic Substances, signed in Vienna in 1988, prohibits sowing, cultivating, harvesting, processing, and marketing of coca leaves. In most countries, possession of coca leaves could lead to arrest and prosecution. This is a contentious issue, some people, especially in the Third World, holding the viewpoint that the traditional usage of coca leaf in South America represents a cultural heritage that does not deserve the condemnation that properly is assigned to the abuse of cocaine. As part of the "War on Drugs," huge efforts and expenditures are underway to eradicate coca plantations, and there are well-intentioned crusades to encourage the planting of replacement crops, which are not as attractive to peasants in extreme poverty.

CULINARY PORTRAIT

Consumption of coca leaves is not practiced in Western culture. However, numerous people have consumed extracts from the plant, albeit without knowing it. Coca-Cola is the world's most popular soft drink. In 1866, John Styth Pemberton of Atlanta, Georgia, introduced his coca-fortified soft drink, Coca-Cola. "Coke" did not become completely cocaine free until 1929. In modern times, de-cocainized coca leaves have been used as a flavoring agent by the Coca-Cola Company.

In the South American Andes, peasants have traditionally sucked wads of leaves, keeping them in their cheeks for hours. Chalk or ash is usually added to increase the alkalinity and dissolve the alkaloids into the saliva. Studies have shown that chewing or sucking 100 g (3½ oz.) of coca leaves satisfies the daily dietary allowance for calcium, iron, phosphorus, and vitamins A, B_2, C, and E. This contribution of vitamins has been shown to be desirable in the starch-heavy diet of the highland South American Indians. Coca leaves do not produce the strong euphoria associated with the abuse of cocaine.

In Peru and Bolivia, "tea" made from the coca leaf is widely available. (Some purists insist that the word "tea" be reserved for Chinese tea (*Camellia sinensis* (L.) Kuntze), and that the word "tisane" be used for all other herbal teas.) Public markets not only sell loose coca leaves, but grocery stores often sell commercial tea bags made with coca. Hotels and airports commonly offer complementary coca tea to tourists, who are often informed that the beverage will counter the fatiguing effects of altitude. Most tourists who try coca tea report that they experienced no obvious effects. During a 1993 visit to Bolivia, Pope John Paul II (1920–2005), is said to have consented to drink coca tea, a mark of respect for the culture of the indigenous population. In the Western World, coca leaf is illegal. Nevertheless, coca tea is available on the Internet, as well as a wide variety of other coca culinary products.

CULINARY VOCABULARY

- *Maté de coca* is the name of the hot, clear coca tea widely available in Peru and Bolivia. This is similar to green Asian tea, but with less color and flavor.
- *Harina de coca* ("coca flour" or "coco powder"), a preparation of powdered coca leaves, is marketed in Peru as well as on the Internet, often as an alleged tonic.
- A *cocada* is a Peruvian term for a period of coca consumption. The Indigenous People of the high Sierra Mountains of Peru are famous for their ability to travel rapidly along mountain paths with heavy burdens, sustained only by occasional chews of coca. They often measure the journey not in hours but by the *cocada*, the time spent chewing one quid of coca, often approximately 1 hour. A traveler will cover about 3 km (1.9 miles) on level ground, or 2 km (1.2 miles) going uphill.

CURIOSITIES OF SCIENCE AND TECHNOLOGY

- Anthropologists have determined that coca was an integral part of the lives of Quechua Indians in South America. They threw coca into the air when a pack train was ready to depart to enlist the help of the mountain gods to ensure a safe trip. Where particularly dangerous passes were present, stones were piled up as a sacred alter against which quids of coca could be thrown as an offering (these are now generally replaced by crosses). Coca was offered to the earth mother to ensure good crops and before inserting the corner stones of a new house. Coca was also offered to a girl's parents to obtain their consent for marriage. In graves where mummies are found, coca is present, often in the mouth of the deceased, to provide strength for the journey in the afterlife.
- Suggestive of use as an anesthetic, the ancient Incas used coca on patients undergoing trephination (an operation in which a hole was bored into the cranium).
- In 1860, Angelo Mariani introduced "Vin Mariani" a wine fortified with coca. Mariani became wealthy selling this drink, which was promoted by such notable persons as Sarah Bernhardt, Queen Victoria of England, Thomas Edison, and Pope Leo the XIII.
- In the nineteenth century, cocaine was used as a "cure" for morphine addiction. However, cocaine is extraordinarily addictive. Starving experimental animals have been observed to choose cocaine in preference to food. Bolivia's yearly income from cocaine exports has been estimated to be about a trillion dollars. North American cocaine dealers make ten times more than the Bolivian producers do. The peasants who grow the coca receive less than 1.5% of the value for which cocaine is sold in the United States.
- COCA has been widely used as an acronym. For examples: Counseling on Cocaine Abuse, Collaborative Objects Coordination Architecture, Center on Contemporary Art, Clearinghouse on Computer Accommodation, Concordia Old Collegians Association, Council of Ontario Construction Associations, Council of Owners and Construction Associates, and Canadian Organization of Campus Activities.
- The larva of a butterfly, *Eloria noyesi*, inhabits coca producing areas and apparently exclusively feeds on coca leaves. It is said to be capable of consuming more than 50 leaves in its 1 month of existence and has been used as a natural biocontrol agent in efforts to eliminate cultivation of coca.
- In recent years, there have been proposals to use a fungus, *Fusarium oxysporum*, as a means of eliminating or at least controlling undesirable plants, especially drug plants. Such proposals have included the possibility of genetically engineering the fungus to specifically attack given plant species, particularly coca, opium poppy, and marijuana. *Fusarium* species are naturally occurring fungi with variants that can cause wilt in numerous plant species. Government-backed research in this area has been conducted since the early 1980s, mostly in secret, and there have been unverified claims that fungi have already been released in attempts to control coca. Environmentalists have objected, arguing that this biological weapon could upset the ecology of a region, endanger food crops and wildlife, and potentially have disastrous unforeseen consequences.
- Heinz Brücher, (1916–1991), an expatriate German botanist, was murdered in Argentina, allegedly the victim of a burglar. At that time, he was working on a viral disease (the Estalla virus) to eradicate the coca plant, thus challenging the interests of the cocaine trade in the Andes, and there is suspicion that the cocaine barons arranged his assassination. Although many consider eliminating plant species to be a special branch of biological warfare, nevertheless Brücher was seemingly a respectable and accomplished researcher. However, a report (*Plant Genetic Resources Newsletter*, vol. 129, pp. 54–57, 2002) revealed that in the 1940s he led a special Nazi troop with the purpose of stealing valuable seeds for breeding purposes from the conquered nations of Europe, a phenomenon now called "biopiracy" (see Chapter 1), and another example of questionable scientific activity.

FIGURE 28.2 A nineteenth-century plantation of coca (*Erythroxylum coca*). (From Jackson, 1890.)

- Why do coca plants produce cocaine? Cocaine is a natural pesticide defense, which acts to poison the nervous systems of many insects that try to feed on the plants.

KEY INFORMATION SOURCES

Acock, M.C., Lydon, J., Johnson, E., and Collins, R. 1996. Effects of temperature and light levels on leaf yield and cocaine content in two *Erythroxylum* species. *Ann. Bot.* 78:49–53.

Bohm, B.A., Ganders, F.R., and Plowman, T. 1982. Biosystematics and evolution of cultivated coca (Erythroxylaceae) *Erythroxylum* spp. *Syst. Bot.* 7:121–133.

Chung, R.C.K., and Brink, M. 2001. *Erythroxylum* P. Browne. In *Plant resources of South-East Asia. 12(1). Medicinal and poisonous plants 1.* Edited by L.S. de Padua, N. Bunyapraphatsara, and R.H.M.J. Lemmens. Backhuys Publishers, Leiden, the Netherlands. pp. 258–262.

Cubas, H.C. 1996. *Commercializing coca, possibilities and proposals.* Catholic Institute for International Relations, London. 40 pp.

Duke, J.A., Aulik, D., and Plowman, T. 1975. Nutritional value of coca. *Bot. Mus. Leafl. Harvard Univ.* 24(6):113–119.

Ganders, F.R. 1979. Heterostyly in *Erythroxylum coca* (Erythroxylaceae). *Bot. J. Linn. Soc.* 78:11–20.

Gentner, W.A. 1972. The genus *Erythroxylum* in Colombia. *Cespedesia.* 1:481–554.

Johnson, E.L. 1995. Content and distribution of *Erythroxylum coca* leaf alkaloids. *Ann. Bot.* 76:331–335.

Johnson, E.L., and Foy, C.D. 1996. Biomass accumulation and alkaloid content in leaves of *Erythroxylum coca* and *Erythroxylum novogranatense* var. *novogranatense* grown in soil with varying pH. *J. Plant Physiol.* 149:444–450.

Johnson, E.L., Saunders, J.A., Mischke, S., Helling, C.S., and Emche, S.D. 2003. Identification of *Erythroxylum* taxa by AFLP DNA analysis. *Phytochemistry*, 64:187–197.

Karch, S.B.1998. *A brief history of cocaine.* CRC Press, Boca Raton, FL. 202 pp.

Karch, S.B. 2003. *A history of cocaine: the mystery of coca java and the Kew plant.* Royal Society of Medicine Press, London. 224 pp.

Manuél, C. 1977. *The coca cultivator's handbook.* Leaf Press, Ukiah, CA. 67 pp.

Martin, R.T. 1970. The role of coca in the history, religion, and medicine of South American Indians. *Econ. Bot.* 24:422–438.

Mortimer, W.G. 2000. *History of coca: "the divine plant" of the Incas.* University Press of the Pacific, Honolulu, HI. 576 pp. [Reprint of 1901 publication].

Nathanson, J.A., Hunnicutt, E.J., Kantham, L., and Scavone, C. 1993. Cocaine as a naturally occurring insecticide. *Proc. Nat. Acad. Sci. (USA)*, 90:9645–9648.

Pacini Hernandez, D., and Franquemont, C. (Eds.). 1986. *Coca and cocaine: effects on people and policy in Latin America: proceedings of the conference, the coca leaf and its derivatives—biology, society and policy*. Cultural Survival, Cambridge, MA. 169 pp.

Plowman, T. 1976. Orthography of *Erythroxylum* (Erythroxoxylaceae). *Taxon*, 25:141–144.

Plowman, T. 1979. Botanical perspectives on coca. *J. Psychedelic Drugs*, 11:103–117.

Plowman, T. 1981. Amazonian coca. *J. Ethnopharmacol*. 3:195–225.

Plowman, T. 1982. The identification of coca (*Erythroxylum* species): 1860–1910. *Bot. J. Linn. Soc.* 84:329–353.

Plowman, T. 1984. The ethnobotany of coca (*Erythroxylum* spp., Erythroxylaceae). *Adv. Econ. Bot.* 1:106–111.

Plowman, T., and Hensold, N. 2004. Names, types, and distribution of neotropical species of *Erythroxylum* (Erythroxylaceae). *Brittonia*, 56:1–53.

Plowman, T., and Rivier, L. 1983. Cocaine and cinnamoylcocaine content of *Erythroxylum* species. *Ann. Bot.* 51:641–659.

Rivier, L. (*Ed.*). 1981. Coca and cocaine (proceedings, symposium, Quito, Ecuador, 1979). *J. Ethnopharmacol.* 3:106–379.

Stephen-Hassard, Q.M. 1970. Sacred plant of the Incas. *Pacific Discovery*, 23(5):26–30.

U.S. Department of Justice, Drug Enforcement Administration. 1991. *Coca cultivation and cocaine processing: an overview*. U.S. Department of Justice, Drug Enforcement Administration, Office of Intelligence, Washington, DC. 15 pp.

Weil, A.T. 1981. The therapeutic value of coca *Erythroxylon* in contemporary medicine. *J. Ethnopharmacol.* 3:367–376.

Specialty Cookbooks

Note: Coca, unless de-cocainized, is illegal in most Western jurisdictions.

Prosper, M. 1977. *The new Larousse gastronomique*. Crown Publishers, New York. 1064 pp. + plates. (Translation of the original French text, first published in 1938, and periodically revised. This edition has a recipe for "Zabaglionhe" or "Sabayon," an Italian cream mousse, made with coca paste. More recent recipes for "Zabaglione" omit the coca.)

29 Coco de Mer (Double Coconut)

Family: Arecaceae (Palmae; palm family)

NAMES

Scientific Name: *Lodoicea maldivica* (J.F. Gmel.) Pers. (*L. callypige* Comm. ex J. St.-Hil., *L. sechellarum* Labill.)

- The name "coco de mer" is derived from the French *coco*, coconut + *de*, of + *mer*, sea. Until the Seychelles were discovered in 1742 by the French explorer Captain Lazare Picault, it was widely believed that the coco de mer lived beneath the sea, and sailors reported how the plants magically vanished when they dove for the fruits. The Dutch (German-born) botanist Georg Eberhard Rumphius (1628–1702), who in 1680 first described the seeds while he was in Europe, stated that the plants grew in the sea.
- The coco de mer is also known as double coconut, sea coconut, Seychelles nut, Seychelles palm, Seychelles nut palm, and coco fesse. (Seychelles is pronounced say-SHELZ or say-SHELLS.)
- Coco de mer is called double coconut because of the resemblance of the nut to two joined coconuts. Although the coconut is also a palm, it is not closely related.
- The names Seychelles nut and Seychelles nut palm reflect the natural geographical distribution of the coco de mer.
- Centuries ago, coco de mer nuts were thought to be the fruit of the tree of knowledge and, reflecting the legendary wisdom of Solomon, the name "coco-de-Solomon" was used.
- "Coco fesse" is French for buttocks coconut, an allusion to the similarity of the double coconut to a woman's pelvis.
- The genus name *Lodoicea* has been explained as commemorating King Louis XV of France (1710–1774; "Louis" can be Latinized as Ludovicus or Lodoicus), but this is not correct. *Lodoicea* was formally proposed as a genus name in 1805 by the botanist Jean Henri Jaume Saint-Hilaire (1772–1845), who adopted the name from another botanist, Philbert Commerson (1727–1773), a crew member on a historic voyage to the Seychelles. Commerson likened the form of the coco de mer seed to Laodice, a lady in classical Greek mythology. Laodice was the most beautiful daughter of Priam, the king of Troy at the time of the Trojan War. It was said that when Troy had fallen she was swallowed up by a chasm in the earth (authorities disagree on whether the war described in Homer's epic *Iliad* actually happened, some arguing that it represented several major historical battles). *Lodoicea* is the Latinized form of Laodice.
- *Maldivica* in the scientific name *L. maldivica* was based on the mistaken belief that the nuts came from the Maldive Islands in the Indian Ocean (the first nuts discovered were washed up on the beaches of the Maldives). The old name "Maldive nut" also reflects this error.

PLANT PORTRAIT

The coco de mer is a palm tree with separate male and female plants, the males reaching a maximum height of 30 m (100 ft.), while females grow no taller than 25 m (82 ft.). The trees have a crown of heavy fan-shaped leaves, with leaf blades often over 6 m (20 ft.) long and 3.7 m (12 ft.) wide (the leaves of young plants are larger, and can reach a length of 14 m or more than 45 ft.). Trunks do not develop until the plants are about 15 years of age. The plants are very slow growing, and female trees do not bear fruit until they are at least 40 years old. A female tree may produce 25 to 35 fruits at one time, in various stages of development. The seeds are the largest of all plants. The seeds are enclosed in a hard shell resembling a pair of coconuts joined in the middle (or, in the words of the prestigious *New Royal Horticultural Society Dictionary of Gardening*, "the rear of a large woman"). One to three seeds may be present in each fruit. The fruits take 6 or 7 years to mature and can weigh up to 32 kg (70 lb), although they typically weigh about 20 kg (44 lb.) when fresh. By comparison, record-size coconuts only weigh about 3 kg (6½ lb.). The fruits sometimes exceed 50 cm (20 in.) in length and 1 m (about a yard) in circumference. The male trees develop an impressive, long, thick flowering stalk that may reach 2 m (6½ ft.) in length.

The species is native to the Seychelles, an archipelago nation with a chain of over 100 islands spread over 500,000 square kilometers (200,000 square miles) of the Indian Ocean, off the east coast of Africa, 900 km (560 miles) northeast of Madagascar. (The Seychelles were part of the British Empire until 1976 when it gained independence. The population is about 73,000.) The trees now grow wild on only two small islands, Praslin (pronounced Prah-lin, measuring at the widest point 12.9 × 1.6 km or 8 × 1 miles) and the nearby Curieuse (3.2 × 1.6 km or 2 × 1 miles, named for the ship whose captain discovered the island in 1768). Curieuse was once covered with coco de mer trees. In an attempt to harvest them the island was set ablaze in 1771, and virtually all were destroyed. The island, which is now almost deserted, revegetated with a wide variety of species. The island of Praslin today is the main sanctuary for the coco de mer. Praslin's Vallée de Mai has thousands of coco de mer palms, some of which are claimed to be nearly 900 years old. However, most coco de mer trees are estimated to live no longer than 200 to 400 years.

The shells of coco de mer were discovered floating or cast up by the sea for centuries before their sources was known, and it was believed that they possessed magical properties. Potentates particularly interested in their alleged aphrodisiac properties paid huge sums for the nuts. On the Maldive Islands, the king had exclusive rights to all double coconuts washed ashore. In the East Indies, a hand was often cut off anyone found to have not turned in a newly discovered nut to the local ruler.

Unlike the coconut, which has spread thousands of miles by floating on the sea, the double coconut is generally too dense to float, and occasional nuts with sufficient air spaces to float cannot survive long sea journeys and remain viable. Because the plant grows far from other lands, it has not been distributed by the sea. The coco de mer is rarely cultivated but is occasionally found in very large public greenhouses. Today, the Seychelles government carefully regulates the harvest of double coconuts and sells them at authorized outlets, so that visitors to the Seychelles are usually able to purchase one (although the weight can be a problem in luggage).

CULINARY PORTRAIT

The young flesh of the coco de mer is jellylike, sweet and very tasty, and is considered a delicacy when the fruits are 10 to 12 months old. After that, the interior becomes very hard. The fruit is extremely rare. Only a few thousand are harvested annually, and the coco de mer is normally available as food only to the extremely wealthy. Indeed, the coco de mer has been called "billionaire's dessert." "Millionaire's salad" is served on the Seychelles, but it does not cost a million, and is not made with coco de mer, but the "heart" (growing point) of another palm tree. (See Peach Palm; the name millionaire's salad is based on the idea that harvesting the growing point kills the tree, which only a millionaire can afford.) The queen of Portugal so enjoyed a preserve made from unripe coco de mer nuts that she ordered a shipload of these fruits every year, a luxury that is no longer possible.

Culinary Vocabulary

- Begging bowls were occasionally made out of the shells of coco de mer and were termed *keshkul* in Persian. Such bowls were carried suspended by a chain from the shoulders of beggars called *fuqara* (plural of *faqir*). Certain of these beggars called "the paupers of God" solicited food, which was placed in the keshkul, until it was eventually filled with different kinds of food. As a heritage of this practice, there evolved a food dish, known as keshkul-e-fuqara (literally "beggar's bowl"), consisting of milk, almonds, and other nuts, which is popular in the Middle East, including Afghanistan and Iran.

CURIOSITIES OF SCIENCE AND TECHNOLOGY

- Before the Seychelles became known as the source of the coco de mer, the rare shells of the nuts were made into polished vessels for the aristocracy, who believed that using them to contain food and beverages ensured that they would not be poisoned. On the Seychelles, shells were used as water pitchers and serving platters, but today these are collector's items.
- The leaves of immature coco de mer palms are often huge—sometimes capable of shading a two-storey house. However, they are not the largest leaves in the plant kingdom, a distinction that belongs to raffia palms (of the genus *Raphia*), the leaves of which sometimes reach 25 m (82 ft.) in length.

FIGURE 29.1 Coco de mer (*Lodoicea maldivica*). (From Harter, 1988.) Fruit at left, male inflorescence at right.

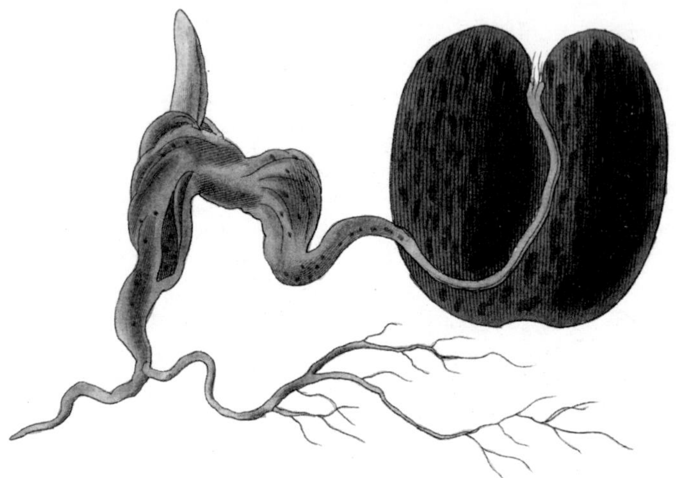

FIGURE 29.2 Coco de mer (*Lodoicea maldivica*). (From Curtis, 1827, vol. 54, plate 2738.) This illustration shows a seedling attached by its "umbilical cord" to the parent double coconut.

- By comparison with the females, male coco de mer are taller, more common, much less likely to be toppled in hurricanes, and have been reported to live longer. This apparent relative "weakness" of females is due to the burden of carrying as much as 250 to 500 kg (550–1100 lb.) of fruit at maturity. By contrast, the sexual part of the male (i.e., their flowering parts) has been estimated to weigh less than 60 kg (132 lb.).
- At the bottom of its trunk, the coco de mer develops extremely dense but flexible wood that allows the tree to rock from side to side during heavy windstorms, and not be toppled over as would happen with most other trees.
- Double coconut nuts that can be used to produce trees, and indeed living potted trees of the double coconut, are extremely difficult to obtain. The Seychelles does not permit unprocessed coco de mer to be exported without a special license, and because of its rarity, the double coconut has been banned from international trade. A germinated nut was presented for sale at a recent palm show in Fort Lauderdale, Florida, for $1,700.00.
- The flesh (endosperm) of the coco de mer fruit matures to become extremely hard, and it is then sometimes used as "vegetable ivory" for carvings. In earlier times, the material was used for buttons.
- Coco de mer is not always one-seeded and two-lobed. Rarely, nuts contain two seeds and are four-lobed, or have three seeds and are four-to six-lobed. One tree in the Vallée de Mer is known to commonly produce four-lobed fruit.
- In the Seychelles, dominoes is a national pastime, and the spots on the game pieces are often made with the "ivory" (white solidified flesh) from coco de mer nuts.
- Why is the seed of the coco de mer so huge? Plant species often produce many small seeds of which only a few will survive (e.g., epiphytic orchids of the tropical rain forests have numerous seeds the size of dust particles), or only a few large seeds whose survival is made more likely by having a large supply of food. The coco de mer has adopted the latter strategy, but in an extreme manner. After a period of up to 18 months, the young seedling emerges from the huge fruit at the end of a special stalk that transports it often as much as 10 m (33 ft.) away where it takes root, but it continues to draw nourishment through the stalk (a sort of umbilical cord) for as long as 3 or 4 years. This example of maternal nursing of seedlings for a very extended period is paralleled by the situation of nursing human infants, sometimes for as long a period.

KEY INFORMATION SOURCES

Akhtar, M.S., Khan, Q.M., and Khaliq, T. 1987. Pharmacological screening for hypoglycemic activity of *Asparagus racemosus* roots and *Lodoicea sechellarum* fruits in rabbits. *J. Pharm.* 8:63–70.

Deb, D.B. 1982. Observation on double coconut in the Indian Botanic Garden. *Bull. Bot. Surv. India*, 24(1/4):177–178.

Edwards, P.J., Kollmann, J., and Fleischmann, K. 2002. Life history evolution in *Lodoicea maldivica* (Arecaceae). *Nordic J. Bot.* 22:227–237.

Fleischmann, K., Edwards, P.J., Ramseier, D., and Kollmann, J. 2005. Stand structure, species diversity and regeneration of an endemic palm forest on the Seychelles. *Afr. J. Ecol.* 43:291–301.

Gerlach, J. 2003. Pollination in the coco-de-mer, *Lodoicea maldivica. Palms*, 47:135–138.

Henderson, A. 2002. Reply to Edwards, P., J. Kollmann & K. Fleishcmann. 2003. Life history evolution in *Lodoicea maldivica* (Arecaceae). *Nordic J. Bot.* 22:418.

Krochmal, A., and Krochmal, C. 1978. Double coconut: the largest seed. *Garden J.* 2(3):22–25.

Krochmal, A., and Krochmal, C. 1979. The double coconut. *Bull. Pacific Trop. Bot. Gard.* 9(2):31–33.

Lionnet, G. 1986. *Coco de mer: the romance of a palm.* L'Île aux images, Bell Village, Île Maurice, Seychelles. 95 pp.

Lucas, S.A. 1982. The coco-de-mer palm at Pacific Tropical Botanical Garden. *Bull. Pacific Trop. Bot. Gard.* 12(2):25–28.

Savage, A.J.P., and Ashton, P.S. 1983. The population structure of the double coconut and some other Seychelles palms. *Biotropica*, 15:15–25.

Schlieben, H.J. 1972. Sea-coconut—the romantic history of an oilpalm. *Nat. Mus.* 102:281–291 (in German).

Silverton, J. 1987. Possible sexual dimorphism in the double coconut: reinterpretation of the data of Savage and Ashton. *Biotropica*, 19:282–283.

Small, E., and Catling, P.M. 2003. Blossoming treasures of biodiversity: 8. Coco de mer (*Lodoicea maldivica*)—What do you do with a 70 pound coconut? *Biodiversity*, 4(2):25–28.

Sneed, M.W. 1976. In quest of the big seed (with observations along the way). *Principes*, 20(1):11–23.

Swabey, C. 1970. The endemic flora of the Seychelle Islands and its conservation. *Biol. Conserv.* 2:171–178.

Specialty Cookbooks

Not unexpectedly because there is no commercial supply, no cookbooks with recipes for coco de mer were found. Aside from species that are near extinction, coco de mer is one of the least available foods in the world. It is certainly the most exotic food plant mentioned in this book. In theory, coconut recipes could be adapted to coco de mer.

30 Culantro

Family: Apiaceae (Umbelliferae; carrot family)

NAMES

Scientific Name: *Eryngium foetidum* L.

- The English name "culantro" is the Spanish word by which the plant is commonly known in Central America. Culantro was named for its similarity to the cilantro form of coriander (i.e., varieties of *Coriandrum sativum* L. selected for edible leaves), the name culantro simply having arisen as a variant of cilantro. Both species are in the carrot family.
- About 70 foreign names have been recorded for culantro. The name recao, by which the plant is widely known in Central America, is also sometimes used in North America.
- Culantro is often confused with cilantro or coriander. Culantro is also known as false, long, Java, Mexican, Puerto Rican, spiny, serrated, saw tooth, and wide-leaved, coriander.
- The names "saw tooth" and "saw leaf herb" reflect the serrated, sawlike edge of the leaves.
- The species is known as fitweed in Guyana, a name based on its supposedly anticonvulsant (i.e., countering fits) property.
- Other English names for culantro: black benny, shado beni (shado benni), spiritweed, stinkweed.
- The genus name *Eryngium* is based on the Greek *eryingion*, the name applied to sea holly (*Eryngium* species).
- *Foetidum* in the scientific name *E. foetidum* is Latin for stinking or bad smelling (accounting for why the plant is sometimes called stinkweed). As with cilantro and coriander, the smell is sometimes equated to a crushed bedbug.

PLANT PORTRAIT

Culantro is a biennial, tap-rooted herb with a basal rosette (i.e., concentric ring at ground level) of leaves about 4 cm (1½ in.) wide and up to 30 cm (1 ft.) long. The leaves have spiny or finely saw-toothed margins. The tiny pale blue or greenish-white flowers are borne in a dense cylindrical head, at the base of which are five to seven spiny-margined leafy bracts that eventually reflex downward. The species is native to continental tropical America and the West Indies and is now cultivated in tropical regions of Asia and Africa, where it has become established as a weed. It can be grown as an annual in most of the United States and southern Canada, and in similar temperate regions.

Culantro is used in dishes throughout the Caribbean, Latin America, and Asia, particularly in India, Korea, and Singapore, but is relatively unknown in Western nations, including most of North America. In 1975, there was a large input of immigrants from Vietnam to the United States, and they brought some of their plants with them. In Texas, the Vietnamese community cultivated culantro (as *ngò gai*, literally "spiny coriander"). Considerable quantities of culantro are exported from Puerto Rico and Trinidad to meet the demands of West Indian, Latin American, and Asian immigrant communities in large cities of the United States, Canada, and the United Kingdom.

Culantro is widely used as a traditional herbal medicine in the Developing World. It is used as a tea for flu, colds, diabetes, constipation, fevers, chills, pneumonia, vomiting, diarrhea, constipation, and malaria.

CULINARY PORTRAIT

Culantro has a similar odor and flavor to that of cilantro, but is more pungent. Because of the similarities, the leaves are used interchangeably with cilantro in many food preparations, leading to confusion between the two. The flavor of culantro is also reminiscent of rau ram (see Chapter 96). Culantro leaves are widely used as a food flavoring and seasoning herb for meat and many other foods. Young leaves can be used as a garnish. Foods often seasoned with culantro include vegetable and meat dishes, chutneys, preserves, sauces, and snacks. Small quantities are sufficient to provide a pungent, unique aroma and characteristic flavor. In Asia, culantro is commonly used with or in place of cilantro in soups, noodle dishes, and curries. In Latin America, culantro is most frequently used in salsa. Those unfamiliar with culantro may sparingly add washed and chopped leaves up to the point that the taste seems to stop improving the flavor of a dish. Fresh leaves can be used in salads and chopped leaves in soups, stews, and curries. It has been suggested that in recipes calling for traditional cilantro, when substituting culantro the amount should be cut in half because culantro is much stronger. Culantro, unlike cilantro, retains its flavor and color after drying and so can be dried and stored. Although the plant is grown for its edible leaves, the roots are also sometimes consumed, especially as a condiment in soups and meat dishes.

CULINARY VOCABULARY

- *Sofrito* and *recaio* are mixtures of seasonings containing cilantro that are widely used in rice dishes, stews, and soups. Sofrito made particularly with culantro, cilantro, and chilies is often sold in West Indian markets of some large North American cities.

CURIOSITIES OF SCIENCE AND TECHNOLOGY

- Candied roots of the European *Eryngium mauritanum* L. became very popular as "kissing comfits" in the eighteenth century, when the candy was considered to be an aphrodisiac.
- The Miskito Indians of eastern Nicaragua rub crushed culantro leaves on the faces of people thought to have become possessed by supernatural beings. Patients are said to revive quickly due to the strong smell of the plant.
- In India, culantro root is eaten raw to treat scorpion stings.

FIGURE 30.1 Culantro (*Eryngium foetidum*). Left: flowering plant by B. Brookes. Right: fruiting plant by B. Flahey.

KEY INFORMATION SOURCES

Bergh, M.H. van den. 1999. *Eryngium foetidum* L. In *Plant resources of South-East Asia. 13. Spices*. Edited by C.C. de Guzman and J.S. Siemonsma. Backhuys, Leiden, the Netherlands. pp. 121–123.

Cardozo, E., Rubio, M., Rojas, L.B., and Usubillaga, A. 2004. Composition of the essential oil from the leaves of *Eryngium foetidum* L. from the Venezuelan Andes. *J. Essent. Oil Res.* 16:33–34.

Garcia, M..D., Saenz, M.T., Gomez, M.A., and Fernandez, M.A. 1999. Topical antiinflammatory activity of phytosterols isolated from *Eryngium foetidum* on chronic and acute inflammation models. *Phytotherapy Res.* 13:78–80.

Ignacimuthu, S., Arockiasamy, S., Antonysamy, M., and Ravichandran, P. 1999. Plant regeneration through somatic embryogenesis from mature leaf explants of *Eryngium foetidum*, a condiment. *Plant Cell Tissue Organ Cult.* 56:131–137.

Leclerq, P.A., Dung, N.X., Lo, V.N., and Toanh, N.V. 1992. Composition of the essential oil of *Eryngium foetidum* L. from Vietnam. *J. Essent. Oil Res.* 4:423–424.

Martins, A.P., Salgueiro, L.R., Proenca-da-Cunha, A., Vila, R., Canigueral, S., Tomi. F., and Casanova, J. 2003. Essential oil composition of *Eryngium foetidum* from S. Tome e Principe. *J. Essent. Oil Res.* 15:93–95.

Mohammed, M., and Wickham, L.D. 1995. Postharvest retardation of senescence in shado benni (*Eryngium foetidum*, L.) plants. *J. Food Quality*, 18:325–334.

Mohamed-Yasseen, Y. 2002. In vitro regeneration, flower and plant formation from petiolar and nodal explants of culantro (*Eryngium foetidum* L.). *In Vitro Cell Dev. Biol. Plant.* 38:423–426.

Morean, F. 1988. *Shado-beni: a popular Caribbean seasoning herb and folk medicine*. Trinidad Guardian, Port-of-Spain, Trinidad and Tobago. 18 pp.

Pino, J.A., Rosado, A., and Fuentes, V. 1997. Chemical composition of the seed oil of *Eryngium foetidum* L. from Cuba. *J. Essent. Oil Res.* 9:123–124.

Pino, J.A., Rosado, A., and Fuentes, V. 1997. Composition of the leaf oil of *Eryngium foetidum* L. from Cuba. *J. Essent. Oil Res.* 9:467–468.

Ramcharan, C. 1999. Culantro: a much utilized, little understood herb. In Perspectives on new crops and new uses. Edited by J. Janick. ASHS Press, Alexandria, VA. pp. 506–509.

Ramcharan, C. 2000. The effect of ProGibb sprays on leaf and flower growth in culantro (*Eryngium foetidum* L.). *J. Herbs Spices Med. Plants*, 7:59–63.

Reddy, C.S., Bhanja, M., and Raju,V.S. 2002. On the occurrence of *Eryngium foetidum* L. (Apiaceae) in Karnataka, with a note on its distribution and economic importance. *J. Econ. Taxon. Bot.* 26:199–200.

Saenz, M.T., Fernandez, M.A., and Garcia, M.D. 1997. Antiinflammatory and analgesic properties from leaves of *Eryngium foetidum* L. (Apiaceae). *Phytother. Res.* 11:380–383.

Sankat, C.K., and Maharaj, V. 1992. Preservation and processing of the shado beni (*Eryngium foetidum* L.) through dehydration. *Proceedings of the Sixth Annual Seminar on Agricultural Research: Sustainable Agriculture (Couva, Trinidad and Tobago, 3–4 Nov. 1992)*. National Institute of Higher Education (Research, Science and Technology), Trinidad and Tobago. pp. 282–298.

Sankat, C.K., and Maharaj, V. 1996. Shelf life of the green herb 'shado beni' (*Eryngium foetidum* L.) stored under refrigerated conditions. *Postharvest Biol. Technol.* 7:109–118.

Santiago-Santos, L.R., and Cedeno-Maldonado, A. 1991. Effect of light intensities on the flowering and growth of spiny coriander (*Eryngium foetidum* L.). *J. Agric. Univ. Puerto Rico*, 75:383–389 (in Spanish, English summary).

Wong, K.C., Feng, M.C., Sam, T.W., and Tan, G.L. 1994. Composition of the leaf and root oils of *Eryngium foetidum* L. *J. Essent. Oil Res.* 6:369–374.

Yasseen, M.Y. 2002. Plant regeneration from leaf blade, pedicle and shoot tip of culantro, *Eryngium foetidum* L., a medicinal plant. *J. Agric. Sci. Mansoura Univ.* 28:885–892.

Specialty Cookbooks

Díaz de Villegas, J.L. 2004. *Puerto Rico: grand cuisine of the Caribbean*. La Editorial de la Universidad de Puerto Rico, San Jaun, Puerto Rico. 292 pp.

Espinoza-Abrams, T. 2004. *Nicaraguan cooking: my grandmother's recipes*. Xlibris Corporation, Bloomington, IN. 146 pp.

McCausland-Gallo, P. 2004. *Secrets of Columbian cooking*. Hippocrene Books, New York. 251 pp.

31 Cumin and Black Cumin

This chapter features

Cumin (*Cuminum cyminum*)
Black cumin (*Nigella sativa*)

CUMIN

Family: Apiaceae (Umbelliferae; carrot family)

NAMES

Scientific Name: *Cuminum cyminum* L. (*Cuminum odorum* Salisb.)

- The English word "cumin" and the genus name *Cuminum* trace to classical names for the plant, including the Latin *cuminum* and the Greek *kýminon*, both probably of Semitic origin (similar ancient words: Hebrew *kammon*, Egyptian *kamnini*, Akkadian *kamunu*).
- Cumin is sometimes found under the old-fashioned spelling cummin.
- Recommended pronunciation of cumin: COME-in (other variations: KUH-mihn, KYOO-mihn, KOO-mihn).
- Cumin is also called Roman caraway. The name Roman caraway alludes to the facts that the Romans used ground cumin seed in the same way that we use pepper, and that caraway (*Carum carvi* L.) has largely replaced cumin in western nations. While Roman caraway is cumin, Roman cumin is caraway!
- The cultivated cumin variety 'Black' is well known as "black cumin," or occasionally as "black caraway." The seeds are darker, often rather small and sweet smelling, with a more complex flavor than ordinary cumin. Its taste is said to fall between that of cumin and caraway. Unfortunately, this cultivar is sometimes confused with *Nigella sativa*, also called black cumin, and discussed below. Cumin is sometimes called "white cumin" to distinguish it from "black cumin," that is, *N. sativa*. Confusion is also possible with *Bunium persicum*, an aromatic herb marketed in the Middle East as black cumin (for additional information, see Black cumin section).
- *Cyminum* in the scientific name *C. cyminum* is based on the Greek *kyminon*, cumin seed.

PLANT PORTRAIT

Cumin is a herbaceous plant, 20 to 80 cm (8–31 in.) tall, with threadlike leaves and white, purple or rose flowers. Commercial plants typically grow to a height of approximately 50 cm (20 in.) in flower, and produce many branches. Wild plants are not definitely known to exist, but the species likely originated in the region of the Mediterranean and the Near East, perhaps in Egypt or southeastern

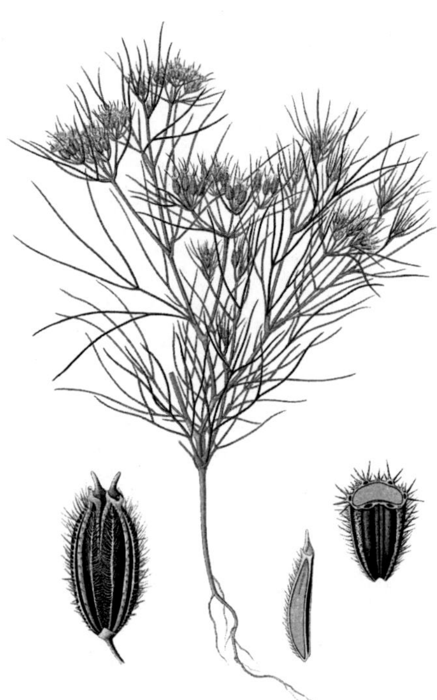

FIGURE 31.1 Cumin (*Cuminum cyminum*). (From Köhler, 1883–1914.)

Europe. Cumin from Iran, India, and the Middle East differs in seed color, quantity of essential oil, and flavor. Cumin seeds are thin, elongated-oval, yellowish brown, approximately 6 mm (1/4 in.) long, and ridged. The seeds resemble those of caraway and so are sometimes confused with the latter. Caraway seeds are more curved than those of cumin. Cumin was cultivated in antiquity. It was included in a list of medicinal plants recorded in ancient Egypt in 1550 BC. In the first century, it was referred to by the Roman scholar Pliny (23–79) as the best appetizer of all condiments. During the Middle Ages, cumin was popular in Europe. In 1419, it became a taxable commodity in England. Usage of cumin declined in Europe as the similar tasting caraway became more popular. In the last 300 years, cumin has been largely discarded from European cooking and is now chiefly used in Indian cooking. Cumin is cultivated primarily in China, India, Morocco, Cyprus, Turkey, Iran, and southern Russia. It is extensively consumed in India, North Africa, the Middle East, Latin America (especially Mexico), and the United States (where it is used particularly in Hispanic and Asian foods). In addition to the culinary uses noted in the next section, the oil of cumin is used commercially in perfumery.

Culinary Portrait

Cumin seeds impart a strong, characteristic odor that has been described as strongly penetrating, irritating, fatty, overpowering, curry-like, heavy, spicy, warm, and persistent, even after drying. The aroma has been compared with the bed-bug odor of coriander, disliked by many. The smell has also been described by some as "resembling musty caraway" and "like dirty socks," but those accustomed to cumin find the plant quite agreeable. More charitably, the flavor has been characterized as slightly bitter, sharp, warm, pungent, and persistent. The flavor is indeed not unlike caraway, although slightly bitter and very hot. People often develop a taste for cumin, provided that it is used sparingly at first. Cumin seeds are available in whole and ground form, the former retaining flavor longer. The seeds require lengthy cooking and for this reason are often sold in small quantities in powdered form, which makes it easier to add the right amount. To bring out the flavor, experienced cooks may crush and roast the seeds before using them, or sauté them briefly in fat or oil and then crush them. In India, cumin is generally dry roasted to bring out its flavor. Cumin is a key ingredient of commercial curry and chili powders. The spice is also added to soups, stews, meats, pickles, and sauerkraut. Cumin complements cheddar cheese canapés, salad dressings, chutneys, devilled eggs, chicken, lamb, and vegetables. It adds zest to mild starchy foods like rice, potatoes, and bread. The seeds from this plant are an important ingredient in Mexican, Middle Eastern, Indian, and, more recently, American cuisine. The Dutch, French, and Swiss use it to flavor cheese; the Germans use it to flavor breads and sauerkraut; and the Spaniards combine it with cinnamon and saffron in stews. In Eastern Europe, cumin is often used in breads, delicatessen meats, and some cheeses. It is an essential ingredient and predominant flavor in most Egyptian, Indian, and Turkish curries and is often used in couscous dishes. The combination of ground cumin and coriander leaves is responsible for the characteristic smell of much Indian food. Cumin is also important in cooking throughout Latin America. It is used extensively in Mexican foods. In some Arab countries, a liquid paste made of ground cumin, pepper, and honey is considered to be an aphrodisiac. In Texas, it is an ingredient of chili con carne.

The oil of cumin is an essential ingredient of kummel liqueurs and some German baked goods. It is also added to soups and stews and makes pungent mango pickles and green mango chutney. The oil is also used in condiments, sausages, meat sauces and cheeses.

Culinary Vocabulary

- Kümmel is a caraway- or cumin-flavored liqueur. The name is derived from the German word for cumin, which in turn comes from the Latin *cuminum.*
- "Cumin water" (*jira pani*) is a tart, cumin-flavored drink of northern India.

Curiosities of Science and Technology

- As early as 5000 BC, the ancient Egyptians preserved the bodies of their kings by mummifying them in cumin, anise, and marjoram.
- Roman scholar Pliny (23–79) recommended smoking cumin seeds to cultivate "an impressive scholastic pallor." Students were said to consume cumin to acquire paleness to make it appear that they had spent long hours in study.
- The heads of villains who were hanged, drawn, and quartered in old times were sometimes parboiled with salt and cumin seed (to keep them from putrefying too quickly) before they were exhibited in public to discourage others from carrying out the same crimes.
- Cumin acquired some interesting symbolic meanings in language idioms. In Spanish, something of trifling importance is likened to a cumin seed. The Mediterranean saying "cumin splitter" refers to people who are so greedy that they would take the time to split a cumin seed to retain part of one. The expression traces to the ancient Greeks and Romans, and became popular with English writers through the seventeenth to the early nineteenth centuries. The phrase is comparable with the modern English "skinning a flint," referring to a "skinflint," that is, a very stingy person.
- "Quitrent" was a form of tax paid by large landowners to the king in England during the early Middle Ages in lieu of providing soldiers for service in the national army. This was often paid in spices, and the custom has persisted until modern times. In 1937, King George VI was presented with 1 lb. of cumin as the quitrent owed by the Duchy of Cornwall.
- The seeds of cumin were once an ingredient in snuff tobacco.
- Cumin is widely used in curry and chili powders, and its high content of cuminaldehyde contributes to the burning sensation of these condiments. As cuminaldehyde is practically insoluble in water, drinking iced water does not cool the burning sensation caused by eating a hot curry or chili-flavored dish. Drinking alcohol (e.g., beer) or whole milk, both of which dissolve cuminaldehyde, should ease the stinging in the mouth from eating too much curry or chili.

Key Information Sources

Amin, G. 2001. Cumin. In *Handbook of herbs and spices*. Edited by K.V. Peter. Woodhead Publishing, Cambridge, U.K. pp. 164–167.

Badr, F.H., and Georgiev, E.V. 1990. Amino acid composition of cumin seed (*Cuminum cyminum* L.). *Food Chem.* 38:273–278.

Baser, K.H.C., Kürkçüoglu, M., and Özek, T. 1992. Composition of the Turkish cumin seed oil. *J. Essent. Oil Res.* 4:133–138.

Behera, S., Nagarajan, S., and Jagan-Mohan-Rao, L. 2004. Microwave heating and conventional roasting of cumin seeds (*Cuminum cyminum* L.) and effect on chemical composition of volatiles. *Food Chem.* 87:25–29.

Bera, M.B., Shrivastava, D.C., Singh, C.J., Kumar, K.S., and Sharma, Y.K. 2001. Development of cold grinding process, packaging and storage of cumin powder. *J. Food Sci. Technol.* 38:257–259.

Boxer, M., Roberts, M., and Grammar, L. 1997. Cumin anaphylaxis: a case report. *J. Allergy Clin. Immunol.* 99:722–723.

British Standards Institution. 1990. *Herbs and spices ready for food use. Part 5. Specification for dried cumin (whole and ground).* British Standard 7087. British Standards Institution, London. 5 pp.

Brody, J.E. 1998. Adding cumin to the curry: a matter of life and death. *J. Med. Food.* 1:57–60.

Chattapadhyay, D., and Sharma, A.K. 1990. Chromosome studies and estimation of nuclear DNA in different varieties of *Cuminum cyminum* L. and *Carum copticum* Benth. and Hook. *Cytologia*, 55:631–637.

El-Ballal, A.S.I. 1987. Cryptic polymorphism of sex expression in cumin (*Cuminum cyminum*, L.). *Acta Hortic.* 208:197–207.

El-Sawi, S.A., and Mohamed, M.A. 2002. Cumin herb as a new source of essential oils and its response to foliar spray with some micro-elements. *Food Chem.* 77:75–80.

Fageria, N.K., Bajpai, M.R., and Parihar, R.L. 1972. Effect of nitrogen, phosphorus and potassium fertilization on yield and yield attributing characters of cumin crop. *Z. Pflanzenernahr. Bodenk,* 132:30–37.

Hemavathy, J., and Prabhakar, J.V. 1988. Lipid composition of cumin, *Cuminum-cyminum* L., seeds. *J. Food Sci.* 53:1578–1579.

Jansen, P.C.M. 1999. *Cuminum cyminum* L. In *Plant resources of South-East Asia. 13. Spices*. Edited by C.C. de Guzman and J.S. Siemonsma. Backhuys, Leiden, the Netherlands. pp. 108–111.

Perveen, S., and Zakaullah. 1979. Diseases of cumin *Cuminum cyminum*—a review. *Pakistan J. Forest.* 29:195–198.

Shah, J.J., and Unnikrishnan, K. 1969. The shoot apex and the ontogeny of axillary buds in *Cuminum cyminum* L. *Aust. J. Bot.* 17:241–253.

Sharma, R.K., Bhati, D.S., and Sharma, M.M. 1999. Cumin. In *Tropical horticulture*, vol. 1. Edited by T.K. Bose, S.K. Mitra, A.A. Farooqui, and M.K. Sadhu. Naya Prokash, Calcutta, India. pp. 740–745.

Shetty, R.S., Singhal, R.S., and Kulkarni, P.R. 1994. Antimicrobial properties of cumin. *World J. Microbiol. Biotech.* 10:232–233.

Singh, K.K., and Goswami, T.K. 1996. Physical properties of cumin seed. *J. Agric. Eng. Res.* 64:93–98.

Singh, K.K., and Goswami, T.K. 1997. Studies on grinding of cumin seed. *J. Food Process. Eng.* 22:175–190.

Singh, K.K., and Goswami, T.K. 2000. Thermal properties of cumin seed. *J. Food Eng.* 45:181–187.

Tassan, C.G., and Russell, G.F. 1975. Chemical and sensory studies on cumin. *J. Food Sci.* 40:1185–1188.

Tawfik, A.A., and Noga, G. 2001. Priming of cumin (*Cuminum cyminum* L.) seeds and its effect on germination, emergence and storability. *J. Appl. Bot.* 75:216–220.

Wijesekera, R.O.B., Nethsingha, C., and Paskaranathan, U. 1975. *Cumin.* Ceylon Institute of Scientific and Industrial Research, Colombo, Sri Lanka. 23 pp.

Specialty Cookbooks

Gandhi, J. 2002. *Indian flavor: curry leaves, cumin seeds, and the spice of healthy cooking*. Laurel Glen Pub., San Diego, CA. 127 pp. [Also published as Gandhi, J. 2002. *Curry leaves and cumin seeds*. New Holland, London. 127 pp.]

Ramos-Elorduy, 1998. J. *Creepy crawly cuisine: the gourmet guide to edible insects*. Park Street Press Rochester, VT. 150 pp. [Presents numerous recipes with cumin, which apparently goes well with insects.]

BLACK CUMIN

Family: Ranunculaceae (buttercup family)

Names

Scientific Name: *Nigella sativa* L.

- "Black cumin" is so named for the blackness of the seeds and their cumin-like appearance.
- Black cumin is also known as black caraway, black seed, devil-in-the-bush (devil in a bush), fennel flower (fennel-flower), fitch, love-in-a-mist, nigella, nutmeg flower, Roman coriander, and small fennel.
- As pointed out earlier, black cumin can be confused with cumin, *Cuminum cyminum*. Particularly confusing, the cultivated cumin variety 'Black' is also commonly called black cumin.
- The Middle Eastern *Bunium persicum* (Boiss.) B. Fedtsch. is also known as black cumin (as well as black caraway). *Bunium persicum* grows wild in the Middle East but is of limited importance.
- Fennel flower and small fennel are names that could lead to confusion with fennel, *Foeniculum vulgare* Mill. In fact, the resemblance of the finely cut foliage of black cumin to that of fennel is why black cumin in often called fennel flower.
- The old name Roman coriander reflects the popularity of black cumin in cooking among the ancient Romans.
- The alternative name nutmeg flower reflects similarity with the strong, agreeable aromatic odor of nutmeg.
- "Kalongi," a Middle Eastern name (Marathi and Punjabi, *kalongi*; Hindi, *kalounji*) meaning "black seed," is frequently used as a synonym for black cumin and is sometimes equated with "onion seed"; rarely, "onion seed" is also used in English to mean black cumin. The name kalongi is commonly used in Indian and Iranian cooking, and spice under this name is sometimes found in Indian and some Middle Eastern markets in Western countries.
- *Nigella damascena* L. is a beautiful garden ornamental known as "virgin in the green" in Europe and "love-in-a-mist" in North America. The latter name is also applied to black cumin. The name "love-in-a-mist" is a reference to the bluish flowers that are almost hidden among the tangled leaves of the plant.
- The genus name *Nigella* is a diminutive of *niger* (black) for the blackness of the seeds (based on the feminine of the Latin *nigellus*, blackish).
- *Sativa* in the scientific name *N. sativa* is Latin for sown or cultivated.

Plant Portrait

Black cumin is native to central and southern Europe, North Africa, and western Asia. It typically grows 30 to 45 cm (12–18 in.) or occasionally to 60 cm (2 ft.) or more in height and produces whitish-blue flowers 2 to 3 cm (about an inch) across. The triangular, dull-black seeds are the economic part of the plant. The seeds are 2–3 mm (1/16–3/32 in.) long and have a rough surface. They are produced in odd, toothed, or horned fruits. The Sanskrit name indicates that black cumin was used in India at a very early date. It was also known to the ancient Hebrews, Greeks, and Romans. The seeds preceded black pepper as a major spice of the Near East and were once used as a substitute for pepper. Black cumin is used as a spice in Egypt, the Middle East, India, and to a small extent in Europe. It is grown in the Mediterranean area and in the Orient, whereas cultivation in Europe has

FIGURE 31.2 Black cumin (*Nigella sativa*). Left: plant and cross-sectioned fruit. (From Köhler, 1883–1914.) Upper right: seed; lower right: fruit with horns, by M. Jomphe.

become insignificant. This spice is a minor condiment, much more popular in Asia and southern Europe than in North America.

Culinary Portrait

The seeds of black cumin are used for seasoning and are used in spice mixes. The odor of crushed seeds has been described as like lemons with a faint suggestion of carrots while the taste is strong, pungent, peppery, rather oddly aromatic, nutty, and said to be like a cross between poppy seeds and pepper. The seeds are added to curries, pickles, cheeses, eggs, fish, poultry, meats, game, pickles, conserves, fruit pies, and confections, particularly cookies, rolls, and bread. The seeds are often found atop Jewish rye bread and pumpernickel and indeed are often sprinkled on top of baked goods in the Middle East. The seeds are also used to flavor a variety of vegetable dishes.

Culinary Vocabulary

- French cooks use black cumin seeds under the name quatre épices or toute épice. However, in French, the name *quatre épices* (or *quatre-épices*) as well as the name *toute épice* (*toute-épice*, *toute-épices*) usually designate allspice (which see). Also, quatre épices designates a standard French "four-spice" blend, based on pepper, usually with nutmeg, cloves, and ginger, sometimes with cinnamon or allspice, and commonly used in charcuterie and in dishes requiring long simmering, such as stews. (Despite the name, quatre épices often has more than four spices.)
- *Choereg* are semi-sweet rolls widely found in Middle Eastern breakfasts; black cumin is an essential ingredient.

Curiosities of Science and Technology

- The translation of names of plants in ancient documents, particularly religious literature, is often difficult. For example, it has been contended that both cumin and black cumin are referred to in Isaiah 28:27: " ... but the fitches are beaten out with a staff, and cummin with a rod." The translation of the Hebrew word *ketyaeh* or *ketzach* as "fitches" instead of vetches (species of *Vicia*) in places in the Authorized Version of the Bible has led to confusion. In this stanza, however, it has been argued that fitches mean black cumin. In other translations, the fitches of this extract have been rendered "caraway" or "dill."
- Charlemagne, 742–814, was a Frankish king (the Franks were a Germanic people) who came to be Emperor of the West. The "Capitulare" was an edict from Charlemagne to all those governed within his realm, which included a list of approximately 90 plants to be grown wherever possible. Black cumin (then known as "gith") was among the edible plants that were ordered to be planted.
- The folk usage of plants as medicinals often traces to ancient times, and once a plant gains a reputation for having curative powers, it often continues to be used, sometimes down to the present. The Roman surgeon and herbalist Dioscorides (50–100) thought that black cumin mixed with vinegar cured dog and crocodile bites. In more recent periods, black cumin seed has been extensively used in folk medicine in Europe and Asia, as a tonic, to increase flow of body fluids, including menstrual flow, lactation, and discharge of pus and urine, and to treat nausea, vomiting, intestinal worms, constipation, headache, blood poisoning, enlarged liver, fever, chest complaints, and other problems. The seeds were once held in the mouth to treat toothache. It was said that Malaysians possessed enough faith in the power of black cumin seed to add it to almost anything. Black cumin is still used in folk medicine in several countries of the Old World.
- Black cumin seeds have been stored with clothing to repel insects.
- Black cumin seeds are believed to be among the earliest of space-traveling herbs, accomplished during the voyage of the Russian satellite Sputnik III in 1958.

Key Information Sources

Ahmed, N.U., and Haque, K.R. 1986. Effect of row spacing and time of sowing on the yield of black cumin (*Nigella sativa*). *Bangladesh J. Agric.* 11:21–24.

Al-Ghamdi, M.S. 2001. The anti-inflammatory, analgesic and antipyretic activity of *Nigella sativa*. *J. Ethnopharmacol.* 76:45–48.

Al-Jassir, M.S., and Rezk, M.A. 1992. Seed anatomy of black cumin, *Nigella sativa* L. *Bull. Fac. Agric. Cairo Univ.* 43:505–514.

Babayan, V.K., Koottungal, D., and Halaby, G.A. 1978. Proximate analysis, fatty acid and amino acid composition of *Nigella sativa* L. seeds. *J. Food Sci.* 43:1314–1315, 1319.

Boselah, N.A.E. 1995. Seed germination of *Nigella sativa* L. *Ann. Agric. Sci. Moshtohor*, 33:793–800.

Burits, M., and Bucar, F. 2000. Antioxidant activity of *Nigella sativa* essential oil. *Phytother. Res.* 14:323–328.

D'Antuono, L.F., Moretti, A., and Lovato, A.F.S. 2002. Seed yield, yield components, oil content and essential oil content and composition of *Nigella sativa* L. and *Nigella damascena* L. *Ind. Crop Prod.* 15:59–69.

Ghosh, D., Roy, K., and Mallik, S.C. 1981. Effect of fertilizers and spacing on yield and other characters of black cumin (*Nigella sativa* L.). *Indian Agric.* 25:191–197.

Hajhashemi, V., Ghannadi, A., and Jafarabadi, H. 2004. Black cumin seed essential oil, as a potent analgesic and antiinflammatory drug. *Phytother. Res.* 18:195–199.

Hitchings, C., and Bird, J. 1993. *Nigella sativa*. *Herbarist*, 59:20–32.

Ipor, I.B., and Oyen, L.P.A. 1999. *Nigella sativa* L. In *Plant resources of South-East Asia. 13. Spices*. Edited by C.C. de Guzman and J.S. Siemonsma. Backhuys, Leiden, the Netherlands. pp. 148–151.

Jünemann, M., and Luetjohann, S. 1998. *The three great healing herbs: tea tree, St. Johns wort, and black cumin: important herbs for health and wellness and as basics in your herbal first aid kit*. Lotus Light, Twin Lakes, WI. 146 pp.

Menounos, P., Staphylakis, K., and Gegiou, D. 1986. The sterols of *Nigella sativa* seed oil. *Phytochemistry*, 25:761–763.

Mozaffari, F.S., Ghorbanli, M., Babai, A., and Sepehr, M.F. 2000. The effect of water stress on the seed oil of *Nigella sativa* L. *J. Essent. Oil Res.* 12:36–38.

Nergiz, C., and Ötle , S. 1993. Chemical composition of *Nigella sativa* L. seeds. *Food Chem.* 48:259–161.

Schleicher, P., and Saleh, M. 2000. *Black cumin: the magical Egyptian herb for allergies, asthma, and immune disorders*. Healing Arts Press, Rochester, VT. 90 pp.

Shah, S., and Ray, K.S. 2003. Study on antioxidant and antimicrobial properties of black cumin (*Nigella sativa* Linn). *J. Food Sci. Technol.* 40:70–73.

Sorvig, K. 1983. The genus *Nigella*. *Plantsman*, 4:229–235.

Steinmann, A., Schaetzle, M., Agathos, M., and Breit, R. 1997. Allergic contact dermatitis from black cumin (*Nigella sativa*) oil after topical use. *Contact Dermatitis*, 36:268–269.

Takruri, H.R.H., and Dameh, M.A.F. 1998. Study of the nutritional value of black cumin seeds (*Nigella sativa* L). *J. Sci. Food Agric.* 76:404–410.

Üstun, G., Kent, L., Çekin, N., and Civelekoglu, H. 1990. Investigation of the technological properties of *Nigella sativa* (black cumin) seed oil. *J. Am. Oil Chem. Soc.* 67:958–960.

Zaoui, A., Cherrah, Y., Mahassini, N., Alaoui, K., Amarouch, H., and Hassar, M. 2002. Acute and chronic toxicity of *Nigella sativa* fixed oil. *Phytomedicine*, 9:69–74.

Specialty Cookbooks

Luetjohann, S. 1998. *The healing power of black cumin: a handbook on Oriental black cumin oils, their healing components, and special recipes*. Lotus Light Pub., Twin Lakes, WI. 156 pp.

32 Cycads

There are approximately 250 species of cycads, classified into three families, with a total of 11 genera, each genus with a restricted geographical range.

Family: *Cycadaceae*

Genus *Cycas* (approximately 90 species, chiefly Australian and Indo-Chinese).

Family: *Stangeriaceae*

Genus *Stangeria* (1 species, in Southern Africa)
Genus *Bowenia* (2 species, in eastern Queensland, Australia)

Family: *Zamiaceae*

Genus *Dioon* (11 species, in Mexico, Honduras, and Nicaragua)
Genus *Encephalartos* (62 species, in Africa)
Genus *Macrozamia* (38 species, in Australia)
Genus *Lepidozamia* (2 species, eastern Australia)
Genus *Ceratozamia* (at least 16 species, in Mexico, Guatemala, and Belize)
Genus *Microcycas* (1 species, in Cuba)
Genus *Zamia* (more than 50 species, in South, Central, and North America)
Genus *Chigua* (2 species in Columbia, South America)

NAMES

- The genus name *Cycas* is based on the Greek *koikas*, a kind of palm. The species are often known as sago or false sago.
- In China, *Cycas* is known as iron-tree and Phoenix-tail canna, and its seeds are referred to as phoenix eggs (according to legend, the phoenix bird is reborn eternally from flame, and the name phoenix egg refers to the plant's ability to survive fire).
- *Cycas* was used as a food source during the economic depression of 1958–1962 in China, at which time it was known as "western rice." In the province of Guangxi, it was also known as "immortal rice."
- The genus name *Stangeria* commemorates Dr. Max Stanger, Surveyor General of Natal province, South Africa.
- The genus name *Bowenia* commemorates Sir George Ferguson Bowen (1821–1899), first governor of Queensland, Australia. The species are often called Byfield fern.
- The genus name *Dioon* is based on the Greek *dis*, two + *oon*, eggs, referring to the paired seeds.
- The genus name *Encephalartos* is based on the Greek *en*, in + *cephale*, head + *artos*, bread, referring to the flour obtained from the trunks of some species by indigenous people.
- The genus name *Macrozamia* is based on the Greek *makros*, large + *Zamia*, the genus of cycads.

- The genus name *Lepidozamia* is based on the Greek *lepidos*, a scale + *Zamia*, the genus of cycads. The "scale" in the name is a reference to the scalelike leaf bases that clothe the short stout stem.
- The genus name *Ceratozamia* is based on the Greek *ceratos*, a horn + *Zamia*, the genus of cycads, referring to the horned seed-bearing parts of the plant.
- The genus name *Microcycas* is based on the Greek *micro*, small, + Cycas, referring to the appearance of the plant being like a small *Cycas*.
- The genus name *Zamia* is based on the Greek *rom* the Greek *azaniae* a pine cone.
- The genus name *Chigua* is based on a Spanish version of the indigenous Indian name for cycads in Panama and Colombia.

PLANT PORTRAIT

Cycads are primitive palm-like or fern-like trees or shrubs and constitute one of the several fundamental groups of plants that reproduce by seeds (by contrast, plants that reproduce by spores include ferns, mosses, algae, and others). Most seed plants today belong to the Angiosperms, estimated to have between 300,000 and 400,000 species, but there are also four small, ancient groups of plants that reproduce by seeds. In addition to angiosperms and cycads, there are the Gnetophyta (genus *Welwitschia*), ginkgo (see Chapter 39; this is a group with just one living species), and the conifers (pines, spruces, etc.). Cycads are known to have lived more than 250 million years ago, before dinosaurs roamed the earth. The Jurassic Period (208 to 146 million years ago) is sometimes called the "Age of Cycads" because they were so common then. Cycads (or at least their ancient extinct relatives) were the main feature of the dinosaurs' landscape. Petrified cycads are frequently found in the same rocks as dinosaur bones, and the plants probably were eaten by some of the herbivorous dinosaurs. Although abundant all over the world in past times, the cycads are now greatly reduced in both numbers and distribution. All cycads are tropical or subtropical. Cycad plants are long lived and slow growing.

There are both male and female trees, and these have a stout, generally unbranched trunk, and a crown of large, evergreen, leathery, fernlike leaves that are compound (i.e., split into leaflets), and often with stiff sharp spines. Most species are less than 3 m (10 ft.) tall, and many are less than 2 m (6½ ft.). Some plants of *Zamia pygmaea* Sims are less than 25 cm (10 in.) tall, whereas the tallest cycad, *Lepidozamia hopei* Regel, grows to 18 m (59 ft.). The stem or trunk of cycads is soft, made up mostly of starch-rich storage tissue, and sometimes the stem is underground and tuberous. The seeds are organized into cones. Cycad seeds are large, with a fleshy outer coat over a hard, stony layer. The roots are unusual. They have the appearance of coral, and so are called coralloid roots. They are contractile, that is, they contract as the plant grows, drawing the sensitive growing point of young plants below the soil surface, which provides protection against drought and fires that are a frequent feature of many cycad habitats. The roots are the host of primitive bacteria which capture atmospheric nitrogen, providing this nutrient which is deficient in the soils of many cycad habitats. These bacteria are photosynthetic, requiring light to live, so the roots grow up out of the soil, rather than down into it, thus exposing the bacteria to the light they need.

Cycads have been used as a food source virtually everywhere that they occur. In Australia where there are many species, cycads, particularly of the genus *Cycas*, have been consumed for more than 4000 years. *Cycas* starch is eaten in some parts of Yunnan province, China, and beer is also made from the starch. In South Africa, the stem of *Encephalartos* is used to make a crude bread, called Hottentot or Kaffir bread by the Dutch settlers (the plant was known as bread tree), and the starch has been employed for brewing beer, and for making porridge. The seeds of the Mexican cycad *Dioon* are an important source of nutrition for some indigenous peoples. Because they grow slowly, cycads are not cultivated commercially for food, although they are very popular house and garden plants. A factory was set up in New South Wales Australia to produce burrawang starch (burrawang is *Macrozamia communis* L.A.S. Johnson) commercially for use as an adhesive paste, but this

FIGURE 32.1 Chestnut dioon (*Dioon edule*). (From Gartenflora, 1881, vol. 30.)

FIGURE 32.2 Suurberg cycad (*Encephalartos longifolius*). (From Curtis, 1856, vol. 82, plate 4903.)

FIGURE 32.3 Burrawong (*Macrozamia spiralis*). (From Gartenflora, 1874, vol. 23.)

venture proved to be inviable. (The names burrawang and burrawong are applied to several species of *Macrozamia*.)

Cycads contain neurotoxins or nerve poisons that have caused serious poisoning of people and livestock. It is remarkable that many indigenous peoples in different parts of the world have independently discovered how to detoxify cycads. For example, native Africans remove the starchy pith from the trunk and bury it in the ground for 2 months while it ages and the poisons are deactivated. The Seminole Indians of Florida and the South Pacific Islanders repeatedly soak the starch in water to leach out the toxins. Even with such knowledge, however, poisoning occurs. It is strongly suspected that "Guam disease" (a combination of the symptoms of Alzheimer's disease, Parkinson's disease, and motor neuron disease) on the island of Guam is due to eating improperly processed seeds of *Cycas*. (Research by Cox et al. (2003) suggested that consumption of fruit bats that have eaten cycad seeds has been responsible for the illness.) Grazing by livestock on leaves of certain species of *Macrozamia* in Australia and *Zamia* in the Americas has resulted in partial or total paralysis of the hind legs, a condition known as the "wobbles" or "staggers."

Cycads have acquired cultural and symbolic significance. In China, a cycad in bloom is a sign of luck and happiness. In the past, cycads symbolized nobility and authority, and even today they are planted in front of government buildings and banks and displayed at important occasions. By tradition, cycads are planted in front of graves, and cycad fronds are sometimes inserted in the door after a funeral to expel evil. Cycads are always planted or cultivated outside of Buddhist temples as they are said to reflect the austerity and constancy of Buddhism and to signify immortality.

In addition to being collected for food and for ornament, cycad leaves are popular for funeral wreaths and Palm Sunday arrangements because they stay green and fresh-looking long after they have been cut from the plant. Unfortunately, because of habitat destruction and overcollecting, all species of cycads are endangered. To help conserve them, cycads are covered by the import/export restrictions of the Convention on International Trade in Endangered Species.

CULINARY PORTRAIT

Cycads contain a large amount of starch in their roots, stem (all cycads contain 20%–60% starch in the central pith), and seeds, and probably starch can be obtained from all cycad species after appropriate detoxification. The plants have been used as a source of food starch particularly during times when normal food supplies have been curtailed, that is, as a "famine food." However, cycad starch has little commercial value.

Preparation of edible starch from cycads requires special techniques to neutralize toxins that are present. To get rid of, or inactivate the poisonous constituents, cycad tissues are processed by a combination of some or all of washing, fermentation, cooking, and aging.

Young cycad leaves are cooked and eaten in Malaysia and Indonesia. They are said to be tender, with a flavor combining cabbage and asparagus.

Culinary Vocabulary

- "Arrowroot starch" is generally obtained from the underground stems of species of *Maranta* (see Chapter 7). However, "Florida arrowroot starch," sometimes also called sago starch, comes from a cycad, *Zamia pumila* L. (*Zamia floridana* A. DC.), called wild sago, comtie, and coontie. It grows in Florida and southern Georgia. In the nineteenth century, there was a short-lived industry based on production of Florida arrowroot starch, used to make baby food, spaghetti, biscuits, and laundry starch.
- "Sago starch" is generally obtained from sago palm (which see) as well as other palms. However, some cycads (including species of *Cycas*, *Zamia*, and *Macrozamia*) are sometimes called sago palms (although they are not true palms) and are used to produce sago starch. Starch is also extracted from the trunk of *Cycas revoluta* Thunb., which is called sago palm and king sago palm, and is widely grown as an ornamental (indeed it is the most common species of cycad in northern greenhouses and nurseries).
- *Sotetsu* is the name under which cycad starch is marketed in Japan. *Nari-miso* is miso made from *Cycas revoluta* (miso is a traditional Japanese fermented product made from rice, barley and/or soybean, and salt). The safety of such preparations has been under investigation.

CURIOSITIES OF SCIENCE AND TECHNOLOGY

- Four thousand years ago, the Etruscans of Italy used fossilized cycad trunks as funeral monuments, placing them on top of their tombs.
- The Taino Indians of the Caribbean Islands offered zamia bread to Christopher Columbus and his men when they arrived in the New World, but the Spanish explorers were repelled by the method of preparation. The Indians left grated balls of the flour in the sun for 2 to 3 days, allowing them to turn black and become infested with worms. These were interpreted as signs that the otherwise poisonous preparations had become safe to eat.
- Coal miners in the eastern United States once used fossilized trunks of cycads that they found as doorstops and good luck charms.
- Gender-linked traditions of consuming cycads are known from Australia. During certain ceremonies, a special cycad cake was made only for men. Conversely, fighting men were prohibited from eating a special bread made from cycad starch.
- Some cycads are known to have lived for more than 1000 years. A 1000-year-old cycad is celebrated on a commemorative stamp in Japan.
- English explorer Captain James Cook (1728–1779), among many other accomplishments, charted the coasts of Australia. Not knowing how the aborigines detoxified cycad seeds, his crew suffered greatly from eating the seeds.

- Florida arrowroot (*Zamia pumila*), the only species of cycad native to the United States, is poisonous until boiled. The Seminole Indians of Florida learned how to prepare non-poisonous "seminole bread" from the tree. During the American Civil War, some Federal soldiers died after eating the starchy roots, not having become acquainted with the careful washing treatment used by the Seminoles.
- South African General J.C. Smuts (1870–1950) during 1901 and 1902 commanded Boer guerrilla forces acting against the British. While camped in the Zuurberg mountains, food supplies ran low, and he and his men turned to eating fresh seeds of the cycad known as "'Hottentot's bread." Soon they were groaning and retching on the ground in agony. Smuts ordered that the men who were most affected were to be tied to their saddles, and he and his men went deep into the mountains to escape the British. They were lucky to survive, as cases of death from consuming cycads that have not been properly processed are widely recorded.
- A seed cone of the South African cycad *Encephalartos caffer* (Thunb.) Lehm. (the source of "Kaffir bread") weighed 42 kg (93 lb.).
- The most famous fossil cycad locality in the United States is near Minnekahta, in the Black Hills of South Dakota. An entire fossil cycad forest once thrived there, and many specimens were dug up and sent to museums around the world. At one time, the U.S. government designated the area as "Fossil Cycad National Monument." However, vandalism and illegal collecting denuded the area so badly that it was closed as a national monument.
- Plants are not "hot blooded" like some animals, but nevertheless some plants actually generate heat, mostly in reproductive organs. Some cycads can maintain their mature male cones 20°C (36°F) or more above the ambient temperature. This helps evaporate chemicals that attract certain insects (which pick up the pollen), but humans consider the smell to be foul. In Guam, it is considered antisocial to have such bad-smelling male cycads in one's garden.

KEY INFORMATION SOURCES

Brenner, E.D., Stevenson, D.W., and Twigg, R.W. 2003. Cycads: evolutionary innovations and the role of plant-derived neurotoxins. *Trends Plant Sci.* 8:446–452.

Castillo, G.C., and Rico, G.V. 2003. The role of macrozamin and cycasin in cycads (Cycadales) as antiherbivore defenses. *J. Torrey Bot. Soc.* 130:206–217.

Chamberlain, C.J. 1919. *The living cycads.* The University of Chicago Press, Chicago, IL. 172 pp.

Chen, C.J. 1999. *Biology and conservation of cycads: proceedings of the fourth international conference on cycad biology held in Panzhihua, Sichuan, China, 1–5 May 1996.* International Academic Publishers, Beijing. 415 pp.

Cox, P.A., Banack, S.A., and Murch, S.J. 2003. Biomagnification of cyanobacterial neurotoxins and neurodegenerative disease among the Chamorro people of Guam. *Proc. Natl. Acad. Sci U.S.A.* 100:13380–13383.

Dehgan, B. 1999. Propagation and culture of cycads: a practical approach. *Acta Hortic.* 486:123–131.

Donaldson, J.S. (for the International Union for the Conservation of Nature and Natural Resources). 2003. *Cycads.* Status survey and conservation action plan. Gland, Cambridge, U.K. 86 pp.

Gilks, C.F. 1988. Cycads and sago. *Lancet*, 1(8578):181–182.

Gutiérrez, M.M., and Paull, R.E. 2008. *Zamia chigua*, chigua. In *The encyclopedia of fruit & nuts.* Edited by J. Janick and R.E. Paull. CABI, Wallingford, Oxfordshire, U.K. p. 921.

Hill, K.D., Chase, M.W., Stevenson, D.W., Hills, H.G., and Schutzman, B. 2003. The families and genera of cycads: a molecular phylogenetic analysis of Cycadophyta based on nuclear and plastid DNA sequences. *Int. J. Plant Sci.* 164:933–948.

Holtzman, G. 1991. Cycads: the "dinosaurs" of plants. *Bull. Natl. Trop. Bot. Gard.* 21(4):17–19.

Jones, D.L. 2002. *Cycads of the world: ancient plants in today's landscape.* 2nd ed. Smithsonian Institution Press, Washington, D.C. 346 pp.

Keng, H. 1972. Cycad seed as food in Malaya. *Malayan Nat. J.* 25:101–103.

Kira, K., and Miyoshi, A. 2000. Utilization of sotetsu (*Cycas revoluta*) in the Amami Islands. *Res. Bull. Kagoshima Univ. For.* 28:31–37.

Laqueur, G.L. 1977. Oncogenicity of cycads and its implications. *Adv. Mod. Toxicol.* 3:231–261.

Lindstrom, A.J. 2008. Cycadaceae, *Cycas* spp., cycads. In *The encyclopedia of fruit & nuts*. Edited by J. Janick and R.E. Paull. CABI, Wallingford, Oxfordshire, U.K. pp. 314–319.

Norstog, K., and Nicholls, T.J. 1997. *The biology of the cycads*. Comstock Pub. Associates, Ithaca, NY. 363 pp.

Rai, H.S., O'Brien, H.E., Reeves, P.A., Olmstead, R.G., and Graham, S.W. 2003. Inference of higher-order relationships in the cycads from a large chloroplast data set. *Mol. Phylogenet. Evol.* 29:350–359.

Read, R.W., and Solt, M.L. 1986. *Bibliography of the living cycads (annotated).* Harold L. Lyon Arboretum, Honolulu, HI. 166 pp.

Schneider, D., Wink, M., Sporer, F., and Lounibos, P. 2002. Cycads: their evolution, toxins, herbivores and insect pollinators. *Naturwissenschaften*. 89:281–294.

Stevenson, D.W. (Ed.). 1987. *The biology, structure, and systematics of the Cycadales. Proceedings of the Symposium CYCAD 87, Beaulieu-sur-Mer, France, April 17–27, 1987.* New York Botanical Garden, Bronx, NY. 210 pp.

Stevenson, D.M. 1992. A formal classification of the extant cycads. *Brittonia*, 44:220–223.

Tang, W. 1995. *Handbook of cycad cultivation and landscaping*. W. Tang, Miami, FL. 33 pp.

Walters, T., and Osborne, R. (Eds.). 2004. *Cycad classification:concepts and recommendations*. CABI Pub., Wallingford, Oxfordshire, U.K. 267 pp.

Whitelock, L.M. 2002. *The cycads*. Timber Press, Portland, OR. 374 pp.

Whiting, M.G. 1963. Toxicity of cycads. *Econ. Bot.* 17:270–302.

Whiting, M.G. 1989. Neurotoxicity of cycads: an annotated bibliography for the years 1829–1989 (annotated). *Lyonia*, 2:201–270.

Specialty Cookbooks

Weil, C. 2006. *Fierce food: the intrepid diner's guide to the unusual, exotic, and downright bizarre*. Penguin Group, New York. 218 pp.

33 Durian

Family: Bombacaceae (bombax family)

NAMES

Scientific Name: *Durio zibethinus* L.

(The name is often cited as "*Durio zibethinus* J. Murr." J.A. Murray used the name in 1784, but the name is correctly attributed to Linnaeus, who used it earlier.)

- The English name "durian" is the Malay and Indonesian word for the fruit and is based on the Malay *duri*, thorn, for the thorns on the fruit.
- Suggested pronunciation: DOOWR-ee-uhn.
- The durian is known as "king of tropical fruits" in Southeast Asia, where it is greatly loved. The tasty flesh is one justification for this name, but another reason is because the durian is reminiscent of the thorny thrones of old Asian kings.
- Another Asian name for the durian is "skunk of the orchard." A similar sentiment is expressed in the self-explanatory German name *stinkfrucht* as well as the Dutch name *stinkvrucht.*
- The genus name *Durio* is based on the Malayan name (see above), essentially meaning spiny-fruited plants.
- *Zibethinus* in the scientific name of the durian is based on the Italian *zibetto*, a reference to the Indian civet cat (*Viverra zibetha*), a creature well known for its musty smell. The notorious odor of the durian has given rise to the unflattering terms "civet cat tree" and "civet fruit" in India.

PLANT PORTRAIT

The durian is a large, tropical, evergreen fruit tree, 27 to 45 m (90–148 ft.) tall, with a trunk 0.5 to 2.5 m (20 in.–8 ft.) in diameter, and prominent buttresses developing at the base of the trunk. It is believed to be native to Borneo and Sumatra and was spread to other tropical Asian countries, including India, Sri Lanka, Myanmar, Thailand, Malaysia, Indonesia, Philippines, and Papua New Guinea. Commercial production is concentrated in Thailand, Indonesia, Malaysia, and Philippines, and the tree is also grown in Australia. Thailand is the largest producer and exporter. The durian is the most important native fruit of Southeastern Asia and neighboring islands. There are hundreds of varieties. The fruit is usually round or oval, 15 to 30 cm (6–12 in.) in length, 12.5 to 20 cm (5–8 in.) in diameter, and weights from 1 to 8 kg (2–18 lb.). The thick, tough skin is usually green, yellow, or yellowish green and is covered with hard, sharp thorns. Handling the fruit without gloves can be painful. The durian is highly esteemed by those who know it well, although it smells bad. To those unfamiliar with the fruit, the smell may be overpoweringly objectionable. The fruit has approximately five segments separated by an inedible white membrane. Each section contains one to several seeds (the best varieties have aborted seeds) covered with custard-like flesh. The adherent flesh is technically termed an aril. The arils differ in color, thickness, texture, aroma, and taste and are typically creamy white, yellowish, pinkish, or orange. The seeds are edible, chestnut-like, 2 to 6 cm (3/4–2¼ in.) long, glossy, and red-brown.

CULINARY PORTRAIT

Various authors have described the smell of the durian by the following phrases: abominable; a mixture of old cheese and onions, flavored with turpentine; rotting fish; unwashed socks; a city dump on a hot summer's day; an unbearable stench; rotten onions with Limburger cheese and low-tide seaweed; French custard passed through a sewer pipe; like sitting on the toilet eating your favorite ice cream; like eating pudding in an outhouse; smells like compost. It has been said that "no other food smells so bad or tastes so good as the durian," delectable and detestable at the same time. According to a popular saying in Singapore, durians have the smell from hell and the taste from heaven. However, once people actually become acquainted with eating the fruit, the objectionable odor generally becomes much less noticeable. Moreover, many people on first eating a really good durian wonder what the fuss was all about. It has been described as like a delicious, sensuous banana pudding with a touch of butterscotch, vanilla, peach, pineapple, strawberry, and almond flavors and a surprising twist of garlic or onion.

The durian is a most unusual tropical fruit combining a delicious flavor with an unpleasant odor, much like certain aged cheeses. Durian fruit is mostly eaten fresh but can be made into a variety of products, including candy, baked goods, milk shakes, ice cream, custards, and jam. The fruits are preserved as powder, paste, chips, and as frozen purée. The freshly fallen fruits are less pungent and may taste best. Dried durian flesh is used to flavor confectionery, pastry, sherbet, and soft drinks. In Asia, durian often accompanies sticky rice, and in China, it is frequently served with pastries. Durian is relatively rare outside Southeast Asia and is a somewhat expensive export when available. Indeed, durians are among the world's most expensive fruits. However, it is often available in ethnic markets of large cities.

Durian seeds may be roasted or baked and consumed like nuts, or sliced and fried, or mixed with sugar to make a sweetmeat, or simply crushed to make a confection.

Tips: Pick comparatively light fruit with a large, solid stalk. Avoid fruits with holes. When shaken, the seeds in a good durian should move. The fruit is mature when its middle exudes a strong, but not sour smell, and especially when an inserted knife comes out sticky. Durians split along lines of natural weakness, which are faintly visible among the spines. To open the fruit, insert a stout knife into such a line. Durian does not store well for long. Without refrigeration, the fruit has a shelf life of 2 to 5 days. The flesh tastes best after being well chilled in a refrigerator, and the cold also serves to reduce the odor. Wrap portions of durian well to prevent the odor from permeating other foods.

Culinary Vocabulary

- *Tempoya* is the name of fermented durian in Indonesia where this is a popular side dish and is also used as a tart sauce for shellfish. Fermented durian may remain palatable for up to a year.
- "Durian leather" is a dried preparation of the fruit flesh. Fruit "leathers" are prepared for a number of the fruits examined in this book. The word leather is used because of the similarity to leather prepared from animal hides.

CURIOSITIES OF SCIENCE AND TECHNOLOGY

- Durian's odor of rotten eggs is due to sulfides, the same compounds that are in part responsible for the bad smell of some cheeses.
- After eating durians, "durian breath" has been said to linger for up to 6 hours, and to be so bad it has been ranked as more unpleasant than garlic.
- Despite the odor of the fruit, a toothpaste flavored with durian has been marketed for durian fanciers.

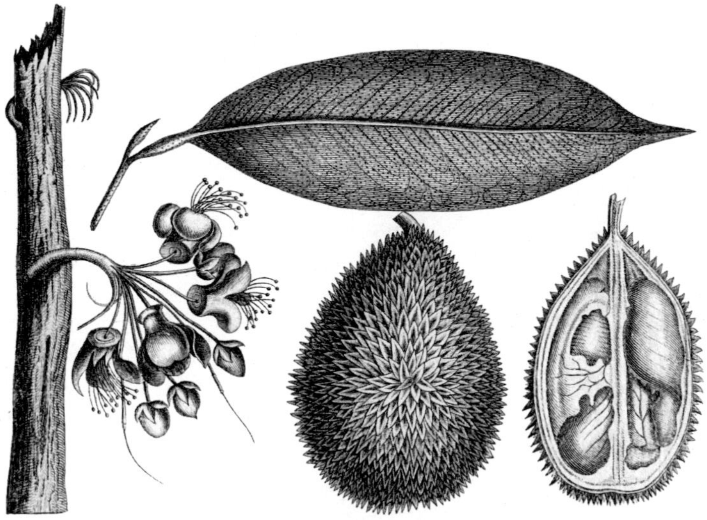

FIGURE 33.1 Durian (*Durio zibethinus*). (From Lamarck and Poiret, 1744–1829, plate 641.)

FIGURE 33.2 An old durian tree in the Malay Peninsula. (From Fairlie, 1893.)

- Durians are generally not on the menus of hotels or airlines servicing Westerners in Southeast Asia, in the belief that many customers would be offended by the smell. In Singapore, the fruit is banned from taxis, buses, ferries, and the jets of Singapore Airlines. In Singapore's spotless subway system, the authorities have posted no-durian signs, showing a durian set in a circle with a red slash through the center.
- In Malaysia, it is believed that the durian produces a type of heat, which when consumed in excess can cause disease, especially sore throat, headache, or fever. To avoid this possibility,

many people follow a simple ritual after devouring the durian. They pour water into the husk, occasionally add a pinch of salt, and drink it.

- Durian plantations are often robbed in Malaysia, and armed men are employed to keep a vigil over the fruit.
- Caution is necessary when approaching a durian tree during the ripening season. Durian fruits fall when ripe, and because they are covered with hard spines and are large, they have been known to seriously injure people. Workers harvesting durians are advised to wear helmets. In some plantations, the fruit is tied to the tree branches from which they hang to prevent them from falling—to keep them from being bruised rather than to prevent them from bruising people.
- In addition to bees, pollinators of the durian include bats, spiders, and ants.
- Orangutans live in the evergreen tropical forests of Sumatra and Borneo in Indonesia. Their favorite fruit is the durian. The feeding style of orangutans seems wasteful—taking a few bites and dropping the rest—but some seeds from the fallen fruit will grow to provide future fruit trees.
- Every culture has at least one food that is an acquired taste, often not attractive for anyone not brought up on it. The durian is an obvious example. Other examples are as follows (with apologies to the offended): lutefisk in Norway (cod fish soaked in lye), fish maw (gut) in China, haggis in Scotland (sheep's stomach, stuffed with oatmeal and steamed), powsowdie in Scotland (a whole boiled sheep's head served in its broth, containing the brains, that was popular with Mary Queen of Scots), vegemite in Australia (made of yeast extract culled from brewery wastes, and quite salty; essentially the same as marmite in New Zealand and the United Kingdom), maggot cheese in Sardinia (cheese covered with cloth is left out so flies will lay their eggs in it, the maggots are allowed to hatch, then spread on bread), menudo in Mexico (soup of boiled tripe: the stomach lining of a cow), drunken shrimp in China (live shrimp swimming in a bowl of rice wine are captured with chopsticks, and the heads bitten off), fish eyes in southeast Asia, gorgonzola in Italy (ripe stinky cheese), spruce beer in Canada, and chiterlings ("chitlins") in the southern United States (fried hog intestines).
- Alfred Russel Wallace (1823–1913), was a brilliant English naturalist who competed with Charles Darwin for the honor of publishing the first book on the role of natural selection in biological evolution. Well acquainted with eating unfamiliar foods, Wallace had no difficulty adjusting to the durian. After eating his first durian in Borneo, he wrote "The more you eat of it the less you feel inclined to stop." He also commented that "the taste of durian is worth the travel to the Far East."
- The botanist E.J.H. Corner proposed a controversial theory known as the Durian Theory, which claims that flowering plants arose from a primitive ancestor with fruit characteristics much like those of the durian tree (see the following two publications: Corner, E.J.H. 1949. The durian theory or the origin of the modern tree. *Ann. Bot. New Ser.*13: 367–414. Corner, E.J.H. 1954. The evolution of tropical forests. In *Evolution as a process.* Edited by J. Huxley, A.C. Hardy, and E.B. Ford. Allen and Unwin, London. pp. 34–46).
- An Asian treatment for rheumatism consists of rubbing the body with a spiny-fruited durian. Strange as this may seem, such application of irritation to the skin is often medically effective.

KEY INFORMATION SOURCES

Abidin, M.Z. 1992. *Technology for commercial production of durian.* Malaysian Agricultural Research and Development Institute, Kuala Lumpur, Malaysia. 7 pp.

Baldry, J., Dougan, J., and Howard, G.E. 1972. Volatile flavoring constituents of durian. *Phytochemistry,* 11:2081–2084.

Brooncherm, P., and Siriphanich, J. 1991. Postharvest physiology of durian pulp and husk. *Kasetsart J.* 25:119–125.

Brown, M.J. 1997. Durio: *a bibliographic review*. International Plant Gene Resources Institute, Office for South Asia, New Delhi, India. 188 pp.

Duarte, O., and Paull, R.E. 2008. *Durio zibethinus*, durian. In *The encyclopedia of fruit & nuts*. Edited by J. Janick and R.E. Paull. CABI, Wallingford, Oxfordshire, U.K. pp. 176–182.

Husin, A., and Abidin, M.Z. 1998. Durian. In *Tropical and subtropical fruits*. Edited by P.E. Shaw, H.T. Chan, Jr., and S. Nagy. Agscience, Auburndale, FL. pp. 261–289.

Irwandi [no initials], and Che-Man, Y.B. 1996. Durian leather: development, properties and storage stability. *J. Food Qual.* 19:479–489.

Kanzaki, S., Yonemori, K., Sugiura, A., and Subhadrabandhu, S. 1998. Phylogenetic relationships of the common durian (*Durio zibethinus* Murray) to other edible fruited *Durio* spp. by RFLP analysis of an amplified region of cpDNA. *J. Hortic. Sci. Biotech.* 73:317–321.

Ketsa, S. 1997. Durian. In *Postharvest physiology and storage of tropical and subtropical fruits*. Edited by S.K. Mitra. CABI, Wallingford, U.K. pp. 323–334.

Ketsa, S., and Daengkanit, T. 1998. Physiological changes during postharvest ripening of durian fruit (*Durio zibethinus* Murray). *J. Hortic. Sci. Biotech.* 73:575–577.

Ketsa, S., and Pangkool, S. 1995. The effect of temperature and humidity on the ripening of durian fruits. *J. Hortic. Sci.* 70:827–831.

Kostermans, A.J.G.H. 1958. The genus *Durio* Adans. (Bombac.). *Reinwardtia*, 4(3):357–460.

Lee, D. 1985. The durian, a most magnificent and elusive fruit. *Fairchild Trop. Garden Bull.* 40(2):19–27.

Lim, T.K., and Luders, L. 1998. Durian flowering, pollination and incompatibility studies. *Ann. Appl. Biol.* 132:151–165.

Lim, T.S. 1990. *Durian: diseases and disorders*. Tropical Press, Kuala Lumpur, Malaysia. 95 pp.

Malo, S.E., and Martin, F.W. 1979. *Vegetables for the hot humid tropics: Part 7. The durian*. Mayaguez Institute of tropical Agriculture, Mayaguez, Puerto Rico. 16 pp.

Mansfield, J.G., and Dostie, B. 1995. *Control of production patterns in tropical fruit: Part 4. Durian*. Report to Rural Industries Research and Development Corporation, Australia. 64 pp.

Morton, J. 1987. Durian. In *Fruits of warm climates*. Creative Resource Systems, Winterville, NC. pp. 287–291.

Naf, R., and Velluz, A. 1996. Sulphur compounds and some uncommon esters in durian (*Durio zibethinus* Murr.). *Flavour Fragrance J.* 11:295–303.

Nantachai, S. (Ed.). 1994. *Durian: fruit development, postharvest physiology, handling and marketing in ASEAN*. ASEAN Food handling Bureau, Kuala Lumpur, Malaysia. 156 pp.

Ramlie, P. 1973. *Bibliography on durian, mango and rambutan in Indonesia*. Departemen Pertanian, Lembaga Perpustakaan Biologi dan Pertanian, Bibliotheca Bogoriensis, Bogor, Indonesia. 46 pp.

Salakpetch, S., Chandraparnik, S., Hiranpradit, H., and Punnachit, U. 1992. Source–sink relationship affecting fruit development and fruit quality in durian, *Durio zibethinus* Murr. *Acta Hortic.* 321:691–694.

Soegeng-Reksodihardjo, W. 1962. The species of *Durio* with edible fruits. *Econ. Bot.* 16:270–282.

Soepadmo, E., and Eow, B.K. 1976. The reproductive biology of *Durio zibethinus* Murr. *Garden's Bull.* 29:25–33.

Subhadrabandhu, S., and Ketsa, S. 2001. *Durian: king of tropical fruit*. CABI Publishing, Wallingford, U.K. 178 pp.

Subhadrabandhu, S., and Schneemann, J.M.P. 1991. *Durio zibethinus* Murray. In *Plant resources of South-East Asia. 2. Edible fruits and nuts*. Edited by E.W.M. Verheij and R.E. Coronel. Pudoc, Leiden, the Netherlands. pp. 157–161.

Valmayor, R.V., Coronel, R.E., and Ramirez, D.A. 1965. Studies of floral biology, fruit set and development in durian. *Philippine Agriculturists*, 48:355–359.

Watson, B.J. 1983. Durian (*Durio zibethinus* Murr.). In *Tropical tree fruits for Australia*. Edited by P.E. Page. Queensland Department of Primary Industries, Brisbane, Queensland, Australia. pp. 45–50.

Weenen, H., Koolhaas, W.E., and Apriyantono, A. 1996. Sulfur-containing volatiles of durian fruits (*Durio zibethinus* Murr.). *J. Agric. Food Chem.* 44:3291–3293.

Wong, K.C., and Tie, D.Y. 1995. Volatile constituents of durian (*Durio zibethinus* Murr.). *Flavour Fragrance J.* 10:79–83.

Specialty Cookbooks

Alejandro, R.J., and Fernandez, D.G. 1999. *Food of the Philippines: authentic recipes from the pearl of the Orient*. Periplus, Hong Kong. 120 pp.

Boetz, M., Christie, S., Thompson, D., and Simons, J. 2007. *Modern Thai food: 100 simple and delicious recipes from Sydney's famous Longrain Restaurant*. Periplus, Hong Kong. 176 pp.

Boutenko, S., and Boutenko, V. 2008. *Fresh: the ultimate live-food cookbook*. North Atlantic Books, Berkeley, CA. 216 pp.

Gates, S. 2005. *Gastronaut: adventures in food for the romantic, the foolhardy, and the brave*. Harcourt, Orlando, FL. 257 pp.

Payne, S., and Payne, W.J.A. 1979. *Cooking with exotic fruit*. B.T. Batsford Ltd., London. 144 pp.

34 Epazote

Family: Chenopodiaceae (goosefoot family)

NAMES

Scientific Name: *Chenopodium ambrosioides* L.

- The name "epazote" is derived from the Nahuatl words *epatl*, skunk, and *tzotl*, sweat, a reference to the rank smell. Nahuatl is the language of the Amerindian peoples of southern Mexico and Central America, spoken by the Aztecs before the Spanish conquest, and still a minority language in Mexico. Suggested pronunciations: eh-pah-soh-teh, e-pah-ZOH-teh.
- The names of some Mexican species unrelated to epazote nevertheless have epazote in their name. *Epazote de zorrillo* refers to *Teloxys graveolens* (Willd.) W. A. Weber, a medicinal plant of Mexico. *Epazote zorrillo* (*Chenopodium graveolens* Willd., skunk saltwort) is another medical species.
- Epazote is also known as American goosefoot, American wormseed, Baltimore wormseed, bitter weed, Californian spearmint (in New Zealand), demi-god's food, epasote, herb sancti Mariae, Indian wormweed, Jesuit's tea, Mexican goosefoot, Mexican tea, mouse food, pazote, Spanish tea, skunkweed, West Indian goosefoot, worm bush, worm grass, wormseed, wormseed goosefoot, wormseed oil plant, and wormweed.
- "Hedge mustard" is used as a name for epazote but is well established for the common weed *Sisymbrium officinale* (L.) Scop. "Stinkweed" is another epazote name that is much more commonly used for other species, for example, *Thlaspi arvense* L. and *Pluchea camphorata* (L.) DC.
- Rarely, the name "ambrosia" has been used for epazote, but ambrosia is almost always used for *Chenopodium botrys* L., a Eurasian plant that is weedy in North America, and is occasionally used as a culinary herb. Ambrosia is also known as Jerusalem oak. Rarely, epazote has been called Jerusalem oak, Jerusalem bush, Jerusalem tea, and Jerusalem parsley.
- For pharmaceutical purposes, epazote is often known as wormseed. Extracted oil has been used as an anthelmintic (a medicine for controlling internal parasites) for many years. In the early 1900s, epazote oil was one of the major anthelmintics used to treat ascarids and hookworms in humans, cats, dogs, horses, and pigs. Wormseed oil from epazote was replaced with other, more effective and less toxic anthelmintics in the 1940s. In a few areas in Latin America, epazote is still used to treat worm infections in livestock.
- The name Baltimore wormseed arose from the fact that the Baltimore Maryland area was the center of production of wormseed oil in North America for over a century.
- The name Mexican tea is sometimes applied to *Ephedra nevadensis* S. Watson, a native of the Southwestern United States and Mexico, more commonly called Mormon tea.
- The genus name *Chenopodium* is based on the Greek *chen*, goose, and *pous*, foot, an allusion to the shape of the leaves.
- *Ambrosioides* in the scientific name *C. ambrosioides* is Latin for ambrosia-like, referring to the strong odor. According to the Greek mythology, ambrosia (Greek for "not mortal") was a nourishment reserved for the Olympic gods.

PLANT PORTRAIT

Epazote, a native of tropical and subtropical America, is an annual or a perennial depending on its situation. It has become naturalized in warm regions throughout the world. In North America, it grows along roadsides and in waste and cultivated grounds as far north as southern Canada. In the southwestern United States, it thrives along stream beds. The plants grow from 0.5 to 1.5 m (12–5 ft.) tall. Large clusters of small, yellowish-green flowers are produced in midsummer to early fall, followed by small brownish-black seeds. Plants that are cultivated to make tea and soup are sometimes put in *C. ambrosioides* var. *ambrosioides*, whereas those cultivated for use as a vermifuge and source of wormseed oil are sometimes placed in var. *anthelminticum* (L.) A. Gray. The species is strongly aromatic. Wormseed oil, the chief product of interest of the plant, is produced mostly in the leaves and the seeds, the latter normally furnishing more than half of the plant's yield. The leaves are conspicuously dotted beneath with oil glands. Epazote was used for centuries as a vermifuge in the Americas and was also used for the same purpose in post-Columbian times in Europe and Asia.

CULINARY PORTRAIT

The availability of epazote as a wild plant has made it a "poor man's herb" for centuries in Mexico, enhancing otherwise common peasant fare. It has been described as "the most Mexican of the culinary herbs." Its rather objectionable odor has been compared with turpentine and petroleum, with hints of citrus, savory, or mint. The taste is medicinal or "antiseptic" and pungently camphoraceous or "musty smelling" and requires habituation. Epazote leaves are used to flavor corn, black beans, mushrooms, fish, soups, stews, chili sauces, shellfish, and freshwater snails. The leaves are also used to brew a tea, and the tender leaves are sometimes used as a potherb. Epazote added to cooking beans is said to reduce the flatulence caused by the latter. In Mexico where the herb is well known, epazote is commonly added to bean dishes to avoid socially embarrassing consequences. Refried beans made with epazote are a Mexican specialty, but when served in Tex-Mex-style restaurants outside of Mexico and the southern United States, this dish almost never contains epazote. Outside of Mexico, epazote has some popularity as a tea in the West Indies, United States, southern France, and Germany. Because epazote is so strongly flavored, it has been recommended that it be added only during the last 15 minutes of cooking so that the food will not become bitter. It has also been advised that this herb be used sparingly.

Epazote contains pharmacologically active principles and cases of poisoning have been recorded. Although widely used as a culinary herb in Mexico, its consumption could be hazardous. If eaten at all, it should only be consumed in very limited quantities, and rarely. The culinary use of epazote outside of the Mexican and Latino communities, where it represents an acquired taste, is limited by its toxicity, peculiar taste, and unappealing aroma. The herb is included in recipes in hundreds of Western cookbooks, but not frequently, often optionally, and in small amounts. Epazote can often be obtained from Latino groceries, both fresh and dried (1 tablespoon (15 mL) of fresh, chopped material is equivalent to 1 teaspoon (5 mL) of dried epazote).

Culinary Vocabulary

- *Sopa de flor de calabaza* (pronounced soh-pay day floor day cah-lah-bah-zah) is a soup prepared in Central and South America from squash blossoms, flavored with epazote, onions, and pepper.

CURIOSITIES OF SCIENCE AND TECHNOLOGY

- In New Mexico, suppositories of dried pulverized leaves of epazote, ground spearmint, and salt have been used as a remedy for appendicitis.

FIGURE 34.1 Epazote (*Chenopodium ambrosioides*). (From Hallier, 1880–1888, vol. 9, plate 870.)

- One of the many health problems treated with an epazote folk remedy is athlete's foot. It has been found that epazote indeed inhibits fungi that cause this disease, confirming the wisdom of this herbal remedy.
- The popular use of epazote in New York City by Latinos has led to it becoming somewhat weedy there, and it is found growing in cracks of sidewalks and in Central Park.
- Epazote is sometimes used to repel pests. It has been used to deter insects from eating stored grains. Indochinese farmers mix the fruiting stalks with fertilizer to discourage insect larvae in their crops. In Brazil, the plant is used as a household insect repellent and insecticide.

KEY INFORMATION SOURCES

Bahrman, N. 1984. Biometrical analysis of the quantitative variability of some morphological characters in *Chenopodium* [*C. ambrosioides* and other species]. *Rev. Gen. Bot.* 91:153–161 (in French, English summary).

Coile, N.C., and Artaud, C.R. 1997. Chenopodium ambrosioides *L., (Chenopodiaceae): Mexican tea, wanted weed?* Department of Agriculture and Consumer Services, Division of Plant Industry, Gainesville, FL. 6 pp.

Gadano, A., Gurni, A., Lopez, P., Ferraro, G., and Carballo, M. 2002. In vitro genotoxic evaluation of the medicinal plant *Chenopodium ambrosioides* L. *J. Ethnopharmacol.* 81:11–16.

Gupta, D., Charles, R., Mehta, V.K., Garg, S.N., and Kumar, S. 2002. Chemical examination of the essential oil of *Chenopodium ambrosioides* L. from the southern hills of India. *J. Essent. Oil Res.* 14:93–94.

Jimenez-Osornio, F.M.V.Z.J., Kumamoto, J., and Wasser, C. 1996. Allelopathic activity of *Chenopodium ambrosioides* L. *Biochem. Syst. Ecol.* 24:195–205.

Kishore, N., Chansouria, J.P.N., and Dubey, N.K. 1996. Antidermatophytic action of the essential oil of *Chenopodium ambrosioides* and an ointment prepared from it. *Phytother. Res.* 10:453–455.

Kliks, M.M. 1985. Studies on the traditional herbal anthelmintic *Chenopodium ambrosioides*: ethnopharmacological evaluation and clinical field trials. *Soc. Sci. Med.* 21:879–886.

Logan, M.H., Gwinn, K.D., Richey, T., Maney, B., and Faulkner, C.T. 2004. An empirical assessment of epazote (*Chenopodium ambrosioides* L.) as a flavoring agent in cooked beans. *J. Ethnobiol.* 24:1–12.

MacDonald, D., VanCrey, K., Harrison, P., Rangachari, P.K., Rosenfeld, J., Warren, C., and Sorger, G. 2004. Ascaridole-less infusions of *Chenopodium ambrosioides* contain a nematocide(s) that is(are) not toxic to mammalian smooth muscle. *J. Ethnopharmacol.* 92:215–221.

Olajide, O.A., Awe, S.O., and Makinde, J.M. 1997. Pharmacological screening of the methanolic extract of *Chenopodium ambrosioides. Fitoterapia*, 68:529–532.

Onocha, P.A., Ekundayo, O., Eramo, T., and Laakso, I. 1999. Essential oil constituents of *Chenopodium ambrosioides* L. leaves from Nigeria. *J. Essent. Oil Res.* 11:220–222.

Pare, P.W., Zajicek, J., Ferracini, V.L., and Melo, I.S. 1993. Antifungal terpenoids from *Chenopodium ambrosioides. Biochem. Syst. Ecol.* 21:649–653.

Pino, J.A., Marbot, R., and Real, I.M. 2003. Essential oil of *Chenopodium ambrosioides* L. from Cuba. *J. Essent. Oil Res.* 15:213–214.

Sagrero-Nieves, L., and Bartley, J.P. 1995. Volatile constituents from the leaves of *Chenopodium ambrosioides* L. *J. Essent. Oil Res.* 7:221–223.

Su, H.C. 1991. Toxicity and repellency of *Chenopodium* oil to four species of stored-product insects. *J. Entomol. Sci.* 26:178–182.

Tapondjou, L.A., Adler, C., Bouda, H., and Fontem, D.A. 2002. Efficacy of powder and essential oil from *Chenopodium ambrosioides* leaves as post-harvest grain protectants against six stored product beetles. *J. Stored Prod. Res.* 38:395–402.

Umemoto, K. 1978. Essential oil of *Chenopodium ambrosioides* L. containing (–)-pinocarveol as a major component. *J. Agric. Chem. Soc. Jpn.* 52:149–150 (in Japanese, English summary).

Weiland, G.S., Broughton, L.B., and Metzger, J.E. 1935. Wormseed oil production. *Univ. Md. Agric. Exp. Stn. Bull.* 384:315–335.

Specialty Cookbooks

Note: Epazote is most likely to be encountered in recipes for Mexican food.

Balmuth, D.L. 1996. *Herb mixtures & spicy blends.* Storey Publishing, Pownall, VT. 156 pp.

Hutson, L. 1987. *The herb garden cookbook.* Texas Monthly Press, Austin, TX. 278 pp.

Jamison, C. 1993. *Texas home cooking.* Harvard Common Press, Boston, MA. 592 pp.

Wilder, J. 2008. *The great chiles rellenos book.* Ten Speed Press, Berkeley, CA. 144 pp.

Wise, V., and Hoffman, S. 1990. *The well-filled tortilla cookbook.* Workman, New York. 301 pp.

35 Feijoa

Family: Myrtaceae (myrtle family)

NAMES

Scientific Name: *Acca sellowiana* (O. Berg) Burret [the species has recently been renamed as *A. sellowiana,* and most literature still uses the older scientific name *Feijoa sellowiana* (O. Berg) O. Berg]

- The English word "feijoa" and the genus name *Feijoa* in which the species was formerly placed commemorate João da Silva Feijó (1760–1824), a Brazilian soldier and naturalist who became director of the Natural History Museum in San Sebastian, Brazil.
- Pronunciation: fay-YOH-ah, fay-JOH-ah, or fee-JOH-ah.
- Feijoa is also called pineapple guava (the taste is reminiscent of pineapple) and is sometimes mislabeled as guava in supermarkets. It has been called Brazilian guava, fig guava, guavasteen (in Hawaii), and New Zealand banana (in New Zealand).
- The genus name *Acca* was coined in 1855 by the German botanist Otto Carl Berg (1815–1866), a specialist on the myrtle family. The name has been said to be based on a South American (especially Peruvian) common name for the plant. Acca is also a Roman mythological figure, and it has been suggested that the genus may have been named for her. According to the story, the twin babies Romulus and Remus were nourished by a wolf. They were discovered in the wolf's lair by a shepherd named Faustulus, who took them to his home to be raised by his wife Acca. The twins are said to have founded Rome (approximately 754 BC), although Romulus, not entirely an admirable figure, killed his brother and to cure a chronic shortage of females and populate the city kidnapped numerous Sabine women (the Sabines were an ancient people in Italy). Acca became important in Roman mythology, although there is a less than complementary legend about her: she is said to have acquired wealth as a courtesan.
- *Sellowiana* in the scientific name *A. sellowiana* commemorates Friedrich Sellow (1789–1831), a German gardener and botanical explorer who collected widely, including specimens of feijoa in 1815 from the province of Rio Grande do Sul in southern Brazil.

PLANT PORTRAIT

Feijoa is a subtropical, evergreen, bushy shrub, 1 to 6 m (3–20 ft.) tall. The species is native to the mountains of southern Brazil, northern Argentina, western Paraguay, and Uruguay. It was consumed in South America before European colonization and was taken to Europe in the nineteenth century. There are now about a dozen cultivated varieties. The fruit is variable in shape—round, oblong, ovoid, or pear shaped, 4 to 6 cm (1½–2½ in.) long and 2.8 to 5 cm (1–2 in.) wide. The fruit surface is coated with whitish hairs until maturity, ripening to dull-green or yellow-green, sometimes with a red or orange caste. The rind or skin is 1 to 1.6 cm (3/8–5/8 in.) thick. Feijoas are strongly perfumed. The flesh is thick, white or creamy, granular (gritty, like pears), and watery, with a translucent, slightly gelatinous central pulp containing very small black seeds. The feijoa has been extensively cultivated in New Zealand and is sometimes commercially grown in South American countries and in the Caribbean area. It has been planted in Florida, California, and Hawaii, and there is a commercial crop in California. In cooler climates, feijoa is more commonly cultivated for its attractive foliage and bright red flowers than for its fruit.

CULINARY PORTRAIT

Feijoas are mature when slightly soft, and the jelly-like sections in the center are clear. They are unripe when the jellied sections are white (at which stage they are sourish or bitter) and overripe when the fruits are turning brown, at which time the flavor becomes unpleasant. It can be difficult to tell whether a fruit is ripe or not without first cutting it open. The fruit is ripe when the flesh has the consistency of a ripe peach and can easily be scooped out with a spoon. The taste is sweet or subacid and has been compared with combinations of pineapple and guava, pineapple and strawberry, and quince, pineapple, and mint. To prevent bruising, the fruit should be handled gently, like ripe peaches. Before eating, the bitter skin should be peeled off. To prevent the flesh from turning brown, peeled fruit should be dipped in water containing fresh lemon juice (orange and pineapple juice may be substituted; a weak salt solution will also work, but unnecessary salt is undesirable). Ripe fruit may be stored in a refrigerator for a few days.

The flesh and pulp (along with the seeds which are so small that they are hardly noticeable when eaten) are consumed raw as a dessert or in salads. Feijoas are also cooked in puddings, pastry fillings, fritters, dumplings, fruit sponge cake, pies, and tarts; made into preserves, chutney, jam, jelly, relish, sauce, and sparkling wine; and used as flavoring for ice cream, yogurt, soft drinks, and, more recently, teas and chips for breakfast cereals. Feijoa goes well with bananas and apples and can be interchanged with these fruits in most recipes.

Culinary Vocabulary

- Although feijoa trees are native to Brazil, feijoa has nothing to do with feijoada, the national dish of Brazil. The Brazilian word *feijoa*, literally translated, means "bean," and beans are the basis of feijoada. Feijoada has been described as a culinary adventure of epic proportions because preparation is often elaborate, often requiring a day or more. According to legend, feijoada evolved over many decades as plantation owners in Brazil fed their slaves the remains of meat animals, such as pig ears and ox tongue, and in time such materials were creatively integrated into a fabulous meat and bean stew served over vegetables or white rice. More probably, feijoada originated from a Portuguese dish, during the time that Brazil was a colony of Portugal.

CURIOSITIES OF SCIENCE AND TECHNOLOGY

- In a suitable climate, an average adult feijoa plant produces about a thousand fruits annually.
- The thick, spicy petals of feijoa flowers are sweet because of the presence of sucrose, that is, table sugar. In their native habitat, the flowers are pollinated by birds, which get dusted by pollen while feeding on the petals (the flowers do not produce nectar, which is normally what attracts birds to plants). The petals are also sometimes eaten fresh by humans, where living plants available. The fleshy petals can be added to salads and desserts. If picked carefully, the remaining flowers can still develop into fruits.
- "Forever Feijoa" is a New Zealand breakfast cereal, which has been described as "A light, satisfying cereal bursting with real feijoa flavor. This wonderful cereal combines wheat and bran flakes, puffed wheat and oatbran sticks, with real freeze dried feijoa pieces."
- In the United States, the feijoa has had its greatest commercial success in California. Although many of the food plants noted in this book have cities names after them, feijoa has achieved at least some recognition: Feijoa Avenue, in Lomita, California.

FIGURE 35.1 Feijoa (*Acca sellowiana*). (From Curtis, 1897, vol. 123, plate 7620.)

KEY INFORMATION SOURCES

Ackerman, M. 1993. Improving of the keeping quality of feijoa fruit after harvest. *Hassadeh*, 74(1):55–59.

Azam, B., Lafitte, F., Obry, F., and Paulet, J.L. 1981. The feijoa in New Zealand. *Fruits*, 36:361–384 (in French).

Bailey, F.L. 1952. Culture of feijoa trees. *N. Z. J. Agric.* 84:291–293, 295–296.

Bowman, C.E. 1984. The feijoa in central California. *J. California Rare Fruit Growers*. 16:32–43.

Cacioppo, O. 1988. *La feijoa* [Feijoa]. Mundi-Prensa, Madrid, Spain. 85 pp.

Crooks, M.R. 1978. Sparkling wine from feijoas. *N. Z. J. Agric.* 136(8):85.

Dettori, M.T., and Palombi, M.A. 2000. Identification of *Feijoa sellowiana* Berg accessions by RAPD markers. *Sci. Hortic.* 86:279–290.

Ducroquet, J.P.H.J., and Hickel, E.R. 1997. Birds as pollinators of feijoa (*Acca sellowiana* Berg). *Acta Hortic.* 452:37–40.

Forte, V. 1993. *The cultivation of feijoia: a plant with ornamental fruit.* 2nd ed. Edagricole, Bologna, Italy. 43 pp (in Italian).

Giacometti, D., and Lleras, E. 1994. Subtropical Myrtaceae. In *Neglected crops. 1492 from a different perspective.* Edited by J.E. Hernández-Bermejo and J. Léon. Food and Agriculture Organization of the United Nations, Rome, Italy. pp. 229–237.

Harman, J.E. 1987. Feijoa fruit growth and chemical composition during development. *N. Z. J. Exp. Agric.* 15:209–216.

Kirkpatrick, S. 1988. Is feijoa the next kiwi fruit? *California Grower*, 12(2):49–50.

Klein, J.D., and Thorp, T.G. 1987. Feijoas—postharvest handling and storage of fruit. *N. Z. J. Exp. Agric.* 15:217–221.

Macpherson, N. 1952. Feijoas have a variety of culinary uses. *N. Z. J. Agric.* 84:336–337.

Mattos, J.R. 1969. The genus *Feijoa*. *Arquivos Botanica Estado Sao Paulo*, 4(4–6):263–267.

Morton, J. 1987. Feijoa. In *Fruits of warm climates.* Creative Resource Systems, Winterville, NC. pp. 367–370.

Nagy, S. 1998. Feijoa. In *Tropical and subtropical fruits.* Edited by P.E. Shaw, H.T. Chan, Jr., and S. Nagy. Agscience, Auburndale, FL. pp. 506–520.

Nodari, R.O., Guerra, M.P., Meler, K., and Ducroquet, J.P. 1997. Genetic variability of *Feijoa sellowiana* germplasm. *Acta Hortic.* 452:47–52.

Patterson, K.J. 1990. Effects of pollination on fruit set size and quality in feijoa *Acca sellowiana*. *N. Z. J. Crop Hortic. Sci.* 18:127–132.

Popenoe, F.W. 1912. *Feijoa sellowiana*: its history, culture and varieties. *Pomona Coll. J. Econ. Bot.* 11:217–242.

Ryerson, K.A. 1933. *Feijoa sellowiana*. *Natl. Hortic. Mag.* 12:240–245.

Sharpe, R.H., Sherman, W.B., and Miller, E.P. 1994. Feijoa history and improvement. *Proc. Annu. Meet. Fla. State Hort. Soc.* 106:134–139.

Shaw, G.J., Ellingham, P.J., and Birch, E.J. 1983. Volatile constituents of feijoa—headspace analysis of intact fruit *Feijoa sellowiana*, a native of Paraguay, Brazil, Uruguay and Argentina. *J. Sci. Food Agric.* 34:743–747.

Shaw, G.J., Allen, J.M., Yates, M.K., and Franich, R.A. 1990. Volatile flavour constituents of feijoa (*Feijoa sellowiana*): analysis of fruit flesh. *J. Sci. Food Agric.* 50:357–361.

Sweet, C. 1990. Feijoa or pineapple guava? *California Grower*, 14(5):30–32.

Swift, J.F. 1984. Feijoa cultivation expands in California. *J. Calif. Rare Fruit Growers*, 16:44–46.

Thomson, P.H. 1970. Pineapple guava. *California Rare Fruit Growers Newsl.* 2(2):2–4.

Thorp, G. 2008. *Acca sellowiana*, feijoa. In *The encyclopedia of fruit & nuts*. Edited by J. Janick and R.E. Paull. CABI, Wallingford, Oxfordshire, U.K. pp. 526–533.

Thorp, T.G., and Bieleski, R.L. 2002. *Feijoas: origins, cultivation and uses*. David Bateman Ltd., Auckland, New Zealand. 87 pp.

Specialty Cookbooks

Drabble, W., and Jenkins, B. 1999. *The feijoa recipe book*. Stonepress, Norsewood, New Zealand. 64 pp.

New Zealand Feijoa Growers Association. 1997. *Feijoa recipes*. NZ Feijoa Growers Assoc., Ngaio, Wellington, New Zealand. 32 pp.

Stone, M. 1993. *Feijoa recipes*. M. Stone, Dannervirke, New Zealand. Unpaginated.

36 Fenugreek

Family: Fabaceae (Leguminosae; pea family)

NAMES

Scientific Name: *Trigonella foenum-graecum* L.

- The name "fenugreek" is derived through the French *fenugrec* from the Latin *fenum graecum*, literally "Greek hay," the classical Roman name of the plant (for additional information, see derivation of *foenum-graecum,* below).
- Pronunciation: FEHN-yoo-greek or FEN-oo-greek.
- Fenugreek is also known by the names billy-goat clover, billy-goat hay, camel grass, common fenugreek, fenugrec, goat's horn, Greek clover, Greek hay, Greek hayseed, and hay fenugreek.
- The genus name *Trigonella* comes from the Greek *treis*, three, and *gonu*, angle or corner, a reference to the triangular appearance of the flowers of some of the species. As well as being the name of a genus of plants, *Trigonella* is also an old genus name for some sea shell (mollusk) species, currently placed in *Scrobicularia*.
- *Foenum-graecum* in the scientific name *T. foenum-graecum* is Latin for Greek hay, a reference to the ancient use of the plant in Greece to scent hay.

PLANT PORTRAIT

Fenugreek is an annual herb growing to a height of 30 to 60 cm (1–2 ft.). The petals of the flowers are usually pale yellow, occasionally white, and sometimes with lilac at the base. The pods are very long and narrow and usually curved. Before ripening, the pod is green or reddish, and at maturity, it is straw colored or light brown. The pod contains 10 to 20 seeds. Fenugreek grows wild in southern Europe and Asia. This ancient crop plant is known from an archaeological location in Iraq dated at 4000 BC as well as from several Early Bronze Age sites of the Near East. The antiquity of fenugreek is also evidenced by the many words for it in various languages of old derivation: Arabic, Indian (Sanskrit), Latin, and classical Greek. It was grown in ancient Egypt, Greece, and Rome. The seeds of fenugreek were used for medicinal and culinary purposes by the Egyptians, Greeks, and Romans, and these uses have persisted down to modern times. The hay was used to promote animal health, a form of veterinary practice in very early times that has survived to the present. Fenugreek is cultivated today in all Mediterranean countries, the Near East and Middle East, northeastern Africa to Ethiopia, Arabia, Turkey, India, China, and Japan. It is grown to a minor extent in Argentina, Canada, and the United States. Fenugreek seed is a source of the steroid diosgenin, which is used in making the female contraceptive pill and synthetic sex hormones. (Diosgenin, mostly obtained from tubers of certain species of South American yams of the genus *Dioscorea*, is the starting compound for over half of the total steroid production by the pharmaceutical industry.) Drugs from the seeds are now used both in human and veterinary medicine. Today, fenugreek is a multipurpose crop, harvested as a spice, a tea, a vegetable (the leaves are occasionally consumed in India), a forage, a dye plant, and a starter material in the production of steroidal hormones.

CULINARY PORTRAIT

Fenugreek is employed principally for its seeds, used as flavoring. The taste and odor of fresh seeds are somewhat reminiscent of lovage and celery. Ground seed has a very strong, maple sweetness, spicy, somewhat meat-like (also described as "farinaceous" or "mealy"), but bitter. The taste is acquired, and most Westerners find fenugreek seeds too bitter to use directly as a spice. The aroma is comparable with burnt sugar but is appetizing and pleasing. Roasting the seeds requires experience—if overheated, they become extremely bitter. The seeds are best known for their use as a constituent of curry powders and spice mixtures and as a condiment. They are also an ingredient of the Jewish sweet dish Halva and are often found in mango pickles, chutney, stews, and soups. Fenugreek seeds or extracts have been used in syrups, pickles, baked goods, candy, condiments, chewing gums, soft drinks, gelatins, pudding, ice cream, icings, cheeses, and some prepared meat seasonings. In India and in other eastern countries, roasted seeds are used as a coffee substitute and in mixtures with wheat and millet as bread. In the Western World, the ground seeds are employed, with other aromatic substances, to produce artificial maple flavoring. The flavor of fenugreek is also useful for blends that mimic caramel, vanilla, butterscotch, rum, and licorice. The seeds are sometimes sprouted and consumed like alfalfa sprouts, especially in Africa and India.

Culinary Vocabulary

- In Hindi, the language of northern India, and the language with an especially great range of fenugreek recipes, both fenugreek seeds and greens are called *methi* (pronounced MEHT-hee).

CURIOSITIES OF SCIENCE AND TECHNOLOGY

- In many Oriental countries, fenugreek seeds were consumed for the purpose of giving "a captivating buxom plumpness" to the female human form. In North Africa and the Middle East, the seeds are also eaten to stimulate breast milk. Several current commercial products based on fenugreek claim to be able to increase the size of breasts. The chemical diosgenin in fenugreek seeds is similar to the female sex hormone estrogen, and this encourages the body to retain water. Just as "The Pill" causes water retention and bloating in some women, perhaps the fenugreek chemical may have caused some women to develop Rubenesque proportions.
- In ancient Egypt, fenugreek was one of the components of *kuphi* ("holy smoke"), an incense used in fumigation and embalming.
- The Romans laid siege to Jerusalem from the years 66 to 70. The Jewish defenders of Jerusalem added fenugreek to the boiling oil and water that was poured on the Roman troops, on ladders, as they attempted to scale the city walls. Mixed with water, ground fenugreek becomes gelatinous and slippery and would have made it more difficult for the Romans to keep their footing.
- In the Middle Ages, fenugreek was used as a cure for baldness in men, and in modern Java, it has continued to be used in hair tonics. Male pattern baldness is related to the activity of the hormone testosterone, and as noted above, fenugreek seeds have hormonal activity.
- Fenugreek increases the flow of milk in cows but imparts its flavor to the milk. In India, women believe that fenugreek increases milk production in recent mothers.
- In India, fenugreek plants are added to stored grain to repel insects.

FIGURE 36.1 Fenugreek (*Trigonella foenum-graecum*). (From Köhler, 1883–1914.) Seeds at lower left by M. Jomphe.

KEY INFORMATION SOURCES

Bajaj, M., Aggarwal, P., Minhas, K.S., and Sidhu, J.S. 1993. Effect of blanching treatments on the quality characteristics of dehydrated fenugreek leaves. *J. Food Sci. Technol.* 30:196–198.

Basch, E., Ulbricht, C., Kuo, G., Szapary, P., and Smith, M. 2003. Therapeutic applications of fenugreek. *Altern. Med. Rev.* 8:20–27.

Basu, S. 2009. *Production of high quality fenugreek* (Trigonella foenum-graecum). VDM Verlag, Saarbrücken, German. 176 pp.

Billaud, C., and Adrian, J. 2001. Fenugreek: composition, nutritional value and physiological properties. *Sci. Alim.* 21:3–26.

Blank, I., Lin, J., Devaud, S., Fumeaux, R., and Fay, L.B 1997. The principal flavor components of fenugreek (*Trigonella foenum-graecum* L.). In *Spices, flavor chemistry and antioxidant properties.* Edited by S.J. Risch and C.T. Ho. American Chemical Society, Washington, DC. pp. 12–28.

Fazli, F.R.Y., and Hardman, R. 1968. The spice fenugreek (*Trigonella foenum-graecum* L.): its commercial varieties of seed as a source of diosgenin. *Trop. Sci.* 10(2):66–77.

Girardon, P., Bessière, J.M., Baccou, J.C., and Sauvaire, Y. 1985. Volatile constituents of fenugreek seeds. *Planta Med.* 51:533–534.

Hooda, S., and Jood, S. 2003. Effect of soaking and germination on nutrient and antinutrient contents of fenugreek (*Trigonella foenum-graecum* L.). *J. Food Biochem.* 27:165–176.

Icon Health Publications. 2004. *Fenugreek: a medical dictionary, bibliography, and annotated research guide to Internet references.* ICON Health Publications, San Diego, CA. 128 pp.

Ismail, I. A. 1996. Changes in some nutrients of fenugreek seeds after boiling. *Bull. Nutr. Inst. Egypt.* 16:78–87.

Lawrence, B. M. 1987. Progress in essential oils. Fenugreek extract. *Perf. Flav.* 12(5):60.

Mansour, E.H., and El-Adawy, T.A. 1994. Nutritional potential and functional properties of heat-treated and germinated fenugreek seeds. *Lesbensm. Wiss. Technol.* 27:578–572.

Mazza, G., Di-Tommaso, D., and Foti, S. 2002. Volatile constituents of Sicilian fenugreek (*Trigonella foenum-graecum* L.) seeds. *Sci. Aliments*, 22:249–264.

Mir, P.S., Mir, Z., and Townley-Smith, L. 1993. Comparison of the nutrient content and in situ degradability of fenugreek (*Trigonella foenum-graecum*) and alfalfa hays. *Can. J. Anim. Sci.* 73:993–996.

Mir, Z., Acharya, S.N., Mir, P.S., Taylor, W.G., Zaman, M.S., Mears, G.J., and Goonewardene, L.A. 1997. Nutrient composition, in vitro gas production and digestibility of fenugreek (*Trigonella foenum-graecum*) and alfalfa forages. *Can. J. Anim. Sci.* 77:119–124.

Mir, Z., Mir, P.S., Acharya, S.N., Zaman, M.S., Taylor, W.G., Mears, G.J., MeAllister, T. A., and Goonewardene, L.A. 1998. Comparison of alfalfa and fenugreek (*Trigonella foenum-graecum*) silages supplemented with barley grain on performance of growing steers. *Can. J. Anim. Sci.* 78:343–349.

Petropoulos, G.A. 2002. *Fenugreek: the genus* Trigonella. Taylor & Francis, New York. 200 pp. [Medicinal aspects].

Rouk, H.F., and Mangesha, H. 1963. *Fenugreek* (Trigonella foenum-graecum *L.*). *Its relationship, geography and economic importance*. Experiment Station Bulletin No. 20. Imperial Ethiopian College of Agriculture and Mechanical Arts, Dire Dawa, Ethiopia. 8 pp.

Sauvaire, Y., Baccou, J.C., and Besancon, P. 1976. Nutritional value of the properties of the fenugreek (*Trigonella foenum-graecum* L.). *Nutr. Rep. Int.* 14:527–537.

Sauvaire, Y., Petit, P., Baissac, Y., and Ribes, G. 2000. Chemistry and pharmacology of fenugreek. In *Herbs, botanicals & teas*. Technomic, Lancaster, PA. pp. 107–129.

Sharma, R.K., Bhati, D.S., and Jain, M.P. 1999. Fenugreek. In *Tropical horticulture*, vol. 1. Edited by T.K. Bose, S.K. Mitra, A.A. Farooqui, and M.K. Sadhu. Naya Prokash, Calcutta, India. pp. 746–751.

Sheoran, R.S., Sharma, H.C., Panuu, P.K., and Niwas, R. 2000. Influence of sowing time and phosphorus on phenology, thermal requirement and yield of fenugreek (*Trigonella foenum-graecum* L.) genotypes. *J. Spices Arom. Crops*, 9(1):43–46.

Singh, J., Gupta, K., and Arora, S.K. 1994. Changes in the anti-nutritional factors of developing seeds and pod walls of fenugreek (*Trigonella foenum-graecum* L.). *Plant Foods Hum. Nutr.* 46:77–84.

Sinskaya, E.N. 1950. Fenugreek—*Trigonella* L. In *Flora of cultivated plants of the USSR*, vol. 13. Edited by E.N. Sinskaya. Ministry of Agriculture of the USSR, Leningrad. pp. 503–518. (Transl. from Russian, 1961, by Israel Program for Scientific Translations, Jerusalem).

Slinkard, A.E. 2002. *Breeding fenugreek for industrial uses*. University of Saskatchewan, Saskatoon, SK. 15 pp.

Soskov, Yu.D., and Bairamov, S.S. 1990. Subspecies of fenugreek (*Trigonella foenum-graecum*). *Nauchno-tekhn. Byull. Inst. Rastenievod.* 198:23–26 (in Russian).

Tawan, C.S., and Wulijarni-Soetjipto, N. 1999. *Trigonella foenum-graecum* L. In *Plant resources of South-East Asia. 13. Spices*. Edited by C.C. de Guzman and J.S. Siemonsma. Backhuys, Leiden, the Netherlands. pp. 225–228.

Tiran, D. 2003. The use of fenugreek for breast feeding woman. *Comp. Ther. Nurs. Midwifery*, 9:155–156.

Toghrol, F., and Pourebrahimi, M. 1976. Estimation of vitamin C in fenugreek, coriander and ribes. *Plant Foods for Man*, 2:1–5.

Udayasekhara, R.P., Sesikeran, B., Srinivasa Rao, P., Nadamuni Naidu, A., Vikas Rao, V., and Ramachandran, E.P. 1996. Short term nutritional and safety evaluation of fenugreek. *Nutr. Res.* 16:1495–1505.

Specialty Cookbooks

Note: Cookbooks specialized on foods of India are the best sources of recipes using fenugreek. For examples, see Chapter 8.

37 Galangal

Family: Zingiberaceae (ginger family)

NAMES

Scientific Names

- Greater galangal—*Alpinia galanga* (L.) Sw. (*Languas galanga* (L.) Stuntz)
- Lesser galangal—*Alpinia officinarum* Hance (*Languas officinarum* (Hance) Farw.)
- Kaempferia galangal—*Kaempferia galanga* L.

- The name "galangal" is based on the Arabic (variously rendered *hulunj, khalanjān*, khulendjan, and khalangian), from Chinese (Mandarin) *Gāoliáng jiāng*, a kind of ginger, based on *Gāoliáng*, an area in Guangdong province + *jiāng*, ginger. Galangal is sometimes spelled galangale, the Middle English spelling (but see "galangale," discussed below).
- Suggested pronunciation: guh-LANG-guhl.
- Greater galangal is also known as big galanga, Java galangal, Java ginger, Laos ginger, galangal major, greater galangal, greater galanga, Siamese galangal, Siamese ginger, and Thai ginger.
- Lesser galangal is also known as aromatic ginger, China root, colic root, Chinese ginger, East Indian catarrh root, East Indian root, galanga, galangal, galanga cardamom, ginger of the Man-tzu, lesser galangal root, and small galangal.
- Kaempferia galangal, like the other two species, is also known as galanga and galangal. It is often mistakenly called both lesser galangal and greater galangal.
- The galangals should not be confused with galingale, species of the sedge genus *Cyperus*, especially *C. longus* L. of Eurasia. *Cyperus longus* was used extensively in medieval England as a spice. The aromatic root of this species led to the Chinese word for ginger (which as explained above is the root of the word galangal) also being used as the root of galingale. Chufa, a species of *Cyperus*, is known as galingale.
- The genus name *Alpinia* commemorates Italian botanist Prosper (or Prospero) Alpino (1533–1616 or 1617).
- The genus name *Kaempferia* commemorates the German botanist and physician Engelbert Kaempfer (1651–1716), the first European scientist to describe many Japanese plants, including ginkgo (see Chapter 39). The genus *Kaempfera* of the Verbenaceae (vervain) family was also named after him, although today this name is usually encountered as a misspelling of *Kaempferia*.
- *Officinarum* in the scientific name *A. officinarum* is Latin for "of the shops" and is indicative of a plant that is (or was) used medicinally and so was sold for this purpose in medicinal shops.

PLANT PORTRAIT

Greater galangal is a perennial herb, 1.2 to 2.1 m (4–7 ft.) high. Its blade-like leaves are 25 to 60 cm (10–24 in.) long, 5 to 15 cm (2–6 in.) wide, and occur in two ranks. The flowers are greenish white with a dark-red veined tip. The fruits are red berries. The underground stems (rhizomes) are orange, pale reddish, or brown and ringed at intervals by the yellowish remnants of atrophied leaf bases. The interior is pale yellow or white and is hard and woody. This tropical species is native to Southeast Asia, probably southern China. It is cultivated in Indochina, Thailand, Malaysia, and Indonesia, and the spice made from the rhizome is very popular in Southeast Asia. Galangal is much used in Thailand as well as Malaysia, Indonesia, Cambodia, Vietnam, and Southern China. This spice is not well known in Western countries, although it was valued in Europe in the early Middle Ages.

Lesser galangal is native to China, where it is cultivated mainly on the southeast coast. It is also grown in India and the rest of Southeast Asia. Lesser galangal, as implied by the name, is smaller than greater galangal, generally less than 1.5 m (5 ft.) in height, and rarely more than 1 m (3¼ ft.) high. The species is reminiscent of an iris. It is more of a subtropical plant than the tropical greater galangal. The leaves are long and slender, roughly half the dimensions of greater galangal. The flowers are small, white with red streaks. The rhizomes are reddish brown on the outside, approximately 2 cm (3/4 in.) in diameter, with somewhat orange or red-brown flesh. These are more pungent than those of greater galangal and are similarly ringed. Lesser galangal is also used as a spice, but to a lesser extent, and mainly in Indochina and Indonesia. However, it is used as a medicinal herb to a much greater extent. Lesser galangal is rarely marketed in North America.

Kaempferia galangal is similar in appearance to lesser galangal. Its rhizomes are reddish with a white interior. The plant is widely cultivated in Southeast Asia but is much less encountered than the above two species, both as a spice and medicine. It is used as a spice mainly in Southeast Asia.

CULINARY PORTRAIT

Greater galangal is by far the principal species encountered as "galangal" in food. Fresh galangal is preferred in Asia, often to flavor vegetables and meats. Fresh galangal is widely used in all the cuisines of Southeast Asia, particularly in seafood and chicken dishes. Like ginger, galangal is a "de-fisher" and so is frequently used in fish and shellfish recipes, often with garlic, ginger, chili pepper, lemon, and/or tamarind. Thin slices are often added to Thai soups with shreds of lemon grass and lime leaves. Galangal is frequently used in the often searingly hot Indonesian cookery. The spice is usually finely cut or chopped for use in stir fries. Ground fresh galangal, often combined with onion, garlic, chili pepper, and ginger, is often used in curry pastes. Dried galangal (powder or slices approximately 3 mm [1/8 in.] thick) is more spicy than the fresh spice, resembling a blend of ginger, pepper, and cinnamon. It has been described as like ginger, but with a much more mellow taste. In Malaysia, galangal fruits are sometimes substituted for cardamom, and the flowers are occasionally consumed in salads. In Russia, galangal has been used to flavor alcoholic beverages, such as the liqueur "nastoika" as well as vinegar. The Tatars of Russia once added galangal to their tea. In Western countries, galangal may be available in the frozen food section of Asian food stores, as a powder in the spice sections of specialty food stores, and as a cut and sifted herb from bulk herb suppliers. Slices retain their quality longer than powdered material, but both forms should be stored in an airtight container and used within a short period. When unavailable, a combination of four parts of powdered ginger and one part of powdered cinnamon or cardamom can be substituted.

Lesser galangal has a much stronger, hotter, gingery flavor than greater galangal and when employed is usually added in smaller amounts than greater galangal. Both the pungent aroma and the flavor have been described as sickly sweet. The peppery rhizome is sometimes available sliced or chopped in jars.

CURIOSITIES OF SCIENCE AND TECHNOLOGY

- St. Hildegard (1098–1179), also known as Hildegard of Bingen and Hildegardis de Pinguia, was an abbess who established a convent and a Benedictine nunnery near the Rhine River. She was one of the most remarkable women the world has ever known, becoming an adviser to popes, kings, and various dignitaries. Hildegard extolled the virtues of galangal in her book *Natural Science.*

- Galangal (both greater and lesser) was used as an aphrodisiac during the Middle Ages in western Europe as well as in Asia. John Gerard (1545–1612), English botanist and barber-surgeon, in his much-cited 1597 book *The Herball, or generall historie of plantes* wrote "they conduce to venery [sex], and heate the too cold reines [loins]." Greater galangal is made into a hallucinogenic drink in New Guinea, which is said to have aphrodisiac properties. In the Jamu medicine of Indonesia, galangal is added to all preparations used as

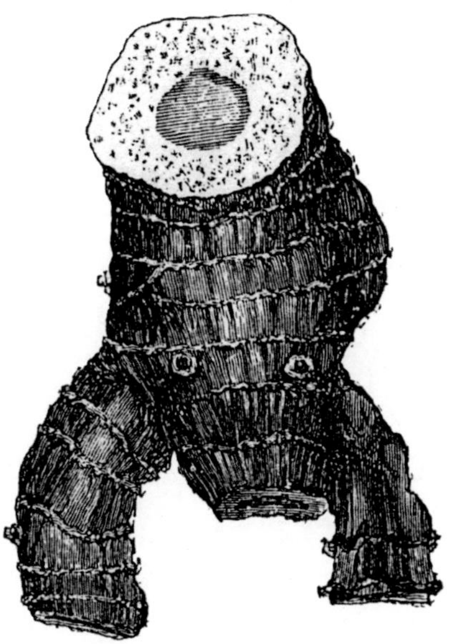

FIGURE 37.1 Root of greater galangal (*Alpinia galanga*). (From Dujardin-Beaumetz and Egasse, 1889.)

FIGURE 37.2 Lesser galangal (*Alpinia officinarum*). (From Köhler, 1883–1914.)

FIGURE 37.3 Kaempferia galangal (*Kaempferia galanga*). (From Linnaeus, 1737.)

aphrodisiacs. Although this widespread usage is very suggestive that galangal actually can increase libido, it should be noted that spicy plants almost everywhere have reputations as aphrodisiacs.

- In some Arab countries, ground galangal has been used as snuff.
- Arabs once fed galangal to horses as a stimulant.
- In India, galangal is used as a body deodorizer and halitosis remedy.

KEY INFORMATION SOURCES

General References

Amit, T., Pant, A.K., Mengi, N., Patra, N.K., and Tewari, A. 1999. A review on *Alpinia* species: chemical, biocidal and pharmacological aspects. *J. Med. Aromat. Plant Sci.* 21:1155–1168.

Ibrahim, H. 2001. *Alpinia* Roxb. In *Plant resources of South-East Asia. 12(2). Medicinal and poisonous plants 2*. Edited by J.L.C.H van Valkenburg and N. Bunyapraphatsara. Backhuys Publishers, Leiden, the Netherlands. pp. 52–61.

Smith, R.M. 1990. *Alpinia* (Zingiberaceae): a proposed new infrageneric classification. *Edinb. J. Bot.* 47:1–75.

Greater galangal (*Alpinia galanga*)

Anand, P.H.M., and Hariharan, M. 1997. In vitro multiplication of greater galangal (*Alpina galanga* (Linn.) Willd). A medicinal plant. *Phytomorphology*, 47:45–50.

Charles, D.J., Singh, N.K., and Simon, J.E. 1992. The essential oil of *Alpinia galanga* Willd. *J. Esent. Oil Res.* 4:81–82.

Cheah, P.B., and Hasim, N.H.A. 2000. Natural antioxidant extract from galangal (*Alpinia galanga*) for minced beef. *J. Sci. Food Agric.* 80:1565–1571.

Jirovetz, L., Buchbauer, G., Shafi, M.P., and Leela, N.K. 2003. Analysis of the essential oils of the leaves, stems, rhizomes and roots of the medicinal plant *Alpinia galanga* from southern India. *Acta Pharm.* 53(2):73–81.

Joy, P.P., Thomas, J., Mathew, S., and Skaria, B.P. 2001. Influence of harvest duration on the yields of rhizome, root, shoot and their oils in *Alpinia galanga*. *J. Med. Aromat. Plant Sci.* 23:341–343.

Pooter, H.L. de, Omar, M.N., Coolsaet, B.A., and Schamp, N.M. 1985. The essential oil of greater galanga (*Alpinia galanga*) from Malaysia. *Phytochemistry*, 24:93–96.

Raina, V.K., Srivastava, S.K., and Syamasunder, K.V. 2002. The essential oil of 'greater galangal' (*Alpinia galanga* (L.) Willd.) from the lower Himalayan region of India. *Flavour Fragrance J.* 17:358–360.

Rangsiruji, A., Newman, M.F., and Cronk, Q.C.B. 2000. Origin and relationships of *Alpinia galanga* (Zingiberaceae) based on molecular data. *Edinb. J. Bot.* 57:9–37.

Scheffer, J.J.C., and Jansen, P.C.M. 1999. *Alpinia galanga* (L.) Willd. In *Plant resources of South-East Asia. 13. Spices*. Edited by C.C. de Guzman and J.S. Siemonsma. Backhuys, Leiden, the Netherlands. pp. 65–68.

Yang, X., and Eilerman, R.G. 1999. Pungent principal of *Alpinia galangal* (L.) Swartz and its applications. *J. Agric. Food Chem.* 47:1657–1662.

Lesser Galangal (*Alpinia officinarum*)

Inoue, T., Shinbori, T., Fujioka, M., Hashimoto, K., and Masada, Y. 1978. Studies on the pungent principle of *Alpinia officinarum* Hance. *Yakugaku Zasshi*, 98:1255–1257.

Ly, T.N., Yamauchi, R., and Kato, K. 2001. Volatile components of the essential oils in galanga (*Alpinia officinarum* Hance) from Vietnam. *Food Sci. Technol. Res.* 7:303–306.

Kaempferia Galangal (*Kaempferia galanga*)

Ibrahim, H. 2001. *Kaempferia* L. In *Plant resources of South-East Asia. 12(1). Medicinal and poisonous plants 1*. Edited by L.S. de Padua, N. Bunyapraphatsara and R.H.M.J. Lemmens. Backhuys Publishers, Leiden, the Netherlands. pp. 331–335.

Maheswarappa, H.P., Nanjappa, H.V., and Hegde, M.R. 2000. Influence of agronomic practices on growth, productivity and quality of galangal (*Kaempferia galanga* L.) grown as intercrop in coconut garden. *J. Plantation Crops*, 28:72–81.

Maheswarappa, H.P., Nanjappa, H.V., and Hegde, M.R. 2000. Dry-matter production and accumulation in different parts of galangal (*Kaempferia galanga*) as influenced by agronomic practices when grown as an intercrop in coconut garden. *Indian J. Agron.* 45:698–706.

Maheswarappa, H.P., Nanjappa, H.V., Hegde, M.R., and Prabhu, S.R. 1999. Influence of planting material, plant population and organic manures on yield of East Indian galangal (*Kaempferia galanga*), soil physico-chemical and biological properties. *Indian J. Agron.* 44:651–657.

Specialty Cookbooks

Poth, S., and Sauer, G. 2000. *The spice lilies: eastern secrets to healing with ginger, turmeric, cardamom and galangal*. Healing Arts Press, Rochester, VT. 128 pp.

38 Ginger

Family: Zingiberaceae (ginger family)

NAMES

Scientific Name: *Zingiber officinale* Roscoe (*Amomum zingiber* L.)

- The English word "ginger" is derived from the Latin word for the plant and spice, *zingiber*, through the Greek *zingiberi*, and ultimately back to the Sanskrit *singabera*, explained below.
- Ginger is also found under the names Canton ginger, common ginger, East Indian ginger, Jamaica ginger (Jamaican ginger), red ginger, stem ginger, and true ginger.
- The word "gingerly," meaning carefully, has nothing to do with ginger.
- Mioga ginger (*Zingiber mioga* (Thunb.) Roscoe) is a native of China and perhaps Japan, which is grown in more temperate climates than ginger. Its underground parts are used as a source of a ginger with a bergamot-like flavor (sometimes called "Japanese ginger"). In California, some mioga ginger has been grown for the Japanese ethnic community, the sprigs eaten fresh as part of a Buddhist tradition.
- Wild ginger (*Asarum canadense* L.) is a native of North America growing from Manitoba to New Brunswick, south to Kansas and North Carolina. Its underground parts have the taste of ginger and have been used by Native Americans for flavor and for medicinal purposes.
- Several exotic ornamental species in the ginger family have ginger in their names, including red ginger (*Alpinia purpurata* (Vieill.) K. Schum.), shell ginger (*A. zerumbet* (Pers.) B.L. Burtt & R.M. Sm.), torch ginger (*Etlingera elatior* (Jack) R.M. Sm. (*Nicolaia elatior* (Jack) Horan.)), white ginger (*Hedychium coronarium* J. König), and two species called yellow ginger (*H. flavescens* Carey ex Roscoe, *H. flavum* Roxb.).
- The genus name *Zingiber* comes from the Sanskrit word *shriingabera*, meaning "horn shaped" (i.e., like a deer's antlers), in reference to the branching shape of the underground stems.
- *Officinale* in the scientific name *Z. officinale* is Latin for "of the shops," a phrase that was used in past centuries to indicate plant species used in medicine (and therefore sold in medicinal shops).

PLANT PORTRAIT

Ginger is a herbaceous perennial, producing reed-like stalks 30 to 150 cm (1–5 ft.) high from a thick, creeping, jointed rhizome (underground stem). Flowering stalks and nonflowering, leafy stems emerge separately from the rhizome (flowering stalks are often not produced in some climates). Varieties vary in rhizome pungency and color, with white, pale yellow, yellow, and red predominating. The area of origin is uncertain, likely tropical Asia or Indonesia. Today, the major producing areas are the West Indies (especially Jamaica), India, West Africa (especially Sierra Leone), Malaysia, Sri Lanka, Fiji, Hawaii, Japan, and Queensland (Australia). Other areas of cultivation include China, Egypt, Nigeria, Mexico, Hawaii, and Central and tropical South America (especially Brazil). Ginger from Jamaica is considered by some to be of the highest quality. Ginger is not cultivated as a crop in North America.

Ginger is one of the oldest and most widely consumed spices and has been extensively spread since ancient times. It was one of the first spices to reach Europe from Asia and was originally thought to be an Arabian product because it was transported by Arabian traders. The ancient Romans distributed ginger products throughout their empire. Ginger was mentioned by the Chinese philosopher Confucius (551–479 BC). Italian traveler Marco Polo (ca. 1254–1324), was the first European to record ginger growing in China. In fourteenth-century England, it was the second most common spice after pepper.

Ginger has a long history of use as a medicinal plant. It was used in the Middle Ages to treat the Black Death and may have been of some help because it can promote sweating. The claim has been made that ginger is an ingredient in almost half of all Oriental herbal medicines. It is considered quite useful in preventing the effects of sea sickness and motion sickness. Indeed, Chinese sailors chewed ginger to prevent sea sickness well before medical science had demonstrated the effectiveness of this treatment.

The familiar commercial ginger comes from the fleshy rhizome, called "root" in the commercial trade. The so-called "stem ginger" is taken from the new growth of rhizomes, highly regarded for tenderness because of lack of fibers. ("Root ginger," by contrast, is harvested from older, more mature portions of rhizome.) Ginger from different countries is often very characteristic because of a combination of the variety grown, cultivation techniques, climate, and mode of preparation. Thus, different forms of commercial ginger are obtained from Jamaica, Sierra Leone, Nigeria, India (Cochin ginger), and Japan. The essential oil from ginger is used in cosmetics such as men's toilet lotions, pharmaceuticals, and perfumes and for culinary purposes as recorded below.

CULINARY PORTRAIT

Ginger is an extraordinary spice. It rounds out the flavor of some foods, accents others, and contributes a unique freshness. Ginger's pungency increases with the application of heat. The range of foods with which ginger can be used, as noted below, is exceptional. Commercial products include

FIGURE 38.1 Ginger (*Zingiber officinale*). (From Köhler, 1883–1914.)

fresh rhizomes, dried ground ginger, and candied ginger. The fresh root is an important component of Asian and Oriental cooking. The taste of fresh material is considered milder than that of the dried product. In New Delhi, India, shops sell 20-year-old pickled ginger called *gillori*, made of large wafer-thin slices rolled into little triangles around a stuffing of spices and pinned with a clove. After scraping the skin away, fresh ginger may be preserved in strong spirit or sherry. Ginger is also canned, crystallized in sugar, or preserved in syrup, a practice originating in ancient China. Stem ginger is the form preferred for the manufacture of preserved or crystallized ginger because it is less fibrous than older parts of the rhizome. The pink-colored young rhizomes are milder than the mature rhizome as they contain less gingerol, which makes ginger "hot." In Japan, the young rhizomes are marinated briefly in rice vinegar and sugar and are known as red-pickled ginger or *beni-shoga*. The rhizomes may be washed and dried with the skin left on; the resulting product is relatively dark and is known as "green" or "black" ginger (although the outer flesh is actually brown). This is sometimes parboiled, skinned, and bleached white. Peeled ginger is called white ginger. A fine powdering of lime may be added, and this preparation is highly regarded. In cooking, dried ginger is "bruised" by hitting with a rolling pin or hammer to open the fibers and allow the flavor to be released.

In the West, dried powdered ginger has traditionally been the most commonly marketed form, although fresh ginger root is now widely available in supermarkets. The powdered product is of lower quality and it may have been adulterated. Fresh ginger "root" (rhizome) should have a firm skin; wrinkled skin indicates the roots are too old. The volatile essential oil that provides the special flavor is often lost on drying. Fresh ginger is peeled and either thinly sliced or ground to a pulp for inclusion in curries, meat, poultry, and fish dishes. The flesh just underneath the skin is youngest and most delicate, and the root should be peeled carefully to preserve this portion ("spring ginger," often imported from Asia in the autumn, does not require peeling). Shrimp, scampis, or mussels are delicious stirred in a pan with a mixture of ginger, soy sauce, and brown sugar. Ginger is commonly added to Chinese stir fries and vegetable dishes and is also popular in Japanese cooking. In Japan, ginger is traditionally served with sushi and sashimi. "Pink ginger," commonly served in Japanese restaurants as a garnish, can be made by removing the skin of very young ginger roots, slicing the ginger very thin, dipping it in lemon juice (which turns the ginger pink), and seasoning with salt. Ginger is a component of many types of baked goods including gingerbread, cakes, gingersnaps and other cookies, pumpkin pie, and puddings. It is also added to creams, sweet sauces, ginger beer, ginger wine, cordials, soft drinks, punches, jam, and candies. As a table condiment, it is sprinkled on fruits, particularly apples, bananas, melons, peaches, and pears. Ginger is delicious on ice cream or made into a syrup for use in barbecue sauce, baked apples, fruit compôte, and ice cream. It can also be added to soups, macaroni and cheese, mincemeat, pickles, sweet rolls, and fried chicken. The taste of winter squash, pumpkin, sweet potatoes, and carrots is particularly enhanced by ginger. Young, spicy ginger shoots are also eaten as a potherb or puréed and used in sauces and dips. Ginger becomes milder with cooking but turns bitter if burned.

Commercially, ginger or one of its extracts is combined with products similar to many of those mentioned above. Ginger is often added to prepared meats such as sausage, knockwurst, bologna, salami, and pressed ham. The essential oil is occasionally used in flavoring drinks, cookies, and desserts. Ginger extracts are used to flavor soft drinks such as ginger ale, ginger beer, and gingerade.

Ginger can be stored fresh in a refrigerator for about a week if wrapped in plastic to protect it against excess humidity. For longer storage, place the piece of rhizome in a small jar of sherry, and refrigerate. It will apparently keep almost indefinitely this way. The rhizome can be frozen. However, it should be grated for use in recipes while still frozen because thawed ginger root becomes too mushy to grate. The rhizomes can be stored for a year by steeping them in a solution of salt, acetic acid, and potassium metabisulphite and then kept in clean jars. This does not require refrigeration. Pieces can be removed from the rhizomes as needed.

Culinary Vocabulary

- The thick, somewhat flattened rhizomes of ginger are called "hands" in the ginger trade because they commonly divide into finger-like projections. These are also called "races" in the trade, based on the Portuguese-Spanish *raices*, roots.
- "Subgum ginger" (also known as ginger pickles and mixed Chinese pickles) is a sweet, tangy combination of preserved ginger, fruit, vegetables, and spices, sold in jars and cans, and used as a relish.
- "Crystallized ginger" in British English is equivalent to "candied ginger" in American English. The British "gingernuts" means "ginger snaps" in North America. The British "root ginger" is expressed as "ginger root" in America.
- A "Moscow Mule" (said to have the kick of a mule) is a cocktail made of vodka and a squeeze of lemon or lime juice topped with ginger beer or ginger ale. The beverage has been credited with popularizing vodka in the 1950s in the United States. Rather than originating from Russia, in fact the Moscow Mule was a marketing ploy of John G. Martin of Heublein Inc., an East Coast spirits and food distributor, best known at the time for A-1 Steak Sauce. In 1939, Martin persuaded the company to purchase the Smirnov (anglicized to "Smirnoff") vodka distillery, an event which became known as "Martin's Folly" until the clever businessman collaborated in 1946 in the invention and marketing of the new cocktail.
- A "Horse's Neck" is a tall glass of ginger ale, served with lemon peel over ice or the same thing with bourbon, whiskey, or gin. (The expression has also been used for a combination of moonshine and dry, hard cider.)
- A "shandy gaff" is a bottle of ginger beer mixed with a pint of ale.
- "The Ginger champagne" cocktail, like ginger beer and ginger ale, is a ginger-flavored "mocktail" (nonalcoholic beverage prepared like a cocktail).
- A "gingersnap" is a cookie made from ginger and molasses. "Snap," an English word dating from 1842, probably is based on the idea that something is done easily (from the German or Middle Dutch *snappen*, to seize quickly).
- "Maupygernon" is chicken smothered in ginger and cloves, a very old dish but not one familiar to most people. In 1954, Sir Robert Perkins delivered a speech in the British House of Commons, attacking the kitchen committee that ran the parliamentary restaurant for not having it on the menu for the last 300 years.

CURIOSITIES OF SCIENCE AND TECHNOLOGY

- After large meals, the classical Greeks ate ginger wrapped in bread as a digestive aid. Gradually, the herb was put in the bread itself, and thus gingerbread was created.
- The rulers of ancient Rome liked the population to believe that it was not taxed. In fact, ginger was very expensive (costing Roman citizens about 15 times as much as pepper) simply because the Romans arranged to collect a very heavy tax in Alexandria, Egypt, which was technically outside of Roman territory, so that the fiction of no taxes could be claimed.
- Ginger has often been used as a bribe. In medieval Venice, spice importers bribed the Master of the Treasury annually with a standardized list of gifts, which included 0.45 kg (1 lb.) each of pepper, cinnamon, and ginger, to allow their merchandise to enter Europe at reasonable customs rates. The House of Burgesses was the lower house of the legislature in colonial Virginia. It started in1619 as the first representative government body in America, consisting of a governor and his council, and six representatives, who were called "Burgesses." Members in the early period of the system were elected by the existing Burgesses. The House of Burgesses is considered by historians to have paved the way for fuller democracy. In 1757, George Washington, the future president of the United States,

lost an election to the House, receiving only 40 votes, but the next year he was elected with a vote of 310. Ginger cookies were among the bribes that politicians used to persuade the voters of Virginia that they deserved to be elected.

- In the thirteenth century, a pound of ginger was worth a sheep.
- In Medieval England, early forms of gingerbread were as popular a present as a box of chocolates is today. Gingerbread slabs presented as presents frequently had the heads of cloves (see Chapter 27) coated with gold paint and driven in like nails to form a fleur-de-lys pattern or had some other artistic gilded applications.
- Ginger has been regarded as an aphrodisiac since ancient times. Greek philosopher Aristotle (384–322 BC), noted that ginger was useful in his day for the same purpose that Viagra is used today. The aphrodisiac qualities were alluded to in *The Thousand and One Nights*. Ginger was purported to be so potent that after simply rubbing one's hand with ginger, just touching a woman was sufficient to win her heart. In Japan, especially thick rhizomes of ginger are believed to be especially potent. The common use of ginger in men's toilet water may stem from the traditional belief that ginger is an aphrodisiac.
- Most heat-producing herbs and spices have reputations as aphrodisiacs. Not only ginger but also black pepper, chili peppers, horseradish, and the mustards are notable for producing heat in the body following ingestion, or even (in the case of the mustards and horseradish) producing heat when simply combined with water.
- Queen Elizabeth I, (1553–1603), is credited with inventing the gingerbread man. She is said to have had her cooks prepare little ginger-flavored cakes shaped like portraits of people familiar to her.
- During the American Revolution, 1773–1765, hard ginger cookies were part of the standard rations of the American soldiers.
- During the nineteenth century, English taverns kept ground ginger that thirsty customers could sprinkle on top of their beer or ale and then stir into their drinks with a red hot poker.
- Canada Dry Ginger Ale was invented by a Canadian, John A. McLaughlin, in 1907.
- Ginger added to the Japanese delicacies sushi and sashimi (raw or slightly cooked fish) may be useful in suppressing the dangerous larvae of the nematode *Anisakis*. The Chinese believe that ginger is an antidote to shellfish poisoning and therefore often season fish and seafood with it.
- Contact with fresh ginger root stimulates circulation, and the fresh root can be applied to the cheeks as a natural "rouge."
- The importance of a plant is often indicated by the extent to which its name has influenced language, and this is true for ginger. During the Middle Ages, ginger was considered so important that the street where traders sold spices in Basel, Switzerland, was called Imbergasse, meaning "Ginger Alley." The name Ginger was commonly applied to redheaded females in Elizabethan England (the period of Elizabeth I) because of their supposed hot temperaments. "Gingerbread-work" refers to fanciful shapes and ornate carvings used to decorate furniture, buildings, and other objects. Gingerbread in this sense was based on the observation of very fancy gingerbread baked goods. Gingerbread work (or just "gingerbread") in architecture was once generally considered gaudy, tawdry, and superfluous, and so the word has been (and continues to be) used pejoratively. However, the elaborate gingerbread on very old houses and objects is now valued. Nevertheless, "gingerbready" means tawdrily showy and overly ornamented.

KEY INFORMATION SOURCES

Akhila, A., and Tewari, R. 1984. Chemistry of ginger: a review. *Curr. Res. Med. Aromat. Plants*, 6:143–156.

Awang, D.V.C. 1992. Ginger. *Can. Pharm. J.* 125:309–311.

Badshah, Z., and Badshah, K. 1979. Diseases of ginger—a review. *Pak. J. For.* 29:110–117.

Bone, M.E., Wilkinson, D.J., Young, J.R., McNeil, J., and Sharlton, S. 1990. Ginger-root—a new antiemetic. The effect of ginger root on postoperative nausea and vomiting after major gynaecological surgery. *Anesthesia*, 45(8):669–671.

Branney, T.M.E. 2005. *Hardy gingers: including* Hedychium, Roscoea, *and* Zingiber. Timber Press, Portland, OR. 268 pp.

Cude, K. 1985. For a lively spice—grow ginger. *California Rare Fruit Growers Newsl.* 17(3):10–15.

Ernst, E., and Pittler, M.H. 2000. Efficacy of ginger for nausea and vomiting: a systematic review of randomized clinical trials. *Br. J. Anaesth.* 84:367–371.

Govindarajan, V.S. 1982. Ginger—chemistry, technology, and quality evaluation. Part 1. *CRC Crit. Rev. Food. Sci. Nutr.* 17:1–96.

Holtman, S., Clarke, A.H., Schereer, H., and Hohn, M. 1989. The anti-motion sickness mechanism of ginger. *Acta Otolaryngol.* 108:168–174.

Kikuzaki, H. 2000. Ginger for drug and spice purposes. In *Herbs, botanicals & teas.* Edited by G. Mazza and B.D. Oomah. Technomic, Lancaster, PA. pp. 75–105.

Leverington, R.E. 1983. Ginger. In *Handbook of tropical foods.* Edited by H.T. Chan, Jr. Marcel Dekker Inc., New York. pp. 297–350.

Peter, K.V., and Kandiannan, K. 1999. Ginger. In *Tropical horticulture*, vol. 1. Edited by T.K. Bose, S.K. Mitra, A.A. Farooqui, and M.K. Sadhu. Naya Prokash, Calcutta, India. pp. 671–682.

Ravindran, P.K., and Babu, K.N. 2004. *Ginger: the genus* Zingiber. CRC Press, Boca Raton, FL. 384 pp.

Ravindran, P.N., Sasikumar, B., George, J.K., Ratnambal, M.J., Babu, K.N., Zachariah, J.T., and Nair, R.R. 1994. Genetic resources of ginger (*Zingiber officinale* Rosc.) and its conservation in India. *Plant Genet. Resour. Newsl.* 98:1–4.

Rodriquez, D.W. 1971. *Ginger, a short economic history.* Agricultural Information Service, Ministry of Agriculture and Fisheries, Kingston, Jamaica. 35 pp.

Schulick, P. 2000. *Ginger: common spice & wonder drug.* 3rd ed. Hohm Press, Prescott, AZ. 166 pp.

Seelig, R.A., and Bing, M.C. 1990. Ginger root. In *Encyclopedia of produce* [Loose-leaf collection of individually paginated chapters]. Edited by R.A. Seelig and M.C. Bing. United Fresh Fruit and Vegetable Association, Alexandria, VA. 7 pp.

Sethi, V., and Anand, J.C. 1982. Preserving ginger all the year round. *Indian Hortic.* 27(3):13.

Sutarno, H., Hadad, E.A., and Brink, M. 1999. *Zingiber officinale* Roscoe. In *Plant resources of South-East Asia. 13. Spices.* Edited by C.C. de Guzman and J.S. Siemonsma. Backhuys, Leiden, the Netherlands. pp. 238–244.

Van Beek, T. A. 1991. Special methods for the essential oil of ginger. In *Modern methods of plant analysis. New Series. Vol. 12. Essential oils and waxes.* Edited by H.F. Linskens and J.F. Jackson. Springer Verlag. pp. 79–97.

Vasala, P.A. 2001. Ginger. In *Handbook of herbs and spices.* Edited by K.V. Peter. Woodhead Publishing, Cambridge, U.K. pp. 195–206.

Wolff, X.Y., Astuti, I.P., and Brink, M. 1999. *Zingiber* G.R. Boehmer. In *Plant resources of South-East Asia. 13. Spices.* Edited by C.C. de Guzman and J.S. Siemonsma. Backhuys, Leiden, the Netherlands. pp. 233–238.

Specialty Cookbooks

Ager, A. 1976. *The ginger cookbook.* Vantage Books/Pinner, London. 144 pp.

Conrad, M., and MacDonald, H. 1997. *The joy of ginger: a winning selection of taste-tingling recipes.* Nimbus, Halifax, NS. 120 pp.

Cost, B. 1989. *Ginger east to west: the classic collection of recipes, techniques, and lore.* Revised edition. Addison-Wesley, Reading, MA. 209 pp.

Gordon-Smith, C., Merrell, J., and Petersen-Schepelern, E. 1998. *Basic flavorings. Ginger.* Courage Books, Philadelphia, PA. 64 pp.

Lorenz. 1996. *Ginger: a book of recipes.* Lorenz, New York. 62 pp.

Pappas, L.S. 1996. *Ginger.* Chronicle Books, San Francisco, CA. 71 pp.

Seeley, C. 1977. *Ginger up your cookery.* Hutchinson, London. 96 pp.

39 Ginkgo

Family: Ginkgoaceae (ginkgo family)

NAMES

Scientific Name: *Ginkgo biloba* L.

- The English name ginkgo and the genus name *Ginkgo* are based on the Chinese ideogram yin-kuo (*yin*, silver + *hing*, apricot), "silver apricot," a reference to the appearance of the seeds. This was translated into the Japanese ideogram *ginkō*, which was in turn translated into the Roman alphabet by Engelbert Kaempfer (1651–1716), a German physician and traveler who studied for some time in Japan. The presence of the second g in ginkgo has been debated as possibly the result of an erroneous transliteration into Roman script, or based on an old Japanese pronunciation (Japanese ideograms often have several pronunciations). In any event, the very awkward spelling in English provides no clue to the pronunciation of the word, which is now widely "GIN-koe," with a hard "g" (i.e., not JIN-koe, which is also encountered). The pronunciations GING-ko and JING-ko are also sometimes heard.
- Ginkgo is occasionally spelled gingko.
- Plural of ginkgo: ginkos, ginkgoes (more frequent).
- The original common name of ginkgo in China was "duck's foot" (*ya chio*), an allusion to the shape of the leaf. Later, the plant became known as "silver apricot," the name that was eventually transformed, as noted earlier, to ginkgo.
- Ginkgo is also known as the maidenhair tree because of the resemblance of the leaves to those of maidenhair ferns (species of the genus *Adiantum*).
- One of the old nicknames in China for ginkgo is "tree of grandfather and grandson," a clever way of saying that when a man plants the slow-growing tree, nuts will not become available before his grandson is born. (Trees usually start to produce nuts at 15–20 years of age, with abundant production after approximately 35 years.)
- An occasional name for ginkgo is "golden fossil tree," references to its yellow autumn foliage and its existence since ancient times.
- In 1771, the English botanist J.E. Smith decided that the genus name *Ginkgo* was "uncouth and barbarous," and he renamed it *Salisburia adiantifolia* (to honor the botanist William Salisbury), a name that was used for some time in literature of the period. However, botanical names are governed by an international code that does not permit anyone to change a scientific name merely because he or she does not like it.
- *Biloba* in the scientific name *G. biloba* means two-lobed, referring to the typical appearance of the leaves.

PLANT PORTRAIT

Ginkgo is a deciduous tree that can grow more than 35 m (115 ft.) in height. The leaves are variable in shape. The leaf blades are typically fan shaped, but not always with two prominent lobes. Ginkgo trees are long lived, and ancient specimens are often found beside Buddhist and Taoist temples and shrines, where the trees are considered to be sacred. (In former times, temples were commonly built beside wild trees, and it is possible that very old trees on temple grounds represent the remnants of ancient forests.) The species may still grow wild in a small area of southeast China, in remote mountain valleys of Zhejiang province. It is there that the oldest trees are found, although it is uncertain whether there are any genuinely wild trees, as they may all have been planted. It has been estimated that there are approximately 100 trees that are more than 1000 years old in China, a few of which may be 3000 years of age. Written records indicate that ginkgo has been cultivated in China for more than 2000 years. From China, it was spread to Japan about a thousand years ago, to Europe in 1727 (in the Netherlands; one of the original trees is still alive), to the United States in 1784 (the original tree is said to have lived for approximately 200 years), and to other temperate areas of the world. It has been claimed that ginkgo now grows in every temperate country of the world.

Ginkgo has both male and female trees. In Asia, female trees are preferred because they produce the nuts used as food. When the fruits ripen they fall to the ground and the fleshy shell bursts open, emitting an extremely disagreeable smell. In Western countries, male trees are overwhelmingly preferred because they are grown as ornamentals not as food plants, and the putrid odor of the fruits (which of course occur only on the females) can be avoided.

The fruits are the size of small plums. They contain a large solitary creamy-white kernel, covered by a thin layer of brown skin, within a hard shell, which is covered by a fleshy outer layer that is bright orange, yellow, or orange-yellow when ripe. The edible seeds are called "ginkgo nuts" and are commercially produced in orchards. In the late 1990s, China annually produced approximately 6000 tons of nuts from approximately 800,000 trees. Occasional female plants are present in North America, and people of Asian background make a point of harvesting the nuts when they fall in the autumn. Although squirrels will collect the fruits, they are not strongly attracted to them, so they accumulate under the trees. Oils in the fresh pulp contain urushiol, which is responsible for poison oak and poison ivy contact dermatitis. Handling the fresh fruits can produce severe dermatitis in many individuals, so rubber or latex gloves should be worn.

The ginkgo family (Ginkgoaceae) originated in the Permian geological period (200–225 million years ago), becoming most diverse in the Jurassic (approximately 100 million years ago), with at least six genera and 12 species. Fossils of species of the genus *Ginkgo* have been dated to 200 million years. Several species of *Ginkgo* once occupied North America, including the apparent ancestor of *G. biloba*. Ginkgoes disappeared from North America approximately 7 million years ago and from Europe approximately 2.5 million years ago. Charles Darwin (1809–1882), the foremost student of biological evolution, used the phrase "living fossil" to designate species that have survived from the distant past. The odd fish known as the coelacanth is possibly the best example of an animal considered to be a living fossil. Of all of the world's plants, none has a stronger claim to the designation of living fossil than ginkgo. Fossils show that *G. biloba* is virtually the same plant that was present when dinosaurs ruled the earth 125 million years ago.

Although ginkgo comes from China, and the Chinese have used the plant for medicine for centuries, ginkgo has become more widely used for medicinal purposes in the West than in its homeland. It is used to treat a wide range of ailments, including asthma, tinnitus (ringing in the ears), hearing loss, cardiovascular disease, and impotence. Ginkgo is now very well known in the treatment of severe memory loss, a usage that has been validated scientifically. It is thought that ginkgo acts by increasing blood flow and oxygen flow to the brain. In Germany, millions of prescriptions for ginkgo are filled annually, and ginkgo is the best-selling plant-based medicine in Europe. Ginkgo is among the 10 most popular "dietary supplements" in the United States. Substances called ginkgolides are thought to be particularly important for medicinal purposes. Advertisements for nonprescription, over-the-counter products suggest it may improve memory for normal people but is not at all clear that there is any truth to this, and millions of people may be wasting their money buying such ginkgo products.

With the rise of medicinal usage of ginkgo leaf extracts in Western culture, trees began to be cultivated in orchards for harvest of leaves in Europe (starting in 1982 in Bordeaux, France) and in China and the United States in the 1990s. The world's largest plantation, occupying almost 5000 ha (12,000 acres), is in Sumpter, South Carolina, and is owned by a German company. The South Carolina plants are cultivated in extremely high density (40 cm or 1.3 ft. apart) and kept very short to

FIGURE 39.1 Ginkgo tree (*Ginkgo biloba*). (From Flora and Sylva, 1904, vol. 2.)

FIGURE 39.2 Ginkgo (*Ginkgo biloba*). Left: female flowering branch. (From Siebold and Zuccarini, 1870.) Right: male flowering branch. (From Jacquin, 1819.)

make mechanical harvesting of the leaves easy. Most of the ginkgo leaf used to manufacture ginkgo medicinal preparations comes from the South Carolina farm. The extracts are prepared in Germany and are the main source of ginkgo medicinal preparations in North America.

Gingkos are widely planted as ornamentals, particularly as park and street trees. They are hardy and extremely resistant to diseases, insects, and pollution (except for acid rain). Ornamental forms with yellow leaves and a selection with yellow/green variegated leaves are available. Some cities

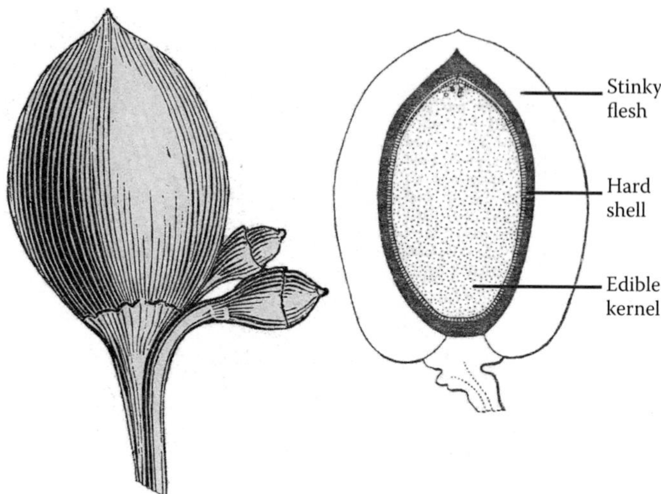

FIGURE 39.3 Ginkgo (*Ginkgo biloba*). Left: mature fruit with two immature fruit attached. (From Loudon, 1844.) Right: anatomy of the fruit. (Based on Engler and Prantl, 1889–1915.)

have forbidden the planting of female trees because of the noxious smell of the fruit and associated problems, including dermatitis, toxicity, and even damage to paint on cars parked below the trees.

CULINARY PORTRAIT

To prepare the nuts, the seeds are taken out of their fleshy shells, soaked to facilitate removal of their skins, and dried. The seeds may be roasted as snacks, or boiled for use in soups, stir fries, and other dishes. The nuts do not store well and are canned in China and Japan. Canned nuts, with the outer fleshy layer removed, are available in Western countries. Once the can is open, the nuts need to be kept covered in water in the refrigerator and will last for only a few days. The canned nuts can be boiled and used in soups, stews, stir-fried dishes, tempura, and other dishes. Some authorities consider the canned nuts to be inferior to fresh nuts, but the latter are rarely available outside of Asia.

Ginkgo nuts are a semiluxury food in China, Korea, and Japan, typically served at weddings, banquets, and other important social functions. When boiled or roasted, they have a moist, tender, fleshy texture, and a pleasantly mild, resinous flavor that has been compared with Swiss cheese, roasted chestnuts, and almonds. The resinous quality is not always appreciated by those unaccustomed to it. The nuts are added to soups and to vegetables, fish, pork, seafood, and poultry dishes. In Japan, ginkgo nuts are commonly grilled or boiled in *chawan-mushi*, a pot-steamed egg dish, or in *nabe-ryori*, a sort of Japanese fondue. Grilled nuts are also often eaten as an accompaniment to sake. The Japanese commonly consume the nuts as a dessert fruit.

A toxic substance (antipyridoxine) occurs in the fresh seeds and has poisoned people consuming the nuts both for medicinal and food purposes. The toxin has an antivitamin effect, and it is dangerous for people suffering from vitamin deficiency to eat large quantities of fresh nuts. However, the nuts mature to some extent after falling from the tree, lowering the level of the toxin, and dried or processed ginkgo nuts used in cooking are generally considered safe. In China, visitors unfamiliar with eating ginkgo nuts have been warned not to eat more than seven at one time, and children especially are not allowed to eat very many nuts. A report in the *Journal of the American Medical Association* in 2001 recommended that patients refrain from taking ginkgo before surgery as it reduces the ability of the blood to clot, which could increase the risk of excessive bleeding during operations.

Culinary Vocabulary

- In Western countries, ginkgo nuts are often sold in ethnic markets as "silver almonds" and "white nuts."

CURIOSITIES OF SCIENCE AND TECHNOLOGY

- During the age of the dinosaurs, a war for survival took place between the huge reptiles and trees on which many of the giant plant eaters fed. The foliage and fruits had to be more than 24 m (80 ft.) to be sure of escaping from the tallest dinosaurs, and ginkgo was often able to grow to 30 m (100 ft.) or more, which was probably one reason why it is still alive. An exceptionally tall tree in Korea was measured at 60 m (197 ft.).
- Ginkgo Petrified Forest near Ellensburg in the state of Washington is estimated to be 15 million years old.
- Very old ginkgo trees often develop "air roots," curious peg-like structures that grow from the trunk or lower branches, take root when they contact soil, and produce new branches. The air roots may be 5 m (16 ft.) long and 30 cm (1 ft.) in diameter. These represent a sort of insurance policy for the plant because when the old plant dies, the air roots turn into independent plants, which are clones of the old tree.
- Male plants are normally necessary to provide pollen before female ginkgo plants will set fruit. In old China, to save the space that a male tree would take, a single branch from a male tree was sometimes grafted onto a female tree, this providing sufficient pollen for the entire female tree. For many years in Western countries, female trees were extremely rare, and often far from male plants that could provide pollen, so horticulturists wishing to produce seeds used the same trick.
- In Chinese gingko orchards, one male plant was commonly planted for every 100 females. More recent practice recommends that three to five male plants be planted for every 100 females.
- Animals, including humans, have sperm that swim to reach and fertilize eggs. In plants, sperm also swim, but only in the more primitive classes of plants, such as algae, mosses, and ferns. In the flowering or seed plants, nonswimming sperm are delivered to the egg by a tube that grows from the pollen grain, penetrates the stigma, and grows directly to the egg. Ginkgo has swimming sperm, reflecting its primitive nature.
- Freshly fallen ginkgo fruits smell like rancid butter and Limburger cheese, and with good reason: the same chemical, butyric acid, is responsible. The odor has been variously described as like vomit, disagreeable, evil, offensive, disgusting, repulsive, nauseating, and abominable. Likely during the age of the dinosaurs, the foul smell of the fallen fruit served to attract animals that would distribute the seeds. It is possible that dinosaurs were attracted by the smell, ate the fruits, and the seeds passed through their guts to be deposited in a huge heap of dung that would have provided nutrients for the growing seedlings, much as happens today for some plants that use elephants to distribute their seeds. Crows and raccoons have been observed to be attracted to smelly ginkgo fruits, and birds are thought to be the most important distributors of ginkgo seeds in modern times. As is well known, dinosaurs are the ancestors of birds (birds are said to be flying dinosaurs), and the birds may simply be playing the same role in eating and distributing ginkgo seeds that the dinosaurs carried out.
- Butyric acid, as noted above, provides not only the foul smell of ginkgo, but also of Limberger cheese and rancid butter. People who eat meat do not notice that they give off the odor of butyric acid, which is present in animal fat; those who eat little meat, such as the Japanese, do notice the odor, and in the nineteenth century they dubbed Westerners *bata-kusai*, "butter-stinkers," because of the characteristic smell. (Prejudices based on

food intake work both ways between the Japanese and Westerners: the considerable fish diet of Japanese has been associated by some with a "fishy odor.")

- Approximately 80% of North Americans are sensitive to poison ivy or poison oak (several species of *Rhus*), most people becoming sensitized in late childhood. Once sensitized, there is an increased likelihood that they will also experience dermatitis from contact with ginkgo (Epstein, W.L. 1974. Poison oak and poison ivy dermatitis: an occupational problem. *Cutis*, 13:544–548).
- Dry leaves of ginkgo, collected in the autumn, when they have developed a beautiful yellow color, have been used traditionally as bookmarks in Japan, not just because they are attractive but also because they are believed to discourage booklice and silverfish that feed on the starch in the book bindings.
- Ginkgo has a strong reputation in Chinese folklore for preventing hangovers. Japanese scientists found some proof for this (reported in *Chem. Pharm. Bull.* 37:155–159, 1989): they discovered substances in the seeds that cause faster elimination of alcohol from the blood.
- The ginkgo-toothed whale, also known as the ginkgo-toothed beaked whale (*Mesoplodon ginkodens*), is so-named because the species has teeth that are shaped like ginkgo leaves. Only mature males have functional teeth, consisting of a single pair, each tooth approximately 10 cm (4 in.) wide, in the middle of the lower jaw. The teeth seem to be barely usable for catching prey and could be weapons to intimidate other males. Prey may be seized and disabled using the hard edges of the mandibles. This extremely rare whale was not named until 1958. It has not been observed swimming at sea and is known only from a dozen or so beached or netted specimens. It is less than 5 m (16 ft.) in length, less than 1500 kg (3300 lb.), and inhabits tropical to warm temperate waters of the Pacific and Indian Oceans.

KEY INFORMATION SOURCES

Arenz, A., Klein, M., Fiche, K., Groß, J., Drewke, C., Hemscheidt, T., and Leistner, E. 1996. Occurrence of neurotoxic 4'-O-methylpyridoxine in *Ginkgo biloba* leaves, ginkgo medications and Japanese ginkgo food. *Planta Med.* 62:548–551.

Becker, L.E., and Skipworth, G.B. 1975. Ginkgo-tree dermatitis, stomatitis, and proctitis. *J. Am. Med. Assoc.* 231:1162–163.

Beek, T.A. van. (Ed.). 2000. Ginkgo biloba. Harwood Academic, Amsterdam, the Netherlands. 548 pp.

Davies, J.R. 1999. Ginkgo biloba *(In a nutshell, healing herbs series).* Element Books Ltd., Boston, MA. 64 pp.

DeFeudis, F.W. 1991. Ginkgo biloba *extract (EGb 761): pharmacological activities and clinical applications.* Elsevier, Amsterdam, the Netherlands. 187 pp.

DeFeudis, F.W. 1998. Ginkgo biloba *extract (EGB 761): from chemistry to the clinic.* Ullstein Medical, Wesbaden, Germany. 401 pp.

Del Tredici, P. 1991. Ginkgos and people: a thousand years of interaction. *Arnoldia,* 51(2):2–15.

Del Tredici, P. 1992. Natural regeneration of *Ginkgo biloba* from downward growing cotyledonary buds (basal chichi). *Am. J. Bot.* 79:522–530.

Ernst, E., and Pittler, P.H. 1999. *Ginkgo biloba* for dementia: a systematic review of double-blind, placebo-controlled trials. *Clinical Drug Investig.* 17:301–308.

Foster, S. 1999. Ginkgo biloba. Revised edition. Botanical Series, American Botanical Council, Austin, TX. 8 pp.

Franklin, A.H. 1959. Ginkgo biloba L. historical summary and bibliography. *Va. J. Sci.* 10:131–176.

Gold, P.E., Cahill, L., Wenk, G.L., Loftus, E.F., McDaniel, M.A., Maier, S.F., Einstein, G.O., and Ceci, S.J. 2002. Ginkgo biloba: *a cognitive enhancer?* Blackwell Publishing, Malden, MA. 38 pp.

Hori, T., Ridge, R.W., Tulecke, W., Del Tredici, P., Trémouillaux-Guiller, J., and Tobe, H. 1997. Ginkgo biloba—*a global treasure. From biology to medicine.* Springer-Verlag, Tokyo. 427 pp.

Horsch, S., and Walther, C. 2004. *Ginkgo biloba* special extract EGb 761 in the treatment of peripheral arterial occlusive disease (PAOD): a review based on randomized, controlled studies. *Int. J. Clin. Pharmacol. Ther.* 42:63–72.

Huh, H., and Staba, E.J. 1992. The botany and chemistry of *Ginkgo biloba* L. *J. Herbs Spices Med Plants.* 1(1/2):91–124.

Laurain-Mattar, D. 2004. *Ginkgo biloba*: a medicinal and ornamental tree of great interest. In *Biotechnology of medicinal plants: vitalizer and therapeutic*. Edited by K.G. Ramawat. Science Publisher Inc., Enfield, NH. pp. 143–177.

Mazza, G., and Oomah, B.D. 2000. Chemistry, pharmacology and clinical applications of St. John's wort and Ginkgo biloba. In *Herbs, botanicals & teas*. Edited by G. Mazza and B.D. Oomah. Technomic, Lancaster, PA. pp. 131–176.

Michel, P.F. 1985. Ginkgo biloba, *l'arbre qui a vaincu le temps*. Félin. Paris. (Also World Wide Fund for Nature/Kiron, 1999). 108 pp.

Packer, L., and Christen, Y. (Eds.). 1998. Ginkgo biloba *extract (EGb 761), lessons from cell biology. Proceedings of the International Symposium, Montignac, France, September 16–17,* 1997. Elsevier, Amsterdam, the Netherlands. 174 pp.

Papadopoulos, V., Drieu, K., and Christen, Y. (Eds.). 1997. *Adaptive effects of* Ginkgo biloba *extract (EGb 761). Proceedings of the International Symposium, Ottrott, France, October 25–26, 1996.* Elsevier, Paris. 151 pp.

Small, E., and Catling, P.M. 2002. Blossoming treasures of biodiversity: 7. *Ginkgo biloba*—brain food from a living fossil. *Biodiversity*, 4(1):29–31.

Smith, P.F., Maclennan, K., and Darlington, C.L. 1996. The neuroprotective properties of the *Ginkgo biloba* leaf: a review of the possible relationship to platelet-activating factor (PAF). *J. Ethnopharmacol.* 50:131–139.

Soholm, B. 1998. Clinical improvement of memory and other cognitive functions by *Ginkgo biloba*: review of relevant literature. *Adv. Ther.* 15:54–65.

Van Beek, T. (Ed.). 2000. *Ginkgo biloba. Part 12. Medical and aromatic plants—industrial profiles*. Harwood Academic Publishers, Amsterdam, the Netherlands. 552 pp.

Van Beek, T.A., Morazzoni, E.B.P., and Peterlongo, F. 1998. *Ginkgo biloba* L. *Fitoterapia*, 69:195–244.

Wilson, J.C., Altland, J.E., Sibley, J.L., Tilt, K.M., and Foshee, W.G. III. 2004. Effects of chilling and heat on growth of *Ginkgo biloba* L. *J. Arboriculture*, 30:45–51.

Xing, S.Y., Huangpu, G.Y., Zhang, Y.H., Hou, J.H., Sun, X., Han, F., and Yang, J. 1997. Analysis of the nutritional components of the seeds of promising ginkgo cultivars. *J. Fruit Sci.* 14:39– 41.

Zhou, Z., and Zheng, S. 2003. The missing link in *Ginkgo* evolution. *Nature*, 423:821–822.

Specialty Cookbooks

So, Y.-K. 2006. *Classic Chinese Cookbook*. DK Publishing, New York. 224 pp.

40 Goji

Family: Solanaceae (potato family)

NAMES

Scientific Name: *Lycium barbarum* L.

- "Goji" is based on one of the Chinese names for the plant, *gouqi*. Although the name goji most often refers to *L. barbarum*, it has also been used for some other species of *Lycium*.
- *Lycium barbarum*, now most commonly called goji, has been known by several English names, including: (Barbary) wolf berry (or wolfberry), Barbary box thorn, Chinese wolf berry, Christmas berry, common matrimony vine, Duke of Argyll's tea tree, and (Barbary) matrimony vine.
- Several names that are used for *L. chinense* Mill., including Chinese box thorn, Chinese matrimony vine, and Chinese wolfberry, are also sometimes applied to *L. barbarum*.
- The name "Duke of Argyll's tea tree" commemorates Archibald Campbell (1682–1761), Scottish nobleman, politician, lawyer, and soldier, who became the third Duke of Argyll in 1743. He was the first governor of the Royal Bank of Scotland, and his portrait has appeared on the front of Royal Bank of Scotland banknotes since 1987. The Duke was an enthusiastic gardener who imported large numbers of exotic species of plants for his estate. When he died, many of these were moved to the Princess of Wales' new garden at Kew, which later became the world-famous Kew Gardens. Some of the Duke's original trees are still alive at Kew. It appears that in the 1930s the Duke received a shipment of plants of goji and common tea (*Camellia sinensis* (L.) Kuntze) with their labels mixed up, and this contributed to goji receiving the name "Duke of Argyll's tea tree" rather than the fact that goji was used for tea.
- The origin of the name "wolfberry" is obscure. It has been suggested that the genus name *Lycium* is reminiscent of the Greek word *lycos*, wolf, and this stimulated the English name. In any event, other plant species are known as wolfberry, particularly *Symphoricarpos* species, including *S. albus* (L.) Blake and *S. occidentalis* Hook.
- The genus (Latin) name *Lycium* is a reference to the ancient country of Lycica in Asia Minor, while *barbarum* in the name *L. barbarum* is thought to refer to Barbary, an old name for part of northern Africa where the species is not native (frequently botanists of the past adopted such geographically based words without knowing the native home of the species).

PLANT PORTRAIT

Goji is a shrub or small tree, typically 0.8 to 2 m (2–6½ ft.) tall, but the viny (arching or climbing) branches can extend to 6 m or 19 ft. The plants can form large, tangled clumps. The older branches are thorny at the nodes. The leaves are up to 7 cm (3 in.) long and 3.5 cm (1.4 in.) wide (they tend to be much smaller on older branches) and are arranged alternately or in fascicles. The lavender, purple, or pinkish flowers are solitary or in small clusters. The flower is funnel shaped, the closed funnel portion 8 to 10 mm (approximately 1/3 in.) long, and the widest part of the flower up to 1.5 cm (0.6 in.) across. The oblong or egg-shaped berries are up to 2 cm (0.8 in.) in length, usually red (fresh red berries tend to become blackish as they dry), and contain 4 to 20 yellow-brown seeds approximately 2 mm (less than 1/10th in.) long.

There is considerable confusion in the literature regarding the separation of *L. barbarum* and *L. chinense*. It is obvious that some suppliers are quite uncertain of the identity of the material they are supplying as "goji" (in fact, both species are known by this name). The most recent authoritative guide to the two species is by Zhi-yun et al. (1994).

Lycium barbarum is a native of China. It also occurs in the provinces adjacent to Mongolia and is widely cultivated and naturalized elsewhere in Asia and Europe. It is grown in northern and Southern China as well as in Mongolia and Tibet, both as a food and medicinal plant (in Asian culture, often no distinction is made between foods and medicines). Recently, it has been experimentally introduced elsewhere as a medicinal crop. It has been cultivated as an ornamental in North America, where it has escaped to become naturalized in waste areas and thickets, especially in the northeast. It has also become established in southeast Australia, from hedges established in agricultural areas. Fruit-based goji commodities in the commercial trade appear to be primarily from *L. barbarum*.

Lycium chinense is also a native of China but additionally is native to eastern Asia; it has been introduced as an ornamental elsewhere, often becoming naturalized, as in North America. The *Flora of China* indicates that *L. chinense* is present in most of the provinces of China and also occurs in Japan, Korea, Nepal, Pakistan, Taiwan, and Europe. *Lycium chinense* is widely cultivated in China as a medicinal plant; the young leaves are also eaten as a vegetable, and the seed oil used as a lubricant and for cooking.

In China, goji berries (both from *L. barbarum* and *L. chinense*) have been used in traditional folk medicines to treat a variety of ailments, including inflammations, skin irritations, nosebleed, aches, and pains, and visual and sexual dysfunction. The berries have been used in tonics, particularly to strengthen muscle and bone and treat problems attributed to malfunctions of the kidneys and liver. A tonic tea is also made from the foliage. Root bark from *L. chinense*, harvested in the winter and

FIGURE 40.1 Goji (*Lycium barbarum*), branch with flowers and fruits, with a long-sectioned fruit at lower right. (From Thomé, 1885.)

dried for later use, has also been used extensively in Chinese herbal medicine (it is unclear to what extent *L. barbarum* is used in the same way). It has been employed to control cough, fever, high blood pressure, diabetes, dizziness, lumbago, impotence, menopausal complaints, eye problems, internal bleeding, nosebleed, and tuberculosis. Bark solutions are applied externally to treat skin rashes.

In addition to food and medicinal use, *Lycium* species (both *L. barbarum* and *L. chinense*) have often been grown as ornamentals, frequently as hedges. Variegated forms with gold-edged leaves are available.

CULINARY PORTRAIT

Goji berries are used as an ingredient and as a condiment in traditional Asian cooking. The fruits have been partly dried in the sun to produce a somewhat chewy material, and the juice has been extracted for use as a beverage. Goji berries have an attractive taste, often compared with cranberry, and a mixture of strawberry and raspberry. The fruit has been said to have a mild, sweet, licorice flavor. Young branches and leaves of *L. chinense* are sometimes lightly cooked and used as a vegetable and flavoring in Asian cuisine. The leaves are especially used in Chinese soups made with pork or liver as well as for flavoring rice (the taste has been described as somewhat cress-like and peppermint-like). The extent to which *L. barbarum* branches are used in the same way is unclear.

Beverages made with goji (often mixtures with much cheaper fruit juices) are perhaps the most popular goji product sold in Western countries. Wine and beer have been made with goji but are little known. Packages of dried fruits are now available in Western supermarkets, sometimes sold as trendy trail mixes. Teas, confections, and cereals also feature goji. Extracts in capsule form are also sold as "nutraceuticals" (nutritional supplements).

Goji food products are not cheap. A 1 L (32 oz.) bottle of juice costs approximately $13.00 (about the cost of a good bottle of red wine, which coincidentally is also known to be very rich in antioxidants); a 0.45 kg (1 lb.) package of dried berries costs $15.0 to –$22.00.

CAUTION

Free radicals are highly reactive "bad" chemical fragments produced as by-products of such metabolic functions of the body as breathing, digesting, and exercising. Free radicals can impair cell function and are believed to be harmful, increasing the risk of cancer, heart disease, other diseases and premature aging. There are also "good" chemicals called antioxidants that occur naturally, and because they can disarm damaging free radicals, they seem to be useful in the body's fight against diseases. Thousands of antioxidants occur naturally in foods, and several vitamins are in fact antioxidants. In the 1990s, some studies suggested that antioxidants can prevent or slow the development of cancer and other diseases by protecting cells against damage by free radicals. This led to the widespread use of consumption of supplements containing antioxidants, and today 10% to 20% of adults in North America and Europe take antioxidant supplements, particularly vitamins A, C, and E. However, some studies have suggested that there may be little benefit in antioxidant supplements (see Chapter 1). Many researchers have concluded that antioxidants work best when they are consumed in food rather than pills. The reasons for this are not clear; it has been suggested that other factors in foods, perhaps in interaction with antioxidants, are responsible for beneficial effects.

Several preliminary studies have been conducted in Asia, suggesting medicinal values of goji, particularly because of the presence of high levels of antioxidants. The widespread marketing claims that goji promotes youthfulness, good health, and sex drive need to be demonstrated experimentally. The U.S. Food and Drug Administration has warned some goji purveyors that their advertisements suggesting that goji prevents, mitigates, treats, or cures disease are claims that can only be legally made for drugs subject to governmental approval.

Warning

Goji is a species in the Solanaceae, and because toxic species are known in the family, some caution should be exercised. The toxin atropine has been found at low concentrations in goji berries. In North America, calves and sheep have been poisoned by eating large amounts of material (likely mostly foliage) growing around houses. It has been suggested that only the fully ripe fruits should be eaten, as unripe berries could be poisonous. For those who have taken a fancy to the fruit, so long as small amounts are consumed, there probably is not cause for concern. More study of the composition and effects of goji is desirable.

Culinary Vocabulary

- "Tibetan goji" and "Himalayan goji" are expressions used in the health food trade to indicate the sources of the goji used, with the suggestion that they are superior.
- "Superfruit" is a marketing term used in the food and beverage industry since the early twenty-first century to indicate fruits that are very high in nutrients and antioxidants. There are no quantitative criteria for defining superfruits, and the term is being applied mainly for marketing purposes, often with exaggerated and unfounded claims of health benefits. Goji is one of the most prominent of superfruits.

CURIOSITIES OF SCIENCE AND TECHNOLOGY

- According to legend, during the Tang Dynasty (around AD 800), a Buddhist temple in the Himalayas relied on a well surrounded by goji plants, which regularly dropped their berries into the water. As a result of drinking the fortified water, the monks lived to a ripe old age, never developing a white hair or losing a tooth. Such legends have been used by purveyors of goji to suggest that it has remarkable health-giving properties.
- Chinese scholar Li Quing Yuen, (1678–1930), allegedly lived to the age of 252 years by consuming goji berries daily (not quite as long as Methuselah who, according to the King James Bible (Genesis 5:27), lived to the age of 969 years). Professor Yuen is said to have outlived 23 wives. This is another or the old stories that are repeated by some goji sellers to indicate that it is health promoting.
- In nineteenth-century London, goji fruit was tied as beads around the necks of teething children to ease their pain.
- Chinese herbal medicine is based on the concept of *yin* and *yang* forces of Daoist herbal theory. Yang represents masculinity, strength, and heat, and yin by contrast is feminine, mild, and cold. In China, tonic herb teas made from goji fruit are frequently used to combat yin problems with the liver (such as blurry vision, dizziness, and headache) and kidneys (such as lumbago). Other yin problems that might be treated with goji tonics include sore back and legs, abdominal pain, impotence, and nocturnal emission.
- Herbs are often used medicinally, by both professionals (herbal practitioners) and lay people, and sometimes the use of a particular herb is dangerous when combined with a conventional (Western) medical treatment. A case has been reported where goji tonic interfered with the use of the blood anticoagulant warfarin (Lam, A.Y., Elmer, G.W., and Mohutsky, A. 2001. Possible interaction between warfarin and *Lycium barbarum* L. *Ann. Pharmacother.* 35: 1199–2101.).

KEY INFORMATION SOURCES

Note: Some of the recent titles listed claiming extraordinary health benefits for goji are included simply to indicate the recent trend to publish uncritical books on the subject.

Gross, P.M., Zhang, R., and Zhang, S. 2006. *Wolfberry: nature's bounty of nutrition and health.* Booksurge, Charleston, SC. 260 pp.

Cheng, K.-T., Chang, H.-C., Huang, H., and Lin, C.-T. 2000. RAPD analysis of *Lycium barbarum* medicine in Taiwan market. *Bot. Bull. Acad. Sin.* 41:11–14.

Haegi, L. 1976. Taxonomic account of *Lycium* (Solanaceae) in Australia. *Aust. J. Bot.* 24:669–679.

Jamin, E. 2009. Superfruits: are they authentic? *Fruit Processing*, 19(4):170–175.

Mindell, E. 2005. *Goji, the Himalayan health secret.* 2nd ed. Momentum Media, Lake Dallas, TX. 91 pp. [A particularly uncritical, partisan, promotional analysis by a medicinal food purveyor, illustrative of the hyperbole associated with goji].

Small, E., and Catling, P.M. 2007. Blossoming treasures of biodiversity: 23. Goji (*Lycium barbarum*): fountain of youth? *Biodiversity*, 8(1):27–35.

Wong, C.-C., Li, H.-B., Cheng, K.-W., and Chen, F. 2006. A systematic survey of antioxidant activity of 30 Chinese medicinal plants using the ferric reducing antioxidant power assay. *Food Chem.* 97:705–711.

Young, Y., Lawrence, R., and Schreuder, M. 2006. *Ningxia wolfberry: the ultimate superfood: how the Ningxia wolfberry and four other foods help combat heart disease, cancer, chronic fatigue, depression, diabetes and more.* Essential Publishing, Orem, UT. 266 pp.

Yu, M.-S., Leung, S.K.-Y., Lai, S.-W., Che, C.M., Zee, S.Y., So, K.F., et al. 2005. Neuroprotective effects of anti-aging oriental medicine *Lycium barbarum* against β-amyloid peptide neurotoxicity. *Exp. Gerontol.* 40:716–727.

Zhang, K.Y.B., Leung, H.W., Yeung, H.W., and Wong, R.N.S. 2000. Differentiation of *Lycium barbarum* from its related *Lycium* species using random amplified polymorphic DNA. *Planta Med.* 67:379–381.

Zhi-yun, Z., An-ming, L., and D'Arcy, W.G. 1994. *Lycium.* In *Flora of China*, vol. 17. Edited by (co-chairs of the editorial committee) W. Zhengyi and P.H. Raven. Missouri Botanical Garden, St. Louis, MO, and Science Press, Beijing. pp. 301–304.

Specialty Cookbooks

Note: The claims in cookbooks that goji has miraculous health benefits should be taken with a grain of salt.

Campbell, S. 2008. *Goji berry cookbook.* Sherri Campbell, Inc., Swampscott, MA. 248 pp.

Hoffmann, C. 2007. *Goji berry: fruits of paradise.* Woodland Publishing, Orem, UT. 32 pp.

Null, G. 2006. *Gary Null's power foods: the 15 best foods for your health.* NAL Hardcover, 336 pp.

Wexler, B. 2007. *Superfruits: power-up you health with pomegranate, acai, gac, mangosteen, and goji.* Woodland Pub., Orem, UT. 38 pp.

41 Grass Pea

Family: Fabaceae (Leguminosae; pea family)

NAMES

Scientific Name: *Lathyrus sativus* L.

- The "pea" in the name "grass pea" is based on resemblance to the pea plant. The "grass" in the name is based on the leaflets, which are long and grass shaped.
- Grass pea (grasspea) is also known as blue vetchling, chickling pea, chickling vetch, dog-tooth pea, grass peavine, Indian pea, khesari, Riga pea, and wedge peavine.
- The genus name *Lathyrus* is based on the ancient Greek name for the plant, combining *la*, very, and *thoures*, a stimulant. The seeds were said to have excitant or stimulant properties.
- *Sativus* in the scientific name *L. sativus* is Latin for cultivated or sown.

PLANT PORTRAIT

Grass pea is an annual herb with white, pink, red, purple, or blue flowers, growing as a suberect, creeping, or climbing vine, typically from 0.6 to 1 m (2–3 ft.) in length, but forms are known that reach 9 m (30 ft.). The pods are flat, 2.5 to 5 cm (1–2 in.) in length, with three to six seeds. The seeds are often white, brownish-gray, or light cream in color and may also be speckled with black. The seeds are distinctively wedge shaped, and range in size from 3 to 7 mm (1/8–1/4 in.) in diameter. The exact area of origin of grass pea is uncertain, probably the Mediterranean area and/or western Asia. The use of grass pea for human food may date back 8000 years in the region of the Balkan Peninsula. From the eastern Mediterranean, grass pea was taken to Europe (where it may well have been one of the first domesticated crops), Africa, and Asia. It has been grown for thousands of years in parts of Europe, North Africa, and Asia, but at present is a major food crop of India, Pakistan, Bangladesh, Nepal, and Ethiopia. Grass pea is raised by several hundred million farmers in the Indian subcontinent and sub-Saharan Africa. In India, 2 million ha (5 million acres) of grass pea are under cultivation. Grass pea is also widely used as a fodder crop in parts of south-central Europe as well as Africa and Asia. The plant is exceptionally able to withstand drought and as a result is often the cheapest or only food available to the poor. Moreover, the seeds contain approximately 25% protein and are often the only protein available in poor regions. This is the source of an extreme health problem called lathyrism, described in the next section.

CULINARY PORTRAIT

Grass pea seed is roasted, made into soup, stews, and gruel, and prepared as paste balls. Flour or meal from grass pea can be used in cooking or to make bread. In Asia, the leaves are sometimes eaten as a pot herb and the immature pods boiled as a vegetable. Grass pea is not presently a food of Western nations, but should scientists currently attempting to create completely nontoxic lines succeed, grass pea would almost certainly become an important Western crop as it produces large amounts of good-tasting seeds under climatic and soil conditions that very few other crops can tolerate.

Lathyrism is caused by an amino acid in the seeds of grass pea. (There is no simple name for this toxic amino acid. It has been called beta-N-oxalyamino-L-alanine, beta-N-oxalyl-L-alpha, beta-diaminopropionic acid, and L-3-oxalyamino-2-amino-propionic acid.) Some 20 amino acids are the basis of proteins that are necessary for human life, but this amino acid is different. It is a neurological poison, destroying nerves, resulting in irreversible crippling lameness, in cattle as well as in humans. Usually when grass pea has made up more than one-third of the diet for 3 or 4 months, lathyrism has resulted. Lathyrism is said to strike below the belt. When too much grass pea is consumed, both legs become paralyzed and degenerate permanently. Continued consumption can result in convulsions and finally death. The onset of the disease is often sudden, and people between 20 and 29 years of age have been the predominant group affected. For reasons that are not fully understood, paralysis is more common among males than females. According to Stewart (2009), "in places like Ethiopia and Afghanistan … the high-protein pea is typically reserved for men to give them strength so that they can feed their families. Instead, it has the opposite effect, reducing them to crawling on their knees." At least 100,000 people in developing countries are believed to suffer from the disease. An old practice in India called "lagua" bonded landless laborers to wages of grass pea seed, but this practice has been stopped. The amount of poison can be reduced by soaking or boiling with changes of water. Baking, roasting, and fermenting also reduce the toxin. Although these methods can reduce the neurotoxin content by more than 90%, the nutritional quality is also lowered and some water-soluble vitamins are lost. Tragically, often poor people simply lack enough water and fuel to carry out such treatments to make the grass pea less poisonous. The poisonous amino acid may be present in a concentration of up to 2.5% in the seeds. Water stress can double the toxin level, whereas salinity in the soil may reduce the toxin level in the seeds. It has been suggested that human consumption is considered to be safe at levels less than 0.2% of the toxin. Low-toxin varieties (less than 0.05%) have been selected recently and are now being distributed in hopes of considerably reducing the problem of lathyrism.

Culinary Vocabulary

- The popular ornamental "sweet pea" (*Lathyrus odoratus* L.), a relative of grass pea, has inedible, somewhat poisonous seeds. Unfortunately, the expression "sweet peas" is often used for the pods and seeds of the common pea (*Pisum sativum* L.), misleading people into believing that the sweet pea plant is edible.

CURIOSITIES OF SCIENCE AND TECHNOLOGY

- Lentil (*Lens culinaris* Medik.) is the most important food species of the pea family in the Old World. Grass pea often appears in lentil crops as a weedy contaminant. This observation has led to the theory that after lentil was domesticated, grass pea seeds evolved to closely mimic the size, shape, and color of lentil seeds, so that grass pea would survive among lentils and be distributed by humans as a contaminant of the lentil crop wherever it was grown.
- Most of our modern crops have been selected from wild ancestors that have poisonous constituents to protect themselves against animals and microorganisms. Usually during domestication, most if not all of the natural toxic chemicals have been removed by selection, and this is beneficial for people, but less so for the plants, which often have to be protected with artificial pesticides. One of the reasons that toxic plants such as grass pea are still cultivated for food in developing countries is that they thrive without synthetic pesticides, which are too expensive in poor regions to buy and use. For example, cassava (*Manihot esculenta* Crantz), one of the most important food plants in the world, has varieties that must be peeled and well cooked, otherwise they are toxic, producing poisonous hydrogen cyanide. Although varieties sold in Western supermarkets are very low in natural toxicity, varieties often raised in poor tropical countries have high amounts of the toxin

because they are much more resistant to decay. If not properly prepared, such high-toxin varieties of cassava produce "konzo disease," marked by either acute poisoning with severe sudden illness and death, or a later onset of a suddenly appearing paralysis (the name konzo was assigned to the disease in the first report from the southern Bandundu region in Zaïre). Curiously, the clinical symptoms of neurolathyrism are identical to those of konzo.

FIGURE 41.1 Grass pea (*Lathyrus sativus*). (From Curtis, vol. 3, 1790, plate 115.) Note opened pod showing seeds.

FIGURE 41.2 Low-toxin seeds of grass pea, bred in Canada. (Courtesy of C.G. Campbell.)

FIGURE 41.3 Farmer crippled by lathyrism, in the grass pea crop that crippled him. The severity of the disease is often judged by whether one or two canes are used. (Courtesy of C.G. Campbell.)

KEY INFORMATION SOURCES

Abegaz, B.M., Haimanot, R.T., Palmer, V.S., and Spencer, P.S. (Eds.). 1994. Nutrition, neurotoxins, & lathyrism: the ODAP challenge. *Proceedings of the Second International* Lathyrus/*Lathyrism Conference in Ethiopia Under the International Network for the Improvement of* Lathyrus sativus *and the Eradication of Lathyrism.* Third World Medical Research Foundation, New York. 139 pp.

Asfaw, T., Asgelil, D., and Bekele, H. 1994. Genetics and breeding of grasspea. In *Cool-season food legumes of Ethiopia.* Edited by T. Asfaw, B. Geletu, M.C. Saxena, C. Mohan, and M.B. Solh. International Center for Agricultural Research in the Dry Areas, Aleppo, Syria. pp. 183–195.

Barrow, M.V., Simpson, C.F., and Miller, E.J. 1974. Lathyrism: a world review. *Q. Rev. Biol.* 49:102–128.

Campbell, C.G. 1997. *Grass pea,* Lathyrus sativus *L.* International Plant Genetic Resources Institute, Rome, Italy. 92 pp.

Campbell, C.G., Mehra, R.B., Agrawal, S.K., Chen, Y.Z., Abdel Moneim, A.N., Khawaja, H.I.T., et al. 1994. Current status and future strategy in breeding grasspea (*Lathyrus sativus* L.). In *Expanding the production and use of cool season food legumes.* Edited by F.J. Muehlbauer and W.J. Kaiser. Kluwer Academic Publishers, Dordrecht, the Netherlands. pp. 617–630.

Croft, A.M., Pang, E.C.K., and Taylor, P.W.J. 1999. Molecular analysis of *Lathyrus sativus* L. (grasspea) and related *Lathyrus* species. *Euphytica,* 107:167–176.

Erskine, W., Smartt, J., and Muehlbauer, F.J. 1994. Mimicry of lentil and the domestication of common vetch and grass pea. *Econ. Bot.* 48:326–332.

Getahun, H., Lambein, F., Vanhoorne, M., and Van der Stufyft, P. 2003. Food-aid cereals to reduce neurolathyrism related to grass-pea preparations during famine. *Lancet,* 362 (9398):1808–1810.

Hanbury, C.D., and Siddique, K.H.M. 1997. *Adaptation of* Lathyrus sativus *and* L. cicera *to Western Australian Mediterranean-type environments.* Centre for Legumes in Mediterranean Agriculture, Nedlands, WA, Australia. 10 pp.

Islam, M.Z., Islam, M.Q., and Ahmed, M.U. 1986. *Abstract bibliography of* Lathyrus sativus. National Agricultural Library and Documentation Centre, Bangladesh Agricultural Research Council, Dhaka, Bangladesh. 61 pp.

Jackson, M.T., and Yunus, A.G. 1984. Variation in the grasspea (*Lathyrus sativus* L.) and wild species. *Euphytica*, 33:549–559.

Jansen, P.C.M. 1989. *Lathyrus sativus* L. In *Plant resources of South-East Asia. 1. Pulses*. Edited by L.J.G. van der Maesen and S. Somaatmadja. Pudoc, Leiden, the Netherlands. pp. 50–51.

Kearney, J., and Smartt, J. 1995. The grasspea. In *Evolution of crop plants*. 2nd ed. Edited by J. Smartt and N.W. Simmonds. Longman Scientific & Technical, Burnt Mill, Harlow, Essex, U.K. pp. 266–270.

Kislev, M.E. 1989. Origins of the cultivation of *Lathyrus sativus* and *L. cicera* (Fabaceae). *Econ. Bot.* 43:264–270.

Kupicha, F.K. 1983. The infrageneric structure of *Lathyrus*. *Notes R. Bot. Gard. Edinb.* 41:209–244.

Raloff, J. 2000. Detoxifying desert's manna. Farmers need no longer fear the sweet pea's dryland cousin. *Sci. News*, 158(5):74–76.

Small, E., and Catling, P.M. 2004. Blossoming treasures of biodiversity 14. Grass pea (*Lathyrus sativus*). Can a last resort food become a first choice? *Biodiversity*, 5(4):29–32.

Smartt, J., Kaul, A., Araya, W.A., Rahman M.M., and Kearney, J. 1994. Grasspea (*Lathyrus sativus* L.) as a potentially safe food legume crop. In *Expanding the production and use of cool season food legumes*. Edited by F.J. Muehlbauer and W.J. Kaiser. Kluwer Academic Publishers. Dordrecht, the Netherlands. pp. 144–155.

Spencer, P.S. (Ed.). 1989. *The grass pea: threat and promise. Proceedings of the International Network for the Improvement of* Lathyrus sativus *and the Eradication of Lathyrism and recommendations of the International INILSEL Coordination Committee*. Third World Medical Research Foundation, New York. 244 pp.

Spencer, P.S., and Palmer, V.S. 2003. Lathyrism: aqueous leaching reduces grass-pea neurotoxicity. *Lancet*. 362 (9398):1775–1776.

Stewart, A. 2009. *Wicked plants. The weed that killed Lincoln's mother & other botanical atrocities*. Algonquin Books of Chapel Hill, Chapel Hill, NC. 236 pp.

Yunus, A.G., and Jackson, M. 1991. The gene pools of the grasspea (*Lathyrus sativus* L.). *Plant Breeding*. 106:319–328.

Yusuf, H.K.M., and Lambein, F. (Eds.). 1995. Lathyrus sativus *and human lathyrism: progress and prospects, from international collaborations. Proceedings of the second international colloquium on* Lathyrus/*Lathyrism, Dhaka, December 10–12, 1993*. University of Dhaka, Dhaka, Bangladesh. 288 pp.

Specialty Cookbooks

Della Crosse, J., and Rizzo, J. 2002. *Umbria: regional recipes from the heartland of Italy*. Chronicle Books, San Francisco, CA. 161 pp.

Scully, T. (Trans.). 2008. *The opera of Bartolomeo Scappi (1570): L'Arte et prudenze d'un maestro cuoco (the art and craft of a master cook)*. University of Toronto Press, Toronto, ON. 776 pp.

42 Guarana

Family: Sapindaceae (soapberry family)

NAMES

Scientific Name: *Paullinia cupana* Kunth

- The name guarana (often spelled guaraná) is from the Spanish and Portuguese *guaraná*, which is based on a Tupi Indian word said to mean liana creeper or climbing plant, descriptive of the guarana plant. The name has also been attributed to the Guarani Indians of South America, who have consumed the plant since pre-Columbian times.
- The genus name *Paullinia* has been claimed to commemorate either C.F. Paullini, a German medical botanist who died in 1712, or Simon Paulli (1603–1680) and his son, Christian Frances (1643–1742), who were Danish botanists. The founder of biological nomenclature, Carl Linnaeus, used the name in the eighteenth century, and it is often difficult to interpret the source of his names.
- *Cupana* in the scientific name *P. cupana* is based on the Latin *cupa*, container, especially for containing wine, in allusion to the beverage use of the plant.

PLANT PORTRAIT

Guarana is an evergreen, high-climbing, woody vine, with or without tendrils, which wraps around the tall trees of its native Amazonian rain forest. In the lush, Brazilian Amazon, it often grows to 12 m (39 ft.) in height. However, when cultivated, it becomes a sprawling shrub about 2 m (6½ ft.) tall, typically extending over an area with a diameter of approximately 4 m (13 ft.). The small, pear-shaped, leathery fruit (about 7 mm or 1/4 in. long) is a yellow or orange-red, three-sided capsule. When ripe, it becomes partially open, usually exposing one (or up to three) black or greenish seeds, which are covered at the base with a white fleshy material (technically called an aril). When European explorers first reached South America, they observed guarana being cultivated. According to one missionary, certain Indian tribes valued it in the same way as Europeans valued gold. The indigenous people of the Amazon rain forest used crushed guarana seed as a beverage and a medicine. Guarana was used to decrease fatigue, to reduce hunger, and to treat diarrhea, arthritis, hangovers from alcohol abuse, and headaches related to menstruation. In the Western World, guarana is marketed mostly as a herbal or dietary supplement, shamelessly claimed to greatly boost energy, sex drive, mental abilities, and athletic performance, to stimulate weight loss, to remove pain (or "produce a buzz"), and to help stop addiction to smoking. The vast majority of guarana is grown in a small area in northern Brazil, with initial attempts underway to cultivate the plant commercially in other countries. It is uncertain when exports began to the Old World, but the use of guarana in Europe was documented in 1775. In Brazil, guarana owes much of its popularity to the stimulation produced by its high caffeine content and the widely held belief in its rejuvenating and aphrodisiacal properties. Several of the most popular commercial brands of guarana preparations contained very little or no guarana until recent Brazilian legislation, aimed at supporting growers, made the inclusion of at least a small amount (60 mg/L) compulsory.

The average American drinks a gallon of soft drinks a week, and carbonated beverage manufacturers are constantly introducing new choices in attempts to hold or increase their market share. In 1995, Pepsico introduced Josta, a mahogany-hued fizzy drink made with guarana, offering 59 mg of caffeine and 44 g of sugars in a 12-oz. can. Pepsi made no medical claims, but on the panther-emblazoned, jungle-themed cans was the message: "Guarana grows deep within the jungle. For centuries, ancient tribes believed that guarana released raw, primal power. Now the legend of guarana has been captured in the potent flavor of Josta. Unleash it." Josta had limited market success in the United States and Brazil and was discontinued in 1999. Part of its failure in Brazil has been attributed to the word Josta resembling the Portuguese word Bosta, which means "excrement."

The Coca-Cola company has attempted to win a share of the guarana market in Brazil by introducing its own guarana brands. However, patriotic Brazilian consumers have been less than rapturous about "gringo guaranas." Coke's fiercest Brazilian rival, Guarana Antarctica (AmBev), ridiculed Coke's new brands by accusing the company of "guarana envy." Nevertheless, Coca-Cola's "Kuat" has had some success in Brazil.

CULINARY PORTRAIT

Roasted seeds of guarana are used mainly to produce an interesting, cola-type soft drink sometimes called "Brazilian cocoa," which since the 1940s has been considered to be Brazil's national drink (although it is second to coffee in popularity). The amber-colored preparation is refreshingly mild, fruity in flavor, and usually is readily accepted by those who try it for the first time. The seeds contain 2.7% to 3.5% caffeine, and the resulting beverages typically have more caffeine than coffee. The traditional method of using guarana is to separate the contents of the seeds and immerse the material in water to form a paste. Sticks are made from these, dried over a slow fire, and smoked for 1 month. To prepare a guarana beverage, part of the stick is simply grated and mixed in water. In modern times, guarana is more widely available as carbonated drinks, although sticks and powder are also sold for personal preparation. In North America, guarana is most likely to be found in stores specializing in Brazilian and Latin foods and in new soft drink introductions. Guarana is occasionally marketed in bubble gum, ice cream, and candy/energy bars.

Because of the high caffeine content, guarana may cause insomnia, trembling, anxiety, palpitations, and increased frequency of urination. It has been recommended that guarana should be avoided during pregnancy and breast-feeding.

Culinary Vocabulary

- Brazil is the world's third-largest soft drink market (after the United States and Mexico), and "guarana" is the most popular "flavor." However, in Brazil the term guarana is often used much like the word "cola" in Western nations to refer to carbonated soda beverages in general.
- Guarana beverages are sometimes called "Zoom" for their stimulating effect.

CURIOSITIES OF SCIENCE AND TECHNOLOGY

- In the Amazon of Brazil, guarana is used as a stimulant by hunters to keep awake and attentive as well as to suppress appetite.
- The stems, leaves, and roots of guarana and other species of *Paullinia* have been placed in streams to kill or stun fish in South America so that they can be easily captured.
- A stimulant paste of guarana has been smoked by Indians of Brazil. In modern times, guarana is occasionally used to flavor cigarettes.
- The high caffeine content of guarana seeds would seem to discourage birds from collecting and dispersing them, but in fact the seeds are dispersed by birds. The seed kernel (embryo and cotyledons) and the seed coat do accumulate considerable caffeine, but the fleshy material at the base of the seed (i.e., the aril) has virtually none, instead containing glucose, fructose, and sucrose up to almost 70% of the dry weight of the aril, thus providing a sweet snack that attracts the birds.

FIGURE 42.1 Guarana (*Paullinia cupana*). (From Köhler, 1883–1914.)

KEY INFORMATION SOURCES

Anonymous. 1999. Light relief. (Kava kava and guarana.). *Funct. Foods Nutraceut.* 3(12):17–19.

Ashurst, P.R. (Ed.). 2005. *Chemistry and technology of soft drinks and fruit juices*. 2nd ed. Wiley-Blackwell, Ames, IA. 392 pp.

Avato, P., Pesante, M.A., Fanizzi, F.P., and Santos, C.A. de M. 2003. Seed oil composition of *Paullinia cupana* var. *sorbilis* (Mart.) Ducke. *Lipids*, 38:773–780.

Baghkhani, L., and Jafari, M. 2002. Cardiovascular adverse reactions associated with guarana: is there a causal effect? *J. Herb. Pharmacother.* 2:57–61.

Bartsch, A. 1996. Guarana—an overall view. Dragoco Report Flavoring Info. *Service*, 1996(3):111–126.

Baumann, T.W., Schulthess, B.H., and Hanni, K. 1995. Guarana (*Paullinia cupana*) rewards seed dispersers without intoxicating them by caffeine. *Phytochemistry*, 39:1063–1070.

Bempong, D.K., and Houghton, P.J. 1992. Dissolution and absorption of caffeine from guarana. *J. Pharm. Pharmacol.* 44:769–771.

Benoni, H., Dallakian, P., and Taraz, K. 1996. Studies on the essential oil from guarana. *Z. Lebensm. Unters. Forsch.* 203:95–98.

Beck, H.T. 1990. A survey of the useful species of *Paullinia* (Sapindaceae). *Adv. Econ. Bot.* 8:41–56.

Clarke, A. 2000. Guarana past its peak but holds new promise. *New Nutrition Business*, 5(3):24–25.

Corrêa, M.P.F., and Santos, V.C. dos. 1983. *Guaraná: resumos informativos 11 (Guarana: informative summaries 11)*. 2nd ed. Empresa Brasileira de Pesquisa Agropecuária, Brasília, Brazil. 124 pp (in Portuguese).

Erickson, H.T. 2008. *Paullinia cupana*, guarana. In *The encyclopedia of fruit & nuts*. Edited by J. Janick and R.E. Paull. CABI, Wallingford, Oxfordshire, U.K. pp. 817–820.

Erickson, H.T., Corrêa, M.P.F., and Escobar, J.R. 1984. Guarana (*Paullinia cupana*) as a commercial crop in Brazilian Amazonia. *Econ. Bot.* 38:273–286.

Escobar, J.R., Correa, M.P.F., and Aguilera, F.P. 1984. Floral structures, flowering and controlled pollination techniques in guarana. *Pesqui. Agropecu. Bras.* 19:615–622 (in Portuguese, English summary).

Henman, A.R. 1982. Guarana (*Paullinia cupana* var *sorbilis*): ecological and social perspectives on an economic plant of the central Amazon basin. *J. Ethnopharmacol.* 6:311–338.

Henman, A. 1986. *O guaraná: sua cultura, propriedades, formas de preparação e uso*. 2nd ed. Global/Ground, São Paulo, Brazil. 77 pp. (in Portuguese)

Houghton, P. 1995. Herbal products. 7. Guarana. *Pharm. J.* 254:435–436.

Lleras, E. 1994. Species of *Paullinia* with economic potential. In *Neglected crops: 1492 from a different perspective*. Edited by J.E. Hernández-Bermejo and J. Léon. Food and Agriculture Organization of the United Nations, Rome, Italy. pp. 223–228.

Mattei, R., Dias, R.F., Espinola, E.B., Carlini, E.A., and Barros, S.B.M. 1998. Guarana (*Paullinia cupana*): toxic behavioral effects in laboratory animals and antioxidant activity in vitro. *J. Ethnopharmacol.* 60:111–116.

Meurer, G.B., Berkov, A., and Beck, H. 1998. Theobromine, theophylline, and caffeine in 42 samples and products of guarana (*Paullinia cupana*, Sapindaceae). *Econ. Bot.* 52:293–301.

Prance, G.T. 2000. *Paullinia cupana* Kunth. In *Plant resources of South-East Asia. 16. Stimulants*. Edited by H.A.M. van der Vossen and M. Wessel. Backhuys, Leiden, the Netherlands. pp. 99–102.

Santa-Maria, A., Lopez, A., Diaz, M.M., Munoz-Mingarro, D., and Pozuelo, J.M. 1998. Evaluation of the toxicity of guarana with in vitro bioassays. *Ecotoxicol. Environ. Saf.* 39:164–167.

Sato, Y., Terazawa, M., Kainuma, K., and Tezuka, S. 1988. On drowsiness preventive gum that contains guarana extract. *Food Industry*, 31(16):64–74.

United Kingdom, Advisory Committee on Novel Foods & Processes. 1996. *Report on guarana infusate*. U.K. Advisory Committee on Novel Foods & Processes, Ergon House/Nobel House London. 8 pp.

Van Straten, M. 1994. *Guarana: the energy seeds and herbs of the Amazon Rainforest*. C.W. Daniel Co., Woodstock, NY. 152 pp.

Walker, T.H., Chaar, J.M., Mehr, C.B., and Collins, J.L. 2000. The chemistry of guarana: guarana, Brazil's super-fruit for the caffeinated beverages industry. In *Caffeinated beverages: health benefits, physiological effects, and chemistry*. Edited by T.H. Parliament and C.-T. Ho. American Chemical Society, Washington D.C. pp. 305–314.

Weckerle, B., Richlin, E., and Schreier, P. 2002. Authenticity assessment of guarana products (*Paullinia cupana*) by caffeine isotope analysis. *Dtsch. Lebensmitt. Rundsch*, 98:122–124.

Weckerle, C.S., and Rutishauser, R. 2005. Gynoecium, fruit and seed structure of Paullinieae (Sapindaceae). *Bot. J. Linn. Soc.* 147:159–189.

Specialty Cookbooks

Hamilton, C. 2005. *Brazil: a culinary journey*. Hippocrene Books, New York. 205 pp.

Idone, C. 1995. *A cook's tour: Brazil*. Clarkson Potter, New York. 240 pp.

Perron, K., and Dembecki, S. 2004. *Jamba juice power! Smoothies and juices for mind, body, and spirit*. Avery, New York. 239 pp.

43 Guava

Family: Myrtaceae (myrtle family)

NAMES

Scientific Name: *Psidium guajava* L.

- The English word "guava" originated through the Spanish *guayaba* from South American Indian language names for the fruit, such as the Tupi *guayava*. *Guajava* in the scientific name *P. guajava* has the same derivation.
- Guava is also known as apple guava, common guava, lemon guava, and yellow guava.
- Although *P. guajava* is the predominant species grown for guava fruits, there are several other species of *Psidium* with edible fruit that are called guava. Most notable of these is *P. cattleianum* Sabine, a native of Brazil, which has two subgroupings: *P. cattleianum* var. *littorale* (Raddi) Fosberg (*P. littorale* Raddi), known as Chinese strawberry guava, strawberry guava, yellow Cattley guava and yellow strawberry guava, and *P. cattleianum* var. *cattleianum* (*P. littorale* var. *longipes* (O. Berg) Fosberg), known as Cattley guava, purple guava, purple strawberry guava, red strawberry guava, and strawberry guava.
- Pineapple guava is an unrelated species (see Chapter 35).
- The genus name *Psidium* is based on *psidion*, the Greek name for pomegranate.

PLANT PORTRAIT

The guava is a tropical to subtropical evergreen shrub or tree to 10 m (33 ft.) high. The fruit has a strong, sweet, musky odor when ripe. Guava fruit is apple or pear shaped, 2.5 to 10 cm (1–4 in.) in diameter, with floral remnants at the tip. The skin of the immature fruit is green and ripens to light-yellow, often with pinkish shading. Under the thin skin, there is a layer of granular flesh, 3 to 13 mm (1/8–1/2 in.) thick, which is white, yellowish, pink, or reddish, acidic or sweet, and juicy. Inside this fleshy layer is the central pulp, which is the same color, juicy, and usually filled with very hard, yellowish, small seeds, or occasionally lacking seeds. The fruits generally weigh 100 to 450 g (1/4–1 lb.) but sometimes are 900 g (2 lb.) or more. The guava is thought to be native from southern Mexico to Central America. At the time of the Spanish conquest, it was found from Peru to Mexico. Early Spanish and Portuguese colonizers brought it to the East Indies and Guam, and it became a crop in Asia and Africa. India and Mexico are by far the world's largest producers. Guava is now also cultivated in Hawaii, southern Florida, and (to a lesser extent) California. There are more than 150 cultivated varieties. The guava is so important in hot areas of the world that it is commonly called "the apple of the tropics." In temperate areas, guavas are luxury items which need to be rapidly transported to stores because they do not keep well.

CULINARY PORTRAIT

Two classes of fruit have been recognized: low-acid, "sweet," dessert types are consumed fresh, whereas "sour," high-acid kinds are used for processing. The sweet types of fruit tend to be larger and white fleshed, whereas the sour kinds are often red fleshed. A third class, consisting of rather bland fruits (low in both acids and sugars), is also sometimes recognized. Guava fruit for fresh consumption is sweet, sometimes slightly acidic. Some people find the strong musky odor to be unpleasant (this tends to be pronounced in ripe, low-acid kinds). Raw guavas are eaten out of hand (along with the small, hard, edible seeds), added to salads, deseeded and sliced as dessert (often with cream or sugar). Usually the fruit is cooked, which eliminates the smell. Stewed guava shells (guava halves with the central seed pulp removed, strained, returned to the shells, and cooked) are popular. In Mexico, sweet potato and guava are a favorite combination. Guava paste, jelly, jam, and juice are widely marketed. The paste tends to crystallize and should be eaten as fresh as possible; it is firm enough to be sliced with a knife, and is often consumed with fresh cream cheese. The juice is often present in cocktail fruit juice mixtures. Sour types of guava are considered highly desirable for jelly because of the resulting distinctive flavor as well as the high pectin and acid content. Jelly can be made from sweet types but is often inferior. Preserved purée is used in pies, cakes, puddings, sauces, ice cream, yogurt, tapioca, jam, butter, marmalade, chutney, relish, catsup, and other products. Some guava is also dehydrated and powdered for commercial use. The best kind to purchase is smooth and unblemished, neither very soft nor very hard. Unripe fruit is too astringent to eat, whereas overripe guava has an unappealing odor. Ripe fruit yields to gentle pressure and may be stored for several days in a refrigerator. The fruit is somewhat laxative.

Culinary Vocabulary

- A "guava colada," naturally enough, is a (usually alcoholic) cocktail featuring guava juice and other ingredients, such as rum, coconut cream, and a slice of kiwi or pineapple.

CURIOSITIES OF SCIENCE AND TECHNOLOGY

- Mahatma Gandhi, (1869–1948), Indian nationalist and spiritual leader, is reputed to have protected his family's guava tree by wrapping a cloth around each fruit when he was a child.
- An average guava provides more than an adult's daily requirement for vitamin C. Guava powder rich in vitamin C was commonly added to Allied armed forces military rations during the Second World War.
- Ripe guavas have traditionally been used as a laxative. By contrast, unripe guavas tend to be constipating.
- In Central and South America, guava leaves are sometimes chewed to relieve toothache.
- In several countries of Asia, the leaves of guava are used as a source of black dye, used to color various textiles.

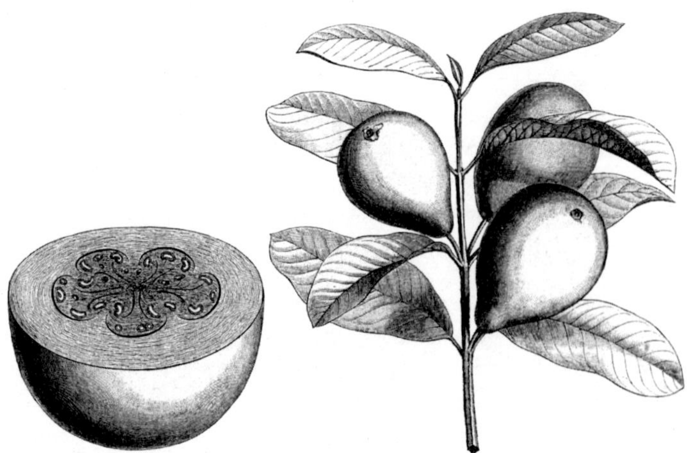

FIGURE 43.1 Guava (*Psidium guajava*). (From Engler and Prantl, 1889–1915.)

KEY INFORMATION SOURCES

Ali, Z., and Lazan, H. 1997. Guava. In *Postharvest physiology and storage of tropical and subtropical fruits*. Edited by S.K. Mitra. CAB International, Wallingford, U.K. pp. 145–165.

Batten, D.J. 1984. Guava (*Psidium guajava*). In *Tropical tree fruits for Australia*. Info. Series Q183018. Edited by P.E. Page. Queensland Dep. Primary Industry, Brisbane, Australia. pp. 113–120.

Boyle, F.P. 1957. *Commercial guava processing in Hawaii*. Hawaii Agricultural Experiment Station, Honolulu, HI. 30 pp.

Campbell, C.W. 1963. Promising new guava varieties. *Proc. Fla. State. Hort. Soc.* 76:363–365.

Chan, H.T., Jr. 1983. Guava. In *Handbook of tropical foods*. Edited by H.T. Chan, Jr. Marcel Dekker Inc., New York. pp. 351–360.

El Baradi, T.A. 1975. Guava. Review article. *Abstr. Trop. Agr.* 1(3):9–16.

Ellshoff, Z.E. 1995. *Annotated bibliography of the genus* Psidium, *with emphasis on* P. cattleianum *(strawberry guava) and* P. guajava *(common guava), forest weeds in Hawai'i*. Cooperative National Park Resources Studies Unit, University of Hawai'i at Manoa, Honolulu, HI. 102 pp.

Gandhi, S.R. 1952. *The guava in India*. Indian Council of Agricultural Research, New Delhi, India. 18 pp.

Gibbs, W.M., Wycocki, A.F., and Olexa, M.T. 2000. *The marketing of a lesser-known Florida fruit: B. The case of guava at a Florida packer shipper*. Institute of Food and Agricultural Sciences, Food and Resource Economics Department, University of Florida, Gainesville, FL. 20 pp.

Hawaii Economic Development Division. 1981. *Hawaii's guava industry*. Hawaii Economic Development Division, Honolulu, HI. 72 pp.

Jain, N.L., and Barker, D.H. 1966. Preparing beverages from guava. *Indian Hort.* 11(1):5–7.

Jaiswal, V.S., and Amin, M.N. 1992. Guava and jackfruit. In *Biotechnology of perennial fruit crops*. Edited by F.A. Hammerschlag and R.E. Litz. CAB International, Wallingford, U.K. pp. 421–431.

Khalifa, A.H., El-Dengawy, R.A., and Ramadan, B.R. 1998. Composition and utilization of guava seeds. *Assiut J. Agric. Sci.* 29:11–18.

Khan, M.I.H., and Ahmad, J. 1985. A pharmacognostic study of *Psidium guajava* L. *Int. J. Crude Drug Res.* 23:95–103.

Lazan, H., and Ali, Z.M. 1998. Guava. In *Tropical and subtropical fruits*. Edited by P.E. Shaw, H.T. Chan, Jr., and S. Nagy. Agscience, Auburndale, FL. pp. 446–485.

Lim, T.K., and Khoo, K.C. 1990. *Guava in Malaysia: production, pests, and diseases*. Tropical Press, Kuala Lumpur, Malaysia. 260 pp.

Menzel, C.M. 1985. Guava: an exotic fruit with potential in Queensland. *Qld Agric. J.* 111(2):93–98.

Misra, A.K., and Prakash, O. 1994. *Guava diseases: an annotated bibliography, 1907–1990*. Bisben Singh Mahendra Pal Singh, Dehra Dun, India. 132 pp.

Mitra, S.K., and Bose, T.K 1985. Guava. In *Fruits of India, tropical and subtropical*. Edited by T.K. Bose. Naya Prokash, Calcutta, India. pp. 277–297.

Mitra, S.K., and Bose, T.K. 1999. Guava. In *Tropical horticulture*, vol. 1. Edited by T.K. Bose, S.K. Mitra, A.A. Farooqui, and M.K. Sadhu. Naya Prokash, Calcutta, India. pp. 297–307.

Morton, J. 1987. Guava. In *Fruits of warm climates*. Creative Resource Systems, Winterville, NC. pp. 356–363.

Paull, R.E., and Bittenbender, H.C. 2008. *Psidium guajava*, guava. In *The encyclopedia of fruit & nuts*. Edited by J. Janick and R.E. Paull. CABI, Wallingford, Oxfordshire, U.K. pp. 541–549.

Prasad, A., and Shukla, J.P. 1979. Studies on the ripening and storage behaviour of guava fruits (*Psidium guajava*). *Indian J. Agric. Res.* 13:39–42.

Rathore, D.S. 1976. Effect of season on the growth and chemical composition of guava (*Psidium guajava* L.) fruits. *J. Hortic. Sci.* 51:41–47.

Ruehle, G.D. 1948. The common guava—a neglected fruit with a promising future. *Econ. Bot.* 2:306–325.

Ruehle, G.D. 1959. *Growing guavas in Florida*. Florida Agricultural Extension Service, Gainesville, FL. 20 pp.

Scott, F.S. 1958. *Commercial uses and consumer preferences for Hawaiian guava products; a guide to market development*. Agricultural Economics Bulletin 13. Hawaii Agricultural Experiment Station, Honolulu, HI. 15 pp.

Shigeura, G.T., and Bullock, R.M. 1983. *Guava (Psidium guajava L.) in Hawaii: history and production*. Hawaii Institute of Tropical Agriculture and Human Resources, College of Tropical Agriculture and Human Resources, University of Hawaii at Manoa, Honolulu, HI. 28 pp.

Singh, B.P., Kalra, S.K., and Tandon, D.K. 1990. Behavior of guava cultivars during ripening and storage. *Haryana J. Hortic. Sci.* 19:1–6.

Smith, K.L. 1957. *Growing and preparing guavas*. Florida Department of Agriculture, Tallahassee, FL. 48 pp.

Soetopo, L. 1991. *Psidium guajava* L. In *Plant resources of South-East Asia. 2. Edible fruits and nuts*. Edited by E.W.M. Verheij and R.E. Coronel. Pudoc, Leiden, the Netherlands. pp. 266–270.

South Africa Institute for Tropical and Subtropical Crops. 2003. *Cultivating guavas*. Revised edition. South Africa. Institute for Tropical and Subtropical Crops. South Africa. South Africa Department of Agriculture, Pretoria, South Africa. 12 pp.

Yadava, U.L. 1996. Guava production in Georgia under cold-protection structure. In *Progress in new crops*. Edited by J. Janick. ASHS Press, Arlington, VA. pp. 451–457.

Wilson, C.W. 1980. Guava. In *Tropical and sub-tropical fruits: composition, properties, and uses*. Edited by S. Nagy and P.E. Shaw. AVI, Westport, CT. pp. 279–299.

Specialty Cookbooks

Hobert, I., and Tietze, H.W. 2001. *Guava as medicine: a safe and cheap form of food therapy*. Subang Jaya, Selangor Darul Ehsan, Pelanduk, Malaysia. 102 pp.

Thursby, I.S. 1932. *The goodly guava*. Bulletin 70. Cooperative Extension Work in Agriculture and Home Economics, Agricultural Extension Division, University of Florida, Tallahassee, FL. 32 pp.

44 Gum Arabic

Family: Fabaceae (Leguminosae; pea family)

NAMES

Scientific Names: *Acacia* Species

- Senegal gum arabic—*A. senegal* (L.) Willd. (*Acacia verek* Guill. & Perr.)
- Shittimwood—*A. seyal* Delile
- Babul gum arabic—*A. nilotica* (L.) Willd. ex Delile (*A. arabica* (Lam.) Willd.)

- The name "gum arabic" was derived from the shipping of this gum to Europe from Arabian ports in former times. Although "Arabic" deserves to be capitalized, and "gum Arabic" is often encountered, "gum arabic" is the predominant spelling. Gum arabic is sometimes called acacia and gum acacia and less often Turkey gum.
- Senegal gum arabic, the principal species of gum arabic, is also known as acacia Senegal tree, gum arabic tree, Senegal gum, and Sudan gum arabic.
- Shittimwood is also known as shittah tree, thirsty thorn (often rendered "thirty thorn"), whistling tree, and white whistling wood. The name shittimwood traces to biblical Hebrew for the wood of the tree (*shittium*, plural *shitta*).
- Babul gum arabic is known as babul, babul acacia, black thorn (not to be confused with a fruit tree called "black thorn" but more often "sloe," *Prunus spinosa* L.), Egyptian acacia, Egyptian mimosa, Egyptian thorn, Indian gum arabic tree, Nile acacia, thorn, prickly acacia, scented thorn, suntwood, thorn mimosa, and thorny acacia.
- The genus name *Acacia* (pronounced AH-kay-sha) is based on the Greek *akakia*, the ancient Greek name for an Egyptian species. The Greek word was based on *ake*, *akis*, tip, thorn, or sharp point and is related to the Akkadian (old language of Mesopotamia) *kakkum*, thorn, and Hebrew *hoah*, thorn, all in recognition of the thorniness of the plants. *Acacia* is a huge genus, some authorities recognizing more than 1000 species.
- *Senegal* in the scientific name *A. senegal* of course refers to the country of Senegal.
- *Seyal* in the scientific name *A. seyal* is derived from an Arabic word for "torrent" used for the species in Egypt and denotes association with water courses. (The word seyal should not be confused with the fact that the species is "Sahelian," i.e., belonging to the Sahel. The Sahel is a geographic zone along the southern fringes of Africa's Sahara desert, generally understood to comprise seven countries [from east to west]: Chad, Niger, Mali, Burkina Faso, Mauritania, Senegal, and the Gambia. The area is a semiarid transition zone from the southern Sahara to the savanna lands of west and central Africa.)
- *Nilotica* in the scientific name *A. nilotica* is Latin for "of the Nile (river area)."

PLANT PORTRAIT

Senegal gum arabic is by far the primary source of gum arabic. It is a spiny, deciduous shrub or tree up to 20 m (66 ft.) tall, with a flat to rounded crown. The bark is typically yellow-brown and smooth on younger trees, changing to dark gray, gnarled and cracked on older trees. The branchlets have thorns. The leaves are made up of numerous small leaflets in 7 to 25 pairs along the leaf midrib. The flowers are white to yellowish, in spikes 5 to 12 cm (2–4¾ in.) long. The seeds pods are 7.5 to 18 cm (3–7 in.) long and 1 to 3.4 cm (0.4–1.3 in.) wide, light brown or gray, papery or woody, with five to six (rarely as many as 15) greenish-brown seeds. Senegal gum is widespread in tropical Africa from Mozambique and Zambia to Somalia, Sudan, Ethiopia, Kenya, and Tanzania. It is cultivated in Nigeria, India, Pakistan, Australia, and South America. The main producing and exporting countries in the "gum belt" include Cameroon, Chad, Mali, Nigeria, and Sudan. Sudan dominates the world gum trade with a market share of approximately 60%. In Sudan, usually around mid-October, gum arabic specialists make cuts in the bark of the trees, and gum subsequently exudes where the bark has been cut. Six weeks later, the first gum collection is made, and up to three further collections are carried out at 3-week intervals. Substitutes such as modified starches, other gums, and sugars made from microorganisms threaten the market for gum arabic.

Shittimwood is the second most important species used as a source of gum arabic. It is a thorny tree 3 to 12 m (10–39 ft.) in height. It has bright yellow flowers in heads 10 to 13 mm (4–5 in.) across and pods that are 7 to 20 cm (2¾–8 in.) long, 0.5 to 0.9 cm (1/5–1/3 in.) in diameter, with six to nine olive or olive-brown seeds. The species is widespread in tropical Africa, ranging northward to Sudan and Somalia. Its gum, which is inferior to that of Senegal gum arabic, is known as gum talha. The gum is not approved for food use in the United States and Europe, and it is mainly used in nonfood products.

Babul gum arabic is a spiny tree, 2.5 to 14 m (8–46 ft.) tall. The flowers are bright yellow, in fluffy heads 6 to 15 mm (1/5–2/3 in.) in diameter. The pods are 8 to 17 cm (3–6 2/3 in.) long (sometimes as long as 24 cm or 9 in.) and 1.3 to 2.2 cm (1/2–7/8 in.) broad, gray to black, with approximately 12 blackish-brown seeds. The species is native form Egypt south to Mozambique and Natal and has been introduced to India, Arabia, and other parts of Africa. Its gum is also of lower quality than that of Senegal gum arabic and is mostly used in the countries where the trees occur.

Gum arabic has been an article of commerce since 4000 BC and was widely used in ancient Egypt in the preparation of ink, water colors, and dyes. It is used today for culinary purposes as detailed in the next section. It is also used, particularly as an emulsifier, thickener, and stabilizer, in pharmaceuticals and cosmetics, in the printing industry (e.g., in inks and pigments and as a sensitizer for lithographic plates), and in the textile industry (e.g., as mordants in calico printing and to give body in finishing silk and rayon fabrics). Many cough drops and syrups use gum arabic because of its demulcent or soothing characteristics. Gum arabic is available in the form of white or

FIGURE 44.1 Senegal gum arabic (*Acacia senegal*). (From Köhler, 1883–1914.)

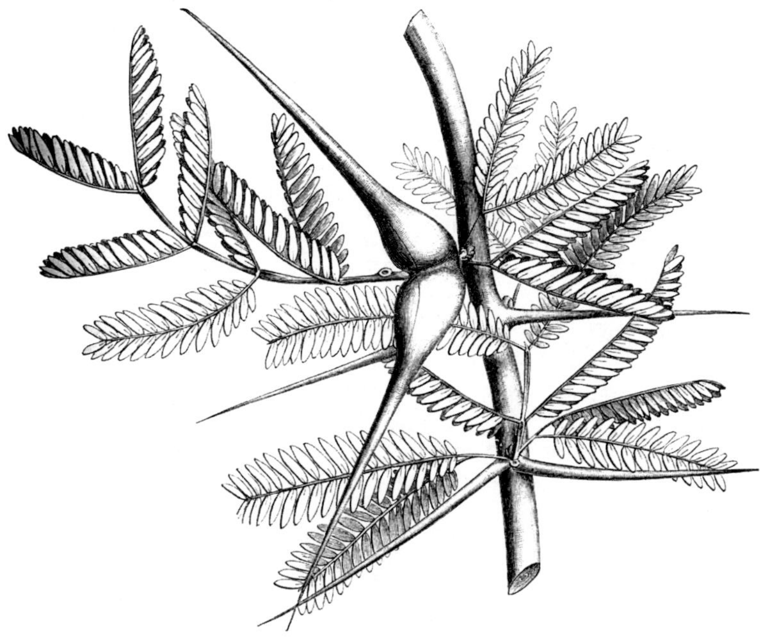

FIGURE 44.2 Shittimwood (*Acacia seyal*). (From Engler and Prantl, 1889–1915.)

FIGURE 44.3 Babul gum arabic (*Acacia nilotica*). (From Baillon, 1876–1892.)

FIGURE 44.4 Acacia woodland in Ethiopia. (From Engler and Prantl, 1889–1915.)

yellowish-white spheroidal tears of varying size, angular fragments, flakes, granules, and powder. The gum is most often presented to industry in a spray-dried form. The gum arabic is heated, dissolved in water (it is almost completely soluble in an equal volume of water), and sprayed through a nozzle, causing the droplets to dry and fall into a powder of uniform particles. Of all the natural gums, only gum arabic is sufficiently soluble in water to make spray-drying an economical option, and so other gum powders are produced simply by pulverization.

In 1997, because of political friction, the United States prohibited trade in gum arabic with Sudan for a period, although the ban was subsequently rescinded. Following the September 11, 2001, terrorist attacks on the United States, particularly against the World Trade Center in New York, stories circulated on the Internet and in some newspapers that Osama bin Laden's wealth, which contributed to the attacks, was based in part on his holdings in a Sudanese company that produced gum arabic. It has been reported that bin Laden divested himself of the holdings in 1996 when Sudan ordered him out of the country, and that in recent times he was not involved in the gum arabic trade. Efforts to boycott Sudanese gum arabic caused difficulties for Sudan, which depends on its exports of gum arabic, and produces approximately three-quarters of the world supply.

CULINARY PORTRAIT

"Gums," whether natural or manufactured, are of great value to the food industry. Principal natural gums include gum arabic, discussed in this chapter, gum tragacanth (from species of the legume genus *Astragalus*), guar gum (obtained from the legume *Cyamopsis tetragonoloba* (L.) Taub or from still another legume—carob; see Chapter 19), and from seaweeds. All gums are carbohydrates, made up of many sugar molecules linked into chains. In water, these long chains make the liquid viscous. The best gum for thickening, stabilizing, and emulsifying is gum arabic. Gum arabic is not normally purchased by most consumers, and few appreciate its important contributions to numerous prepared foods that are regularly eaten.

Gum arabic is the oldest and best known of the edible gums. It is used in foods as a suspending or emulsifying agent, bulk-forming agent, stabilizer, adhesive or film-forming agent, and flavor fixative and to prevent crystallization of sugar. Gum arabic is used in numerous processed foods, including candy, snack foods, alcoholic and nonalcoholic beverages (particularly soft drinks), baked goods, frozen dairy desserts, gelatins, and puddings, imitation dairy products, breakfast cereals, and fats and oils. For example, working as an emulsifier, gum arabic helps prevent fruit particles in soft drinks from falling to the bottom, and it keeps fats evenly distributed throughout food preparations, particularly in caramels and toffees. Gum arabic is a basic ingredient of such familiar foods as chewing gum, marshmallow, and liquorice. The superior film-forming ability of gum arabic makes it ideal for protecting flavor components from oxidation, evaporation, and absorption of moisture from the air. In brewing, gum arabic is used as a foam stabilizer and agent to promote adhesion of foam to glass. It acts as a clarity stabilizer in the chemical treatment of wines. Gum arabic itself is odorless and has a bland taste.

Culinary Vocabulary

- "Gum drops" are small candies made of sweetened, colored, and flavored gum arabic (hence the name gum drop) and often coated with sugar (gelatin and other cheaper ingredients are now usually substituted for gum arabic, except in the Middle East).
- "Acacia honey," which has been dubbed "the Rolls-Royce of honeys," is a pale yellow or amber honey with a delicate taste. China is the major source for acacia honey, which is also produced in California and elsewhere. This gourmet honey crystallizes very slowly with age, remaining liquid far longer than most other honeys. Acacia honey is often not produced from species of *Acacia*. Depending on country, the honey may be made from the nectar of other tree legumes, particularly the black locust, *Robinia pseudoacacia* L.
- *Eau parfumée* (French for "perfumed water") is a Moroccan beverage prepared by placing a vessel of water (often orange flavored) over smoldering gum arabic ashes, resulting in the scent being captured by the water.
- "Gum syrup" or "gum," as widely used by professional bartenders in preparing some cocktails, is essentially sugar and water with the addition of gum arabic for extra body and a pleasing "mouth feel."

CURIOSITIES OF SCIENCE AND TECHNOLOGY

- It was once believed that a man could live on nothing but 170 g (6 oz.) of gum arabic a day. However, gum arabic's nutritional value is roughly comparable with sugar and does not provide a balanced or sustainable source of nutrition.
- Egyptians of 2000 BC used gum arabic as a suspending and binding agent for paint pigments. The survival of some Egyptian wall paintings for 4000 years bears testimony to the value of gum arabic.
- According to Greek historian Herodotus (fifth century BC), in ancient Egypt cargo boats were constructed from acacia trees and were fitted with sails of papyrus.
- Gum arabic used to be given intravenously to counteract low blood pressure after surgery and to treat swelling associated with kidney infection, but it was found that this caused kidney and liver damage as well as allergic reactions, and such usage has been abandoned.
- Because of the presence of an enzyme in gum arabic, it is not suitable for use as a component of pharmaceuticals with certain other ingredients, including cod liver oil (it quickly reduces the vitamin A content), morphine (which it destroys), and strongly alcoholic liquids.
- In Exodus 3, Moses saw a "burning bush," which although on fire was not being consumed by the flames ("The Lord appeared unto him in a flame of fire out of the midst of a bush"). Some authorities believe that this was the thorny *A. nilotica* (the original Hebrew meant "thorny bush"), while the "fire" was a reddish parasitic mistletoe, *Loranthus acaciae* Zucc., that occurs on the plant. Both of these plants are sometimes found in gardens of biblical plants.

- Ronald Reagan (1911–2004), 40th president of the United States, was no friend of trees. In 1966 as governor of California, he opposed the expansion of Redwood National Park, declaring "A tree is a tree. How many more do you have to look at?" In 1981, he asserted that "Trees cause more pollution than automobiles do." Reagan's claim was ridiculed, but by 1995 scientists were aware that evaporation of chemicals from the leaves of trees can contribute to air pollution (although by no means as significantly as automobiles). In particular, plants produce large quantities of the chemical isoprene (3-methyl 3-butadiene), a volatile compound that can contribute to the formation of photochemical smog. It has been suggested that isoprene protects plants against heat shock at ambient temperatures above 35°C (95°F), and much greater amounts are produced at higher temperatures. Unfortunately, chemicals that evaporate from plants can combine with man-made pollutants, chiefly fumes from vehicle exhaust and smokestacks, and especially in hot weather the result is the formation of smog. Among the most important contributors to atmospheric isoprene are species of *Acacia* and *Eucalyptus*, which are very widely used in tropical areas for reforestation. However, even cutting lawns releases significant amounts of isoprene from the grass.
- Although spiders have been known occasionally to eat nectar and pollen, virtually all 40,000 species have been considered to be carnivores. *Bagheera kiplingi* is the only species that is predominantly vegetarian. This jumping spider is found in Central America, including Mexico, Costa Rica, and Guatemala. It is a specialist feeder on Acacia trees that produce "Beltian bodies" at their leaf tips. Ants are known to guard the trees against predators, rewarding the ants with the nutritious, protein- and fat-rich Beltian bodies (this is a textbook example of mutualism—an association between species to their mutual benefit). The spiders are adept at avoiding the ants, and sometimes 90% of their food consists of Beltian bodies. They will, however, steal larvae from the ants and cannibalize young of their own species.

KEY INFORMATION SOURCES

Adamson, A.D. 1974. *The market for gum arabic*. Report No. G87. Tropical Products Institute, London. 99 pp.

Anderson, D.M.W. 1986. *Evidence for the safety of gum arabic (*Acacia senegal *(L.) Willd.) as a food additive—A brief review*. *Food Addit. Contam.* 3:225–230.

Anderson, D.M.W. 1991. Food safety assurance: revised specification for gum arabic. *Br. Food J.* 93:2, 20–24.

Anderson, D.M.W. 1993. Some factors influencing the demand for gum arabic (*Acacia senegal* (L.) Willd.) and other water-soluble tree exudates. *For. Ecol. Manage.* 58:1–18.

Anderson, D.M.W., and Eastwood, M.A. 1989. The safety of gum arabic as a food additive and its energy value as an ingredient: a brief review. *J. Hum. Nutr. Diet.* 2:137–144.

Awouda, E.H.M. 1974. *Production and supply of gum arabic*. E.H.M. Awouda, Khartoum, Sudan. 39 pp.

Ayoub, S.M.H. 1983. Algicidal properties of *Acacia nilotica*. *Fitoterapia*, 53:175–178.

Barbier, E.B. 1990. *The economics of controlling degradation: rehabilitating gum arabic systems in Sudan*. London Environmental Economics Centre, London. 28 pp.

Beshai, A.A. 1984. The economics of a primary commodity: gum arabic. *Oxf. Bull. Econ. Stat.* 46:371–381.

Blunt, H.S. 1926. *Gum arabic, with special reference to its production in the Sudan*. Oxford University Press, London. 45 pp.

Cheema, M.S.Z.A., and Qadir, S.A. 1973. Autecology of *Acacia senegal* (L.) Willd. *Vegetatio*, 27:131–162.

Chikamai, B.N., Casadei, E., Coppen, J.J.W., Abdel-Nour, H.O., and Cesareo, D. 1996. *A review of production, markets and quality control of gum arabic in Africa*. Food and Agriculture Organization, Rome, Italy. 191 pp.

Fagg, C.W., and Allison, G.E. 2004. *Acacia senegal and the gum arabic trade: monograph and annotated bibliography*. Oxford Forestry Institute, University of Oxford, Oxford, U.K. 261 pp.

Flowerman, P.M. 1985. *Marketing Sudanese gum arabic in the U.S.A.: facts and options*. [Submitted by Checchi and Company to] U.S. Agency for International Development, Khartoum, Sudan. 216 pp.

Glicksman, M. 1983. Gum arabic (gum acacia). *Food Hydrocoll.* 2:7–29.

Hanson, J.H. 1992. Extractive economies in a historical perspective: gum arabic in West Africa. *Adv. Econ. Bot.* 9:107–114.

Informatics Inc. 1972. *GRAS (generally recognized as safe) food ingredients: gum arabic*. Informatics Inc., Rockville, MD. 86 pp.

Islam, A.M., Phillips, G.O., Sljivo, A., Snowden, M.J., and Williams, P.A. 1997. A review of recent developments on the regulatory, structural and functional aspects of gum arabic. *Food Hydrocoll.* 11:493–505.

Karg, C. 2000. The nutritional benefits of gum arabic (acacia gum). *Nutraceuticals Now*, (Winter):44–45.

Krochta, J.M., Baldwin, E.A., and Nisperos-Carriedo, M.O. (*Eds.*). 1994. *Edible coatings and films to improve food quality*. CRC, Boca Raton, FL. 392 pp.

Lister, P.R., Holford, P., Haigh, T., and Morrison, D.A. 1996. *Acacia* in Australia: ethnobotany and potential food crop. In *Progress in new crops*. Edited by J. Janick. ASHS Press, Alexandria, VA. pp. 228–236.

Menzies, A.R., Osman, M.E., Malik, A.A., and Baldwin, T.C. 1996. A comparison of the physicochemical and immunological properties of the plant gum exudates of *Acacia senegal* (gum arabic) and *Acacia seyal* (gum tahla). *Food Addit. Contam.* 13:991–999.

Mocak, J., Jurasek, P., Phillips, G.O., Varga, S., Casadei, E., and Chikemai, B.N. 1998. The classification of natural gums. X. Chemometric characterization of exudate gums that conform to the revised specification of the gum arabic for food use, and the identification of adulterants. *Food Hydrocoll.* 12:141–150.

Mugah, J.O., Chikamai, B.N., Mbiru, S.S., and Casadei, E. (Eds.). 1988. *Conservation, management and utilisation of plant gums, resins and essential oils. Proceedings of a Regional Conference for Africa held in Nairobi, Kenya 6–10 October 1997.* Food and Agriculture Organization, Forestry Department, Rome, Italy. 125 pp.

Osman, M.E., Williams, P.A., Menzies, A.R., and Phillips, G.O. 1993. Characterization of commercial samples of gum arabic. *J. Agric. Food Chem.* 41:71–77.

Pande, M.B., Talpada, P.M., Patel, J.S., and Shukla, P.C. 1981. Note on the nutritive value of babul (*Acacia nilotica* L.) seeds (extracted). *Indian J. Anim. Sci.* 51:107–108.

Phillips, G.O. 1998. Acacia gum (gum arabic): a nutritional fibre: metabolism and calorific value. *Food Addit. Contam.* 15:251–264.

United Nations. 1983. *The gum arabic market and the development of production*. U.N. Sudano-Sahelian Office, International Trade Centre UNCTAD/GATT, New York. 170 pp.

United States National Toxicology Program. 1982. *NTP technical report on the carcinogenesis bioassay of gum arabic (CAS No. 9000-01-5) in F344 rats and B6C3F1 mice (feed study)*. U.S. Department of Health and Human Services, Public Health Service, National Institutes of Health, Bethesda, MD. 124 pp.

Wiliams [sic], P.A., and Phillips, G.O. 2001. Gum arabic. Production, safety, and physiological effects, physicochemical characterization, functional properties, and food applications. In *Handbook of dietary fibre*. Edited by S.S. Cho and M.L. Dreher. Marcel Dekker, New York. pp. 675–693.

Specialty Cookbooks

Culinary Institute of America. 2004. *Baking and pastry: mastering the art and craft*. Wiley, New York. 880 pp.

Fal, F. 2002. *Food of Morocco: authentic recipes from the North Africa coast*. Periplus Editions, Boston, MA. 120 pp.

Farah, M. 2001. *Lebanese cuisine: more than 200 simple, delicious, authentic recipes*. Four Wall, Eight Windows, New York. 256 pp.

Mazouz, M. 2005. *The momo cookbook: a gastronomic journey through North Africa*. Simon & Schuster, London. 225 pp.

Salloum, M. 1992. *A taste of Lebanon: cooking today the Lebanese way*. Interlink Publishing, New York. 190 pp.

45 Hemp (Hempseed)

Family: Cannabaceae (hemp family)

NAMES

Scientific Name: *Cannabis sativa* L.

- The origin of the English word "hemp" is obscure. It appears to have arisen from the old Latin *hanapus* and the Old High German *hanaf*, referring to a bowl or basket. This corresponds with the English "hamper," meaning a hemp bag or wicker basket. It has been contended that both the genus name *Cannabis* and the word hemp are based on a language of Central Asia or the Near East (see below). Although "hemp" seems to be quite unrelated to "cannabis," it may have resulted from a process called the Gothonic sound shift, whereby there is a substitution of h for k and of f or p for b in Teutonic languages. Dozens of plants have "hemp" in their names, but *Cannabis* is the "true hemp."
- The earliest name for the cannabis plant is the Sanscrit *sana*, a hollow reed-like plant or cane. Corresponding to this is the Persian *canna* and *kannap*, hence the Arabian *cannab*, a small reed or cane, the Greek *kanna* and *kannabis*, a reed and anything made from it, and the Latin *cannabis*, from *canna*, a reed or cane, which led to the genus name *Cannabis*.
- *Sativa* in the scientific name is Latin for cultivated or sown.
- *Cannabis sativa* has been divided into two subspecies: *C. sativa* subsp. *sativa* (appropriately referred to as hemp) and *C. sativa* subsp. *indica* (Lam.) E. Small & Cronq. (appropriately referred to as marijuana). ("Marijuana" is more often spelled this way today. In the past, "marihuana" was more frequent. Both spellings are correct.) Although "hemp" and "marijuana" have been used interchangeably, such usage is undesirable. Most plants of subspecies *sativa* (hemp) have limited intoxicant potential and have characteristics suiting them for production of fiber from the stem and/or oil from the seeds; most plants of subspecies *indica* (marijuana) have considerable intoxicant potential and limited ability to produce commercial amounts of fiber and seeds. This chapter is about hemp, not marijuana.
- Italicized, *Cannabis* refers to the biological name of the plant (only one species of this genus is commonly recognized, *C. sativa*). Nonitalicized, "cannabis" is a generic abstraction, widely used as a noun and adjective, and commonly (often loosely) used both for cannabis plants and/or any or all of the intoxicant preparations made from them.

PLANT PORTRAIT

Cannabis sativa is an extraordinary, multipurpose plant. Its stem produces a durable fiber for textiles and many other applications, its reproductive parts (flowers and fruits) provide intoxicating drugs, and its edible seeds are the source of a multipurpose oil. However, varieties that are useful for producing fiber and oil are quite different from intoxicant varieties. This annual species, originally from Eurasia, varies greatly in appearance, depending on variety and how it is cultivated. Plants taller than 6 m (20 ft.) have often been recorded, although typically cultivated plants range from 1 to 4 m (3 to 13 ft.) in height. The sexes are separated on different plants, except for certain cultivated varieties in which male and female flowers occur on the same plant. Male plants are taller but more delicate in appearance, and they die after shedding their pollen.

Hemp is one of the world's most ancient crop plants and was valued by the Chinese 8500 years ago. Historically, the plant was used mostly as a fiber plant. Hemp is one of the oldest sources of textile fiber, and hempen cloth aged 6000 years has been found. During the era of sailing ships, *Cannabis* was considered to provide the very best canvas, and indeed this word is derived from *Cannabis*. Hemp was introduced to western Asia and Egypt and subsequently to Europe somewhere between 1000 and 2000 BC. Cultivation in Europe became widespread after AD 500. Hemp was brought to South America in 1545, in Chile, and to North America in Port Royal, Acadia, in 1606. It was widely grown in North America until the early part of the present century, followed by a brief revival during the Second World War. Until the beginning of the nineteenth century, hemp was the leading cordage fiber. Until the middle of the nineteenth century, hemp rivaled flax as the chief textile fiber of vegetable origin and indeed was described as "the king of fiber-bearing plants—the standard by which all other fibers are measured." The majority of all twine, rope, ship sails, rigging, and nets

FIGURE 45.1 Hemp (*Cannabis sativa*). (From Baillon, 1876–1892.) Left: female plant; right: male plant.

FIGURE 45.2 Seeds of hemp (*Cannabis sativa*), with a match for scale. (Photo by E. Small.)

FIGURE 45.3 Some North American foods fortified with hemp seed and/or hemp seed oil. From left to right: beer, salad dressing, energy bars, "ice cream," chips. (Photo by E. Johnson.)

up to the late nineteenth century was made from hemp fiber. The popularity of fiber usage of hemp declined in the late nineteenth and early twentieth centuries. Numerous countries forbade the cultivation of *C. sativa* during the twentieth century because of concern about its narcotic content.

Earliest reference to narcotic use of *C. sativa* appears to date to China of 5 millennia ago, but it was in India over the last millennium that cannabis consumption became more firmly entrenched than anywhere else in the world. Not surprisingly, the most highly domesticated drug strains were selected in India. Although *Cannabis* has been extensively used as a narcotic for thousands of years in India, the Near East, parts of Africa, and other Old World areas, such widespread use simply did not develop in temperate countries, where fiber hemp was raised. The use of *Cannabis* as a recreational inebriant in sophisticated, largely urban settings is substantially a twentieth-century phenomenon. Marijuana, a preparation of leaves and flowers of intoxicant varieties, has become the most widely used illegal drug in the world. Hashish is a relatively pure preparation of the resinous secretions of intoxicant varieties of the plant.

Cannabis drug preparations have been employed medicinally in folk medicine since antiquity and were extensively used in Western medicine between the middle of the nineteenth century and the Second World War, particularly as a substitute for opiates. Medical use declined with the introduction of synthetic analgesics and sedatives, and there is very limited authorized medical use today. However, there has recently been a great upsurge of interest in using marijuana for treatment of various ailments, especially for the nausea and loss of appetite that accompanies radiation treatment and chemotherapy for cancer.

The use of *Cannabis* for seed oil began at least 3000 years ago. Until 1800, hemp oil was the most widely burned lighting oil in the world. Hemp oil is a drying oil that has occasionally been used in paints and varnishes and in the manufacture of soap. In the 1990s, European firms introduced lines of hemp oil-based personal care products, including soaps, shampoos, bubble baths, and perfumes. Hemp oil is now marketed throughout the world in a range of body care products, including creams, lotions, moisturizers, and lip balms. In Germany, a laundry detergent manufactured entirely from hemp oil has been marketed.

Today, varieties of *C. sativa* low in the intoxicating compound tetrahydrocannabinol can be grown under license for production of nonnarcotic materials (primarily fiber and oilseed) in many countries, with the notable exception of the United States.

CULINARY PORTRAIT

Hempseed was one of the grains of ancient China, although there was limited subsequent direct use of hempseed as food by humans. In the past, the seeds were more likely to have been used as wild bird and poultry feed than as culinary items. Today, there has been a great resurgence of interest in using hempseed as human food, especially in North America. A principal reason for this is that efficient methods have been found to hull the seeds, that is, remove the outer hard shell. In the past, hempseed had to be eaten whole, or was only partly hulled, and so was rather gritty. A great difficulty under current regulations in most Western countries is that hempseed has to be sterilized (to prevent plants from being grown from the seed). Once sterilized, the seeds go rancid in a few weeks, unless refrigerated. The oil also goes rancid quickly, unless kept in the dark and refrigerated. Nevertheless, dozens of food items are now currently available in North America, made from hemp seeds. Hemp seeds, oil, and flour are currently being added to many foods. Hemp seeds have an attractive nutty taste and are now marketed in many forms, often mimicking familiar foods or incorporated into them. Those marketed in North America included nutritional (granola-type) bars, "nut butters," bread, pretzels, tortilla chips, cookies, yogurts, pancakes, porridge, fruit crumble, frozen dessert ("ice cream"), pasta, burgers, pizza, salt substitute, salad dressings, mayonnaise, "cheese," and beverages ("milk," "lemonade," and "coffee nog"), including alcoholic beverages (beer and wine). Hemp seed is considered to be quite nutritious. In most Western countries, there are stringent limits to the amount of tetrahydrocannabinol that is permissible in human foods, and indeed negligible levels are normally present. This is comparable with the permitted levels of opiate chemicals in poppy seeds used as food (see Chapter 71). As well as marketed food products, individuals can purchase hemp oil and hulled hemp seed for use in cookery. Hemp oil is unsuitable as a frying oil (the smoke point, i.e., the temperature when smoke is produced, is too low).

CULINARY VOCABULARY

- "Hemp nut" is hulled hemp seed; that is, the inedible shell covering has been removed. ("HempNut" is a claimed trade mark. "Hemp Hearts," the phrase also trade-marked, is occasionally encountered.)

CURIOSITIES OF SCIENCE AND TECHNOLOGY

- The oldest surviving paper is more than 2000 years of age, comes from China, and was made from hemp fiber. Egyptian papyrus sheets that might be thought to be an older form of paper are not "paper" as this term is understood by experts, because the fiber strands are woven, not "wet-laid." Until the early nineteenth century, hemp and flax were the chief paper-making materials. Wood-based paper came into use when mechanical and chemical pulping was developed in the mid-1800s in Germany and England. Today, more than 90% of paper is made from wood pulp.
- The Hmong are one of China's largest minority ethnic groups (they are known as the Miao in China) and are also found in Thailand, Laos, and Vietnam. By tradition, they are laid to rest in hemp garments (a common practice in China). Hemp must be used for funeral dress or the ancestors will refuse the dead's soul in the afterworld. Each son and daughter must give their departed parent hemp trousers or a hemp skirt to be worn in the coffin. Depending on the number of children, the deceased may be buried in as many as a dozen sets of clothing.

- The first and second drafts of the American Declaration of Independence were written on hemp paper. The final version was copied onto animal parchment and signed on August 2, 1776. The Magna Carta and the King James Bible were also written on hemp paper.
- The paintings of Rembrandt (1606–1669), Vincent Van Gogh (1853–1890), and Thomas Gainsborough (1727–1788) were painted primarily on hemp canvas, often with hemp-based paint.
- In colonial America, citizens of several colonies were required by law to grow hemp.
- In 1682, the Virginia legislature made hemp fiber legal tender for up to one quarter of all debts. Similar laws were enacted in Maryland in 1683 and Pennsylvania in 1706. By 1810, hemp was Kentucky's major crop and was also used as money.
- Wood-based paper came into use when mechanical and chemical pulping were developed in the mid-1800s in Germany and England. Today, at least 95% of paper is made from wood pulp. Before then, paper was made from rags, most commonly hemp rag.
- The 1892 World's Fair in Chicago featured hundred of architectural "marble" columns that were actually made up of hemp and plaster of Paris. Today, especially in France, similar cement-like materials made of hemp and plaster of Paris are being used in house construction.
- The first diesel engine was designed to run on vegetable oils, one of which was hemp oil.
- In 1941, Henry Ford (1863–1947), built a car body from a mixture of plant resins, including hemp, and demonstrated that it could run on fuel made from plants, including hemp.
- In the 1990s, the European Union provided a subsidy of more than $1000 per hectare (approximately $400/acre) for farmers who grew hemp. In a classic example of how good-intentioned legislation can be abused, thousands of hectares of hemp were grown in Spain, the subsidies were collected, and all of the hemp was simply burned in the fields.

KEY INFORMATION SOURCES

Baxter, B., and Scheifele, G. 2000. *Growing industrial hemp in Ontario*. Ontario Ministry of Agriculture, Food and Rural Affairs, Toronto, ON. 10 pp.

Bócsa, I., and Karus, M. 1998. *The cultivation of hemp: botany, varieties, cultivation and harvesting*. Hemptech, Sebastopol, CA. 184 pp.

Boyce, S.S. 1900. *Hemp (*Cannabis sativa*). A practical treatise on the culture of hemp for seed and fiber with a sketch of the history and nature of the hemp plant.* Orange Judd Company, New York. 112 pp.

Ceapoiu, N. 1958. *Hemp, monographic study*. Bucharest. Editura Academiei Republicii Populare Rominae, Bucharest, Romania. 734 pp (in Romanian).

Clarke, R.C. 1977. *The botany and ecology of* Cannabis. Pods Press, Ben Lomond, CA. 57 pp.

Conrad, C. 1997. *Hemp for health: the medicinal and nutritional uses of* Cannabis sativa. Healing Arts Press, Rochester, VT. 264 pp.

Grotenhermen, F., and Russo, E. (Eds.). 2002. Cannabis *and cannabinoids: pharmacology, toxicology, and therapeutic potential*. Haworth Integrative Healing Press, New York. 439 pp.

Hemptech. 1995. *Industrial hemp: practical products—paper to fabric to cosmetics*. Hemptech, Ojai, CA. 48 pp.

Hillig, K.W., and Mahlberg, P.G. 2004. A chemotaxonomic analysis of cannabinoid variation in *Cannabis* (Cannabaceae). *Am. J. Bot.* 91:966–975.

Jones, K. 1995. *Nutritional and medicinal guide to hemp seed*. Rainforest Botanical Laboratory, Gibsons, BC. 60 pp.

Joyce, C.R.B., and Curry, S.H. (Eds.). 1970. *The botany and chemistry of* Cannabis. J. & A. Churchill, London. 217 pp.

Kriese, U., Schumann, E., Weber, W.E., Beyer, M., Bruehl, L., and Mattaeus, B. 2004. Oil content, tocopherol composition and fatty acid patterns of the seeds of 51 *Cannabis sativa* L. genotypes. *Euphytica*, 13:339–351.

Leizer, C., Ribnicky, D., Poulev, A., Dushenkov, S., and Raskin, I. 2000. The composition of hemp seed oil and its potential as an important source of nutrition. *J. Nutraceut. Funct. Med. Foods*, 2(4):35–53.

Matthews, P. 2003. *Cannabis culture*. Bloomsbury, London. 276 pp.

McPartland, J.M., Clarke, R.C., and Watson, D.P. 2000. *Hemp diseases and pests: management and biological control*. CABI Publ., New York. 251 pp.

Meijer, E.P.M. de, and Keizer, L.C.P. 1996. Patterns of diversity in *Cannabis*. *Genet. Resour. Crop Evol.* 43:41–52.

Meijer, E.P.M. de, Bagatta, M., Carboni, A., Crucitti, P., Moliterni, V.M.C., Ranalli, P., and Mandolino, G. 2003. The inheritance of chemical phenotype in *Cannabis sativa* L. *Genetics*, 163:335–346.

Montford, S., and Small, E. 1999. A comparison of the biodiversity friendliness of crops with special reference to hemp (*Cannabis sativa* L.). *J. Int. Hemp Assoc.* 6:53–63.

Nova Institute. 1995. Bioresource hemp. In *Proceedings of the Symposium, Frankfurt am Main, Germany, March 2–5, 1995*. 2nd ed. Distributed by Hemptech, Ojai, CA. 626 pp. (contributions in English and German).

Nova Institute. 1997. Bioresource hemp 97. In *Proceedings of the Symposium, Frankfurt am Main, Germany, February 27–March 2, 1997*. Distributed by Hemptech, Sebastopol, CA. 699 pp (contributions in English and German).

Ranalli, P. (Ed.). 1998. *Advances in hemp research*. Food Products Press (of Haworth Press), London. 272 pp.

Robson, P. 2001. Therapeutic aspects of cannabis and cannabinoids. *Br. J. Psychiatry*, 178:107–115.

Roulac, J. 1997. *Hemp horizons: the comeback of the world's most promising plant*. Chelsea Green Pub., White River Junction, VT. 211 pp.

Schreiber, G. 2002. *The hemp handbook*. Revised edition. Fusion, London. 173 pp.

Sherman, C., Smith, A., and Tanner, E. 1999. *Highlights: the illustrated history of* Cannabis. Ten Speed Press, Berkeley, CA. 159 pp.

Small, E. 1979. *The species problem in* Cannabis, *science and semantics*. Corpus, Toronto, ON. 2 vols.

Small, E. 1995. Hemp. In *Evolution of crop plants*. 2nd ed. Edited by J. Smartt and N.W. Simmonds. Longman Scientific & Technical, Burnt Mill, Harlow, Essex, U.K. pp. 28–32.

Small, E. 2004. Narcotic plants as sources of medicinals, nutraceuticals, and functional foods. In *Proceedings of the International Symposium on the Development of Medicinal Plants, Hualien District Agricultural Research and Extension Station, Haulien, Taiwan, August 24–25, 2004*. Edited by F.-F. Hou, H.-S. Lin, M.-H. Chou, and T.-W. Chang. pp. 11–67.

Small, E. 2007. *Cannabis* as a source of medicinals, nutraceuticals, and functional foods. In *Advances in medicinal plant research*. Edited by S.N. Acharya and J.E. Thomas. Research Signpost, Transworld Research Network, Trivandrum, Kerala, India. pp. 1–39.

Small, E., and Cronquist, A. 1976. A practical and natural taxonomy for *Cannabis*. *Taxon*, 25:405–435.

Small, E., and Marcus, D. 2002. Hemp: a new crop with new uses for North America. In *Trends in new crops and new uses*. Edited by J. Janick and A. Whipkey. ASHS Press, Alexandria, VA. pp. 284–326.

Specialty Cookbooks

Note: The use of psychoactive *Cannabis* material is illegal under most circumstances, and cookbooks focused on recipes in this regard are not listed; the use of hemp seed products is authorized in many countries.

Benhaim, B. 2000. *H.E.M.P. Healthy eating made possible*. Fusion Press, London. 332 pp.

Cicero, D., Czartoryski, K., Gruber, S., and Lipp, M. 2002. *The Galaxy Global Eatery hemp cookbook*. Frog, Ltd., Berkeley, CA. 312 pp.

Dalotto, T. 1999. *The hemp cookbook: from seed to shining seed*. Healing Arts Press, Rochester, VT. 184 pp.

Hiener, R., and Mack, B. 1999. *The hemp cookbook*. Ten Speed Press, Berkeley, CA.143 pp.

Krieger, M.C.R., and Krieger, G.W. 2000. *Cooking for life: recipes with cannabis butter: research—in search of wellness*. Lark Pub., Calgary, AB. 136 pp.

Leson, G., Pless, P., and Roulac, J. 1999. *Hemp foods & oils for health: your guide to cooking, nutrition, and body care*. Hemptech, Sebastopol, CA. 62 pp.

Rose, R., Mars, B., and Pirello, C. 2004. *The hempnut cookbook: ancient foods for a new millennium*. Book Publishing Company, Summertown, TN. 180 pp.

Suzanne, K. 2009. *Krysten Suzanne's ultimate raw vegan hemp recipes: fast & easy raw food hemp recipes for delicious soups, salads, dressings, bread, crackers, butter, spreads, dips, breakfast, lunch, dinner & desserts*. Green Butterfly Press, Scottsdale, AZ. 116 pp.

Woodland Publishing. 2005. *Healthy recipes: delicious recipes for using hemp foods*. Woodland Publishing, Orem, UT. 38 pp.

46 Horseradish Tree

Family: Moringaceae (horseradish tree family)

NAMES

Scientific Name: *Moringa oleifera* Lam. (*M. pterygosperma* Gaertn., *M. aptera* Gaertn.)

- The horseradish tree is so-named for the very close similarity of taste of the condiment prepared from its roots to true horseradish (from *Armoracia rusticana* P. Gaertn., B. Mey. & Scherb.).
- Other names for the horseradish tree include ben tree, ben-oil tree, benzolive tree, cabbage tree, kelor tree, moringa, radish tree, oil of ben tree, and West Indian ben. The many names with "ben" reflect the fact that ben oil (or behen oil) is also obtained from African species of *Moringa*, where the Ben or Benas, an African Bantu-speaking people north of Lake Nyasa, are located.
- The very long woody pods are called "drumsticks" in India. The name was first introduced by the British in India because of the resemblance of the pods to drumsticks. The tree accordingly acquired the name drumstick tree.
- In the Philippines, the leaves of the horseradish tree are cooked and fed to babies, resulting in the name "mother's best friend." Eating the leaves is thought to increase a nursing mother's milk production.
- The species is commonly known in the Philippines as *malunggay*, and this name is often used as an English word.
- In Mexico, the horseradish tree is called *coatli*.
- In the African country of Burkina Faso, the Moré language name for the tree is *arzam tigha*, the tree of paradise. Another colorful African name, from Nigeria, is the Yoruba language *idagbo monoye*, the tree that grows crazily.
- Because of its exceptional resistance to drought, the horseradish tree is called *nebeday* in West Africa, the name apparently derived from the English "never die."
- The genus name *Moringa* is based on a Malayan name (*muringo*) for the plant, said to indicate the long, thin shape of the pods. In Dravidian (a group of languages spoken mostly in India and nearby countries), the horseradish tree is called *morunga*.
- *Oleifera* in the scientific name *M. oleifera* is Latin for oil bearing. The mature seed kernels contain an average of approximately 40% oil, the chief economic product of the species.

PLANT PORTRAIT

The horseradish tree (which is sometimes a large shrub) is commonly 5 to 10 m (16–33 ft.) in height, occasionally reaching 15 m (49 ft.). The trunk is slim, ranging from 10 to 25 cm (4–12 in.). The leaves may be evergreen, depending on climate, but in many regions the plant conserves water by shedding its foliage during the dry season. The leaves are 20 to 50 cm (8–20 in.) long and are repeatedly dissected like those of many ferns. White or cream colored flowers 2.5 cm (1 in. wide) are displayed in large clusters. The fruits are pod-like, borne singly or in pairs, and at maturity are 2 cm (4/5 in.) wide, 25 to 45 cm (10–18 in.) long, up to 120 cm or 4 ft. in some cultivated varieties, three sided or nearly cylindrical, nine ribbed, brown, and somewhat woody. When fully mature, the fruits split open into three valves. Inside is a single row of round or triangular brown seeds each with three white papery wings, embedded in dry, white tissue. This tropical tree is apparently native only to restricted areas in the southern foothills of the Himalayas, notably in northern India and Pakistan, and is now cultivated in all tropical countries, particularly in Asia, Africa, tropical regions of the Americas, and Oceania.

Characterized by some as one of the world's most valuable multipurpose trees, the horseradish tree has been dubbed "the miracle tree." A pale yellow, nondrying oil in the seeds has a mild, nutty flavor. The chief commercial use of the horseradish tree is as a source of this oil, known as ben oil, ben nut oil, and behen oil. It is used for light lubrication (as in watches and other delicate machinery), perfumery, artwork, and cosmetics as well as for culinary purposes. This seed oil has also been used for illumination as it burns without producing smoke. Additionally, it is used as a hair dressing and for making soap. Ben oil is also obtained from other species of *Moringa*. The horseradish tree is widely grown as an ornamental in many tropical countries and is sometimes planted in southern California and southern Florida. Seeds are available from some horticultural supply sources, and the horseradish tree is sometimes grown as a potted plant. It can be grown as an annual in cold areas.

Hunger in regions of the world, especially Africa, is due in part to climates that are inhospitable for the cultivation of food plants, and innovative solutions are being sought. The horseradish tree is considered to be particularly useful in addressing this problem because it is extremely fast growing (up to 4 m or 13 ft. in its first year), tolerates extreme drought, and the leaves and fruit can be harvested and stored as nutritious food for the off-season when little food is otherwise available.

CULINARY PORTRAIT

In India, Malaysia, the Philippines, and tropical Africa, the horseradish tree is prized for its young edible fruits, seeds, leaves, flowers, roots, and seed oil. The seed oil is slow to turn rancid and can be used in salads. Tender young plants (less than 50 cm or 20 in. tall) and young leaves and flowers are cooked and consumed in salads, soups, and sauces, or as greens. The foliage contains raffinose and stachyose, which can produce flatulence. Slim young pods are cooked and eaten like spinach. In India and in the Philippines, the still green but almost mature pods are peeled, stewed or boiled, and added to curries. These so-called "drum sticks" are commonly cut into sections and canned for export, and they are often available in Oriental markets in North America. The pods are best for human consumption when they can be easily broken, without visible strings of fiber. If left to become mature, the tough pod coat is discarded while the inner contents are consumed (the contents of stewed fruits are simply sucked out and the pod thrown away). The drumsticks are said to have a taste like asparagus and are eaten much like green beans. This usage as a vegetable has given rise to the proposal that the species belongs to a new category of edible plants, "vegetable trees." The immature seeds are used in India and elsewhere much like green peas. A white, bitter hull around the green seeds is removed before cooking. The mature seeds are fried and taste like peanuts. Mature seeds are also ground to a powder for use in seasoning sauces.

The roots are pungent like horseradish and are much used in India as a substitute for horseradish. In India, plants as small as 60 cm (2 ft) in height are pulled up, the root bark scraped off, the inner root ground up, and vinegar and salt added to make a popular condiment. The root bark must be completely removed as it contains potent, toxic constituents (indeed, the root bark has been used to induce abortion, often killing the expectant mother). Cautions have been made that even when the condiment is properly prepared, eating it in excess may be harmful. As well, it has been recommended that toxicity studies should be conducted on other parts of the plant, as the effects of long-term consumption are not well known.

Culinary Vocabulary

- In India, the edible young pods ("drumsticks") are known as *susumber*, and this term will be encountered in restaurants specializing in Indian cuisine.

CURIOSITIES OF SCIENCE AND TECHNOLOGY

- Arab women in the Sudan discovered that the muddy water of the Blue Nile could be clarified by swirling it in a container with horseradish tree seeds. In the Nile Valley, the name of the tree is *Shagara al Rauwaq*, "tree for purifying." The seeds are now used as an agent to purify surface water from rivers, ponds, water holes and shallow wells in rural areas of the Tropics. Fecal matter is common in the water of these regions, and this method of removal is important for disease control. Chemicals in the seeds make traces of silt and clay settle out as effectively as the alum (aluminum sulfate) that city water departments use.
- In some parts of Nigeria, crushed horseradish tree leaves are used for scrubbing cooking utensils and for cleaning walls.
- In Java, pet turtle doves are bathed in horseradish tree leaf juice in the belief that this enhances their singing.
- The long thin pods of the horseradish tree have a reputation as an aphrodisiac because of their fancied phallic appearance (although known to exceed a meter or yard in length). In southern India, the pods are fed to bridegrooms to promote their fertility.
- In Ayurveda, the ancient traditional medicinal system of India, the leaves of the horseradish tree are used to prevent approximately 300 diseases. The tree is employed to treat such ailments as rheumatism, gout, and high blood pressure.
- The fruits, seeds, flowers, and particularly the leaves of the horseradish tree are all considered to be nutritious vegetables. The leaves have been said to contain seven times the concentration of vitamin C as oranges, four times the calcium in milk, four times the vitamin A in milk, three times the potassium in bananas, and twice the protein in milk.
- Because of its nutritional and medicinal values, hospitals in Senegal have planted horseradish trees in their yards. In particular, small quantities of horseradish leaf powder have proven to greatly benefit children suffering from malnutrition, who have been brought to the hospitals.

FIGURE 46.1 Horseradish tree (*Moringa oileifera*). (From Lamarck and Poiret, 1744–1829, plate 337.)

KEY INFORMATION SOURCES

Bennett, R.N., Mellon, F.A., Foidl, N., Pratt, J.H., Dupont, M.S., Perkins, L., and Kroon, P.A. 2003. Profiling glucosinolates and phenolics in vegetative and reproductive tissues of the multi-purpose trees *Moringa oleifera* L. (horseradish tree) and *Moringa stenopetala* L. *J. Agric. Food Chem.* 51:3546–3553.

Bond, R.E. 1985. The horseradish tree. *California Rare Fruit Growers Newsl.* 17(4):14–16.

Caceres, A., Freire, V., Giron, L.M., Aviles, O., and Pacheco, G. 1991. *Moringa oleifera* (Moringaceae): ethnobotanical studies in Guatemala. *Econ. Bot.* 45:522–523.

Coote, H.C., Stewart, M., and Bonongwe, C. 1997. *The distribution, uses, and potential for development of Moringa oleifera in Malawi.* Forestry Research Institute of Malawi, Zomba, Malawi. 40 pp.

Dahot, M.U. 1988. Vitamin contents of the flowers and seeds of *Moringa oleifera. Pak. J. Biochem.* 21:21–24.

Dalla Rosa, K.R. 1993. Moringa oleifera: *a perfect tree for home gardens.* Agroforestry Information Service, Paia, Hawaii. 2 pp.

D'Souza, J., and Kulkarni, A.R. 1993. Comparative studies on nutritive values of tender foliage of seedlings and mature plants of *Moringa oleifera* Lam. *J. Econ. Taxon. Bot.* 17:479–485.

Dutt, B.S.M., Narayana, L.L., Radhakrishnaiah, M., and Nageshwar, G. 1984. Systematic position of *Moringa. J. Econ. Taxon. Bot.* 5:577–580.

Foidl, N., and Paull, R.E. 2008. *Moringa oleifera*, moringa or horseradish tree. In *The encyclopedia of fruit & nuts.* Edited by J. Janick and R.E. Paull. CABI, Wallingford, Oxfordshire, U.K. pp. 509–512.

Folkard, G., and Sutherland, J. 1996. *Moringa oleifera*: a tree and a litany of potential. *Agroforestry Today.* 8(3):5–8.

Fuglie, L.J. (Ed.). 2001. *The miracle tree: the multiple attributes of moringa.* Church World Service, Dakar, Senegal. 172 pp.

Fuglie, L.J. 1999–2003. *The miracle tree:* Moringa oleifera, *natural nutrition for the Tropics.* Church World Service, New York. 63 pp.

Indian Council of Forestry Research and Education. 1994. Moringa oleifera. Indian Council of Forestry Research and Education and Forest Research Institute, Dehra Dun, India. 10 pp.

Juliani, H.R., Fonseca, Y., Acquaye, D., Malumo, H., Malainy, D., and Simon, J.E. 2009. Nutritional assessment of Moringa (*Moringa* spp.) from Ghana, Senegal and Zambia. In *African natural plant products: new discoveries and challenges in chemistry and quality.* Edited by H.R. Juliani, J.E. Simon, and C.-T. Ho. American Chemical Society, Washington, DC. pp. 469–484.

Jyothi, P.V., Atluri, J.B., and Reddi, C.S. 1990. Pollination ecology of *Moringa oleifera* Moringaceae. *Proc. Indian Acad. Sci. Plant Sci.* 100:33–42.

Makkar, H.P.S., and Becker, K. 1996. Nutritional value and antinutritional components of whole and ethanol extracted *Moringa oleifera* leaves. *Animal Feed Sci. Technol.* 63:211–228.

Makkar, H.P.S., and Becker, K. 1997. Nutrients and antiquality factors in different morphological parts of the *Moringa oleifera* tree. *J. Agric. Sci.* 128:311–322.

Morton, J.F. 1991. The horseradish tree, *Moringa pterygosperma* (Moringaceae)—a boon to arid lands? *Econ. Bot.* 45:318–333.

Mughal, M.H., Saba, Srivastava, P.S., and Iqbal, M. 1999. Drumstick (*Moringa pterygosperma* Gaertn.): a unique source of food and medicine. *J. Econ. Taxon. Bot.* 23:47–61.

Mulavi, G.M., Sprent, J.I., Soranzo, N., Provan, J., Odee, D., Folkland, G., et al. 1999. Amplified fragment length polymorphism (AFLP) analysis of genetic variation in *Moringa oleifera* Lam. *Mol. Ecol.* 8:463–470.

Ndabigengesere, A., and Narasiah, K.S. 1996. Influence of operating parameters on turbidity removal by coagulation with *Moringa oleifera* seeds. *Environ. Technol.* 17:1103–1112.

Ndabigengesere, A., and Narasiah, K.S. 1998. Quality of water treated by coagulation using *Moringa oleifera* seeds. *Water Res.* 32:781–791.

Ndabigengesere, A., and Narasiah, K.S. 1998. Use of *Moringa oleifera* seeds as a primary coagulant in wastewater treatment. *Environ. Technol.* 19:789–800.

Oliveira, J.T.A., Silveira, S.B., Vasconcelos, I.M., Cavada, B.S., and Moreira, R.A. 1999. Compositional and nutritional attributes of seeds from the multiple purpose tree *Moringa oleifera* Lamarck. *J. Sci. Food Agric.* 79:815–820.

Olson, M.E. 2002. Combining data from DNA sequences and morphology for a phylogeny of Moringaceae (Brassicales). *Syst. Bot.* 27:55– 73.

Olson, M.E., and Carlquist, S. 2001. Stem and root anatomical correlations with life form diversity, ecology, and systematics in *Moringa* (Moringaceae). *Bot. J. Linn. Soc.* 135:315–348.

Palada, M.C. 1996. Moringa (*Moringa oleifera* Lam.): a versatile tree crop with horticultural potential in the subtropical United States. *HortScience*, 31:794–797.

Parrotta, J.A. 1993. Moringa oleifera *Lam.: resed, horseradish tree, Moringaceae, horseradish-tree family*. International Institute of Tropical Forestry, U.S. Department of Agriculture, Forest Service, Río Piedras, Puerto Rico. 6 pp.

Peter, K.V. 1987. Drumstick (*Moringa oleifera* Lam.)—a multipurpose perennial Indian vegetable tree of considerable medicinal value. In *Medicinal and poisonous plants of the Tropics. Proceedings of Symposium 5-35 of the 14th International Botanical Congress, Berlin, July 24–August 1, 1987*. Edited by A.J.M. Leeuwenberg. Pudoc, Wageningen, the Netherlands. pp. 124–127.

Polprasid, P. 1993. *Moringa oleifera* Lamk. In *Plant resources of South-East Asia. 8. Vegetables*. Edited by J.E. Siemonsma and K. Piluek. Pudoc Scientific Publishers, Wageningen, the Netherlands. pp. 213–215.

Ramachandran, C., Peter, K.V., and Gopalakrishnan. P.K. 1980. Drumstick (*Moringa oleifera*): a multipurpose Indian vegetable. *Econ. Bot.* 34:276–283.

Rao, N.V., Avita, S., and Inamdar, J.A. 1983. Studies on the Moringaceae. *Feddes Repert.* 94:213–223.

Ray, A.L. 1970–1979? *The horseradish tree: a valuable source of nutritious food*. Grace Mountain Mission, Port-au-Prince, Haiti. 13 pp.

Ronse-Decraene, L.P., Laet, J. de., and Smets, E.F. 1998. Floral development and anatomy of *Moringa oleifera* (Moringaceae): what is the evidence for a Capparalean or Sapindalean affinity? *Ann. Bot.* 82:273–284.

Shindano, J., and Kasase, C. 2009. Moringa (*Moringa oleifera*): a source of food and nutrition, medicine and industrial products. In *African natural plant products: new discoveries and challenges in chemistry and quality*. Edited by H.R. Juliani, J.E. Simon, and C.-T. Ho. American Chemical Society, Washington, DC. pp. 421–467.

Shukla, S., Mathur, R., and Prakash, A.O. 1988. Antifertility profile of the aqueous extract of *Moringa oleifera* roots. *J. Ethnopharmacol.* 22:51–62.

Verdcourt, B. 1985. A synopsis of the Moringaceae. *Kew Bull.* 40:1–34.

Villasenor I.M., Lim-Sylianco, C.Y., and Dayrit, F. 1989. Mutagens from roasted seeds of *Moringa oleifera*. *Mutat. Res.* 224:209–212.

Specialty Cookbooks

Fuglie, L.J. 2001. Moringa preparations. In *The miracle tree: the multiple attributes of moringa*. Edited by L.J. Fuglie. Church World Service, Dakar, Senegal. pp. 145–152.

Holst, S. 2000. *Moringa: nature's medicine cabinet*. Sierra Sunrise Books, Sherman Oaks, CA. 122 pp.

47 Jackfruit

Family: Moraceae (mulberry family)

NAMES

Scientific Name: *Artocarpus heterophyllus* Lam. (*A. integer* or *A. integrifolius* of some authors)

- The "jack" in "jackfruit" is derived from the Portuguese *jaca*, which in turn comes from the Malayalam *cakkai* or *chakka*, the name of the fruit. (Malayalam is the Dravidian language of Kerala, southwest India, closely related to the Tamil language.)
- See Chapter 13 for the derivation of the genus name *Artocarpus*.
- *Heterophyllus* in the scientific name *A. heterophyllus* is based on Greek meaning "diversely-leaved," a reference to the unusual and variable leaves of the jackfruit tree.

PLANT PORTRAIT

The jackfruit is an evergreen tree, 9 to 21 m (30–70 ft.) tall. It is thought to have originated in the rainforests of the Western Ghats (a mountain chain of southwestern India) and is adapted to grow only in humid tropical and near-tropical climates. The ancient Romans were aware of the jackfruit as a result of their expeditions. Jackfruit is an important food crop in the tropics and is cultivated in India, Burma, Sri Lanka (Ceylon), southern China, Malaya, East Indies, Philippines, central and eastern Africa, Surinam, and Australia. The fruit is most popular in India and Sri Lanka. A sticky, white latex is present in all parts of the plant. The fruit is the largest of all edible tree fruits: 20 to 90 cm (8–36 in.) long and 15 to 50 cm (6–20 in.) wide and weighing 4.5 to 20 kg (10–44 lb.) or sometimes as much as 50 kg (110 lb.). (Strictly speaking, the fruit is a "multiple fruit" or "aggregate fruit," composed of many individual fruits.) The inedible rind of the fruit is green or yellowish-green, turning yellow when ripe, made up of numerous hard, cone-like, spiny points attached to a thick, rubbery, pale yellow or whitish wall. The interior of the fruit is made up of large bulbs of yellow, banana-flavored flesh, among narrow ribbons of thin, tough material and a central, pithy core. Each bulb encloses a white-membrane-covered, light-brown pit or seed, 2 to 4 cm (3/4–1½ in.) long. There may be 100 to 500 pits in a single fruit. In India, a good yield is 150 large fruits per tree annually, with large mature trees sometimes producing as many as 500 fruits.

CULINARY PORTRAIT

The jackfruit is not to the taste of Westerners unfamiliar with it but is quite pleasant when cooked and made into a vegetable, chutney, jam, jelly, custard, or a dessert-like ice cream. In Asia, unripe fruit pulp, seeds, and flowers are used in curried dishes or baked and served as a vegetable, and while ripe, raw fruits are used in desserts, sometimes with added sugar. In India, jackfruit is used in fruit salad, the pulp is preserved in sugar syrup, flakes of pulp are fried and consumed like potato chips, and the juice and flesh are used in various prepared dishes. When fully ripe, the unopened jackfruit emits a strong disagreeable odor, resembling that of decayed onions, while the pulp of the opened fruit smells like pineapple and banana. Jackfruits turn brown and deteriorate quickly after ripening but can be kept for a month in cold storage. The ripe fruit is somewhat laxative, and consuming large amounts will cause diarrhea.

The raw seeds are indigestible. However, these can be eaten when boiled or roasted (the skin of the seeds is inedible), and in tropical countries they are used in various culinary dishes. In India, the seeds are dried and ground into a flour, which is used in food preparations. Roasted pits are said to be reminiscent of roasted chestnuts in size and taste. Canned seeds are sometimes available. Tender young leaves and male flower clusters are sometimes roasted and consumed as vegetables. In India, jackfruit leaves are used as a food wrap and as spoons or cups in some villages. Plates are occasionally made by stitching the leaves together.

Jackfruit continues to ripen after harvest from the tree, like bananas. Because the fruits are so large, they are often sold in pieces. These should have a strong aroma but should not smell fermented. Jackfruit removed from its rind can be kept in a refrigerator for several days. Gummy, sticky, white latex from the fruit can accumulate on hands unless they are first rubbed with salad oil. To prevent the sticky sap from adhering to a knife used to open a whole jackfruit, an oiled knife can be used. Experienced chefs often cut the fruit in quarters lengthwise, discard the inedible core, separate the small "fruitlets," and remove the seed in each of these.

Culinary Vocabulary

- *Halo-halo* is a Filipino milk shake, typically prepared with strips of jackfruit, young coconut, and sweet red beans.

CURIOSITIES OF SCIENCE AND TECHNOLOGY

- The fruit of the jackfruit tree is unusual in being borne on the main branches and the trunks, occasionally even from surface roots of the tree. This makes sense, however, because the smaller branches and twigs would not be strong enough to bear the enormous weight of the huge fruits. (Trees that bear fruits on their trunks are termed "cauliflorous," a word that has nothing to do with cauliflower.)

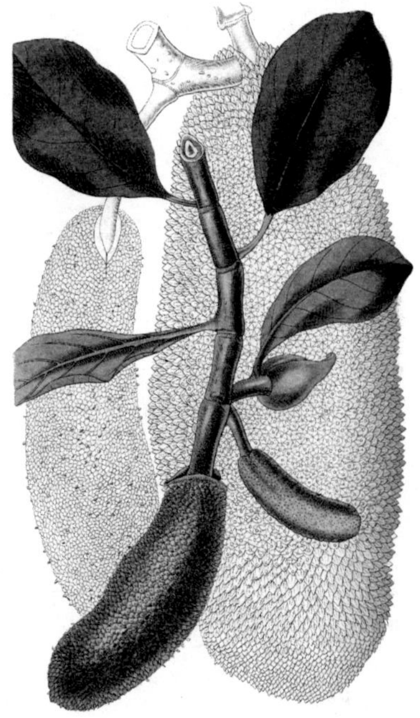

FIGURE 47.1 Jackfruit (*Artocarpus heterophyllus*). (From Curtis, 1828, vol. 55, plate 2833.)

- Heated latex from the jackfruit is used as a household cement for mending chinaware and earthenware and to caulk boats and holes in bucket.
- The wood of the jackfruit tree resembles mahogany and is extremely resistant to termites. Palaces have been built of jackfruit wood in Bali (an Indonesian island of the eastern end of Java), and the limited supply was once reserved for temples in Indochina.
- In religious ceremonies in Malabar (a coastal region of southwestern India on the Arabian Sea), dried branches of the jackfruit tree are used to produce fire by friction.
- A rich yellow dye is obtained from the wood of the jackfruit tree and is used for coloring silk and the cotton robes of Buddhist priests in Sri Lanka and Burma.
- In China, jackfruit is used to treat hangovers.
- It has been said that jackfruit should not be prepared or consumed with alcohol, as the combination has an unpleasant effect on some people. The same contention has been advanced about the durian (see Chapter 33), and it is unclear whether or not this is simply a myth.
- Appropriately, the largest of all tree fruits, the jackfruit, is a favorite food of the largest of terrestrial animals, the elephant.

KEY INFORMATION SOURCES

Acedo, A.L. 1992. *Jackfruit biology, production, use, and Philippine research*. Forestry/Fuelwood Research and Development Project, Winrock International Institute for Agricultural Development, Arlington, VA. 51 pp.

Berry, S.K., and Kalra, C.L. 1988. Chemistry and technology of jack fruit (*Artocarpus heterophyllus*)—a review. *Indian Food Packer*, 42(3):62–76.

Campbell, R.J., and El-Sawa, S.F. 1999. New jackfruit cultivars for commercial and home garden use in Florida. *Proc. Annu. Meet. Fla. State Hort. Soc.* 111:302–304.

Campbell, R.J., and Ledesma, N. 2003. *The exotic jackfruit: growing the world's largest fruit*. Fairchild Tropical Garden, Coral Gables, FL. 71 pp.

Campbell, R.J., El-Sawa, S.F., and Eck, R. 1998. *The jackfruit (*Artocarpus heterophyllus *Lam.)*. Fairchild Tropical Garden, Miami, FL. 23 pp.

Che-Man, Y.B., and Sanny, M.M. 1996. Stability of jackfruit leather in different packaging materials. *ASEAN Food J.* 11(4):163–168. [Jackfruit "leather" is dehydrated fruit, reminiscent of jerky, for eating.]

Che-Man, Y.B., and Sin, K.K. 1997. Processing and consumer acceptance of fruit leather from the unfertilised floral parts of jackfruit. *J. Sci. Food Agric.* 75:102–108.

Corner, E.J.H. 1939. Notes on the systematy and distribution of Malayan phanerogams, II. The jack and the chempedak. *Garden's Bulletin Straits Settlements*, 10:56–81.

Dutta, S. 1956. Cultivation of jack fruit in Assam. *Indian J. Hortic.* 13:189–197.

Hossain, M., Haque, A., and Hossain, M. 1979. Nutritive value of jackfruit. *Bangladesh J. Agric.* 4:9–12.

Jarrett, F.M. 1959. Studies in *Artocarpus* and related genera. III. A revision of *Artocarpus* subgenus *Artocarpus*. *J. Arnold Arb.* 40:329–334.

Kader, A.A. 2001. Jackfruit: recommendations for maintaining postharvest quality. *Perishables Handling Quart.* 106(May):11–12.

Kanzaki, S., Yonemori, K., Sugiura, A., and Subhadrabandhu,S. 1997. Phylogenetic relationships between the jackfruit, the breadfruit and nine other *Artocarpus* spp. from RFLP analysis of an amplified region of cpDNA. *Sci. Hortic.* 70:57–66.

Lin, C.N., Lu, C.M., and Huang, P.L. 1995. Flavonoids from *Artocarpus heterophyllus*. *Phytochemistry*, 39:1447–1451.

Maia, J.G.S., Andrade, H.A., and Zoghbi, M. das G.B. 2004. Aroma volatiles from two fruit varieties of jackfruit (*Artocarpus heterophyllus* Lam.). *Food Chem.* 85:195–197.

Mitra, S.K. 1999. Jackfruit. In *Tropical horticulture*, vol. 1. Edited by T.K. Bose, S.K. Mitra, A.A. Farooqui, and M.K. Sadhu. Naya Prokash, Calcutta, India. pp. 359–362.

Moncur, M.W. 1985. Floral ontogeny of the jackfruit, *Artocarpus heterophyllus* Lam. (Moraceae). *Aust. J. Bot.* 33:585–593.

Morton, J.F. 1965. The jackfruit (*Artocarpus heterophyllus* Lam.): its culture, varieties, and utilization. *Proc. Fla. State Hort. Soc.* 78:336–344.

Morton, J. 1987. Jackfruit. In *Fruits of warm climates*. Authored by J.F. Morton. Creative Resource Systems, Winterville, NC. pp. 58–64.

Naik, K.C. 1957. *The jackfruit in India*. Indian Council of Agricultural Research, New Delhi, India. 16 pp.

Rahman, M.A., Nahar, N., Mian, A.J., and Mosihuzzaman, M. 1999. Variation of carbohydrate composition of two forms of fruit from jack tree (*Artocarpus heterophyllus* L.) with maturity and climatic conditions. *Food Chem*. 65:91–97.

Samaddar, H.N. 1985. Jackfruit. In *Fruits of India, tropical and subtropical*. Edited by T.K. Bose. Naya Prokash, Calcutta, India. pp. 487–497.

Schnell, R.J., Olano, C.T., Campbell, R.J., and Brown, J.S. 2001. AFLP analysis of genetic diversity within a jackfruit germplasm collection. *Sci. Hortic*. 91:261–274.

Selvaraj, Y., and Pal, D.K. 1989. Biochemical changes during the ripening of jackfruit *Artocarpus heterophyllus* L. *J. Food Sci. Technol*. 26:304–307.

Soepadmo, E. 1991. *Artocarpus heterophyllus* Lamk. In *Plant resources of South-East Asia. 2. Edible fruits and nuts*. Edited by E.W.M. Verheij and R.E. Coronel. Pudoc, Leiden, the Netherlands. pp. 86–91.

Thomas, C.A. 1980. Jackfruit, *Artocarpus heterophyllus* (Moraceae), a source of food and income. *Econ. Bot*. 34:154–159.

Wong, K.C., Lim, C.L., and Wong, L.L. 1992. Volatile flavour constituents of chempedak (*Artocarpus polyphema* Pers.) fruit and jackfruit (*Artocarpus heterophyllus* Lam.) from Malaysia. *Flavour Fragrance J*. 7:307–311.

Wuthrich, B., Borga, A., and Yman, L. 1997. Oral allergy syndrome to jackfruit (*Artocarpus integrifolia*). *Allergy*, 52:428–431.

Yong, H.S. 1979. Sukun, Nangka and Cempedak [breadfruit, jackfruit and champedak] from Mutiny on the Bounty, the fruit with the strongest and richest smell. *Malaysian Panorama*, 9(2):18–23.

Specialty Cookbooks

Hutton, W., and Matsuhisa, N. 2002. *Tropical Asian cooking: exotic flavors from Equatorial Asia*. Periplus, Singapore. 192 pp.

Johari, H. 2000. *Ayurvedic healing cuisine. 200 vegetarian recipes for health, balance, and longevity*. Healing Arts Press, Rochester, VT. 264 pp.

Rajah, C.S., and Thompson, D. 2008. *Heavenly fragrance: cooking with aromatic Asian herbs, fruits, spices and seasonings*. Periplus, Boston, MA. 256 pp.

Trang, C. 2003. *Essentials of Asian cuisine: fundamentals and favourite recipes*. Simon & Schuster, New York. 608 pp.

Varadarajan, V. 2008. *Samayal: the pleasures of South Indian vegetarian cooking*. 5th ed. Orient Enterprises, Chennai, India. 164 pp.

48 Japanese Vegetables

This chapter features

Gobo (*Arctium lappa*)
Garland chrysanthemum (*Glebionis coronaria*)
Mitsuba and honewort (*Cryptotaenia* species)
Water dropwort (*Oenanthe javanica*)

Most Japanese cuisine is dominated by seafood, coupled with plant foods such as rice, soy, and tea that are quite familiar to people in the Western World. Two exotic flavoring herbs especially found in Japan, perilla and wasabi, are treated separately in this book. Most Japanese vegetables are variations of Chinese and Western vegetable species. For example, the "daikon," often considered the most common of all Japanese vegetables, is a giant radish. The four plants highlighted in this chapter are decidedly unfamiliar to most English-speaking people and are among the most important vegetables of Japan. Although none is of dominating importance, collectively they well represent the special vegetable cuisine of Japan.

GOBO (BURDOCK)

Family: Asteraceae (Compositae; sunflower family)

Names

Scientific Name: *Arctium lappa* L. (*Lappa major* Gaertn.)

- "Gobo" is a Japanese word for burdock that has been adopted in English.
- Pronunciation: GOH-boh.
- The "bur" in "burdock" is said to come from the French *bourre*, derived from the Latin *burra*, a lock of wool, alluding to bits of wool often found attached to burs in sheep pastures. "Bur" is a general word for fruits that have barbed, pointed, or rough outgrowths so that they can cling to the fur or hair of animals and be distributed. The "dock" in "burdock" is an old English name for sorrel (*Rumex* species, including the edible herbs garden sorrel (*R. acetosa* L.) and French sorrel (*R. scutatus* L.)).
- Gobo is also known as great burdock and edible burdock.
- Another name, "elephant ears," reflects the large leaves of the plant.
- As a weed or medicinal plant, there are many names for burdock, most of which are no longer used or represent confusion with similar species. In particular, cocklebur (species of *Xanthium*, especially *X. strumarium* L.) is often confused with burdock. Some old colorful names include beggar's buttons, love leaves, and happy major.

FIGURE 48.1 *Arctium lappa*. Left: edible roots of the cultivated variety (gobo). (From Vilmorin-Andrieux, 1885.) Center and right: wild variety (burdock). (From Oeder et al., 1761–1883, plate 642.)

- The genus name *Arctium* is based on the Latin name *arction*, derived from the earlier Greek name for the plant, *arktion*, which is from *arktos*, a bear, an allusion to the rough nature of the bur.
- *Lappa* in the scientific name *A. lappa* is a Latin word for bur.

Plant Portrait

Gobo is a biennial herbaceous root vegetable grown as an annual. It is a domesticated form of burdock, largely selected in Japan, although the Chinese have also bred superior varieties. In its first year, the plant produces a rosette of large, wavy leaves on long leaf stalks, from which it develops a branched flowering stalk with purple flowers in the second year. The brown-skinned, white-fleshed roots are harvested in the first year of growth, after 2 or 3 months in the ground, before they branch and become woody. Compared with the wild burdock, gobo roots are longer, smoother, and more refined in taste. Gobo is widely used as a vegetable in Japan, where it is known both as gobo and *takinogawa*. In Japan, gobo is highly prized as a delicacy. Under exceptional conditions, the roots can grow to 1.2 m (4 ft.) in length and become the size of a baseball bat. Marketed roots vary from 10 to 35 mm (0.4–1.4 in.) in thickness and are often 60 cm (2 ft.) or more in length. The wild ancestor of the cultivated plant is native to Eurasia and has been introduced along roadsides, thickets, fields, and waste places in North America. In North America, gobo roots are cultivated as a minor crop to supply Asian grocery outlets. Artfully packed boxes of long gobo roots, reminiscent of boxes of cigars, are frequently found in Japanese markets, whether in Japan or elsewhere.

Burdock roots were regarded as an important herbal medicine in the Middle Ages, for example, finding use as a blood purifier and as a treatment for skin diseases such as eczema, boils, and ringworm. Poultices were used to reduce swelling from wasp and bee stings. An infusion (boiled tea) was thought to cure gout and diabetes. Herbal practitioners today sometimes use the root for a variety of medical conditions, and burdock preparations are sold as over-the-counter health supplements.

Culinary Portrait

The roots of gobo are crisp in texture, with a sweet, mild flavor that has been said to be oyster-like, similar to salsify (the root vegetable *Tragopogon porrifolius* L.), and also somewhat like artichoke. Gobo is valued for its high fiber, high vitamin B, low calories, and blandness (hence useful for texture in cooking). Before cooking, the outer layer is scraped off, down to the white fiber underneath (the roots should not be peeled, because the finest flavor is closest to the skin). The roots are usually boiled in two changes of water to improve the vegetable's color and flavor. Another technique, intended to reduce the slight bitter aftertaste, is to soak the roots in salty water for 5 to 10 minutes before cooking. If not consumed promptly, the roots should be dipped in water to which vinegar or lemon juice has been added to prevent discoloration. When cooked the roots are crunchier than carrots and slightly stringy, but not unpleasantly so. Because the flesh remains slightly fibrous after cooking, it is often grated, finely sliced, or diced. Gobo can be used in Oriental-style stir-fries or added to vegetable dishes, stews, soups, and meat dishes. Gobo is also pickled and processed into a paste, tea, or soft drink. The roasted root can taste surprisingly like meat and has been made into "vegetable hamburger."

The best roots are young and tender, approximately 46 cm (18 in.) long, and no wider than 2.5 cm (1 in.). Thin, old roots tend to be coarse and woody. Canned gobo is inferior, although sources of fresh roots are difficult to find in the Western World. The roots should not be washed until just before they are cooked. They may be stored in paper (not plastic) bags in the refrigerator for up to 2 weeks (alternatively, wrap in paper towels to absorb any moisture and place in perforated plastic bags) but should be consumed as soon as possible because they lose much of their flavor with time.

From a quarter to almost half of burdock roots may be made up of inulin, a naturally occurring, indigestible sweet starch-like substance. Inulin-containing plants are thought to have some use in regulating hypoglycemia and prediabetic conditions. Too much inulin consumption leads to fermentation in the gut and indigestion for some people.

The young leaves and very young shoots of gobo can be used in soups and stews, and the leaves can be used in the same way as leaf vegetables such as spinach and cabbage. The young leafstalks are occasionally peeled before consumption, and the young floral stalks are also sometimes eaten.

Although gobo has been selected as a root vegetable, the wild form of the plant, burdock, has also occasionally been eaten, perhaps most notably in Scotland. Scots also used the wild plant to prepare burdock ale and burdock wine. Young leaves of burdock are sometimes consumed as a salad in Scandinavia and Japan. In the past, burdock stalks were cut before flowering, stripped of their rinds and eaten like asparagus or in a salad. Such stalks were also once candied, like angelica. Sprouts produced from the roots in the spring were eaten raw or cooked.

Culinary Vocabulary

- One of the most popular of gobo dishes is *kinpira gobo*, in which gobo and carrots are shred into thin strips and stir-fried; they may then be glazed with soy and sesame sauce and topped with sugar and sake or other flavorful additives.

Curiosities of Science and Technology

- Burdock was often used medicinally in the past, although most of the usages seem improbable today. In Ireland, a poultice of burdock was used to cure ringworm, but it was considered necessary to apply the material on the opposite side of a river from the patient, or the "worm" (actually a fungus) would simply move to a safer location on the person. In nineteenth-century Arkansas, children were given a necklace of burdock roots to protect them against enchantment. A similar type of neckware, a few burdock seeds in a bag carried around the neck, was used to relieve rheumatism. Babies with colic were also treated with a burdock necklace. An old American treatment for aching feet: take portions of the blade of a large burdock leaf, place them on a hot shovel to soften, and attach to the hollow of the feet by a cloth bandage. To cure fever in New England, burdock leaves were bandaged to a patient, the leaf tips pointing down to either the wrists or the ankles, in the belief that the leaves would absorb the fever and it would run out at the points.
- Some Japanese farmers grow gobo in boxes that they build on top of the ground and fill with soil. At harvest time, they dismantle the boxes, which saves the work of digging deep into soil. This technique also serves to keep gobo from spreading.

- Burdock owes its spread to the hooked prickles on the burs, which are fruits containing seeds. The burs adhere to almost anything. Shakespeare had Pandarus say in *Troilus and Cressida* "They are burs, I can tell you, they'll stick where they are thrown."
- The Swiss engineer George de Mestral, (1907–1990), had a deep respect for nature as the ultimate engineer. He often took long walks through the autumn countryside with his dog, and both would invariably return covered in burs. De Mestral inspected the prickly fruits under a microscope and noticed hundreds of tiny hooks that would grab onto anything passing by. He experimented with plastic models that mimicked the bur's action and produced Velcro in 1948. Velcro consists of two strips of nylon fabric, one with thousands of small hooks, the other with small loops. When the two strips are pressed together, they form a strong bond, which can be detached and reattached thousands of times. The word "Velcro" is made from the terms velvet (a fabric made of loops) and crochet (French for "hook"). The patent for the invention ran out in the 1980s, and the brand became a generic term for an endless number of copies. Velcro (excuse the pun) has attached itself to everything, including shoes, jackets, pants, and wallets, and has replaced a large percentage of zippers and buttons.

Key Information Sources

Asano, N., and Matsuzawa, Y. 1977. The cultivation method of edible burdock (*Arctium lappa* L.) by autumn sowing. *Bull. Ibaraki Agric. Exp. Stn.* 18:79–90 (in Japanese).

Douglas, M.H., Burgmans, J.L., Burton, L.C., and Smallfield, B.M. 1992. The production of burdock (*Arctium lappa* L.) root in New Zealand—a preliminary study of a new vegetable, vol. 22. *Proceedings of the Annual Conference of the Agronomy Society of New Zealand*, Lincoln University, Canterbury, New Zealand. pp. 67–70.

Duistermaat, H. 1996. Monograph of *Arctium* L. (Asteraceae). Generic delimitation (including *Cousinia* Cass. p.p.), revision of the species, pollen morphology, and hybrids. *Gorteria Suppl.* 3:1–143.

Gross, R.S., Werner, P.A., and Hawthorn, W.R. 1980. The biology of Canadian weeds. 38. *Arctium minus* (Hill) Bernh. and *Arctium lappa* L. *Can. J. Plant Sci.* 60:621–634.

Han, J.S., Cheigh, M.J., Kim, S.J., Rhee, S.H., and Park, K.Y. 1996. A study on wooung (burdock, *Arctium lappa*, L.) kimchi—changes in chemical, microbial, sensory characteristics and volatile flavor components in wooung kimchi during fermentation. *J. Food Sci. Nutr.* 1:30–36.

Hasama, W. 1990. Occurrence and control of diseases and pests of edible burdock in Kyushu district. *Plant Protect.* 44:446–449 (in Japanese).

Hawthorn, W.R., and Hayne, P.D. 1978. Seed production and pre-dispersal seed predation in the biennial composite species *Arctium minus* and *Arctium lappa*. *Oecologia*, 34:283–296.

Honda, K., and Ishikawa, M. 1974. Studies on the cultivation of edible burdock (*Arctium lappa* L.). 1. Life history and absorption of inorganic elements. *Bull. Ibaraki Agric. Exp. Stn.* 15:104–112 (in Japanese).

Honda, K., and Ishikawa, M. 1976. Suitable soil conditions for edible burdock (*Arctium lappa* L.) culture. *Agric. Hortic.* 51:669–673 (in Japanese).

Ishimaru, M., Kagoroku, K., Chachin, K., Imahori, Y., and Ueda, Y. 2004. Effects of the storage conditions of burdock (*Arctium lappa* L.) root on the quality of heat-processed burdock sticks. *Sci. Hortic.* 101:1–10.

Liu, L., Gao, W.D., and Zhang, J.H. 1997. Techniques for preventing edible burdock from browning. *Food Machinery*. 1:27–28 (in Chinese).

Moriizumi, S., and Osaki, K. 1984. Studies on the relation between soil hardness and the underground part shape of root vegetables. II. On the root shape of edible burdock. *Gakujutsu Hokoku Sci. Rep. Fac. Agric. Ibaraki Univ. Ibaraki-ken.* 32:53–60 (in Japanese, English summary).

Nagai, T., Imamura, H., and Kiriyama, S. 1980. Dietary fiber breads containing gobo residue, gobo holocellulose, and konjac powder. *Cereal Chem.* 57:307–310.

Nishida, T. 1998*a*. Germination and growth properties of edible burdock. *Bull. Hokkaido Prefect. Agric. Exp. Stn.* 74:43–51 (in Japanese, English summary).

Nishida, T. 1998b. Nutritive properties and optimum nitrogen fertilizer amount of edible burdock. *Bull. Hokkaido Prefect. Agric. Exp. Stn.* 74:53–61 (in Japanese, English summary).

Porter, B., and Barl, B. 2000. *Burdock production in Saskatchewan*. Saskatchewan Agriculture and Food, Regina, SK. 2 pp. [Burdock considered as a medicinal plant.]

Rhoads, P.M., Tong, T.G., Banner, W.R., and Anderson, R. 1984. Anticholinergic poisonings associated with commercial burdock root *Arctium lappa* tea. *J. Toxicol. Clin. Toxicol.* 22:581–584.

Rodriguez, P., Blanco, J., Juste, S., Garces, M., Perez, R., Alonso, L., and Marcos, M. 1995. Allergic contact dermatitis due to burdock (*Arctium lappa*). *Contact Dermatitis*, 33:134–135.

Schmelzer, G.H., and Horsten, S.F.A.J. 2001. *Arctium lappa* L. In *Plant resources of South-East Asia No. 12(2). Medicinal and poisonous plants 2*. Edited by J.L.C.H van Valkenburg and N. Bunyapraphatsara. Backhuys Publishers, Leiden, the Netherlands. pp. 78–82.

Takei, Y., Yutani, Y., and Moteki, M. 1995. Effects of the seeding time and the covering time in the plastic-tunnel to the growth of the edible burdock. *Res. Bull. Gunma Hortic. Exp. Stn.* 1:1–6 (in Japanese, English summary).

Tatokoro, Y., Nakajima, A., and Kazama, Y. 1977. Prevention of discoloration of sliced edible burdock (*Arctium lappa* L). *J. Food Sci. Tech.* 24:643–644 (in Japanese, English summary).

Wu, M.C. 1995. Studies on browning reaction of frozen edible burdock. *Food Sci.* 22:308–316 (in Chinese, English summary).

Specialty Cookbooks

Barber, K. 2007. *The Japanese kitchen*. Kyle Books, London. 240 pp.

Moromito, M. 2007. *Moromito: the new art of Japanese cooking*. DK Publishing, New York. 272 pp.

Tsuji, S. 2007. *Japanese cooking: a simple art*. Kodansha International, New York. 507 pp.

GARLAND CHRYSANTHEMUM (CHOP SUEY GREEN)

Family: Asteraceae (Compositae; sunflower family)

NAMES

Scientific Name: *Glebionis coronaria* (L.) Cass. ex Spach (*Chrysanthemum coronarium* L. var *spatiosum* L.H. Bailey, *C. spatiosum* (L.H. Bailey) L.H. Bailey).

On the basis of the work by Bremer and Humphries (1993) and Watson et al. (2000), the species highlighted in this chapter was transferred from *Chrysanthemum* to *Glebionis*. However, as is evident from the literature cited at the end of this presentation, most people still call the species *Chrysanthemum coronarium*.

- The name chrysanthemum is based on an old Greek name, *chryanthemon*, made up of *khrusos*, gold + *anthemon*, flower, that is, "golden flower."
- The word "garland" in "garland chrysanthemum" is based on use in preparing garlands, as noted later. Since at least Roman times, the flowers were woven into garlands that were worn on the head and around the neck.
- "Chop suey green" (= chop suey greens) is an American-coined name for garland chrysanthemum based on "chop suey." "Chop suey" is a name of Chinese origin: Cantonese *shap sui* or *tsap seui*, from the Mandarin *tsa sui*, "odds and ends." It is a stir-fried dish of such mixed vegetables as bean sprouts, bamboo shoots, water chestnuts, onions, celery, mushrooms, a starchy soy sauce, and small pieces of beef, pork, chicken, shrimp, or fish, served over rice. ("Americanized chop suey" is a New England dish made of ground beef, noodles, and tomato sauce.) According to widely circulated mythology, chop suey was created in America by Chinese immigrants and laborers in the mid- to late nineteenth century. In fact, chop suey is a dish that comes from Toisan, a rural district of Canton. Many early immigrants to California came from the region, which is probably why California is most frequently mentioned as the place where chop suey originated. There does not seem

to be a particular culinary reason why the name chop suey is associated with the garland chrysanthemum.

- Garland chrysanthemum—both the species and the vegetable—are also known as cooking chrysanthemum, edible chrysanthemum, edible-leaved chrysanthemum, chrysanthemum greens, and Japanese greens.
- The Chinese name tangho is encountered as an English name for garland chrysanthemum and is also used in several foreign languages. The species is known as *shungiku* or *shiyungiki* in Japan, *tanghoe* in China, and *fior d'oro* in Italy.
- Ornamental and weedy forms of the species are known as crown daisy as well as garland chrysanthemum.
- The genus name *Glebionis* is based on the Latin *gleba*, soil + *ionis*, characteristic of, perhaps an allusion to the association of the weedy species with agriculture.
- *Coronarium* in the scientific name *G. coronarium* means "pertaining to a crown, wreath, or garland."

Plant Portrait

Garland chrysanthemum is an annual herbaceous plant grown for its edible leaves. In flower, it grows up to 1 m (about a yard) in height. The lower leaves are shallowly divided, and the upper leaves on the flowering stem are especially divided into fine segments. The vegetable form of the species has fleshier leaves than the kinds cultivated as ornamentals, a large terminal leaf lobe, and much smaller lobes near the base. The flower heads resemble small chrysanthemum flowers. The species is native to the Mediterranean and southwestern Europe. It has been found as an occasional garden escape in North America, and in the San Diego region of California it is an invasive weed. Varieties grown as garden ornamentals usually lack the somewhat resinous flavor that is favored in varieties used for culinary purposes in Asia.

FIGURE 48.2 Garland chrysanthemum (*Glebionis coronaria*). Left: flowering shoot of the edible variety, by B. Flahey. Upper right: rosette of the edible variety, by M. Jomphe. Lower right (left to right): flower head of ornamental variety, from Nicholson (1885–1889); flower head of edible variety, by B. Flahey. Note the doubled flowers (i.e., with numerous petals) in the ornamental variety.

Garland chrysanthemum and indeed chrysanthemums in general have acquired considerable symbolic value in Asia. The "chrysanthemum throne" refers to the monarchy in Japan. A chrysanthemum with 16 "petals" is the traditional emblem of the Mikado (monarch). Garland chrysanthemum has been interpreted as the national flower of Japan since the fourteenth century. In the Far East, chrysanthemum symbolizes purity, perfection, and long life. In China, garland chrysanthemum is a symbol of a life of ease and joviality and is particularly appreciated by older adults as a symbol of mature beauty.

Culinary Portrait

Garland chrysanthemum is grown for its edible leaves in China, Japan, and Taiwan. The leaves are generally used in combination with other foods, and indeed this vegetable is best combined with other greens or vegetables. Sometimes the leaves are used as an edible garnish. The leaves have a characteristic odor of chrysanthemums and a distinct, tangy taste. Young material is pleasantly mild, older leaves are stronger in taste, and the foliage becomes bitter as the plants produce flowers. The spicy flavor may seem rather strong to those unaccustomed to it. Overcooking results in a bitter flavor, and the greens should be added to stir-fry dishes toward the end of the cooking cycle so that they are heated for only a few minutes. Steaming or cooking briefly in a small amount of water, as with spinach, is also recommended. The tender shoots are used in stir-fries or steamed and are added to flavor soups, meat, and fish dishes. In Japan, garland chrysanthemum is typically present in the well-known sukiyaki (beef, vegetables, and yam noodles cooked in broth) and in miso soup. In Kyoto, Japan, chrysanthemum fritters are a specialty.

The flower heads are also eaten in Japan, although they are bitter (it is preferable to consume just the petals). It has been suggested that the flowers be dipped in boiling, salted water before serving, and that dried petals should be soaked in water before adding to other ingredients. Petals of garland chrysanthemum are a component of a distinctive Japanese pickle called *kikumi*. In Japan, garland chrysanthemum is used in some culinary ceremonies, consistent with the view that it has health-giving properties. The petals are soaked in sake, considered to be another life extender, and eaten at a chrysanthemum festival. The Japanese custom of dipping flowers in sake to begin a meal is also reputed to give health and long life. The high regard in which garland chrysanthemum is held in Japan is also reflected by its presence as an ingredient in a special dish served as part of the tea ceremony.

In Western countries, the most likely place garland chrysanthemum can be purchased is in Oriental produce markets. It is best to buy material with the stem attached to the leaves, as the leaves will last longer. Flowers should not be on the plant (indicating over-maturity), and the leaves should be bright green. Garland chrysanthemum can be stored by wrapping in damp paper towels inside a plastic bag in a refrigerator for up to a week. The leaves may wilt but can still retain good flavor. This vegetable can be grown in the home garden and in containers in bright windows for a fresh supply.

Garland chrysanthemum is a member of the sunflower family, which includes ragweed, notorious for causing hay fever. Rarely, people who are allergic to ragweed exhibit "cross sensitivity," becoming allergic to other members of the sunflower family, and is wise to be aware of this phenomenon.

Culinary Vocabulary

- A "fire pot" is basically an elaborate Chinese chafing dish used to cook food at the table. Often made of brass, it consists of a pot in which a charcoal or alcohol fire is built in a central well that is vented up through the center of the pot, so that the container around the central heated area holds gently simmering stock into which diners dip bits of meat, poultry, seafood, or vegetables until cooked to their taste. A fondue pot can serve much the same purpose. In some special Chinese banquets, edible chrysanthemum petals are tossed over the broth as a decorative garnish, and this has led to the alternative name "chrysanthemum pot."

- "Chrysanthemum tea" (*chiu hwa*, pronounced tchee-oo har) is a Chinese blend of chrysanthemum blossoms and tea. Reminiscent of jasmine tea, it is sweetened with rock candy and served after a meal along with pastries.
- A "chrysanthemum cut" is an ornamental method of preparing a root vegetable, traditionally carried out by holding a portion of vegetable upright with chopsticks held parallel to each other, slicing vertically between the chopsticks, not quite to the base, while very slowly rotating the material through 180°, producing fine strands that radiate outward from the center, reminiscent of a daisy flower.

Curiosities of Science and Technology

- Garland chrysanthemum was once regarded as an important medicinal plant in Japan. It was a chief ingredient in a formula for promoting longevity, returning gray hair to its original color and replacing lost teeth, so that "an old man of 80 would become like a boy again."
- The outer floral petals (technically, the ray petals, i.e., the large petals of the small flowers at the outer edge) of the flower heads are traditionally used to garnish Cantonese snake meat dishes in China.
- Words often are coined by analogy with familiar objects, and in Asia, where chrysanthemums have a very long history, a canine came to be called the "chrysanthemum dog." The dog in question is the shih tzu, one of the most distinctive of the toy breeds. The name originated because its facial hair grows in all directions, resembling the petals of a chrysanthemum.

Key Information Sources

Alvarez, C.P.P., and Pascual-Villalobos, M.J. 2003. Effect of fertilizer on yield and composition of flowerhead essential oil of *Chrysanthemum coronarium* (Asteraceae) cultivated in Spain. *Ind. Crop. Prod.* 17:77–81.

Alvarez-Castellanos, P.P., Bishop, C.D., and Pascual-Villalobos, M.J. 2001. Antifungal activity of the essential oil of flowerheads of garland chrysanthemum (*Chrysanthemum coronarium*) against agricultural pathogens. *Phytochemistry*, 57:99–102.

Bremer, K., and Humphries, C.J. 1993. Generic monograph of the Asteraceae–Anthemideae. *Bull. Nat. Hist. Mus. London Bot.* 23:71–77.

Chiang, M.H., and Park, K.W. 1994. Effects of temperature, light and mechanical treatment on the seed germination of *Chrysanthemum coronarium* L. *J. Korean Soc. Hortic. Sci.* 35:534–539 (in Korean, English summary).

Ching, M.H., and Park, K.W. 1993. The effect of daylength, shading and irrigation on the flowering of *Chrysanthemum coronarium* L. *J. Biol. Prod. Facilities Environ. Control.* 2:136–146 (in Korean, English summary).

Dasuki, U.A., and Bergh, M.H. van den. 1993. *Chrysanthemum coronarium* L. In *Plant resources of South-East Asia. 8. Vegetables.* Edited by J.E. Siemonsma and K. Piluek. Pudoc Scientific Publishers, Wageningen, the Netherlands. pp. 140–142.

Flamini, G., Cioni, P.L., and Morelli, I. 2003. Differences in the fragrances of pollen, leaves, and floral parts of garland (*Chrysanthemum coronarium*) and composition of the essential oils from flowerheads and leaves. *J. Agric. Food Chem.* 51:2267–2271.

Gill, B.S., and Gupta, R.C. 1978. Structural hybridity in *Chrysanthemum coronarium* L. *Perspect. Cytol. Genet.* 3:523–530.

Hara, T., Inno, Y., and Nakamura, T. 1997. Changes of qualities and chemical contents in garland chrysanthemum (*Chrysanthemum coronarium* L.) after harvest. *Bull. Osaka Agric. Res. Center*, 33:8–12 (in Japanese).

Kasahara, K., and Nishibori, K. 1995. The suppressing effect of garland chrysanthemum on the odor of Niboshi [boiled and dried sardines or anchovies] soup stock. *Fisheries Sci.* 61:672–674.

Mochizuki, M., Kosuge, E., and Cho, O. 1985. *The cultivation of leafy vegetables*. No-san-gyoson Bunka Kyokai, Showa 60, Tokyo, Japan. 237 pp (in Japanese).

Morishita, M., Kida, K., and Yamada, K. 1987. Selection of variety and environmental control for growth of garland chrysanthemum in summer. *Bull. Osaka Agric. Res. Center*, 24:67–75 (in Japanese).

Park, K.W. 1980. *Chrysanthemum coronarium* as a vegetable culture. *Gemuese*, 16:398, 400 (in German).

Perez, M.P., amd Pascual-Villalobos, M.J. 1999. Effects of the essential oil of flower heads of *Chrysanthemum coronarium* L. on white fly and stored product pests. *Investigacion Agraria. Produccion Proteccion Veg.* 14:249–258 (in Spanish, English summary).

Schmoll, M. 1984. *Chrysanthemum coronarium* 'Chopsuey Green'. *Garten Organisch*, 1984(1):6 (in German).

Senatore, F., Rigano, D, de Fusco, R., and Bruno, M. 2004. Composition of the essential oil from flowerheads of *Chrysanthemum coronarium* L. (Asteraceae) growing wild in Southern Italy. *Flavour Fragrance J.* 19:149–152.

Soreng, R.J., and Cope, E.A. 1991. On the taxonomy of cultivated species of the *Chrysanthemum* genus-complex (Anthemideae; Compositae). *Baileya*, 23:145–165.

Strother, J.L. 2006. *Glebionis*. In *Flora of North America north of Mexico*, vol. 19. Edited by Flora of North America Editorial Committee. Oxford University Press, Oxford, U.K. pp. 554–555.

Sulas, L., Re, G.A., Molle, G., and Ligios, S. 1999. Chrysanthemum coronarium *L.: a new pasture species for Mediterranean forage systems*. Centre International de Hautes Etudes Agronomiques Mediterranéennes, Zaragoza, Spain. [vol. 39]. 296 pp.

Tamaki, K. 1986. Measurement of weight growth of *Chrysanthemum coronarium* cultivated in hydroponic culture. *Environ. Control Biol.* 24:87–93 (in Japanese, English summary).

Turland, N.J. 2004. Proposal to conserve the name *Chrysanthemum coronarium* [=*Glebionis coronaria*] (Compositae) with a conserved type. *Taxon*, 53:1072–1074.

Valente, M.E., Borreani, G., Caredda, S., Cavallarin, L., and Sulas, L. 2003. Ensiling forage garland (*Chrysanthemum coronarium* L.) at two stages of maturity and at different wilting levels. *Anim. Feed Sci. Technol.* 108:181–190.

Vogel, g. 1992. Culture of *Chrysanthemum coronarium* as vegetable. 3. *Gartenbau Magazin*, 1(9):54–55 (in German).

Wang, H., Ye, X.Y., and Ng, T.B. 2001. Purification of chrysancorin, a novel antifungal protein with mitogenic activity from garland chrysanthemum seeds. *Biol. Chem.* 382:947–951.

Watson, L.E., Evans, T.M., and Boluarte, T. 2000. Molecular phylogeny of the tribe Anthemideae (Asteraceae), based on chloroplast gene *ndh*F. *Mol. Phylogenet. Evol.* 15:59–69.

Yamamoto, F., Tsuruoka, M., and Tanaka, A. 1993. Effects of planting densities and picking node orders of main stem on yield and quality of garland chrysanthemum. *Bull. Chiba Hortic. Exp. Stn.* 15:17–24 (in Japanese, English summary).

Yamasaki, M., and Tsuji, H. 2001. The effect of high temperature stress on the occurrence of dead heart before and during flower bud differentiation in garland chrysanthemum (*Chrysanthemum coronarium* L.). *J. Jpn. Soc. Hortic. Sci.* 70:496–500 (in Japanese).

Yulian, Fujime, Y., and Okuda, N. 1996. Effects of daylength and temperature on capitulum initiation and development of garland chrysanthemum (*Chrysanthemum coronarium* L.). *Environ. Control Biol.* 34:21–28.

Zheng, C.H., Kim, T.H., Kim, K.H., Leem Y.H., and Lee, H.J. 2004. Characterization of potent aroma compounds in *Chrysanthemum coronarium* L. (garland) using aroma extract dilution analysis. *Flavour Fragrance J.* 19:401–405.

Specialty Cookbooks

Harrington, G. 2009. *Growing Chinese vegetables in your own back yard: grow 40 vegetables and herbs in gardens and pots*. Storey Publishing, North Adams, PA. 216 pp.

Larkom, J. 2008. *Oriental vegetables: the complete guide for the gardening cook*. Revised edition. Kodansha America, New York. 232 pp.

MITSUBA AND HONEWORT

Family: Umbelliferae (Apiaceae, carrot family)

NAMES

Scientific Names: *Cryptotaenia* species

- Mitsuba—*C. japonica* Hassk. (*C. canadensis* (L.) DC. var. *japonica* (Hassk.) Makino)
- Honewort—*C. canadensis* (L.) DC. (*Deringa canadensis* (L.) Kuntze)

- Mitsuba is the Japanese name of this plant, simply carried over into English. Mitsuba is Japanese for three leaflets, a reference to the fact that the leaves are frequently split into three leaflets. The word is sometimes translated into English as "trefoil," Latin for three-leaved. Unfortunately, "trefoil" is rather ambiguous; it is applied to a number of genera, especially *Trifolium* (better known as clover).
- The word honewort, literally "hone plant," is the name of several unrelated species of the carrot family, particularly the European *Sison amomum* L., which was used to cure a swelling (especially a hard swelling in the cheek) called a hone.
- Honewort has also been called Canada honewort, Canadian honewort, and white chervil.
- Mitsuba is also known as Japanese parsley, a name that is also used for water dropwort (discussed later) and occasionally for coriander (*Coriandrum sativum* L.).
- Additional names for mitsuba are Japanese honewort and Japanese wild chervil. The name "mountain celery" is occasionally encountered in English translations of Japanese publications.
  - The Chinese name *ya er qin* for mistuba translates as "duck celery."
  - The genus name *Crypotaenia* comes from the Greek *cryptos*, hidden, and *taenia*, a fillet, referring to concealed oil tubes in the fruits.
  - *Japonica* and *canadensis* in the scientific names of mitsuba and honewort, respectively, mean "Japanese" and "of Canada."

FIGURE 48.3 Mitsuba (*Cryptotaenia japonica*), by S. Rigby.

FIGURE 48.4 Honewort (*Cryptotaenia canadensis*). Left by B. Flahey, right from U.S. Department of Agriculture Forest Service Illustration Collection.

Plant Portrait

Mitsuba is a succulent perennial herb of southeastern Asia, which looks somewhat like parsley and smells rather like celery. It is cultivated in Japan, Korea, China, Taiwan, and Indonesia, where it is a well-known vegetable and condiment. Mitsuba will often be found in Chinese and Japanese markets in North America and is grown commercially for the Oriental market near Los Angeles in plastic houses. However, it is largely unknown in most of North America, although sold by several seed supply houses in Canada and the United States. It is also little known in most of Europe. The plant varies from 20 to 125 cm (8–20 in.) in height and is grown particularly for its long, slender, hollow leafstalks, and for its leaf blades. Mitsuba develops long, stringy roots from a thick, creeping rootstock. Cultivar 'Atropurpurea' of *C. japonica* is a purple-leaved ornamental variety. Ornamental forms of edible plants often lack attractive taste, and one supplier of this plant commented on its Web site, "Edible but little point! Good for flower arranging."

Honewort is almost identical to mitsuba but is a wild North American herb found in rich woods and thickets and on the banks of streams from western New Brunswick to Manitoba, south to Georgia, Alabama, Arkansas, and Texas. Some authorities combine the two species into one, under the name *C. canadensis*.

Culinary Portrait

Mitsuba is widely used as a garnish and flavoring in Japanese cuisine. The very distinct, pungent flavor is reminiscent of celery but has been described as a blend of parsley, celery, and angelica, or somewhat between celery and sorrel. All parts of mitsuba are edible. The leaves and leafstalks are eaten raw in salads and sandwiches, boiled or fried and consumed alone as a vegetable, used as a potherb, and added to soups, egg dishes, tempura, and a variety of fried foods. They can also simply be used as a garnish. The leaves become bitter and lose their fragrance when simmered too long or subjected to too much heat, so mitsuba should be no more than lightly parboiled or very gently stir-fried in a scant amount of oil. The roots are occasionally fried and eaten. Seedlings can be used as sprouts, added to soups, sandwiches, and salads.

In the late spring and early summer, while it is still young and tender, wild plants of honewort are collected for use as food in North America. Much like its Japanese counterpart, it is used for soup, as a potherb, in salads, as a root vegetable, and as a seasoning. Young leaves, stems, and flowers can be boiled for use as a potherb or chopped like parsley for addition to salads and green soups. Honewort has occasionally been harvested in North America and sold as mitsuba.

There is a report (Makino, 1989) of contact dermatitis resulting from handling mitsuba, but this seems to be a very rare occurrence.

Culinary Vocabulary

- *Shabu-shabu* (also spelled *syabu-syabu*) is a traditional "hot pot" country dish of Japan, composed of thinly sliced meat (usually beef) and vegetables, often served with dipping sauces and mitsuba. Shabu-shabu translates as "swish-swish" based on the mode of preparation: the slices of meat and vegetables are swished back and forth in boiling water or broth.
- *Cha-wan-musha* or *chawan-mushi*, Japanese egg custard, is also often accompanied by mitsuba.

Curiosities of Science and Technology

- In addition to the North American *C. canadensis* and the Asian *C. japonica*, there are many other examples of plant species of eastern North America that have very closely related species in eastern Asia. Indeed, some 34 herbaceous genera of plants are known to exhibit this pattern. The very similar deciduous forests of eastern Asia and eastern North America seem to be the major survivors of an ancestral forest that formed a continuous band around the northern hemisphere 15 to 20 million years ago. This was fragmented by climatic cooling, uplift of mountains, and continental glaciations.
- Mitsuba is an important and widely used medicinal herb in China, used to treat colds, diarrhea, rheumatism, painful menstruation, and certain glandular conditions.

Key Information Sources

Abe, K., Sakimoto, M., and Miyazaki, A. 1994. Effects of cultivation methods and cultivar variations on characteristic aroma of Japanese honewort (*Cryptotaenia japonica* Hassk.). *J. Jpn. Soc. Hortic. Sci.* 62:903–908 (in Japanese).

Andreev, G.N., and Kostolomov, M.N. 1979. First discovery of *Cryptotaenia japonica* Hassk. (Apiaceae) in the continental part of the Soviet Far East. *Bot. Zh. Akad. Nauk. S.S.S.R.* 64:1034–1038 (in Russian).

Baskin, J.M., and Baskin, C.C. 1988. The ecological life cycle of *Cryptotaenia canadensis* (L.) DC. (Umbelliferae), a woodland herb with monocarpic ramets. *Am. Midl. Nat.* 119:165–173.

Eckenbach, U., Lampman, R.L., Seigler, D.S., Ebinger, J., and Novak, R.J. 1999. Mosquitocidal activity of acetylenic compounds from *Cryptotaenia canadensis*. *J. Chem. Ecol.* 25:1885–1893.

Follett, J.M. 1990. Mitsuba and shungiku—two traditional Japanese vegetables. *Hortic. N. Z.* 1(2):22–25.

Hara, S., and Ito, K. 1979. Studies on the systematized water culture of "mitsuba" (*Cryptotaenia japonica*) using the belt type medium. *Bull. Osaka Agric. Res. Center.* 16:1–11 (in Japanese).

Hardway, T.M., Spalik, K., Watson, M.F., Katz-Downie, D.S., and Downie, S.R. 2004. Circumscription of Apiaceae tribe Oenantheae. *S. Afr. J. Bot.* 70:393–406.

Hawkins, T.S., Baskin, J.M., and Baskin, C.C. 2005. Life cycles and biomass allocation in seed- and ramet-derived plants of *Cryptotaenia canadensis* (Apiaceae), a monocarpic species of eastern North America. *Can. J. Bot.* 83:518–528.

Ikeda, H., Yoshida, Y., and Osawa, T. 1985. Effects of ratios of NO_3/NH_4 and temperature of the nutrient solution on growth of Japanese honewort, garland chrysanthemum and Welsh onion. *J. Jpn. Soc. Hortic. Sci.* 54:58–65 (in Japanese, English summary).

Izaki, M., Hara, H., Igawa, H., Okazawa, N., and Kurihara, M. 1988. Studies on cooperation phenomenon in garden crop community: 1. The present phenomenon in *Cryptotaenia japonica* Hassk. *Sci. Rep. Fac. Agric. Ibaraki Univ.* (36):19–24 (in Japanese).

Kami, T., Otaishi, S., Hayashi, S., and Matsuura, T. 1969. A study on low-boiling chemical constituents of *Cryptotaenia japonica* Hassk. *Agric. Biol. Chem.* 33:1717–1722.

Kanzaki, T. 1989. Contact dermatitis due to *Cryptotaenia japonica* Makino. *Contact Dermatitis*, 20:60.

Lee, M.S. 1987. Volatile flavor components of *Artemisia selengensis* and *Cryptotaenia japonica*. *Korean J. Food. Sci. Technol.* 19:279–284 (in Korean).

Li, P.P. 1999. Light and temperature requirements for hydroponic culture of *Cryptotaenia japonica*. *China Vegetables*, 1:39–40 (in Chinese).

Matsumura, S., Ikeya, T., Ozawa, H., and Aoyama, S. 1984. Development of the devices for continuous production of vegetables by water culture and year-round cultivation of Japanese honewort with the use of them in glasshouse. *Bull. Farms Tokyo Univ. Agric. Technol.* 11:19–30 (in Japanese, English summary).

Mizuno, N., and Hiraoka, T. 1978. Studies on water culture of Mitsuba (*Cryptotaemia japonica* Hassk.). *Bull. Agric. Res. Inst. Kanagawa Prefect.* 120:15–19 (in Japanese, English summary).

Miyazawa, M., Okamura, S., Okuno, Y., and Morii, S. 1999. Components of the essential oil of *Cryptotaenia japonica* Hassk. 'Itomitsuba' for Japanese food. *Flavour Fragrance J.* 14:273–275.

Nakamura, S., and Enohara, N. 1980. Germination improvement of vegetable seeds using polyethylene glycol. I. Eggplant, *Cryptotaenia japonica* and carrot. *J. Jpn. Soc. Hortic. Sci.* 48:443–452 (in Japanese, English summary).

Okude, T., and Hayashi, S. 1970. Sesquiterpene constituents of the essential oil of mitsuba (*Cryptotaenia japonica* Hassk.). *Chem. Soc. Jpn. Bull.* 43:2984–2985.

Omori, S., and Sugimoto, M. 1976. Studies on etiolating in mitsuba (*Cryptotaenia japonica*). 2. Effects of temperature, day-length, sowing time and nutritional condition on the bolting in mitsuba. *Bull. Kanagawa Hortic. Exp. Stn.* 23:43–48 (in Japanese, English summary).

Small, E., and Catling, P.M. 2001. Poorly known economic plants of Canada—31. Honewort, *Cryptotaenia canadensis* (L.) DC. *Bull. Can. Bot. Assoc.* 34(4):44–46.

Spalik, K., and Downie, S.R. 2007. Intercontinental disjunctions in *Cryptotaenia* (Apiaceae, Oenantheae): an appraisal using molecular data. *J. Biogeogr.* 34:2039–2054.

Tanno, L. 1988. Differentiation of anthocyanin pigmentation in *Cryptotaenia japonica* Hassk. Species. *Bull. Inst. Agric. Res. Tohoku Univ.* 39:29–35.

Yamauchi, N., Yoshimura, M., Shono, Y., and Kozukue, N. 1995. Chlorophyll degradation in mitsuba leaves during storage. *J. Jpn. Soc. Food Sci. Technol.* 42:709–714.

Yoshimura, M., Shono, Y., and Yamauchi, N. 1995. Influence of vacuum packaging and processing temperature on the quality change of mitsuba. *J. Jpn. Soc. Food Sci. Technol.* 42:588–593 (in Japanese).

Specialty Cookbooks

Kazuko, E. 2002. *Masterclass in Japanese cooking*. Pavilion, London. 192 pp.

Kijima, N. 2006. *Healthy & tasty sushi rolls and onigiri*. Shufunotomo Company, Tokyo, Japan. 62 pp.

Staff, J. 2009. *Kawaii bento boxes: cute and convenient Japanese meals on the go*. Japan Publications Trading, Tokyo, Japan. 81 pp.

Tohata, A. 2001. *Japanese vegetable cooking*. Graph-sha, Tokyo, Japan. 64 pp.

Weinstein, K. 2007. *Japanese kosher cooking: shushi, sushi and more*. KTAV Publishing House, Jersey, NJ. 193 pp.

Yagihashi, T., and Salat, H. 2009. *Takashi's noodles*. Ten Speed Press, Berkeley, CA. 168 pp.

WATER DROPWORT

Family: Apiaceae (Umbelliferae; carrot family)

Names

Scientific Name: *Oenanthe javanica* (Blume) DC. (*O. stolonifera* (Roxb.) C.B.Clarke)

- "Dropwort" is an old European name, applied to various species. The "drop" in dropwort may have originated from the perceived appearance of the small tubers of one species, hanging like drops of water by slender stems. "Wort" in "dropwort" is an old English word meaning plant (it has nothing to do with warts). The "water" in water dropwort reflects its aquatic habitat.

- Water dropwort is also called Chinese celery (a name that has been applied to Chinese varieties of common celery, *Apium graveolens* L.), Japanese parsley (a name also used for mitsuba (discussed above) and coriander (*Coriandrum sativum* L.)), Indian pennywort, Javan waterwort, Korean watercress, Vietnamese celery, Vietnamese parsley (also occasionally applied to rau ram, discussed in Chapter 96), water celery, and water parsley.
- Wild species of *Oenanthe* are called "water dropwort," and many of these are very poisonous. Some of these may be encountered as cultivated plants. Do not consume a "water dropwort" unless absolutely certain that it is edible.
- The aquatic herb *Oxypolis filiformis* (Walt.) Britt. is also called water dropwort. This species is also in the carrot family and is native to the southeastern United States, Bahamas, and Cuba. It is not edible.
- The name "water celery" is also applied to eelgrass (species of the aquatic genus *Vallisneria*).
- The name "water parsley" is also applied to some quite poisonous species of *Sium*, another genus in the carrot family.
- In Asian markets, water dropwort will be found under local language names. The Korean name of water dropwort is *minari*. The Japanese name is *seri*. The Chinese is *shuî qín* or *shui ching* (Mandarin) and *shuí kan* (Cantonese). The Thai is *ak chi lawm*. The Vietnamese is *rau càn* (*cân nuóc*).
- According to one explanation, the genus name *Oenanthe* is based on the Greek *oinanthe*, flowers of the grape vine, formed from *ainos*, wine + *anthos*, a flower, from the wine-like scent of the flowers. A second interpretation is that *oenanthe* is an ancient Greek name for some other plant, which was taken up to designate the present genus.
- There is a bird genus named *Oenanthe*, which includes the wheatear (*Oenanthe oenanthe*), a small thrush of most northern regions. Zoological nomenclature is independent of botanical nomenclature, so some plant names overlap with animal names. Other examples are as follows: *Arenaria* is a genus name for sandworts (plants) and turnstones (birds), *Aster* for asters (plants) and starfish, and *Prunella* for species of self-heal (plants) and accentors (birds).
- *Javanica* in the scientific name *O. javanica* is Latin for Javanese (Java is an Indonesian island).

Plant Portrait

Water dropwort is a robust, perennial, aquatic herb which spreads by runners like watercress and is considered to be an invasive water weed. The plant resembles celery. It produces erect, slender, hollow stems 10 to 150 cm (4 in. to 5 ft.) tall, with roots often developed at the lower nodes, where leaves arise. The roots are white, long, and threadlike. The leaves are divided like celery, up to 30 cm (1 ft.) long, with an odor resembling carrot tops. The tiny, fragrant, white or greenish-yellow flowers are in large umbels (flat-topped clusters of flowers with the flower stalks arising at one point) 5 to 8 cm (2–3 in.) across. The species is native from Japan to Australia. The plant grows wild in freshwater marshes and swampy fields and along ditches, canals, and streams. It is cultivated in Asia and in Hawaii. Also, this Asian aquatic leafy vegetable has recently been introduced to the United States by Vietnamese immigrants, and it is being grown as a specialty item in scattered locations across North America. Water dropwort has been grown experimentally in Florida and has been cultivated in Ontario, Canada, by immigrants of Asian origin. It is said to be naturalized as far north as British Columbia.

Water dropwort is cultivated as an ornamental plant in water gardens and in fish pools (koi are said to "eat it like candy"). The cultivar 'Flamingo' has leaves that are variegated with pink and cream and is widely grown. It is a low-growing, fast-spreading ground cover that does very well in shade. The plant is quite edible, and the divided pink leaves look attractive in salads.

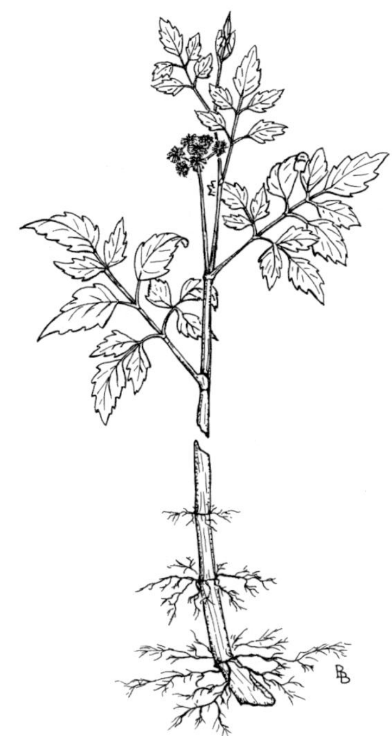

FIGURE 48.5 Water dropwort (*Oenanthe javanica*), by B. Brookes.

Culinary Portrait

Water dropwort is a popular Asian flavoring herb which is considered by many to be extremely delicious. The tops are eaten raw in salads or used as a garnish like parsley. The young stems and leaves are also steamed with rice or boiled and chopped as greens. The flavor has been described as a cross between parsley and carrot tops. The leaves and shoots are best for eating in the fall and winter when they are tender and mild flavored. The young leaves are milder than the mature ones, but both are suitable for use in all types of salad. These tender winter greens are a favorite in Japan for sukiyaki. Leaves and stems, chopped finely, combine well with other flavors in salads and in soups or other savory dishes. The distinct celery flavor makes this herb an excellent addition to baked meats and stuffing for poultry. The thin, cord-like roots are prized in Japanese cooking.

Culinary Vocabulary

- "Seven herbs rice gruel" or *Nana-kusa-gayu* (*nana* means seven, *kusa* means herbs or grasses in Japanese) is customarily consumed on January 7 in Japan. The practice is said to have started during the reign of Emperor Daigo in the early tenth century. The number seven was considered lucky in ancient Japan, and it was believed that eating this dish kept one in good health for the rest of the year. In modern times, preassembled seven grass rice gruel is commonly purchased. The seven herbs that were added to rice porridge varied somewhat depending on district. Typically, they were seri (water dropwort), nazuna (shepherd's purse, *Capsella bursa-pastoris* (L.) Medik.), gogyo (cudweed, *Gnaphalium multiceps* Wall.; the Japanese name means "mother and children herb"), hakobera (chickweed, *Stellaria media* (L.) Vill.), hotoke-no-za (henbit, *Lamium amplexicaule* L.), suzuna (turnip tops), and suzushiro (daikon, i.e. Japanese radish). Kudzu (*Pueraria montana* (Lour.) Merr.) is sometimes substituted for one of these species.

Curiosities of Science and Technology

- Aquatic plants that remove nutrients from the water are called filter plants or biofilter plants. Removal of nutrients is desirable in many circumstances to prevent excessive growth of algae and deterioration of water quality. Water dropwort is considered to be an excellent filter plant.
- A relative of water dropwort, Pacific water parsley (*Oenanthe sarmentosa* K. Presl. ex DC.), is native to the California Coastal Ranges, the Sierra Nevada Mountains, and north to British Columbia. It has traditionally been collected as a wild food by American Indians. Its black tubers are said to have a "cream-like taste," and the leaves and stems are also edible, tasting somewhat like celery. Unfortunately, some similar looking species are poisonous, and indeed *O. sarmentosa* itself is recorded as having been used as a poison by some Indian groups (the reason for this conflicting information is unclear). The plant is considered as useful as a filtering agent because it forms floating colonies, cleansing water of various chemicals.
- Another relative of water dropwort, *Oenanthe crocata* L., known as (hemlock) water dropwort, is a common plant in England, which has been said to have been responsible for more fatal accidental poisonings in the country than any other plant.
- *Oenanthe crocata*, discussed earlier, has been proposed as the plant responsible for what was termed "sardonic smile" by the ancient Greek poet Homer (eighth or ninth century BC; his work may represent the contributions of several people). According to Homer, people were executed by giving them an intoxicating potion that made them smile as they were being killed. *Oenanthe crocata* has chemicals capable of forcing facial muscles into a grimace. (Reference: Appendino, G., Pollastro, F., Verotta, L., Ballero, M., Romano, A., Wyrembek, P., et al. 2009. Polyacetylenes from Sardinian *Oenanthe fistulosa*: a molecular clue to *risus sardonicus*. *J. Nat. Prod.* 72: 962–965.)

Key Information Sources

An, W.B., and Lee, B.Y. 1991. Basic studies on the development of hydroponic system of water dropwort, *Oenanthe stolonifera* DC. 2. Optimal composition of macronutrients. *J. Korean Soc. Hortic. Sci.* 32:425–433 (in Korean, English summary).

Ashraf, M., Aziz, J., Mahmood, S., and Bhatty, M.K. 1979. Studies on the essential oils of the Pakistani species of the family Umbelliferae. XXI. *Oenanthe javanica*, DC. (surkhai) seed oil. *Pak. J. Sci. Ind. Res.* 22:82–83.

Choi, D.J., Kim, C.B., Lee, S.H., Yoon, J.T., Choi, B.S., and Kim, H.K. 2000. Effects of precooling and packaging film materials on quality of water dropwort (*Oenanthe stolonifera* DC.) at low temperature storage. *J. Korean Soc. Hortic. Sci.* 41:379–382 (in Korean).

Dai, Q., Cai, S., and Zhang, X. 1998. Purification of gold-containing wastewater by *Oenanthe javanica* and accumulation of gold in it. *Chinese J. Appl. Ecol.* 9:107–109 (in Chinese, English summary).

Haeng, J.R., Moo, S.K., and Ok, J.C. 1995. A study on the volatile constituents of the water dropwort (*Oenanthe javanica* DC)—according to extraction methods, parts and heating methods. *J. Korean Soc. Food Sci.* 11:386–395 (in Korean, English summary).

Huh, M.K., Choi, J.S., Moon, S.G., and Huh, H.W. 2002. Genetic diversity of natural and cultivated populations of *Oenanthe javanica* in Korea. *J. Plant Biol.* 45:83–89.

Hwang, J.M., and Park, Y.M. 1998. Growth and composition of fall season water dropwort under various cultivation systems. *J. Korean Soc. Hortic. Sci.* 39:657–660 (in Korean, English summary).

Kim, B.W., Lee, B.Y., and Kim, K.D. 1987. Studies on seed propagation method of *Oenanthe stolonifera* DC. 1. Flowering habit, seed structure, and seed development of *O. stolonifera*. *Agric. Res. Seoul Nat. Univ.* 12:15–20 (in Korean, English summary).

Kim, Y.O., and Park, Y.J. 1995. Study on the nutrient composition of hydroponic water dropwort. *J. Korean Soc. Food Nutr.* 24:1016–1019.

Lee, J.G., and Lee, B.Y. 2002. Growth and nutrient-water uptake characteristics of *Oenanthe stolonifera* DC. as affected by the concentrations of nutrient solution in closed hydroponic system. *J. Korean Soc. Hortic. Sci.* 43:582–586 (in Korean).

Lee, J.G., and Lee, B.Y. 2003. Growth of water dropwort (*Oenanthe stolonifera*) and changes of macroelement concentration as affected by nutrient solution composition in closed hydroponic system. *J. Korean Soc. Hortic. Sci.* 44:23–27 (in Korean).

Mitchell, M.I., and Routledge, P.A. 1978. Hemlock water dropwort [*Oenanthe crocata*] poisoning—a review. *Clin. Toxicol.* 12:417–426.

Morton, J.F., and Snyder, G.H. 1979. Trial of water celery [*Oenanthe javanica*] as an aquatic flavoring herb for Everglades farmlands. *Proc. Fla. State Hort. Soc.* 91:301–305.

Mun, S.I., Joh, Y.G., and Ryu, H.S. 1990. Protein and amino acid composition of water dropwort. *J. Korean Soc. Food Nutr.* 19:133–142 (in Korean, English summary).

Nagase, H., and Abe, S. 1998. Seed germination and year-round cultivation of water dropwort from summer to fall. *Bull. Oita Prefect. Agric. Res. Center*, 28:57–81 (in Japanese, English summary).

Park, J.C., Ha, J.O., and Park, K.Y. 1996. Antimutagenic effect of flavonoids isolated from *Oenanthe javanica*. *J. Korean Soc. Food Sci. Nutr.* 25:588–592 (in Korean, English summary).

Park, J.C., Yu, Y.B., Lee, J.H., and Kim, N.J. 1994. Studies on the chemical components and biological activities of edible plants in Korea. 6. Anti-inflammatory and analgesic effects of *Cedrela sinensis*, *Oenanthe javanica* and *Artemisia princeps* var. *orientalis*. *J. Korean Soc. Food Nutr.* 23:116–119 (in Korean, English summary).

Sasmitamihardja, D. 1993. *Oenanthe javanica* (Blume) DC. In *Plant resources of South-East Asia. 8. Vegetables*. Edited by J.E. Siemonsma and K. Piluek. Pudoc Scientific Publishers, Wageningen, the Netherlands. pp. 220–222.

WonHo, S., and HyungHee, B. 2005. Identification of characteristic aroma-active compounds from water dropwort (*Oenanthe javanica* DC.). *J. Agric. Food Chem.* 53:6766–6770.

Yang, S.Y., Jeong, Y.G., and Yang, W.M. 1989. Studies on the major characteristics of the Korean water-dropwort (*Oenanthe javanica* DC.) for selection of superior strains. *J. Korean Soc. Hortic. Sci.* 30:180–186 (in Korean, English summary).

Xia, Q., and Li, L. 2000. Contact dermatitis owing to *Oenanthe javanica* Blume DC.: a case report. *J. Clin. Dermatol.* 29:233.

Specialty Cookbooks

Heiter, C. 2007. *The sushi book*. ThingsAsian Press, San Francisco, CA. 269 pp.

Kim, E. 2008. *Cooking Korean food with Maangchi: books 1 & 2*. CreateSpace, Scotts Valley, CA. 118 pp.

Invernizzi, L., and Hutton, W. 2000. *The food of Malaysia: authentic recipes from the crossroads of Asia*. Periplus Editions, Singapore. 144 pp.

Sugimoto, T., Iwatate, M., Trotter, C., and Kawana, M. 2006. *Shunju: new Japanese cuisine*. Periplus, Singapore. 272 pp.

49 Jicama

Family: Fabaceae (Leguminosae; pea family)

NAMES

Scientific Name: *Pachyrhizus erosus* (L.) Urb. (*P. angulatus* Rich. ex DC., *P. bulbosus* Kurz) (See below for additional species sometimes cultivated as "Jicama")

- The word "jicama" is from American Spanish *jícama*, based on the Nahuatl (Aztec language) *xicamatl*. In Spanish, *jícama* is used for any edible root.
- Pronunciation: HEE-kah-mah, HICK-a-mah, or hee-ca-MA.
- Jicama is also called chop suey bean, Mexican potato, Mexican turnip, Mexican water chestnut, Mexican yam bean, potato bean, short-podded yam bean, and yam bean.
- The names Chinese potato (used also for other root vegetables, including Chinese arrowhead, *Sagittaria sagittifolia* L., and cinnamon vine, *Dioscorea batatas* Decne) and Chinese turnip are sometimes encountered but are inappropriate because jicama is not native to China.
- In Ecuador and Peru, *Smallanthus sonchifolius* (Poepp. & Endl.) H. Rob. (*Polymnia sonchifolia* Poepp. & Endl.), an edible-tuber plant of the sunflower family that is usually known as aricoma and yacón, is often called jicama.
- In addition to *P. erosus*, some other species of *Pachyrhizus* are also known as jicama and yam bean, and the roots are used similarly, although these are much less important. *Pachyrhizus tuberosus* (Lam.) Spreng., a native of the Amazon Basin, produces a larger root and is cultivated in South America. *Pachyrhizus ahipa* (Wedd.) Parodi is cultivated in northwestern Argentina, Bolivia, and Peru.
- As noted above, several alternative names for jicama have "yam bean" in their names. The African yam bean is *Sphenostylis stenocarpa* (Hochst. ex A. Rich.) Harms and is occasionally cultivated in West Africa. It is also a member of the pea family.
- The genus name *Pachyrhizus* is based on the Greek *pachy*, thick + *rhiz*, root, a reference to the tuberous roots.
- *Erosus* in the scientific name *P. erosus* means having an irregularly toothed or apparently gnawed margin, descriptive of the leaves.

PLANT PORTRAIT

Jicama is a perennial, climbing, tropical vine growing to a length of 6 m (20 ft.) or more. The species is native to Central America and was introduced to Southeast Asia via the Philippines in the seventeenth century. Subsequently, it was spread across Asia, particularly in China and the Pacific. It is commonly consumed only in the Philippines, southern China, Mexico, and other parts of Latin America but recently has been acquiring popularity in North America. Jicama has been grown in its native Mexico for centuries, and Mexico is the main exporter. It has not been successfully developed as a crop in the United States, although attempts to do this are underway.

The root is large (10–15 cm or 4–6 in. in diameter) and turnip shaped (typically like a turnip that has been slightly flattened on both ends), with thin, light brown skin and white crunchy flesh. The roots may weigh up to 23 kg (50 lb.), but when large they become woody. Those in stores weigh 1 to 3 kg (2.2–6.5 lb.). Roots marketed in North America often have four lobes and occasionally have dwarf tubers attached, whereas those grown in Asia usually produce a single, smoothly rounded tuber. Jicama is available year-round in North America, particularly from November through May, in Mexican markets and many large supermarkets, often sold cut in pieces wrapped in plastic.

There are two cultivated forms of jicama: jicama de agua and jicama de leche. The latter has an elongated root and milky juice. The agua form has a root that varies between top shaped and flattened-turnip shape. Its watery juice is translucent, and it is the preferred form in markets.

By starting the plant indoors, jicama can be grown outdoors as far north as southern Canada, at least as an attractive ornamental. It blooms profusely, producing white to lavender flowers that resemble sweet peas and large heart-shaped leaves. (When grown for food, the flowers are pinched off to promote better root development.) Five to 9 months is required to produce good-sized roots, and the plants are killed by frost, so growing jicama as a vegetable can be done only in milder areas of the United States and Europe. Although the starchy root is edible, the above-ground parts of the plant are toxic. Those attempting to grow it should take care that no humans or animals mistakenly eat the poisonous, narcotic leaves and seeds. The young pods are sometimes eaten as a vegetable in the tropical areas where jicama is cultivated but are probably best avoided because of their content of rotenone (see the following sections).

CULINARY PORTRAIT

The taste of jicama has been compared with water chestnut, apple, and pear. Numerous recipes are available on the Web, which is not surprising because the vegetable has the versatility of potatoes with the added feature that it can be eaten raw. The best roots to purchase are medium to small (larger ones are less juicy and more fibrous) and are clean, firm, and without evidence of decay. Jicama has crisp, sweet, nutty flesh that is adapted to raw usage in salads, appetizers, and hors d'oeuvres. It can be steamed, baked, boiled, mashed, or fried and is used in soups, stews, and stir-fries. If cooked, light cooking or adding to the cooking pot later than the other ingredients has been recommended to preserve the crispy texture, although the flesh retains its crispness rather well when cooked. The thin skin should be peeled away before use. Because of its crisp texture, white flesh, and blandness, jicama can be substituted for the more expensive water chestnuts and bamboo shoots. It has the advantage over turnips of not discoloring when exposed to air. Jicama can be sliced, diced, or cut into strips for use as a garnish, in salads, and with dips. It is often served as a snack sprinkled with lime or lemon juice or a dash of chili powder. The crisp flesh may be served slightly chilled to enhance its taste. This vegetable can be stored like potatoes in a cool, dry area for several weeks (or up to 2 months according to some authorities). The tubers can also be placed in a plastic bag in the refrigerator for up to 2 weeks. With longer storage, conversion of starch in the root to sugar can occur.

Culinary Vocabulary

- *Popiah* are spring rolls, popular in Taiwan, Singapore, Malaysia, and parts of China. Although many vegetables can be used as the filling, jicama is frequently the principal one used.

CURIOSITIES OF SCIENCE AND TECHNOLOGY

- The toxin in jicama leaves and seeds has been identified as rotenone, and there are large amounts. Rotenone is a commercial insecticide, obtained as a natural plant extract from certain members of the pea family (notably derris, *Derris elliptica* (Wall.) Benth.), which are cultivated for the purpose, especially in the Malay Peninsula and Indonesia. Not surprisingly, jicama suffers from few pests. The presence of the chemical is a strategy of the plants to protect themselves from insects. Because it is a natural (not a synthetic or man-made) insecticide, it has been rather popular with some who advocate "organic" cultivation techniques, especially for fruits and vegetables. However, in 2005, the U.S. Department of Agriculture removed it from its list of approved agents for organic production because of concern about its safety. Rotenone is also used to control lice, ticks, and warble flies on animals and to eliminate undesired fish in fish management. It has to be used with care because it can be toxic to humans and pets. Children need to be kept

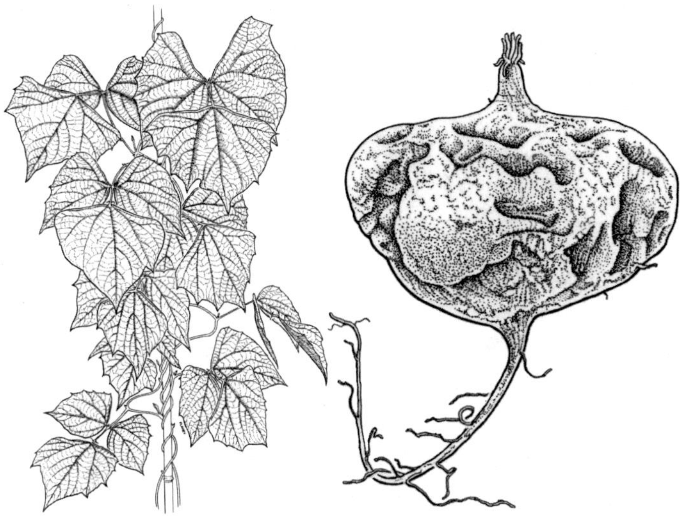

FIGURE 49.1 Jicama (*Pachyrhizus erosus*). Left: portion of vegetative vine; right: root. Artist: B. Flahey.

FIGURE 49.2 Jicama (*Pachyrhizus erosus*), flowering branch and fruit. (From Hooker's *Icones Plantarum,* third series, 1886–1887, plate 1842.)

away from jicama plants and seeds because of the possibility of ingestion and consequent poisoning.

- The poisonous seeds of jicama were used medicinally by the ancient Aztecs.
- The fiber in the tough vines of jicama has been used as cord and for fish nets.
- Rotenone is an effective fish poison, and this fact has been exploited as a means of incapacitating fish so that they can be captured. The roots of jicama, and indeed of other plants that contain rotenone, were crushed by indigenous peoples and placed in water where fish were swimming. Rotenone is much more easily absorbed through the gills of fish than through the skin or intestinal mucosa of humans, so the fish that have been trapped in this manner are only slightly poisonous to humans.

KEY INFORMATION SOURCES

Alvarenga, A.A., and Vaalio, I.F.M. 1989. Influence of temperature and photoperiod on flowering and tuberous root formation of *Pachyrhizus tuberosus*. *Ann. Bot.* 64:411–414.

Aquino-Bolanos, E.N., Cantwell, M.I., Peiser, G., and Mercado-Silva, E. 2000. Changes in the quality of fresh-cut jicama in relation to storage temperatures and controlled atmospheres. *J. Food Sci.* 65:1238–1243.

Bergsma, K.A., and Brecht, J.K. 1992. Postharvest respiration, moisture loss, sensory analysis and compositional changes in jicama (*Pachyrrhizus erosus*) roots. *Acta Hortic.* 318:325–332.

Bruton, B.D. 1983. Market and storage diseases of jicama. *J. Rio Grande Valley Hortic. Soc.* 36:29–34.

Cantwell, M. 2000. Produce facts: jicama. *Perishables Handling Quart.* 103(Aug.):17–18.

Cantwell, M., Orozco, W., Rubatzky, V., and Hernandez, L. 1992. Postharvest handling and storage of jicama roots. *Acta Hortic.* 318:333–343.

Clausen, R.T. 1944. *A botanical study of the yam beans* (Pachyrrhizus). Memoir 264. Agricultural Experiment Station, Cornell University, Ithaca, NY. 38 pp.

Cotter, D.J., and Gomez, R.E. 1979. Daylength effect on root development of jicama (*Pachyrhizus erosus* Urban). *Hortic. Sci.* 14:733–734.

Dabydeen, S., and Sirju-Charran, G. 1990. The developmental anatomy of the root system in yam bean, *Pachyrhizus erosus* Urban. *Ann. Bot.* 66:313–320.

Fernandez, M.V., Warid, W.A., Loaiza, J.M., and Montiel, A. 1997. Developmental patterns of jicama (*Pachyrhizus erosus* (L.) Urban) plant and the chemical constituents of roots grown in Sonora, Mexico. *Plant Foods Human Nutr.* 50:279–286.

Fine, A.J. 1991. Hypersensitivity reaction to jicama *Pachyrhizus* yam bean. *Ann. Allergy*, 66(2):173–174.

Grum, M., Halafihi, M., Stölen, O., and Sørensen, M. 1994. Yield performance of yam bean in Tonga, South Pacific. *Exp. Agr.* 30:67–75.

Hansberry, R., Clausen, R.T., and Norton, L.B. 1947. Variations in the chemical composition and insecticidal properties of the yam bean. *J. Agric. Res.* 74:55–64.

Johnson, H., Jr., Orozco, W., Cantwell, M., and Hernandez, L. 1991. Jicama, yam bean. Crop Sheet SMC-020. In *Specialty and minor crops handbook*. Edited by C. Myers. The Small Farm Center, Division of Agriculture and Natural Resources, University of California, Oakland, CA. 4 pp.

Juarez, M.S., and Paredes-Lopez, O. 1994. Studies on jicama juice processing. *Plant Foods Human Nutr.* 46:127–131.

Lynd, J.Q., and Purcino, A.A.C. 1987. Effects of soil fertility on growth, tuber yield, nodulation, and nitrogen fixation of yam bean (*Pachyrhizus erosus* (L.) Urban) grown on a typic eutrustox. *J. Plant Nutr.* 10:485–500.

Mercado-Silva, E., and Cantwell, M. 1998. Quality changes in jicama roots stored at chilling and nonchilling temperatures. *J. Food Qual.* 21:211–221.

Paull, R.E., and Chen, N.J. 1988. Compositional changes in yam bean during storage. *HortScience.* 23:194–196.

Paull, R.E., Chen, N.J., and Fukuda, S.K. 1988. Planting dates related to tuberous root yield, vine length, and quality attributes of yam bean. *HortScience*, 23:326–329.

Porterfield, W.M. 1939. The yam bean as a source of food in China. *N. Y. Bot. Gard. J.* 40:107–108.

Santos, A.C.O., Cavalcanti, M.S.M., and Coelho, L.C.B.B. 1996. Chemical composition and nutritional potential of yam bean seeds (*Pachyrhizus erosus* L. Urban). *Plant Foods Hum. Nutr.* 49:35–41.

Schroeder, C.A. 1968. Sociological aspects of the jicama in Mexico. *Ethnos*, 33:78–89.

Schroeder, C.A. 1968. The jicama, a root crop from Mexico. *Proc. Am. Soc. Hortic. Sci.* 11:65–71.

Sinha, R.P., Prakash, R., and Haque, M.F. 1977. Genetic variability in yam bean (*Pachyrhizus erosus* Urban). *Trop. Grain Legume Bull.* 7:21–23.

Sørensen, M. 1988. A taxonomic revision of the genus *Pachyrhizus* Rich. ex DC. nom. cons. *Nordic J. Bot.* 8:167–192.

Sørensen, M. 1996. *Yam bean:* Pachyrhizus *DC.* International Plant Genetics Resources Institute, Rome, Italy. 141 pp.

Sørensen, M., and Hoof, W.C. van. 1996. *Pachyrhizus erosus* (L.) Urban. In *Plant resources of South-East Asia. 9. Plants yielding non-seed carbohydrates.* Edited by M. Flach and F. Rumawas. Backhuys Publishers, Leiden, the Netherlands. pp. 137–141.

Sørensen, M., Estrella, J.E., Hamann, O., and Ríos, S.A.R. (Eds.). 1998. *Proceedings of the second international symposium on tuberous legumes, Celaya, Mexico, August 5–8, 1996.* Mackeenzie (printer), Copenhagen, Denmark. 566 pp. (Includes numerous contribution on jicama.)

Sørensen, M., Doygaard, S., Estrella, J.E., Kvist, L.P., and Nielsen, P.E. 1997. Status of the South American tuberous legume *Pachyrhizus tuberosus* (Lam.) Spreng. *Biodivers. Conserv.* 6:1581–1625.

Sørensen, M., Grum, M., Paull, R.E., Vaillant, V., Venthou-Dumaine, A., and Zinsou, C. 1993. Yam bean (*Pachyrhizus* species). In *Underutilized crops: pulses and vegetables.* Edited by J.T. Williams. Chapman & Hall, London. pp. 59–102.

Tapia, C., and Sørensen, M. 2003. Morphological characterization of the genetic variation existing in a Neotropical collection of yam bean, *Pachyrhizus tuberosus* (Lam.) Spreng. *Genet. Res. Crop Evol.* 50:681–692.

Venthou-Dumaine, A., Vaillant, V., Soerensen, M., Estrella, J.E., Hamann, O.J., and Ruiz, S.A.R. (Eds.). 1998. *Proceedings of the 2nd international symposium on tuberous legumes* 566 pp. (Includes numerous contributions on jicama.)

Specialty Cookbooks

Gabilondo, A. 1986. *Mexican family cooking.* Ballantine Books, New York. 385 pp.

Hoyer, D. 2005. *Culinary Mexico: authentic recipes and traditions.* Gibbs Smith, Salt Lake City, UT. 240 pp.

Poore, M. 2001. *1000 Mexican recipes.* Wiley, New York. 656 pp.

50 Jujube

Family: Rhamnaceae (buckthorn family)

NAMES

Scientific Name: *Ziziphus jujuba* Mill. (*Z. sativa* Gaertn., *Z. spinosa* (Bunge) Hu ex F.H. Chen, *Z. zizyphus* (L.) Meikle)

- The word "jujube" is based on the Middle Latin *jujuba*, an alteration of the Latin *zizyphum*, from the Greek *zizyphon*. The word refers to the plant or its fruit. Jujube also has the meaning "a fruit-flavored gumdrop or lozenge" because this confection was originally made from jujube fruit. Candy jujubes are now made with gum arabic (see Chapter 44) and flavoring.
- Pronunciations: JEW-joob, JUH-ju-bee, JUH-jew-bee.
- Other names for the jujube are ber, common jujube, Chinese jujube, Chinese date, and red date.
- The Indian jujube, *Z. mauritiana* Lam., is an evergreen tree adapted to warmer climates than the deciduous jujube tree described here. It is cultivated primarily in India and like other cultivated species of *Zizyphus* produces less desirable fruit. Often called a "poor man's fruit," attempts are underway to breed better varieties. The literature at times uses the name *Z. jujuba* when in fact *Z. mauritiana* is being discussed.
- *Jujubinus* is a genus of molluscs.
- The genus name *Ziziphus* is from the Arabic *zizouf*, the name of the lotus fruit of antiquity.

PLANT PORTRAIT

The jujube is a shrub or small tree to 9 m (30 ft.) in height, occasionally to 12 m (40 ft.), with small, glossy, deciduous leaves and drooping branches growing in zigzag fashion. Many of the plants have spiny young branches (in Asia, some spineless varieties are grown, but are typically grafted onto rootstocks from spiny trees). The species is native to temperate Asia and is now fairly commonly grown in India, Russia, the Middle East, southern Europe, and especially China. The jujube has been cultivated for more than 4000 years in China, where the crop has been most extensively developed, with more than 400 known varieties. As late as the middle of the twentieth century, the jujube was the most common fruit tree in China (followed by persimmons). Today, the area devoted to growing jujube in China (290,000 ha or 720,000 acres) is equivalent to the area of Florida on which citrus is grown. During the Roman era, jujube trees were established throughout southern Europe and North Africa. Jujube trees first reached the United States in 1837 and were planted in North Carolina. They are attractive ornamentals and are often seen in the southern United States. Most specimens cultivated in Europe and North America have not been propagated from superior trees, the fruits are generally no larger than an olive, and like apples on apple trees grown from seeds, the fruit is inferior, although nevertheless occasionally consumed. The fruit of the species is egg shaped or pear shaped, ranges in size from that of a cherry to a plum, and has dark red, thin skin that matures to black, accompanied by crisp, whitish, sweet flesh (with more than 20% sugar) reminiscent of an apple but not as juicy. An elongated pit is present. The fruit will dry like figs if left on the tree, in which case the smooth skin becomes wrinkled. Jujube fruit has never become popular outside of Asia, and indeed in Western countries most people encountering the word "jujube" think of the fruit-flavored gumdrops that have been losing popularity.

CULINARY PORTRAIT

Fresh or dried jujubes are eaten out of hand or cooked. The taste of the dried fruit has been compared with a combination of apples and prunes. Most fruits are sweet, but some tart-fruited forms are known in China. In China, jujubes are eaten fresh, dried, smoked, candied, pickled, made into a spread or sauce, prepared as a kind of raisin bread, and boiled with rice. They are also used to impart a subtly sweet taste to braised and steamed dishes and soups. The Chinese also make a genuine jujube candy, unlike the often rubbery jujubes that have been sold as confectionary. Jujubes in Chinese jujube confectionary have slash marks, made to facilitate their stewing. Dried jujubes can be used wherever recipes call for raisins or dates. For sweet dishes, these are preferably soaked in cold water for several hours before use. Cooked fruit may be used to make jam or compôte or used in stews, stuffings, and soups. Juice has been expressed from the fruit, and alcoholic beverages produced by fermentation. Ripe fruits may be stored at room temperature for about a week, whereas tree-dried fruit stores indefinitely. Fresh jujubes are occasionally available in Western countries, usually imported from China, and these can be stored for up to 2 weeks in the refrigerator.

Culinary Vocabulary

- The dried fruit of jujube is commonly called "Chinese date" and "red date" and although unrelated to the true date, the fruits are similar in appearance, texture and flavor, although not as sweet.

CURIOSITIES OF SCIENCE AND TECHNOLOGY

- "Lotus-eaters" were a people described in *The Odyssey* of the Greek epic poet Homer (approximately 850 BC), a story about a long, adventurous journey. The lotus-eaters lived on "lotus" fruit as their staple, and this left them in a permanent state of lethargy. Ulysses, the hero of the tale, sent three of his crew into the country of the lotus-eaters to investigate the people, and they were given lotus to eat, causing them to lose all interest in continuing the journey. Ulysses had to drag the men back to his ship and tie them up until they returned to their senses. The term lotus-eater is now used to refer to people who are idle, lazy daydreamers without clear objectives and accomplishments. Scholars have concluded that the jujube is the most likely plant that Homer had in mind when he used the word lotus.
- In ancient Egypt, jujubes were offered as tributes to the dead at funerals.
- The Indian jujube is used in Ethiopia to stupefy fish.
- The Indian jujube is used in India to produce shellac, made from an excretion of the female lac insect (the scale insect *Kerria lacca* (Kerr.), more commonly known as *Laccifer lacca* Kerr.), which is allowed to feed on the tree.
- The 22,000-page Oxford English dictionary is arguably the greatest work of scholarship ever produced. It has been claimed that the only deliberately humorous entry is "jubjub," a word that appeared in Lewis Carroll's "Through the Looking-Glass" (1871), a word that some etymologists have proposed was inspired by "jujube."

FIGURE 50.1 Jujube (*Ziziphus jujuba*), from Nicholson (1885–1889). Long section of fruit at bottom right is from Lamarck and Poiret (1744–1829, plate 185).

KEY INFORMATION SOURCES

Abbas, M.F. 1997. Jujube. In *Postharvest physiology and storage of tropical and subtropical fruits*. Edited by S.K. Mitra. CAB International, Wallingford, Oxon, U.K. pp. 405–415.

Cheng, G., Bai, Y., Zhao, Y., Tao, J., Liu, Y., Tu, G., Ma, L., Liao, N., and Xu, X. 2000. Flavonoids from *Ziziphus jujuba* Mill var. *spinosa*. *Tetrahedron*, 56(45):8915–8920.

Ciminata, P. 1996. The Chinese jujube, *Zizyphus jujuba*. *WANATCA Yearbook* (*Australia*), 20:34–36.

Crawford, M. 2002. The jujube: *Ziziphus jujuba*. *WANATCA Yearbook* (*Australia*), 26:37–42.

Dutta, S. 1954. Jujubes of Assam. *Indian J. Hortic*. 11:53–56.

Juhaeti, T. 2003. *Zizyphus* Miller. In *Plant resources of South-East Asia, Vol. 12(3). Medicinal and poisonous plants 3*. Edited by R.H.M.J. Lemmens and N. Bunyapraphatsara. Backhuys Publishers, Leiden, the Netherlands. pp. 423–424.

Kader, A.A. 2000. Produce facts: Chinese jujube (Chinese date). *Perishables Handling Quart*. 101(Feb.):13–14.

Kirkbride, J.H., Jr., Wiersema, J.H., and Turland, N.J. 2006. Proposal to conserve the name *Ziziphus jujuba* against *Z. zizyphus* (Rhamnaceae). *Taxon*, 54:1049–1050.

Lanham, W.B. 1926. *Jujubes in Texas*. Texas Agricultural Experiment Station, College Station, TX. 28 pp.

Latiff, A.M. 1991. *Ziziphus mauritiana* Lamk. In *Plant resources of South-East Asia, Vol. 2. Edible fruits and nuts*. Edited by E.W.M. Verheij and R.E. Coronel. Pudoc, Leiden, the Netherlands. pp. 310–312.

Lyrene, P. 1978. Jujubes in southwest Alabama. *North American Pomona*, 11:136–137.

Lyrene, P.M. 1979. The jujube tree (*Zizyphus jujuba* Lam). *Fruit Var. J*. 33:100–104.

Lyrene, P.M. 1983. Flowering and fruiting of Chinese jujubes in Florida. *HortScience*, 18:208–209.

Lyrene, P. 2008. *Zizyphus jujube*, Chinese jujube. In *The encyclopedia of fruit & nuts*. Edited by J. Janick and R.E. Paull. CABI, Wallingford, Oxfordshire, U.K. pp. 615–617.

Lyrene, P. 2008. *Zizyphus mauritiana*, Indian jujube. In *The encyclopedia of fruit & nuts*. Edited by J. Janick and R.E. Paull. CABI, Wallingford, Oxfordshire, U.K. pp. 617–619.

Lyrene, P.M., and Crocker, T.E. 1979. *The Chinese jujube. Fruit crops facts sheet*. Agricultural Extension Service, University of Florida, Gainesville, FL. 3 pp.

Meyer, R., and Chambers, R.C. 1998. *Jujube primer & source book*. California Rare Fruit Growers, Inc./ Fullerton Arboretum, California State University, Fullerton, CA. 267 pp.

Mitra, S.K. 1999. Ber. In *Tropical horticulture*, vol. 1. Edited by T.K. Bose, S.K. Mitra, A.A. Farooqui, and M.K. Sadhu. Naya Prokash, Calcutta, India. pp. 344–350.

Noel, D. 1995. *Ziziphus*: jujube family species, distribution, exploitation. *WANATCA Yearbook (Australia)*, 19:14–20.

Outlaw, W.H., Jr., Zhang, S., Riddle, K.A., Womble, A.K., Anderson, L.C., Outlaw, W.M., et al. 2002. The jujube (*Ziziphus jujuba* Mill.), a multipurpose plant. *Econ. Bot.* 56:198–200.

Pareek, O.P. 1983. *The ber*. ICAR, New Delhi, India. 71 pp.

Shimizu, H.H. 1990. *Ziziphus jujuba*. *Public Garden*, 5(1):39–41.

Sweet, C. 1985. Large market potential seen for the Chinese date (jujube). *California Grower*, 9(12):41–43, 48.

Teaotia, S.S., and Chauhan, R.S. 1963. Flowering, pollination, fruitset and fruit-drop studies in ber (*Zizyphus mauritiana* Lamk.). I. Floral biology. *Punjab Hort. J.* 3:58–70.

Thomas, C.C., and Church, C.G. 1924. *The Chinese jujube*. U.S. Department of Agriculture, Washington, DC. 31 pp.

Yamdagni, R. 1985. Ber. In *Fruits of India, tropical and subtropical*. Edited by T.K. Bose. Naya Prokash, Calcutta, India. pp. 520–536.

Specialty Cookbooks

Grillon, O. 1911. *Modern practical gum work manual. Contains the best and most modern processes ... for the manufacturing of gum drops, soft and hard, superior and ordinary licorice, jujube paste ...* New York. 140 pp.

Hache, V., and Li-Ping, X. 1995. Jujube—not only in China a fruit with a high potential. *Fruit Processing*. 5(2a): 36–37. (Deals with manufacture of dried fruit, candied jujube, jujube marmalade, jujube juice, jujube wine, jujube brandy, and jujube vinegar.)

Lee, C.H.-J. 2009. *Quick and easy Korean cooking: more than 70 everyday recipes*. Chronicle Books, San Francisco, CA. 168 pp.

Miller, G.B. 1984. *The thousand recipe Chinese cookbook*. Simon & Schuster, New York. 926 pp.

Park, A. 2007. *Discovering Korean cuisine: recipes from the best Korean restaurants in Los Angeles*. Dream Character, Lomita, CA. 170 pp.

Wakiya, Y., Bouley, D., Matsuhisa, N., and Masashi, K. 2008. *Haute Chinese cuisine from the kitchen of Wakiya*. Kodansha International, New York. 192 pp.

51 Kava

Family: Piperaceae (pepper family)

NAMES

Scientific Name: *Piper methysticum* G. Forst.

- Kava is a Tongan word, meaning acrid, and is used both for the plant and for the intoxicating beverage prepared from it. The name kava means bitter in various Polynesian languages, in reference to the bitter taste of the beverage. The word kava may be used in Polynesia to designate properties of food and drink. In Hawaii, kava may mean bitter, sour, sharp, or pungent; in the Marquesas, it may mean bitter, sour, or sharp; and in Tahiti, it may mean bitter, sour, acid, acrid, salty, sharp, or pungent.
- Kava is also known as kava kava (kavakava), kava pepper, and kawa pepper.
- In Hawaii, in Tahiti, in the Marquesas, and in some other Polynesian areas, kava is called ava (awa), a Polynesian word also used to designate various properties of food and drink such as bitter, sharp, or pungent.
- In Fiji, the name yanggona or yangona is used.
- The genus name *Piper* (as well as the English word pepper) is derived from the Sanskrit name *pippali* or *pippalii*, transferred to the Greek *péperi* and Latin *piper*, all referring to some type of pepper (not necessarily the common pepper, *Piper nigrum* L.).
- *Methysticum* in the scientific name *P. methysticum* is the Latin transcription of the Greek *methustikos*, intoxicant.

PLANT PORTRAIT

Kava is a shrub, generally approximately 2 m (6½ ft.) tall, but in favorable circumstances it grows as high as 6 m (20 ft.). There are separate male and female plants, but the females tend to be sterile (unable to produce seeds), and as male plants are common, the species is usually propagated vegetatively. The stems are green to black, succulent, with strongly swollen nodes reminiscent of bamboo stems. The leaves are heart shaped, 15 to 28 cm (6–11 in.) long, and when held up to light appear dotted because of the presence of oil-containing glands. The species is thought to have originated in Melanesia and is common in Polynesia. Kava is believed to have been cultivated on islands in the South Pacific for more than 3000 years. A drink made from the kava plant was the beverage of choice for the royal families of the South Pacific, and indeed kava has been called "the South Pacific's most revered herb." The first Europeans to observe the plant and its ritualistic consumption by natives of Oceania were Dutch explorers Jacob Le Maire and William Schouten in 1616. It was noted that indigenous people chewed or pounded the rhizome (underground stem, usually called a "root") and mixed it with water to produce a brownish, often bitter brew, which they then consumed for its psychoactive properties. Kava's active chemicals are called kavalactones and are concentrated in the "roots." The drink is still consumed in Western Polynesia, especially in Samoa, Tonga, and most of Melanesia including Fiji, although missionaries reduced its use considerably. Kava was also traditionally used as a medicine for various ailments, and has been used as a sedative and aphrodisiac.

CULINARY PORTRAIT

Although kava is sometimes considered to be a "drug," it is more properly viewed as a traditional social beverage, much like wine. Indeed, in many of the Oceanic islands where kava was used, alcoholic beverages were unknown. As with alcoholic beverages, heavy consumption of kava can cause intoxication and loss of coordination. Although kava is not consumed for its nutritional content, the same is true for several other social beverages such as coffee and tea, so this herb is legitimately within the realm of food plants. Many procedures have been used to prepare kava traditionally. In the "Tonga method," the rhizome is chewed by young men or women until it is fine and fibrous, then soaked in water, and after a period of time decanted and drunk. In the "Fiji method," which is currently the most widespread, the rhizome is first mechanically pulverized, followed by the extraction of the residue with water.

Among herbal supplements, kava ranked ninth in sales at the beginning of the twenty-first century in the United States, so that is almost entirely in this form that North Americans are consuming the plant. Some information on its effects is therefore in order. When drunk or eaten in its paste form, the tongue and inside of the mouth go numb, as with a shot of novocaine. However, commercial preparations (tablets, capsules, extracts, tinctures, drinks, and tea bags) are usually diluted to prevent this. Kava reduces anxiety much like the well-known Valium and is a potent muscle relaxant. It promotes relaxation and sociability, but its effects are different from those produced by either alcohol or synthetic tranquillizers. It does not result in a hangover, and even more significant, it does not cause dependency or addiction. Not surprisingly, in recent times, kava has become extremely popular and has been widely marketed specifically to reduce anxiety, depression, and insomnia.

In 2002, the U.S. Food and Drug Administration cautioned that there is potential risk of severe liver injury from the use of dietary supplements containing kava. Recent reports from health authorities in Germany, Switzerland, France, and the United Kingdom linked kava use to at least 25 cases of liver toxicity, including hepatitis, cirrhosis, and liver failure. Canada banned kava in 2002. Kava products are also banned in Singapore and Germany (at one time there was an attempt to produce a kava-based soft drink in Germany), and some countries have also either banned the sale of kava or taken measures to restrict the market. It has been noted that liver damage appears to be rare. Those with liver disease or liver problems or taking drugs that can affect the liver should discuss the problem with their health care practitioner before using kava. Kava should not be consumed along with alcohol. Use of kava is discouraged during pregnancy and breast feeding and by those with Parkinson's or clinical depression. Children should not be given kava. Prolonged use can result in yellow coloration of skin, nails, and hair, allergic skin reactions, and visual and equilibrium disturbances. In Europe, it has been recommended that kava not be consumed for longer than 3 months without medical advice and that driving and operating machinery during consumption should be avoided.

CULINARY VOCABULARY

- A "kava bowl" has long been used ritualistically in the South Pacific and Hawaii to drink prepared kava. It was traditionally a coconut shell or carved wooden bowl in the past, but in modern times plastic bowls are often used. Antique kava bowls are valuable, and rather cheap replicas are commonly marketed at expensive prices. Victor Jules Bergeron, Jr. (1902–1984), better known as "Trader Vic," was a well-known owner of a chain of Polynesian-style restaurants. He was famous for inventing (allegedly in some cases) well-known cocktails such as the Mai Tai. One of his cocktails was the Kava Bowl, a rum cocktail made without kava and served in a rather gaudy bowl that is now a collector's item.
- "Kava colada" is a beverage described in Kilham (2001) as a blend of one tray of frozen pineapple-juice cubes, a can of coconut milk, 1 oz. of fluid kava extract, and a little honey and vanilla (serves six).

FIGURE 51.1 Kava (*Piper methysticum*). (From Delessert, 1820–1846.)

FIGURE 51.2 "Young Girls Preparing Kava Outside of the Hut Whose Posts Are Decorated with Flowers," painted about 1891 by the American artist and writer John LaFarge (1835–1910).

CURIOSITIES OF SCIENCE AND TECHNOLOGY

- The daughters of Polynesian chiefs were often enlisted to prepare kava by chewing the roots for hours. The desire to escape from this fatiguing duty is thought to have motivated many of them to elope.
- Former First Ladies Claudia "Lady Bird" Johnson (1912–2007), wife of the 36th president, Lyndon Baines Johnson, and Hillary Rodham Clinton (1947–), wife of the 42nd president, William Jefferson Clinton; U.S. Senator and Secretary of State, and Pope John Paul II (1920–2005), drank kava during welcoming ceremonies in the Pacific, respectively, in 1966, 1992, and 1986. Queen Elizabeth II of Great Britain (1926–) and members of the Royal Family have consumed kava during visits to Fiji on a number of occasions.
- Like kava kava, ylang ylang (a volatile oil from the ylang ylang tree, *Cananga odorata* (Lam.) Hook. f. & Thomson) is a plant product the name of which repeats the same word. Indeed, ylang ylang has been called "the kava kava of essential oils" because it is used in aromatherapy to reduce insomnia, nervous tension, anxiety, and stress, uses that are the same as for kava kava.

KEY INFORMATION SOURCES

Anke, J., and Ramzan, I. 2004. Kava hepatotoxicity: are we any closer to the truth? *Planta Med.* 70:193–196.

Basch, E., Hammerness, P., Sollars, D., Basch, S., Boon, H., Ulbricht, C., et al. 2002. Kava monograph: a clinical decision support tool. *J. Herb. Pharmacother.* 2(4):65–91.

Bilia, A.R., Gallori, S., and Vincieri, F.F. 2002. Kava-kava and anxiety: growing knowledge about the efficacy and safety. *Life Sci.* 70(22):2581–2597.

Brown, J.F. 1989. *Kava and kava diseases in the South Pacific*. Australian Centre for International Agricultural Research, Canberra, ACT, Australia. 70 pp.

Brunton, R. 1989. *The abandoned narcotic: kava and cultural instability in Melanesia*. Cambridge University Press, Cambridge, U.K. 219 pp.

Cox, P.A., and O'Rourke, L. 1987. Kava (*Piper methysticum*, Piperaceae). *Econ. Bot.* 41:452–454.

Davis, R.I., and Brown, J.F. 1999. *Kava (*Piper methysticum*) in the South Pacific: its importance, methods of cultivation, cultivars, diseases and pests*. Australian Centre for International Agricultural Research, Canberra, ACT, Australia. 32 pp.

Dietlein, G., and Schroeder B.D. 2003. Doctors' prescription behaviour regarding dosage recommendations for preparations of kava extracts. *Pharmacoepidemiol. Drug Saf.* 12:417–421.

Fackelmann, K. 1992. Pacific cocktail: the history, chemistry and botany of the mind-altering kava plant. *Sci. News*, 141(26):424–425.

Gatty, R. 1956. Kava: Polynesian beverage shrub. *Econ. Bot.* 10:241–249.

Kilham, C. 2001. *Psyche delicacies: coffee, chocolate, chiles, kava, and cannabis, and why they're good for you*. Rodale, Emmaus, PA. 221 pp.

Kunisaki, J., Araki, A., and Y Sagawa, Y. 2003. *Micropropagation of `awa (kava,* Piper methysticum*)*. University of Hawai'i at Manoa, Honolulu, HI. 11 pp.

Lebot, V., and Cabalion, P. 1988. *Kavas of Vanuatu: cultivars of Piper methysticum Forst.* South Pacific Commission, Noumea, New Caledonia. 191 pp.

Lebot, V., and Levesque, J. 1989. The origin and distribution of Kava (*Piper methysticum* Forst. f., Piperaceae): a phytochemical approach. *Allertonia*, 5(2):223–281.

Lebot, V., and Levesque, J. 1996. Evidence for conspecificity of *Piper methysticum* Forst. f. and *Piper wichmannii* C. DC. *Biochem. Syst. Ecol.* 24:775–782.

Lebot, V., and Levesque, J. 1996. Genetic control of kavalactone chemotypes in *Piper methysticum* cultivars. *Phytochemistry*, 43:397–403.

Lebot, V., Aradhya, M.K., and Manshardt, R.M. 1991. Geographic survey of genetic variation in kava, *Piper methysticum* Forst. f., and *Piper wichmannii* C. DC. *Pac. Sci.* 45(2):169–185.

Lebot, V., Merlin, M.D., and Lindstrom, L. 1992. *Kava: the Pacific drug*. Yale University Press, New Haven, CT. 255 pp.

Lebot, V., Johnston, E., Zheng, Q.Y., McKern, D., and McKenna, D.J. 1999. Morphological, phytochemical, and genetic variation in Hawaiian cultivars of 'awa (kava, *Piper methysticum*, Piperaceae). *Econ. Bot.* 53:407–418.

Loew, D., and Franz, G. 2003. Quality aspects of traditional and industrial kava extracts. *Phytomedicine*, 10:610–612.
Nveenimo, T., and Ngere, O. 1991. *Kava. A potential cash crop for the Papua New Guinea highlands*. Lowlands Agricultural Research Station, Keravat, Papua New Guinea. 12 pp.
Onwueme, I.C. 2000. *Piper methysticum* G. Forster. In *Plant resources of South-East Asia, vol. 16. Stimulants*. Edited by H.A.M. van der Vossen and M. Wessel. Backhuys, Leiden, the Netherlands. pp. 106–108.
Onwueme, I.C., and Papademetriou, M.K. 1997. *The kava crop and its potential*. Food and Agriculture Organization Regional Office for Asia and the Pacific, Bangkok, Thailand. 46 pp.
Pittler, M..H., and Ernst, E. 1999. Efficacy of kava extract for treating anxiety: systematic review and meta-analysis. *J. Clin. Psychopharmacol.* 20:84–89.
Pollock, N.J. 2000. Kava. In *The Cambridge world history of food*. Edited by K.F. Kiple and K.C. Ornelas. Cambridge University Press, Cambridge, U.K. pp. 664–671.
Prescott, J., and McCall, G. 1988. *Kava: use and abuse in Australia and the South Pacific. Proceedings from the Symposium on Kava, University of New South Wales, Nov. 11, 1988.* National Drug and Alcohol Research Centre, University of New South Wales, Kensington, N.S.W., Australia. 58 pp.
Simeoni, P., and Lebot, V. 2002. Identification of factors determining kavalactone content and chemotype in kava (*Piper methysticum* Forst. f.). *Biochem. Syst. Ecol.* 30:413–424.
Singh, Y.N. 1986. *Kava: a bibliography*. Pacific Information Centre, University of the South Pacific, Suva, Fiji. 111 pp.
Singh, Y.N. 2003. *Kava: the genus kava*. Taylor & Francis, New York. 192 pp.
Singh, Y.N. 2004. *Kava: from ethnology to pharmacology*. CRC Press, Boca Raton, FL.167 pp.
Singh, Y.N., and Blumenthal, M. 1997. *Kava: an overview*. HerbalGram, 39:33–55.
Steinmetz, E.F. 1960. Piper methysticum: *kava-kawa-yaqona, famous drug plant of the South Sea Islands*. E.F. Steinmetz, Amsterdam, the Netherlands. 46 pp.
Tyler, V.E. 1999. Herbs affecting the central nervous system. In *Perspectives on new crops and new uses*. Edited by J. Janick. ASHS Press, Alexandria, VA. pp. 442–449.
Whitton, P.A., Lau, A., Salisbury, A., Whitehouse, J., and Evans, C.S. 2003. Kava lactones and the kava-kava controversy. *Phytochemistry*, 64:673–679.
Xuan, T.D., Yuichi, O., Junko, C., Eiji, T., Hiroyuki, T., Mitsuhiro, M., et al. 2003. Kava root (*Piper methysticum* L.) as a potential natural herbicide and fungicide. *Crop Prot.* 22:873–881.

Specialty Cookbooks

N.B. Note the information above that discourages consumption of kava because of its toxic properties.

Collections of Recipes from Atkins and Others. 2009. *125 vegetarian recipes (vegan recipes, kava kava, weight loss)*. C.M. Harbin/Amazon Digital Services. Kindle book. 167 KB.

52 Khat

Family: Celastraceae (staff-tree family)

NAMES

Scientific Name: *Catha edulis* (Vahl) Forssk. ex Endl. (*Celastrus edulis* Vahl)

- The English word "khat" and the genus name *Catha* are based on the Arabic word for the plant and drug, *qatt*. This in turn has been said to have been derived from the Arabic *waqtalqaylula*, meaning the time of rest and relaxation.
- Pronunciation: cot; often mispronounced as "cat" in North America.
- Khat is also spelled chat, gat, ghat, kat, quat, qat, and tschat. It is also known as Abyssinian tea, African salad, African tea, Arabian tea, Bushman's tea, cafta, catha, flower of paradise, Somali tea, and by the foreign names kus-es-salahin, miraa (mirra) and tohai.
- The synthetic drug methcathinone, a white chunky powder known in the United States by the street name "cat," is very dissimilar to khat but could be confused with it because both are narcotics.
- *Edulis* in the scientific name *C. edulis* is Latin for edible.

PLANT PORTRAIT

Khat is an evergreen shrub or tree, mostly growing 2 to 7 m (6½–23 ft.) in height, occasionally to 15 m (50 ft.), rarely to 25 m (82 ft.). It is usually cut back in cultivation to form a shrub, making it easier to harvest the twigs and leaves, the economically important parts of the plant. The leaves and sometimes also the young twigs are consumed as a masticatory (a chewed substance) for a stimulant effect. The leaves are up to 10 cm (4 in.) long, resemble those of basil (*Ocimum basilicum* L.), and emit a strong, sweetish smell. They are crimson-brown and glossy when young, becoming yellow-green and leathery as they age. The young branches near the top of the plant produce the best material, but the middle and lower leaves and stems are also used. In wetter regions, khat shoots are tender and the soft green twigs are also consumed. In dryer regions, the twigs are woody and bitter, and only the leaves are used. The youngest leaves are usually quite weak in narcotic effect in comparison with the more mature ones. At the retail level, khat is sold in leafy bundles, wrapped in banana leaves or plastic to preserve freshness. Each bundle contains the quantity of leaves and stem tips that most consumers chew in 1 day.

The species is native to northeastern Africa and was probably domesticated in the Ethiopian highlands, although some have suggested an independent domestication in Yemen. Khat is cultivated primarily in countries in East Africa and the Arabian Peninsula. In Yemen, it has been estimated that more than half of the arable land is now used to cultivate khat, which has replaced other crops, particularly coffee. Khat is Ethiopia's second largest crop, after coffee. It is also grown in Kenya, Somaliland, Tanzania, Madagascar, and Uganda. Ethiopia and Kenya are principal exporters. The taste for khat is acquired and is almost completely restricted to people who have been culturally habituated to it. Khat is consumed in many Old World countries, including Kenya, Malawi, Uganda, Tanzania, Arabia, the Congo, Zimbabwe, Zambia, Madagascar, South Africa, Somalia, Ethiopia, and Yemen. The species is also grown as an ornamental in warm areas, such as Florida and California.

Khat is mostly used by males in friendly social situations, typically consumed during much of the evening and night. The harvesting, dealing, and sorting were traditionally almost entirely left to

FIGURE 52.1 Khat (*Catha edulis*). (From Richard, 1847.)

women. In more recent times, women have taken up holding their own sessions, separately from the men. In the Old World, depending on the country, 10% to more than 60% of women now consume khat. In Yemen, it has been reported that women's sessions typically have dancing and music and are often more lively than men's meetings. On a world basis, it has been estimated that 5 to 10 million people use khat, consuming about 5 million kg (11 million pounds) of leaf material daily.

Khat chewing is generally believed to have originated and spread from Yemen. A study of ancient Egyptian hieroglyphics has suggested that khat was used for ritual religious purposes. Khat has certainly been used at least since the thirteenth century as a recreational and religious drug in Eastern Africa, the Arabian Peninsula, and throughout the Middle East. It is traditionally consumed in many Muslim countries where alcohol is forbidden by religious law. However, until the late 1970s, khat chewing was largely a weekend habit of the rich. More recently, khat was taken up by a large proportion of Yemen's population (more than 90% of the male adults) as well as by many in Ethiopia, Somalia, and Djibouti.

During the 1980s, a flood of refugees from the Horn of Africa entered the United States, Canada, Australia, and various West European countries and brought with them their custom of chewing khat. As the leaves become old or dry, their stimulatory ability is reduced. Because only fresh khat has a strong effect, fresh material has often been transported into many countries by plane, especially for emigrants to Britain and North America habituated to its use. Because khat in leaf form starts to lose its potency after 48 hours, it is frequently shipped to North America on Thursdays, Fridays, and Saturdays for weekend use. In North America, the use of khat is most popular among immigrants from Yemen and the East African nations of Somalia and Ethiopia.

The stimulating ability of khat is due mostly to the alkaloid cathinone, which is closely related to and as potent as amphetamine. Cathine, a milder form of cathinone, is also present. Cathinone, the most potent active principle of khat, is chemically unstable; it has a half-life potency of only 1½ hours, and 2 to 3 days after harvest cathine is the only significant stimulant remaining, explaining why khat users prefer to chew only fresh leaves and shoots. (Deep freezing has been reported to prolong the potency of khat for months.) When chewed in moderation, khat alleviates fatigue (or increases energy), reduces appetite, and increases alertness. Approximately 30% of users have reported becoming sleepy rather than excited. Self-esteem and the ability to communicate are usually increased, typically taking the form of users embarking on long speeches in the belief that they are treating their listeners to pearls of knowledge.

Khat users can develop ill effects similar to those associated with amphetamines, including mental or emotional problems such as increased irritability, aggression, depression, anxiety, and paranoid delusions. Frequent use may be associated with loss of appetite, digestive difficulties, cancer of the mouth, heart disease, and loss of sexual potency. Where khat use is prevalent, it has been claimed that the semidrugged condition of many workers slows economic production. Khat is usually not considered a physically addictive drug, but in 1973, the World Health Organization listed it as a "dependence producing drug," that is, one for which a strong craving has developed.

Although the use of khat is predominantly by Muslims, less than half of 1% of the world's Muslims use the drug. Although Muslim religious leaders are divided on whether the use of khat contravene's the Koran's general injunction against the use of intoxicants, the World Islamic Conference for the Campaign Against Alcohol and Drugs, which met at Medina, Saudi Arabia in May, 1983, issued the following resolution:

> After reviewing reports submitted to the Conference on the health, psychological, ethical, behavioral, social and economic damages resulting from khat, the Conference judges khat to be a drug prohibited by religion and accordingly the Conference recommends to Islamic states to apply punishment of the basis of Islamic Shari'ah [canon law] against any person who plants this tree and markets or consumes khat.

Most Arab countries have outlawed khat. The Ethiopian and Somali governments have prohibited khat by law, but in practice this has not reduced consumption or production. Fresh khat is also illegal and classified as a narcotic in the United States (where it is a Schedule I narcotic, like marijuana and peyote), Canada, and much of Europe, although most of this prohibition is quite recent. Although khat often is still smuggled in because of the cost (often \$25–\$50 for an evening's supply for one person), many immigrants habituated to khat have turned to beer, which is cheaper and more available. As with other strongly desired drugs that society has made illegal, a Mafia-like control has developed over production and distribution.

A 1973 estimate suggested that more than 4 billion hours of work a year were lost in Yemen as a result of khat chewing and, whether true or not, khat has a reputation for encouraging laziness. In 1967, the Marxist government of South Yemen attempted to do away with khat because of this belief. With wide resistance to a total ban, the government imposed a heavy tax. Surprisingly, the people paid the tax and kept on chewing. By 1985, khat ranked first among taxes on agricultural products and second among all excise duties in increasing revenue. A 1992 report of the Yemen Times suggested that the value of khat to the Yemeni economy was twice the value from cultivation of all other crops. The World Bank has estimated that an astounding one-quarter of the national income of Yemen is spent on khat.

CULINARY PORTRAIT

Khat is rich in vitamin C, but like coffee and tea, is used primarily as a stimulant. A single mature leaf is said to produce an effect like a strong cup of coffee, but with accompanying euphoria. Several leaves are picked from a young fresh branch and chewed slowly. After the first few leaves are well chewed, they are kept in the side of the cheek, and additional leaves are added and chewed, all the while swallowing the juice. This process is continued until the cheek becomes uncomfortably full, after which the material is discarded. Generally, a mouthful of khat is retained 15 to 45 minutes to ensure the maximum extraction of alkaloids. Between 100 and 200 g (3½–7 oz.) of leaves are typically chewed more than 3 or 4 hours. Chewing khat leaves produces a strong aroma and generates intense thirst, and drinking water, coffee, or cola drinks is frequent. Stimulation can occur within 15 minutes of chewing. The peak "high" is reached in the third hour, and effects can remain up to

24 hours. Following the high, a slight depression or melancholy commonly sets in and remains for a few hours.

Khat is occasionally ground and taken as juice, a custom that is widespread in Ethiopia. Sometimes a paste to be chewed is made of khat leaves, water, and sugar or honey and may be flavored with herbs. Sometimes an infusion of fresh khat in water or milk is made and sweetened with honey. Khat is rarely added to tea or smoked in combination with tobacco. Khat is sometimes dried and consumed as a pleasantly stimulating tea, but the effect from dried material is much milder than chewing fresh herb. The laws against dried khat are usually less harsh than those against fresh khat, or there might not be any laws against dried khat, as in tea, depending on country.

Culinary Vocabulary

- A *chobe khat* is a long, notched stick traditionally used in Afghanistan by bakers to keep track of the number of breads baked for a household for a week, after which payment is expected. It has nothing to do with khat.

CURIOSITIES OF SCIENCE AND TECHNOLOGY

- Potency, flavor, and tenderness of the leaves of khat differ according to growing conditions, soil, and other variables, and experienced khat users, it has been claimed, can distinguish not only the geographical area of origin but even the field in which the product originated.
- In the Alamaya district of Harer, Ethiopia, a kind of khat called *kuda* results from damage by a leafhopper (a species of *Empoasca*). The insect injects saliva containing a toxin into the leaf tissue during feeding, resulting in a milky taste that is preferred by connoisseurs of this material, which is sold at a premium price.
- Yemeni Jews have used khat since at least the seventeenth century. In a poetic play by author Sholem bin Joseph al-Shibezi (1619–1686), there is a dialogue between coffee and khat. The play continues to be performed in Arabic by Yemeni jews in Israel. Almost all Jews who lived in Yemen migrated to Israel, and they brought their interest in cultivating khat with them. A khat-based frozen concentrate, "Pisgat," was manufactured and sold in Israel as a health food, its maker claiming that 2 tablespoons is equivalent to the effect achieved by several hours of chewing fresh khat.
- Khat is exceptionally high in ascorbic acid (vitamin C), and the practice of chewing it is believed to supply users with some of their daily requirements. Curiously, ascorbic acid has been reported to act as an antidote to amphetamine-like substances such as are present in khat.
- Agents at international airports usually do not need trained dogs to detect the odor of khat ("like old grass clippings") being smuggled in suitcases. The pungent smell is sufficient and leads to numerous seizures.
- KHAT has been used as an acronym for Kids & Hospitals Against Trauma.

KEY INFORMATION SOURCES

Al-Bekairi, A.M., Abulaban, F. S., Qureshi, S., and Shah, A.H. 1991. The toxicity of *Catha edulis*, khat, a review. *Fitoterapia*, 62:291–300.

Al-Meshal, I.A., Ageel, A.M., Parmar, N.S., and Tariq, M. 1985. *Catha edulis* (khat): use, abuse and current status of scientific knowledge. *Fitoterapia*, 3:131–152.

Al-Meshal, I.A., Ageel, A.M , Tariq, M., and Parmar, N.S. 1983. Gastric anti-ulcer activity of khat, *Catha edulis*. *Res. Commun. Subst. Abuse*, 4:143–150.

Al-Motarreb, A., Baker, K., and Broadley, K.J. 2002. Khat: pharmacological and medical aspects and its social use in Yemen. *Phytother. Res.* 16:403–413.

Balint, G.A., Ghebrekidan, H., and Balint, E.E. 1991. *Catha edulis*, an international socio-medical problem with considerable pharmacological implications. *East Afr. Med. J.* 68:555–561.

Brooke, C. 2000. Khat. In *The Cambridge world history of food*. Edited by K.F. Kiple and K.C. Ornelas. Cambridge University Press, Cambridge, U.K. pp. 671–684.

Crombie, L., Crombie, W.M.L., and Whiting, D.A. 1990. Alkaloids of khat (*Catha edulis*). *Alkaloids*, 39:139–164.

Dagne, D. (Ed.). 1984. *Proceedings international symposium on khat (*Catha edulis*), Addis Ababa, Ethiopia, Dec 15, 1984: chemical and ethnopharmacological aspects*. Natural Products Research Network for Eastern and Central Africa, Addis Ababa, Ethiopia. 89 pp.

Elhag, H.M., and Mossa, J.S. 1996. *Catha edulis* (khat): in vitro culture and the production of cathinone and other secondary metabolites. *Medicinal and Aromatic Plants*, 9:76–86.

Elhag, H., Mossa, J.S., and El-Olemy, M.M.. 1999. Antimicrobial and cytotoxic activity of the extracts of khat callus cultures. In *Perspectives on new crops and new uses*. Edited by J. Janick. SHS Press, Alexandria, VA. pp. 463–466.

Gebissa, E. 2004. *Leaf of Allah: khat & agricultural transformation in Harerge, Ethiopia 1875–1991*. Ohio University Press, Athens, OH. 210 pp.

Geisshusler, S., and Brenneisen, R. 1987. The content of psychoactive phenylpropyl and phenylpentenyl khatamines in *Catha edulis* Forsk. of different origin. *J. Ethno. Pharmacol.* 19:269–277.

Halbach, H. 1972. Medical aspects of the chewing of khat leaves. *World Health Organ. Bull.* 47:21–29.

Kalix, P. 1990. Pharmacological properties of the stimulant khat. *Pharmacol. Ther.* 48:397–416.

Kalix, P. 1991. The pharmacology of psychoactive alkaloids from *Ephedra* and *Catha*. *J. Ethno. Pharmacol.* 32:201–208.

Krikorian, A.D. 1984. Kat and its use: an historical perspective. *J. Ethnopharmacol.* 12:115–178.

Krikorian, A.D. 1985. Growth mode and leaf arrangement in *Catha edulis* (kat). *Econ. Bot.* 39:514–521.

Maitai, C.K. 1996. Catha edulis *(miraa): a detailed review focusing on its chemistry, health implication, economic, legal, social, cultural, religious, moral aspects and its cultivation*. National Council for Science and Technology, Nairobi, Kenya. 52 pp.

Nelhans, B. 1974. *Khat, a stimulating drug in Eastern Africa and the Arabian peninsula*. University of Gothenburg, Göteborg, Sweden. 68 pp.

Pantelis, C., Hindler, C.G., and Taylor, J.C. 1989. Use and abuse of khat (*Catha edulis*): a review of the distribution, pharmacology, side effects, and a description of psychosis attributed to khat chewing. *Psychol. Med.* 19:657–668.

Paris, R., and Moyse, H. 1948. Abyssinian tea (*Catha edulis* Forsk., Celastraceae). A study of some samples of varying geographical origin. *Bull. Narc.* 10:29–34.

Randrianame, M., Szendrei, K., Tongue, A., and Shahandeh, B. (Eds.). 1983. *The health and socio-economic aspects of khat use. International Conference on Khat, Antananarivo, Madagascar, Jan. 17–21, 1983.* International Council on Alcohol and Addictions, Lausanne, Switzerland. 251 pp.

Revri, R. 1983. Catha edulis *Forsk.: geographical dispersal, botanical, ecological and agronomical aspects with special reference to Yemen Arab Republic*. Institut fur Pflanzenbau und Tierhygiene in den Tropen und Subtropen der Universität, Göttingen, Germany. 157 pp.

Sheikh, M.H. (Ed.). 1984. *Studies on khat: its social, economic and health effects*. Scientific Research Bureau of the SRSP, Central Committee and National Committee for the Eradication of Khat, Mogadishu, Somalia. 148 pp.

Small, E. 2004. Narcotic plants as sources of medicinals, nutraceuticals, and functional foods. In *Proceedings of the International Symposium on the Development of Medicinal Plants, Hualien District Agricultural Research and Extension Station, Haulien, Taiwan, August 24–25, 2004*. Edited by F.-F. Hou, H.-S. Lin, M.-H. Chou, and T.-W. Chang. pp. 11–67.

Tariq, M., Ageel, A.M., Parmar, N.S., and Al-Meshal, I.A. 1984. The pharmacological investigation of the Saudi Arabia variant of *Catha edulis*. *Fitoterapia*, 55:195–200.

United Nations. 1980. Special issue devoted to *Catha edulis* (khat). *Bull. Narc.* 32(3):1–99.

Specialty Cookbooks

Note: Khat is illegal in most Western nations.

Bladholm, L. 2000. *The Indian grocery store demystified*. Renaissance Books, Los Angeles, CA. 265 pp. [Cites "khat mix snack."]

53 Kiwi

Family: Actinidiaceae (kiwi family)

NAMES

Scientific Names: *Actinidia* Species

- Kiwi—*A. deliciosa* (A. Chev.) C.F. Liang & A.R. Ferguson (*A. chinensis* Planch. var. *deliciosa* (A. Chev.) A. Chev.)
- Kiwi—*A. chinensis* Planch.
  Note: Most literature treats kiwifruit as one species, *A. deliciosa*. However, Liang and Ferguson (1986) assigned some cultivated varieties marketed today to a second species, *A. chinensis* Planch. According to Ferguson (2008), the cultivar 'Hayward', supplying 80% of the world market, belongs to *A. deliciosa*, whereas approximately 8% of marketed fruit (currently from China and New Zealand) is from *A. chinensis*. The two species have different wild geographical ranges in China, but in the marketplace the distinctions are of interest mainly to fruit specialists.
- Grape kiwi—*A. arguta* (Siebold & Zucc.) Planch. ex Miq.

- The name "kiwi" (pronounced KEE wee) was adopted in New Zealand about 1953. The fuzziness of the kiwi fruit, reminiscent of the fuzziness of the flightless kiwi bird, New Zealand's national bird, is said to have been the inspiration for the plant name (although it has also been claimed that the fruit was named for its resemblance to the egg of the bird). The name kiwi has proven to be a much more attractive word for marketing purposes than the older English name "Chinese gooseberry" (first recorded in New Zealand in 1925; said to have been based on the resemblance of the seeds to gooseberry seeds). The name Chinese gooseberry is a translation of a Chinese phrase that has been used for the kiwi and also for the carambola (see Chapter 18). Possibly even less attractive are the alternative Chinese names *hou tao*, "monkey peach," *yang t'ao*, "goat peach," and *orr mei*, "goose fruit." The French initially had another rather questionable name for kiwi fruit when they first became popular: *souris végetales*, "vegetable mice."
- The kiwi bird's name (and hence the name of the kiwi fruit) comes from the Maori language and imitates the cry of the male kiwi bird during the mating season.
- The popular cultivar 'Hayward' is named for Hayward Wright, who selected it in Avondale, New Zealand around 1924.
- The grape kiwi is also known as the baby kiwi, bower actinidia, bower vine, hardy kiwi, Siberian kiwi, strawberry peach, tara, and tara vine.
- The genus name *Actinidia* is based on the Greek *aktis*, a ray, an allusion to the styles (part of the female flowers) that radiate like the spokes of a wheel.
- *Deliciosa* in the scientific name *A. deliciosa* means delicious.
- *Chinensis* in the scientific name *A. chinensis* means "of China."
- *Arguta* in the scientific name *A. arguta* means sharply toothed or serrated, which is descriptive of the leaves.

PLANT PORTRAIT

The kiwi is a large, woody, perennial deciduous vine growing up to 10 m (33 ft.) in length, with male and female flowers on separate plants. The plants look similar to a grapevine but have a rather

FIGURE 53.1 Kiwi (*Actinidia deliciosa*). (From Curtis, 1913, vol. 139, plate 8538.) Fruit at bottom right is from Hooker (1836–1900, vol. 16, plate 1593).

FIGURE 53.2 Grape kiwi (*Actinidia arguta*). (From Curtis, 1895, vol. 121, plate 7497.) Fruit cluster at bottom left is from Bailey (1900–1902).

treelike appearance. A mature vine can produce more than 90 kg (200 lb.) of fruit. The kiwi is one of the very few temperate fruit crops to have been domesticated in the twentieth century. In the early 1900s, it was still just a wild plant in its native region of southwestern China, but by 1970, it had been developed into a new fruit crop in New Zealand. The fruit has become so strongly associated with that country that "Kiwi" became a (generally affectionate) informal term for a New Zealander, much used by New Zealanders themselves. Te Puke, in the center of New Zealand's kiwi producing region, calls itself "The Kiwifruit Capital of the World." The main producers of kiwis are New Zealand, China, Chile, and Italy, with significant crops also in France, Spain, the United States, Australia, Japan, Korea, and other countries. 'Hayward' is the main cultivar grown to date. 'Hayward' fruits are fuzzy, brown, oblong in shape, averaging 6.6 cm (2⅝ in.) long, 5.4 cm (2⅛ in.) wide, with flesh that is tart-sweet and tastes like a combination of citrus, melon, and strawberry. When fruit are cut crosswise, the emerald-green flesh has a ring of small 'black' edible seeds. In addition to their beautiful appearance, excellent flavor and texture, kiwis are high in vitamin C and retain 90% of this vitamin after 6 months of storage.

There are many other species of *Actinidia* that show great promise for development as new fruits. Best of these is the grape kiwi, which has been esteemed in Asia for many years. Small commercial plantings of it have been established in Canada, France, Germany, Italy, New Zealand, and the United States. The fruit of this species is approximately the size of a large grape—approximately 3 cm (1¼ in.) long, 2 cm (3/4 in.) in diameter. Approximately 15 grape kiwi fruits weigh as much as one regular kiwi fruit. The grape kiwi fruit is completely hairless, so the skin is consumed with the fruit. The fruits are greenish-yellow in color and acidic until ripe. When ripe, they are very sweet and juicy and the flavor is considered to be better than that of the familiar kiwifruit. This type of kiwi can be easily grown in areas that are too cold for regular kiwifruit cultivation and is claimed to survive winter temperatures of –32 °C (–25 °F). The grape kiwi has been touted as having 20 times more vitamin C than citrus fruit and capable of producing up to 45 kg (100 lb.) of fruit per vine. 'Issai' is a self-fruitful variety, so male plants are not necessary. However, for most grape kiwi plants sold to homeowners, a male plant needs to be present for the female to set fruit (generally one female kiwi will suffice for a family). It should be kept in mind when adding this vine to a garden that the plants often take several years to mature, need a substantial trellis for support, and usually do not bear fruit until they are 5 to 9 years old.

CULINARY PORTRAIT

Kiwi is used predominantly as a fresh dessert fruit. The brown, fuzzy skin is edible but is usually removed. The fruit can be sliced in half and the flesh scooped out with a spoon, but a kiwi is often peeled and eaten directly or sliced and served with cream. Slices of fresh kiwi improve the taste of cereals, ice cream, and sherbet. The flesh is sweet, juicy, and slightly acidic. Fruit salads are enhanced both in flavor and attractiveness by the brilliant green color of diced or sliced kiwi segments. Kiwi complements meat, poultry, and fish and is used in a wide range of prepared desserts, including cakes and pies. Preserved whole fruit or fruit segments in glass or cans is available, and wine, liquor, jam, and marmalade made from kiwi are also occasionally sold. Sauces (including sweet-and-sour types) and soups are also prepared using kiwi, but generally the fruit should not be cooked for a long time to preserve the color and delicate flavor. The fruits become sweeter as they ripen, at which stage they yield to light pressure of the fingers. For many people, kiwi is somewhat laxative. The fruits are high in oxalates, which can cause problems for susceptible people. A small percentage of people are allergic to kiwifruit, some getting blisters from placing their lips on the fuzzy skin of the kiwi.

Kiwi fruits contain enzymes (particularly actinidin) that can act as a meat tenderizer. Meat can be improved in texture and flavor by mashing a few slices of kiwi with a fork and letting it sit in the meat for at least 20 minutes before cooking. The enzymes of unripe kiwis that are peeled will even tend to tenderize the fruit itself! When used in fruit salads, it may be wise to add kiwi at the last minute to prevent softening of the other fruits. The enzymes prevent gelatin from setting and makes milk products go off (yogurt and ice cream are not affected), unless the kiwi is lightly cooked first. If uncooked kiwi is used in milk products, these should be served soon to prevent deterioration.

Culinary Vocabulary

- In New Zealand, kiwi is used as the topping in one of the country's pièces de résistance, "Pavlova" (named after Russian ballerina Anna Pavlova (1881–1931), the most celebrated dancer of her time), a delightful meringue shell filled with whipped cream and fruit (usually kiwi, passion fruit, and pineapple). The Pavlova is also a popular dessert in Australia.

CURIOSITIES OF SCIENCE AND TECHNOLOGY

- Large supermarket kiwi fruits contain more than 1000 seeds, but they are so small they are not detectable when eating them.
- In China, both paper (from the bark and leaves) and pencils (from the bark) have been made from kiwi.
- Before the Chinese gooseberry became the kiwi, in the early 1950s New Zealand exporters found that the fruit was being refused entry to the United States because the "Chinese" in the name suggested it was from China, and the United States at the time was enforcing importation bans on selected fruit from Communist countries.
- The green color of the flesh of most marketed fruits of kiwi is due to the presence of chlorophyll. Chlorophyll, the main chemical responsible for photosynthesis in green plants, is normally not present in significant amounts in the flesh of mature fruits.
- A substance irresistible to cats and other felines including lions and tigers has been obtained from several species of *Actinidia*, particularly *A. polygama* (Siebold & Zucc.) Maxim., which is grown as an ornamental called silvervine and cat powder.
- The several species of kiwi birds are all flightless natives of New Zealand, with furlike feathers. They burrow in the ground and are largely nocturnal.
- The kiwi is the only bird in the world that has its nostrils at the end of its beak and one of the few to have a highly developed sense of smell. A kiwi literally sniffs out its food and also uses its bill also to smell danger.
- Male kiwi birds are wimps. The female kiwi is bigger than the male and occasionally kicks her smaller partner when warding off his unwanted advances. During courting, the male follows the female about, grunting at her, but if she gets mad he runs away. In the North Island, the male brown kiwi (a different species from the common kiwi) does most of the egg incubating. Not surprisingly, the female bird (as well as the male) has long whiskers. Males do not have roving eyes and stay faithful to the females for life (sometimes as long as 30 years).
- Pregnancy is an ordeal for female kiwi birds because they lay a huge egg in proportion to their body size. A pregnant female's belly bulges so much that it touches the ground. The egg makes up 15% to 20% of her body weight, and she has to walk with her legs wide apart to accommodate it. A female will sometimes stand up to her belly in cold water when she is heavy with egg—to soothe the inflamed stretched skin and to temporarily relieve her of the weight she is carrying. Just before it is laid, the egg is so huge it almost fills the kiwi's whole body, leaving little room for food in her stomach, and she has to fast for the 2 or 3 days that precede laying.

KEY INFORMATION SOURCES

Astridge, S.J. 1975. Cultivars of Chinese gooseberry *(Actinidia chinensis)* in New Zealand. *Econ. Bot.* 29:357–360.

Bailey, F.L., and Topping, E. 1951. *Chinese gooseberries, their culture and uses*. Bulletin No. 349. New Zealand Department of Agriculture, Wellington, New Zealand. 23 pp.

Brown, S.H., and Beutel, J.A. 1996. *The kiwifruit: a home gardener's guide*. Revised edition. University of California, Division of Agriculture and National Resources, Oakland, CA. 14 pp.

Costa, G. 1999. Kiwifruit orchard management: new developments. *Acta Hortic.* 498:111–119.

Ferguson, A.R. 1984. Kiwifruit: a botanical review. *Hortic. Rev.* 6:1–64.

Ferguson, A.R. 1990. Botanical nomenclature: *Actinidia chinensis*, *Actinidia deliciosa*, and *Actinidia setosa*. In *Kiwifruit: science and management*. Edited by I.J. Warrington and G.C. Weston. Ray Richards Publishers and New Zealand Society for Horticultural Science. Auckland, New Zealand. pp. 36–57.

Ferguson, A.R. 1990. Kiwifruit (*Actinidia*). *Acta Hortic.* 290:603–653.

Ferguson, A.R. 1990. Kiwifruit (*Actinidia*). In *Genetic resources of temperate fruit and nut crops*. Edited by J.N. Moore and J.R. Ballington, Jr. International Society for Horticultural Science, Wageningen, the Netherlands. pp. 601–653.

Ferguson, A.R. 1990. Kiwifruit management. In *Small fruit crop management*. Edited by G.J. Galletta and D.G. Himelrick. Prentice Hall, Englewood Cliffs, NJ. pp. 472–503.

Ferguson, A.R. 1995. Kiwifruit. In *Evolution of crop plants*. 2nd ed. Edited by J. Smartt and N.W. Simmonds. Longman Scientific & Technical, Burnt Mill, Harlow, Essex, U.K. pp. 1–4.

Ferguson, A.R. 1999. New temperate fruits: *Actinidia chinensis* and *Actinidia deliciosa*. In *Perspectives on new crops and new uses*. Edited by J. Janick. ASHS Press, Alexandria, VA. pp. 342–347.

Ferguson, A.R. 2008. Actinidiaceae: *Actinidia deliciosa*: kiwifruit. In *The encyclopedia of fruit & nuts*. Edited by J. Janick and R.E. Paull. CABI, Wallingford, Oxfordshire, U.K. pp. 1–7.

Ferguson, A.R., and Huang, H.W. 2007. Genetic resources of kiwifruit: domestication and breeding. *Hortic. Rev.* 33:1–121.

Ferguson, A.R., Seal, A.G., and Davison, R.M. 1990. Cultivar improvement, genetics and breeding of kiwifruit. *Acta Hortic.* 282:335–347.

Ferguson, A.R., Seal, A.G., McNeilage, M.A., Fraser, L.G., Harvey, C.F., and Beatson, R.A. 1996. Kiwifruit. In *Fruit breeding: Vol. 2. Vines and small fruits*. Edited by J. Janick and J.N. Moore. Wiley, New York. pp. 371–417.

Fletcher, W.A. 1971. *Growing Chinese gooseberries*. A.R. Shearer, Government Printer, Wellington, New Zealand. 37 pp.

Freye, H.B. 991. Life-threatening anaphylaxis to kiwifruit and prevalence of kiwifruit hypersensitivity in the United States. *Am. J. Rhinol.* 5(2):51–55.

Giulivo, C., and Ferguson, A.R. (Eds.). 1990. *International Symposium on Kiwifruit, Padova, Italy, October 14–17, 1987*. International Society for Horticultural Science, Wageningen, the Netherlands. 446 pp.

Hasey, J.K. 1994. *Kiwifruit growing and handling*. University of California, Division of Agriculture and Natural Resources, Oakland, CA. 134 pp.

Hopping, M.E. 1986. Kiwifruit. In *CRC handbook of fruit set and development*. Edited by S.P. Monselise. CRC Press, Boca Raton, FL. pp. 217–232.

Huang, H. (Ed.). 2003. *Proceedings of the Fifth International Symposium on Kiwifruit, Wuhan, P.R. China, September 15–20, 2002*. International Society for Horticultural Science, Leuven, Belgium. 548 pp.

Hung, H., and Ferguson, A.R. 2001. Kiwifruit in China. *N. Z. J. Crop Hortic. Sci.* 29:1–14.

Johnson, D.M., Hanson, C.A., and Thomson, P.H. 1988. *Kiwifruit handbook*. Bonsall Publications, Bonsall, CA. 106 pp.

Li, H.L. 1952. A taxonomic review of the genus *Actinidia*. *J. Arnold Arbor.* 33:1–61.

Liang, C.F., and Ferguson, A.R. 1986. The botanical nomenclature of the kiwifruit and related taxa. *N. Z. J. Bot.* 24:183–184.

Morton, J. 1987. Kiwifruit. In *Fruits of warm climates*. Creative Resource Systems, Winterville, NC. pp. 293–300.

Organisation for Economic Co-operation and Development. 1992. *Kiwis = Kiwifruit*. O.E.C.D., Washington, DC. 69 pp.

Perera, C.O., Young, H., and Beever, D.J. 1998. Kiwifruit. In *Tropical and subtropical fruits*. Edited by P.E. Shaw, H.T. Chan, Jr., and S. Nagy. Agscience, Auburndale, FL. pp. 336–385.

Retamales, J. (Ed.). 1999. *Fourth International Symposium on Kiwifruit, Santiago, Chile, January 11–14, 1999*. International Society for Horticultural Science, Leiden, the Netherlands. 370 pp.

Sale, P.R. 1985. *Kiwifruit culture*. 2nd ed. Government Printer, Wellington, New Zealand. 96 pp.

Sfakiotakis, E., and Porlingis, J. (Eds.). 1997. *Third International Symposium on Kiwifruit, Thessaloniki, Greece, September 19–22, 1995*. International Society for Horticultural Science, Leiden, the Netherlands. 2 vols.

Strik, B., 2004. *Growing kiwifruit*. Revised edition. Oregon State University, Corvallis, OR. 23 pp.

Thompson, A.K. 1982. *The storage and handling of kiwifruit*. Tropical Products Institute, London. 11 pp.

Warrington, I., and Hewett, E.W. (Eds.). 1992. *The Second International Symposium on Kiwifruit, February 18–21, 1991. Massey University, Palmerston North, New Zealand.* International Society for Horticultural Science, Wageningen, the Netherlands. 2 vols.

Warrington, I.M., and Weston, G.C. (Eds.). 1990. *Kiwifruit: science and management*. New Zealand Society for Horticultural Science, Auckland, New Zealand. 576 pp.

Specialty Cookbooks

Armstrong, A. 1968. *Kiwi cook book*. Seven Seas Pub., Wellington, New Zealand. 59 pp.

Beutel, M. 1975. *Kiwifruit recipes*. Kiwi Growers of California, Gridley, CA. 64 pp.

Beutel, M. 1982. *Kiwifruit cook book*. Revised edition. Kiwi Growers of California, Carmichael, CA. 63 pp.

Brennan, B., McGeough, B., and Smith, E. 1989. *Kiwifruit collection: cookbook*. Island Directories, Sidney, BC. 93 pp.

Bilton, J. 1990. *New Zealand kiwifruit cookbook*. Revised edition. I. Holt, Auckland, New Zealand. 72 pp.

Caplan, K. 2001. *The purple kiwi cookbook*. Frieda's, Los Angeles, CA. 127 pp.

R & R Publications. 1993. *101 kiwifruit ideas*. R & R Pub., Epping, N.S.W., Australia. 112 pp.

54 Kumquat

Family: Rutaceae (rue family)

NAMES

Scientific Names: *Fortunella* Species

- Nagami (or oval) kumquat—*F. margarita* (Lour.) Swingle (oblong fruits, approximately 3 × 4 cm (1¼ × 1½ in.), with acid juice).
- Marumi (or round) kumquat—*F. japonica* (Thunb.) Swingle (round fruits, approximately 3 cm (1¼ in.) in diameter, with acid juice).
- Meiwa kumquat—*F. crassifolia* Swingle (a hybrid between the previous species; round fruits, approximately 4 cm (1½ in.) in diameter, with sweet rind and almost juiceless pulp).

- The name kumquat (also spelled cumquat and comquot) arose from the Chinese (Cantonese) name *cam kwat* (*kin kü*), meaning "golden orange."
- The genus *Fortunella* is named after Robert Fortune (1812–1880), Scottish horticulturist and plant explorer for the Royal Horticultural Society, who introduced the kumquat to Europe in the mid-nineteenth century.
- In the scientific names of the three species of kumquat, *margarita*, *japonica*, and *crassifolia* are Latin for pearly, Japanese, and thick leaved, respectively.

PLANT PORTRAIT

Kumquat trees are small, subtropical, evergreen trees or shrubs 2.4 to 4.5 m (8–15 ft.) tall, which are somewhat hardier than citrus species and will hybridize readily with them. There are several commercial hybrids between citrus species and the kumquat, including the limequat, orangequat, lemonquat, and citrangequat (a hybrid of the orange, kumquat, and the trifoliate orange, *Poncirus trifoliata* (L.) Raf.). The calmondin is a cross of kumquat and the mandarin orange. Kumquat fruits are citrus-like, deeply colored, round, or distinctly oval, 2.5 to 5 cm (1–2 in.) long and frequently about half as wide. The pulp is scant, in three to six segments, not very juicy, acid to subacid, and contains small, pointed seeds or sometimes none. Kumquats are thought to be native to China, where they have probably been cultivated for thousands of years. Candied kumquats are quite popular during the Chinese Spring Festival as the Chinese name means gold, symbolizing a prosperous start for the New Year. The fruit first appeared in the West in the seventeenth century, and the trees have been grown in Europe and North America since the mid-nineteenth century mainly as ornamentals. Commercial cultivation is carried out in China, Japan, Taiwan, southern Europe, the United States, Puerto Rico, Guatemala, Surinam, Colombia Brazil, Australia, Israel, and South Africa. In the United States, kumquats are grown mainly in California, Florida, and Texas, mostly in home gardens, but a few trees are often found in citrus orchards, especially of growers catering to the gift package trade. Clusters of kumquats are frequently placed in gift packages for their ornamental effect, and often kumquats are sold still attached to a branch with its decorative green leaves.

CULINARY PORTRAIT

Kumquats, which have a pungent, orange-ginger, bittersweet but pleasant taste, are eaten as a fresh fruit, generally as a dessert, but also in salads. The golden-yellow to reddish-orange rind of kumquats is thin, tender, and edible, so the whole fruit may be eaten out of hand. It has been suggested that before eating a kumquat, the fruit be washed well, then rolled between the fingers to release the essential oils in the rind, improving the aroma. The rind of the kumquat is sweet, whereas the rather dry flesh is very tart, and accordingly kumquats should not be peeled. Very ripe fruit can be sliced and added raw to salads or used as a garnish. Served as a dessert, kumquats are sometimes rolled in sugar. Halved kumquats cooked in syrup and presented with ice cream make an excellent dessert. However, the kumquat is usually cooked, candied, or pickled whole or used in preserves, jellies, marmalades, chutneys, relishes, stuffings, cakes, pies, muffins, sweet-and-sour sauces, cocktails, and many other confections. Kumquat goes well with fish, poultry, and lamb. In China, whole kumquats are sometimes preserved in sugar syrup and served as desserts. When purchasing kumquats, it is best to choose fruit that is plump and firm, not shriveled. The fruit is more perishable than oranges because of the thin skin but can be stored for up to 2 weeks in a refrigerator.

Culinary Vocabulary

- What is "citrus?" Botanically, the "citrus family" is the Rutaceae, which includes many plants that are not "citrus species." Kumquats have been called "the little gems of the citrus family" and "the smallest fruits of the citrus family," which are accurate statements insofar as the family affiliation is concerned. However, it is frequently claimed that kumquats are citrus species, which is not quite correct. Strictly, species that produce citrus fruits are in the genus *Citrus*, and this includes oranges, grapefruits, lemons, limes, citrons, and many other lesser known fruit. As noted earlier, there are many hybrids of the kumquat genus *Fortunella* and *Citrus* species, and whether such hybrids are "citrus species" is a semantic question because one or more parents are citrus species and one or more parents are not. Some botanists once considered kumquats to be *Citrus* species, and some think that kumquats should be reunited in the same genus with *Citrus*, so it is possible that one day kumquats will accurately be described as true citrus.
- A "mojito" is a traditional, famous Cuban highball made of rum (or vodka), sugar (or cane juice), lime, sparkling, water, ice, and mint. "Kumquat mojitos," which replace the limes with kumquats, have become quite popular.

CURIOSITIES OF SCIENCE AND TECHNOLOGY

- Kumquats are unusual fruits in that the peel is sweeter than the pulp.
- By grafting, it is possible to produce a citrus tree that produces lemons, limes, oranges, tangerines, kumquats, and grapefruits growing on different branches.
- According to *The Book of Lists* (1990s edition), the 10 ugliest-sounding words in English, excluding indecent words, are fructify, kumquat, quahog, crepuscular, kakkak, gargoyle, cacophonous, aasvogel, brobdingnagian, and jukebox.
- Pangrams are sentences containing all the letters of the alphabet. Examples:
  - J. Hoefler cabled: "puzzling over waxy kumquats."
  - Sixty-five wildly panting fruitflies gazed hungrily at the juicy bouncing kumquats.
- For some applications (e.g., in preparing particular dishes in high-end restaurants), kumquats, which have two to five seeds, are deseeded by hand, a tedious process. 'Nordmann Seedless' is a seedless mutation of the Nagami variety, which was discovered by George Otto Nordmann in 1965 in DeLand, Florida. Other seed mutations are known, and in time breeders are likely to produce seedless varieties that will replace the currently marketed types.

FIGURE 54.1 Marumi (round) kumquat (*Fortunella japonica*). (From Siebold and Zuccarini, 1835–1870, plate 15.)

FIGURE 54.2 Nagami (oval) kumquat (*Fortunella margarita*). (From Gartenflora, 1882, vol. 31.)

KEY INFORMATION SOURCES

Badr, F.H. 2002. Chemical and polysaccharides composition of kumquat fruits. *Zagazig J. Agric. Res.* 28:413–423.

Bonian, L. 1986. The main commercial strains of Ning Buo kumquat (*Fortunella* Swingle). *Acta Agric. Univ. Zhejiangensis*, 12:203–211 (in Chinese, English summary).

Cappello, C., Bovalo, F., Tonarelli de Rossin, G., Retamar, J.A., and Talenti, C.J. 1982. Researches on Argentine kumquats of the *Fortunella margarita* genus: 2. The juice. *Essenze Derivati Agrumari*, 52(1):73–78 (in Italian).

Cappello, C., Calvarano, I., Tonarelli de Rossin, G., and Retamar, J.A. 1982. Researches on Argentine kumquats of the *Fortunella margarita* genus: 1. Essential oil. *Essenze Derivati Agrumari*, 52:67–72 (in Italian).

Chalutz, E., Lomaniec, E., and Waks, J. 1989. Physiological and pathological observations on the post-harvest behaviour of kumquat fruit. *Trop. Sci.* 29:199–206.

Dabhade, R.S., and Khedkar, D.M. 1981. Kumquat: good for processing. Physical properties and chemical composition. *Indian Food Pack.* 36(4):29–31.

Fantz, P.R. 1988. Nomenclature of the Meiwa and Changshou kumquats, intrageneric hybrids of *Fortunella. HortScience*, 23:249–250.

Fujikawa, K., Kono, A., Okurano, H., Matsushima, K., Shinohara, K., and Tokito, T. 1997. Effects of temperature on flowering and fruitset of kumquat (*Fortunella crassifolia*) growing in a greenhouse. *Bull. Kagoshima Fruit Tree Exp. Stn.* 1:21–25 (in Japanese, English summary).

Hall, D.J. 1986. Use of postharvest treatments for reducing shipping decay in kumquats. *Proc. Annu. Meet. Fla. State Hortic. Soc.* 99:108–112.

Hiep, N.T., and Jansen, P.C.M. 1991. *Fortunella* Swingle. In *Plant Resources of South-East Asia. Vol. 2. Edible fruits and nuts*. Edited by E.W.M. Verheij and R.E. Coronel. Pudoc, Leiden, the Netherlands. pp. 169–171.

Ives, V. 1986. Kumquats. *Pacific Hortic.* 47(4):28–30.

Ives, V. 1987. Kumquats: easy to grow, tough to pick. *California Grower*, 11(6):20–21.

Koyasako, A., and Bernhard, R.A. 1983. Volatile constituents of the essential oil of kumquat. *J. Food Sci.* 48:1807–1812.

Morton, J. 1987. Kumquat. In *Fruits of warm climates*. Creative Resource Systems, Winterville, NC. pp. 182–185.

Porras-Castillo, I., Gonzalez-Benavente, A., and Garcia-Lidon, A. 1994. Ornamental uses of round kumquat (*Fortunella* sp. L.). *Agricola Vergel*, 154:526–533 (in Spanish).

Rahman, M.M., and Nito, N. 1994. Phylogenetic relationships in the kumquat (*Fortunella*) as revealed by isozyme analysis. *Sci. Hortic.* 57:17–28.

Raman, H., and Dhillon, B.S. 1994. Isozyme polymorphism and phylogenetic relationship among species of *Citrus*, *Fortunella* and *Poncirus*. *Crop Improvement*, 21:37–43.

Swingle, W.T. 1915. A new genus, *Fortunella*, comprising four species of kumquat oranges. *J. Wash. Acad. Sci.* 5(5):166–176.

Umano, K., Hagi, Y., Tamura, T., Shoji, A., and Shibamoto, T. 1994. Identification of volatile compounds isolated from round kumquat (*Fortunella japonica* Swingle). *J. Agric. Food Chem.* 42:1888–1890.

Ye, Y.M. 1985. The status of *Fortunella* genetic resources in China. *Fruit Var. J.* 39(2):17–20.

Specialty Cookbooks

Greey, M. 1999. *Get fresh! How to cook a kumquat and other useful tips for more than 100 fruits and vegetables*. Macmillan, Toronto, ON. 260 pp.

55 Lemongrass

Family: Poaceae (Gramineae; grass family)

NAMES

Scientific Name: *Cymbopogon citratus* (DC. ex Nees) Stapf (*Andropogon citratus* DC. ex Nees, *A. nardus* var. *ceriferus* Hack., *A. roxburghii* Nees)

- The name "lemongrass" accurately reflects the lemon taste of this member of the grass family. The spellings lemon grass and lemongrass are both frequently encountered.
- The English name "lemongrass" is the same in French, Spanish, and Italian, and almost the same (lemongras) in German.
- Lemongrass is also known as citronella root, citron grass, fever grass, sere grass, and West Indian lemongrass.
- The name "fever grass" arose in southern India and Ceylon because of the alleged curative properties of infusions made with it. In Southeast Asia and western Africa, fever (lemon) grass is used to combat malaria. Natives of Guatemala have used lemongrass in folk medicine for fever, grippe, flatulence, and low blood pressure. Lemongrass is one of the most used plants in Brazilian folk medicine, the leaves used in a tea for the treatment of nervous and gastrointestinal disturbances as well as for fever.
- The genus name *Cymbopogon* is based on the Greek *kumbe*, boat + *pogon*, beard. The spathe (a sort of leaf covering the flowering branch of grasses) is boat shaped, whereas the many awns (bristles among grass flowers) give the appearance of a beard.
- *Citratus* in the scientific name *C. citratus* is Latin for lemony.

PLANT PORTRAIT

Lemongrass is a perennial, tall, tropical grass growing in dense clumps or tufts 70 to 150 cm (28–59 in.) or more in height. The plants seldom flower under cultivation. The slightly enlarged, bulbous, juicy base is the most palatable portion. All parts are strongly lemon scented. Lemongrass likely originated in southern India, Ceylon, or perhaps Indonesia. It is often available in Asian markets, including those of North America, particularly Thai, Vietnamese, Cambodian, and Laotian. Lemongrass is widely grown as a culinary herb but is raised mostly as a source of essential oil. It is cultivated in tropical to warm temperate areas, including southern and eastern Asia, western Africa, the West Indies, Central America, Brazil, and Australia. It is grown on a vast scale in India and on a very minor scale in the southern parts of California and Florida.

In addition to lemongrass, several other species of *Cymbopogon* have minor culinary uses. These include the following.

- *Cymbopogon flexuosus* (Nees ex Steud.) J.F. Watson is known as East Indian lemongrass, Malabar (lemon) grass, and Cochin grass. This is native to India and cultivated selections are grown in India, Southeast Asia (including China), equatorial Africa, Guatemala, and Paraguay. Steam distillation produces an essential oil called Indian lemongrass oil, Cochin lemongrass oil, and Malabar oil. This is one of the world's major essential oils. Food manufacturers add East Indian lemongrass oil to beverages, ice cream, candy, baked goods, gelatin desserts, and chewing gum.

- *Cymbopogon martinii* (Roxb.) J.F. Watson is native to southern Asia. Variety *martinii* is known as palmarosa grass, East Indian geranium, Turkish geranium, and rosha grass. It is cultivated in India, East Asia, Angola, and Brazil for its essential oil called palmarosa oil, rusa oil, East Indian geranium oil, and Turkish geranium oil. The oil contains the chemical geraniol and is used to flavor ice cream, gelatin desserts, chewing gum, and bakery products.
- *Cymbopogon nardus* (L.) Rendle is called Ceylon citronella grass, Ceylon lemongrass, citronella, mana grass, lenabatu grass, nard grass, and nardus grass. It is native to North Africa and India. Two types of plant are recognized: Java and Ceylon. Culinary uses are limited, but the leaves are sometimes used to flavor soups, cooked fish, and curry and to prepare tea, and the essential oil components are used as flavoring. This species is a major source of citronella, used in candles to ward off mosquitoes and other insects.

CULINARY PORTRAIT

The taste of lemon seems universally attractive to various cultures, and different plants have furnished the aroma to particular societies. Throughout Southeast Asia, where lemons were not available historically, lemongrass was and remains the dominant source of lemon flavoring.

In Southeast Asia, the base and lower shoots of the plant are eaten. This herb is used to impart its distinctly lemon-like aroma and flavor to food. The most tender part is the swollen base of the leafy shoots, generally the bottom 6 cm (2½ in.), and this should be cut off and peeled. The finely chopped basal portions of the stems can be eaten raw or cooked with fish, sauces, marinades, stir-fries, and curries. Lemongrass is particularly compatible with garlic, shallots, ginger, coconut, and both (hot) chili peppers and (mild) green peppers. The taste combines well with cilantro in flavoring fish, shellfish, chicken, and pork. Raw pieces are added to salads, particularly those containing cold meats or fish. The flavor is sufficiently versatile to be used in many parts of a meal from soup to tea. Teas are usually steeped from the dried herb, which imparts less flavor than the fresh leaves. The fresh bulbous base and leaves make the best tea. The tough, straw-like, outer leaves may be tied in a loop and cooked with rice, fruit, and vegetable dishes, sauces, soups, and stews to provide flavor. However, these leaves need to be removed before serving, remaining fibrous after cooking, and in general the tougher parts of lemongrass should not be chewed or swallowed. This does not apply to the young, tender, basal portions, which can be used as described above. The leaves can be placed in the cavity of fish or chicken and steamed for a subtle flavor. The leaves do not become bitter with long cooking. Leaves and stems should be pounded lightly with a heavy object to help release the volatile oils. The tender hearts of the young shoots are served as a vegetable with rice. Fresh lemongrass can be stored for up to 2 months in a refrigerator. For most culinary applications, dried lemongrass should be soaked in water for 30 minutes before using, and the dried form should be used only when fresh herb is unavailable.

Sliced and dried or powdered lemongrass is sold in many ethnic markets. As with numerous herbs that are predominantly used fresh, the taste is not as good, and indeed dried material is unsuitable for many recipes. One teaspoon of powder is approximately equal to one stalk of fresh lemongrass.

Lemongrass oil is widely used as a flavoring and aroma ingredient in food and drinks. Lemongrass oil is used in alcoholic and nonalcoholic drinks, frozen dairy desserts, candy, baked goods, gelatins, puddings, meat, and meat products.

A rust fungus (*Puccinia nakanishikii*) is sometimes found on lemongrass, producing orange-colored fungal bodies in the crown of the plant where the leaf blades converge. Any material with fungus growing on it should be discarded.

Culinary Vocabulary

- "Trahanas" are a fermented combination of milk and cereal. Lemongrass is substituted for yogurt in the preparation of "nistisemos trahanas" (fasting trahanas), eaten during religious holidays in Greece and Turkey, where the regular type of trahanas made with animal milk would be sacrilegious.

CURIOSITIES OF SCIENCE AND TECHNOLOGY

- Lemongrass is one of many plants capable of producing hydrocyanic acid (prussic acid), and plants with this capability are said to be cyanogenetic (cyanogenic). This gas is so toxic that inhalation of a small amount may produce death in a few minutes. Fortunately, the gas is not produced normally by such plants but is chemically combined in the form of a cyanogenetic glucoside. However, this glucoside is often associated with an enzyme capable of decomposing it and freeing the hydrocyanic acid, and this can occur in a bruised or wilting plant, or much more dangerously in the alimentary tract of an animal (even without the enzyme). Cases of poisoning of stock animals from hydrocyanic plants are occasionally reported. In some plants, this potential for poisoning animals protects the plants against being consumed. There does not seem to be any literature suggesting that the cyanogenetic potential of lemongrass poses a hazard for people. To those alarmed by this poisonous potential of lemongrass, note that several commonly consumed plants, including rice, oats, sugar cane, corn, lima bean, cherries, and apple, are also cyanogenetic.
- The rhizome (underground stem) of lemongrass is chewed by Latin Americans until it is frayed, then used as a toothbrush.
- Cocaine is produced from the coca plant (*Erythroxylum coca* Lam.), which is treated in this book (see Chapter 28). Lemongrass is one of the crops being promoted as a substitute for coca in the Huallaga Valley, the main coca-producing region of Peru.
- In Australia, lemongrass is cultivated mainly for adding aroma to dishwashing liquids.

FIGURE 55.1 Lemongrass (*Cymbopogon citratus*), by B. Flahey.

KEY INFORMATION SOURCES

Abegaz, B., Yohannes, P.G., and Dieter, R.K. 1983. Constituents of the essential oil of Ethiopian *Cymbopogon citratus* Stapf. *J. Nat. Prod.* 46:424–426.

Anon. 1952. Lemon grass paper. *Chemurg. Dig.* 11(10):16.

Beech, D.F. 1977. Growth and oil production of lemongrass (*Cymbopogon citratus*) in the Ord Irrigation Area, Western Australia. *Aust. J. Exp. Agric. Anim. Husb.* 17:301–307.

Blake, S.T. 1974. *Revision of the genera* Cymbopogon *and* Schizachyrium *(Gramineae) in Australia.* Queensland Herbarium Contributions, 17. Department of Primary Industries, Brisbane, Australia. 70 pp.

Carlini, E.A., Contar, J.D., Silva-Filho, A.R., Da Silverira-Filho, N.G., Frochtengarten, M.L., and Bueno, O.F. 1986. Pharmacology of lemongrass (*Cymbopogon citratus* Stapf.). I. Effects of teas prepared from the leaves on laboratory animals. *J. Ethnopharmacol.* 17:37–64.

Elson, C.E., Underbakke, G.L., Hanson, P., Shrago, E., Wainberg, R. H., and Qureshi, A.A. 1989. Impact of lemongrass oil, an essential oil, on serum cholesterol. *Lipids*, 24:677–679.

Guha, S.R.D., Pant, R., and Karira, B.G. 1973. Chemical pulps for writing and printing papers from *Cymbopogon citratus* (DC.) Stapf (West Indian lemon grass). *Indian For.* 99:717–720.

Gupta, B.K., and Jain, N. 1979. Contribution to the bibliography of *Cymbopogon* grasses. *Indian J. For.* 2:71–96, 99–117.

Jones, M.A., and Arrillaga, N.G. 1950. *The production of lemon-grass oil.* Federal Experiment Station in Puerto Rico, Mayagüez, Puerto Rico. 41 pp.

Kulkarni, R. N., and Ramesh, S. 1992. Development of lemongrass clones with high oil content through population improvement. *J. Essent. Oil Res.* 4:181–186.

Kumar, S. 2000. Cymbopogon: *the aromatic grass: monograph.* Central Institute of Medicinal and Aromatic Plants, Lucknow, India. 380 pp.

Leite, J. R., Seabra, M. L., Maluf, E., Assolant, K., Suchecki, D., Tufik, S., et al. 1986. Pharmacology of lemongrass (*Cymbopogon citratus* Stapf.). III. Assessment of eventual toxic, hypnotic and anxiolytic effects on humans. *J. Ethnopharmacol.* 17:75–83.

Lorenzetti, B.B., Souza, G.E., Sarti, S.J., Filho, D.S., and Ferreira, S.H. 1991. Myrcene mimics the peripheral analgesic activity of lemongrass tea. *J. Ethnopharmacol.* 34:43–48.

Prasad, L.K., and Mukherji, S.R. 1980. Effect of nitrogen, phosphorus, and potassium on lemon grass—*Cymbopogon citratus*, fertilizers. *Indian J. Agron.* 25:42–44.

Rajendrudu, G., and Rama Das, V.S. 1983. Interspecific differences in the constituents of essential oils of *Cymbopogon*. *Proc. Indian Acad. Sci.* 92:331–334.

Rao, B.L., and Sobti, S.N. 1987. Breeding of a high oil yielding lemongrass for flavour industry. *Indian Perfum.* 31(1):32.

Robbins, S.R.J. 1982. *Selected markets for the essential oils of lemongrass, citronella and eucalyptus.* Tropical Products Institute, London. 90 pp.

Singh, N., Luthra, R., and Sangwan, R.S. 1989. Effect of leaf position and age on the essential oil quantity and quality in lemongrass (*Cymbopogon flexuosus*). *Planta Med.* 55:254–256.

Soenarko, S. 1977. The genus *Cymbopogon* Sprengel (Gramineae). *Reinwardtia*, 9(Part 3):225–375.

Stapf, O. 1906. The oil grasses of India and Ceylon. (*Cymbopogon*, *Vetiveria* and *Andropogon* spp.). *Kew Bull.* 8:313–364.

Valencia, J., and Myers, C. 1991. Lemon grass. Crop Sheet SMC-037. In *Specialty and minor crops handbook.* Edited by C. Myers. The Small Farm Center, Division of Agriculture and Natural Resources, University of California, Oakland, CA. 2 pp.

SPECIALTY COOKBOOKS

Hutton, W., Solomon, C., Kawana, M., Ong, C., and Ong, M. 2004. *Green mangoes and lemon grass: Southeast Asia's best recipes from Bangkok to Bali.* Periplus, North Clarendon, VT. 223 pp.

Kotylo, J.M. 2002. *The everything Thai cookbook: from pad thai to lemongrass chicken skewers, 300 tasty, tempting Thai dishes you can make at home.* Adams Media, Avon, MA. 287 pp.

Poladitmontri, P., and Lew, J. 2002. *Thai cuisine: lemon grass cookbook.* Joie Inc., New York. 88 pp.

Tholstrup, M., Read, M., and Cazals, J. 2001. *Lemongrass and lime: new Vietnamese cooking.* Ten Speed Press, Berkeley, CA. 191 pp.

56 Lemon Verbena

Family: Verbenaceae (vervain family)

NAMES

Scientific Name: *Aloysia citriodora* Paláu (*A. triphylla* (L'Hér.) Britton, *Lippia triphylla* (L'Hér.) Kuntze, *L. citriodora* Kunth, *Verbena triphylla* L'Hér.)

- "Verbena" in the name "lemon verbena" reflects the fact that the species was once placed in the genus *Verbena* (it does look like vervains, i.e., species of *Verbena*). The "lemon" in the name points out its extremely lemony smell.
- Lemon verbena is also known as cidron, herb Louisa, lemon-scented verbena, lemon vervain, limonetto, real vervain, sweet verbena, sweet-scented verbena, verbena, and verveine citronella.
- In 1864, the Emperor Maximilian (1832–1867), emperor of Mexico during 1864–1867, and his wife, the empress Carlotta, planted lemon verbena in the royal gardens of Montezuma's palace in Mexico, naming it "yerba Louisa" after Carlotta's mother, the queen of Belgium. *Yerba Louisa* should not be confused with *hierba luisa*, a Spanish name used in South America to designate lemongrass (see Chapter 55).
- The genus name *Aloysia* commemorates Maria Luisa (often "Maria Louisa;" 1751–1819), wife of King Charles IV (1748–1819) of Spain. Her complete influence over him led to disastrous policies, and he was forced to abdicate in 1808.
- *Citriodora* in the scientific name *A. citriodora* is Latin for citrus-odored or lemon-scented.

PLANT PORTRAIT

Lemon verbena is a deciduous shrub which can be semievergreen in warm climates. The species is native to Argentina, Chile, Peru, and Uruguay. It grows as tall as 5 m (16 ft.) in climates with mild winters but is usually 1 to 1.5 m (3–5 ft.) high in northern gardens or 0.5 to 1 m (20–39 in.) when grown in pots. The leaves grow in whorls of three or four; they are dotted with oil glands on the underside and pleasantly aromatic, with an intense lemon scent (so much so that the plant has been described as "lemon perfume"). Filmy spikes of small white and purple flowers are produced in summer and autumn. Lemon verbena was taken to Spain by the Spanish conquistadors in the 1700s and became naturalized in parts of Europe. It has been cultivated in the Mediterranean region since the end of the eighteenth century. About the same time, it was introduced into North America by a New England sea captain who brought it from Chile. The species is found as an escape in South Africa and parts of Asia where it is sometimes cultivated for its essential oil.

Lemon verbena is of minor economic importance, although commercial herb tea mixtures often contain it. Culinary use of this herb is most developed in Latin America. The species is widely cultivated in home gardens in Mediterranean Europe, Africa, India, and other mild climates. France consumes several hundred tons annually, most of it imported. Lemon verbena is grown commercially in North Africa (Algeria and Tunisia), South Africa, and parts of Asia, and occasionally in East Africa and Chile. The species is cultivated as a garden ornamental in warm climates, such as in the southern United States. An ornamental variegated form is available.

The essential oil has been used to make flavoring for soft drinks and liqueurs and has been used in perfume, toilet water, and eau de cologne. The oil has also been used to make a "health tea" in Italy, France, and Spain. The oil is expensive, however, and has largely been replaced by other, cheaper essential oils with similar properties.

CULINARY PORTRAIT

Lemon verbena is grown mainly for its foliage, which is employed for culinary purposes. The leaves can be used fresh or dried to impart a sweet lemon-lime flavor and aroma. Many think that the foliage of lemon verbena has the most delicious fragrance of all the lemon-scented herbs. The sharp, long-lasting fragrance has also been described as "more lemony than lemons themselves," so that when it is substituted for other lemon herbs, only half the usual amount needs to be used. The leaves are especially desirable for making an excellent tea, used alone or in combination with other herbs, including common black or Chinese tea. Whole leaves are used to add lemon flavor to cold drinks, fruit cups, sauces, dressings, marinades, salads, jellies, soups, omelets, vegetables, fish, chicken, and meat dishes. Leaves are placed at the bottom of a cake pan to flavor a cake while baking and removed before serving. Occasionally, the young leaves are eaten like spinach. Fresh leaves are tough and should be minced finely if they are to be consumed rather than used simply to impart their flavor. Dried leaves can easily be crumbled finely.

The simplest way of getting material of this outstanding culinary herb is to grow a pot of it. In cold climates, the potted plant can be moved indoors. Although the plant typically reacts to being moved by dropping all of its leaves, it will generally recover (the soil should be kept on the dry side until the foliage begins to grow again). Leaves can be harvested at any time but are at their most flavorful just as the flowers are coming into bloom. The mature plants are preferable because the lemon fragrance becomes stronger with age. Harvested leaves can be dried in a dark, warm, airy place. The dried leaves retain their flavor and scent for years if stored in a cool dark place in an airtight jar. Leaves can also be frozen in ice cubes trays to make flavorful decorative ice cubes (a technique that can be used with other lemony herbs). The leaves can be fairly tough and when used in cooking they should be removed before serving. The main problem in using lemon verbena is simply that the aroma can be overpowering, so the herb should be consumed in moderation.

Culinary Vocabulary

- "Yerba Louisa" tea, made with lemon verbena and sweetened with honey, is a popular drink in parts of Europe.

CURIOSITIES OF SCIENCE AND TECHNOLOGY

- There are numerous nonfood medicinal uses for lemon verbena in Middle American countries. The plant is used as an antidote for bites of poisonous animals in Argentina. Other ailments treated with lemon verbena include asthma, colds, fever, flatulence, headache, migraines, nausea, rabies, stomach cramps, and vertigo. Tea made from lemon verbena is fairly widely used in South America to relieve acidity, indigestion, and flatulence, and as a stimulant for lethargy and depression.
- The dried leaves of lemon verbena can be added to a potpourri for scent. Although long lasting, the scent often fades in approximately 6 months and should be replenished.
- To freshen a house, lemon verbena leaves can be placed in the vacuum cleaner so that the scent will be blown out.

FIGURE 56.1 Lemon verbena (*Aloysia citriodora*). (From Curtis, 1797 vol. 11, plate 367.)

KEY INFORMATION SOURCES

Armada, J., and Barra, A. 1992. On *Aloysia* Palau (Verbenaceae). *Taxon*, 41:8–90.

Bellakhdar, J., Idrissi, A.I., Canigueral, S., Iglesias, J., and Vila, R. 1994. Composition of lemon verbena (*Aloysia triphylla* (L'Herit.) Britton) oil of Moroccan origin. *J. Essent. Oil Res*. 6:523–526.

Carnat, A., Carnat, A.P., Fraisse, D., and Lamaison, J.L. 1999. The aromatic and polyphenolic composition of lemon verbena tea. *Fitoterapia*, 70:44–49.

Casadoro, G., and Rascio, N. 1982. Glands of *Lippia triphylla* (L'Her.) O. Kuntze. *Cytobios*, 35(138):85–94.

Crabas, N., Marongiu, B., Piras, A., Pivetta, T., and Porcedda, S. 2003. Extraction, separation and isolation of volatiles and dyes from *Calendula officinalis* L. and *Aloysia triphylla* (L'Her.) Britton by supercritical CO_2. *J. Essent. Oil Res*. 15:272–277.

El-Hamidi, A., Ahmed, S.S., and Shaarawy, F. 1983. *Lippia citriodora* grown in Egypt. A new crop under development. *Acta Hortic*. 132:31–33.

Fitz-Gerald, M.E. 1942. Lemon verbena. *Herbarist*, 8:36–37.

Gonzalez, D.B.C. 1999. *Essential oil content and organoleptic evaluation of different clones of lemon verbena (*Aloysia triphylla* (L'Herit.) Britt.)*. Thesis. Talca University, Talca, Chile. 47 pp (in Spanish, English summary).

Kim, N.S., and Lee, D.S. 2004. Headspace solid-phase microextraction for characterization of fragrances of lemon verbena (*Aloysia triphylla*) by gas chromatography-mass spectrometry. *J. Sep. Sci*. 27:96–100.

Lamaison, J.L., Petitjean-Freytet, C., and Carnat, A. 1993. Le verbascoside, composé phénolique majeur des feuilles de frêne (*Fraxinus excelsior*) et de verveine (*Aloysia triphylla*). *Plant. Med. Phytother*. 26(3):225–233.

Montes, M., Valenzuela, L., Wilkomirsky, T., and Arrivé, M. 1973. Composition of essential oil of *Aloysia triphylla*. *Planta Med*. 23:119–124 (in French).

Nakamura, T., Okuyama, E., Tsukada, A., Yamazaki, M., Satake, M., Nishibe, S., et al. 1997. Acteoside as the analgesic principle of cedron (*Lippia triphylla*) a Peruvian medicinal plant. *Chem. Pharm. Bull*. 45:499–504.

Ozek, T., Kirimer, N., Baser, K.H.C., and Tumen, G. 1996. Composition of the essential oil of *Aloysia triphylla* (L'Herit.) Britton grown in Turkey. *J. Essent. Oil Res*. 8:581–583.

Perrira, C.G., and Meireles, M.A.A. 2007. Evaluation of global yield, composition, antioxidant activity and cost of manufacturing of extracts from lemon verbena (*Aloysia triphylla* (L'Herit.) Britton) and mango (*Mangifera indica* L.) leaves. *J. Food Process Eng.* 30:150–173.

Silva Brant, R. da, Brasil Pereira Pinto, J.E., Vilela Bertolucci, S.K., Silva, A. da, and Brant Alburquerque, C.J. 2009. Lemon verbena's *Aloysia triphylla* (L'Herit.) Britton (Verbenaceae) essential oil content in different harvest periods and post-harvesting process. *Cien. Agrotecnol.* 33(Special Issue):2065–2068 (in Portuguese, English summary).

Skaltsa, H., and Shammas, G. 1988. Flavonoids from *Lippia citriodora. Planta Med.* 54:465.

Stefanini, M.B., Ming, L.C., Uesugi, G., and Figueiredo, R.O. de. 2004. Influence of IBA and boric acid on rooting of stem-cuttings of *Aloysia triphylla* (L'Herit.) Britton. *Acta Hortic.* 629:329–332.

Van Brunt, E.R. 1972/1973. Lemon-verbena. *Plants Gard.* 28(4):56–57.

Vogel, H., Silva, M.L., Razmilic, I. 1999. Seasonal fluctuation of essential oil content in lemon verbena (*Aloysia triphylla*). *Acta Hortic.* 500:75–79.

Wannmacher, L., Fuchs, F.D., Paoli, C.L., Fillman, H.S., Gianlupi, A., Lubianca, N.J.F., et al. 1990. Plants employed in the treatment of anxiety and insomnia. II. Effect of infusions of *Aloysia triphylla* on experimental anxiety in normal volunteers. *Fitoterapia*, 61:449–453.

Specialty Cookbooks

Brown, K., and Pollack, J. 1999. *Herbal teas: 101 nourishing blends for daily health & vitality.* Storey Publishing, Pownal, VT. 153 pp.

Chiarello, M., and Petzke, K. 2006. *Michael Chiarello's flavored oils and vinegars: 100 recipes for cooking with infused oils and vinegars.* Chronicle Books, San Francisco, CA. 192 pp.

Perry, S. 1995. *The book of herbal teas: a guide to gathering, brewing, and drinking.* Chronicle Books, San Francisco, CA. 120 pp.

Platt, E.S. 2002. *Lemon herbs: how to grow and use 18 great herbs.* Stackpole Books, Mechanicsburg, PA. 132 pp.

Saffi, T. 2001. *Healthy teas: green, black, herbal, fruit.* Periplus, Boston, MA. 112 pp.

Schlosser, K.K. 2007. *The Herb Society of America's essential guide to growing and cooking with herbs.* Louisiana State University Press, Baton Rouge, LA. 349 pp.

Zak, V. 2009. *20,000 secrets of tea: the most effective ways to benefit from nature's healing herbs.* Dell, New York. 272 pp.

57 Loofah

Family: Cucurbitaceae (gourd family)

NAMES

Scientific Names: *Luffa* Species

- Angled loofah—*L. acutangula* (L.) Roxb.
- Smooth loofah—*L. aegyptiaca* Mill. (*L. cylindrica* M. Roem.)

- The English name "loofah" and the genus name *Luffa* are based on the Arabic *louff*, the name of *L. aegyptiaca*.
- Angled loofah is also known as angled luffa, California okra, Chinese okra, ribbed gourd, ribbed loofah, ridged gourd, silky gourd, silk squash, strainer vine, and vegetable gourd. In China, it is called sing-kwa, cee-kwa, or cee gwa. The name cee-kwa is often encountered also as the name in English.
- The young immature fruit of angled luffa is called Chinese okra because of the okra-like shape and external appearance of the tender ridges, combined with the popularity of the vegetable in China.
- Smooth loofah is commonly known as vegetable sponge and sometimes as sponge gourd because it is the main source of bathroom loofah sponges, as noted in the next section.
- Smooth loofah is also called dishcloth gourd, dishrag gourd, loofah, and smooth luffa. In China it is called sze-kwa.
- *Aegyptiaca* in the scientific name *L. aegyptiaca* means Egyptian, the supposed homeland of the species when it was described.
- *Acutangula* in the scientific name *L. acutangula* is Latin for sharp angled, in reference to the surface of the gourds.

PLANT PORTRAIT

Angled loofah is an annual, herbaceous vine with lobed cucumber-like leaves, which climbs by tendrils. When crushed, the leaves give off a rank odor. The species has been cultivated since ancient times in hot, humid, tropical areas of Asia. Cultivated forms probably originated in India and today are raised in tropical Asia and parts of the Caribbean. Angled loofah is grown less than smooth loofah, as described in the next section, although its fruits are superior in taste. In parts of the tropics, individual plants produce 15 to 20 fruits. The fruits are cylindrical, generally growing to 30 to 60 cm (1–2 ft.) in length. Approximately 10 sharp, elevated ridges run the length of the pods.

Smooth loofah is a vigorous, climbing, annual plant, similar in appearance to angled loofah. The fruits are shaped like cucumbers but larger, 30 to 60 cm (1–2 ft.) in length and 10 to 12.5 cm (4–5 in.) thick. The exterior is green, sometimes mottled, and smooth with longitudinal lines. The site of domestication is unknown, although India has been suggested. Domesticated smooth loofah was known in China in AD 600 and is now found in many tropical regions.

The main area of cultivation of loofahs for food purposes is Asia, but these interesting gourds are occasionally grown and marketed as specialty vegetables in North America. They are also raised in home gardens. These vigorous plants are noted for their zucchini-like vegetable, and for sponges, as noted in the following.

The interiors of angled loofah and smooth loofah are cucumber-like when immature but develop into a network of fibers surrounding a large number of flat, blackish seeds. Mature fruits can be peeled and the fibrous interior used as a sponge or scrub brush after separation of the skin, flesh, and seeds. Smooth loofah is superior to angled loofah for such purposes and indeed is mainly grown to obtain loofah sponges, material for filters, and stuffing for pillows, saddles, and slippers. Normally, mature fruits are left on the vine to dry, after which the thin outer skin is removed. The fruit is then soaked in running water for several days, and the softer tissue is removed. After additional soaking and drying, the seeds are shaken out and the remains of the gourd are bleached chemically or by the sun then marketed. Loofah sponges are widely sold in the cosmetic and bath sections of stores and are popular because of their gently exfoliating effect on the skin. They are natural and biodegradable and therefore attractive to the environmentally conscientious consumer.

CULINARY PORTRAIT

Although some British colonists in India found the taste inferior to that of vegetables familiar in England, Indians consider angled loofah to be one of the best of indigenous vegetables and use it extensively in curries. The fruits of angled loofah are harvested when immature (10–15 cm or 4–6 in. in length), at which time they are dark green and have tender ridges. Normally, the ridges are cut off the angled fruits with a vegetable peeler to prepare them for eating. The skin can be left on young fruits (the rind of older fruits is bitter, and at maturity is almost papery). The fruits are cooked in vegetable dishes, stews, and soups. The young fruits are cooked like okra or summer squash and can also be sliced and used in salads in place of cucumbers. Young leaves, blossoms, and seeds are also edible and can be simmered until tender then added to vegetable dishes. In China and Japan, the fruit is sliced, and the dried slices are kept for later use.

Smooth loofah is used for culinary purposes like angled loofah. The immature fruits, leaves, flower buds, and seeds can be cooked and eaten in vegetable dishes. The Burmese consider it to be a delicious vegetable. The unripe fruit has been used to make pickles by the Arabians.

The seeds of various other species of *Luffa* have been used as a drastic purgative (to induce vomiting). They are toxic and should not be eaten by humans, pets, or livestock. By contrast, the seeds of pumpkins, another species in the gourd family, are quite edible.

Culinary Vocabulary

- In India, where smooth loofah is often cultivated for food, it is called "ghia."

CURIOSITIES OF SCIENCE AND TECHNOLOGY

- *Luffa* gourds are among the many, varied types that are found, mainly in the gourd family, the Cucurbitaceae. Gourds were so important to Haitians that in the early 1800s they were made the national currency of Haiti. To this day, the standard Haitian coin is called a "gourde."
- Both male and female flowers occur on the loofah species described here, but male flowers greatly outnumber the female flowers.
- Smooth loofah vines are often found growing up telephone poles in Honduras, the gourds frequently hanging down from the telephone wires.
- Before the Second World War, loofah sponges were used as filters in U.S. navy steam engines.
- Loofah sponges are made into a remarkable variety of consumer goods, including dolls, hats, slippers, and toys.

FIGURE 57.1 Angled loofah (*Luffa acutangula*). (From Gartenflora, 1899, vol. 48.)

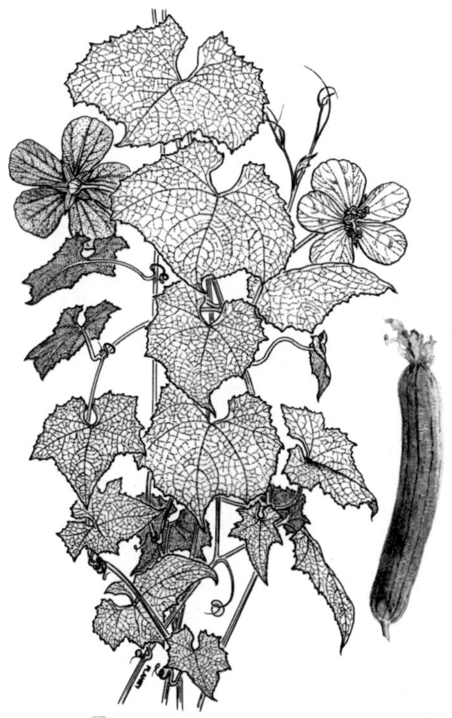

FIGURE 57.2 Smooth loofah (*Luffa aegyptiaca*), by B. Flahey. Fruit by B. Brookes.

KEY INFORMATION SOURCES

Cox, L. 1982. The amazing luffa plant. *Small Farmer's J.* 6(2):54–55.

Davis, J.M. 1994. Luffa sponge gourd production practices for temperate climates. *HortScience*, 29:263–266.

Davis, J.M., and DeCourley, C.D. 1993. Luffa sponge gourds: a potential crop for small farms. In *New crops*. Edited by J. Janick and J.E. Simon. Wiley, New York. pp. 560–561.

Dillon, B.A. 1979. How to harvest and prepare luffa gourd sponges. *Flower Garden* (Southern Edition), 23(9):8–9.

Dutt, B., and Roy, R.P. 1971. Cytogenetic investigations in Cucurbitaceae. 1. Interspecific hybridizations in *Luffa*. *Genetica*, 42:139–156.

Dutt, B., and Roy, R.P. 1990. Cytogenetics of the Old World species of *Luffa*. In *Biology and utilization of the Cucurbitaceae*. Edited by D.M. Bates, R.W. Robinson, and C. Jeffrey. Cornell University Press, Syracuse, NY. pp. 134–140.

Fisher, B. 1974. Sponges on a vine: the luffa plant makes a handy household or bathing sponge—and it's delicious eating when harvested young. *Org. Gard. Farming*, 21(6):58–59.

Hamid, S., Salma, Sabir, A.W., and Khan, S.A. 1985. Cultivation conditions and physico-chemical properties of *Luffa acutangula* var. *acutangula* seed oil. *Pak. J. Sci. Ind. Res.* 28(2):119–122.

Heiser, C.B., Jr. 1979. *The gourd book*. University of Oklahoma Press, Norman, OK. 248 pp.

Heiser, C.B., Jr. 1989. Domestication of Cucurbitaceae: *Cucurbita* and *Lagenaria*. In *Foraging and farming: the evolution of plant exploitation*. Edited by D.R. Harris and G.C. Hillman. Unwin Hyman, London. pp. 471–480.

Heiser, C.B., and Schilling, E.E. 1988. Phylogeny and distribution of *Luffa* (Cucurbitaceae). *Biotropica*, 20:185–191.

Heiser, C.B., and Schilling, E.E. 1990. The genus *Luffa*: a problem in phytogeography. In *Biology and utilization of the Cucurbitaceae*. Edited by D.M. Bates, R.W. Robinson, and C. Jeffrey. Cornell University Press, Ithaca, N.Y. pp. 120–133.

Huyskens, S., Mendlinger, S., Benzioni, A., and Ventura, M. 1993. Optimization of agrotechniques in the cultivation of *Luffa acutangula*. *J. Hortic. Sci.* 68:989–994.

Iyer, K.R.P.H., Subramanian, R.B., and Inamdar, J.A. 1991. Structure, ontogeny and biology of nectaries in *Luffa acutangula*. *Indian Bot. Contactor*, 8(3):137–142.

Jansen, G.J., Gildemacher, B.H., and Phuphathanaphong, L. 1993. *Luffa* P. Miller. In *Plant resources of South-East Asia. Vol. 8. Vegetables*. Edited by J.E. Siemonsma and K. Piluek. Pudoc Scientific Publishers, Wageningen, the Netherlands. pp. 194–197.

Johnson, W.B. 1984. Luffa—*vegetable sponge or dishcloth gourd: how to grow and prepare it for use*. FS Cooperative Extension Service, New Brunswick, NJ. 2 pp.

Joshi, S.S., and Shrivastava, R.K. 1978. Amino acid composition of *Luffa cylindrica* and *Luffa acutangula* seeds. *J. Inst. Chem.* 50(2):73–74.

Kalloo, G. 1993. Loofah, Luffa spp. In *Genetic improvement of vegetable crops*. Edited by G. Kalloo and B.O. Bergh. Pergamon Press, New York. pp. 265–266.

Kamel, B.S., and Blackman, B. 1982. Nutritional and oil characteristics of the seeds of angled luffa *Luffa acutangula*. *Food Chem.* 9:277–282.

Martin, F.W. 1979. *Sponge and bottle gourds,* Luffa *and* Lagenaria. Agricultural Research (Southern Region), Science and Education Administration, U.S. Department of Agriculture, Mayagüez Institute of Tropical Agriculture (Puerto Rico), New Orleans, LA. 19 pp.

Okusanya, O.T. 1983. The mineral nutrition of *Luffa aegyptiaca*. *Can. J. Bot.* 61:2124–2132.

Paris, H.S., and Maynard, D.N. 2008. *Luffa cylindrica*, sponge gourd; *Luffa acutangula*, angled luffa. In *The encyclopedia of fruit & nuts*. Edited by J. Janick and R.E. Paull. CABI, Wallingford, Oxfordshire, U.K. pp. 302–305.

Porterfield, W.M., Jr. 1955. Loofah, the sponge gourd. *Econ. Bot.* 9:211–223.

Sahni, G.P., Singh, R.K., and Saha, B.C. 1987. Genotypic and phenotypic variability in ridge gourd. *Indian J. Agric. Sci.* 57:666–668.

Schilling, E.E., and Heiser, C.B. 1981. Flavonoids and the systematics of *Luffa*. Biochem. *Syst. Ecol.* 9:263–265.

Singh, B.P. 1991. Interspecific hybridization in between New and Old-World species of *Luffa* and its phylogenetic implication. *Cytologia*, 56:359–365.

Singh, R.D., and Singh, J.P. 1998. Improvement and cultivation: *Lagenaria* and *Luffa*. In *Cucurbits*. Edited by N.M. Nayar and T.A. More. Science publishers, Enfield, NH. pp. 199–203.

Trivedi, R.N., and Roy, R.P. 1976. Interspecific hybridization and amphidiploid studies in the genus *Luffa*. *Genet. Iber.* (Spain), 28:83–106.

Specialty Cookbooks

Chang, N. 2001. *My student's favorite Chinese recipes*. Travelling Gourmet, Wappingers Falls, NY. 192 pp.
Gelle, G.G. 2008. *Filipino cuisine: recipes from the islands*. Museum of New Mexico Press, Santa Fe, NM. 280 pp.
Jelani, R. 2003. *Healthy Asian vegetarian dishes: your guide to the exciting world of Asian vegetarian cooking*. Periplus, Singapore. 128 pp.
Kraft, K., and Kraft, P. 1977. *Exotic vegetables: how to grow and cook them*. Walker and Company, New York. 116 pp.
Pathak, J. 2007. *Taste of Nepal*. Hippocrene, New York. 470 pp.

58 Loquat

Family: Rosaceae (rose family)

NAMES

Scientific Name: *Eriobotrya japonica* (Thunb.) Lindl.

- The name loquat is derived from the Chinese (Cantonese) name for the plant, *lōkwat*, literally "rush-orange," so named because it grows well in marshy soil among "rushes."
- In its native homeland, southern China, one of the names of the loquat is *pipa*, after the lute; the Chinese version of the musical instrument is alleged to resemble in shape the leaves and/or the fruits of the loquat (pear-shaped leaves and pear-shaped fruits often occur on loquat trees and indeed are reminiscent of the outline of Chinese lutes).
- The loquat is also known as Japanese medlar, Japanese plum, and May apple, names that are best reserved for other plants. "Medlar" is usually applied to *Mespilus germanica* L., an uncommon fruit tree (see Chapter 62). "May apple" is the usual name of *Podophyllum peltatum* L., a potentially toxic herb of eastern North America, known for its medicinal value. "Japanese plum" is the usual name of *Prunus salicinia* Lindl., a species grown to a small extent for its fruit.
- The genus name *Eriobotrya* is derived from the Greek for "wooly cluster," an allusion to the appearance of the young cluster of fruits (some varieties have fuzzy-skinned fruits).
- *Japonica* in the scientific name means Japanese, although in fact the homeland of the loquat is China.

PLANT PORTRAIT

The loquat is an evergreen, large shrub or moderately sized tree 6 to 9 m (20–30 ft.) tall that bears its fruits in clusters of 4 to 30. The fruits are oval, rounded, or pear shaped, 2.5 to 7 cm (1–2.8 in.) long, with smooth or downy, yellow to orange, sometimes reddish, tough, plum-like skin, and white, yellow, or orange succulent, sweet to acid pulp. The fruit usually has three to five (sometimes up to ten) brown seeds typically approximately 1.5 cm (5/8 in.) long, which make up about a quarter of the weight of the whole fruit. The fruit develops in the early spring. Loquats are native to southeastern China and are adapted to a subtropical to mild-temperate climate. The loquat has been grown from antiquity in Japan and northern India. Today it is cultivated extensively in the Mediterranean basin (Spain, Algeria, Turkey, Israel, and Italy), Japan, China, to some extent in India and Brazil, and in a more limited fashion in Chile and the United States (mostly California and Florida). Loquats are popular landscape plants in the southernmost parts of the United States, and dwarf cultivars are available for small lots. The largest producers are Spain, Algeria, Japan, China, and Brazil. There are more than 800 varieties. Two groups are recognized, Chinese and Japanese. Chinese varieties have slender leaves and fruit that is pear shaped or nearly round, with thick, orange skin and dark-orange flesh, not very juicy, subacid, but of distinct flavor, with small numerous seeds, and of good keeping quality. Japanese varieties have broad leaves and fruit that is pear shaped or long oval, the skin is usually pale-yellow, the flesh whitish, very juicy, acid but otherwise not distinct in flavor, with few but large seeds, and only fair keeping quality. For the fresh market, loquats are picked before full maturity; otherwise, they are too acid. After harvest, the fruit keeps for 10 days at room temperature and for 60 days in cool storage. However, after removal from storage, the shelf-life may be only 3 days. Cold storage of loquats in polyethylene bags alters the flavor of the fruit and promotes internal browning and decay.

CULINARY PORTRAIT

Depending on variety, the flesh of the loquat may be firm or soft, very juicy or only moderately so, or quite acidic or not (acidity is higher in unripe fruit). The taste varies from sourish to sweet. The taste of loquat has been said to be reminiscent of a remarkable range of fruits, including apples, apricots, cherries, grapes, kumquats, pears, pineapples, plums, and strawberries. Inedible seeds are usually present. Fresh loquats are uncommonly available in Western countries. These delicate fruits bruise easily and are difficult to ship, so that in many areas canned loquats preserved in syrup are more easily obtained. Where they are available, peeled and seeded fruits are eaten fresh, often in fruit salads. In Japan, loquats are often served with chicken dishes. The fruit can also be cooked and served with cream, stewed, candied, or preserved. Loquats are also used in gelatin desserts, pie filling, sauce, syrup, jam, jelly, and liqueur.

The seeds and peel of the loquat contain amygdalin, which is easily converted into hydrocyanic (prussic) acid. (Seeds of many edible fruits of the rose family have this capacity to produce poisonous cyanide.) There have been instances of poisoning in poultry from ingestion of loquat seeds, which should be discarded. Sometimes loquat is consumed with the peel on, and the seed (whole or ground) is used as a spice, but these practices are not recommended because of the toxic potential.

CURIOSITIES OF SCIENCE AND TECHNOLOGY

- When in blossom, loquat flowers have a sweet, penetrating scent. Unfortunately, this results in some individuals suffering headaches when too close to a loquat tree in bloom.
- Loquat wood is prized by stringed instrument craftspeople, especially violin makers.
- In Japan, paper bags are sometimes placed around growing loquats to produce unblemished specimens, which may be sold in white boxes like candy.

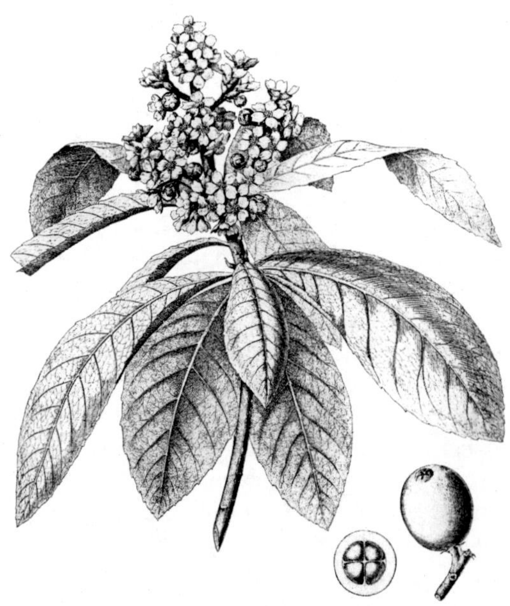

FIGURE 58.1 Loquat (*Eriobotrya japonica*). (From Edwards and Lindley, 1815–1847, vol. 5.)

KEY INFORMATION SOURCES

Badenes, M.L., Martinez-Calvo, J., and Llacer, G. 2000. Analysis of a germplasm collection of loquat (*Eriobotrya japonica* Lindl.). *Euphytica*, 114:187–194.

Blumenfeld, A. 1980. Fruit growth of loquat. *J. Am. Soc. Hortic. Sci.* 105:747–750.

Caldeira, M.L., and Crane, J.H. 2000. Evaluation of loquats (*Eriobotrya japonica* (Thunb.) Lindl.) at the Tropical Research and Education Center, Homestead. *Proc. Annu. Meet. Fla. State Hort. Soc.* 112:187–190.

California Rare Fruit Growers. 1985. Loquat. *California Rare Fruit Growers Yearbook*, 17:1–39.

Chachin, K., and Hamauzu, Y. 1997. Loquat. In *Postharvest physiology and storage of tropical and subtropical fruits*. Edited by S.K. Mitra. CABI, Wallingford, Oxon, U.K. pp. 397–403.

Chen, Z., Lin, S., and Liu, Q. 1991. Loquat (*Eriobotrya japonica* Lindl). *Biotechnol. Agric. For.* 16:62–75.

Condit, I.J. 19115. The loquat. *Bull. Calif. Agric. Exp. Stat.* 250:251–284.

Ding, C., Chen, Q., Sun, T., and Xia, Q. 1995. Germplasm resources and breeding of *Eriobotrya japonica* Lindl. in China. *Acta Hortic.* 403:121–126.

Ding, C.K., Chachin, K., Hamauzu, Y., Ueda, Y., and Imahori, Y. 1998. Effects of storage temperatures on physiology and quality of loquat fruit. *Postharvest Biol. Technol.* 14:309–315.

Ding, C.K., Chachin, K., Ueda, Y., Imahori, Y., and Wang, C.Y. 2002. Modified atmosphere packaging maintains postharvest quality of loquat fruit. *Postharvest Biol. Technol.* 24:341–348.

Donan, J.F. 1985. Yes, Virginia, the loquat is a member of the rose family. *California Rare Fruit Growers*, 17:1–22.

Ezzat, A.H., Rokbah, A.M., and Khalil, F.A. 1972. Seasonal changes of the loquat fruit. *Agric. Res. Rev.* 50(4):33–38.

Fererra, J. 1996. Nispero, medlar or loq-what? *Fresh Produce J.* 1996(May 17):25–26.

Froehlich, O., and Schreier, P. 1990. Volatile constituents of loquat *Eriobotrya japonica* Lindl. fruit. *J. Food Sci.* 55:176–180.

Goubran, F.H., and El-Zeftawi, B.M. 1986. Induction of seedless loquat. *Acta Hortic.* 179:381–384.

Hiep, N.T., and Verheij, E.W.M. 1991. *Eriobotrya japonica* (Thunb.) Lindley. In *Plant resources of South-East Asia. Vol. 2. Edible fruits and nuts*. Edited by E.W.M. Verheij and R.E. Coronel. Pudoc, Leiden, the Netherlands. pp. 161–164.

Hirai, M. 1980. Sugar accumulation and development of loquat fruit. *J. Jpn. Soc. Hortic. Sci.* 49:347–353.

Kader, A.A. 1999. Loquat. Recommendations for maintaining postharvest quality. *Perishables Handling Quart.* 100:19–20.

Khan, B.M., Shahid, M., and Chaudhry, M.I. 1986. Effect of honey bee pollination on the fruit setting and yield of loquat. *Pak. J. For. Peshawar*, 36(2):73–77.

Kumar, R. 1976. Induction of seedlessness in loquat. *Indian J. Hortic.* 33:26–32.

Lin, S. 2008. *Eriobotrya japonica*, loquat. In *The encyclopedia of fruit & nuts*. Edited by J. Janick and R.E. Paull. CABI, Wallingford, Oxfordshire, U.K. pp. 642–651.

Lin, S., Sharpe, R.H., and Janick, J. 1999. Loquat: botany and horticulture. *Hortic. Rev.* 23:179–231.

Martinez-Calvo, J., Badense, M.L., Llacer, G., Bleiholder, H., Hack, H., and Meier, U. 1999. Phenological growth stages of loquat tree (*Eriobotrya japonica* (Thunb.) Lindl.). *Ann. Appl. Biol.* 134:353–357.

Morton, J. 1987. Loquat. In *Fruits of warm climates*. Creative Resource Systems, Winterville, NC. pp. 103–108.

Pathak, R.K. 1999. Loquat. In *Tropical horticulture*, vol. 1. Edited by T.K. Bose, S.K. Mitra, A.A. Farooqui, and M.K. Sadhu. Naya Prokash, Calcutta, India. pp. 386–389.

Pathak, R.K., and Gautam, H.O. 1985. Loquat. In *Fruits of India, tropical and subtropical*. Edited by T.K. Bose. Naya Prokash, Calcutta, India. pp. 549–558.

Pathak, R.K., and Hari, O.G. 1985. Loquat. In *Fruits of India. Tropical and subtropical*. Edited by T.K. Bose. Naya Prokash, Calcutta, India. pp. 548–558.

Randhawa, G.S., and Singh, R.K.N. 1970. *The loquat in India*. Indian Council of Agricultural Research, New Delhi, India. 64 pp.

Sawyer, P., Houghton, P., and Manuel, L. 1984. Loquats: a literature search. *California Rare Fruit Growers Yearbook*, 17:23–33.

Schroeder, C.A. 1996. The loquat: a fruit of quality. *WANATCA Yearbook* (Australia), 20(0):24–29.

Shaw, P.E. 1980. Loquat. In *Tropical and subtropical fruits: composition, properties and uses*. Edited by S. Nagy and R.E. Shaw. AVI Publishing Company, Westport, CT. pp. 479–491.

Tilsher, W.G. 1976. The mouth-watering loquat. *Org. Gard. Farm.* 23(4):136–137.

Topuz, A., and Ozdemir, F. 1999. Determination of some physical and chemical properties of loquat and possibilities of processing into marmalade, nectar and canned fruit. *Fruit Processing*, 9:398–401.

Vilanova, S., Badenes, M.L., Martinez-Calvo, J., and Llacer, G. 2001. Analysis of loquat germplasm (*Eriobotrya japonica* Lindl) by RAPD molecular markers. *Euphytica*, 121:25–29.
Zappi, D., and Turner, J. 2001. *Eriobotrya japonica* Rosaceae. *Curtis's Bot. Mag.* 18(2):108–113.

Specialty Cookbooks

Donan, J., Hand, D., and Houghton, P. 1985. Loquat and other Rosaceae recipes. *J. Calif. Rare Fruit Grow.* 17:35–39.
Ziedrich, L. 2009. *The joy of jams, jellies, and other sweet preserves: 200 classic and contemporary recipes showing the fabulous flavours of fresh fruits*. Harvard Common Press, Boston, MA. 432 pp.

59 Lychee, Longan, and Rambutan

Family: Sapindaceae (soapberry family)

The lychee, longan, and rambutan are extremely similar fruits of the soapberry family, used interchangeably, and so are discussed together in this chapter.

This chapter features

Lychee (*Litchi chinensis*)
Longan (*Dimocarpus longan*)
Rambutan (*Nephelium lappaceum*)

LYCHEE

NAMES

Scientific Name: *Litchi chinensis* Sonn.

- The English name "lychee" (also spelled litchi, litchee, leechee, and lichi) and the genus name *Litchi* are based on Mandarin Chinese *lìzh*, which is composed of *lì,* litchi + *zh*, twig. The Chinese characters in the name indicate that the fruit should be harvested with the twig attached.
- The tough skin led to a frequent name for the fruit, "lychee nut" (also sometimes "Chinese nut"), although in no sense is it a nut.
- *Chinensis* in the scientific name *L. chinensis* is Latin for "of China," indicating the native home of the lychee.

PLANT PORTRAIT

The lychee is an evergreen fruit tree growing to a height of 7.6 to 30 m (25–100 ft.). It is native to southern China, where it may have been harvested since 1500 BC. The production of superior types of lychee was a matter of great family pride and local rivalry in China, where the fruit is esteemed. Many varieties were named after the families that produced them. According to legends, some ancient Chinese devotees of the fruit consumed a thousand of them daily. Lychees are given as good luck gifts during the Chinese New Year, and in southern China the fruit is considered to be a symbol of love. The tree is adapted to tropical and subtropical regions. Major producers today include China, India, Pakistan, Bangladesh, Burma, Taiwan, Thailand, Mauritius, Japan, Philippines, Australia, Israel, Madagascar, Brazil, and South Africa. Lychee trees were started in Hawaii in 1873, Florida in 1883, and California in 1897. There is some commercial cultivation in Florida. A good tree can produce 91 to 136 kg (200–300 lb.) of fruit in a single year. Freshly picked lychees keep their color and quality only 3 to 5 days at room temperature but nevertheless have been successfully shipped by air from Florida to markets throughout the United States and Canada. The sweet, fragrant, musky, delicious fruits are borne in large clusters of up to 30, hanging at the ends of branches, appearing rather like bunches of strawberries. The fruits are round, heart-shaped or egg-shaped, typically 2.5 cm (1 in.) wide and 4 cm (1½ in.) long, deep pink, brownish-yellow, or bright red or purple-red, with pearl white, translucent pulp. Red fruit is most often marketed, these rather suggestive of a large, luscious grape or strawberry. The thin but tough, scaly, leathery skin may have tiny warts and peels readily from the pulp in fresh fruit. Dried fruit have shriveled brown flesh reminiscent of a raisin. Some varieties have a small amount of clear, delicious juice immediately under the skin, but this is considered undesirable as they leak juice when the skin is broken. Preferred varieties are drier and tend not to drip while the fruit is being eaten. There is usually one inedible, nonclinging seed, approximately 20 mm (3/4 in.) long, but in some desirable fruit, this is small or undeveloped.

Culinary Portrait

Lychee pulp when fresh is sweet, acidic, and slightly musty in flavor. The taste has been described as a combination of strawberry, rose, and muscat grape. The flavor varies with degree of maturity: unripe fruit are gelatinous and rather bland, and overripe fruits have lost much of their flavor. Lychees are best harvested at full maturity as they will not ripen afterward, but because the fruit does not withstand shipping well, rather brownish, tasteless lychees are often sold in Western supermarkets. Fresh fruit can be shelled and pitted and served in both vegetable and fruit salads. If cooked, this should be done very briefly (by adding the fruit at the end of the cooking of other ingredients) to preserve the delicate flavor. The fruit may be used to flavor rice, vegetables, stuffings, and sauces. The fruit also goes well with poultry, shellfish, and pork and blends admirably with sweet-and-sour sauces. In Chinese cooking, lychees are often combined with fish or meat. Fresh fruit can be stored for up to 3 weeks in a refrigerator (wrapped in a paper towel, placed in a perforated plastic bag) but is best consumed very promptly. The fruit can simply be frozen in its shell. Lychees make excellent preserves. Canned lychees are common and are nearly as good as the fresh fruit, which is available only in season. The dried fruits are eaten out of hand like raisins and are said to resemble very sweet Muscat grapes. Dried fruit usually cannot be substituted for fresh or canned lychee that is required in recipes.

FIGURE 59.1 Lychee (*Litchi chinensis*). (From Bailey, 1916.)

Culinary Vocabulary

- A "lychee martini" has acquired some popularity in Asian restaurants in North America. A sample recipe: combine 60 mL (2 fluid oz.) of vodka, 60 mL of lychee liqueur, and a splash of lychee juice with ice and shake in a cocktail shaker; strain into a frozen martini glass, and garnish with fresh or canned lychee.
- The atrophied seeds in some lychee fruits are called "chicken tongue." "Chicken-tongue lychees" have undeveloped seeds and so are considered desirable. The phrase "chicken tongue" originated in China and has been carried over to English.

Curiosities of Science and Technology

- The Sapindaceae family is named for one of its genera, *Sapindus*, the soapberry, some species of which are grown as ornamentals in the southern United States. The fruits are rich in foaming chemicals called saponins and are used as soap in some countries.
- Chinese (T'ang) Emperor Hsuan Tsung (712–756) organized a "pony express" to carry fresh lychees from tropical south China to the northern court for Lady Yang Kuei Fei.
- One of the earliest books providing information on growing crops was written in 1056 and was about lychees.
- Lychees were once used as a form of money in China.
- In 2002, a single lychee weighing approximately 14 g (1/2 oz.) sold for approximately $101,000.00 at an auction in the southern Chinese city of Zengcheng. This particular fruit was considered extremely valuable because it came from an ancient tree, more than 400 years old, called Xiyuangualu, which produces only a few dozen fruits each year. During the Quing Dynasty (1644–1911), under the emperors Qianlong and Jiaqing, lychees from this tree served as tributes to the imperial court and were dubbed "the king of fruit."

Key Information Sources (for Additional References, see Longan Section)

Anonymous. 1985. *An album of Guangdong litchi varieties in full color*. Guangdong Province Scientific Technology Commission, Guangdong, China. 78 pp.

Batten, D. 1984. *Lychee varieties*. Department of Agriculture, New South Wales, Sydney, Australia. 15 pp.

Bhatia, B. S., Mehra, G.L., and Nair, K.G. 1963. Dehydration of litchi (*Litchi chinensis*). *Food Sci.* 12:313–315.

Cavaletto, C.G. 1980. Lychee. In *Tropical and subtropical fruits: composition, properties and uses*. Edited by S. Nagy and R.E. Shaw. AVI Publishing Company, Westport, CT. pp. 469–478.

Galán Saúco, V., and Menini, U.G. 1989. *Litchi cultivation*. Food and Agriculture Organization Plant Production and Protection Paper No. 83. FAO, Rome, Italy. 136 pp (in Spanish).

Greer, N. 1990. *Growing lychee in south Queensland*. Queensland Department of Primary Industries, Brisbane, Australia. 44 pp.

Joubert, A.J. 1986. Litchi. In *CRC handbook of fruit set and development*. Edited by S.P. Monselise. CRC Press, Boca Raton, FL. pp. 233–246.

Langdon, R.D. 1969. The lychee. *California Rare Fruit Growers Yearbook*, 1:1–8.

Maiti, S.C. 1985. Litchi. In *Fruits of India, tropical and subtropical*. Edited by T.K. Bose. Naya Prokash, Calcutta, India. pp. 388–408.

Menzel, C.M. 1983. The control of floral initiation in lychee: a review. *Sci. Hortic*. 21:201–215.

Menzel, C.M. 1984. The pattern and control of reproductive growth in lychee: a review. *Sci. Hortic.* 22:333–345.

Menzel, C.M. 1985. Propagation of lychee, a review. *Sci. Hortic*. 25:31–48.

Menzel, C.M. 1991. *Litchi chinensis* Sonn. In *Plant resources of South-East Asia. 2. Edible fruits and nuts*. Edited by E.W.M. Verheij and R.E. Coronel. Pudoc, Leiden, the Netherlands. pp. 191–195.

Menzel, C.M., and Simpson, D.R. 1987. Lychee nutrition: a review. *Sci. Hortic*. 31:195–224.

Menzel, C.M., and Simpson, D.R. 1990. Performance and improvement of lychee cultivars, a review. *Fruit Var. J*. 44:197–215.

Menzel, C.M., and Simpson, D.R. 1991. A description of lychee cultivars. *Fruit Var. J*. 45:45–56.

Menzel, C.M., and Simpson, D.R. 1994. Lychee. In *Handbook of environmental physiology of fruits crops. Vol. 2. Subtropical and tropical crops*. Edited by B. Schaffer and P.C. Anderson. CRC Press, Boca Raton, FL. pp. 123–144.

Menzel, C.M., Watson, B.J., and Simpson, D.R. 1988. The lychee in Australia. *Qld. Agric. J.* 114:19–27.

Mitra, S.K. 1999. Litchi. In *Tropical horticulture, vol. 1*. Edited by T.K. Bose, S.K. Mitra, A.A. Farooqui, and M.K. Sadhu. Naya Prokash, Calcutta, India. pp. 319–330.

Mo, B.C. (Ed.). 1992. *Litchi high production cultivation and technology*. Guangxi Science and Technology Publishers, Guangxi, China. 97 pp (in Chinese).

Morton, J. 1987. Lychee. In *Fruits of warm climates*. Creative Resource Systems, Winterville, NC. pp. 249–259.

Pandey, R.M. 1989. *The litchi*. Publications and Information Division, Indian Council of Agricultural Research, New Delhi, India. 80 pp.

Paull, R.E., and Chen, N.J. 1987. Effect of storage temperature and wrapping on quality characteristics of litchi fruit. *Sci. Hortic.* 33:223–236.

Singh, L.B., and Singh, U.P. 1954. *The litchi*. Superintendent, Printing and Stationery, Lucknow, India. 86 pp.

Underhill, S.J.R., Coates, L.M., and Salco, Y. 1997. Litchi. In *Postharvest physiology and storage of tropical and subtropical fruits*. Edited by S.K. Mitra. CABI, Wallingford, U.K. pp. 191–208.

Villiers, E.A. de. 2002. *The cultivation of litchi*. Agricultural Research Council, Nelspruit, South Africa. 213 pp.

Young, T.W. 1966. A review of the Florida lychee industry. *Proc. Fla. State Hort. Soc.* 79:395–398.

Zee, F. 1999. *Growing lychee in Hawaii*. Cooperative Extension Service, University of Hawaii at Manoa, College of Tropical Agriculture and Human Resources, Honolulu, HI. 8 pp.

Zee, F., and Paull, R.F. 2008. *Litchi chinensis*, litchi. In *The encyclopedia of fruit & nuts*. Edited by J. Janick and R.E. Paull. CABI, Wallingford, Oxfordshire, U.K. pp. 799–808.

Specialty Cookbooks

See Longan section. Lychee and longan are interchangeable in recipes.

LONGAN

Names

Scientific Name: *Dimocarpus longan* Lour. (*Euphoria longan* (Lour.) Steud., *E. longana* Lam.)

- The name "longan" (as well as the word *longan* in the scientific name *D. longan*) is derived from the Chinese (Mandarin) *lóng*, dragon + *yǎn*, eye. "Lungan" is a variation of the name sometimes encountered. Rarely, "langngan" is encountered.
- The longan is also called cat's eye, dragon's eye (dragon eye), and eyeball. The "eye" in these names is due to the appearance of the seed, which is black, reddish-brown, or brown, with a circular white spot giving it the appearance of an eye.
- The genus name *Dimocarpus* is based on the Greek *dimorphos*, two formed (i.e., with two shapes), and *karpos*, fruit.

Plant Portrait

The longan is a tropical to subtropical evergreen tree, usually growing 9 to 12 m (30–40 ft.) in height, rarely, as high as 40 m or 131 ft., with fruits in drooping clusters. Longan fruits are very similar to lychees but slightly smaller and smoother-skinned. Because the fruit is not as desirable as the lychee, it is grown because it tolerates cooler and less exacting conditions, and in the case of China, because it is used medicinally as well as for food. The fruits are spherical or ovoid, 1.3 to 3.8 cm (1/2–1½ in.) in diameter, with a thin, brittle or leathery, yellow-brown to light reddish-brown, somewhat pebbled (but almost smooth) rind, and a single large brown seed. Between the rind and the seed is a gelatinous, pleasantly flavored, whitish, translucent, sweet,

FIGURE 59.2 Longan (*Dimocarpus longan*). (From Curtis, 1844, vol. 70, plate 4096.)

musky pulp, which is the edible portion. The pulp separates readily from the rind and the seed. The species is native to southern China (also, according to some authorities, Myanmar (Burma), southwest India, Sri Lanka, and the Indochinese peninsula) and is commonly grown in Thailand (where it is the country's largest fruit export), Cambodia, Laos, Vietnam, Taiwan, and India, and to some extent in Australia, Kenya, and South Africa. Thailand, China, and Taiwan are the main centers of commercial production. The longan was introduced to the United States in 1903, and there are now small commercial crops in Hawaii, California, and Florida. In China and in the United States, the similar lychee is more successful. Indeed, in China, the longan is known as the "little brother" of the lychee, reflective of its more limited importance. There are numerous cultivated varieties, approximately 40 of which are commonly grown. In China, the hard lumber of the tree is used for furniture and construction.

Culinary Portrait

The cherry-sized longan fruit is primarily eaten fresh as a dessert or snack (discarding the peel and seed, both of which are inedible) but is also canned and dried. Fresh longans are uncommonly available in Western countries, but canned, pitted fruit in syrup, and dried fruit are often sold. Longans can be used fresh in salads, baked in pies as filling, poached, and added to stir-fries, marinades, and sauces. They complement rice, vegetables, and other dishes. In China, longans are frequently used to flavor sweet soups and in sweet-and-sour dishes. Dried fruit, commonly called nuts, look like large raisins and are eaten like them. Liqueurs and nonalcoholic beverages are also prepared. The longan is invariably compared with its close, very similar relative, the lychee, and is interchangeable in recipes, although the lychee is considered tastier and better textured. However, longan fruits are milder in flavor and less acidic. The taste, said to be like a semisweet, musky white grape, is acquired, and demand outside of Asia tends to be from ethnic Asians. Once removed from the tree, the fruit will not increase in sweetness. The fruit has a short shelf life (several days) at room temperature and can be placed in plastic bags in the refrigerator for 5 to 7 days or simply frozen (longans can be frozen in their shells).

Culinary Vocabulary

- *Naam lumyai* (lumyai = longan) is a popular longan beverage in Thailand.

Curiosities of Science and Technology

- In the old days in Vietnam, longans were among the food items reserved as tributes for the Kings.
- As noted earlier, the longan is used medicinally in Asia. In southern Asia, longan fruit flesh is regarded as an antidote for poisons. In Vietnam, the "eye" of the longan seed is simply pressed against snakebites in the belief that it will absorb the venom.
- The seeds of longan are rich in saponins, foaming chemicals often used to prepare soap, as noted earlier. In China, the seeds have been used to prepare hair shampoo.

Key Information Sources

Achariyaviriya, A., Soponronnarit, S., and Tiansuwan, J. 2001. Study of longan flesh drying. *Dry Technol.* 19:2315–2329.

Campbell, CW., and Malo, S.E. 1981. Evaluation of the longan as a potential crop for Florida. *Proc. Fla. State Hort. Soc.* 94:307–309.

Chattopadhyay, P.K., Ghosh, A., and Hasan, M.A. 1994. The less-known fruit, longan. *Indian Hortic.* 39(1):49–50.

Chian, G.Z., Hee, Y.Q., Lin, Z.Y., Liu, M.Y., and Feng., X.B. 1996. *Longan high production cultivation techniques.* Guangxi Science Technology Publishers, Nanking, China. 90 pp (in Chinese).

Choo, W.K., and Ketsa, S. 1991. *Dimocarpus longan* Lour. In *Plant resources of South-East Asia. Vol. 2. Edible fruits and nuts.* Edited by E.W.M. Verheij and R.E. Coronel. Pudoc, Leiden, the Netherlands. pp. 146–151.

Choo, W.K., and Yuen, G.Y. 1992. The diversity of *Dimocarpus longan* subspecies *malesianus* variety *malesianus* in Sarawak. *Acta Hortic.* 292:29–39.

Diczbalis, Y.A. 2002. *Longan: improving yield and quality: a report for the Rural Industries Research and Development Corporation.* Rural Industries Research and Development Corp., Barton, ACT, Australia. 59 pp.

Groff, G.W. 1921. *The lychee and lungan.* Orange Judd, New York. 188 pp.

Huang, H., and Menzel, C. (Eds.). 2001. *Proceedings of the first international symposium on litchi and longan, Guangzhou, China, 16–19 June, 2000.* International Society for Horticultural Science, Leuven, the Netherlands. 446 pp.

Huang, J.S., Xu, X.D., Zheng, S.Q., and Xu, J.H. 2001. Selection for aborted-seeded longan cultivars. *Acta Hortic.* 558:115–118.

Jiang, Y.M., Zhang, Z.Q., Joyce, D.C., and Ketsa, S. 2002. Postharvest biology and handling of longan fruit (*Dimocarpus longan* Lour.). *Postharvest Biol. Technol.* 26:242–252.

Kader, A.A. 2001. Longan: recommendations for maintaining postharvest quality. *Perishables Handling Quart.* 106:13–14.

Lai, Z.X., Chen, C.L., and Chen, Z.G. 2001. Progress in biotechnology research in longan. *Acta Hortic.* 558:137–141.

Leenhouts, P.W. 1971. A revision of *Dimocarpus* (Sapindaceae). *Blumea*, 19:113–131.

Liu, X.H., and Ma, C.L. 2001. Production and research of longan in China. *Acta Hortic.* 558:73–82.

Marcelina, E. 1997. A look at lychees and longans: Southern Florida growers are marketing these tropical fruit treats. *AgVentures*, 2(5):41–44.

McConchie, C.A., Vithanage, V., and Batten, D.J. 1994. Intergeneric hybridisation between litchi (*Litchi chinensis* Sonn.) and longan (*Dimocarpus longan* Lour.). *Ann. Bot.* 74:111–118.

Menzel, C. 1989. The subtropical longan: an Australian view. *California Grower*, 13(7):20, 22–23.

Menzel, C., Mitra, S.K., and Waite, G.K. 2003. *Litchi and longan: botany, cultivation and uses.* CABI, Wallingford, U.K. 400 pp. (2005 edition, same publisher, 336 pp.)

Menzel, C.M., Watson, B.J., and Simpson, D.R. 1988. Longans—a place in Queensland's horticulture? *Qld. Agric. J.* 115:251–265.

Morton, J. 1987. Longan. In *Fruits of warm climates*. Creative Resource Systems, Winterville, NC. pp. 259–262.

Phillips, R.L., Campbell, C.W., and Malo, S.E. 1986. *The longan. Fruit Crops Facts Sheet*. University of Florida, Agricultural Extension Service, Gainesville, FL. 2 pp.

Queensland Department of Primary Industries. 1995. *Longan and lychee*. Queensland Department of Primary Industries, North Region, Townsville, Queensland, Australia. 37 pp.

Stanley, D. 1998. Florida growers like lychees and longans. *Agric. Res*. 46(1):8–9.

Tongdee, S.C. 1997. Longan. In *Postharvest physiology and storage of tropical and subtropical fruits*. Edited by S.K. Mitra. CAB International, Wallingford, Oxon, U.K. pp. 335–345.

Welch, T., and Ferguson, J. 1996. *Fourth national lychee seminar including longans theme: from Marcot to market proceedings (Yeppoon, Qld.)*. Australian Lychee Growers' Association, Maroochydore, Queensland, Australia. 164 pp.

Winston, E.C., O'Farrell, P.J., and Young, K.E. 1993. Yield and fruit quality of longan (*Dimocarpus longan* Lour.) cultivars on the Atherton Tableland of tropical north Australia. *Fruit Var. J*. 47:153–160.

Wongsiri, S., Thapa, R., and Kongpitak, P. 1998. Longan: a major honey plant in Thailand. *Bee World*, 79(1):23–28.

Yen, C.-R., and Paull, R.E. 2008. *Dimocarpus longan*, longan. In *The encyclopedia of fruit & nuts*. Edited by J. Janick and R.E. Paull. CABI, Wallingford, Oxfordshire, U.K. pp. 795–799.

Zee, F.T.P., Chan, H.T., Jr., and Yen, C.-R. 1998. Lychee, longan, rambutan and pulasan. In *Tropical and subtropical fruits*. Edited by P.E. Shaw, H.T. Chan, Jr., and S. Nagy. Agscience, Auburndale, FL. pp. 290–335.

Specialty Cookbooks

Bartell, K.H. 2001. *The best of Taiwanese cuisine: recipes and menus for holidays and special occasions*. Hippocrene, New York. 122 pp.

Choi, T.T., Isaak, M., and Holzon, H. von. 2005. *Authentic recipes from Vietnam*. Periplus, Singapore. 112 pp.

Moromito, M. 2007. *The new art of Japanese cooking*. DK Publishing, New York. 272 pp.

Nguyen, A., and Beisch, L. 2006. *Into the Vietnamese kitchen: treasured foodways, modern flavors*. Ten Speed Press, Berkeley, CA. 352 pp.

Periplus Editors. 2003. *Fabulous Asian homestyle recipes: nutritious meals in minutes*. Periplus, Singapore. 96 pp.

Solomon, C. 2006. *The complete Asian cookbook*. Tuttle Publishing, Boston, MA. 512 pp.

Tan, T., Tan, C., Thompson, D., and Ho, E. 2003. *Shiok! Exciting tropical Asian flavors*. Periplus, Singapore. 192 pp.

Wibisono, D., Wong, D., and Tettoni, L.I. 2005. *Authentic recipes from Singapore: 63 simple and delicious recipes from the tropical island city-state*. Periplus, Singapore. 112 pp.

RAMBUTAN

Names

Scientific Name: *Nephelium lappaceum* L. (*Euphoria nephelium* DC., *Dimocarpus crinita* Lour.)

- The name "rambutan" is based on the Malay word *rambut*, hair, for the hair-like covering of the fruit.
- The rambutan is sometimes encountered under the nicknames "hairy lychee" and "hairy cherry."
- The genus name *Nephelium* is an old name tracing to classical Greek and Roman words for the burdock (*Arctium* species), applied because of the burr-like fruits.
- *Lappaceum* in the scientific name *N. lappaceum* is Latin for burr-like, descriptive of the appearance of the fruit.

Plant Portrait

The rambutan is a tropical, evergreen tree growing 15 to 25 m (50–80 ft.) in height. The species is native to Malaysia and Indonesia. It is grown throughout southern China, the Indo-Chinese region, and the Malay Archipelago, and to a lesser extent in India, Brunei, Vietnam, Singapore, Kampuchea, Burma, Sri Lanka, Central America, northern Australia, Hawaii, Mauritius, and Madagascar. Major producers are Thailand, Indonesia, Malaysia, and the Philippines. Exports are primarily from Malaysia and Thailand. There are more than two dozen cultivated varieties. Most rambutan trees propagated from seeds are not true to type and the fruits are usually sour. Half of the trees produced from seeds may be male plants, which do not bear fruit. To avoid these difficulties, orchards are established from cuttings of trees that are known to produce good fruit.

A relative of the lychee (see Lychee section), the rambutan fruit has been described as a "warty chestnut" and a "small hedgehog." The fruit is ovoid, or ellipsoid, 3.4 to 8 cm (1⅓–3¼ in.) long (about the size of an apricot), and varies in color through shades of pink, red, orange, yellow, yellowish-brown, maroon, and purple. Red fruit are the most popular. The rind is thin, leathery, and covered with tubercles, each of which has a soft, fleshy, red, pink, or yellow spine 0.5 to 2 cm (1/5–3/4 in.) long. The rind is removable by a twist of the hands. Inside is a thin layer of flesh around a slightly flattened, large seed, 2.5 to 3.4 (1–1⅓ in.) long. There are both "freestone" varieties (in which the pit adheres to the flesh) and "clingstone" varieties (in which the pit is free from the flesh). The crispy flesh is white or rose-tinted, translucent, juicy, and acid to sweet. There may be one or two small undeveloped fruits close to the stem of a mature fruit. Drying causes a darkening of the peel and the soft hairs, giving the fruit an unattractive appearance, although they remain edible.

Culinary Portrait

Rambutans are frequently consumed like lychees, for which they may be substituted in recipes. The fruit is eaten primarily fresh, out of hand. The flavor ranges from sweet, mild, and fragrant to slightly sour or acid, depending on variety. To prepare the fruit for eating, cut through the outer skin with a sharp paring knife and discard the tough, papery skin (cases of poisoning are known from

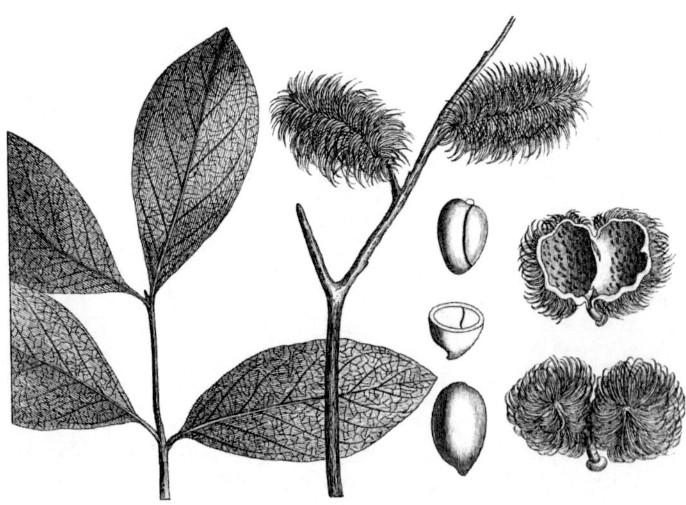

FIGURE 59.3 Rambutan (*Nephelium lappaceum*). (From Lamarck and Poiret, 1744–1829, plate 764.)

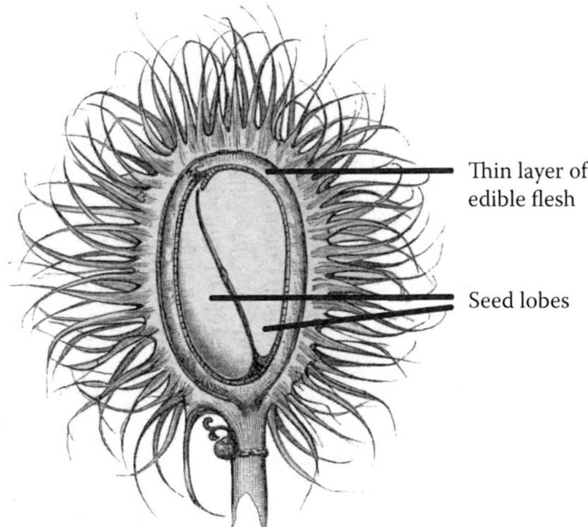

FIGURE 59.4 Sectioned fruit of rambutan (*Nephelium lappaceum*). (From Engler and Prantl, 1889–1915.)

eating this fruit wall, although it is used medicinally). Hold the rambutan with the fingers. If the fruit is a clingstone variety, do not bite too deeply as the flesh will come away with the tough skin of the seed attached; the whole fruit can be popped into the mouth and chewed around the seed. The fruit is also peeled and used in salads and dessert toppings and goes well with ice cream. A novel way of serving a rambutan is to remove only the top half of the shell, and present it in the bottom part of the shell, like an egg in an egg cup. The peeled fruits can be stewed as dessert, made into preserves and jams, and canned. Cooked rambutan may accompany meat and vegetable dishes or made into stuffing, and rambutan sauce is excellent for chicken and seafood. The seeds are sometimes roasted and eaten, although they are bitter and reputedly poisonous and narcotic. The fruits are quite perishable. They can be stored briefly in the refrigerator but should be consumed as soon as possible.

Culinary Vocabulary

- Although in Western culture the term "boutique" refers to a small shop or area within a large department store, in Southeast Asia the word is used for roadside stalls that sell produce and other goods. Rambutans are a particularly prominent offering at these boutiques when the fruit comes into season.

Curiosities of Science and Technology

- An average rambutan tree can produce between 5000 and 6000 fruits annually, although trees producing 10,000 fruits have been observed.
- Young rambutan shoots are used to dye silk green, while a dye obtained from the fruit is used to color silk black.
- The seeds of rambutan are rich in saponins, foaming chemicals often used to prepare soap, as noted earlier. A fatty tallow from rambutan seeds has been used to make candles.
- Colorful rambutans are often used in decorative floral bouquets and arrangements. Bouquets are made by arranging and tying clusters of fruits together along a stalk.
- Thailand, the world's largest producer of rambutan, holds an annual Rambutan Fair in August in Surat Thani Province.

Key Information Sources

Abidin, M.I.B.Z. 1990. Rambutan. In *Cultivation of tropical fruits*. Edited by M.I.B.Z. Abidin. Hi-Tec Enterprise, Kuala Lumpur. Malaysia. pp. 34–42.

Almeyda, N., Malo, S.E., and Martin, F.W. 1979. *Cultivation of neglected tropical fruits with promise: part 6. The rambutan*. Science and Education Administration, US Department of Agriculture, New Orleans, LA. 14 pp.

Bernardo, A.M.N., Garcia, R.N., and Mendoza, E.M.T. 1998. Identification of rambutan (*Nephelium lappaceum* Linn.) cultivars using isozyme electrophoretic patterns. *Philipp. J. Crop Sci.* 23:127–131.

Chin, H.F. 1975. Germination and storage of rambutan (*Nephelium lappaceum* L.) seeds. *Malays. Agric. Res.* 4:173–180.

Diczbalis, Y.A., and Paull, R.E. 2008. *Nephelium lappacium*, rambutan. In *The encyclopedia of fruit & nuts*. Edited by J. Janick and R.E. Paull. CABI, Wallingford, Oxfordshire, U.K. pp. 809–816.

Gan, Y.Y., Tan, S.G., Ahmad, A.R., and Idris-Abdol. 1983. Phylogenetic studies on rambutans (*Nephelium lappaceum*). In *Biochemistry in agriculture: proceedings of the ninth Malaysian Biochemical Society Conference*. Edited by Marziah-Mahmood [no initial]. Universiti Pertanian Malaysia, Serdang, Selangor, Malaysia. pp. 63–65.

Kader, A.A. 2000. Rambutan. Recommendations for maintaining postharvest quality. *Perishables Handling Quart.* 104(Nov.):15–16.

Kelso, J.M., Jones, R.T., and Yunginger, J.W. 1998. Anaphylaxis after initial ingestion of rambutan, a tropical fruit. *J. Allergy Clin. Immunol.* 102:145–146.

Lam, P.F., and Kosiyachinda, S. (Eds.). 1987. *Rambutan, fruit development, postharvest physiology and marketing in ASEAN*. ASEAN Food Handling Bureau, Kuala Lumpur, Malaysia. 82 pp.

Landrigan, M., Morris, S.C., and McGlasson, B.W. 1996. Postharvest browning of rambutan is a consequence of water loss. *Am. Soc. Hortic. Sci.* 121:730–734.

Landrigan, M., Sarafis, V., Morris, S.C., and McGlasson, W.B. 1994. Structural aspects of rambutan (*Nephelium lappaceum*) fruits and their relation to postharvest browning. *J. Hortic. Sci.* 69:571–579.

Leenhouts, P.W. 1986. A taxonomic revision of *Nephelium* (Sapindaceae). *Blumea*, 31:373–436.

Lim, A.L. 1984. The reproductive biology of rambutan, *Nephelium lappaceum* L. (Sapindaceae). *Gard. Bull. Singapore*, 37(pt. 2):181–192.

Mendoza, D.B., Jr., Pantastico, E.B., and Javier, F.B. 1972. Storage and handling of rambutan *(Nephelium lappaceum* L.). *Philipp. Agric.* 4:322–332.

Morton, J. 1987. Rambutan. In *Fruits of warm climates*. Creative Resource Systems, Winterville, NC. pp. 262–265.

Nakasone, H.Y., and Paull, R.E. 1998. Litchi, longan, and rambutan. In *Tropical fruits*. CABI, Wallingford, U.K. pp. 173–207.

O'Hare, T.J. 1995. Postharvest physiology and storage of rambutan. *Postharvest Biol. Technol.* 6:189–199.

O'Hare, T.J., Prasad, A., and Cooke, A.W. 1994. Low temperature and controlled atmosphere storage of rambutan. *Postharvest Biol. Technol.* 4:147–157.

Ong, P.K.C., Acree, T.E., and Lavin, E.H. 1998. Characterization of volatiles in rambutan fruit (*Nephelium lappaceum* L.). *J. Agric. Food Chem.* 46:611–615.

Salma, I. 1986. *Rambutan (Nephelium lappaceum L.) clones and their classification*. Report No. 107. Malaysian Agricultural Research and Development Institute, Kuala Lumpur, Malaysia. 33 pp.

Seibert, B. 1991. *Nephelium* L. In *Plant resources of South-East Asia. 2. Edible fruits and nuts*. Edited by E.W.M. Verheij and R.E. Coronel. Pudoc, Leiden, the Netherlands. pp. 233–235.

Tindall, H.D., Menini, U.G., and Hodder, A.J. 1994. *Rambutan cultivation*. Food and Agriculture Organization Plant Production and Protection Paper No. 121. FAO, Rome, Italy. 182 pp.

Torres, J.P., Dionido, P.B., and Zamora, A. 1962. Selection and propagation of rambutan in Oriental Mindoro, Philippines. *Araneta J. Agric.* 9:146–160.

University of the Philippines at Los Banos. 1982. *An abstract bibliography of post harvest handling of mango and rambutan*. ASEAN Post-harvest Horticulture Training and Research Center, University of the Philippines at Los Banos, College, Laguna, Philippines. 121 pp.

Valmayor, R.V., Mendoza, D.B., Jr., Aycardo, H.B., and Palencia, C.O. 1970. Growth and flowering habits, floral biology and yield of rambutan (*Nephelium lappaceum* Linn.). *Philipp. Agric.* 54:359–374.

Watson, B.J. 1988. Rambutan cultivars in north Queensland. *Qld. Agric. J.* 114:37–41.

Welzen, P.C. van, and Verheij, E.W.M. 1991. *Nephelium lappaceum* L. In *Plant resources of South-East Asia. Vol. 2. Edible fruits and nuts*. Edited by E.W.M. Verheij and R.E. Coronel. Pudoc, Leiden, the Netherlands. pp. 235–240.

Zee, F. 1993. Rambutan and pili nuts: potential crops for Hawaii. In *New Crops*. Edited by J. Janick, and J.E. Simon. Wiley and Sons, New York. pp. 461–465.

Specialty Cookbooks

Solomon, C. 1998. *Charmaine Solomon's encyclopedia of Asian food*. Periplus, Editions, Boston, MA. 480 pp.

60 Macadamia Nut

Family: Proteaceae (protea family)

NAMES

Scientific Names: *Macadamia* Species

- Smooth-shell(ed) macadamia nut—*M. integrifolia* Maiden & Betche (*M. ternifolia* of some authors; trees cultivated for nuts under the name *M. ternifolia* are *M. integrifolia*).
- Rough-shell(ed) macadamia nut—*M. tetraphylla* L.A.S. Johnson (also referred to as *M. ternifolia* by some authors).
- For the derivation of the English name "macadamia nut," see information on the genus name *Macadamia* in the next section. Although the word "macadamia" is based on the name of a person, the initial letter is usually not capitalized.
- As the common name indicates, the seed coat of smooth-shell (or smooth-shelled) macadamia nut is smooth. Despite its name, the seed coats of some varieties of the rough-shell (or rough-shelled) macadamia are smooth, although other varieties have rough and pebbled seed coats.
- *Macadamia integrifolia* is known as Australian nut, macadamia nut (occasionally shortened to macnut), and Queensland nut. Australian hazel (in recognition of the similarity of the nut to a large hazel nut) is an occasional name.
- *Macadamia tetraphylla* is known as Australian nut, macadamia nut, rough-shell, rough-shell macadamia nut, and Queensland nut.
- The name "Queensland nut" reflects the primary source of the nuts, Queensland, Australia's second largest state. "Australian nut" and "Australian hazel" of course acknowledge that the nuts came originally from Australia.
- Australian aborigines had different common names for macadamia nuts, including boombera, burrawang, and kindal kindal. The European settlers of Australia also had many early names for the nuts, some of which are still used today. Some of the odder names include Bauple nut (after Mt. Bauple, north of Gympie, where possibly the largest number of native trees grew), and Mullimbimby nut.
- Baron Ferdinand von Mueller (1840–1896), considered Australia's foremost botanist, and Walter Hill (1820–1904), Director of the Botanic Gardens at Brisbane, Australia, found wild *M. integrifolia* in 1857. The species had been collected earlier, but no one had appreciated that it was very different from known species. Von Mueller proposed the plant as a new genus, which he named *Macadamia* in honor of his friend, John Macadam (1827–1865), Scottish-born Australian chemist, health officer, and politician; no relation to John Loudon McAdam, 1756–1836, the British engineer who introduced improved roadways built of crushed stone known as "macadamized" roads. John Macadam, after whom macadamia nuts are named, is believed to have never seen nor tasted a macadamia nut. According to one account, he died at sea at the age of 38 years, after fracturing his ribs and developing pleurisy while on a rough voyage from Australia to New Zealand in 1865. Another account states that he died of overwork and self-neglect. He was a doctor of medicine but did not practice, and his medical knowledge did not save him.
- *Integrifolia* in the scientific name *M. integrifolia* is Latin for entire leaved, a botanical way of saying that the edges of the leaves are smooth, not toothed. In fact, whereas *M.*

integrifolia tend to have fewer teeth on the leaf edges than does *M. tetraphylla*, *M. integrifolia* leaves often have some teeth.
- *Tetraphylla* in the scientific name *M. tetraphylla* is Latin for four-leaved, reflecting the frequent occurrence of the leaves of this species in whorls of four on the stems. By contrast, *M. integrifolia* tends to have three leaves coming from the nodes of a stem.

PLANT PORTRAIT

Macadamia nuts are obtained from two species of evergreen trees of the genus *Macadamia*, and their hybrids. Smooth-shelled macadamia nut is native to the southeastern part of the state of Queensland where it grows in the rain forests and close to streams. Rough-shelled macadamia nut is native to southeastern Queensland and northeastern New South Wales and grows in rain forests, in moist places, and along stream banks. At the point where the two species meet, there are types that appear to be natural hybrids, and hybrids have in fact been produced in cultivation. The trees grow as tall as 20 m (66 ft.), but normally not much more than half this height. Their glossy, dark-green leaves resemble holly and are sometimes used to make Christmas wreaths. The fruits are borne profusely in long, grape-like open clusters, and fall to the ground (at least in most areas where they are grown) when mature. The fruit is made up of three parts, a fleshy husk, enclosing a spherical hard shell, inside of which is the kernel or macadamia nut. The hulled nut is approximately 2.5 cm (1 in.) in diameter. Nuts are not produced until the fourth or fifth year, and trees can live for hundreds of years. The bearing life of macadamia trees is unknown, but some of the earliest planted trees are still producing after a century. Mature trees can produce 27 to 68 kg (60–150 lb.) of unshelled nuts annually.

The husk and the shell account for most of the macadamia fruit's weight. The husk is approximately 50% of the weight of the fruit. The average kernel recovery rate from "in-shell nuts" (without the husk) is 25% to 35%. This means that the edible kernel is only 15% of the whole macadamia fruit. Not surprisingly, attempts have been made to find uses for the shells and husks. The shells can

FIGURE 60.1 Smooth-shelled macadamia nut (*Macadamia integrifolia*). (From Turner, 1893a.)

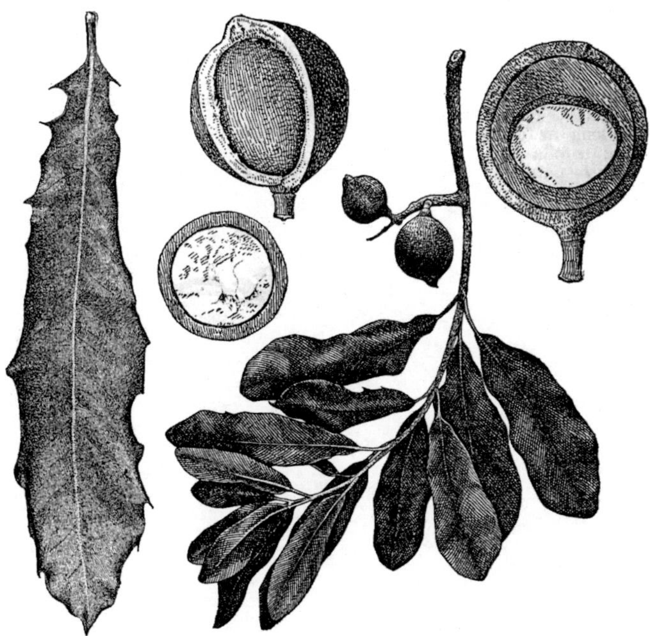

FIGURE 60.2 Smooth-shelled macadamia nut (*Macadamia integrifolia*). (From Bailey, 1916.)

be used as mulch, fuel for processing macadamia nuts, planting medium for culture of plants, manufacturing plastics, and as a substitute for sand in the sand-blasting process. The husks are used as mulch or composted for fertilizer.

Up to the middle of the nineteenth century, the nuts were known only to Australian aboriginals, who have gathered them since ancient times. The first macadamia tree to be cultivated, a smooth-shelled type, was planted in 1858 and is thought to still be alive. Commercial orchards were established in Australia in 1888. Macadamia was first grown in Hawaii in 1882 (once again the original tree is still alive). Improved varieties are grown today. Hawaii and Australia are the largest producers. There is also some minor production in California and in numerous semitropical and tropical countries, especially in Latin America (notably Guatemala and Brazil) and Africa (notably South Africa, Malawi, and Kenya). The quality of the kernels of rough-shell macadamia is more variable than that of the smooth-shell macadamia. However, some rough-shell nuts are as good as those of the smooth-shell nuts, and in Australia, there is some mixing of the two types of nuts. In Hawaii, only the smooth-shell type is grown, whereas in California, the rough-shell type has proven to be more suitable for the climate. However, most commercial production in the world uses the smooth-shell macadamia nut.

Australian Aborigines collected macadamia nuts for sale or barter among themselves and with European settlers. The Aborigines extracted oil from the nuts for use as a liniment base, as a cosmetic oil, and as an aid in decorating the face and body. Copying the knowledge of the aborigines, oil is expressed to some extent from macadamia nuts not good enough to sell and used by the cosmetics industry, especially in Japan, to produce soaps, sunscreens, and shampoos.

CULINARY PORTRAIT

The famous American plant breeder and horticulturist Luther Burbank (1849–1926), proclaimed the macadamia to be the perfect nut. It is a high-calorie, gourmet delicacy, ranking as one of the world's most expensive and tasty nuts. The meat is white, crisp, and slightly sweet, somewhat reminiscent of hazelnut. Rich in oil, the nuts produce excellent nut butter. The nuts are generally cooked

FIGURE 60.3 Macadamia nuts (*Macadamia integrifolia*). (Photograph by Eric Johnson.)

before eating, either dry-roasted or cooked in oil. They are often marketed in vacuum-packed glass jars, and purchasing such containers is advisable to ensure freshness. Macadamia nuts are consumed mostly as snacks, typically as hors d'oeuvres with cocktails or other beverages. Chocolate-covered macadamia nuts have become very popular in recent years. The nuts are occasionally used for baking and toppings, especially for ice cream, and can be used to flavor curries, salads, stews, vegetables, rice, and pastries but are expensive for such purposes. Macadamia candy, jam, cakes, nut meal, and soups have been produced and exported from Hawaii. This premium nut is sometimes distributed as a treat to airline passengers. Macadamia nuts are high in calories because of the very high fat content—typically approximately 80%—higher than any other commercial nut. They spoil less quickly than other fatty nuts, but once shelled, the kernels will become rancid. To preserve their quality, the nuts can be stored (sealed to prevent absorption of odors) in a refrigerator or even better in a freezer. Because the shells are so hard, macadamia nuts are seldom marketed whole, although special nutcrackers have been developed and a hammer can be used (see below for additional information).

Culinary Vocabulary

- "Pupu" (pu pu, pronounced as "POO-poo") is a Hawaiian term for various appetizers or hors d'oeuvres, either hot or cold. Not surprisingly in view of its importance as a crop in Hawaii, macadamia nuts are a common form of pupu in Hawaii. A "pupu platter" (or pu pu platter) is a tray with various hors d'oeuvres, commonly available at Chinese and other Asian restaurants in North America, sometimes with a lighted burner in the center for heating the offerings.

CURIOSITIES OF SCIENCE AND TECHNOLOGY

- When macadamia nuts were first collected, they were thought to be poisonous, the aborigines said to have provided this information (which was likely applicable to another tree species). Walter Hill (noted above as a co-collector of the macadamia nut in 1857), Director of the Brisbane Botanic Gardens, gave some seeds to a young assistant to crack out of their shells in a vice. To his horror, Hill found his assistant eating some of the kernels. Several days later, when the lad's health remained sound and his assistant stated that the nuts were outstanding, Hill tasted a kernel and became the first European to realize the commercial potential of macadamia nuts.

- Macadamia is the only major horticultural crop to be developed from the native Australian flora, a country with an impressive 25,000 native plant species, more than in North America north of Mexico.
- Flowers of macadamia trees are carried on flowering branches (technically, *racemes*), each often with several hundred flowers. However, not more than approximately 10 of the flowers on a raceme will develop into mature nuts. The tremendous apparent overproduction of flowers by many trees seems extremely wasteful but may be explained in various ways, for example, as a means of creating a large floral display to attract pollinators so that at least some of the flowers will be fertilized.
- Smaller macadamia nuts have been fashioned into buttons in Australia.
- During the Second World War, food packets were sent to troops around the world and sometimes they contained macadamias. In many cases, rats would attack the packs, and it is said that the macadamias were always eaten first. It was reported that some knowledgeable rats even learned to gnaw through tin lids to get the delectable treats.
- Macadamia nuts are extremely difficult to break open. It takes 140 kg/cm^2 (more than 2000 lb./sq. in.) of compression to crack the macadamia shell. That is why they are seldom sold in grocery stores (except sometimes where they are grown). Factories have special equipment to husk and shell macadamias. Aborigines of Australia were skilled in cracking macadamia nuts. Their main method required a hard rock with a depression, which would hold the nut. A flat, wedge-shaped stone was then placed on top of the nut, and this was struck with a heavy stone. At some of the aboriginal feasting grounds, the early European settlers found large heaps of shells. Those who have tried this method say that it does less damage than a hammer.
- It can be very difficult to date Australian trees like macadamia nut by counting growth rings in the trunk, as is commonly done with temperate region deciduous trees. Because of the climate, most of Australia's native trees are evergreen and there is continuous growth all year so that there are no well-defined annual growth rings.
- Macadamia is the most important tree crop in Hawaii (and the third largest crop), contributing \$300 million each year to the economy and employing an estimated 2000 people. Although macadamia nuts are native to Australia, they first developed as a major commercial nut in Hawaii, after they were taken there. For many years, there was intense resentment in Australia that foreigners had in effect "stolen" their crop, although in the last decade Australia has considerably increased its market share of the world macadamia market. Most of the crops grown in all countries of the world are in fact not native but have been acquired or "stolen" from other countries. Although this is of historical interest, more relevant is the continuing effort to use wild plants, especially from tropical regions, as material with which to develop new profitable crops or drugs. Many developing countries deeply object to the idea that their present genetic resources could be taken by foreigners who are unwilling to share in the profits (see the discussion of "biopiracy" in Chapter 1).

KEY INFORMATION SOURCES

Allan, P. 2001. *Illustrated guide to identification of macadamia cultivars in South Africa.* South African Macadamia Growers' Association and University of Natal, Pietermaritzburg, South Africa. 38 pp.

Bell, H.F.D. 1996. Breeding and selecting new varieties of macadamia. *WANATCA Yearbook*, 20:13–18.

Bittenbender, H.C. (Ed.). 1993. *Proceedings of the first international macadamia research conference, Kona Hilton, Kailua-Kona, Hawaii, July 28–30, 1992*. Hawaii Cooperative Extension Service, College of Tropical Agriculture and Human Resources, University of Hawaii at Manoa, Honolulu, HI. 141 pp.

Cavaletto, C.G. 1980. Macadamia nut. In *Tropical and subtropical fruits: composition, properties and uses*. Edited by S. Nagy and R.E. Shaw. AVI Publishing Company, Westport, CT. pp. 542–561.

Cavaletto, C.G. 1983. Macadamia nuts. In *Handbook of tropical foods*. Edited by H.T. Chan, Jr. Marcel Dekker Inc., New York. pp. 361–397.

Fouras, D. 1973. *The Australian macadamia nut industry: a review of the current situation and future market prospects*. Queensland Department of Primary Industries, Marketing Services Branch, Brisbane, Australia. 24 pp.

Hamilton, R.A., and Fukunaga, E.T. 1959. *Growing macadamia nuts in Hawaii*. Hawaii Agricultural Experiment Station, Honolulu, HI. 51 pp.

Hamilton, R.A., and Ito, P.J. 1984. *Macadamia nut cultivars recommended for Hawaii*. Hawaii Institute of Tropical Agriculture and Human Resources, College of Tropical Agriculture and Human Resource, University of Hawaii at Manoa, HI. 7 pp.

Hamilton, R.A., Ito, P.J., and Chia, C.L. 1983. *Macadamia: Hawaii's dessert nut*. Circular 485. Revised edition. University of Hawaii, Honolulu, HI. 13 pp.

Hancock, W.M. 1991. *Macadamia reference manual: Macadamia integrifolia*. Tree Nut Authority, Blantyre, Malawi. 136 pp.

Hartung, M.E., and Storey, W.B. 1939. The development of the fruit of *Macadamia ternifolia*. *J. Agric. Res.* 59:397–406.

Huett, D.O. 2004. Macadamia physiology review: a canopy light response study and literature review. *Aust. J. Agric. Res.* 55:609–624.

Johnson, L.A.S. 1954. *Macadamia ternifolia* and a related new species. *Proc. Linn. Soc. N. S. W.* 79:15–18.

Joubert, A.J. 1986. Macadamia. In *CRC handbook of fruit set and development*. Edited by S.P. Monselise. CRC Press, Boca Raton, FL. pp. 247–252.

Leigh, D.S., Trochoulias, T., and Austen, V.C. 1971. Propagation of macadamia; a review of investigations in New South Wales. *Agric. Gaz. N. S. W.* 82:333–336.

Mason, R.L., and McConachie, I. 1994. A hard nut to crack. A review of the Australian macadamia nut industry. *Food Aust.* 46:466–471.

McGregor, A. 1991. *A review of the world production and market environment for macadamia nuts*. Pacific Islands Development Program, East-West Center, Honolulu, HI. 44 pp.

McHargue, L.T. 1996. Macadamia production in southern California. In *Progress in new crops*. Edited by J. Janick. ASHS Press, Arlington, VA. pp. 458–462.

Nagao, M.A. 2008. *Macadamia integrifolia*, macadamia nut. In *The encyclopedia of fruit & nuts*. Edited by J. Janick and R.E. Paull. CABI, Wallingford, Oxfordshire, U.K. pp. 600–610.

Nagao, M.A., and Hirae, H.H. 1992. Macadamia: cultivation and physiology. *Crit. Rev. Plant Sci.* 10:441–470.

Ohler, J.G. 1969. Macadamia nuts. *Trop. Abstr.* 24:781–791.

Paino, A. 1998. *Growing macadamia nuts*. Agriculture Western Australia, Perth, Western Australia. 43 pp.

Peace, C., Allan, P., Vithanage, V., Turnbull, C., and Carroll, B. 2001. *Identifying relationships between macadamia varieties in South Africa by DNA fingerprinting*. South African Macadamia Growers' Association Yearbook 2001, 9:64–71.

Shigeura, G.T. and Ooka, H. 1984. *Macadamia nuts in Hawaii: history and production*. Research and Extension Series 039. University of Hawaii, Manoa, HI. 91 pp.

Stephenson, R.A., and Trochoulias, T. 1994. Macadamia. In *Handbook of environmental physiology of fruit crops. Vol. 2. Subtropical and tropical crops*. Edited by B. Schaffer and P.C. Andersen. CRC Press, Boca Raton, FL. pp. 147–163.

Stephenson, R.A., and Winks, C.W. 1991. *Macadamia integrifolia* Maiden & Betche. In *Plant resources of South-East Asia. Vol. 2. Edible fruits and nuts*. Edited by E.W.M. Verheij and R.E. Coronel. Pudoc, Leiden, the Netherlands. pp. 195–198.

Storey, W.B. 1965. The ternifolia group of *Macadamia* species. *Pac. Sci.* 19:507–514.

Thomson, P.H. 1980. The macadamia in California. *California Rare Fruit Growers Yearbook*, 12:46–115.

Urata, U. 1954. *Pollination requirements of macadamia*. Hawaii Agricultural Experiment Station, University of Hawaii, Honolulu, HI. 40 pp.

Wagner-Wright, S. 1995. *History of the macadamia nut industry in Hawai'i, 1881–1981: from bush nut to gourmet's delight*. E. Mellen Press, Lewiston, N.Y. 265 pp.

Specialty Cookbooks

Buyers, R. 1982. *The marvellous macadamia nut*. Irena Chalmers Cookbooks, New York. 84 pp.

De Domenico, A. 1979. *Anita De Domenico's macadamia nut recipes from our plantation house*. 3rd ed. Hawaiian Holiday Macadamia Nut Co., Honokaa, HI. 100 pp.

Food Consultants of Hawaii. 1992. *Take home aloha recipes featuring Hawaiian macadamia nut oil*. Food Consultants of Hawaii, Honolulu, HI. 32 pp.

Koy, P. 1959. *Recipes by Paul Koy for Royal Hawaiian macadamia nuts*. Castle & Cooke, Inc., Honolulu, HI. 24 pp.

Macadamia Power Pty. 1982. *Macadamia power in a nutshell*. Macadamia Power Pty., Hamilton, Australia. 120 pp.

Mansfield, L. 1998. *The Mauna Loa macadamia cooking treasury*. Celestial Arts, Berkeley, CA. 168 pp.

61 Mangosteen

Family: Clusiaceae (Guttiferae; garcinia or mangosteen family)

NAMES

Scientific Name: *Garcinia mangostana* L.

- The name "mangosteen" and the word *mangostana* in the scientific name *G. mangostana* are based on the Malay word for the plant, *manggista* or *mangustan* (replaced by *manggis* in modern times).
- The mangosteen is not related to the mango (*Mangifera indica* L.), which is claimed to be the most important fruit produced in the Tropics.
- Suggested pronunciation: MANG-uh-steen.
- The mangosteen is occasionally known as king's fruit (or king of fruits, a name also applied to the mango), mangostan, and purple mangosteen.
- So-called "yellow mangosteens" are species of *Garcinia* that produce yellow fruit (in contrast to the purple fruit of the mangosteen). The yellow fruit is usually extremely sour and is occasionally used for preserves.
- "Black mangosteen" is *Garcinia indica* (Thouars) Choisy, a species native to India, with fruits that are used for flavoring and other purposes. In India (and in English), the plant and the spice produced from it are often called kokam or kokum. The spice is also known as black mangosteen.
- The "button mangosteen" (*Garcinia prainiana* King) is still another edible-fruited species, albeit with very limited commercial value. Its fruit is reminiscent of a tangerine, but the skin is very thin, unlike the mangosteen. The species is cultivated in southeast Asia and sometimes in Florida.
- The genus name *Garcinia* commemorates Laurent Garcin (1683–1751), French botanist, physician, and scholar, who traveled in India and Malaysia.

PLANT PORTRAIT

The mangosteen is a very slow-growing tree reaching 6 to 25 m (20–82 ft.) in height. The leaves are thick, leathery, and evergreen, and the large pink flowers are so attractive that the plant is often cultivated as an ornamental. The species may have originated in the Sunda Islands and the Moluccas or perhaps Malaya. It has been suggested that it was first domesticated in Thailand or Myanmar (Burma). It is extensively cultivated in Thailand, southern Vietnam, Burma, Malaya, Philippines, Indonesia, and Singapore and occasionally in other tropical areas including Ceylon and India. Mangosteens are considered difficult to grow and require a tropical climate. Planted trees have done poorly in Hawaii, California, and Florida. Depending on location and age, trees may produce no fruit, only a few dozen fruit, or as many as 3000 in a year.

The fruit is reminiscent of an orange in size, thickness of the rind, and segmentation, although the purple color is distinctive. The fruit is round, dark-purple to red-purple, and smooth externally, 3.4 to 7.5 cm (1⅜–3 in.) in diameter. The rind is 6 to 10 mm (1/4–3/8 in.) thick and red, while it is purplish-white on the inside. The rind contains bitter, yellow latex and a purple, staining juice that has been used for dyeing in the leather industry. The interior of the fruit is made up of four to eight, fleshy, snow-white, juicy segments. The fruit is seedless or contains one to five edible seeds approximately 2.5 cm (1 in.) long, which cling to the flesh. The rind and a thick reddish membrane enclosing the central white flesh are inedible and constitute approximately three-fourth of the fruit.

CULINARY PORTRAIT

The mangosteen has been acclaimed as exquisitely luscious and delicious, indeed by some as the best of tropical fruits (hence the occasional names "king of fruit," "queen of fruit," and "princess of fruit"). Unlike many other tropical fruits, the mangosteen does not require getting used to and is almost universally liked by people tasting it for the first time. The pulp is soft and melting, reminiscent of a ripe plum, with a sweet, slightly acidic taste. The flavor has been described as a mixture of grape, peach, and apple. Mangosteens are usually eaten fresh as dessert. The fleshy segments are sometimes canned or made into jams, preserves, and juice, but the flavor does not keep well in juice and preserves. Immature fruit have proven much more suitable for canning than mature fruit, the flavor also not keeping well in older fruit. A purée makes an excellent flavoring or topping for ice cream, sherbet, yogurt, cakes, and puddings. Vinegar is prepared from the fruit in Asia. The fruit purchased from supermarkets should be eaten promptly but can be refrigerated for up to a week. For each segment in the fruit, there is a corresponding lobe from the old flower at the apex of the fruit, and it has been recommended that fruits with the highest number of such lobes be chosen because as a general rule the more segments a mangosteen has, the fewer seeds there are. In Asia, oil is extracted from the seeds.

The following procedures have been recommended to open a mangosteen. First fold your hands in front of you with your fingers interlocked. Then pick up the fruit, stem up, between the heels (not the palms) of your hands. Gently but firmly squeeze until the thick peel suddenly splits in half. Another option is to use a sharp knife to cut through the rind approximately 1 cm (1/3 in.) deep all the way around the "equator" (but not through the segments) and then simply lift off the top. Discard the top half of the rind, and lift the segments out with a fork. Some authorities recommend chilling the fruit before eating.

FIGURE 61.1 Mangosteen (*Garcinia mangostana*). (From Curtis, 1855, vol. 81, plate 4847.)

When buying mangosteens, it is best to choose purple-skinned fruits that yield to gentle pressure. Fruits with very hard skin may be overripe. Mangosteens spoil quickly and should be consumed soon after purchase. They may be stored for up to a week in a refrigerator.

CURIOSITIES OF SCIENCE AND TECHNOLOGY

- The mangosteen was prescribed by colonial doctors in Southeast Asia for fevers, dysentery, and other illnesses.
- *Garcinia* species are the source of a yellow gum resin, known as gamboge, obtained from incisions made in the bark. Gamboge has been used as a dye to provide yellow color for the silk robes of Buddhist monks. It is also used as an artist's pigment and medicinally as an expectorant.
- The mangosteen is unusual in that its seeds are not the product of true sexual fertilization but of a process called apomixis. The plants produced from the seeds are virtually identical genetically to the maternal plant. (By contrast, apple seeds are normally capable of producing trees and apples that are different from the tree they came from.) Indeed, there is very little variation among cultivated mangosteen plants; hence, cultivated varieties differ relatively little.

KEY INFORMATION SOURCES

Almeyda, N., and Martin, F.W. 1976. *Cultivation of neglected tropical fruits with promise. Part 1. The mangosteen.* Agricultural Research Service S-155. United States Department of Agriculture, Mayaguez, Puerto Rico. 18 pp.

Barrett, O.W. 1994. The mangosteen and related species. *WANATCA Yearbook*, 18:54–63.

Campbell, C.W. 1967. Growing the mangosteen in southern Florida. *Proc. Fla. State Hort. Soc.* 79:399–400.

Choehom, R., Ketsa, S., and Doorn, W.G. van. 2003. Chilling injury in mangosteen fruit. *J. Hortic. Sci. Biotechnol.* 78:559–562.

Chong, S.T. 1992. Vegetative propagation of mangosteen (*Garcinia mangostana* L.). *Acta Hortic.* 292:73–80.

Gonzalez, L.G., and Anoos, Q.A. 1951. The growth behavior of mangosteen and its graft affinity with relatives. *Philipp. Agric.* 35:379–395.

Idris, S., and Rukayah, A. 1987. Description of the male mangosteen (*Garcinia mangostana* L.) discovered in Peninsular Malaysia. *Short MARDI Res. Bull.* 15:63–66.

Jansen, P.C.M. 1991. *Garcinia* L. In *Plant resources of South-East Asia. 2. Edible fruits and nuts.* Edited by E.W.M. Verheij and R.E. Coronel. Pudoc, Leiden, the Netherlands. pp. 175–177.

Kader, A.A. 2001. Mangosteen: recommendations for maintaining postharvest quality. *Perishables Handling Quart.* 106(May):15–16.

Kanchanapoom, K., and Kanchanapoom, M. 1998. Mangosteen. In *Tropical and subtropical fruits.* Edited by P.E. Shaw, H.T. Chan, Jr., and S. Nagy. Agscience, Auburndale, FL. pp. 191–216.

Ketsa, S., and Koolpluksee, M. 1993. Some physical and biochemical characteristics of damaged pericarp of mangosteen fruit after impact. *Postharvest Biol. Technol.* 2:209–215.

Krishnamurthi, S., and Madhava Rao, V.N. 1965. The mangosteen (*Garcinia mangostana* L.) and its introduction and establishment in peninsular India. *Adv. Agric. Sci.* 1965:401–420.

MacLeod, A.J., and Pieris, N.M. 1982. Volatile flavour components of mangosteen, *Garcinia mangostana*. *Phytochemistry*, 21:117–119.

Mahabusarakam, W., and Wiriyachitra, P. 1987. Chemical constituents of *Garcinia mangostana*. *J. Nat. Prod.* 50:474–478.

Martin, F.W. 1980. Durian and mangosteen. In *Tropical and subtropical fruits: composition, properties and uses.* Edited by S. Nagy and R.E. Shaw. AVI Publishing Company, Westport, CT. pp. 407–414.

Morton, J. 1987. Mangosteen. In *Fruits of warm climates.* Creative Resource Systems, Winterville, NC. pp. 301–304.

Othman, Y., Tindall, H.D., Minini, U.G., and Hodder, A. 1995. *Mangosteen cultivation.* Food and Agriculture Organization of the United Nations, Rome, Italy. 114 pp.

Ramage, C.M., Sando, L., Peace, C.P., Carroll, B.J., and Drew, R.A. 2004. Genetic diversity revealed in the apomictic fruit species *Garcinia mangostana* L. (mangosteen). *Euphytica*, 136:1–10.

Richards, N.J. 1990. Studies in *Garcinia*, dioecious tropical fruit trees, the origin of the mangosteen (*Garcinia mangostana* L.). *Bot. J. Linn. Soc.* 103:301–308.

Rukayah, A., and Zabedah, M. 1992. Studies on early growth of mangosteen (*Garcinia mangtostana* L.). *Acta Hortic.* 292:93–100.

Salakpetch, S., and Paull, R.E. 20008. *Garcinia mangostana*, mangosteen. In *The encyclopedia of fruit & nuts.* Edited by J. Janick and R.E. Paull. CABI, Wallingford, Oxfordshire, U.K. pp. 263–267.

Shrivastava, H.C., Kripal Singh, K., and. Mathur, P.B. 1962. Refrigerated storage of mangosteen *(Garcinia mangostana). Food Sci.* 1962(Aug.):226–228.

Tongdee, S.C., and Suwanagul, A. 1989. Postharvest mechanical damage in mangosteens. *ASEAN Food J.* 4:151–155.

Verheij, E.W.M. 1991. *Garcinia mangostana* L. In *Plant resources of South-East Asia. Vol. 2. Edible fruits and nuts.* Edited by E.W.M. Verheij and R.E. Coronel. Pudoc, Leiden, the Netherlands. pp. 177–181.

Yapwattanaphun, C., Subhadrabandhu, S., Honsho, C., and Yonemori, K. 2004. Phylogenetic relationship of mangosteen (*Garcinia mangostana*) and several wild relatives (*Garcinia* spp.) revealed by ITS sequence data. *J. Am. Soc. Hortic. Sci.* 129:368–373.

Specialty Cookbooks

Gerungan, L. 2008. *The Bali cookbook: over 100 delicious recipes from Bali's most famous chef.* Kyle Cathie, London. 192 pp.

Hawaii Farm Bureau Federation, Namkoong, J., and Matson-Mathes, H. 2006. *The Hawaii Farmers Market cookbook.* Watermark, Honolulu, HI. 182 pp.

Laursen, T.D. 2003. *From Bangkok to Bali in 30 minutes: 175 fast and easy recipes with the lush, tropical flavors of Southeast Asia.* Harvard Common Press, Boston, MA. 288 pp.

Straten, M. van. 2007. *The complete superfoods cookbook: dishes and drinks for energy, detoxing, and healing.* Whitecap, North Vancouver, BC. 256 pp.

Straten, M. van. 2007. *More super juice: juicing for health and healing.* Whitecap, North Vancouver, BC. 144 pp.

Sweetser, W. 2008. *Juice it: 85 deliciously healthy drinks.* New Holland, London. 192 pp.

62 Medlar

Family: Rosaceae (rose family)

NAMES

Scientific Name: *Mespilus germanica* L.

- The origin of the English name "medlar" and the genus name *Mespilus* are obscure. One interpretation is that they are derived from old names for the plant: the Medieval Latin *mespila*, from the classical Greek *mespilē*.
- Pronunciation: medlr.
- The Japanese medlar is the loquat (see Chapter 58). Confusingly, the medlar has been called the "northern loquat."
- The Italian, Mediterranean, Naples, Neopolitan, or Welsh medlar is azarole (*Crataegus azarolus* L.), a minor fruit tree grown in Europe.
- The rock medlar is *Amelanchier ovalis* Medik., a minor fruit tree of North America.
- The Indian medlar, also known as the wild medlar (of South Africa), is *Vangueria infausta* Burch.
- The medlar was given the least complementary name of any edible fruit, indeed possibly of any edible plant: "open-arse," the English name that was common a thousand years ago, and persisted for many centuries. Equally distressing is the French nickname *cul-de-chien*, "dog's arse." There are similar names in German. The basis of these unflattering names is the appearance of the nonstem end of the fruit, which gapes open to expose the fruit's interior. In the time of William Shakespeare (1564–1616), the name was often too embarrassing to mention in public; in *Romeo and Juliet*, instead of uttering "open-arse," Mercutio delicately says "open et cetera." The name is so shocking that discussions of the medlar almost never mention its horrid old name.
- In medieval times in Europe, medlars were called "openers." This is often "explained" today as based on the necessity of leaving the fruits to decay before they could be opened and used as food. In fact, the word is a variation of "openars," derived from "open-arse," as explained above.
- *Germanica* in the scientific name *M. germanica* is the Latinized word for German. It was once thought that the medlar was native to Germany.

PLANT PORTRAIT

The medlar is a small, deciduous, relatively slow-growing tree, 3 to 6 m (1–20 ft.) in height. The plant is long-lived; one English tree recorded as surviving almost 400 years. The tree is quite ornamental, with contorted branches and often even a contorted trunk, a mass of large, very snowy-white or pinkish flowers in the early spring, and beautiful reddish-brown foliage in the fall. Wild plants are typically thorny, whereas cultivated varieties are thornless. The fruit generally resembles large crabapples, greenish-yellow when unripe, with a curious indentation on top which exposes the five seed compartments, each of which contains a single seed. The medlar is such an unusual fruit that it has stimulated some authors to describe it in memorable prose. English garden writer J.C. Woodsford in 1939 called it "a crabby-looking, brownish-green, truncated, little spheroid of unsympathetic appearance." The English novelist and playwright D.H. Lawrence characterized the fruits as "wineskins of brown morbidity." Staub (2007) described the medlar as having "rather controversial physiognomy, looking for all the world like a crabapple into which someone has inserted a small explosive, resulting in a rather provocative opening sporting a tattered skirt." There are several cultivated varieties, including some with large fruits, up to 6.4 cm (2½ in.) in diameter. The species is native to the eastern part of the Mediterranean area and western Asia. It may have been grown by the ancient Assyrians and Babylonians. It was cultivated by the classical Greeks and Romans and was present in orchards all over medieval Europe, when medlars were as common as apples and cherries. Medlars were frequently found for sale in markets and were a basic food resource. The fruit was quite popular in Victorian England. In northern Europe, medlars are usually not picked until after they have been softened by frost, then they are stored until ripe. Medlars are rarely cultivated for fruit today and in North America are obtainable normally only from trees that are planted as garden or ornamental specimens. The fruits are still found in the marketplace in Italy and occasionally in the Caucasus and southern Ukraine.

CULINARY PORTRAIT

It is commonly claimed that in warm climates, as typical of Italy, the medlar's fruits often ripen to the point that they can be eaten raw. However, this appears to be based on confusion with the loquat (see Chapter 58; also see the Names section for similar names applied to the two species). In Italy as well as in cool climates such as that of England, the fruit needs to be "bletted," that is, softened by partial rotting, which sounds disgusting but actually improves the taste considerably (the word "blet" was coined in 1839 by botanist John Lindley). When the fruit is picked in late autumn, it is still very hard and unpalatable and if eaten is quite astringent (puckering the mouth). The fruit is kept in a cool place, often in sawdust, and internal fermentation occurs, the fruit flesh becoming soft like applesauce or stewed apple, aromatic, and brownish, the skin turning dark brown and rather wrinkled. Maturity (i.e., rotting) can be judged by squeezing lightly: the flesh should yield easily under the thick skin. After this treatment, the fruit can be sucked empty, leaving the skin and stones behind (this is recommended as the easiest way to consume the fresh fruit). The mushy pulp can also be scooped out and mixed with cream and/or sugar as a dessert. The flesh is rather coarse (gritty), tart, and acidic, with a distinctive taste that has been described as like apple sauce with considerable cinnamon—appealing to some and not others. The fruit is high in pectin and is used for preserves, jams, jellies, tarts, syrups, and sweet beverages. A fermented liqueur or wine is sometimes made from medlars, and the fruits are occasionally candied, baked, stewed, or roasted. Medlars were a traditional accompaniment of port for dessert and are still recommended with a glass of vintage port.

Culinary Vocabulary

- A famous marmalade-like preserve known in Orléans, France, as *contignac*, was always offered to a French sovereign when he entered the town and indeed was presented to Joan of Arc (ca. 1412–1431), when she led her troops into that city. It has been said that medlars were the foundation of this preserve, although others contend that it was made from quince (*Cydonia oblonga* Mill.).
- The creamy, rotted flesh of the medlar is said to make an excellent "cheese," the term used by virtue of the similarity in texture to some true cheeses. Medlar cheese is similar to lemon curd.

CURIOSITIES OF SCIENCE AND TECHNOLOGY

- Medlar wood is hard and was once used to make spearpoints, cudgels, clubs, and hard sticks for hunting or warfare. The wood was also used for making parts of windmills,

FIGURE 62.1 Medlar (*Mespilus germanica*). (From Morren, 1851–1885, vol. 6.)

FIGURE 62.2 Medlar (*Mespilus germanica*). (From Loudon, 1844, vol. 6.)

especially some of the turning wheels, as well as walking sticks and canes. Medlar wood is prized for making violins.

- It was widely through in past centuries in Europe that if apple or pear trees were grown close to a medlar tree, the fruit would be of rotten quality, and so medlars were planted at some distance from the other trees.

- In past times in England, it was a popular practice to graft different fruits of the rose family, such as medlar, pear, quince, and azarole, onto a hawthorn rootstock, fabricating a tree that produced several different kinds of fruits.
- Medlar grafted onto a hawthorn root often produces much larger fruits than are normally developed.
- In Victorian times in England, it was common to keep medlars in a silver dish of moist sawdust on the sideboard (i.e., dining room furniture with compartments and shelves for holding articles of table service).
- MEDLAR is an acronym for MEDical Literature Analysis and Retrieval and MEDium LARge and an abbreviation of MEchanizing Deduction in Logics of prActical Reasoning.
- The *makila* is a unique walking stick manufactured from local medlar trees by the Basque people of southwestern France. Basque shepherds of the regions always kept a makila with them, considering it as an aid to walking in the treacherous foothills of the Pyrenees Mountains, a weapon, and a mystical aid. A makila has a message inscribed on it in the Basque language and is produced over a decade by artisans who guard the secrets of its construction.

KEY INFORMATION SOURCES

Akhmedov, A.I. 1979. New flavouring extract from wild medlar. *Konservnaya Ovoshchesushil'naya Promyshlennost'*, 7:22–24 (in Russian).

Ayaz, F.A, Huang, H.S., Chuang, L.T., VanderJagt, D.J., and Glew, R.H. 2002. Fatty acid composition of medlar (*Mespilus germanica* L.) fruit at different stages of development. *Ital. J. Food Sci.* 14:439–446.

Ayaz, F.A., Glew, R.H., Huang, H.S., Chuang, L.T., VanderJagt, D.J., and Strnad, M. 2002. Evolution of fatty acids in medlar (*Mespilus germanica* L.) mesocarp at different stages of ripening. *Grasas Aceites*, 53:352–356.

Aydin, N., and Kadioglu, A. 2001. Changes in the chemical composition, polyphenol oxidase and peroxidase activities during development and ripening of medlar fruits (*Mespilus germanica* L.). *Bulgarian J. Plant Physiol.* 27(3–4):85–92.

Baird, J.R., and Thieret, J.W. 1989. The medlar (*Mespilus germanica*, Rosaceae) from antiquity to obscurity. *Econ. Bot.* 43:328–372.

Bernal, A.A. 1976. Hormonal treatments and rapid maturing of the medlar fruit. Summary of three years of experiments. *Technol. Agrar.* 3:133–143 (in Spanish, English summary).

Bostan, S.Z. 2002. Interrelationships among pomological traits and selection of medlar (*Mespilus germanica* L.) types in Turkey. *J. Am. Pomol. Soc.* 56:215–218.

Browicz, K. 1968. Distribution of woody Rosaceae in West Asia. II. On the distribution of *Mespilus germanica* L. *Arbor Kornick.* 13:21–29. (Translated 1974).

Butzke, H. 1986. On the geographical distribution and the habitats of medlar *Mespilus germanica* L. in the western part of north Rhine-Westphalia West Germany and the properties of its wood. *Decheniana*, 139:178–192 (in German).

Byatt, J.I., Ferguson, I.K., and Murray, B.G. 1977. Intergeneric hybrids between *Crataegus* L. and *Mespilus* L.: a fresh look at an old problem. *Bot. J. Linn. Soc.* 74:329–343.

Dincer, B., Colak, A., Aydin, N., Kadioglu, A., and Guner, S. 2002. Characterization of polyphenoloxidase from medlar fruits (*Mespilus germanica* L., Rosaceae). *Food Chem.* 77:1–7.

Glew R.H., Ayaz, F.A., Vanderjagt D.J., Millson, M., Dris, R., and Niskanen, R. 2003. Mineral composition of medlar (*Mespilus germanica*) fruit at different stages of maturity. *J. Food Qual.* 26:441–447.

Glew, R.H., Ayaz, F.A., Sanz, C., VanderJagt, D.J., Huang, H.S., Chuang, L.T., et al. 2003. Changes in sugars, organic acids and amino acids in medlar (*Mespilus germanica* L.) during fruit development and maturation. *Food Chem.* 83:363–369.

Glew, R.H., Ayaz, F.A., Sanz, C., VanderJagt, D.J., Huang, H.S., Chuang, L.T., et al. 2003. Effect of postharvest period on sugars, organic acids and fatty acids composition in commercially sold medlar (*Mespilus germanica* 'Dutch') fruit. *Eur. Food Res. Technol.* 216:390–394.

Jacob, H., and Dietrich, H. 2002. Medlars. An uncommon fruit with special properties. *Kleinbrennerei*, 54(5):4–6 (in German).

Lipp, L.F. 1972. Do you know the medlar? *Plants Gard.* 28(1):34.

Miller, R.H. 1984. The multiple epidermis-cuticle complex of medlar fruit *Mespilus germanica* L. (Rosaceae). *Ann. Bot.* 53:779–792.

Phipps, J.B., O'Kennon, B., and Lance, R. 2003. *Hawthorns and medlars*. Timber Press, Portland, OR. 139 pp.

Phipps, J.B., Weeden, N.F., and Dickson, E.E. 1991. Isozyme evidence for the naturalness of *Mespilus* L. (Rosaceae, subfam. Maloideae). *Syst. Bot.* 16:546–552.

Romero-Rodriguez, A., Simal-Lozano, J., Vazquez-Oderiz, L., Lopez-Hernandez, J., and Gonzalez-Castro, M.J. 2000. Physical, physicochemical and chemical changes during maturation of medlars and persimmons. *Dtsch. Lebensmitt. Rundsch.* 96(4):142–145.

Staub, J.S. 2007. *75 remarkable fruits for your garden*. Gibbs Smith, Salt Lake City, UT. 224 pp.

Webster, T. 2008. *Mespilus germanica*, medlar. In *The encyclopedia of fruit & nuts*. Edited by J. Janick and R.E. Paull. CABI, Wallingford, Oxfordshire, U.K. pp. 674–678.

Specialty Cookbooks

Medieval and Victorian era European cookbooks commonly provided medlar recipes. However, there are almost no modern cookbooks that deal in any substantial way with the medlar, and recipes for "medlar" in cookbooks and on the Web more often than not refer to the Japanese medlar, that is, the loquat. One of the best ways to prepare the fruit is as a jelly, and the same recipes prescribed for quince may be used. In brief, bletted fruit can be chopped, simmered in water for 3 hours, filtered overnight through cheesecloth (to get rid of the seeds and skins), combined with an equal amount of sugar, and simmered until it gels.

63 Melons (Exotic)

Family: Cucurbitaceae (gourd family)

This chapter features

Bitter melon (*Momordica charantia*)
Horned melon, Kiwano (*Cucumis metuliferus*)

BITTER MELON

Names

Scientific Name: *Momordica charantia* L.

- The "bitter" in "bitter melon" is an accurate description, the taste reminiscent of quinine. Whether it is a "melon" is subjective; although melons are usually considered to be dessert plants, the bitter melon is consumed as a vegetable and looks much more like a cucumber.
- Bitter melon is also called African cucumber, ampalaya (a name used in the Philippines), balsam pear, bitter cucumber, bitter gourd, Chinese bitter melon, foo gua (China), karela (carilla, karala, a name used in China, Pakistan, and India and reflecting the Indian region of Kerala), kho-qua (Vietnamese), maiden's blush (a West Indies name), la-kwa, and leprosy gourd. In Cuba, Puerto Rico, and Spanish-speaking tropical America, the names balsamina and cundeamor are encountered.
- The name "leprosy gourd" reflects use in the Orient for the treatment of leprosy.
- Bitter melon is also called balsam apple, a name better reserved for its relative *Momordica balsamina* L., a plant with smaller fruits, which are pickled in India when young and green and cooked in curry dishes and stews when ripe and red.
- The name "alligator pear" has been applied to bitter melon, a phrase that was once used in the United States for the avocado (*Persea americana* Mill.) and is also sometimes used for chayote (see Chapter 23).
- The genus name *Momordica* is based on the Latin *mordica*, bitten, for the seeds, which have a jagged margin as though bitten.
- *Charantia* in the scientific name *M. charantia* is an old plant name used before modern binomial nomenclature started (in 1753), which was simply taken up and applied to the species.

Plant Portrait

Bitter melon is a rank-smelling, fast-growing, trailing or climbing annual vine with thin stems and coiled tendrils. In tropical regions, the vines often grow 6 to 9 m (20–30 ft.) in length. Male and female flowers are borne separately on the same plant, singly in the leaf axils. Male flowers appear first and usually exceed the number of female flowers by approximately 25 to 1. The flowers open

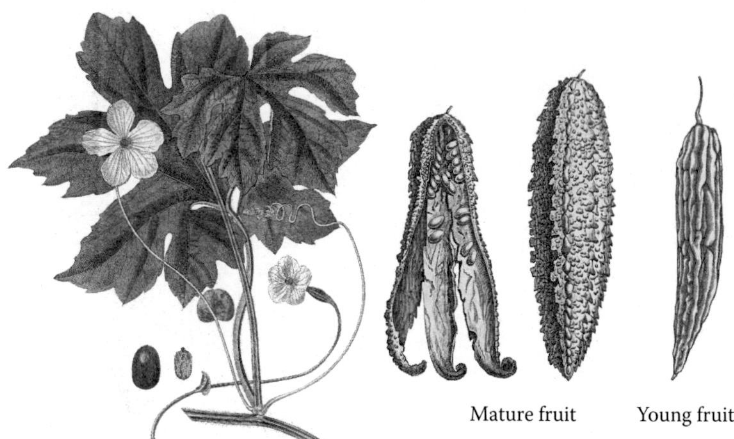

FIGURE 63.1 Bitter melon (*Momordica charantia*). Left: portion of shoot (seed at lower left). (From Curtis, 1824, vol. 51, plate 2455.) Center: mature fruit. (From Lamarck and Poiret, 1744–1829, plate 794.) Right: young fruit by M. Jomphe.

at sunrise and remain open for only 1 day. Fruit production requires insects for pollination. The fruits have a pebbly surface of smooth warts and smooth lengthwise ridges. Immature fruits are light green, oblong, pointed at the blossom end, and have white flesh. Generally, a bitter melon fruit looks like a lime-green, elongated cucumber with a furrowed, convoluted rind. As the fruits begin to mature, the surface gradually turns yellow or orange. At maturity, the melons tend to split open, revealing orange flesh. The fruits vary, depending on variety, from 2.5 to 30 cm (1–12 in.) in length and from 1.3 to 6.4 cm (1/2–2½ in.) in width. The seeds are tan, oval, with a rough etched surface and are covered by a scarlet fleshy growth (technically called an aril), which is eaten by humans and birds.

The area of origin of bitter melon is unknown, although it is best developed in India. It was brought to Brazil in the seventeenth or eighteenth century, possibly from Africa. Bitter melon is predominantly a third-world subsistence crop, with some exports to Europe from tropical countries. The plant is now cultivated throughout Southeast Asia, China, and the Caribbean but is also grown in small acreages in the United States, primarily in California and Florida. Bitter melon is a weed along the Gulf Coast of the United States and is a serious pest in citrus groves in Florida. In temperate areas, including North America and Europe, the plant is cultivated as an ornamental vine.

Bitter melon has been used as a folk medicine in parts of Asia, indeed more extensively than for other edible fruits, both as a "blood tonic" and in the treatment of malaria, colds, liver complaints, and kidney stones. The fruit pulp has been mashed and mixed with olive oil for use as a liniment for piles, burns, and chapped hands. The fruit juice has been given as a substitute for quinine (it is widely misreported that the plant contains quinine) and also has been used as a vermifuge (worm medicine) and remedy for liver and spleen ailments. The fruits, the seeds, and the roots are reported to contain compounds with pharmaceutical effects, including a substance with the clinical properties of insulin. Molecules with insulin-like activity have been found in both the fruits and seeds. In human experiments, the fruit juice was found to significantly improve the glucose tolerance of many patients when given orally. However, toxic effects are associated with ingestion of large doses of the ripe fruits. The fruit is purgative (causing vomiting) and abortifacient (causing abortion). Ingestion of mature fruits has been implicated in cases of poisoning of dogs. The use in folk medicine is considered dangerous. Nevertheless, bitter melon extracts are widely sold, especially on the Web, for self-medication to treat numerous illness, some very serious.

Culinary Portrait

Bitter melon, like other bitter foods, is a taste that has been acquired by many in Asia in particular as well as in tropical countries generally and is considered desirable, at least in food mixtures by those who have become accustomed to it. Bitter melon is rarely eaten alone and often accompanies sea food and pork. In China, which makes most use of bitter melon, it is particularly used in soups and braised dishes. In India, bitter melon is used in curries, as a pickle, and is also boiled and fried and sometimes eaten raw after steeping in water to remove the bitter taste. In North America, bitter melon is grown entirely for its immature, gherkin-like fruits, which are used in Oriental cooking. Varieties grown for Western markets are sometimes much less bitter than those grown in Asia, where bitterness is actually desired. The very young fruits are the least bitter, do not require removal of the seeds, and may be cooked with meat and fish without pretreatment to reduce bitterness. The bitter flavor in both the fruits and leaves is due to momordicine, an alkaloid, which can be reduced somewhat by parboiling or soaking in salt water. Such treatment, it is said, reduces the bitter taste of the fruit to give a flavor comparable with dandelion leaves. Ripe fruits are extremely bitter and, as noted above, have been reported to be toxic to man and animals, so fully ripe fruit is not used for food. The seeds must be removed from semiripe fruits (the spongy center which contains the seeds is simply scooped away), and normally the peel is also removed. Immature fruits are consumed as a vegetable with the peel removed (some cooks retain the rind), either boiled or fried. The flesh tends to turn brown on cooking. The immature fruits are also used in curries and pickles. To reduce the bitterness, the flesh can be soaked in salted water for up to an hour. Canned and fresh bitter melon can be purchased in some ethnic markets in the larger cities of North America. The fruits are sensitive to chilling injury, and commercial growers recommend not storing them below 13°C (55°F), but they can be kept for a few days (up to a week according to some authorities) in a refrigerator, individually wrapped in paper toweling and placed in closed plastic bags or simply placed in perforated plastic bags.

In some countries, the young leaves and flowers are also harvested and used as a potherb. The tender young shoots and leaves are parboiled to leach out the bitterness before being eaten and should be boiled in two changes of water.

Culinary Vocabulary

- "Ilocano" refers to a person and the language of an agricultural people centered in northern Luzon of the Philippines. Although rice and vegetables are the predominant foods, Ilocano cuisine specialties are rather unique, often flavored with fermented fish sauce and/or bitter melon.
- *Goya champura* is Okinawa's signature dish. Composed of stir-fried tofu and eggs, the key ingredient is bitter melon, which gives the preparation is characteristic bitter taste.

Curiosities of Science and Technology

- Certain African tribes include the root of bitter melon in aphrodisiac preparations (although the root is dangerous and has often been used to produce abortion).
- The fruits of bitter melon have been used as a substitute for soap to wash clothes in Columbia and Cuba.
- In Asia and Hawaii, bitter melon fruits on the vine are frequently covered with paper bags, cloths, or leaves to protect them from fruit flies. Bagging is also widely practiced to reduce bitterness. Bagging often produces fruit that is almost white at harvest (some cultivars are naturally white).

Key Information Sources

Abusaleha, and Dutta, O.P. 1994. Performance of bittergourd (*Momordica charantia*) under different training systems. *Indian J. Agric. Sci.* 64:479–481.

Binder, R.G., Flath, R.A., and Mon, T.R. 1989. Volatile components of bittermelon. *J. Agric. Food Chem.* 37:418–420.

Chang, M.K., Chapital, D.C., Conkerton, E.J., Wan, P.J., Vadhwa, O.P., and Spiers, J.M. 1998. Comparison of the fruit from four cultivars of Chinese melon (*Momordica charantia* L.). *Tropic. Sci.* 38:128–133.

Chang, M.K., Conkerton, E.J., Chapital, D.C., Wan, P.J., Vadhwa, O.P., and Spiers, J.M. 1996. Chinese melon (*Momordica charantia* L.) seed: composition and potential use. *J. Am. Oil Chem. Soc.* 73:263–265.

Deshpande, A.A., Venkatasubbaiah, K., Bankapur, V.M., and Nalawadi, U.G. 1979. Studies on floral biology of bitter gourd (*Momordica charantia* L.). *Mysore J. Agric. Sci.* 13:156–159.

Hien, N.H., and Widodo, S.H. 2001. *Momordica* L. In *Plant resources of South-East Asia. Vol. 12(1). Medicinal and poisonous plants 1.* Edited by L.S. de Padua, N. Bunyapraphatsara, and R.H.M.J. Lemmens. Backhuys Publishers, Leiden, the Netherlands. pp. 353–359.

Hunter, J. 1985. *Bitter melon.* Cooperative Extension, University of California, Division of Agricultural and Natural Resources, Berkeley, CA. 4 pp.

Huyskens, S., Mendlinger, S., Benzioni, A., and Ventura, M. 1992. Optimization of agrotechniques for cultivating *Momordica charantia* (karela). *J. Hortic. Sci.* 67:259–264.

Marr, K.L., Mei, X.Y., and Bhattarai, N.K. 2004. Allozyme, morphological and nutritional analysis bearing on the domestication of *Momordica charantia* L. (Cucurbitaceae). *Econ. Bot.* 58:435–455.

Maynard, D.N., and Paris, H.S. 2008. *Momordica charantia*, bitter melon. In *The encyclopedia of fruit & nuts.* Edited by J. Janick and R.E. Paull. CABI, Wallingford, Oxfordshire, U.K. pp. 305–307.

Mohammed, M., and Wickham, L.D. 1993. Extension of bitter gourd (*Momordica charantia* L.) storage life through the use of reduced temperature and polyethylene wraps. *J. Food Qual.* 16:371–382.

Morton, J.F. 1967. The balsam pear—an edible, medicinal and toxic plant. *Econ. Bot.* 21:57–68.

Ng, T.B., Wong, C.M., Li, W.W., and Yeung, H.W. 1986. Insulin-like molecules in *Momordica charantia* seeds. *J. Ethnopharmacol.* 15:107–117.

Prakash, G. 1979. Ecophysiological study of flowering and sex behaviour of *Momordica charantia* L. *Comp. Physiol. Ecol.* 4:280–285.

Raj, N.M., Prasanna, K.P., and Peter, K.V. 1993. Bitter gourd, *Momordica* spp. In *Genetic improvement of vegetable crops.* Edited by G. Kalloo and B.O. Bergh. Pergamon Press, New York. pp. 239–246.

Reyes, M..E.C., Gildemacher, B.H., and Jansen, G.J. 1993. *Momordica* L. In *Plant resources of South-East Asia. Vol. 8. Vegetables.* Edited by J.E. Siemonsma and K. Piluek. Pudoc Scientific Publishers, Wageningen, the Netherlands. pp. 206–210.

Rodriguez, D.B., Raymundo, L.C., Lee, T.-C., Simpson, K.L., and Chichester, C.O. 1976. Carotenoid pigment changes in ripening *Momordica charantia* fruits. *Ann. Bot.* 40: 615–624.

Singh, V.P., Singh, V., Prakash, G., Agrawal, P.K., and Kumar, D. 1978. Induction of sex by seed pre-treatment with cold and growth substances in *Momordica charantia* L. *Incompat. Newsl. (Wageningen)*, 10: 110–122.

Virdi, J., Sivakami, S., Shahani, S., Suthar, A.C., Banavalikar, M.M., and Biyani, M.K. 2003. Antihyperglycemic effects of three extracts from *Momordica charantia. J. Ethnopharmacol.* 88:107–111.

Walters, T.W., and Decker-Walters, D.S. 1988. Balsam-pear (*Momordica charantia*, Cucurbitaceae). *Econ. Bot.* 42:286–288.

Welihinda, J., Karunanayake, E.H., Sheriff, M.H., and Jayasinghe, K.S. 1986. Effect of *Momordica charantia* on the glucose tolerance in maturity onset diabetes. *J. Ethnopharmacol.* 17:277–282.

Williams, J.T., and Ng, N.O. 1976. [Taxonomic] Variation within *Momordica charantia* L., the bitter gourd (Cucurbitaceae). *Ann Bogor.* 6(2):111–123.

Yang, S.-L., and Walters, T.W. 1992. Ethnobotany and the economic role of the Cucurbitaceae of China. *Econ. Bot.* 46:349–367.

Yen, G.C., and Hwang, L.S. 1985. Lycopene from the seeds of ripe bitter melon (*Momordica charantia*) as a potential red food colorant. II. Storage stability, preparation of powdered lycopene and food applications. *J. Chin. Agric. Chem. Soc.* 23:151–161.

Zong, R.J., Morris, L., and Cantwell, M. 1995. Postharvest physiology and quality of bitter melon (*Momordica charantia* L.). *Postharvest Biol. Tech.* 6:65–72.

Specialty Cookbooks

Alejandro, R. 1985. *Philippine cookbook*. Perigree Trade, New York. 256 pp.

Bhargava, P. 2009. *From mom with love. A complete guide to Indian cooking*. Crest Books, Nesconset, NY. 182 pp.

Pharm, M. 1995. *The best of Vietnamese & Thai cooking: favorite recipes from Lemon Grass Restaurant and Cafes*. Prima Lifestyles, Rocklin, CA. 288 pp.

Sacharoff, S.N. 1991. *Flavors of India: vegetarian Indian cuisine*. Book Publishing Co., Summertown, TN. 191 pp.

Yan, M. 2008. *Martin Yan's China*. Chronicle Books, San Francisco, CA. 240 pp.

HORNED MELON (KIWANO)

Names

Scientific Name: *Cucumis metuliferus* E.H. May. ex Schrad.

- The horned melon is most commonly marketed in North America under the trademark name "Kiwano." This name was registered as a trademark of Prinut Inc., and has been owned by M.I. Exotics since 1997. The name kiwano originated in New Zealand and was named similarly to the kiwi fruit (see Chapter 53) because some people thought the flesh was similar. Although both fruits are often imported from New Zealand, the kiwano is a member of the cucumber family, not the kiwi family.
- The horned melon is also known as African horned cucumber, African horned melon, English tomato, hedged gourd, horned cucumber, jelly melon, kiwano melon, melano, and metulon.
- The "horned cucumber" acquired this name from its cucumber taste, coupled with its spines.
- In the southeastern United States, the horned melon is known as the "blowfish fruit." (Blowfish, also known as "fugu," are a popular dish in Japan, despite the possibility of being fatally poisoned if the toxic portions are not removed. When blowfish inflate themselves with water for protection, they are reminiscent of the horned melon, indeed often with spines.)
- The genus name *Cucumis* is the ancient Roman name for the common cucumber.
- *Metuliferus* in the scientific name *C. metuliferus* is Latin for "bearing conical horns," which is descriptive of the fruit.

Plant Portrait

The horned melon is an annual vine, 1.5 to 3 m (5–10 ft.) long, climbing by tendrils. The small, deeply cut, five-lobed leaves are similar to those of the watermelon. The fruits are oblong or ellipsoid, 5 to 13 cm (2–5 in.) long, typically weighing 0.23 to 0.45 kg (1/2–1 lb.). The thick skin is studded with conical spiny protuberances. The fruit has been described as an "oval melon with horns" and a "prickly hand grenade." Fruits are picked immature, when they are a mottled light green. When they mature, they turn a bright yellow-reddish orange on the outside. The flesh of the ripe interior is lime-green, jelly-like, with numerous white, soft, edible seeds. The horned melon is native to the semiarid regions of southern and central Africa, including the Kalahari Desert, mainly in Botswana, South Africa, Namibia, Zimbabwe, Malawi, and Nigeria. It has been known for thousands of years in Africa, where it is eaten today as a local wild food. The fruit has been sold commercially only recently. Part of its success is due to

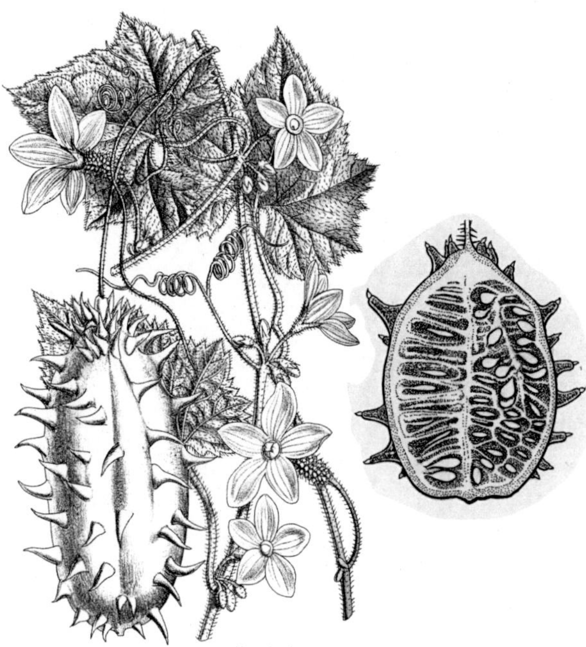

FIGURE 63.2 Horned melon (*Cucumis metuliferus*). (From Curtis, 1911, vol. 137, plate 8385.) The sectioned fruit at right is by B. Flahey.

effective marketing techniques by specialty producers and distributors, particularly Frieda's Finest Produce Specialties Inc. of Los Angeles. Its success is also due to an extremely long shelf life and a very unique and decorative appearance. This fruit is a good example of the significant increase in the importation of exotic fruits and vegetables into North America. It is now available year-round from crops grown in New Zealand and California. It has been cultivated in New Zealand since the 1930s and more recently in California. The horned melon is also grown in Kenya and Israel, and interest in the crop is expanding.

Culinary Portrait

The fruit taste has been described as "a mix of lemon and banana" or "subtle banana-lime." Current varieties have a rather bland taste, which limits the potential of the fruit, and attempts are underway to breed more attractive varieties. The fruit may be sliced open to expose the green cucumber-like gel full of white pips and eaten like watermelon. The seeds and pulp, but not the rind, are eaten raw, generally as a dessert. The flesh can be mixed with sour cream, sugar, cream cheese, yogurt, orange juice, or other ingredients. There are many ways of serving the flesh: drizzled over a smoked chicken salad; added to a fresh fruit salad; poured over honeydew melon or cantaloupe balls; added to soups sauces, sorbets, yogurt, or ice cream; or mixed into alcoholic cocktails. The abundant, large, sticky seeds make the fruit difficult to use, and the pulp can simply be strained for juice. Some people dislike the gelatinous, jelly-like texture, considering it to be slimy; liquefying the pulp or combining it with other ingredients is a way of overcoming this problem. The hard rinds are frequently used as decorative food bowls. Horned melon is sometimes marketed merely as a garnish or for decorative purposes. Although it is usually consumed as a fruit (i.e., for dessert), it is also used in exotic vegetable dishes, for example, added to stir-fries, and it has also been used to flavor meat dishes.

When purchasing horned melon, it is advisable to avoid fruits that are overly soft or tender or have discolored areas. Good-quality horned melon is golden-orange when ripe; often the riper the fruit, the brighter orange it is (yellow fruit is also often available). The fruits have excellent shelf life and can be stored for up to 6 months at room temperature. Horned melon must not be stored in a refrigerator, but it can be chilled before eating to improve its flavor. The fruit should not be stacked as the spines can puncture neighboring horned melons.

Curiosities of Science and Technology

- In its native area, the fruits of the horned melon have not been considered to be of much value. In Nigeria, the fruits are rather bitter and not eaten. In the Kalahari region of South Africa, they are ingested by game animals and, in times of scarcity, are consumed by the Bushmen and fed to cattle. The reluctance to eat the fruits is due to the presence of chemicals called cucurbitacins, compounds that are very toxic to mammals. They have been said to be the most bitter substances known and so deter feeding by animals. Indeed the fruits of many native plants of Africa are quite inedible. When fruits were first imported into Los Angeles in 1986, many buyers were put off by an unpleasant bitter aftertaste, which sometimes persisted for an hour. Of course, horned melons sold in supermarkets today have had most of these compounds removed by selecting nonbitter cultivars.
- Many exotic plants introduced as new crops or ornamentals have proven to be hard to keep under control. The horned melon was introduced to Australia 70 years ago and became a weed there. The plant is also considered to have considerable weedy potential in California, where it is now grown.
- Because of the spines, special measures are necessary to handle the horned melon commercially. Thick-padded gloves are required during harvesting for protection. Before packing, the ends of the spines are blunted by a grinding wheel. The fruits are placed in single-layer trays to avoid their spines puncturing adjacent fruits.

Key Information Sources

Benzioni, A., Mendlinger, S., Ventura, M., and Huskens [sic], S. 1991. The effect of sowing dates and temperatures on germination, flowering and yield of *Cucumis metuliferus*. *HortScience*, 26:1051–1053.

Benzioni, A., Mendlinger, S., Ventura, M., and Huyskens [sic], S. 1993. Germination, fruit development, yield, and postharvest characteristics of *Cucumis metuliferus*. In *New crops*. Edited by J. Janick and J.E. Simon. Wiley, New York. pp. 553–557.

Bon, H. de, and Cottin, R. 1990. Behavior of horned melon (*Cucumis metuliferus*) in Martinique. *PHM Rev. Hortic*. 304:39–43 (in French).

Corral, M.C.S., Lorenzo [sic], J.L.F., and Vasquez, J.P.M. 1991. Kiwano. Experimentation in Galicia (Spain). *PHM Rev. Hortic*. 313:59–63 (in French).

Corral, M.C.S., and Lourenzo [sic], J.L.F. 1990. Comparison of kiwano (*Cucumis metuliferus*) cultivation systems. In *Proceedings Iberian Congress on Horticultural Sciences, Lisbon, Portugal, June 18–21, 1990*. Associacao Portuguesa de Horticultura e Fruticultura and Sociedad Espanola de Ciencias Horticolas, Lisboa, Portugal. pp. 378–383 (in Spanish, English summary).

Fassuliotis, G. 1977. Self-fertilization of *Cucumis metuliferus* Naud. and its cross-compatibility with *Cucumis melo* L. *J. Am. Soc. Hortic. Sci*. 102:336–339.

Helm, M.A., and Hemleben, V. 1997. Characterization of a new prominent satellite DNA of *Cucumis metuliferus* and differential distribution of satellite DNA in cultivated and wild species of *Cucumis* and in related genera Cucurbitaceae. *Euphytica*, 94:219–226.

Marsh, D.B. 1993. Evaluation of *Cucumis metuliferus* as a specialty crop for Missouri. In *New crops*. Edited by J. Janick and J.E. Simon. Wiley, New York. pp. 558–559.

Mendlinger, S., Benzioni, A., Huskens, S., and Ventura, M. 1992. Fruit development and postharvest physiology of *Cucumis metuliferus* Mey., a new crop plant. *J. Hortic. Sci*. 67:489–493.

Morton, J.F. 1987. The horned cucumber, alias “kiwano” (*Cucumis metuliferus*, Cucurbitaceae). *Econ. Bot.* 41:325–327.

Myers, C. 1991. Kiwano, African horned cucumber or melon, jelly melon. Crop Sheet SMC-021. In *Specialty and minor crops handbook*. Edited by C. Myers. The Small Farm Center, Division of Agriculture and Natural Resources, University of California, Oakland, CA. 2 pp.

Norton, J.D., and Granberry, D.M. 1980. Characteristics of progeny from an interspecific cross of *Cucumis melo* with *Cucumis metuliferus*. *J. Am. Soc. Hortic. Sci.* 105:174–180.

Raharjo, S.H.T., and Punja, Z.K. 1993. Plantlet regeneration from petiole explants of the African horned cucumber, *Cucumis metuliferus*. *Plant Cell Tissue Organ Cult.* 32:169–174.

Romero-Rodriguez, M.A., Vazquez-Oderiz, M.L., Lopez-Hernandez, J., and Simal-Lozano, J. 1987. Physical and analytical characteristics of the kiwano. *J. Food Composit. Anal.* 5:319–322.

Sweet, C. 1987. Kiwano: can it make it here in the U.S.? *Calif. Grower*, 11(4):23–24.

Walters, S.A., and Wehner, T.C. 2002. Incompatibility in diploid and tetraploid crosses of *Cucumis sativus* and *Cucumis metuliferus*. *Euphytica*, 128:371–374.

Specialty Cookbooks

Better Homes and Garden Books, and Fuller, K. (Ed.). 2000. *Big book of healthy family dinners*. Better Homes and Gardens, Des Moines, IA. 288 pp.

Editors of Cooking Light Magazine. 2009. *Cooking light: fresh fast food. Over 280 incredibly flavorful 5-ingredient 15-minute recipes*. Oxmoore House, Birmington, AL. 368 pp.

Duby, D., and Duby, C. 2006. *Wild sweets: exotic dessert and wine pairings*. Whitecap, Vancouver, BC. 162 pp.

64 Miracle Fruit

Family: Sapotaceae (sapote family)

NAMES

Scientific Name: *Synsepalum dulcificum* (Schumach. & Thonn.); Daniell (*Richardella dulcifica* (Schumach. & Thonn.) Baehni)

- The name "miraculous berry" was reported by W.F. Daniell in 1852 (in *Pharm. Journal* 11: 445–448), who noted that European travelers and traders had coined the name to describe the extraordinary sweetening ability of the fruit. This name is still often used, but "miracle fruit" is adopted here because it is more widely used and is shorter.
- Miracle fruit is also known as miracle berry.
- Very confusingly and undesirably, the names miracle fruit and miracle berry have been applied to another sweet-fruited, West African plant, *Thaumatococcus daniellii* (Benn.) Benth. ex B.D. Jacks. Its effect on taste is discussed below.
- The name miracle fruit has also been applied to *Gymnema sylvestre* (Retz.) Schult., a tropical Old World medicinal plant. Its effect on taste is discussed below.
- The phrase "miracle fruit" has been applied by vested interests exaggerating the medicinal value of other fruits than *S. dulcificum*. These plants include varieties of blueberry, noni, papaya, and raspberry. Similarly, "miracle berry" has been applied to varieties of blackberry, blueberry, and raspberry.
- The genus name *Synsepalum* is based on the Greek *syn*, together + the Latin *sepalum*, sepal, indicating that the sepals of the flower are highly fused together.
- *Dulcificum* in the scientific name *S. dulcificum* is based on the Latin *dulce*, sweet + *icum*, a suffix indicating exceptional development of the characteristic.

PLANT PORTRAIT

Miracle fruit is native to hot, humid, tropical lowlands of West Africa, including parts of Ghana, Benin, Nigeria, Cameroon, Central African Republic, The Democratic Republic of the Congo (formerly Zaire), and Gabon. The species is now cultivated in tropical regions of Asia, South America, and Florida. It is one of the most interesting of plants, known for its curious sweetening properties. This white-flowered shrub grows up to a height of 5.5 m (18 ft.) in its native habitat but rarely more than 1.5 m (5 ft.) elsewhere. The species typically develops an oval or pyramidal shape. The leaves are evergreen, deep green, papery-leathery, elongated, and grow in a spire-like fashion. Forms of the plant with rather hairy leaves and forms with smooth leaves have been reported. The flowers are brownish-white and small, generally approximately 1 cm (0.4 in.) long. A large plant can produce hundreds of berries. The fruits are dull green when immature, ripening to bright red. They are ellipsoid, approximately 2 to 3 cm (0.8–1.2 in.) long. A single seed is present in the olive-like fruit, surrounded by thin, soft pulp.

The berries contain a glycoprotein (a protein molecule with a carbohydrate component) called miraculin, which is not sweet but changes the flavor of sour acid foods such as lemon and rhubarb and bitter foods such as Brussels sprouts into a delicious sweetness and even removes the bad taste of vinegar. There has been scientific investigation of why miraculin changes taste perception, but a clear explanation is not yet available. It is widely thought that there is a lock-and-key mechanism between taste receptors (primarily on the tongue) and flavor molecules (the shapes of which fit into complementary openings in the receptors). It has been hypothesized that miraculin binds to the tongue's taste buds, distorting the shape of sweetness receptors so that molecules that are normally sour (particularly acids) or bitter are temporarily perceived as sweet.

Ghanaians and other West Africans use the fruits to improve the flavor of maize dishes and beverages such as palm wine or tea; to obscure the sour taste of various food substances, such as lemon, lime, and grapefruit; and even to improve the taste of stale food. An ambitious attempt to market miracle fruit chewing gum and other products in the United States ended in 1974 when the Food and Drug Administration banned the sale of all miracle fruit products (by contrast, in Japan, however, miracle fruit extracts are often added to foods). Nevertheless, such products are advertised and offered from international sources on the Internet. Indeed, miracle fruit has become so popular recently that it has been described as "the most fashionable fad among the Foodie Set." The berries have become the subject of "taste-tripping parties," the guests consuming the fruits and testing the effects on the taste of various foods, such as Brussels sprouts, broccoli, rhubarb, sauerkraut, martinis, oysters, limes, grapefruit juice, and pineapple juice. Under the influence of miracle berry, many familiar foods acquire quite different tastes, for example, beer is said to taste like a chocolate milkshake, lemons like sweet lemonade, goat cheese like cheesecake, vinegar like apple juice, and Tabasco sauce like doughnut glaze. Banana, however, is said to still taste like banana.

Miracle fruit requires a frost-free climate, although mature plants are said to be able to survive light frost. In its native area, it is often grown on farms and near dwellings. Miracle fruit can be grown outdoors in southern Florida and Hawaii and as a container plant in more northerly locations. The plants require an acidic soil with some organic content, and if grown in a pot, a water-soluble, acidic fertilizer is recommended. The plants are slow growing, but berries may be produced in as little as 2 years. The fruits furnished are a marvellous topic of conversation.

CULINARY PORTRAIT

Although miracle fruit is not itself sweet (it is relatively tasteless), when a single fruit is eaten and the fleshy pulp is allowed to coat the taste buds of the tongue and inside of the mouth, the effect is extraordinary. One can eat a slice of lemon or lime without wincing. The attractive aroma and inherent sweetness of the citrus remain but the sourness is almost completely removed. The taste reprogramming effect often remains strong for 30 minutes and detectable for 2 hours. Although fresh berries are very potent, damaged fruit or crude extracts quickly become denatured (chemists have learned how to stabilize the extracted miraculin). Miraculin becomes denatured with heat, and so miracle fruit and its extracts cannot be used in cooking. Cold tends to prolong the life of the berries. The berries should be eaten soon after picking, and even if stored in the refrigerator for a day or so, they often lose their effectiveness. Freeze-dried extracts (claimed to last for a year or so) have been widely marketed in tablets via the Internet (but not legally in many countries).

Miracle berry has been touted by some as useful for health purposes: in diabetic and weight-control diets (as a sugar substitute, by virtue of the ability to make nonsugary foods taste sweet) and in cancer and other therapies where medical treatments have the effect of reducing the taste appeal of foods (chemotherapy, for example, may produce an unpleasant, metallic aftertaste). Miracle berry is used to some extent in Japan in treating diabetes and obesity, but authorized medical use of miracle berry is generally not supported.

CULINARY VOCABULARY

- "Taste-tripping party" (less commonly "flavor-tripping") is a phrase that has become popular in the early twenty-first century. It denotes parties in which guests consume miracle fruits or miracle fruit extracts and afterward sample a variety of foods to experience how their taste perceptions have changed.

WARNING

Although consuming miracle fruit makes very acidic foods like lemons and vinegar palatable, such foods can be damaging to the mouth and alimentary canal. Common sense dictates that such foods should be sampled only in very small amounts.

CURIOSITIES OF SCIENCE AND TECHNOLOGY

- One-thousandth of a gram (approximately 0.000035 oz.) of miraculin is sufficient to make a lemon (without sugar) taste like lemonade.
- In West Africa, the small twigs of miracle fruit have been used as "chew sticks," reminiscent of toothbrushes.
- Like miracle fruit, globe artichoke (*Cynara scolymus* L.) tends to reduce sensitivity to sour taste.
- West Africa has two additional species, noted for their taste-altering proteins found in the fruits. One of these is *Thaumatococcus daniellii*, a herbaceous perennial of Sierra Leone and Guinea; as noted earlier, it is confusingly called miracle fruit or miracle berry, names that are most commonly applied to *S. dulcificum*. Katemfe fruit is a preferable name. Proteins (primarily thaumatin) extracted from its fruit are 2000 to 3000 times sweeter than sugar. Thaumatin has passed extensive safety testing, but the technology for using it in industry has not yet been sufficiently advanced to make it a common food additive. Thaumatin, marketed as Talin by the Talin Food Company, Merseyside, England, is used today for some limited flavor-enhancing applications.
- The other species of West Africa noted for its taste-altering protein is serendipity berry, *Dioscoreophyllum cumminsii* (Stapf) Diels, an interesting plant from West Africa, the fruit

FIGURE 64.1 Branch of miracle fruit (*Synsepalum dulcificum*) with fruits and flowers. (Photo courtesy of Logee's Tropical Plants.)

FIGURE 64.2 Miracle fruit (*Synsepalum dulcificum*) grown as a houseplant. (Photo courtesy of Logee's Tropical Plants.)

of which produces an intensely sweet protein, monellin. This species is grown to some extent in the United States. Natural monellin has not been mass marketed to the food industry.

- *Gymnea sylvestre* (Indian vine) of Eruasia is still another plant with taste-altering proteins, and confusingly it too is called miracle fruit. Like *S. dulcificum*, its action belongs to the "weird but true" category: it inhibits sweetness, so it is possible to suppress the sweet taste of sugar. Although reducing sweetness seems hardly desirable, the ability to taste bitterness is also decreased. Jujube (see Chapter 50) also temporarily reduces sweet sensitivity.
- As is well known, salt in low concentrations enhances the sweetness of sugar (this is why many add salt to grapefruits). Monosodium glutamate is another well-known nonbotanical substance that has the ability to enhance the overall flavor of foods.
- Lettuce has been genetically engineered to produce miraculin (Sun et al., 2006); two lettuce leaves produced as much as did one miracle berry. Tomato plants have also been engineered to produce miraculin (Sun et al., 2007).

KEY INFORMATION SOURCES

Adansi, M.A., and Holloway, H.L.O. 1977. Germination of seeds of the sweet or miraculous berry (*Synsepalum dulcificum* (Schum. & Thonn.) Daniell. *Acta Hortic.* 53:181–182.

Ayensu, E.S. 1972. Morphology and anatomy of *Synsepalum dulcificum* (Sapotaceae). *Bot. J. Linn. Soc.* 65:179–187.

Bartoshuk, L.M., Gentile, R.L., Moskowitz, H.R., and Meiselman, H.L. 1974. Sweet taste induced by miracle fruit (*Synsepalum dulcificum*). *Physiol. Behav.* 12:449–456.

Brouwer, J.N., Glaser, D., Hard-Af-Segerstad, C., Hellekant, G., and Ninomiya, Y. 1983. The sweetness-inducing effect of miraculin; behavioural and neurophysiological experiments in the rhesus monkey *Macaca mulatta*; sweetness-inducing protein from the miracle fruit *Synsepalum dulcificum*. *J. Physiol.* 337:221–240.

Buckmire, R.E., and Francis, F.J. 1976. Anthocyanins and flavonols of miracle fruit, *Synsepalum dulcificum*, Schum. *J. Food Sci.* 41:1363–1365.

Buckmire, R.E., and Francis, F.J. 1978. Pigments of miracle fruit, *Synsepalum dulcificum* Schum, as potential food colorants. *J. Food Sci.* 43:908–911.

Dastoli, F.R., and Harvey, R.J. 1974. Miracle fruit concentrate. In *Symposium: sweeteners*. Edited by G. E. Inglett. AVI Pub. Co., Westport, CT. pp. 204–210.

Gollner, A.L. 2008. *The fruit hunters: a story of nature, adventure, commerce and obsession*. Scribner, New York. 288 pp.

Giroux, E.L., and Henkin, R.I. 1974. Purification and some properties of miraculin, a glycoprotein from *Synsepalum dulcificum* which provokes sweetness and blocks sourness. *J. Agric. Food Chem.* 22:595–601.

Guney, S., and Nawar, W.W. 1977. Seed lipids of the miracle fruit (*Synsepalum dulcificum*). *J. Food Biochem.* 1:173–184.

Hohmann, B. 1978. Microscopic examination of the fruit of *Synsepalum dulcificum* (DC.) Daniell, a berry that changes sour to sweet flavor. *Dtsch. Lebensmitt. Rundsch*, 74:434–438 (in German, English summary).

Hu, J., and Liu F. 1989. Preliminary study on the fruiting practices for potted *Synsepalum dulcificum* Daniell. *Trop. Crop Res.* 3:59–60, 52.

Inglett, G.E. 1970. Natural and synthetic sweeteners [*"Richardella dulcifica," Stevia rebaudiana, Dioscoreophyllum cumminsii*]. *HortScience*, 5:139–141.

Kurihara, Y., and Terasaki, S. 1982. Isolation and chemical properties of multiple active principles from miracle fruit *Richardella dulcifica*. *Biochim. Biophys. Acta.* 719:444–449.

Kurihara, K., and Beidler, L.M. 1968. Taste modifying protein from miracle fruit. *Science*, 161:1241–1243.

Marchal, J., Montagut, G., and Martin-Prevel, P. 1972. Mineral balance of *Synsepalum dulcificum*. *Fruits*, 27:223–225 (in French).

Milhet, Y., and Costes, C. 1984. Biology of two sweetener plants. *Acta Hortic.* 144:77–84 (in French, English summary).

Montagut, G. 1972. Attempts of *Synsepalum dulcificum* culture in Dahomey. *Fruits*, 27:219–221 (in French).

Nakajo, S., Theerasilp, S., Nakaya, K., Nakamura, Y., and Kurihara, Y. 1988. A quantitative enzyme immunoassay for miraculin in *Richardella dulcifica* miracle fruit. *Chem. Senses*, 13:663–670.

Parke, A. 1976. Miracle fruit, *Synsepalum dulcificum*. *Bull. New Series Ministry Agric. Fish.* 64:21–22.

Sun, H.J., Cui, M.L., Ma, B., and Ezura, H. 2006. Functional expression of the taste-modifying protein, miraculin, in transgenic lettuce. *FEBS Lett.* 580(2):620–626.

Sun, H.J., Kataoka, H., Yano, M., and Ezura, H. 2007. Genetically stable expression of functional miraculin, a new type of alternative sweetener, in transgenic tomato plants. *Plant Biotech. J.* 5:768–777.

Theerasilp, S., and Kurihara, Y. 1988. Complete purification and characterization of the taste-modifying protein, miraculin, from miracle fruit. *J. Biol. Chem.* 263(23):11536–11539.

Tripp, N. 1985. The miracle berry. *Horticulture*, 63(1):58–60, 62, 64–72.

Specialty Cookbooks

Miracle fruit is not used in cookery, and recipes for its use in food preparations are not available.

65 Myrrh

Family: Burseraceae (bursera family)

NAMES

Family: Burseraceae (bursera family; sometimes known as the torchwood family, copal family, and incense tree family).

Scientific Names: *Commiphora* species

- Myrrh—*C. myrrha* (Nees) Engl. (*C. molmol* (Engl.) Engl.).
- Abyssinian myrrh—*C. habessinica* (O. Berg) Engl. (*C. abyssinica* (O. Berg) Engl.).

- The English "myrrh" traces back through Latin *murra* and Greek *myrrha*, from a Semitic word like the Hebrew *mōr*, meaning myrrh and bitter. The Assyrian *murru* has also been suggested as the ultimate root of the word myrrh. *Myrrha* in the scientific name *C. myrrha* has the same basis.
- Myrrh is also called African myrrh, herabol (harabol) myrrh, and Somali myrrh.
- Abyssinian myrrh is also called Arabian myrrh and Yemen myrrh.
- Mecca myrrh is *C. gileadensis*, a species usually called "balm of Gilead." The latter name is applied to several trees or shrubs of the genus *Commiphora* but especially refers to *C. gileadensis* (L.) C. Chr. (*C. opobalsamum* (L.) Engl.) of Arabia and Somalia. (The name balm of Gilead is also used for some unrelated plants.)
- The unrelated sweet cicely (*Myrrhis odorata* (L.) Scop.), a culinary herb, is sometimes called British myrrh, garden myrrh, and sweet myrrh.
- The genus name *Commiphora* is based on the Greek *kommi*, gum + *phero*, to bear.
- *Habessinica* in the scientific name *C. habessinica* is a variation of *abyssinica*, Abyssinian.

PLANT PORTRAIT

According to Christian belief, three wise men traveled to Bethlehem to worship the Christ child, bringing gifts of gold, frankincense, and myrrh, the most valuable substances of the Old World. Myrrh is often interpreted as having the greatest value of the time (at least, it sometimes exceeded the value of gold, and in ancient Rome it was once priced at five times that of frankincense). It is mentioned about two dozen times in the Bible and was employed as incense in ceremonies and indeed was used in ancient Egypt in religious rituals. The scriptural myrrh or true myrrh refers to several species of small East African and Arabian trees of the genus *Commiphora*, particularly the two highlighted here. *Commiphora myrrha*, the most important species, is a small tree or large shrub to 3 m (10 ft.) in height, found in northeastern Africa and the Arabian Peninsula. It has whitish-gray bark and branches that end in spines. *C. habessinica* is a shrub or small tree, also found in northeastern Africa and the Arabian Peninsula.

The word myrrh also refers to a yellow to reddish brown, fragrant, bitter gum resin, typically in tear-shaped or irregular masses that have exuded naturally or from incisions made in the bark of the myrrh trees. This dried tree sap was once esteemed for its fragrance for ceremonial, religious, and embalming purposes. In ancient times, it was used as an ingredient of incense, fumigants, and perfumes and as a remedy for numerous infections, including leprosy and syphilis. Myrrh was also recommended for relief from bad breath and for dental conditions. In traditional Chinese medicine,

FIGURE 65.1 Myrrh (*Commiphora myrrha*). (From Köhler, 1883–1914.)

FIGURE 65.2 Abyssinian myrrh (*Commiphora habessinica*). (From Engler and Prantl, 1889–1915.)

it has been used for bleeding disorders and wounds. Myrrh is an important article of commerce in northeastern Kenya where it is locally known by the Somali word *molmol*. At present, myrrh is used in some medicinal preparations (of uncertain usefulness) such as a stimulating tonic; for its antiseptic properties in mouthwashes and toothpastes; as a fragrance in soaps, cosmetics, and perfumes; and as noted below as a flavoring in food products. Bellium is an aromatic gum resin used in medicine and perfumes. It is similar to myrrh and is produced by certain Asian and African shrubs and trees of the genus *Commiphora*.

CULINARY PORTRAIT

Myrrh is a very minor food additive, used commercially to flavor soft drinks, liquor, soups, baked goods, ice cream, candy, honey, and chewing gum. Myrrh was once used to flavor and preserve wines. It imparts a burning, biting, smoky, somewhat bitter taste, and a somewhat musky, incense-like odor that some people describe as medicinal, others as spicy.

CULINARY VOCABULARY

- "Myrrh" in recipes normally refers to the culinary and pot herb *Myrrhis odorata*, noted in the Names section.

CURIOSITIES OF SCIENCE AND TECHNOLOGY

- The ancient Assyrians reportedly treated hangovers with ground swallow beaks and myrrh.
- Myrrh was part of the combat gear of ancient Greek warriors, valued for its antiseptic and anti-inflammatory properties. It was used to clean wounds, to prevent infection, and to stop the spread of gangrene.
- "Wine mingled with myrrh," referred to in the Bible (Mark 15:23), was a Roman preparation given to criminals being executed by crucifixion to intoxicate them for partial relief of their suffering. When the Roman soldiers offered it to Jesus while he was on the cross, "he received it not" (Mark 15:23). There is some indication that myrrh has opiate qualities (*Science News* 149(2): 20, 1996). It has been suggested that the Romans offered vinegar, not wine, to those being crucified, and the alleged intoxicant property of the preparation was from added myrrh, not alcohol.
- "Before a girl's turn came to go in to King Xerxes [King Ahasuerus in Hebrew], she had to complete twelve months of beauty treatments prescribed for the women, six months with oil of myrrh and six with perfumes and cosmetics" (Old Testament, Esther 2:12, pertaining to the harem of beautiful virgins assembled for Xerxes the Great, the king of Persia, who lived approximately 519 to 465 BC). Myrrh has a long history of use for improving skin conditions and today is still found in some skin beauty products.
- In ancient times, myrrh was used in the preservation of mummies. Myrrh was also used to embalm the body of Christ: "And there came also Nicodemus, which at first came to Jesus by night, and brought a mixture of myrrh and aloes, about a hundred pound weight" (John 19:39).
- Hundreds of plants were assigned special meanings in the "Victorian Language of Flowers," popular in Victorian times, and many delighted in sending coded messages by this means, especially for romantic purposes. Myrrh meant "gladness."
- Ivory Soap was originally named "Proctor & Gamble White Soap." In 1879, Harley Proctor found the new name during a reading in church of the 45th psalm of the Bible: "All thy garments smell of myrrh, and aloes, and cassia, out of ivory palaces, whereby they have made thee glad."

KEY INFORMATION SOURCES

Abd El Aleem, H., Youssef, M.A., Abd El Mageed, A.M., Shehata, S.H., and Youssef, M.S. 1993. In vitro and in vivo experimental studies on *Commiphora myrrha*, myrrh as an indigenous fertility regulating agent. *Assiut Vet. Med. J.* 29:227–234.

Allam, A.F., El Sayad, M.H., and Khalil, S.S. 2001. Laboratory assessment of the molluscicidal activity of *Commiphora molmol* (myrrh) on *Biomphalaria alexandrina*, *Bulinus truncatus* and *Lymnaea cailliaudi*. *J. Egypt. Soc. Parasitol.* 31:683–690.

Baser, K.H.C., Demirci, B., Dekebo, A., and Dagne, E. 2003. Essential oils of some *Boswellia* spp., myrrh and opopanax. *Flavour Fragrance J.* 18:153–156.

Brieskorn, C.H., and Noble, P. 1982. The terpenes of the essential oil of myrrh. *World Crops Prod. Util. Descr.* 10:221–226.

Brieskorn, C.H., and Noble, P. 1983. Furanosesquiterpenes from the essential oil of myrrh isolated from *Commiphora molmol. Phytochemistry*, 22:1207–1211.

Dagne, E., Dekebo, A., Desalegn, E., Bekele, T., Tesso, H., and Bisrat, D. 1998. Preliminary report on essential oils from frankincense, myrrh and other plants of Ethiopia. In *Conservation, management and utilisation of plant gums, resins and essential oils (Proceedings, Rome, Italy, 1998).* Edited by J.O. Mugah, B.N. Chikamai, S.S. Mbiru, and E. Casadei. Food and Agriculture Organization, Rome, Italy. pp. 79–84.

Dolara, P., Corte, B., Ghelardini, C., Pugliese, A.M., Cerbai, E., Menichetti, S., and Lo-Nostro, A. 2000. Local anaesthetic, antibacterial and antifungal properties of sesquiterpenes from myrrh. *Planta Med.* 66:356–358.

Gallo, R., Rivara, G., Cattarini, G., Cozzani, E., and Guarrera, M. 1999. Allergic contact dermatitis from myrrh. *Contact Dermatitis*, 41:230–231.

Groom, N. St. J. 1981. *Frankincense and myrrh: a study of the Arabian incense trade*. Longman, New York. 285 pp.

Karamalla, K.A., 1998. The chemical characterisation of myrrh and frankincense and opportunities for commercial utilisation. In *Conservation, management and utilisation of plant gums, resins and essential oils (Proceedings, Rome, Italy, 1998).* Edited by J.O. Mugah, B.N. Chikamai, S.S. Mbiru, and E. Casadei. Food and Agriculture Organization, Rome, Italy. pp. 75–78.

Maradufu, A. 1982. Furanosesquiterpenoids of *Commiphora erythraea* and *Commiphora myrrh*. Naturally occurring insecticides, used on livestock against ticks. *Phytochemistry*, 21:677–680.

Maradufu, A., and Warthen, J.D., Jr. 1988. Furanosesquiterpenoids from *Commiphora myrrh* oil. *Plant Sci.* 57:181–184.

Michie, C.A., and Cooper, E. 1991. Frankincense and myrrh as remedies in children. *J. R. Soc. Med.* 84:602–605.

Morteza-Semnani, K., and Saeedi, M. 2003. Constituents of the essential oil of *Commiphora myrrha* (Nees) Engl. var. *molmol. J. Essent. Oil Res.* 15:50–51.

Rao, R.M., Khan, Z.A., and Shah, A.H. 2001. Toxicity studies in mice of *Commiphora molmol* oleo-gum-resin. *J. Ethnopharmacol.* 76:151–154.

Steyn, M. 2003. *Commiphora: a field guide, Southern Africa.* Marthinus Steyn, Bendor Place, Polokwane, South Africa. 92 pp (in English and Afrikaans).

Tipton, D.A., Lyle, B., Babich, H., and Dabbous, M.K. 2003. In vitro cytotoxic and anti-inflammatory effects of myrrh oil on human gingival fibroblasts and epithelial cells. *Toxicol. In Vitro*, 17:301–310.

Tucker, A.O. 1986. Frankincense and myrrh. *Econ. Bot.* 40:425–433.

Vollesen, K. 1986. *Commiphora*, some thoughts on the classification of an "impossible" genus. In *Research on the Ethiopian flora. Proceedings of the first Ethiopian flora symposium held in Uppsala May 22–26, 1984.* Edited by I Hedberg. Uppsala Univ., Uppsala, Sweden. pp. 204–212.

Walt, J.J.A. van der. 1973. The South African species of *Commiphora. Bothalia*, 11:53–102.

Walt, J.J.A. van der. 1975. The fruit of *Commiphora. Boissiera*, 24a:325–330.

Wiendl, R.M., Muller, B.M., and Franz, G. 1995. Proteoglycans from the gum exudate of myrrh. *Carbohydr. Polym.* 28:217–226.

Zhu, N., Sheng, S., Sang, S., Rosen, R.T., and Ho, C.-T. 2003. Isolation and characterization of several aromatic sesquiterpenes from *Commiphora myrrha. Flavour Fragrance J.* 18:282–285.

Specialty Cookbooks

Antol, M.N. 1996. *Healing teas: how to prepare and use teas to maximize your health.* Avery, Garden City Park, NY. 246 pp.

Chifolo, A.F., and Rayner, W.H. 2006. *Cooking with the bible: biblical foods, feasts, and lore.* Greenwood Press, New York. 416 pp.

Dawson, A.G. 2000. *Herbs: partners in life. Healing, gardening, and cooking with wild plants.* Healing Arts Press, Rochester, VT. 280 pp.

DiMarco, V. 2007. *Egg pies, moss cakes, and pigeons like puffins: eighteenth century British cookery from manuscript sources.* iUniverse, Inc., New York. 487 pp.

66 Neem

Family: Meliaceae (mahogany family)

NAMES

Scientific Name: *Azadirachta indica* A. Juss. (*Antelaea azadirachta* (L.) Adelb., *Melia azadirachta* L., *M. indica* (A. Juss.) Brandis)

- The word "neem" is the Hindi word for the plant, *nīm*, which is derived from the Sanskrit *nimba*.
- The neem tree is also known as holy tree, Indian lilac, margos, margosa tree, nim, nimtree, paradise tree, and white cedar.
- Curry leaf (*Murraya koenigii* (L.) Spreng) is usually known as "cooking neem" in Indian stores, and asking for "neem" in such a store is likely to result in this culinary herb being offered. "Sweet neem" is an alternative name for curry leaf and is also a name used for Siamese neem (*A. indica* var *siamensis* Valeton, also known as "edible neem"), which has red instead of white flowers.
- The Chinaberry, *Melia azedarach* L. (also known as gora neem, Indian bead tree, pride of India, and Texas umbrella tree), was once easily confused with neem when the latter's scientific name was *Melia azadirachta*. This common ornamental of the southern United States is sufficiently toxic that consumption can result in death. Children have died from eating six to eight fruits, and dogs have died from consuming several; birds, however, seem to love the berries. In parts of Florida, the species is now classified as a noxious weed.
- The genus name *Azadirachta* is based on the Persian *azad dhirakat* or *azaddhirakt*, "excellent tree" or "noble tree," referring to the usefulness and the considerable economic importance of the genus.
- *Indica* in the Latin scientific name *A. indica* means Indian.

PLANT PORTRAIT

Neem is a tropical tree, 15 to 20 m (49–66 ft.), or very rarely even 30 m (98 ft.) in height, with a dense, rounded crown that can be 20 m (66 ft.) wide. The trunk diameter sometimes is as wide as 1.8 m (6 ft.). The tree is normally evergreen but will drop its leaves in very dry localities during extreme drought and in areas of frost occurrence. Neem trees commonly live for more than a century but rarely reach 200 years of age. The species is native to east India and Myanmar (Burma) and has been planted in much of Southeast Asia and West Africa. Neem is also grown in Australia and in Central and South America. The somewhat edible fruit are olive-like, up to 2 cm (3/4 in.) long, with one to three seeds, the fruit ripening from green to yellow. A mature tree can produce 50 kg (110 lb.) of fruit annually (as much as 150 kg or 330 lb. in a year by a single tree has been recorded). The seed kernels contain approximately 45% oil. Neem oil is very widely used in southern Asia as a natural insecticide for agricultural crops (to a lesser extent, leaves, powdered seeds, and water extracts of various parts of the plant have also been traditionally used for the same purpose). The oil has sulfur-containing chemicals that cause it to smell like garlic, and it is unsuitable as food. The oil is used to prepare soap, toothpaste, and other products and is used as a lamp fuel. Chemicals called azadirachtins are the main basis of neem's anti-insect activity, and attempts are underway to produce improved insect-control products on the basis of these chemicals. Neem is the leading

FIGURE 66.1 Neem tree (*Azadirachta indica*), by B. Flahey.

FIGURE 66.2 Neem (*Azadirachta indica*), by B. Flahey. Fruiting branch at right, flowering branch at left.

candidate to replace synthetic chemical pesticides with botanical control agents, which are far less toxic to humans and animals in the ecosystem, and because they are biodegradable, they are less persistent in the environment than are synthetic pesticides, which tend to pollute soil and water. Neem is also useful as a source of timber and as a livestock feed and is often used for shade and as an ornamental in tropical and subtropical regions. The myriad uses and considerable potential of the neem have led to its being called a "wonder tree."

Neem is widely used in Ayurvedic medicine, an ancient system developed in India. The word Ayurveda comes from the Sanskrit words *ayus*, life + *veda*, science, that is, "the science of life."

Roots, bark, gum, leaves, flowers, fruit, seed kernels, and seed oil from neem are all used in preparations intended to heal and prevent disease. A very large number of illnesses are treated with neem, so much so that the tree is sometimes called "the village pharmacy." For examples, neem is used in India to treat leprosy, ulcers, diabetes, and skin disorders. It is also used as a birth control agent, specifically as a spermicide.

Neem is considered sacred by Hindus. On the Indian New Year day (Gudipadwa), Hindus eat bitter neem leaves to symbolize acceptance of the bitterness as well as the sweetness of life and bathe in water containing boiled neem leaves in the belief that this will result in freedom from diseases. During the festival of Ghatasthapana ("installation of the sacred pot"), a container is filled with water in which five neem twigs and a coconut are placed, then covered with flowers, worshipped, and presented with sacrifices to avert disease and bad luck.

CULINARY PORTRAIT

Neem is very extensively used in the Developing World for numerous purposes, but the culinary usage is quite minor. Its potential as a human food warrants attention because the tree is commonly available in regions where food is often scarce, and by means of either breeding improved edible varieties or processing, it is possible to extract considerable nutrition from the plant. Methods of preparing edible neem oil have been developed, and attempts at commercialization are underway.

The fruits are eaten in India, but enjoying them requires experience. When completely ripe, the flesh has a pleasant, sweet-sour flavor, but it is not easily separated from the core, which is extremely bitter. For the most part, neem fruits are not currently considered particularly desirable and are mainly consumed as a famine food. The fruit is also made into an alcoholic drink. The seeds and leaves are also extremely bitter. The seed kernels and leaves are occasionally consumed as a bitter spice in some of the hotter Indian foods. After meals, many people eat one- or two-seed kernels in the belief that this will aid digestion and kill bacteria in the mouth. In many parts of India, neem leaves are consumed as a chutney (paste) to enhance health and resistance to disease. The flowers are eaten raw or fried in curries and soups. Because of uncertainties regarding the long-term health effects of consuming neem, the recommendation has been made that the leaves and leaf extracts should not be consumed by people or fed to animals over a long period.

Siamese neem (mentioned above) is known as "edible neem" and "sweet neem" because its young leaves and flowers contain lower amounts of bitter substances than neem and are consumed as a vegetable in Myanmar (Burma).

CULINARY VOCABULARY

- As noted in Names, curry leaf can be confused with neem ("sweet neem" can refer both to edible neem and to curry leaf). In India (and in recipes dealing with Indian food), the phrases (in Hindi, often carried over into English) *meetha neem* and *mitha neem* refer to curry leaves, not to neem.
- In parts of southern India, neem flowers are very popular in *ugadi pachhadi*, a souplike pickle. *Veppampoo rasam*, served by Tamils in India, is another souplike dish based on neem flowers. *Bevina hoovina gojju* is a type of curry prepared with neem blossoms in India.

CURIOSITIES OF SCIENCE AND TECHNOLOGY

- To protect against any lingering infection, the Puranas (Indian sacred texts) urge that neem leaves be chewed after attending a funeral, a practice that continues symbolically in India to signify grief. The Puranas also suggest that neem leaves be strewn as an antiseptic barrier on the threshold of a house where a death has occurred. These recommendations date to times when many people died in epidemics, and such use of neem leaves likely has resulted in saving lives. Inhaling the smoke of burning neem leaves is said to exorcise evil spirits, and once again the antibacterial power of the plant may in fact serve to rid the body of harm.

- Twigs of the neem tree are used daily as a toothbrush in Bangladesh, India, and Pakistan by approximately 600 million people. The end of the twig is chewed to form bristles. It is thought that materials in neem help prevent cavities and gum diseases. A toothpaste is also manufactured using neem. Attempts are underway to market neem toothpastes in the Western World, with essential oils added to mask the natural, bitter, garlic-like taste.
- Although the neem tree is extraordinarily defended against insects by its own natural insecticides, it is attacked by dozens of insects, including several that specialize on it. This is in accord with a principle enunciated by biologist Daniel H. Janzen of the University of Pennsylvania: "for every plant defence there is a predator that has found a way of overcoming it."

KEY INFORMATION SOURCES

Benge, M.D. 1986. *Neem: the cornucopia tree*. Agency for International Development, Washington, D.C. 90 pp.

Catling, P.M., and Small, E. 2003. Blossoming treasures of biodiversity:10. Neem—an economic and environmental wonder plant. *Biodiversity*, 4(4):25–28 + back cover.

Conrick, J. 2001. *Neem, the ultimate herb*. Lotus Press, Twin Lakes, WI. 165 pp.

Gilbert, H. 1986. *The neem tree,* Azadirachta indica, *an inhibitor of insect feeding and growth, 1970–1985: 297 citations*. U.S. Department of Agriculture, National Agricultural Library, Beltsville, MD. 28 pp.

Gilbert, H. 1989. *The neem tree: an inhibitor of insect feeding and growth, January 1982–April 1989: 302 citations*. U.S. Department of Agriculture, National Agricultural Library, Beltsville, MD. 30 pp.

Gupta, B.N., and Sharma, K.K. 1998. *Neem, a wonder tree*. Indian Council of Forestry Research & Education, Dehra Dun, India. 168 pp.

Jacobson, M. 1989. *The neem tree*. CRC Press, Boca Raton, FL. 178 pp.

Koul, O., and Wahab, S. (Eds.). 2004. *Neem: today and in the new millennium*. Kluwer Academic Publishers, Boston, MA. 276 pp.

Koul, O., Isman, M.B., and Ketkar, C.M. 1990. Properties and uses of neem, *Azadirachta indica*. *Can. J. Bot.* 68:1–11.

März, U. 1989. *The economics of neem production and its use in pest control*. Wissenschaftsverlag Vauk Kiel, Germany. 153 pp.

Mitra, C.R. 1963. *Neem*. Indian Central Oilseeds Committee, Hyderabad, India. 190 pp.

Narwal, S.S., Tauro, P., and Bisla, S.S. (Eds.). 1997. *Neem in sustainable agriculture*. Scientific Publishers, Jodhpur, India. 266 pp.

National Research Council (U.S.). 1992. *Neem: a tree for solving global problems*. Report of an Ad Hoc Panel of the Board on Science and Technology for International Development, National Research Council. National Academy Press, Washington, D.C. 141 pp.

Norten, E., and Pütz, J. 2000. *Neem: India's miraculous healing plant*. Healing Arts Press, Rochester, VT. 92 pp.

Pennington, T.D., and Styles, B.T. 1975. A generic monograph of the Meliaceae. *Blumea*, 22:419–540.

Puri, H.S. 1999. *Neem: the divine tree Azadirachta indica*. Harwood Academic Publishers, Amsterdam, the Netherlands. 182 pp.

Randhawa, N.S., and Parmar, B.S. 1996. *Neem*. 2nd ed. New Age International, New Delhi, India. 332 pp.

Read, M.D., and French, J.H. (Eds.). 1993. *Genetic improvement of neem: strategies for the future. Proceedings of an International Consultation Held at Kasetsart University, Bangkok, Thailand, 18–22 January 1993*. Winrock International Institute for Agricultural Development, Forestry/Fuelwood Research and Development Project, Bangkok, Thailand. 194 pp.

Schmutterer, H. (Ed.). 1995. *The neem tree: Azadirachta indica A. Juss. and other meliaceous plants: sources of unique natural products for integrated pest management, medicine, industry, and other purposes*. VCH, New York. 696 pp.

Schmutterer, H., and Ascher, K.R.S. (Eds.). 1983. *Natural pesticides from the neem tree* (Azadirachta indica *A. Juss) and other tropical plants. Proceedings of the Second International Neem Conference, Rauischholzhausen, Federal Republic of Germany, 25–28 May, 1983.* Deutsche Gesellschaft für Technische Zusammenarbeit, Eschborn, Germany. 587 pp.

Schmutterer, H., Ascher, K.R.S., and Rembold, H. 1981. *Natural pesticides from the neem tree* (Azadirachta indica *A. Juss). Proceedings of the First International Neem Conference, Rottach-Egern, Federal Republic of Germany, 16–18 June, 1980.* German Agency for Technical Cooperation, Eschborn, Germany. 297 pp.

Singh, R.P. 1996. *Neem and environment.* Science Publishers, Lebanon, NH. 2 vols.

Singh, R.P., Saxena, R.C., and Hassan, E. (Eds.). 1999. Azadirachta indica *A. Juss. Proceedings International Neem Conference, University of Queensland, 1996.* Science Publishers, Enfield, NH. 322 pp.

Tewari, D.N. 1992. *Monograph on neem* (Azadirachta indica *A. Juss.*). International Book Distributors, Dehra Dun, India. 279 pp.

Venketeshwarlu, B. (Ed.). 1998 *Neem applications in agriculture, health care and environment. Proceedings and Extended Summaries, International Conference on Neem: Setting Goals for a Global Vision (Dec. 20–22, 1996).* Neem Foundation, Mumbai, India. 91 pp.

Vijayalakshmi, K., Shiva, V., and Radha, K.S. 1995. *Neem, a user's manual.* Centre for Indian Knowledge Systems and Research Foundation for Science, Technology, and Natural Resource Policy (New Delhi), Madras, India. 96 pp.

Specialty Cookbooks

Note: In virtually all recipes in cookbooks, the word neem refers to curry leaf; in "recipes" for health-promoting formulations in books dealing with aromatherapy or the application of oils to skin, the word neem refers to neem.

Varadarajan, V. 2008. *Festival Samayal: an offering to the gods.* Orient Enterprises, Chennai, India. 98 pp. [Presents recipe with neem flowers.]

67 Noni (Indian Mulberry)

Family: Rubiaceae (madder family; more familiarly, this is occasionally called the "coffee family")

NAMES

Scientific Name: *Morinda citrifolia* L. (Not *M. citrifolia* Bedd., the correct name of which is *M. tinctoria* Roxb.)

- The name "noni" is the Hawaiian word for the plant, probably acquired from a similar Asian name during the transport of the plant from Asia to Hawaii. The original settlers of Polynesia migrated from Southeast Asia and Indonesia before settling the Polynesian islands from 1000 BC to AD 500. Hawaii was one of the last island groups to be occupied, probably between AD 500 and 700. The name in Southeast Asia is *nhau*, in the islands of the South Pacific the plant is known as *nonu*, and in Samoa and Tonga the name is *nono.*
- "Tahitian noni" is a registered trade mark, but the plant, growing in Tahiti, is nevertheless legitimately called Tahitian noni.
- Next to the word noni, *M. citrifolia* is best known as "Indian mulberry," a name that indicates the resemblance of the aggregated fruits to those of the true mulberry (species of *Morus*). Unfortunately, the name Indian mulberry is applied to some kinds of white mulberry (*Morus alba* L., also known by the old names *Morus indica* L. and *Morus tatarica* L.), and the expression "Indian mulberry" is sometimes used simply to indicate a mulberry species cultivated in or native to India.
- Noni is also known as awl tree, beach mulberry, canary wood, cheesefruit (in Australia), dog dumping (in Barbados), headache tree, hog apple, great morinda, painkiller (in the Caribbean, for its medicinal values), Polynesian bush fruit, rotten cheese plant (for the bad odor of the fruit), starvation fruit, vomit fruit, and wild pine.
- The genus name *Morinda* is from the Greek *moron*, mulberry + *inda*, Indian.
- *Citrifolia* in the scientific name *M. citrifolia* is Latin for lemon-leaved.

PLANT PORTRAIT

Noni is a tropical, evergreen shrub or tree, usually less than 3 m (10 ft.) high but sometimes more than 10 m (33 ft.) in height. The small white flowers are in globose heads approximately 2.5 cm (1 in. long), and these mature into fruit 5 to 10 cm (2–4 in.) long and 5 to 7 cm (2–3 in.) thick. The fruit surface is divided into warty, polygonal, pitted cells. The apparent "fruit" is actually a compound structure composed of numerous, fused, separate tiny fruits (sometimes called "stones" because of the hard covering), each produced from a separate flower, and each containing up to four seeds. Each section or hexagonally marked area on the mature, yellowish-white compound fruit represents the transformation of a flower into an individual tiny fruit. The green immature compound fruit ripens to yellow and white, at which stage the skin becomes translucent, and the yellowish-white flesh is soft and juicy and has an unpleasant, foul odor which has been compared with vomit. The seeds float and are known to be carried unharmed by sea water over very long distances to establish new plants. Accordingly, the plants are mostly found growing near seashores. The species is native

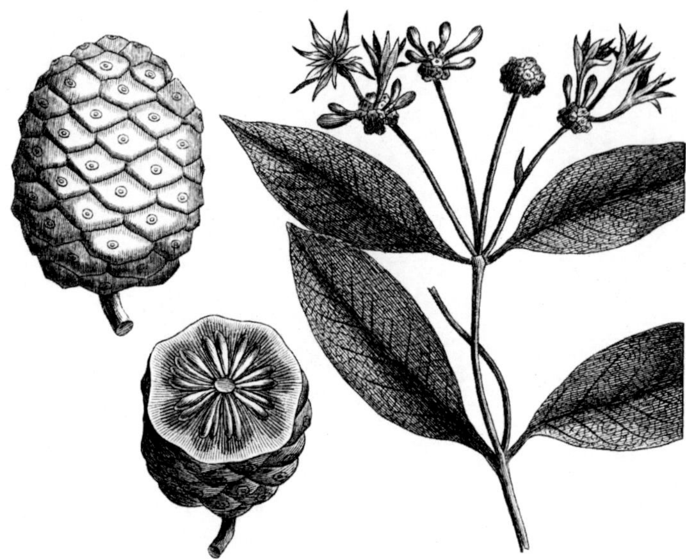

FIGURE 67.1 Noni (*Morinda citrifolia*). (From Lamarck and Poiret, 1744–1829, plate 153.)

to Malaysia, Australia, and Polynesia and is widely naturalized in southern Asia, tropical America, and the West Indies. It has been claimed that noni was originally native to Queensland, Australia, from where it was taken to other locations. The species has been extensively grown in the past in India and elsewhere, not as a food plant but rather for an excellent red dye that can be extracted from the roots of young plants (generally harvested when 3 years old) as well as a yellow dye from the peeled roots. Noni was introduced long ago to Hawaii but lost most of its popularity there as a cultivated plant by the early twentieth century.

In addition to being used as food in the Malay Peninsula and on the Pacific Islands, especially in Hawaii, noni fruit acquired usage as a medicinal plant for diabetes, high blood pressure, cancer, and numerous chronic disorders, although such applications have not been scientifically validated. Nevertheless, in recent years, noni has become a popular folk medicine, with claims that it can help cure high blood pressure and cancers. Numerous companies have taken advantage of noni's cure-all reputation and are now marketing noni extract juice from the ripened fruits or dried extract in capsule form. Cultivation of noni has increased dramatically to meet the increasing market demand for noni products. Noni appears to have some beneficial effects but is being advertised in the tradition of the infamous snake-oil salesman, with obviously exaggerated claims that it will cure almost everything. An Internet search for "noni" reveals thousands of Web sites, almost all of which advertise products alleging remarkable medical virtues, and the superiority of a particular company's offerings. Companies getting their noni from Tahiti claim it is superior to Hawaiian noni and vice versa. Noni products are now sold not only as juice, tea bags, extracts, powders, and capsules but even as facial cleansers, bath gels, and soaps. It has commonly been said that good medicine has to taste bad, and it is widely acknowledged that noni tastes bad. With so many traditional uses for noni, it is likely that there are some medical conditions that would benefit from the plant. A few pharmacological analyses of noni have been conducted to date, but there is a need for much more extensive and objective study. A wide variety of chemical components have been investigated for possible physiological effects, but exact mechanisms of action are not yet known. Most of the claims about scientific research conducted on noni are highly deceptive, and indeed there have been convictions in the United States for fraudulent advertising. With the present state of knowledge, noni should be taken with a grain of salt, and noni juice should not be substituted for any standard medical treatment.

CULINARY PORTRAIT

The fruit stinks and tastes bad (it has been compared with putrid cheese) and, quite sensibly, it was mostly eaten in the past mainly as an emergency or famine food. However, in some Pacific islands, including Raratonga, Samoa, and Fiji, noni fruit was once a staple food of choice, eaten raw or cooked. The fruit has been eaten raw with salt in Indochina and by Australian Aborigines and cooked as a curry. The slightly bitter young leaves were eaten in Java and Thailand, and mature leaves have been wrapped around fish, cooked, and eaten along with the fish. Such usages are still found occasionally but are uncommon. Until recently, most of the world has not been attracted to either the culinary or the medicinal virtues of noni, but rather the plant has been grown mainly as a garden ornamental and dye source of tropical climates. However, because of an explosion of interest in its alleged medicinal virtues in the last decade, noni juice is now being widely marketed. The fruits can hardly be recommended as food. The seeds have been roasted and eaten, but the advice has been offered that they should be avoided. The best way to consume noni fruit is in capsules or in mixed juices.

Noni is high in potassium, and people with kidney problems on potassium-restricted diets should not consume it. There is suspicion that noni may cause liver damage in some individuals (contrarily, there is some evidence that noni has liver-protective properties), and those with liver disease should consult a doctor before taking noni products. Noni has historically been used to induce abortion, and it has been recommended that it should not be consumed during pregnancy or breast feeding.

CURIOSITIES OF SCIENCE AND TECHNOLOGY

- Noni was known as the queen of the "canoe plants," which were sacred plants that Polynesians took with them when settling a new island. In addition to medicinal uses, the Hawaiians used the resilient leaves to wrap food during cooking, dyes from the plant for clothing, and the wood for preparing household objects. Some of the familiar canoe plants brought to Hawaii include bamboo, banana, breadfruit, coconut, sugar cane, sweet potato, turmeric, and yam.
- In French Polynesia, pigs were often allowed to roam free where noni was grown because the smell of the fruits was so bad it was thought that the additional odor from the pigs would not matter. In recent times in Hawaii, numerous trees were cut down because of their rancid smell.
- Hawaiians once placed noni juice on their heads to get rid of lice.
- Noni fruit pulp has been used to clean hair, iron, and steel.
- Noni fruits appear to attract weaver ants (*Oecophylla smaragdina*), which often make remarkable nests out of the living leaves of the plant. By offering such nesting sites to the insects, the plants may in turn benefit as the ants will attack other insects that attempt to feed on the tree. Weaver ants are large, reddish, Old World ants with a fierce bite. They create their nests using chains of workers to pull leaf edges against each other then squeeze their larvae to produce silk to glue the leaf edges together. In China, the ants have deliberately been placed in citrus orchards to protect the fruit against insects. The pupae of these ants are consumed as a delicacy in Southeast Asia.

KEY INFORMATION SOURCES

Aalbersberg, W.G.L., Hussein, S., Subramaniam, S., and Parkinson, S. 1993. Carotenoids in the leaves of *Morinda citrifolia*. *J. Herbs Spices Med. Plants*, 2(1):51–54.

Brett, J.W., Jensen, C.J., and Westendorf, J. 2008. A new vegetable oil from noni (*Morinda citrifolia*) seeds. *Int. J. Food Sci. Tech.* 43:1988–1992.

Dittmar, A. 1993. *Morinda citrifolia* L.: use in indigenous Samoan medicine. *J. Herbs Spices Med. Plants*, 1(3):77–92.

Dixon, A.R., McMillen, H., and Etkin, N.L. 1999. Ferment this: the transformation of noni, a traditional Polynesian medicine (*Morinda citrifolia*, Rubiaceae). *Econ. Bot.* 53:51–68.

Elkins, R. 2002. *The noni revolution: today's tropical wonder that can battle disease, boost energy and revitalize your health*. Woodland Pub., Pleasant Grove, UT. 175 pp.

Fairechild, D.1998. *Noni: aspirin of the ancients*. 3rd ed. Flyana Rhyme, Anahola, HI. 191 pp.

Farine, J.P., Legal, L., Moreteau, B., and Quere, J.L. le. 1996. Volatile components of ripe fruits of *Morinda citrifolia* and their effects on *Drosophila*. *Phytochemistry*, 41:433–438.

Furusawa, E., Hirazumi, A., Story, S., and Jensen, J. 2003. Antitumour potential of a polysaccharide-rich substance from the fruit juice of *Morinda citrifolia* (noni) on Sarcoma 180 ascites tumour in mice. *Phytother. Res*. 17:1158–1164.

Groenendijk, J.J. 1991. Morinda citrifolia L. In *Plant resources of South-East Asia. Vol. 3. Dye and tannin-producing plants*. Edited by R.H.M.J. Lemmens and N. Wulijarni-Soetjipto. Pudoc, Leiden, the Netherlands. pp. 94–96.

Heinicke, R.M. 1985. The pharmacologically active ingredient of noni. *Bull. Pacific Trop. Bot. Gard*. 15(1):10–14.

ICON Health Publications. 2004. *Morinda citrifolia—a medical dictionary, bibliography, and annotated research guide to Internet references*. ICON Health Publications, San Diego, CA. 116 pp.

Levand, O., and Larson, H.O. 1979. Some chemical constituents of *Morinda citrifolia*. *Planta Med*. 36:186–187.

McClatchey, W. 2002. From Polynesian healers to health food stores: changing perspectives of *Morinda citrifolia* (Rubiaceae). *Integr. Cancer Ther*. 1:110–120.

Morton, J.F. 1992. The ocean-going noni, or Indian mulberry (*Morinda citrifolia*, Rubiaceae) and some of its "colourful" relatives. *Econ. Bot*. 46:241–256.

Nelson, S.C. 2001. *Noni cultivation in Hawaii*. U.S. Department of Agriculture, Cooperative Extension Service, College of Tropical Agriculture & Human Resources, Honolulu, Hawaii, and University of Hawaii at Manoa. 4 pp.

Nelson, S.C. (Ed.). 2003. *Proceedings of the 2002 Hawai'i Noni Conference, Hilo Hawaiian Hotel, Hilo, Hawai'i, October 26, 2002*. University of Hawaii at Manoa, Manoa, HI. 50 pp.

Nelson, S.C. 2008. *Morinda citrifolia*, noni. In *The encyclopedia of fruit & nuts*. Edited by J. Janick and R.E. Paull. CABI, Wallingford, Oxfordshire, U.K. pp. 764–767.

Pawlus, A.D., and Kinghorn, D.A. 2007. Review of the ethnobotany, chemistry, biological activity and safety of the botanical dietary supplement *Morinda citrifolia* (noni). *J. Pharm. Pharmacol*. 49:1587–1609.

Potterat, O., and Hamburger, M. 2007. *Morinda citrifolia* (noni) fruit—phytochemistry, pharmacology, safety. *Planta Med*. 73:191–199.

Sang, S., Wang, M., He, K., Liu, G., Dong, Z., Badmaev, V., et al. 2002. Chemical components in noni fruits and leaves (*Morinda citrifolia* L.). In *Quality management of nutraceuticals. Proceedings of a Symposium, Washington, DC, August 2000*. Edited by C.T. Ho and Q.Y. Zheng. American Chemical Society, Washington, DC. pp. 134–150.

Solomon, N. 1998. *Noni: nature's amazing healer*. Woodland Pub., Pleasant Grove, UT. 101 pp.

Solomon, N. 2000. *Tahitian noni juice: how much, how often, for what*. Direct Source Pub., Vineyard, UT. 60 pp.

Solomon, N., and Udall, C. 1999. *The noni phenomenon*. Direct Source Pub., Vineyard, UT. 296 pp.

Starling, S. 2003. Noni begins to bear fruit. *Functional Ingredients*, 2003(June):26–27.

Tap, N., and Bich, N.K. 2003. *Morinda* L. In *Plant resources of South-East Asia. Vol. 12(3). Medicinal and poisonous plants 3*. Edited by R.H.M.J. Lemmens and N. Bunyapraphatsara. Backhuys Publishers, Leiden, the Netherlands. pp. 302–305.

Younos, C., Rolland, A., Fleurentin, J., Lanhers, M.C., Misslin, R., and Mortier, F. 1990. Analgesic and behavioural effects of *Morinda citrifolia*. *Planta Med*. 56:430–434.

Wang, M.Y., West, B.J., Jensen, C.J. Nowicki, D., Chen, S.U., Palu, A.K., et al. 2002. *Morinda citrifolia* (noni): a literature review and recent advances in noni research. *Acta Pharmacol. Sin*. 23:1127–1141.

Wei, G.J., Huang, T.C., Huang, A.S., and Ho, C.T. 2004. Flavor compounds of noni fruit (*Morinda citrifolia* L.) juice. In *Nutraceutical beverages: chemistry, nutrition and health effects; proceedings of a symposium, Chicago, August 2001*. Edited by F. Sahidi and D.K. Weerasinghe. American Chemical Society, Washington, DC. pp. 52–61.

Specialty Cookbooks

Navarre, I. 2005. *101 ways to use noni fruit juice*. Direct Source Publishing, Orem, UT. 445 pp.

Navarre-Brown, I. 2001. *76 ways to use noni fruit juice for your better health: a handbook of oral, topical, and internal applications and procedures*. Pride Pub., Vineyard, UT. 345 pp.

Tsutsumi, C.C. 2003. *101 great tropical drinks: cocktails, coolers, coffees, and virgin drinks*. Island Heritage Publishing, Weipahu, HI. 152 pp.

68 Nutmeg and Mace

Family: Myristicaceae (nutmeg family)

NAMES

Scientific Name: *Myristica fragrans* Houtt.

- The word nutmeg comes from the Latin *nux*, nut, and *muscat*, musky.
- The word mace comes from the Greek *makir*, which was used to denote an Oriental spice, possibly made from the bark of a tree.
- The genus name *Myristica* is apparently derived from the Greek *mýron*, balm or ointment.
- *Fragrans* in the scientific name *M. fragrans* comes from the Latin *fragrare*, smell, reflecting the aroma of the plant.

PLANT PORTRAIT

Two major spices, nutmeg and mace, come from the same tropical, evergreen tree, native to the Moluccas or Spice Islands of Indonesia. Nutmeg and mace were recorded in first-century Rome, and at that time were also well known in the Arab world, but did not become common in Europe until the end of the twelfth century, after the Crusaders became familiar with the spices in the Holy Land. Although cultivated in many tropical countries, most commercial exports of nutmeg and mace come from Indonesia and Grenada (West Indies). Grenada is often called the "Nutmeg Isle." The nutmeg is the national symbol of Grenada and is emblazoned on the country's red, yellow, and green flag. Nevertheless, nutmeg from Granada ("West Indian nutmeg") is often regarded as inferior to Indonesian nutmeg ("East Indian nutmeg").

The trees, which are usually either male or female, grow to a height of 4 to 10 m (13–33 ft.), sometimes as high as 20 m (66 ft.), and can bear fruit for 60 years or longer. The fruit of the nutmeg tree resembles a small peach or apricot. The fleshy outer shell is used locally where the tree is cultivated for making jam. Nutmeg is the single seed within the fruit, and it is surrounded by a hard black shell. Mace is the scarlet, lacy covering that surrounds the hard, black shell enclosing the seed. Exceptional trees may produce more than 10,000 fruits a year, but generally good trees yield 1500 to 2000 annually, typically the equivalent of approximately 4.5 kg (10 lb.) of nutmeg and a much smaller amount of mace (often less than 1/20th they yield of nutmeg). Nutmeg is light brown. White nutmeg has been treated with lime. Orange-yellow mace likely comes from Grenada, whereas orange-red mace is probably from Indonesia.

The Dutch waged a bloody war, including the massacre and enslavement of the inhabitants of the island of Banda (the largest of the Molucca spice islands of Indonesia), to control nutmeg production in the East Indies. In the eighteenth century, the Dutch kept the price of nutmeg artificially high by burning full warehouses of nutmegs in Amsterdam. Nutmegs exported from the islands were dipped in lime to prevent the seed from germinating and allowing competitors to establish trees. The Dutch were partially frustrated in their attempts to keep the seeds under their control by fruit pigeons, which carried the seeds to other islands. The Dutch held control of the Spice Islands until the Second World War, after which the French ambassador to Mauritius transported seedlings to that country.

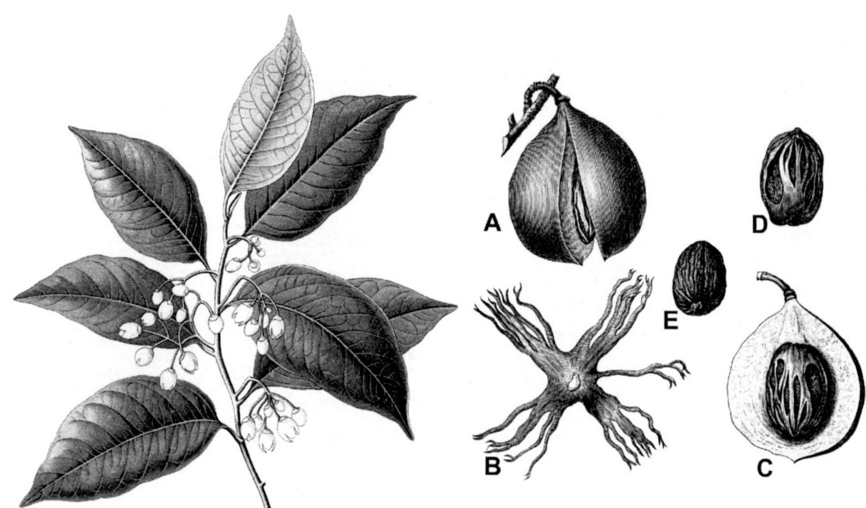

FIGURE 68.1 Nutmeg (*Myristica fragrans*). Left: flowering branch. (From Köhler, 1883–1914.) Right: (From Lamarck and Poiret, 1744–1829, plate 833.) A, opened fruit; B, aril ("mace"); C, split fruit; D, seed surrounded by aril; E, seed ("nutmeg").

CULINARY PORTRAIT

Both nutmeg and mace have an aromatic, sweet, warm, rich flavor, which is said to be more robust in nutmeg. Nutmeg is popular in sweet and savory dishes, baked goods, custards, puddings, meats, sausages, sauces, vegetables, eggs, and beverages. It is essential in bechamel sauce and gives a delicate flavor to cabbage, eggnog, cauliflower, and spinach. Nutmeg is a popular spice in traditional holiday recipes such as pumpkin pie, eggnog, and plum pudding. A survey showed that the five most popular spices during the winter holidays in the United States are, in decreasing order, ground black pepper, ground cinnamon, seasoned salt, garlic power, and ground nutmeg.

Mace is more expensive than nutmeg, and more powerful. Although somewhat less known than nutmeg, mace is used similarly, especially in pie spices and poultry seasonings, and to flavor cakes, cookies, fruit pies, preserves, cheeses, sauces, puddings, and soups. Mace is preferred for use with savory dishes, fish dishes, sauces for fish, and oyster stew. Mace is usually lighter in color than nutmeg and may be used instead of the latter in light-colored dishes where darker spice fragments would detract from appearance. Mace is used in the manufacture of pickles and tomato ketchup.

When ground, nutmeg soon loses the volatile oils that give it flavor and taste, so grating fresh nutmeg is recommended. If a nutmeg grater is unavailable, it is advisable to use the finest blade on a larger hand-held manual grater. To test if nutmegs are fresh and are of good quality, insert a darning needle a centimeter (3/8 in.) into the meat of the nutmeg; if a tiny drop of oil seeps out, the nut is good. Freshly grated nutmeg is best added at the end of the cooking process because heat diminishes the flavor. Whole fresh nutmegs as well as ground nutmeg and mace should be kept in a tightly sealed jar or container in a cool, dark place. Leftover fresh nutmeg should be wrapped tightly to prevent loss of oils. Once ground, nutmeg and mace are best used promptly or stored in a refrigerator.

Nutmeg and mace contain myristicin and elemicin, which are toxic, and accordingly these spices should be consumed only in limited quantities. Myristicin is hallucinogenic. Eating small amounts of nutmeg is harmless, but consuming one to three whole nutmegs (more than 1 teaspoon ground) can cause wild hallucinations, nausea, vomiting, and/or circulatory collapse within 1 to 6 hours. Very large doses can be fatal. Although nutmeg is reputed to counteract stomach distress from gas and to be useful for other ailments, home remedies should be avoided.

Culinary Vocabulary

- "Blade mace" is unground mace, generally available as fragments. It is less frequently encountered than ground mace but is often specified for some recipes—for example, in clear soups and sauces, where nutmeg powder might spoil the appearance.
- The classical French spice mixture *quatre épices* ("four spices") has been described as "fortified black pepper." It contains nutmeg, black pepper, cloves, and ginger (and, optionally, allspice and cinnamon). This is used to flavor meat dishes, especially those cooked or braised for a rather long time such as stews and ragouts, and sometimes also in sausages and pastries.
- A "Yard of Flannel" is a beverage and alleged cure for colds, prepared from ale, eggs, brown sugar, and nutmeg, and served warm.

CURIOSITIES OF SCIENCE AND TECHNOLOGY

- The citizens of medieval Rome set fire to piles of nutmeg on their city streets when the English King Henry VI (1421–1471), visited their city. These nutmeg bonfires gave their city a sweet, welcoming aroma while demonstrating the extravagant wealth of the city-state of Rome.
- In thirteenth-century Europe, half a kilogram (approximately 1 lb.) of nutmeg cost as much as three sheep or a cow. A kilogram of mace (approximately 2 lb.) would buy a cow. In 1393, a pound of nutmeg was worth seven fat oxen in Germany.
- At the height of its value in Europe, nutmeg was carried around by aristocrats as a demonstration of their wealth. Diners would flourish tiny graters and grind their own in fancy restaurants. Personal nutmeg graters became quite fashionable accoutrements, giving rise to intricate designs and shapes made of precious metals. These antique graters are now prized.
- Mace is generally in short supply and is more valuable than nutmeg. During the time that the Dutch controlled the Moluccas (the Spice Islands), a colonial administrator, not realizing that nutmeg and mace came from the same tree, issued an order that the colonists should plant fewer nutmeg trees and more mace trees.
- Nutmeg has been an ingredient of numerous balms and elixirs for centuries. In the eighteenth century, it was considered a cure-all. In England, a nutmeg was often carried in the pocket by those hoping to avoid rheumatism. Because of resemblance in shape to the human brain, nutmeg kernels were particularly used as a remedy for ailments of the brain. Perhaps related to this belief, nutmegs encased in silver were worn at night to induce sleep.
- Connecticut was known as the "Nutmeg State" because Yankee peddlers (mostly from Connecticut) whittled wooden "nutmegs," scented them with nutmeg, and sold them to gullible housewives. Use of the motto was abandoned recently by Connecticut because it did not reflect well on the state. Nicknames for Connecticuters include nutmegs and nutmeggers.
- In 1946, before his conversion to Islam, militant black social reformer Malcolm X (1925–1965), used nutmeg while in jail when his supplies of marijuana ran out. When the authorities became aware of the hallucinogenic use of nutmeg, it was removed from many prison kitchens. United States federal prison regulations now call for nutmeg and mace to be stored in a locked cabinet, along with other items that could be abused, such as alcohol-based flavorings.
- In nutmeg orchards, male trees are typically assigned a harem of 12 female trees.
- Mace, the brand of personal defense chemical sprays used in self-defense and riot control is not derived from the mace of nutmeg fruits. Mace especially uses the chemical capsaicin from the hot chili pepper, but some defensive sprays also use tear gas and other components.

KEY INFORMATION SOURCES

Armstrong, J.E., and Drummond, B.A. 1986. Floral biology of *Myristica fragrans* Houtt. (Myristicaceae), the nutmeg of commerce. *Biotropica*, 18:32–38.

Bavappa, K.V.A., and Ruettimann, R.A. 1981. *Nutmeg: cultivation and processing. Sri Lanka*. Department of Minor Export Crops, Kandy, Sri Lanka. 11 pp.

Brizan, G.I. 1979. *The nutmeg industry, Grenada's black gold*. Government Printing Office, St. George's, Grenada. 75 pp.

Ceylon Industrial Development Board. 1973. *Nutmeg cultivation*. Industrial Development Board of Ceylon, Ceylon. 7 pp.

Ehlers, D., Kirchoff, J., Gererd, D., and Quirin, K.W. 1998. High-performance liquid chromatography analysis of nutmeg and mace oils produced by supercritical CO_2 extraction—comparison with steam-distilled oils—comparison of East Indian, West Indian and Papuan oils. *Int. J. Food Sci. Technol.* 33: 215–223.

Federation of American Societies for Experimental Biology (U.S.A.). 1973. *Evaluation of the health aspects of nutmeg, mace and their essential oils as food ingredients*. FASEB, Bethesda, MD. 18 pp.

Flach, M., and Cruickshank, A.M. 1969. Nutmeg (*Myristica fragrans* Houtt. and *Myristica argentea* Warb.). In *Outlines of perennial crop breeding in the tropics*. Edited by F.P. Ferwerda and F. Wit. Misc. Paper 4, Landbouwhogeschool, Wageningen, the Netherlands. pp. 329–338.

Flach, M., and Willinck, M.T. 1999. *Myristica fragrans* Houtt. In *Plant resources of South-East Asia. Vol. 13. Spices*. Edited by C.C. de Guzman and J.S. Siemonsma. Backhuys, Leiden, the Netherlands. pp. 143–148.

Food and Agriculture Organization of the United Nations. 1994. *Nutmeg and derivatives*. FAO, Rome, Italy. 107 pp.

Food and Drug Research Laboratories (Maspeth, NY). 1972. *GRAS (generally recognized as safe) food ingredients: oil of nutmeg and myristica oil*. Distributed by National Technical Information Service, Springfield, VA. 19 pp.

Forrest, J.E., and Heacock, R.A. 1972. Nutmeg and mace, the psychotropic spices from *Myristica fragrans*. *Lloydia*, 35:440–449.

Gopalakrishnan, M. 1992. Chemical composition of nutmeg and mace. *J. Spices Aromat. Crops*, 1(1): 49–54.

Grover, J.K., Khandkar, S., Vats, V., Dhunnoo, Y., and Das, D. 2002. Pharmacological studies on *Myristica fragrans*: antidiarrheal, hypnotic, analgesic and hemodynamic (blood pressure) parameters. *Methods Find. Exp. Clin. Pharmacol.* 24:675–680.

Inamdar, J.A., Avita (no initial), and Rao, N.V. 1982. Studies in *Myristica fragrans* Houtt. *J. Econ. Taxon. Bot.* 3:869–875.

International Organization for Standardization. 2002. *Nutmeg, whole or broken, and mace, whole or in pieces* (Myristica fragrans *Houtt.*). *Specification*. 2nd ed. International Organization for Standardization, Geneva, Switzerland. 8 pp.

Joseph, J. 1980. The nutmeg *Myristica fragrans*—its botany, agronomy, production, composition, and uses. *J. Plant. Crops*, 8(2):61–72.

Krishnamoorthy, B., and Rema, J. 2001. Nutmeg and mace. In *Handbook of herbs and spices*. Edited by K.V. Peter. Woodhead Publishing, Cambridge, U.K. pp. 238–248.

Mayers, J.M. 1974. *The nutmeg industry of Grenada*. Institute of Social and Economic Research, University of the West Indies, Kingston, Jamaica. 50 pp. (Reprinted 1981).

McKee, L.H., and Harden, M.L. 1991. Nutmeg: a review. *Lebensm. Wiss. Technol.* 24:198–203.

Milton, G. 1999. *Nathaniel's nutmeg, or, the true and incredible adventures of the spice trader who changed the course of history*. Farrar, Straus and Giroux, New York. 388 pp.

Mohana-Rao, P.R. 1974. Nutmeg seed: its morphology and developmental anatomy. *Phytomorphology*, 24:263–273.

Peter, K.V., and Kandiannan, K. 1999. Nutmeg. In *Tropical horticulture, vol. 1*. Edited by T.K. Bose, S.K. Mitra, A.A. Farooqui, and M.K. Sadhu. Naya Prokash, Calcutta, India. pp. 724–730.

Pillai, K.S. 1965. *The nutmeg*. Trivandrum Agricultural Information Service, Department of Agriculture, Kerala, India. 28 pp.

Simpson, G.I.C., and Jackson, Y.A. 2002. Comparison of the chemical composition of East Indian, Jamaican and other West Indian essential oils of *Myristica fragrans* Houtt. *J. Essent. Oil Res.* 14:6–9.

Sinclair, J. 1969. The genus *Myristica* in Malesia and outside Malesia. Government Printing Office, Singapore, Singapore. 540 pp.

Utami, N.W., and Brink, M. 1999. *Myristica* Gronov. In *Plant resources of South-East Asia. Vol. 13. Spices.* Edited by C.C. de Guzman and J.S. Siemonsma. Backhuys, Leiden, the Netherlands. pp. 139–143.

Wilson, T.K., and Maculans, L. 1967. The morphology of Myristicacea. Flowers of *Myristica fragrans* and *M. malabarica*. *Am. J. Bot.* 54:214–220.

Specialty Cookbooks

Peterson, J. 2008. *Sauces: classical and contemporary sauce making*. 3rd ed. Wiley, Hoboken, NY. 612 pp.

Vernazza, M., Tharp, R., and Jaekel, S. 1984. *More than you ever wanted to know about nutmeg*. Surrey Top Inc., U.S. Virgin Islands. 60 pp.

69 Oca

Family: Oxalidaceae (wood sorrel family)

NAMES

Scientific Name: *Oxalis tuberosa* Molina

- The word oca (sometimes spelled occa, and pronounced OH-kah) is Spanish, probably based on the Quechua *oqa*. (The Quechua language was used by the Incas and is widespread among Indian peoples of Peru, Bolivia, Ecuador, Chile, and Argentina.)
- Oca is rarely spelled oka, but this may result in confusion with oka (pronounced o-KAH), a Canadian cow's milk cheese made by Trappist monks at Oka, Quebec.
- Oca is called "yam" in New Zealand but is unrelated to the true yams (*Dioscorea* species). The name "New Zealand yam" is also used.
- Occasionally, the name "oca potato" will be encountered.
- The genus name *Oxalis* is based on the Greek name for sorrel (*Rumex* species), *oxys*, meaning sour, because the oxalic acid content imparts an acid taste to the leaves. Oca has been called "oxalis" in England, although the genus *Oxalis* contains many other species.
- *Tuberosa* in the scientific name *O. tuberosa* is Latin for tuberous, referring to the tubers.

PLANT PORTRAIT

Oca is a perennial, herbaceous plant with underground tubers, which are also called oca. The stems are succulent, growing to a height of 25 cm (10 in.), and the leaves are clover like, with three leaflets. The underground stem (rhizome) is considerably branched, and the tips are swollen into fleshy, firm-textured, floury tubers the size of small potatoes, generally 5 to 7.5 cm (2–3 in.) long and 2.5 to 3.8 cm (1–1½ in.) wide. Some forms of the plant have lost their ability to produce flowers and seeds. In the Andean region, oca is the most well-known root vegetable after the potato. The tuber is unusual in appearance: often brightly colored, waxy, and crinkled. Tuber colors include white, black, yellow, pink, and red. Oca was grown throughout the Incan and Mayan civilizations. It has been consumed in Peru for more than 4000 years and has long been a staple food item of the Andean Indians. The crop is now cultivated from Venezuela to Argentina, typically at altitudes of 3000 to 4000 m (10,000–13,000 ft.). Oca is also grown commercially in Mexico and New Zealand. Nearly all production is for fresh market consumption. Compared with potatoes, which are now cultivated in 130 countries around the world, oca is scarcely known outside of the Andean region. Unsuccessful attempts were made to introduce oca as a crop in Europe during the early part of the nineteenth century, although it was once grown in southern France for pig feed. However, oca has in fact been grown as a garden ornamental in Europe for more than a century, with little realization that the tubers are edible. The cultivation of oca appears to be declining in some areas of the Andes and some traditional methods of preparation may be disappearing, all this reflecting the unfortunate viewpoint that oca is a poor people's food.

CULINARY PORTRAIT

Oca tubers have a tangy, acid, sweet, nutty, mild flavor. The acidic character is due to oxalic acid, which is present in other food plants such as rhubarb and spinach, and sometimes causes health problems by locking up certain minerals needed in the diet. The amounts of oxalic acid in sweet

FIGURE 69.1 Oca (*Oxalis tuberosa*), by B. Flahey.

FIGURE 69.2 Oca tubers. (From Vilmorin-Andrieux, 1885.)

types of oca are not considered sufficient to pose a threat to human health. Like the potato and other Andean root vegetables, oca is eaten both fresh and in a dehydrated state. A dehydrated, freeze-dried form of root vegetables known as "chuño" is widely prepared in high, cold mountains in South America, and this preparation can last several years and serve as a secure food source when necessary. In the Andes, oca is often sun dried for several days to make it sweeter and then parboiled, roasted, or prepared as *pachamanca* (meat roasted in a hole in the ground). New Zealanders often serve oca with their national dish, roast lamb. Oca is also traditionally served in soups and stews and can be boiled and mashed, steamed, baked, fried, or roasted. It has sometimes been pickled in vinegar. Oca is a very versatile vegetable. Peeling is not necessary so long as the roots are trimmed off and the tubers washed to get rid of adhering soil (easily accomplished as the skin is somewhat waxy). In Mexico, oca is commonly sprinkled with salt, hot pepper, and lemon juice and eaten raw. The raw vegetable has a pleasing, crunchy texture and often a lemony taste. The leaves and young shoots are occasionally consumed as a green vegetable (consumption should be limited in view of the oxalic acid content). In North America and Europe, fresh oca is often available as an import from New Zealand, and frozen tubers are occasionally found in specialty or Hispanic stores. Tubers can be purchased from some garden plant sources to grow in home vegetable gardens, but it should be kept in mind that production of tubers tends to start in late summer in north temperate countries, so that harvest needs to be later than for potatoes.

Culinary Vocabulary

- The French expression *à la péruvienne* (pronounced ah lah pay-roo-vee-ahn) refers to a garnish for meat made with oca and sauce.

CURIOSITIES OF SCIENCE AND TECHNOLOGY

- Like many root crops of the Andes, including the potato, a major problem with growing the plants outside of their native range is their natural adaptation to the local Andean climate. In particular, Andean plants tend to form tubers only when the day length is relatively short, and during longer days (as are typical in the temperate regions of the world), the plants fail to produce edible tubers. This was a major problem facing the potato when it was first brought to Europe but in time forms were selected that would produce tubers in northern latitudes. The problem has not yet been completely solved for oca, but likely forms that are currently grown in New Zealand, Mexico, and Chile are suitable for adapting to temperate climates in Europe and North America.
- Species of *Oxalis*, including oca, often have an explosive mechanism for distributing their seeds. When mature, the capsules containing the seeds literally explode, scattering the seeds.
- The leaves of many species of *Oxalis* fold downward, like a regular umbrella, at night or when the light becomes weak. This tends to protect the foliage against rain and insects.
- OCA is an acronym for Orthodontic Centers of America, Orthodox Church in America, Orangedale Community Association, Ohio Counselling Association, Organization of Chinese Americans, Oldsmobile Club of America, Olympic Council of Asia, Ohio Climbers Association, Ontario Cattlemens Association, Ontario Camping Association, Oklahoma Correctional Association, Oklahoma Chess Association, and Oregon Citizens Alliance.
- Among the place names listed for Fresno County, California, are Oxalis Station (named for one of the common *Oxalis* species, wood sorrel) and Silaxo Station (which spells *Oxalis* backward).

KEY INFORMATION SOURCES

Aarbizu, C., and Tapia, M. 1994. Andean tubers. In *Neglected crops. 1492 from a different perspective*. Edited by J.E. Hernández-Bermejo and J. Léon. Food and Agriculture Organization of the United Nations, Rome, Italy. pp. 149–163.

Albihn, P.B.E., and Savage, G.P. 2001. The effect of cooking on the location and concentration of oxalate in three cultivars of New Zealand-grown oca (*Oxalis tuberosa* Mol.). *J. Sci. Food Agric.* 81:1027–1033.

Albihn, P.B.E., and Savage, G.P. 2001. The bioavailability of oxalate from oca (*Oxalis tuberosa*). *J. Urol.* 166:420–422.

Bianco, M., and Sachs, C. 1998. Growing oca, ulluco, and mashua in the Andes: socioeconomic differences in cropping practices. *Agric. Human Values*, 15:267–280.

De Azkue, D., and Martinez, A. 1990. Chromosome number of the *Oxalis tuberosa* alliance (Oxalidaceae). *Plant Syst. Evol.* 169:25–29.

Emshwiller, E. 2002. Ploidy levels among species in the '*Oxalis tuberosa* alliance' as inferred by flow cytometry. *Ann. Bot.* 89:741–753.

Emshwiller, E. 2002. Biogeography of the *Oxalis tuberosa* alliance. *Bot. Rev.* 68:128–152.

Emshwiller, E., and Doyle, J.J. 1998. Origins of domestication and polyploidy in oca (*Oxalis tuberosa*: Oxalidaceae): nrDNA ITS data. *Am. J. Bot.* 85:975–985.

Emshwiller, E., and Doyle, J.J. 2002. Origins of domestication and polyploidy in oca (*Oxalis tuberosa*: Oxalidaceae). 2. Chloroplast-expressed glutamine synthetase data. *Am. J. Bot.* 89:1042–1056.

Emshwiller, E., Theim, T., Grau, A., Nina, V., and Terrazas, F. 2009. Origins of domestication and polyploidy in oca (*Oxalis tuberosa*: Oxalidaceae):3. AFLP data of oca and four wild, tuber-bearing taxa. *Am. J. Bot.* 96:1839–1848.

Flores, T., Alape-Giron, A., Flores-Diaz, M., and Flores, H.E. 2002. Ocatin. A novel tuber storage protein from the Andean tuber crop oca with antibacterial and antifungal activities. *Plant Physiol.* 128:1291–1302.

Hernandez-Lauzardo, A.N., Mendez-Montealvo, G., Velazquez-del-Valle, M.G., Solorza-Feria, J., and Bello-Perez, L.A. 2004. Isolation and partial characterization of Mexican *Oxalis tuberosa* starch. *Starch/Staerke*, 56:357–363.

Kays, S.J., Gaines, T.P., and Kays, W.R. 1979. Changes in the composition of the tuber crop *Oxalis tuberosa* Molina during storage. *Sci. Hortic.* 11:45–50.

King, S.R., and Gershoff, S.N. 1987. Nutritional evaluation of three underexploited Andean tubers: *Oxalis tuberosa* (Oxalidaceae), *Ullucus tuberosus* (Basellaceae), and *Tropaeolum tuberosum* (Tropaeolaceae). *Econ. Bot.* 41:503–511.

Ross, A.B., Savage, G.P., Martin, R.J., and Vanhanen, L. 1999. Oxalates in oca (New Zealand yam) (*Oxalis tuberosa* Mol.). *J. Agric. Food Chem.* 47:5019–5022.

Sangketkit, C., Savage, G.P., Martin, R.J., and Mason, S.L. 2001. Oxalate content of raw and cooked oca (*Oxalis tuberosa*). *J. Food Compost. Anal.* 14:389–397.

Sangketkit, C., Savage, G.P., Martin, R.J., Mason, S.L., and Vanhanen, L. 1999. Oxalates in oca: a negative feature? *Proc. Nutr. Soc. N. Z.* 24:44–50.

Sangketkit, C., Savage, G.P., Martin, R.J., Searle, B.P., and Mason, S.L. 2000. Sensory evaluation of new lines of oca (*Oxalis tuberosa*) grown in New Zealand. *Food Qual. Prefer.* 11:189–199,

Shah, A.A., Stegemann, H., and Galvez, M. 1993. The Andean tuber crops mashua, oca and ulluco: optimizing the discrimination between varieties by electrophoresis and some characters of the tuber proteins. *Plant Var. Seeds*, 6:97–108.

Sperling, C.R., and King, S.R. 1990. Andean tuber crops: worldwide potential. In *Advances in new crops.* Edited by J. Janick and J.E. Simon. Timber Press, Portland, OR. pp. 428–435.

Stegemann, S., Majino S, and Schmiediche, P. 1988. Biochemical differentiation of clones of *Oxalis tuberosa* by their tuber proteins and the properties of these proteins. *Econ. Bot.* 42:37–44.

Trognitz, B.R., and Hermann, M. 2001. Inheritance of tristyly in *Oxalis tuberosa* (Oxalidaceae). *Heredity*, 86:564–573.

Trognitz, B.R., Carrion, S., and Hermann, M. 2000. Expression of stylar incompatibility in the Andean clonal tuber crop oca (*Oxalis tuberosa* Mol., Oxalidaceae). *Sex Plant Reprod.* 13:105–111.

Trognitz, B.R., Hermann, M., and Carrion, S. 1998. Germplasm conservation of oca (*Oxalis tuberosa* Mol.) through botanical seed. Seed formation under a system of polymorphic incompatibility. *Euphytica*, 101:133–141.

Specialty Cookbooks

Note: Oca can generally be cooked like new potatoes, and can be substituted in most potato recipes. Oca cooks more quickly than potatoes and sweet potatoes and should be checked for doneness.

Cotter, D. 2007. *Wild garlic, gooseberries, and me: a chef's stories and recipes from the land.* Collins, London. 318 pp.

Kijac, M.B. 2003. *The South American table: the flavor and soul of authentic home cooking from Patagonia to Rio de Janeiro, with 450 recipes.* Harvard Common Press, Boston, MA. 480 pp.

70 Okra

Family: Malvaceae (mallow family)

NAMES

Scientific Name: *Abelmoschus esculentus* (L.) Moench (*Hibiscus esculentus* L.)

- The name "okra" is of West African origin, similar to the Akan (Twi) word *nkruma*. In the southern United States, slaves from the Gold Coast called okra *nkrumun*.
- Okra is also known as gumbo, gobo, gombo, and lady's finger(s). In the Caribbean, the spelling "ochro" is also used.
- Okra brought to the United States by slaves from Angola was called *ochinggombo* (a word of Bantu origin, the term for okra in the Umbundu language of Angola), later shortened to *ngombo*. It has been suggested that *ngombo* became "gombo" and finally evolved into "gumbo." The word gumbo came to refer not only to okra but also to the stew made with it. Alternatively, it has been suggested that the name "gumbo" is derived from the Portuguese version of the plant's African name. There is still another explanation. The word gumbo has also been used in a mix of French and various African languages spoken by black people in some parts of Louisiana and the French West Indies. It has been suggested that the word may have been derived from *nkombo*, meaning "runaway slave" in the Bantu language of the Kongo people of Zaire.
- "Wild okra" is a name applied to a violet, *Viola palmata* L. of eastern North America, which like true okra is very mucilaginous and is used for thickening soups.
- The genus name *Abelmoschus* is apparently from the Arabic *abulmosk*, father of musk, an allusion to the smell of the seeds.
- *Esculentus* in the scientific name *A. esculentus* means edible, good to eat.

PLANT PORTRAIT

Okra is an annual herbaceous plant, apparently only known in cultivation. The most closely related wild relative is *A. tuberculatus* Pal & Singh of northern India. Other wild relatives occur in Ethiopia, the upper Nile in Sudan, and West Africa. A type of okra called "Guinean okra" has been described from parts of Africa. Okra was used in Egypt from at least AD 1200, and its use spread into the eastern Mediterranean and subsequently to India. Okra has never been popular in Europe, except in Spain, Greece, and the Balkans. African slaves introduced it to the Americas, and the French brought it to the Mississippi delta in the seventeenth century. It is still an important part of Creole cooking of Louisiana. This vegetable was predominantly consumed by the poor for centuries. It has still not escaped from its past reputation as a food for the poor and would otherwise be more popular. Okra is a tropical or subtropical plant requiring warm growing conditions. It is cultivated in the warmer parts of Africa, the Mediterranean, and the Americas. The major production areas in North America are the southeastern United States, Texas, and California. Seed production is concentrated in southwestern Arizona.

Depending on variety, okra grows 1 to 2.4 m (3–8 ft.) tall. The fruit is a long, generally ribbed pod, harvested while still tender and immature. The pods attain lengths of 15 to 20 cm (6–8 in.) and up to 2.5 cm (1 in.) or more in diameter. Smaller pods are more tender and flavorful, the larger pods tending to become woody and tough. The finger-like, slightly curved pods are greenish, occasionally

FIGURE 70.1 Okra (*Abelmoschus esculentus*) in flower. (From Gartenflora, 1894, vol. 43.)

FIGURE 70.2 Okra (*Abelmoschus esculentus*) fruits. (From Vilmorin-Andrieux, 1885.)

reddish, and are filled with seeds that range in color from beige to pinkish beige. The pods are harvested before they are completely ripe as they become tough and fibrous with maturity. In the young pods, the small seeds are still immature and are soft and tender. Newer varieties lack spines on the pods.

CULINARY PORTRAIT

The flavor of okra has been described as between eggplant and asparagus but is rather unique. Okra pods are used fresh as a cooked vegetable and are added to soups, stews, and casseroles. The pods are also canned, frozen, or dehydrated for later use and can be pickled (a common use in the Middle East). Young okra pods can be batter- or stir-fried and are important in Creole cooking for "gumbo" soups. Okra can also be sautéed, braised, boiled, and steamed but does not purée well. In the United States, okra frequently accompanies chicken. In parts of the southern United States, the pods are dried and consumed in the winter. In India, the pods are used in curries and are eaten fresh.

Those unfamiliar with okra may need to get used to the swelling, mucilaginous material. A particularly sticky texture and pasty flavor are sometimes associated with cooked okra, which results from pods being broken during cooking or overcooking. Shunned by many for its gummy, slimy qualities, okra's gumminess is used to advantage to thicken stews and sauces. Okra is often used in catsup, its mucilage serving to make it thick and hard to get out of the bottle.

The sliminess of okra is the reason why many detest this vegetable. A 1974 survey of vegetable preferences by the U.S. Department of Agriculture revealed that okra was one of the three most disliked vegetables. In the Middle East, okra is sometimes soaked in lemon juice or sprinkled with salt for an hour before cooking and then fried whole to avoid the slimy texture. Dried okra lacks the mucilaginous quality of fresh okra. However, rather than trying to change the essential character of okra, for those who just do not like it, it would seem preferable to simply avoid it. Green beans can be substituted in recipes, although the consistency will differ considerably.

Traditionally, gumbo soups were made from filé powder (from sassafras, *Sassafras albidum* (Nutt.) Nees), but okra is now generally preferred as the thickener. When sassafras is used, it means that gumbo (the stew) is made without gumbo (i.e., okra). The advice has been given not to use both okra and filé powder together in gumbo stews, or the stew may have the consistency of glue.

The seeds can be used similarly to dried beans and have been used as a substitute for pearled barley. The seeds were once used as a substitute for coffee. Occasionally, an edible oil is extracted from the seeds, or the seeds are dried and ground into flour. Young okra leaves can be used as cooked greens (the leaves and young shoots are widely consumed throughout tropical Africa). In addition to its use as a vegetable, dehydrated okra is used as a flavoring and emulsifier in food processing, commonly in salad dressing.

When purchasing okra, it is advisable to select bright green, plump, unblemished pods that are 5 to 7.5 cm (2–3 in.) long. Stiff pods should be avoided (overripe pods are not only tougher but will have a very sticky texture). Fresh okra can be refrigerated in a paper bag or wrapped in paper towels inside a perforated plastic bag but should be used within a day or two (although it will keep longer, it loses flavor). In preparation, okra should first be washed under cold running water. Some older varieties have downy pods, and some cooks gently scrub away their surface fuzz with a vegetable brush or paper towel. Frozen okra is generally available and is an excellent substitute for fresh okra. Canned okra is a less desirable choice. Iron or copper cookware will discolor okra, although the taste will not be affected. It may be noted that the water in which okra is cooking has a tendency to boil over very easily. Okra can be cooked uncut, this preventing the mucilage from oozing out. Alternatively, the stem end can be sliced off or the conical portion can be peeled without breaking the surface of the okra. After trimming away the stem end, okra can be sliced crosswise into desired lengths. The sliced okra can be added raw to a bowl of salad greens along with dressing. When used to thicken soups and stews, okra should be added approximately 10 minutes before the end of cooking.

The majority of people are sensitive to okra's small spines, and contact frequently results in a rash or itch which may persist for weeks. Harvesters should wear gloves, long sleeves, and long pants while picking.

Culinary Vocabulary

- *Bhindi* is a Hindi term for okra. In the context of Indian cooking, okra is often called bhindi in English.
- "Lady's fingers" (or ladyfingers) is a term for an okra dish most often encountered in England and in the southern United States. Ladyfingers also refers to a delicate sponge cake served with a topping or ice cream or used as the starting point to prepare more elaborate desserts such as baked Alaska or charlotte. In the past, a number of other species have been called lady's fingers, including types of potato and apple. In Australia, ladyfingers means bananas.
- "Foo-foo" (foofoo, foufou, foutou) is an extremely thick porridge made from ground cassava, corn, rice, and/or yams. It is a staple of Central and West African cuisines. The Caribbean counterpart of foo-foo is coo-coo (coo coo, coocoo; called "funchi" [pronounced "foon-jee"] and "fungi" in the Netherlands Antilles and the Virgin islands), a pasty dish of steamed cornmeal and okra garnished with vegetables and herbs. Different versions of coo coo use as a major constituent any of okra, breadfruit, cassava, coconut, or plantain. The North African counterpart is couscous, a meat or vegetable stew mixed with bits of semolina flour.

CURIOSITIES OF SCIENCE AND TECHNOLOGY

- In the 1800s, slaves from Africa used okra as a part of their diet, and this included use of ground okra seeds as a coffee substitute. During the American Civil War blockades of the 1860s when coffee was scarce, southerners also resorted to making coffee with okra. Okra seeds are still sometimes used as a coffee substitute.
- Okra stems have been used to make paper and cordage.

KEY INFORMATION SOURCES

Ariyo, O.J. 190. Measurement and classification of genetic diversity in okra (*Abelmoschus esculentus*). *Ann. Appl. Biol.* 116:335–341.

Bates, D.M. 1968. Notes on the cultivated Malvaceae. 2. *Abelmoschus*. *Baileya*, 16:99–112.

Charrier, A. 1984. *Genetic resources of the genus Abelmoschus Med. (okra)*. International Board for Plant Genetic Resources Secretariat, Rome, Italy. 61 pp.

Chheda, H.R., and Fatokun, C.A. 1982. Numerical analysis of variation patterns in okra (*Abelmoschus esculentus* (L.) Moench). *Bot. Gaz.* 143:253–261.

Egley, G.H., and Elmore, C.D. 1987. Germination and the potential persistence of weedy and domestic okra (*Abelmoschus esculentus*) seeds. *Weed Sci.* 35:45–51.

Hamon, S., and Sloten, D.H. van. 1995. Okra. In *Evolution of crop plants*. 2nd ed. Edited by J. Smartt and N.W. Simmonds. Longman Scientific & Technical, Burnt Mill, Harlow, Essex, U.K. pp. 350–357.

Hamon, S., Charrier, A., Koechlin, J., and Van Sloten, D.H. 1991. Potential improvement of okra (*Abelmoschus* spp.) through the study of its genetic resources. *Plant Genet. Resour. Newsl.* 86:9–15.

Holbrook, C.R., and Wamble, A.C. 1951. *Processing okra seed for oil recovery*. Texas Engineering Experiment Station, College Station, TX. 11 pp.

International Board for Plant Genetic Resources. 1991. *Report of an international workshop on okra genetic resources, held at the National bureau for Plant Genetic Resources, New Delhi, India, 8–12 Oct. 1990.* International Crop Network Series 5. International Board for Plant Genetic Resources, Rome, Italy. 133 pp.

Joshi, A.B., Gadwal, V.R., and Hardas, M.W. 1974. Okra. In *Evolutionary studies in world crops; diversity and change in the Indian subcontinent*. Edited by J.B. Hutchinson. University Press, Cambridge, U.K. pp. 99–105.

Malik, C.P., Bhalla, P.L., Singh, M.B., and Singh, H. 1985. The biology of okra—a review. In *Widening horizons of plant sciences*. Edited by C.P. Malik. Cosmo Publications, New Delhi, India. pp. 355–390.

Martin, F.W. 1982. Okra, potential multiple-purpose crop for the temperate zones and tropics. *Econ. Bot.* 36:340–345.

Martin, F.W. 1982. A second edible okra species, and its hybrids with common okra. *Ann. Bot.* 50:277–283.

Martin, F.W., and Ruberté, R. 1978. *Vegetables for the hot humid tropics. Part 2. Okra,* Abelmoschus esculentus. Mayaguez Institute of tropical Agriculture, Mayaguez, Puerto Rico. 22 pp.

Martin, F.W., Rhodes, A.M., Ortiz, M., and Diaz, F. 1981. Variation in okra. *Euphytica*, 30:697–705.

Myers, C. 1991. Okra, gumbo. Crop Sheet SMC-026. In *Specialty and Minor Crops Handbook.* Edited by C. Myers. The Small Farm Center, Division of Agriculture and Natural Resources, University of California, Oakland, CA. 4 pp.

Schweers, V.H., and Sims, W. 1976. *Okra production.* Leaflet 2679. Division of Agricultural Sciences, University of California, Berkeley, CA. 5 pp.

Sharma, G. 1993. Okra, *Abemoschus* spp. In *Genetic Improvement of Vegetable Crops.* Edited by G. Kalloo and B.O. Bergh. Pergamon Press, Oxford, U.K. pp. 751–769.

Siemonsma, J.S. 1982. West African okra—morphological and cytogenetical indications of the existence of a natural amphidiploid of *Abelmoschus esculentus* (L.) Moench and *A. manihot* (L.) Medikus. *Euphytica*, 31:241–252.

Siemonsma, J.S. 1993. *Abelmoschus manihot* (L.) Medikus. In *Plant Resources of South-East Asia. Vol. 8. Vegetables.* Edited by J.E. Siemonsma and K. Piluek. Pudoc Scientific Publishers, Wageningen, the Netherlands. pp. 57–60.

Sikes, V.R. 1980. *Keys to profitable okra production for food and seed.* Texas Agricultural Extension Service, College Station, TX. 4 pp.

Specialty Cookbooks

B.F. Trappey's Sons. 1900–1992. *Yam and okra recipes; the key of Creole cooking.* B.F. Trappey's Sons, New Iberia, LA. 16 pp.

Oklahoma State Dept. of Agriculture. 1983. *Oklahoma okra: recipes.* State Department of Agriculture, Marketing Industry Division, Oklahoma City, OK. 6 pp.

Raymond, D. 1987. *Book of eggplant, okra & peppers.* Revised edition. Villard Books, New York. 87 pp.

Raymond, D., and Raymond, J. 1984. *The gardens for all book of eggplant, okra & peppers.* Gardens for All, Burlington, VT. 36 pp.

Steffens, G. McM. 1985. *The National Gardening book of eggplant, okra & peppers.* Revised edition. National Gardening Association, Burlington, VT. 31 pp.

71 Opium Poppy

Family: Papaveraceae (poppy family)

NAMES

Scientific Name: *Papaver somniferum* L.

- The English word "poppy" is derived from the classical Latin name for the plant, *papaver* (see information below about the genus name).
- "Opium" comes from *opion*, a diminutive of *opos*, the Greek word for juice.
- The word "morphine," a constituent of opium, is derived from the Greek god of dreams, Morpheus.
- The opium poppy is also known as common poppy, garden poppy, mawseed, oil poppy, poppy, and white poppy.
- The ripe seeds of the opium poppy are sometimes called maw seed and, as discussed in the next section, are pressed for poppy oil, hence the name oil poppy.
- Poppies often grow as weeds among cultivated crops, and in parts of England, poppy species are sometimes known as "corn rose" (note that "corn" in England generally means the principal cereal of a region, whereas "maize" designates the "corn" of North America).
- The "California Poppy," the state flower of California, is *Eschscholtzia californica* Cham., which is not related to *Papaver*, although it is in the poppy family and is reputed to be mildly narcotic.
- The genus name *Papaver* is based on the classical Latin name for the plant. The word is related to the Latin *buro*, "to burn," and the Hebrew *bera*, "fire" or "burning," presumably because of the blazing red flowers of many of the species.
- *Somniferum* in the scientific name *P. somniferum* is from the Latin *somnus*, sleep + *ferre*, bring or bear, literally "sleep-bearing," and refers to the plant's narcotic qualities.

PLANT PORTRAIT

Most people are unaware that the source of poppy seeds commonly used on bakery products is the opium poppy, exactly the same species that produces opiate drugs such as heroin. Indeed, the very same plants are often the source of both products. Almost all varieties of opium poppy have some opiate drugs, so the cultivation of the plants, either for food or for legitimate drugs, is strictly controlled. The culture of the opium poppy is illegal in many countries, although often legislation is interpreted as permitting the ornamental use in such countries. Indeed, opium poppy cultivars are often marketed and grown as ornamentals in numerous countries. Unlike such narcotic plants as marijuana, cultivation of opium poppy for production of illicit drugs is mostly absent in most countries, and so there has been no significant concern about the cultivation of opium poppy as an ornamental. However, very recently in North America, opium poppies have been illicitly cultivated for production of "doda," a tea made with poppy capsules, and used by citizens of Asian heritage. Although notably lower in the opiate chemicals that are found in opium, doda is now considered to be a significant health and law enforcement issue.

The Oriental poppy (*Papaver orientale* L.) is widely and legally cultivated as an ornamental and has seeds that some think are as tasty as those of *P. somniferum*. However, the opium poppy remains the exclusive commercial source of poppy seed products.

FIGURE 71.1 Opium poppy (*Papaver somniferum*). (From Köhler, 1883–1914.) A seed is at lower right. A cross section of a capsule is at lower left and shows fruit sectors filled with tiny seeds.

The opium poppy is an annual, growing as high as 2 m (6½ ft.), with attractive white, pink, red, bluish, or purple flowers (doubled in some garden forms, i.e., with extra petals) to 10 cm (4 in.) across. The species is believed to grow wild in the Mediterranean region, from the Canary Isles eastwards. It is found as an escape from cultivation in fields, roadsides, and waste places in scattered localities around the world, including North America. The economic part of the plant for culinary purposes is the seed. The kidney-shaped seeds range through white and yellow to slate-blue or black and are tiny. They and the oil expressed from them have virtually no narcotic properties. In mature wild plants, the wind shakes the seeds out of the fruit through pores at the apex. The opium poppy has been selected to produce three kinds of plants: narcotic cultivars, oilseed/condiment cultivars, and ornamentals.

The narcotic effects may have been known to the ancient Sumerians, about 4000 BC, as they had a symbol for it that has been translated as "joy plant." The opium poppy was well established in classical times. It was familiar to ancient Egyptian, Greek, and Roman civilizations and seems to have been used effectively as a therapeutic drug. However, by the time of Mohammed (570–632), the narcotic qualities were also appreciated in Arabia. Islamic traders and missionaries spread the cultivation of the opium poppy to Persia, India, China, and Southeast Asia, where it was used to relieve pain but also began to be used excessively as a habit-forming, destructive narcotic, initially in India and then in China. The drug was first eaten in the Orient but began to be smoked in the seventeenth century in China, which resulted in a tremendous upsurge of use and production. Crude opium has a disagreeable smell and a hot biting taste. The Swiss alchemist Paracelsus (1493–1541), overcame these objectionable features by making up a solution of opium in alcohol. This deadly mixture, known as tincture of opium or laudanum, enslaved numerous people during the following centuries. Portuguese, Dutch, and British merchants trafficked extensively in opium. In the nineteenth century, the British brought about the Opium Wars of 1840 and 1855 to prevent the Chinese from outlawing the opium trade. It has been estimated that in 1886 about a fourth of the Chinese (about 15 million people) were addicted to opium. During Victorian times in England, tincture of opium was readily available and was often administered

FIGURE 71.2 Ornamental garden form of opium poppy (*Papaver somniferum*) with doubled flowers. (From Vilmorin-Andrieux, et al. 1884.)

to teething or upset babies to make them sleep. In the United States, opium preparations became widely available in the nineteenth century, and morphine was used so widely as a painkiller for wounded soldiers during the American Civil War that opium addiction became known as "the army disease" and "soldier's disease."

Crude opium is the hardened milky sap of the unripe fruit. Dried latex is obtained from unripe capsules and is used medicinally as well as for illicit narcotics. The drug opium is a mixture of many constituents, including the alkaloids morphine and codeine. It has traditionally been obtained by making incisions into the nearly ripe poppy capsules 10 to 20 days after flowering. In cooler climates, incisions do not seem to result in good exudation of latex, and mature capsules are simply collected for chemical extraction. Morphine is often extracted from the capsules of oilseed cultivars after the seeds have been harvested, although narcotic cultivars are more productive. Heroin is manufactured from morphine. The major illegal growing areas are in the highlands of mainland Southeast Asia (especially Burma, Laos, Thailand, and adjacent southern China and northwestern Vietnam), in Southwest Asia (notably Pakistan, Iran, and Afghanistan), in Mexico, and to a lesser extent in Lebanon, Guatemala, and Columbia. Limited authorized pharmaceutical production occurs, largely in Europe. Much of the poppy grown for the extraction of opium for medicinal purposes is cultivated in government-regulated farms in India, Turkey, and Tasmania (Australia). The importance of opium poppy derivatives is indicated by the fact that codeine is a constituent of more than 22% of analgesic preparations in the United States. Opium poppy has been used as a tranquilizer and analgesic for millennia. It has also been used as an aphrodisiac and to treat numerous medical conditions and continues to be used in China for a variety of medicinal purposes. Centuries of abuse of the drug as a recreational inebriant and the well-known misery that has resulted suggest that the opium poppy has been more of a curse on humanity than a blessing. However, opium poppy is one of the most important medicinal plants, and scarcely anyone has not experienced pain relief by taking medically prescribed morphine or codeine. Growing poppies for pharmaceuticals in most of the world, including North America, is possible only under license.

The culinary use of the opium poppy is also ancient, like the narcotic use. Many seeds have been found at the sites of ancient domiciles, including Swiss Lake Dwellings of Neolithic Age (at least 4000 years ago). About 2000 BC, the Egyptians cultivated poppy to obtain the edible oil in the seeds. The Classical Greeks also used poppy seeds as a food. Both the ancient Greeks and Romans added the seeds to cakes and bread as a flavoring. By the Middle Ages, the use of poppy seeds as a condiment on bread was well established. Today, poppy seed is legally produced in the Netherlands, Poland, Romania, the Czech Republic, the former Yugoslavia, Russia, India, Iran, Turkey, Argentina, and many Asian and Central and South American countries. Most importation to North America is from Australia and the Netherlands.

Of the weight of opium poppy seeds, 40% to 60% is oil, and the plant is commonly grown as a source of edible and industrial oil. In southern Europe, poppy seed oil is not competitive with olive oil and so is not produced. Artists use poppy seed oil as a drying oil, useful in paints and varnishes. The oil is also a component of some soaps.

CULINARY PORTRAIT

Poppy seeds have a mildly spicy, oily, agreeably hazelnut-like flavor and are used as a condiment on baked goods, including pastry. The pleasantly nutty flavor and crunchy texture are enhanced by toasting or baking. Poppy seeds are used as a topping on rolls, breads, cookies, cakes, and other baked goods or as a garnish. The seeds can also be used in a variety of cooked vegetable dishes and stews and go well with pasta, potato salads, cheese, and marinades. Crushed seeds act as a thickener and are used as a filling for cakes and pastries, cream cheese, and sour cream dips. Poppy seeds are also used to flavor butter and various food preparations.

Poppy oil is a pale yellow, fixed, tasteless, oil, useful as a salad oil as it is less liable to become rancid than olive oil. The oil is also used to make margarine and salad dressing, and is used as a cooking oil.

Culinary Vocabulary

- "Hamantashen" are small, triangular (three-cornered), Jewish pastries with a sweet filling of honey and poppy seeds, puréed prunes, raisins, apricots, or nuts. This confection is generally served during the holiday of Purim, a celebration of an ancient Jewish victory over Haman, minister of king Ahaseurus of Persia who was attempting to exterminate the Jews, described in the Old Testament book of Esther. "Hamantashen" literally means "Haman's pockets." The pastries are said to represent Haman's tricornered hat as well as his bribe-stuffed pockets.
- The "kolache" is a traditional Czech sweet bun made with an egg and yeast dough and filled with preserves, poppy seeds, or cottage cheese.
- "Seven spice powder" is a Japanese spice blend usually made with poppy seed, anise pepper seed, sesame seed, flax seed, rape seed, nori, and dried tangerine or orange peel.

CURIOSITIES OF SCIENCE AND TECHNOLOGY

- An Egyptian text dated at approximately 1300 BC described the medical practice of giving poppy preparations to children to stop them from crying.
- Avicenna, a celebrated Arab philosopher and physician, died of opium intoxication in 1037.
- In his 1822 book *Confessions of an English Opium Eater*, author Thomas DeQuincy (1785–1859), invented the word "tranquilizer" to describe the effect of opium.

- In 1865, opium was grown in the state of Virginia and was used to produce a product with 4% morphine. In 1867, it was grown in Tennessee and in 1873 in Kentucky. During the period, opium (as well as marijuana and cocaine) could be purchased legally over the counter from druggists.
- Floral names were very popular during the late Nineteenth Century in America and England. Favorite names were Rose, Blossom, Daisy, Iris, Pansy, Fern, Viola, Violet, Zinnia, and Poppy.
- The "Victorian language of flowers" was a secret coded language in Victorian times in England, with flowers and plants symbolic of certain messages, so when the flower or plant was mentioned in a letter, those who knew the code could understand the hidden information. "Red poppy" meant "consolation," "scarlet poppy" stood for "fantastic extravagance," and "white poppy" translated as "sleep," "my bane," or "my antidote."
- In the July 1906 issue of Good Housekeeping, a Shaker community member recounted the growing of flowers for opium: "We always had extensive poppy beds and early in the morning, before the sun had risen, the white-capped sisters could be seen stooping among the scarlet blossoms to slit those pods from which the petals had just fallen. Again after sundown they came out with little knives to scrape off the dried juice. This crude opium was sold at a large price and its production was one of the most lucrative as well as the most picturesque of our industries." (The Shakers were a religious sect that came to America from England, practiced celibacy, and since there were no children, went out of existence.)
- "Lettuce opium" or lactucarium is a drug from the dried milky juice of wild and cultivated lettuce species and is known to have been used by the ancient Egyptians. It enjoyed popularity in North America as a sedative and painkiller in the nineteenth century and again as a legal, allegedly mind-altering drug in the mid-1970s. Although lettuce opium has been found to have some sedative components, it is not thought to have any efficacy as a drug.
- Three categories are often used to describe the relative safety of herbs: GRAF—generally recognized as food, GRAS—generally recognized as safe, and GRAP—generally recognized as poisonous. Opium poppy falls into all three categories: the widely eaten poppy seeds are GRAF and usually also considered GRAS, and poppy alkaloids are GRAP.
- Etymologists have established that "to have a yen for" is an expression that arose in the United States and means to have an intense desire or longing for. "Yen" in this expression is based on the Chinese *yen*, opium smoke.
- Alkaloids are plant-produced compounds that contain nitrogen, are generally bitter, and often very toxic. There are thousands of alkaloids, some of which are very important in foods, for example, piperine is responsible for the burning taste of black pepper, and caffeine is a stimulant in coffee and tea. The first alkaloid ever isolated chemically was morphine, the basic alkaloid from opium. This was carried out in 1803 by F.W.A. Sertürner (1793–1841), a pharmacist's assistant in the German town of Paderborn.
- Minute quantities of morphine have apparently been found in lettuce, hay, cow's milk, and human milk. The finding in lettuce is interesting in view of the uses of lettuce opium, but the concentration is too low to result in any effect. The same is true for milk and hay.
- Workers in opium poppy fields assigned the task of incising the capsules with a knife to induce flow of latex sometimes complain of "narcosis" from the fumes. Sometimes masks are worn to prevent excessive inhalation if there is insufficient wind to carry the fumes away.
- Poppy seeds are exceptionally small, approximately 20 million in 1 kg (9 million in 1 lb.).
- Poppy seeds that have been contaminated with traces of the latex may produce a false-positive result on a urine test for drugs, suggesting that the person has recently used morphine or heroin.

- In 2004, the Royal Canadian Mint produced the world's first colored circulation coin. The silver-colored 25-cent piece featuring the red image of a poppy—Canada's flower of remembrance—inlaid in the center over a maple leaf. Thirty million copies of the two-color quarter were distributed to commemorate Canada's 117,000 war dead. In 2007, sensational accounts in the media reported that the unusual quarter led to an American Defense Department false espionage warning that the coin contained nanotechnology-based radio transmitters being used to track people and spy in the United States. The supposed nanotechnology was actually a conventional protective coating applied to prevent the poppy's red color from rubbing off.

KEY INFORMATION SOURCES

Allen, G., and Frappell, B.D. 1970. The production of oil poppies. *Tasmanian J. Agric.* 41(2):89–94.

Andrews, A.C. 1951. The opium poppy as a food and spice in the classical period. *Agric. Hist.* 25:152–155.

Bjerver, K., Jonsson, J., Nilsson, A., Schuberth, J., and Schuberth, J. 1982. Morphine intake from poppy *Papaver somniferum* seed food. *J. Pharm. Pharmacol.* 34:798–801.

Bonicamp, J.M., and Santana, I.L. 1998. Can a poppy seed food addict pass a drug test? *Microchem. J.* 58:73–79.

Chitty, J.A., Allen, R.S., Fist, A.J., and Larkin, P.J. 2003. Genetic transformation in commercial Tasmanian cultivars of opium poppy, *Papaver somniferum*, and movement of transgenic pollen in the field. *Funct. Plant Biol.* 30:1045–1058.

Commonwealth Bureau of Pastures and Field Crops. 1977. *Oilseed poppy* (Papaver somniferum) *bibliography 1966–1977*. Annotated bibliography G478. Commonwealth Bureau of Pastures and Field Crops, Hurley, U.K. 8 pp.

Dittbrenner, A., Lohwasser, U., Mock, H.P., and Borner, A. 2008. Molecular and phytochemical studies of *Papaver somniferum* in the context of infraspecific classification. *Acta Hortic.* 799:81–88.

Duke, J.A. 1973. Utilization of *Papaver*. *Econ. Bot.* 27:390–400.

Duke, J.A., Gunn, C.R., Leppik, E.E., Reed, C.F., Solt, M.L., and Terrell, E.E. 1973. *Annotated bibliography on opium and oriental poppies and related species*. Agricultural Research Service, U.S. Department of Agriculture, Washington, DC. 349 pp.

Eklund, A., and Agren, G. 1975. Nutritive value of poppy seed protein. *J. Am. Oil Chem. Soc.* 52(6):188–190.

Forsyth, A. 1986. Poppy culture. *Harrowsmith*, 65(Jan./Feb.):61–69.

Frick, S., Kramell, R., Schmidt, J., Fist, A.J., and Kutchan, T.M. 2005. Comparative qualitative and quantitative determination of alkaloids in narcotic and condiment *Papaver somniferum* cultivars. *J. Nat. Prod.* 68:666–673.

Grove, M.D., Spencer, G.F, Wakeman, M.V., and Tookey, H.L. 1976. Morphine and codeine in poppy seed. *J. Agric. Food Chem.* 24:896–897.

Husain, A., and Sharma, J.R. (Eds.). 1983. *The opium poppy*. Central Institute of Medicinal & Aromatic Plants, Lucknow, India. 167 pp.

Kapoor, L.D. 1995. *Opium poppy: botany, chemistry, and pharmacology*. Food Products Press, New York. 326 pp.

Lööf, B. 1966. Review article: poppy cultivation. *Commonwealth Agriculture Bureaux Field Crop Abstracts*. 19(1):1–5.

Merlin, M.M. 1984. *On the trail of the ancient opium poppy*. Associated University Presses, London. 324 pp.

Németh, E., and Bernáth, J. 2009. Selection of poppy (*Papaver somniferum* L.) cultivars for culinary purposes. *Acta Hortic.* 826:413–419.

Nyman, U., and Hall, O. 1974. Breeding oil poppy (*Papaver somniferum*) for low content of morphine. *Hereditas*, 76:49–54.

Pelders, M.G., and Ros, J.J.W. 1996. Poppy seeds: differences in morphine and codeine content and variation in inter- and intra-individual excretion. *J. Forensic Sci.* 41:209–212.

Pushpangadan, P., and Singh, S.P. 2001. Poppy. In *Handbook of herbs and spices*. Edited by K.V. Peter. CRC Press, Boca Raton, FL. pp. 261–268.

Ram, M., Ram, D., Singh, S., and Kumar, S. 1999. Cost effective technology for seed production in opium poppy (*Papaver somniferum*). *J. Med. Aromatic Plant Sci.* 21:335–337.

Sangwan, N.K., Dhindsa, K.S., and Gupta, R. 1985. Effect of variety and growing location on the proximate and fatty-acid composition of opium poppy *Papaver somniferum*. *Int. J. Trop. Agric.* 3(1):1–8.

Schweizer, G. 1974. On the anatomy of the poppy seed (*Papaver somniferum* L.). *Berichten Deutsch. Bot.* 49:414–423.

Shaari, K., and Brink, M. 2001. *Papaver* L. In *Plant resources of South-East Asia. Vol. 12(1). Medicinal and poisonous plants 1*. Edited by L.S. de Padua, N. Bunyapraphatsara and R.H.M.J. Lemmens. Backhuys Publishers, Leiden, the Netherlands. pp. 373–379.

Sharma, J.R., Lal, R.K., Gupta, M.M., Verma, R.K., and Misra, H.O. 2002. A novel non-narcotic seed variety Sujata of opium poppy (*Papaver somniferum*). *J. Med. Aromatic Plant Sci.* 24:481–485.

Sharma, J.R., Lal, R.K., Gupta, A.P., Misra, H.O., Pant, V., Singh, N.K., et al. 1999. Development of non-narcotic (opiumless and alkaloid-free) opium poppy, *Papaver somniferum*. *Plant Breed.* 118:449–452.

Singh, H.G. 1982. Cultivation of opium poppy. In *Cultivation and utilization of medicinal plants*. Edited by C.K. Atal and B.M. Kapur. Regional Research Laboratory, Council of Scientific & Industrial Research, Jammu-Tawi, India. pp. 120–138.

Singh, S.P., Shukla, S., and Khanna, R.R. 1995. Opium poppy. In *Advance in Horticulture—medicinal and aromatic plants. Vol. 11*. Edited by K.L. Chadha and R. Gupta. pp. 535–574.

Singh, S.P., Shukla, S., and Singh, N. 1998. Genetic divergence in relation to breeding for fatty acids in opium poppy (*Papaver somniferum* L.). *J. Genet. Breed*, 52:301–306.

Singh, S.P., Khanna, K.R., Shukla, S., Dixit, B.S., and Banerji, R. 1995. Prospects of breeding opium poppies (*Papaver somniferum* L.) as a high-linoleic-acid crop. *Plant Breed*, 114:89–91.

Srinivas, H., and Rao, M.S.N. 1986. Functional properties of poppy *Papaver somniferum* seed meal. *J. Agric. Food Chem.* 34(2):222–224.

Tetenyi, P. 1995. Biodiversity of *Papaver somniferum* L. (opium poppy). *Acta Hortic.* 390:191–201.

Tetenyi, P. 1997. Opium poppy (*Papaver somniferum*): botany and horticulture. *Hortic. Rev.* 19:373–408.

U.S. Department of Justice. 1992. *Opium poppy cultivation and heroin processing in Southeast Asia*. U.S. Department Justice, Drug Enforcement Admin., Washington, DC. 31 pp.

Verma, S., Agarwal, S.K., Singh, S.S., Siddiqui, M.S., and Kumar, S. 1999. Poppy seed: composition and uses. *J. Med. Aromatic Plant Sci.* 21:442–446.

Specialty Cookbooks

Klivans, E. 2005. *Cupcakes*. Chronicle Books, San Francisco, CA. 143 pp.

Ode, K. 2006. *Baking with the St. Paul Bread Club: Recipes, Tips & Stories*. Minnesota Historical Society Press, St. Paul, MN. 156 pp.

72 Palmyra Palm

Family: Arecaceae (Palmae; palm family)

NAMES

Scientific Name: *Borassus flabellifer* L.

- The species described here is known most commonly as "palmyra palm," or simply as "palmyra." The name "palmyra" is based on the Portuguese *palmeira*, and traces to *palma*, the Latin word for palm (of a hand) and palm tree; the tree was named originally for the resemblance of the leaf to the palm of a hand. Occasionally, palmyra is spelled palmyrah.
- Palmyra is also the name of an island in the central Pacific and an ancient city (Tamar) in Syria northeast of Damascus, rather commonly referred to in scripture and in literature. Numerous modern cities are also named Palmyra. The mistaken belief that palmyra palm comes from a place called Palmyra is likely responsible for the frequent capitalization of the p in palmyra (as Palmyra palm).
- Palmyra palm is also known as doub palm, lontar palm, sugar palm, tala palm, toddy palm, and wine palm.
- "Sugar palm" is an ambiguous name as several other palms, some actually used like the palmyra palm to produce sugar, are also called sugar palm.
- "Wine palm" is also ambiguous because the sap of several species of palms called wine palm is used to make palm wine.
- The genus name *Borassus* is based on the Greek *borassos*, referring to the immature flowering top of the date palm.
- *Flabellifer* in the scientific name *B. flabellifer* is Latin for *flabellum*, a small fan + *ifer*, bearing, a poetic way of saying that the tree has fanlike leaves.

PLANT PORTRAIT

The palmyra palm is a large, tropical Asian tree growing up to 30 m (98 ft.) in height, with a trunk up to 0.5 m (1½ ft.) wide at the base. The leaves are leathery, fan-shaped, 1 to 3 m (3–10 ft.) in diameter, with numerous segments. There are separate male and female plants. The fruits look like miniature coconuts, 12 to 15 cm (4¾–6 in.) wide, with a smooth, thin, leathery outer covering that is brown but turns nearly black after harvest. The interior contains a juicy mass of long, tough, coarse, white fibers coated with yellow or orange pulp. There are often two or three seeds in each fruit, embedded in the pulp. When mature, the seed is a solid white kernel resembling coconut meat but much harder. When the fruit is very young, the kernel is hollow, soft like jelly, and translucent like ice. The very young fruit also contains a sweet, watery liquid. Each tree may bear 6 to 15 bunches of fruit yearly, with average annual yield generally less than 100 fruits per tree. The species has been interpreted as growing wild from the Persian Gulf to the Cambodian–Vietnamese border. Alternatively, the palmyra palm is viewed as a cultivated selection from the wild species *Borassus aethiopum* C. Mart. of Africa. The palmyra is commonly cultivated in India, Southeast Asia, Malaysia, and occasionally in other warm regions, including Hawaii and southern Florida. It is not as developed as the world's most important food palms (coconut (*Cocos nucifera* L.), oil palm (*Elaeis guineensis* Jacq.), date (*Phoenix dactylifera* L), and sago palm (*Metroxylon sagu* Rottb.)), with little selection of cultivated varieties.

FIGURE 72.1 Palmyra palm (*Borassus flabellifer*) in Ceylon. (From Marilaun, 1895.)

Palm wine or toddy is an ancient beverage based on collecting and fermenting the sap of a number of different palm species. Tapping the stem, as is done with the sugar maple (*Acer saccharum* Marshall), is the simplest method, but it has been found for many of the species that much more sap can be collected from the flowering tops. The chief product of the palmyra is the sweet sap obtained by tapping the flowering stalk at the top of the plant. Tapping palm trees is carried out throughout tropical regions of the world but is most popular in Asia and Africa. The coconut and the palmyra are principal sugar trees in Asia. In Africa, several palm species are used, including the oil palm. Tapping palms in Latin America and the Caribbean has a long history, but the practice is uncommon today.

The leaves of the palmyra palm are used for house construction (such as for thatching and walls) in some parts of Asia. Fiber from various parts of the palmyra palm is made into brooms, hats, baskets, mats, fans, and other household items.

CULINARY PORTRAIT

Sugar is a chief product of the palmyra palm. The juice is traditionally processed into three types of sugar: liquid sugar (sugar palm syrup, reminiscent of molasses produced from cane and beet sugar), crystalline palm sugar, and block sugar. The most common type consumed in rural areas of Asia is sugar palm syrup, which is typically concentrated to 80% dry matter. Occasionally, the syrup is smoked first, giving the sugar a black color and distinctive flavor. When further evaporated, palm sugar (commonly known as jaggery) is produced. The color of palm sugar generally ranges from pale golden to very dark brown. Palm sugar is thick and crumbly and can be gently melted before adding to sauces or dressings. Palmyra palm sugar is somewhat more nutritious than crude cane sugar, containing 1% protein. Palm sugar is available from Asian food stores. Candy is often made from the sugar.

When fermented, palm sap yields palm wine (palm toddy). Additional fermentation of toddy produces a mild palm vinegar. From toddy, the highly intoxicating drink arrack is prepared by distillation.

In Asia, various parts of the palmyra palm are consumed. Immature seeds are eaten fresh or roasted. The hollow, jellylike young kernels are often sold in markets. These soft-shelled immature seeds are sliced longitudinally to form attractive loops or rings, and these as well as the whole kernels are canned in clear, mildly sweetened water, and exported. The peeled seedlings are eaten fresh or sun-dried,

FIGURE 72.2 Collecting sap from palmyra palm (*Borassus flabellifer*) to prepare toddy (palm wine). (From Baillon, 1876–1892.)

raw, or cooked in various ways. Small fruits are pickled in vinegar. The top part of immature fruit is cooked as a vegetable. The sweet jellylike pulp from the mature fruit is eaten directly (after separation from the wiry fibers) or used in various dishes. The roots are roasted. Edible starch is obtained from palmyra seedlings. Also, the outer portion of the young shoots are boiled, dried, and milled to provide an edible flour. However, some reports have suggested that this flour has toxic properties.

Culinary Vocabulary

- "Jaggery," based on the Hindi word *jagri*, is an unrefined, brown, palm sugar, made from the sap of several palm species, including palmyra palm.
- "Toddy" (rarely "tody") has several meanings, all concerned with sweet and/or alcoholic beverages. The word is a corruption of *taudi*, the Indian name for the sweet sap from palm flowering heads, based in turn on the Sanskrit *told* or *taldi*, referring to palm sap. Palm toddy is wine made from the sap of (primarily) Old World palm species, although the name has also been used for nonalcoholic beverages made from unfermented sweet palm sap. Palm toddy is commonly manufactured in Africa and Southeast Asia. In Western countries, a toddy is typically made with whiskey flavored with citrus fruits and spices. A "hot toddy" refers to a warm drink made with liquor, hot water, sugar, and spices (originally toddy was made cold, but today it is usually heated).
- "Arrack" (arak), based on the Arabic *araq*, sweet juice or liquor, refers to strong spirits distilled chiefly in Asia from fermented juice of fruits, grains, palms, or sugarcane. In the nineteenth century, Ceylon (now Sri Lanka) was noted for palm arrack, and in modern times Indonesia is said to make the best arrack. Fermentation of sap from the palmyra palm is a chief source of palm arrack.

FIGURE 72.3 Palm sap collector. (From Hartwich, 1911.)

CURIOSITIES OF SCIENCE AND TECHNOLOGY

- An ancient Hindu song about the palmyra palm of India enumerates 801 uses for the plant. People in temperate regions are generally aware of the usefulness of the world's most important plant family, the grasses (which includes the cereals and numerous forages that provide the bulk of the food of humans and livestock), but rarely appreciate that in tropical regions the palm family rivals the grass family in importance.
- In parts of Asia, drums have been made from hollowed-out palmyra palm stems.
- Tapping the sweet sap from the flowering stalks at the tops of the tall palmyra palm is an athletic, skilled activity. Cambodian tappers use a single long bamboo pole as a ladder, with the stumpy remnants of the leaf bases at the nodes serving as rudimentary steps for climbing. When the trees are located close to each other, one or two long bamboo poles are used as an aerial bridge to move between the trees. Before commencing to tap a flowering stalk, juice production needs to be stimulated by mildly crushing the tissues, sometimes for a week. Tappers are capable of attending to 20 to 30 palm trees twice a day, provided an assistant is available at the base of the trunk to receive the collected juice (collection is carried out twice daily—morning and afternoon—to limit exposure of the juice to contamination by yeast and other fermenting microorganisms).
- When a new tapper replaces another in the middle of the season, palmyra trees frequently respond by lowering sap production for a day or two, perhaps like milk cows disturbed when a new milker replaces a familiar one.
- Female palmyra palm trees produce approximately 50% more sweet sap than the males.

KEY INFORMATION SOURCES

Anderson, P.H., and Poulsen, E. 1985. Mutagenicity of flour from the palmyrah palm (*Borassus flabellifer*) in *Salmonella typhimurium* and *Escherichia coli*. *Cancer Lett.* 26:113–119.

Ariyasena, D.D., Jansz, E.R., and Abeysekera, A.M. 2001. Some studies directed at increasing the potential use of palmyrah (*Borassus flabellifer* L) fruit pulp. *J. Sci. Food Agric.* 81:1347–1352.

Ariyasena, D.D., Jansz, E.R., Jayesekera, S., and Abeysekara, A.M. 2000. Inhibitory effect of bitter principle of palmyrah (*Borassus flabellifer* L.) fruit pulp on the growth of mice: evidence using bitter and non-bitter fruit pulp. *J. Sci. Food Agric.* 80:1763–1766.

Arseculeratne, S.N., Gunatilaka, A.A.L., and Panabokke, R.G. 1982. Studies on the toxicology of the palmyrah palm (*Borassus flabellifer* L.). I. A bioassay for the neurotoxin. *J. Nat. Sci. Council Sri Lanka,* 10:269–275.

Borin, K. 1998. Sugar palm (*Borassus flabellifer*): potential feed resource for livestock in small-scale farming systems. *World Animal Rev.* 91(2):21–29 (multilingual edition).

Ceylon Institute of Scientific and Industrial Research. 1967. *Products from the palmyrah palm.* Ceylon Institute of Scientific and Industrial Research, Colombo, Ceylon. 51 pp.

Dassanayake, M.D., and Sivakadachchan, B. 1973. Germination and seedling structure of *Borassus flabellifer* L. *Ceylon J. Sci. Biol. Sci.* 10:157–166.

Davis, T.A., and Johnson, D.V. 1987. Current utilization and further development of the palmyra palm (*Borassus flabellifer* L., Arecaceae) in Tamil Nadu State, India. *Econ. Bot.* 41:247–266.

Flach, M., and Paisooksantivatana, Y. 1996. *Borassus flabellifer* L. In *Plant resources of South-East Asia. Vol. 9. Plants yielding non-seed carbohydrates.* Edited by M. Flach and F. Rumawas. Backhuys Publishers, Leiden, the Netherlands. pp. 59–63.

Greig, J.B., Kay, S.J.E., and Bennetts, R.J. 1980. A toxin from the palmyra palm, *Borassus flabellifer*: partial purification and effects in rats. *Food Cosmet. Toxicol.* 18:483–488.

Hussain, M.D., and Hussain, M.I. 1992. Juice harvesting from date and palmyra palm tree in Bangladesh. *Indian J. Agric. Eng.* 2:17–24.

Jeyaseelan, K., and Seevaratnam, S. 1986. Ethanol and biomass from palmyra palm sap. *Biotechnol. Lett.* 8:357–360.

Kovoor, A. 1983. *The palmyrah palm: potential and perspectives.* FAO Plant Production and Protection Paper No. 52, Food and Agriculture Organization, Rome, Italy. 77 pp.

Lubeigt, G. 1977. The sugar palm, *Borassus flabellifer* L., its different products and associated technology in Burma. *J. Agric. Tradit. Bot. Appl.* 24:311–340 (in French, English summary).

Morton, J.F. 1988. Notes on distribution, propagation, and products of *Borassus* palms (Arecaceae). *Econ. Bot.* 42:420–441.

Nath, T.K., Karmakar, N.C., Molla, M.A.H., and Kabir, M.A. 2002. Tal (*Borassus flabellifer* Linn.: Arecaceae): a valuable economic palm of rural Bangladesh. *For. Trees Livelihood,* 12:283–295.

Theivendirarajah, K., and Chrystopher, R.K. 1986. Alcohol from palmyrah palm (*Borassus flabellifer* L.) fruit pulp. *Vingnanam,* 1:44–46.

Theivendirarajah, K., and Chrystopher, R.K. 1986. Chemical analysis of palmyrah palm *Borassus flabellifer* L. wine (toddy). *Vingnanam,* 1:1–7.

Specialty Cookbooks

Note: "Palm sugar" is often specified in recipes in Asian cookbooks, and an Internet search for "palm sugar recipes" will result in thousands of offerings. The use of palm sugar produces a somewhat caramel flavor to dishes. Virtually any recipe calling for sugar can use palm sugar, but health concerns today suggest that sugar intake is advisedly reduced.

73 Passionfruit (Granadilla)

Family: Passifloraceae (passionflower family)

NAMES

Scientific Names: *Passiflora* Species

- "Passionflower" and "passion flower" on the one hand and "passionfruit" and "passion fruit" on the other are widely used alternative spellings.
- The passionflower or "flower of passion" acquired this name because Spanish missionaries to South America, when they became aware of it in the jungles of Brazil in the sixteenth century, thought the flower represented some of the objects associated with the Crucifixion or suffering of Christ. The "passion" in the name is based on the Passion of Christ, the period between the Last Supper and his death. For the Jesuit missionaries, it was the bloom of the mystic St. Francis of Assisi (1181 or 1182 to 1226), founder of the Franciscan orders, who saw a flower in a vision. The 10 apostles who remained faithful to Jesus (leaving out Peter who denied him and Judas who betrayed him) throughout the Passion are symbolized by the five petals and the five petal-like sepals, while a fringe of colored, hair-like filaments above the petals (botanically known as the corona) was thought to represent the crown of thorns that Jesus wore (or by some, considered to symbolize a halo). The three spreading styles atop the stigma represented the three nails by which Christ was attached to the cross. The five hammer-like anthers on top of the stamens were thought by some to represent the hammer used to drive the nails; others interpreted the five stamens as the wounds of Christ (hence the name "flower of the five wounds" sometimes used in Peru, the West Indies, and elsewhere) and interpret the ovary (unfertilized seed case) as the hammer. Some have even interpreted the lobed leaves and long green vines as the hands and whips (scourges) of Christ's prosecutors, the column of the ovary as the pillar of the cross, the white tint of the flower as purity, and the blue tint as heaven. Supposedly 72 corona filaments are present, reflecting the traditional number of thorns in the crown of thorns, and some even saw on the underside of the leaves dark spots representing the 30 pieces of silver and the trinity in the three-lobed leaves. The Jesuit Fathers interpreted the eagerness of the Indians to consume the fruits as a heavenly sign that they were hungering for conversion from heathenism to Christianity and with overbearing religious zeal quickly succeeded in wiping out thousands of years of New World culture.
- The yellow-fruited form of *Passiflora edulis* (known as yellow passionfruit and golden passionfruit) could be confused with yellow granadilla, *P. laurifolia*, the fruits of which are also eaten.
- The name "granadilla" (sometimes spelled grenadilla; pronunciation: grah-nah-DEE-yah) is derived from the Spanish diminutive of *granada*, pomegranate (*Punica granatum* L.), so-named because the Spanish who first viewed them thought they were like little pomegranates.
- *Edulis* in the scientific name *P. edulis* is Latin for eatable or edible.
- The name "Maypop" has been explained on the basis that the plant seemingly pops out of the ground in May. Another explanation is that if one squeezes the fruit, it "may pop" like a balloon. The name Maypop may in fact be derived from a similar Native American word for *Passiflora incarnata*, such as the Algonkian *maricock* or *maracock*, or the Powhatan

FIGURE 73.1 Passionfruit (*Passiflora edulis*). (From Curtis, 1808, vol. 45, plate 1989.)

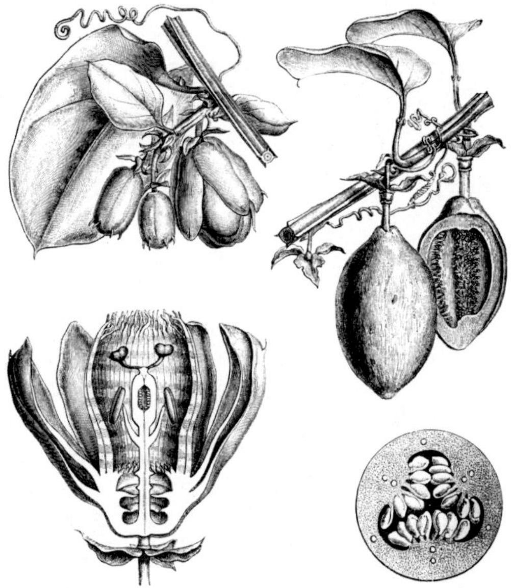

FIGURE 73.2 Wingstem passionflower (*Passiflora alata*). (From Engler and Prantl, 1889–1915.)

mahcawq. It has been suggested that the North American Indians acquired the word from the Tupi Indians of South America, who called the plant *maraca-cui-iba*, the rattle fruit, because seeds in the gourdlike fruits produce a rattling sound when the fruits are dried.

- Local names used historically in the southern United States for the Maypop include apricot, apricot vine, field apricot, ground ivy, holy trinity flower, Mayapple (likely to result in confusion with the true Mayapple, *Podophyllum peltatum* L.), mollie cockle, and molly-pop.

TABLE 73.1
Important Species of *Passiflora* Grown for Fruit

Species	Common Names	Area of Origin	Area of Cultivation
P. edulis Sims	Passionfruit, purple passionfruit, purple granadilla, yellow passionfruit, golden passionfruit	Southern Brazil	Throughout tropics and subtropics
P. quadrangularis L.	Giant granadilla, granadilla	Tropical South America	Tropical lowland gardens
P. ligularis Juss.	Sweet granadilla, sweet passionfruit	Mexico to western Bolivia	Tropics; throughout Andes to South America
P. incarnata L.	Maypop, Maypop passionflower, wild passionflower	Eastern United States	An occasional garden plant in temperate areas
P. alata Curtis	Wingstem passionflower, fragrant granadilla	Peru and Brazil	Brazil
P. laurifolia L.	Water lemon, bell apple (belle apple), yellow granadilla, Jamaica honeysuckle, vinegar pear, yellow granadilla, pomme-de-liane	West Indies and northeastern South America	Widely in Tropics; West Indies, northeastern South America
P. maliformis L.	Sweet calabash, chulupa, sweet cup, conch apple, curuba	West Indies to northern South America	Ecuador, Columbia, West Indian islands
P. tripartita (Juss.) Poir. var. *mollissima* (Kunth) Holm-Niels. & P.M. Jørg. (*P. mollissima* (Kunth) L.H. Bailey)	Tasco, curuba, banana passionfruit	Andes of Venezuela to Bolivia	Throughout Andes, Hawaii, New Zealand

- *Incarnata* in the scientific name of the Maypop, *P. incarnata*, means flesh colored, descriptive of the yellowish fruit.
- The genus name *Passiflora* has the same general basis as the "passion" in passionflower or passionfruit, noted earlier. The name *Passiflora* is specifically from the Latin *passio*, passion + *flos*, flower (Table 74.1).

PLANT PORTRAIT

Most species of *Passiflora* are native to South America, but approximately 20 are found in Asia and Australia. Of the approximately 500 species, about 60 produce edible fruit, of which only 10 to 12 have been commercially exploited for their fruits. The most common types have orange, purple, or yellow skin and are egg- to apple-sized. The fruit has a hard, leathery shell, which encloses many small, dark, crunchy, edible seeds that are enveloped by juicy, tart, gelatinous flesh. The flesh may be yellow, orange, pinkish-green, white, or colorless. Passionfruit is grown widely throughout the tropics and subtropics. Commercial production occurs in South America, the Caribbean, Mexico, Australia, New Zealand, Southeast Asia, Taiwan, India, Africa, the Mediterranean area, Hawaii, and South Africa. There are small crops in California and Florida.

The most important species is *P. edulis*, which is known simply as passionfruit. It is native to South America, from southern Brazil through Paraguay to northern Argentina, and is widely grown, including plantations in Australia, Hawaii, and Mexico, with some cultivation in southern

California. The passionfruit plant is a shallow-rooted, woody, perennial vine up to 15 m (49 ft.) long, which climbs by tendrils, and has large, deciduous, three-lobed, serrated leaves. Its flowers are 5 to 7.5 cm (2–3 in.) wide, with white petals and a fringelike crown. The fruit is nearly round or ovoid, 4 to 7.5 cm (1½–3 in.) wide, with a tough, waxy rind varying from dark-purple with white specks to yellow. The fruit is filled with sacs of orange, pulpy juice and numerous small, hard, dark-brown, yellow, or black pitted seeds. The flavor is acidic. There are two types of plant, one with purple fruit and the other with yellow, and sometimes referred to as forma *flavicarpa*. Both are called granadilla in Spanish. The purple form is also called purple, red, or black granadilla, whereas the yellow type is widely known as yellow passionfruit, or in Australia, golden passionfruit. The yellow form may have arisen as a mutation of the purple form. The yellow type has a more vigorous vine and generally larger fruit than the purple, but the pulp of the purple is less acidic, richer in aroma and flavor, and has more juice. The purple form has black seeds, whereas the yellow has brown seeds. The purple passionfruit is adapted to a subtropical climate, whereas the yellow is adapted to a more tropical climate.

The second most important passionfruit species is the giant granadilla, *P. quadrangulata*, which is often cultivated in tropical lowland gardens. This produces the largest fruit in the genus *Passiflora*. The melon-like fruit is 12 to 15 cm (4¾–6 in.) wide and 10 to 30 cm (8–12 in.) long. The fruit is tasty but somewhat bland.

Another notable passionfruit, the sweet granadilla, *P. ligularis*, is popular in markets near its native distribution, from central Mexico through Central America and western South America, through western Bolivia to south-central Peru. It has also been grown in Hawaii since the nineteenth century. The pulp of the fruit is usually consumed fresh, whereas the strained juice is used in cold drinks and sherbets. Many consider this species to provide the tastiest fruit.

Maypop (*P. incarnata*, also known as wild passionflower and apricot vine) is widely grown in tropical areas as an ornamental, and its fruits are edible (and known as maypops). It is one of the few species of *Passiflora* native to the United States, occurring from Virginia to southern Illinois and southeast Kansas, south to Florida and Texas. The Maypop is a common roadside weed throughout many areas of the southeastern United States, often growing in large masses in ditches and open fields. It can be grown as far north as U.S. Department of Agriculture climate zone 5, which includes parts of southern Canada. The plants are also excellent indoors in containers. The ripe, yellow to yellow-green, egg-shaped fruit usually has delicious pulp (wild fruit sometimes is unpleasant) but is seedy and best made into jelly. Overripe fruit ferments into a foul paste. Extracts from *P. edulis* and *P. incarnata* are commonly used as sedatives in herbal medicine.

CULINARY PORTRAIT

The fruit of the main commercial passionfruit species, *P. edulis*, can be cut lengthwise and the seedy pulp scooped out of the inedible shell with a spoon. The seeds are usually removed by squeezing the pulp through two thicknesses of cheesecloth or pressing the pulp through a strainer. Alternatively, the seeds may simply be eaten along with the pulp. Passionfruit can be prepared as a dessert with cream and sugar or sprinkled with lemon or lime juice. It can also be used in fruit salads and beverages or made into jellies and jams. Passionfruit juice can be cloying alone but is an excellent flavoring even when used in small quantities. It is used in sauces, gelatin desserts, candy, ice cream, yogurt, sherbet, icing, pies, soups, and cocktails. The fresh juice and concentrate are excellent mixers with alcoholic beverages, such as gin, vodka, and rum. The juice is sometimes fermented to make alcoholic beverages. Most people are familiar, often unknowingly, with the flavor of passionfruit from having tasted Hawaiian punch. It has been said that passionfruit juice gives the punch to Hawaiian punch.

Good quality passionfruit has very old-looking, brown, shriveled skin, which may even have some mold growing on it; this is acceptable as long as the skin is not cracked. The underlying shell should remain hard, and soft fruits should be avoided. Smooth-skinned passionfruit generally needs ripening (although some kinds, often called maracuya, which have large yellow-orange fruit with white and gray flesh, are generally eaten while the skin is still smooth). The fruit should not be refrigerated until it is ripe, at which stage it may be stored in a refrigerator for up to a week.

Culinary Vocabulary

- A "Hurricane" is a cocktail made with dark rum, passionfruit flavoring, and citrus juices, served in a hurricane glass (a footed glass that is bulbous at the bottom, tapering to a flaring cylinder at the top, used for blended or frozen tropical and specialty drinks). The origin of the beverage is obscure, but it has long been associated with Pat O'Brien's French Quarter Bar in New Orleans, Louisiana.
- A "Grand Passion" is a cocktail prepared with passionfruit liqueur, gin, and Angostura bitters.

CURIOSITIES OF SCIENCE AND TECHNOLOGY

- Captain James Smith (who brought 105 people to Virginia and founded Jamestown in 1607, the first permanent English settlement in the New World) observed Indians cultivating the Maypop for its fruits.
- In 1865, Queen Victoria (1819–1901) had a wreath of passionflowers sent to the funeral of Abraham Lincoln (1809–1865), 16th president of the United States. The Queen also declared that the passionflower was the ideal flower for funerals. Gravestones with elaborately carved passionflower vines became very popular for the wealthy in England.
- The caterpillars (larvae) of several species of zebra butterflies can only eat passionflower leaves, so if one wants to attract butterflies to a garden, growing the cold-hardy Maypop is a good idea. Several passionflower species have adaptations to prevent damage by the caterpillars. Some plants disguise themselves by producing leaves that do not look like typical passionflower leaves. Some species have nectar-secreting glands to attract wasp and ant parasites of the larvae. Some species produce chemicals that poison the larvae. And still other species produce tiny egg-like particles on the leaves that send the message to the female butterfly that the leaves are already occupied so she should find another plant on which to lay her eggs.
- The Maypop (*P. incarnata*), like some other species of *Passiflora*, has nectar-producing glands on the leaf stalks, which attract ants that in turn have been experimentally demonstrated to protect the plant against attack from other insects. This type of cooperative arrangement, whereby plants produce nectar on secretory glands outside of the flowers to attract ants for protective purposes, is widespread in the plant kingdom. It has been shown that the Maypop and some other species of *Passiflora*, when their leaves have been damaged by insects, seem to have the ability to increase their production of secretory glands to attract more protective ants.

KEY INFORMATION SOURCES

Abeysinghe, A. 1973. *Commercial passion fruit: cultivation, processing, and marketing*. Ministry of Plantation Industries, Government of Sri Lanka, Colombo, Sri Lanka. 29 pp.

Anonymous. 1974. *Passion fruit culture in Hawaii*. University of Hawaii, Cooperative Extension Ser., Honolulu, HI. 35 pp.

Anonymous. 1978. *Passion fruit industry analysis*. College of Tropical Agriculture, University of Hawaii, Manoa, HI. 2 vols.

Bora, P.S., and Narain, N. 1997. Passion fruit. In *Postharvest physiology and storage of tropical and subtropical fruits*. Edited by S.K. Mitra. CAB International, Wallingford, U.K. pp. 375–386.

Casimir, D.J., Keffod, J.F., and Whitfield, F.B. 1981. Technology and flavor chemistry of passion fruit juices and concentrates. In *Advances in food research, vol. 27.* Edited by C.O. Chichester, E.M. Mrak, and G.F. Stewart. Academic Press, New York, N.Y. pp. 243–295.

Chan, H.T., Jr. 1980. Passion fruit. In *Tropical and subtropical fruits: composition, properties and uses.* Edited by S. Nagy and R.E. Shaw. AVI Publishing Company, Westport, CT. pp. 300–315.

Chapman, T. 1967. Passion fruit growing in Kenya. *Econ. Bot.* 17:165–168.

Fouqué, A. 1972. Espèces fruitières d'Amérique tropicale: genre *Passiflora. Fruits*, 27:369–382.

George, A., and Paull, R.E. 2008. *Passiflora edulis*, passsionfruit. In *The encyclopedia of fruit & nuts.* Edited by J. Janick and R.E. Paull. CABI, Wallingford, Oxfordshire, U.K. pp. 586–595.

Gurnah, A.M. 1991. *Passiflora edulis* Sims. In *Plant resources of South-East Asia. Vol. 2. Edible fruits and nuts.* Edited by E.W.M. Verheij and R.E. Coronel. Pudoc, Leiden, the Netherlands. pp. 244–248.

Knight, R.J., Jr. 1969. Edible-fruited passionvines in Florida: the history and possibilities. *Proc. Trop. Reg. Am. Soc. Hort. Sci.* 13:265–274.

Leal, F. 1990. Granadilla. In *Fruits of tropical and subtropical origin: composition, properties and uses.* Edited by S. Nagy. Florida Science Source Inc., Lake Alfred, FL. pp. 322–327.

Lippmann, D. 1978. *Cultivation of Passiflora edulis S. General information on passion fruit growing in Kenia* [sic]. German Agency for Technical Cooperation (Schriftenreihe der GTZ), Eschborn, Germany. 88 pp.

Martin, F.W., and Nakasone, H.Y. 1970. The edible species of *Passiflora. Econ. Bot.* 24:333–343.

McKenzie, D.E., Chalker, F.C., and Buggie, G.J. 1981. *Passionfruit growing.* North Coast Region, Department of Agriculture, Coffs Harbour, N.S.W., Australia. 24 pp.

Menzel, C.M., and Simpson, D.R. 1994. Passionfruit. In *Handbook of environmental physiology of fruit crops. Vol. 2. Subtropical and tropical crops.* Edited by B. Schaffer and P.C. Anderen. CRC Press, Boca Raton, FL. pp. 225–241.

Menzel, C.M., Winks, C.W., and Simpson, D.R. 1989. Passionfruit in Queensland. 3. Orchard management. *Qld. Agric. J.* 115:155–164.

Morton, J. 1987. Passionfruit. In *Fruits of warm climates.* Creative Resource Systems, Winterville, NC. pp. 320–328.

Mott, J. 1969. *The market for passion fruit juice.* Tropical Products Institute, Ministry of Overseas Development. London. 15 pp.

Murray, I.E., Shipton, J., and Whitfield, F.B. 1972. Volatile constituents of passion fruit, *Passiflora edulis. Aust. J. Chem.* 25:1920–1933.

Notodimedjo, S. 1991. *Passiflora quadrangularis* L. In *Plant resources of South-East Asia. Vol. 2. Edible fruits and nuts.* Edited by E.W.M. Verheij and R.E. Coronel. Pudoc, Leiden, the Netherlands. pp. 248–249.

Rupert-Torres, R., and Martin, F.W. 1974. First-generation hybrids of edible passionfruit species. *Euphytica*, 23:61–70.

Sao José, A.R., Ferreira, F.R., and Vaz, R.L. 1991. *A cultura do maracuja no Brasil [The culture of passion-flower in Brazil].* Fundacao de Estudos e Pesquisa em Agronomia, Medicuna Veterinaria e Zootecnica (FUNEP), Jaboticabal, Brazil. 247 pp (in Portuguese).

Seale, P.I., and Sherman, G.D. 1960. *Commercial passion fruit processing in Hawaii.* Hawaii Agricultural Experiment Station, University of Hawaii, Honolulu, HI. 18 pp.

Ulmer, T., and MacDougal, J.M. 2004. *Passiflora: passionflowers of the world.* Timber Press, Portland, OR. 430 pp.

Vanderplank, J. 1996. *Passionflowers and passionfruit.* Revised edition. Cassell, London. 224 pp.

Whittaker, D.E. 1972. Passion fruit: agronomy, processing and marketing. *Trop. Sci.* 14:59–77.

Williams, C.G. 1954. Preparing passion fruit for the fresh fruit market. *Qld. Agric. J.* 78(2):81–88.

Winks, C.W., Menzel, C.M., and Simpson, D.R. 1988. Passionfruit in Queensland. 2. Botany and cultivars. *Qld. Agric. J.* 114:217–224.

Specialty Cookbooks

Gebhard, M.L., and Butler, W.H. 1967. *Pineapples, passion fruit, and poi; recipes from Hawaii.* Charles E. Tuttle Co., Rutland, VT. 127 pp.

Hawaiian Electric Company. 1956. *Reddy's passion fruit recipes.* Hawaiian Electric Company, Honolulu, HI. 9 pp.

Vinacke, W.R. 1955. *Passion fruit recipes for sweetened, frozen, pure passion fruit juice.* Food Processing Laboratory, University of Hawaii, Honolulu, HI. 11 pp.

74 Peach Palm

Family: Arecaceae (Palmae; palm family)

NAMES

Scientific Name: *Bactris gasipaes* Kunth (*Guilielma speciosa* C. Mart., *G. utilis* Oersted, *G. chontaduro* Triana, *G. gasipaes* (Kunth) L.H. Bailey)

- The peach palm was so named by German botanist-explorer Alexander von Humboldt (1769–1859), and is based on the peach-like aroma of the fermented pulp. The fruit itself is not particularly reminiscent of a peach.
- Peach palm is also widely known as pejibaye (palm). This name of Spanish origin appears to have been applied to the peach palm because of the popularity of the fruit in the neighborhood of the Costa Rican city of Pejibaye.
- There are at least 300 different names among the Native American languages of the New World for peach palm.
- The genus name *Bactris* is based on the Greek *baktron*, cane, reflecting the use of the young stems of some young palm stems as walking-sticks.
- *Gasipaes* in the scientific name *B. gasipaes* is derived from the Amerindian name *cochipay* used in the Magdalena River valley of Colombia.

PLANT PORTRAIT

The peach palm is a particularly good example of a palm that furnishes "hearts of palm." There are several thousand species in the palm family, distributed around the tropical world, and all have edible "hearts," although many do not have a pleasant taste and contain chemicals that may be unhealthy to consume. Palm plants develop new leaves at a growing point (meristem) located at the top of the stem. The very young leaves are mostly just leaf stalks packed together and in palms typically form a cylinder of tissue in the center of the uppermost part of the stem. The cylinder of tender young leaves, before they emerge from the center of the stem (and become green), together with the stem tissue just below the growing point, make up the "heart." Only approximately 100 palm species have hearts that are large enough to be worth collecting, and most of these are used in local cuisines as a vegetable.

International trade in palm hearts has been based to a considerable degree on the genus *Euterpe*, which occurs in South and Central America. Brazil began exporting *Euterpe edulis* C. Mart. (known as assai palm) in the 1950s based on wild palms from the Atlantic forests of southern Brazil and soon decimated the wild populations. The industry then shifted to the estuary of the Amazon River to harvest wild *Euterpe oleracea* C. Mart. (manicole palm, also known as assai palm), and the majority of Brazil's exports are now derived from it. (As detailed in Chapter 3, the berries of the plant have become extremely popular as a specialty health food.) Many palm trees die when their growing point is harvested, and it is absolutely irresponsible to threaten species in the wild with extermination merely to eat a very small portion of it, as has occurred in Brazil. (Multistemmed palms can be sustainably collected from the wild, so long as one does not destroy too many of the stems by harvesting their growing points.) There is some export of palm hearts from Asia on the basis of the coconut palm (*Cocos nucifera* L.), which is widely grown, and from *Daemonorops schmidtiana* Becc., which lacks a widespread English common name. Often when palms are cut

FIGURE 74.1 Peach palm (*Bactris gasipaes*). (From Mora-Urpí, 1994.) Bottom left: profile and sections of fruit. Upper right: cluster of fruit. Bottom right: spine-covered internodes of the stem. (Reproduced with permission from the Food and Agriculture Organization of the United Nations.)

down for timber or because of age, the edible heart is harvested. The peach palm is highlighted here because it has the largest potential of dominating the industry.

Sabal palmetto palm, cabbage palm, cabbage palmetto, and swamp cabbage are names for *Sabal palmetto* Schult. & Schult. f. This used to be regarded as a food for poor people in Florida and was widely harvested during the Depression Era. When its delicacy became appreciated, it became known as "millionaires' salad," a term often applied to edible palm hearts (which are often so large that they can furnish a salad for many people). Because so many people were harvesting sabal palmetto, Florida enacted a law to protect it.

The peach palm is a graceful erect tree growing 15 to 30 m (49–100 ft.), with a single slender stem 13 to 30 cm (5–12 in.) thick or, more often, several stems in a cluster. The trunk is armed with concentric rings of stiff, black spines, which occur at alternating nodes, that is, where leaves originated but have fallen off the plant. Spines also occur along the midribs of the leaves. The fruits hang in clusters of 50 to 100 or sometimes as many as 300 and are yellow to orange, scarlet, yellow-and-red, or brownish at first, turning purple when fully ripe. The fruits are ovoid or cylindrical, 2.5 to 5 cm (1–2 in.) long, and have been said to resemble tiny American footballs.

Peach palm is apparently native to Amazonian areas of Colombia, Ecuador, Peru, and Brazil, but it has been cultivated and distributed by Indians from ancient times. Today, it is found throughout Central and South America. Although some other American palms were altered slightly by selection, the peach palm is the only American palm that has been truly domesticated, that is, changed genetically from its wild ancestors. The species was probably domesticated in southwestern Amazonia. Harvesting has greatly reduced wild-growing stands. The peach palm was the most important palm of pre-Columbian America and the main crop of the Amerindians of an extensive territory of the humid tropics. When Columbus arrived in the New World, the Carib Indians were taking full advantage of what they called the pejibaye tree. They housed themselves with its bark, put a roof over their heads with the leaves, consumed the fruits of the mature tree, and ate the center core of the young plants.

FIGURE 74.2 Commercial palm hearts. (Photo by E. Johnson.)

The peach palm is an important backyard tree in much of tropical Latin America and is used as a dietary staple by some Amerindian tribes as well as for livestock food. The fruit is used principally as a cooked snack. It is rich in starch and oil and has been an occasional source of starch. However, the fruit has achieved very limited success to date as an international item of trade (occasionally, peach palm fruit is available in North American supermarkets, but the fruit has a short shelf life). The fruits contain calcium oxalate crystals, which are considered undesirable nutritionally.

The palm heart, in contrast to the fruit, is well known as an export, and its consumption is expanding quickly worldwide. Palm heart production and canning is a rapidly growing industry in Latin America. The peach palm heart has been on international markets since 1978 and is increasing in market share in both Europe and North America. Plantation technology for the peach palm heart is most developed in Costa Rica, Brazil, and Ecuador, but Peru, Bolivia, Colombia, Guyana, Surinam, Venezuela, Panama, Guatemala, and the Dominican Republic are rapidly planting peach palm. Most peach palm stems are spiny, which makes it difficult to extract the heart, but several spineless populations have been found and eventually spineless varieties will dominate the industry. The young plant is harvested at approximately 12 months of age, when the center "cord" is between 1.9 and 2.5 cm (3/4–1 in.) in diameter, and tender. At harvest time, the plant is approximately 1.5 m (5 ft.) tall and has a diameter of 10 cm (4 in.). If harvested at a later date, the center cord develops a "woody" taste as the plant turns into a tree. The center cord is attached to a slightly more fibrous cylindrical base with a larger diameter. The entire cylindrical center cord and the attached base are edible, but the center cord is considered more of a delicacy because of its lower fiber content. In the marketplace, palm hearts compete with several comparable gourmet items, including asparagus, bamboo shoots, artichoke, and certain tubers, but they have their unique appeal.

CULINARY PORTRAIT

Commercial hearts of palm are slender, ivory-colored, delicately flavored, and expensive. They resemble white asparagus, without the tips. Their texture is firm and smooth, and the taste has been compared with olives, asparagus, artichoke, and celery. (Hearts from *Euterpe* palms have a relatively biting (acrid) taste, those from the peach palm have a rather neutral taste, and those from the coconut have a sweet taste, favored in mixed vegetable sautées and curries.) Each stalk is approximately 10 cm (4 in.) long and ranges in diameter from pencil thin to 2.5 to 3.8 cm (1–1½ in.). Canned hearts of palm are packed in water, often with citric acid, and can be found in gourmet markets and many large supermarkets. Once opened, the hearts should be transferred to a nonmetal container with an airtight cover. They can be refrigerated in their own liquid for up to a week. Hearts of palm can be used raw in salads and in main dishes, or deep-fried. They can be boiled and mixed with vegetables and eggs to make a casserole. Central and South Americans like their hearts of palm very soft, and the processors cook the product longer to obtain the softer texture. The majority of Americans like their hearts of palm more crunchy, as they are when newly cut in the field.

The peach palm fruit is consumed several ways by Amerindians: the cooked fruit is used as a vegetable, flour is extracted for bread and confectionaries, a nutritious, slightly alcoholic gruel known as chicha (a word used generally to refer to alcoholic drinks prepared from fermented starchy foods) is prepared, and vegetable oil is extracted. Peach palm fruit is not palatable raw and is usually cooked. Heat is necessary to deactivate two toxic factors, an oxalic acid and a trypsin inhibitor. (Oxalates may interfere with calcium metabolism in the body, especially in a calcium-poor diet. Trypsin inhibitors are proteins, widely distributed in the plant kingdom and in plant foods, that inhibit the ability of digestive enzymes to break down protein, and so can be detrimental.) The fruits have the qualities of a starchy cereal or root crop. Boiled or roasted, they have the texture of new potatoes and the taste of chestnuts. The fruit has been used to make a good poultry stuffing as well as pleasant-tasting nonalcoholic beverages and fermented drinks. Canned fruit is marketed in Central America. The seed kernel is edible.

Culinary Vocabulary

- Palm hearts are often called "cabbages" or "palm cabbages," especially in Europe (but note that the expression "palm tree cabbage" is used for some kales (certain *Brassica* species). The hearts are also occasionally called "palmettos" and "palmitos," especially in the New World.
- *Sopa de palmito e de leite de coco* (pronounced soh-pah duh pahl-mee-two ee duh lay-tuh duh koh-koh) is a cream soup prepared in Brazil from hearts of palm and coconut milk.

CURIOSITIES OF SCIENCE AND TECHNOLOGY

- In its native area, peach palm has been used as a folk remedy for headache and stomach-ache and to expel worms.
- Native Amerindians fashioned sewing needles out of the long spines of peach palm.
- Lowland Indian tribes of South America used the tough, strong, elastic wood of peach palm and some other species of *Bactris* for various weapons and hunting implements, including spears, bows, arrows, blowgun darts, fishing poles, and harpoons.
- The sap of some palm species is rich in sugar, and the trees are tapped to produce "palm sugar," which is often fermented to produce a mildly alcoholic (up to 8%) beverage called toddy, and a potent spirit called arrack. Occasionally, peach palm trunks are tapped, like the sugar maple, to produce a sweet sap that is fermented into a wine. (For additional information, see Chapter 72.)
- Strong alcoholic beverages are sometimes made from the fruit of the peach palm, and these are prohibited in several Central and South American areas (except often for Amerindians for whom the drinks are traditional).
- Peach palm is being promoted as a crop alternative to illegal coca production for cocaine (from *Erythroxylum coca* Lam.) in Bolivia and Peru.
- A fascinating relationship exists between the peach palm and a very small beetle, *Andranthobius palmarum* in Central America. The peach palm tree has both male and

female flowers. Females of the beetle are attracted to the petals of the male flowers by chemical secretions, feed on them, and lay their eggs in the petals. The insect's eggs develop into larvae within the male flowers after they have fallen to the ground, and the larvae migrate to the soil where they pupate. In only 11 days, adults emerge and transfer pollen from tree to tree. This cooperative relationship ensures that pollinators are available to the plant, and food is available to the insect.

- A very few species of plants are capable of producing heat, and in most cases the heat is produced by the flowers—the situation in the peach palm. Sometimes the heat serves to evaporate floral chemicals producing odor that attracts insect pollinators, and this may be the purpose of the heat produced by the peach palm (see above relationship with beetles).
- Because the trunks are generally covered with spines, Amerindians generally collect fruits of the peach palm by cutting off the fruit clusters with a long bamboo pole tipped with a hooked blade and catching the fruit with a net or blanket.
- As noted earlier, peach palm fruits contain toxic factors, and so the fruits must be cooked. It has been observed in Central and South America that most wild animals instinctively do not consume fallen fruits until they have rotted somewhat, which also deactivates the toxic components.

KEY INFORMATION SOURCES

Adin, A., Weber, J.C., Sotelo Montes, C., Vidaurre, H., Vosman, B., and Smulders, M.J.M. 2004. Genetic differentiation and trade among populations of peach palm (*Bactris gassipaes* Kunth) in the Peruvian Amazon—implications for genetic resource management. *Theor. Appl. Genet.* 108:1564–1573.

Almeyda, N., and Martin, F.W. 1980. *The pejibaye*. Agricultural Research (Southern Region), Science and Education Administration, U.S. Department of Agriculture, Mayagüez Institute of Tropical Agriculture (Puerto Rico), New Orleans, LA. 10 pp.

Arkcoll, D.B., and Aguiar, J.P.L. 1984. Peach palm (*Bactris gasipaes* HBK) a new source of vegetative oil from the wet tropics. *J. Sci. Food Agric.* 35:520–526.

Balick, M.J. 1987. The palm heart as a new commercial crop from Tropical America. *Principes*, 20:24–28.

Beach, J.H. 1984. The reproduction biology of the peach or 'pejibaye' palm (*Bactis gasipaes*) and a wild congener (*B. porschiana*) in the Atlantic lowlands of Costa Rica. *Principes*, 28:107–119.

Blaak, G. 1976. Pejibaye. *Abs. Trop. Agric.* 2(9):9–17.

Clement, C.R. 1988. Domestication of the pejibaye palm (*Bactis gasipaes*): past and present. *Adv.* Econ. Bot. 6:155–174.

Clement, C.R. 1989. The potential use of the pejibaye palm in agroforestry systems. *Agroforestry Systems*, 7:201–212.

Clement, C.R. 1990. Pejibaye. In *Fruits of tropical and subtropical origin: composition, properties and uses*. Edited by S. Nagy. Florida Science Source Inc., Lake Alfred, FL. pp. 302–321.

Clement, C.R. 1992. Domesticated palms. *Principes*, 36:70–78.

Clement, C.R. 1995. Pejibaye. In *Evolution of crop plants*. 2nd ed. Edited by J. Smartt and N.W. Simmonds. Longman Scientific & Technical, Burnt Mill, Harlow, Essex, U.K. pp. 383–388.

Clement, C.R., and Arkcoll, D.B. 1989. The pejibaye palm: economic potential and research priorities. In *New crops for food and industry*. Edited by G.N. Wickens, N. Haq, and P. Day. Chapman & Hall, New York. pp. 304–322.

Clement, C.R., and Arkcoll, D.B. 1991. The pejibaye (*Bactris gasipaes* H.B.K., Palmae) as an oil crop: potential and breeding strategy. *Oleagineux*, 46:293–299.

Clement, C.R., and Mora Urpí, J. 1987. The pejibaye (*Bactris gasipaes* H.B.K., Arecaceae): multi-use potential for the lowland humid tropics. *Econ. Bot.* 41:302–311.

Clement, C.R., Manshardt, R.M., DeFrank, J., Zee, F., and Ito, P. 1993. Introduction and evaluation of pejibaye (*Bactris gasipaes*) for palm heart production in Hawaii. In *New crops*. Edited by J. Janick and J.E. Simon. Wiley, New York. pp. 465–472.

Clement, C.R., Manshardt, R.M., Cavaletto, C.G., DeFrank J., Mood, J., Jr., Nagai, N.Y., et al. 1996. Pejibaye heart-of-palm in Hawaii: from introduction to market. In *Progress in new crops*. Edited by J. Janick. ASHS Press, Arlington, VA. pp. 500–507.

Clement, C.R., Weber, J.C., van Leeuwen, J., Domian, C.A., Cole D.M., Arévalo Lopez, L.A., et al. 2004. Why extensive research and development did not promote use of peach palm fruits in Latin America. *Agroforestry Systems*, 61:195–206.

Deenik, J., Ares, A., and Yost, R.S. 2000. Fertilization response and nutrient diagnosis in peach palm (*Bactris gasipaes*): a review. *Nutr. Cycl. Agroecosyst.* 56:195–207.

Ferreira, E. 1999. The phylogeny of pupunha (*Bactis gassipaes* Kunth, Palmae) and allied species. In *Evolution, variation and classification or palms*. Edited by A. Henderson and F. Borchsenius. Memoirs of the New York Botanical Garden 83. New York Botanical Garden Press, Bronx, NY. pp. 225–236.

Henderson, A. 2000. Bactris *(Palmae)*. Flora Neotropical Monograph 79. New York Botanical Garden, New York. 181 pp.

Johannessen, C.L. 1966. Pejibaye palm: yields, prices and labor costs. *Econ. Bot.* 20:302–315.

Johannessen, C.L. 1966. The domestication process in trees reproduced by seed: the pejibaye palm in Costa Rica. *Geogr. Rev*. 56:363–376.

Johannessen, C.L. 1967. Pejibaye palm: physical and chemical analysis of the fruit. *Econ. Bot.* 21:371–378.

Mogea, J.P., and Verheij, E.W.M. 1991. *Bactris gasipaes* Kunth. In *Plant resources of South-East Asia. Vol. 2. Edible fruits and nuts*. Edited by E.W.M. Verheij and R.E. Coronel. Pudoc, Leiden, the Netherlands. pp. 100–104.

Mora-Urpí, J.E. 1984. *The pejibaye palm (*Bactis gasipaes *H.B.K.)*. Food and Agriculture Organization, Rome, Italy. 16 pp.

Mora-Urpí, J. 1994. Pejibaye palm. (*Bactis gasipaes*). In *Neglected crops. 1492 from a different perspective*. Edited by J.E. Hernández-Bermejo and J. Léon. Food and Agriculture Organization of the United Nations, Rome, Italy. pp. 211–221.

Mora-Urpí, J., and Clement, C.R. 1988. Races and populations of peach palm found in the Amazon basin. In *Final report: peach palm* (Bactris gasipaes H.B.K.) *germplasm bank*. Project report. Edited by C.R. Clement and L. Coradin. U.S. Agency for International Development, INPA-CENARGEN, Manaus, Brazil. pp. 78–94.

Mora-Urpí, J., Weber, J.C., and Clement, C.R. 1997. *Peach palm,* Bactris gasipaes *Kunth*. International Plant Genetic Resources Institute, Rome, Italy. 83 pp.

Morton, J. 1987. Pejibaye. In *Fruits of warm climates*. Creative Resource Systems, Winterville, NC. pp. 12–14.

Patiño, V.M. 1992. An ethnobotanical sketch of the palm *Bactris* (*Guilielma*) *gassipaes*. *Principes*, 36:143–147.

Popenoe, W., and Jimenez, O. 1921. The pejibaye, a neglected food plant of tropical America. *J. Hered.* 12:154–166.

Quast, D.G., and Bernhardt, L.W. 1978. Progress in palmito (heart-of-palm) processing research. *J. Food Prot.* 41:667–674.

Rodrigues, D.P., Astolfi Filho, S., and Clement, C.R. 2004. Molecular marker-mediated validation of morphologically defined landraces of pejibaye (*Bactris gassipaes*) and their phylogenetic relationships. *Genet. Res. Crop Evol*. 51:871–882.

Tabora, P.C., Jr., Balick, M.J., Bovi, M.L.A., and Guerra, M.P. 1993. Hearts of palm (*Bactris*, *Euterpe* and others). In *Pulses and vegetables*. Edited by J.T. Williams. Chapman & Hall, New York. pp. 193–218.

Zapata, A. 1972. Pejibaye palm from the Pacific coast of Columbia (a detailed chemical analysis). *Econ. Bot.* 26:156–159.

Zumbado, A.M., and Murillo, R.M. 1984. Composition and nutritive value of pejibaye (*Bactris gasipaes*) in animal feeds. *Rev. Biol. Trop*. 32:51–56.

Specialty Cookbooks

Note: Palm hearts have been characterized as a "national ingredient" of Brazilian food, and Brazilian cookbooks are an excellent source of recipes.

Bogosian, F.S. 2008. *Delicious Brazil: gastronomy and tourism*. Create Space, Scotts Valley, CA. 150 pp.

Botafogo, D. 1993. *The art of Brazilian cookery*. Hippocrene Books, New York. 240 pp.

Calvo, I.M. 1981. *Usos culinarios del chontaduro*. Fundación Educación Superior, Instituto Vallecaucano de Investigaciones Científicas, Cali, Columbia. 73 pp (in Spanish. Presents 40 recipes for peach palm, mostly for baked products such as bread, cake, and cookies).

Mendelson, A., and Martinez, Z. 2004. *Zarela's Veracruz: Mexico's simplest cuisine*. Houghton Mifflin Harcourt, Boston, MA. 400 pp.

75 Pepino

Family: Solanaceae (potato family)

NAMES

Scientific Name: *Solanum muricatum* Aiton

- The name "pepino" is Spanish for cucumber, the vegetable that early Spanish colonists in the New World considered similar. The Spanish word traces to the Latin *pepō*, melon.
- Pronunciation: peh-PEE-noh or puh-PEE-noh.
- Plural: "pepinos" is much more common than "pepinoes."
- The pepino is also known as melon pear, melon shrub, mellow fruit, native cucumber, pear melon, pepion de aqua, pepino dulce, pepino melon, sweet cucumber, sweet melon, sweet pepino, tree melon, and Peruvian pepino.
- The name "pepino dulce" is Spanish for sweet cucumber (*dulce* means sweet).
- The name pepino is also used for *Cucumis anguria* L., the West Indian gherkin, which is mainly used to produce small pickles.
- The name pepino is also sometimes used in South America for the cassabanana (see Chapter 21), a vine grown in Latin America for its cucumber-like fruit.
- The genus name *Solanum* is derived from the Latin name *solanum* for some species of the genus, possibly *Solanum nigrum* L., the common nightshade, a plant with poisonous potential that is nevertheless used medicinally and for food purposes. It has also been suggested that the genus name is based on the Latin *solamen*, meaning quieting, an allusion to the sedative properties of some of the species.
- *Muricatum* in the scientific name *S. muricatum* means having a surface that is rough, with short, hard points (a word based on the rough shell of the mollusk *Murex*). It might be thought that this was descriptive of the fruit of forms of the fruit that were first collected (modern pepino fruit is smooth), but it seems that the word *muricatum* was chosen by the author of the scientific name, Aiton, as a reference to the rough or warty character of the adventitious roots that develop from the base of the stems.

PLANT PORTRAIT

The pepino is a perennial herbaceous bush (usually cultivated as an annual) with a woody base and fibrous roots. Several stems arise from the base, and branches may bend over and take root. The plant grow 60 to 100 cm (2–3 ft.) high and about as wide. Pepinos are close relative of tomatoes and potatoes, and there are similarities in the leaves and flowers. Like tomato bushes, the plants may need to be staked for support. The species is considered to be native to the Andean regions of Colombia, Peru, and Chile, although it is known only in cultivation. The pepino was domesticated in pre-Hispanic times and is pictured on South American pottery thousands of years old. The Spanish colonizers of South America discouraged Native Indians from continuing to cultivate many of their traditional crops because they preferred their familiar European vegetables and fruits. The pepino was one of the crops that were discouraged. The Spanish prohibited its consumption and gave the fruit the pejorative name *mataserrano*, "highlander killer." The plant was originally cultivated along the Andes, from southern Colombia to Bolivia and the Peruvian coast. It is still widely grown in South America and is common in the markets of Columbia, Ecuador, Peru, Bolivia, and

FIGURE 75.1 Pepino (*Solanum muricatum*). Left: (From Bailey, 1900–1902.) Upper right: (From *Gardeners' Chronicle,* March 14, 1903, p. 160.)

Chile. With European colonization, the pepino was introduced into Mexico and Central America. It was taken to New Zealand in 1973, where attractive varieties have been developed. Since 1984, pepinos have been one of New Zealand's most lucrative fruit exports. Pepino is frost sensitive but can be grown for fruit in certain cool subtropical and temperate climates, including the warmest parts of the United States. The plants will not set fruit until night temperatures are above 18°C (65°F). Seeds produce variable fruit, and plants are typically propagated commercially by cuttings to ensure uniformly good fruit. There are about a dozen cultivated varieties, all of which are grown from cuttings. Commercial crops are produced in Chile, New Zealand, Western Australia, and the United States (primarily California, with some cultivation in Hawaii). Pepinos have been successfully marketed in recent years in Japan, Europe, and South America, often packaged attractively for an upscale clientele.

The fruit found today in Western markets is melon-like, 5 to 20 cm (2–8 in.) in diameter. It has a sweet smell, pleasant flavor, and yellow, golden, white or ivory skin, with jagged purple stripes, and white to golden-yellow or orange flesh. Although pepino fruit range in size from that of a plum to that of a large papaya, typically they are approximately 13 cm (5 in.) long. In South America, a much wider variety of fruit colors is available, with purple, gray, and green stripes and with greenish, yellow, salmon, or nearly clear flesh. Fruit shape is also highly variable in South America, ranging from globose to elongated with pointy ends.

CULINARY PORTRAIT

The pepino is generally consumed out of hand, particularly chilled and eaten fresh much like a cantaloupe or other melon. The small melons that are typically marketed are ideal for one person. The skin, the soft seeds, and the flesh are all edible, but the fruit should be peeled before use (the skin of some varieties has a disagreeable flavor). The seeds are generally discarded (easily done because they occur in a cluster at the center). These melons are suitable for fruit salads, appetizers, and as an accompaniment or garnish for meats and vegetables. Fresh pepinos may be flavored with ginger, lemon or lime juice, or liqueurs. Puréed fruit may be added to ice cream, sorbets, and beverages. Unripe fruit is often cooked like squash. Good quality fruit is moderately sweet, refreshing, and juicy, with a taste and aroma like a combination of cantaloupe and honeydew melon. The taste has also been compared with cucumber. In poor varieties, there may be an unpleasant soapy aftertaste. Pepinos are frequently available from specialty produce markets and supermarkets. The best fruits are fragrant and yield slightly to palm pressure (a ripe pepino is as firm as a partially ripe plum). Pepinos need to be harvested carefully as they bruise easily. Finger mark bruises indicate that the fruit was not harvested carefully enough. Ripe fruit are often wrapped individually for protection. From a marketing viewpoint, the purple stripes on the fruit are advantageous in that they tend to mask bruise marks that show up well on golden-yellow parts of the fruit. Pepinos soften as they ripen. The best flavor is developed when they are picked ripe, but as with many exotic fruit that need to be shipped, there is a tendency to pick them on the green side. A deep yellow or gold color indicates that the fruit is fully ripened. The ripe fruit can be stored up to a week in the refrigerator.

Culinary Vocabulary

- A "pepino de comer" (pronounced puh-pee-noh day koh-mehr) is a small, white-skinned, cucumber-flavored pepino that is eaten raw or pickled in vinegar in South American cuisines.

CURIOSITIES OF SCIENCE AND TECHNOLOGY

- Herdsmen of the Virú and Moche valleys of northern coastal Peru take pepinos in knapsacks for eating but also for hydration during long treks through deserts. The pepino is more than 90% water.
- Goiter (enlarged thyroid, visible in the front of the neck) is due to disturbance in iodine balance (typically insufficient iodine in the diet). Pepino has a good content of iodine and has been used in South America to treat the condition.
- Most species develop fruits only after their flowers have been fertilized so that seeds are produced, which combine the characteristics of different kinds of plants. However, sometimes seedless "parthenocarpic" fruits develop without fertilization, and this occurs at times in pepino (similarly, chickens have been selected to produce eggs without fertilization by roosters). Although this is of no value to wild plants, in cultivation it makes the plants valuable to humans, who tend to preserve them by vegetative reproduction (as with seedless bananas). As noted above, cultivated varieties of pepino are propagated by vegetative reproduction. However, pollination greatly increases fruit production so that fertilization in the plant is still regarded as necessary.

KEY INFORMATION SOURCES

Ahumada, M., and Cantwell, M. 1996. Postharvest studies on pepino dulce (*Solanum muricatum* Ait.): maturity at harvest and storage behavior. *Postharvest Biol. Technol.* 7:129–136.

Anderson, G.J., Jansen, R.K., and Kim, Y. 1996. The origin and relationships of the pepino, *Solanum muricatum* (Solanaceae): DNA restriction fragment evidence. *Econ. Bot.* 50:369–380.

Blanca, J.M., Prohens, J.A., Gregory, J., Zuriaga, E., Cañizares, J., and Nuez, F. 2007. AFLP and DNA sequence variation in the Andean domesticate pepino (*Solanum muricatum*, Solanaceae): implications for evolution and domestication. *Am. J. Bot.* 94:1219–1229.

Burge, G.K. 1989. Fruit set in the pepino (*Solanum muricatum* Ait.). *Sci. Hortic.* 41:63–68.

Ercan, N., and Akilli, M. 1996. Reasons for parthenocarpy and the effects of various hormone treatments on fruit set in pepino (*Solanum muricatum* Ait.). *Sci. Hortic.* 66:141–147.

Grigg, F.D.W., Smith, P.R., Stenersen, M.A., and Murray, B.G. 1988. Variable pollen fertility and abnormal chromosome behavior in the pepino (*Solanum muricatum* Ait., Solanaceae). *Sci. Hortic.* 35:259–268.

Hammett, K.R.W., Horner, M.B., Boyd, M., and Irwin, J. 1989. *Bibliography of the pepino:* Solanum muricatum. Division of Horticulture and Processing, Department of Scientific and Industrial Research, Auckland, New Zealand. 14 pp.

Heyes, J.A., Blaikie, F.H., Downs, C.G., and Sealey, D.F. 1994. Textural and physiological changes during pepino (*Solanum muricatum* Ait.) ripening. *Sci. Hortic.* 58:1–15.

Kader, A.A. 2000. Produce facts: pepino. *Perishables Handling Quart.* 101(Feb.):15–16.

Martinez-Romero, D., Serrano, M., and Valero, D. 2003. Physiological changes in pepino (*Solanum muricatum* Ait.) fruit stored at chilling and non-chilling temperatures. *Postharvest Biol. Technol.* 30:177–186.

Morley-Bunker, M.J.S. 1983. A new commercial crop, the pepino (*Solanum muricatum* Ait.) and suggestions for further development. *Annu. Rep. R. N. Z. Inst. Hortic.* 11:8–19.

Murray, B.G., Hammett, K.R.W., and Grigg, F.D.W. 1992. Seed set and breeding system in the pepino *Solanum muricatum* Ait., Solanaceae. *Sci. Hortic.* 49:83–92.

Pluda, D., Rabinowitch, H.D., and Kafkafi, U. 1993. Pepino dulce (*Solanum muricatum* Ait.) quality characteristics respond to nitrogen nutrition and salinity. *J. Am. Soc. Hortic. Sci.* 118:86–91.

Prohens, J., and Nuez, F. 1999. Strategies for breeding a new greenhouse crop, the pepino (*Solanum muricatum* Aiton). *Can. J. Plant Sci.* 79:269–275.

Prohens, J., and Nuez, F. 2001. Improvement of mishqui (*Solanum muricatum*) earliness by selection and ethephon applications. *Sci. Hortic.* 87:247–259.

Prohens, J., Ruiz, J.J., and Nuez, F. 1996. The pepino (*Solanum muricatum*, Solanaceae): a "new" crop with a history. *Econ. Bot.* 50:355–368.

Prono-Widayat, H., Schreiner, M., Huyskens-Keil, S., and Luedders, P. 2003. Effect of ripening stage and storage temperature on postharvest quality of pepino (*Solanum muricatum* Ait.). *J. Food Agric. Environ.* 1:35–41.

Redgwell, R.J., and Turner, N.A. 1986. Pepino (*Solanum muricatum*): chemical composition of ripe fruit. *J. Sci. Food Agric.* 37:1217–1222.

Rodriguez-Burruezo, A., Prohens, J., and Nuez, F. 2003. Wild relatives can contribute to the improvement of fruit quality in pepino (*Solanum muricatum*). *Euphytica*, 129:311–318.

Ruiz, J.J., and Nuez, F. 1997. The pepino (*Solanum muricatum* Ait.): an alternative crop for areas affected by moderate salinity. *HortScience*, 32:649–652.

Ruiz, J.J., and Nuez, F. 2000. High temperatures and parthenocarpic fruit set: misunderstandings about the pepino breeding system. *J. Hortic. Sci. Biotech.* 75:161–166.

Ruiz-Bevia, F., Font, A., Garcia, A.N., Blasco, P., and Ruiz, J.J. 2002. Quantitative analysis of the volatile aroma components of pepino fruit by purge-and-trap and gas chromatography. *J. Sci. Food Agric.* 82:1182–1188.

Shiota, H., Young, H., Paterson, V., and Irie, M. 1988. Volatile aroma constituents of pepino fruit. *J. Sci. Food Agric.* 43:343–354.

Vega, I.S. 1994. Andean fruits. In *Neglected crops. 1492 from a different perspective.* Edited by J.E. Hernández-Bermejo and J. Léon. Food and Agriculture Organization of the United Nations, Rome, Italy. pp. 181–191.

Specialty Cookbooks

Luard, E. 2003. *The Latin American kitchen: a book of essential ingredients with over 200 authentic recipes.* Larel Glen, San Diego, CA. 240 pp.

76 Perilla

Family: Lamiaceae (Labiatae; mint family)

NAMES

Scientific Name: *Perilla frutescens* (L.) Britton (*P. arguta* Benth., *P. ocymoides* L. ("*P. ocimoides*"), *Ocimum frutescens* L.)

- The English name perilla and the genus name *Perilla* are based on a native East Indian name.
- Perilla is also called beef-steak plant (beefsteak plant), Chinese basil, mint perilla, perilla mint, purple mint, purple perilla, rattlesnake weed, shiso, summer coleus, suttso, wild coleus, and wild sesame.
- The leaves of perilla are reminiscent of coleus, hence the names summer coleus and wild coleus. The leaves are also similar to those of basil, hence the name Chinese basil.
- *Frutescens* in the scientific name *P. frutescens* is Latin for "shrubby."

PLANT PORTRAIT

Perilla is an erect, white-, purple-, or red-flowered, bushy annual herb growing 0.5 to 2.0 m (20 in. to 6 ½ ft.) in height. The foliage is strongly aromatic, with a cinnamon scent. The leaves often become purplish at maturity. Leaves of some horticultural selections may be quite purplish, bronze-purple, or green mottled with purple, red, pink, or white, and are sometimes wrinkled. Cultivated varieties differ considerably in appearance and chemistry. The species is indigenous from the Himalayas to eastern Asia, including India, Burma, and China. Perilla seems to have first been cultivated in China and has long been raised in Japan, Korea, Vietnam, Burma, India, Iran, and southern Europe. It has been grown to obtain oil from the seeds at least since the thirteenth century in Korea. Perilla has escaped from cultivation in many areas. The herb was introduced as an ornamental to the United States and escaped to become a troublesome weed in much of the southeastern United States. It has been collected as far north as Ontario in Canada and New England in the United States and as far west as eastern Texas, Oklahoma, and Kansas.

Perilla is a multipurpose plant. The seeds are used as food for birds and humans; the seed oil is used as a fuel, drying oil, and cooking oil; the leaves are used as a potherb, for medicine, and for food coloring and to produce an essential oil for flavoring. The plant is also grown as an aromatic ornamental, honey plant, and condiment.

Perilla seeds are pressed to obtain an oil, and this is the main economic use of the plant. This oil, which makes up approximately 40% of the seeds, is commercially produced in Korea and to a lesser extent in Russia, Cyprus, South Africa, and Austria. Perilla seed oil is a drying oil, resembling linseed oil and tung oil, and when dry provides a hard, tough, brilliant film, useful for waterproofing paper. The oil is used in cooking and to manufacture cheaper lacquer varnishes and in printing ink, dyes, paints, enamels, Japanese oil paper, artificial leather, linoleum, and protective clothing.

An aromatic oil is distilled from the leaves and inflorescences in Japan. This herb oil (in contrast to the seed oil) is largely used as a flavoring and sweetening agent in confectionary and in cosmetics. It is used to sweeten tobacco and flavor sauces, chewing gum, and candy. The oil is strongly antiseptic and is also used in toothpaste, mouthwash, and as an antimildew agent. A nonnutritive

FIGURE 76.1 Perilla (*Perilla frutescens*). (From Curtis, 1823, vol. 50, plate 2395.)

sweetener, much sweeter than sucrose, is extracted from perilla in Japan. Perilla extracts are added to fragrances and can be used in food.

Perilla has many medicinal uses in Asian herbal medicine. It is employed as a sedative, to treat spasms, uterine and head problems, and to increase perspiration. In Korea it is used for colds and coughs. In Vietnam it is used against fever and malaria. In Chinese medicine it is employed alone or in combination with other herbs to treat cholera, colds, influenza, stomach problems, nausea, poisoning, asthma, bronchitis, chest pain, rheumatism, and other ailments. In Chinese folk medicine the seeds are used to treat premature ejaculation.

CULINARY PORTRAIT

Perilla is a well-known condiment in southern Asia, particularly in India, Korea, and Japan. According to Larkcom (2007), "Perilla is the quintessential Japanese herb for seasoning and garnishing." It is especially important in *shisho*, the Japanese national dish, and is also used as a food colorant in Japan. A shiso-flavored beverage, Pepsi Shiso, was released by Pepsi Japan in 2009. The aroma of the herb has been described as pleasant, and the flavor as somewhat cinnamon-like. Seeds and leaves are used as food, to give a reddish color to pickles, as a condiment in foods such as tofu, and as a garnish for tempura (battered vegetables such as potatoes or carrots, or seafood deep-fried in diluted sesame oil). Perilla sprouts are used as a garnish by restaurants and hotels in Japan, particularly with sushi and sashimi (sliced raw fish). Young leaves and tender shoots are deep fried in butter. The leaves are an ingredient in mume (pickled Japanese plum) tea. Flowers are used to make an unusual tempura: cut young flower spikes are dipped in tempura batter and deep fried. Flowers are also occasionally added to soups as a seasoning. The seeds and the flowers can be used in herbal teas. Immature flower clusters are used as a garnish for soups and chilled tofu, whereas older ones are fried. The leaves are also used to flavor dishes in India, China, Korea, and Vietnam. In Hawaii, the Japanese use the young leaves as serving beds for deep-fat-fried shrimp. The leaves can be wrapped around shrimp, fish, chicken, or beef and dipped in spicy sauces. Red perilla is manufactured into a dry powder for use as a dye for fruits, vegetables, rice, and cake mixes. Perilla products are generally available in Oriental food stores in North America that cater to people from Korea and Japan. Seasoned leaves in oil, packaged in small flat tins, are commonly marketed in the United States, as are fresh greens in season, seed oil, pickled plums, plum sauce, and other condiments with added perilla.

Perilla is well known as a cause of poisoning in domestic animals. Although cattle avoid it, perilla has been implicated in cattle poisoning. A chemical called perilla ketone appears to be the primary toxin, the amount increasing notably during fruiting. Both seeds and leaves contain the compound, which has been described as a potent lung toxin. Despite this known toxicity, the plant and its oil continue to be widely used in Asia as human food. However, consumption should be limited, and perilla should be avoided during pregnancy.

Culinary Vocabulary

- The Korean name of perilla is *deulkkae* or *tŭlkkae*, which means "wild sesame." Unfortunately, English versions of Korean cookbooks sometimes translate the term as "sesame," resulting in the use of the wrong flavoring agent. The French for perilla, *sésame sauvage*, could result in the same erroneous interpretation.
- *Shichimi togarashi* (pronounced she-Chee-mee toh-gah-RAH-shee) is a Japanese seven-spice seasoning, used to flavor soups, noodle dishes, meat dishes, fish, and stews. It often contains dried chilies, sesame seeds, mustard, dried seaweeds, and perilla.

CURIOSITIES OF SCIENCE AND TECHNOLOGY

- Perilla was brought to the United States at least by1800 by Japanese and Korean immigrants, and it became common in port towns and nearby coasts where immigrants were common. When Japanese Americans were interned in camps throughout the Midwestern and southern states during the Second World War, the herb began to show up in the woods of those regions, a poignant indicator of the clash of cultures.
- Perilla was once used as a fish poison. Despite this, perilla has a folk reputation as an antidote to fish poisoning. It has been speculated that the widespread use of perilla as a condiment in sushi in Japan has often prevented poisoning, and indeed that unconsciously the Japanese have chosen to use perilla in raw fish products for this reason.
- Many people develop dermatitis from touching perilla, and in Japan often more than half of the workers in perilla harvesting operations develop a rash. Ironically, perilla has been found to be very effective for treating dermatitis.
- One form of the chemical perillaldehyde in perilla, used as a sweetening agent in Japan, is 2000 times as sweet as cane sugar and approximately six times sweeter than saccharin (commonly consumed as "Sweet-n-Low"; saccharin is banned or restricted in many countries).
- In the Washington, D.C., area, perilla is cultivated in window boxes and gardens near Oriental restaurants as edible landscaping.

KEY INFORMATION SOURCES

Anjula Pandey, A., and Bhatt, K.C. 2008. Diversity distribution and collection of genetic resources of cultivated and weedy type in *Perilla frutescens* (L.) Britton var. *frutescens* and their uses in Indian Himalaya. *Genet. Resour. Crop. Evol.* 55:883–892.

Aritomi, M. 1982. Chemical studies on the constituents of edible plants (part 1). Phenolic compounds in leaves of *Perilla frutescens* Britton var. *acuta* Kudo f. *viridis* Makino. *J. Home Econ. Jpn.* 33:353–359.

Aritomi, M., Kumori, T., and Kawasaki, T. 1985. Cyanogenic glycosides in leaves of *Perilla frutescens* var. *acuta*. *Phytochemistry*, 24:2438–2439.

Brenner, D.M. 1993. Perilla: botany, uses and genetic resources. In *New crops—exploration, research, and commercialization*. Edited by J. Janick, and J.E. Simon. J. Wiley and Sons, New York. pp. 322–328.

Fujita, T., Funayoshi, A., and Nakayama, M. 1994. A phenylpropanoid glucoside from *Perilla frutescens*. *Phytochemistry*, 37:543–546.

Guzman, C.C. de, and Siemonsma, J.S. 1999. *Perilla frutescens* (L.) Britton. In *Plant resources of South-East Asia. Vol. 13. Spices*. Edited by C.C. de Guzman and J.S. Siemonsma. Backhuys, Leiden, the Netherlands. pp. 166–170.

Ito, M., and Honda, G. 1996. A taxonomic study of Japanese wild *Perilla* (Labiatae). *J. Phytogeogr. Taxon*. 44:43–52.

Kerr, L. A., Johnson, B. J., and Burrows, G. E. 1986. Intoxication of cattle by *Perilla frutescens* (purple mint). *Vet. Hum. Toxicol*. 28:412–416.

Kimura, M., Ishii, M., Yoshimi, M., Ichimura, M., and Tomitaka, Y. 2000. Essential oils and glandular trichomes of perilla seedlings as affected by light intensity. *Acta Hortic*. 515:219–225.

Larkcom, J. 2007. *Oriental vegetables: the complete guide for the gardening cook*. Revised edition. Kodansha, New York. 232 pp.

Lee, J.-I., Han, E.-D., Park, H.-W., and Bang, J.-K. 1989. Review of the research results on *Perilla* and its prospects in Korea. In *Proceedings of the National Symposium on Oil Crop Production and Its Utilization*. Crop Experiment Station, Rural Development Administration, Suwon, Korea. pp. 40–52 (in Korean, English summary).

Lee, J.K., and Ohnishi, O. 2001. Geographic variation of flowering response to daylength in *Perilla frutescens* var. *frutescens* in East Asia. *Korean J. Crop Sci*. 46:395–400.

Lee, J.K., and Ohnishi, O. 2003. Genetic relationships among cultivated types of *Perilla frutescens* and their weedy types in East Asia revealed by AFLP markers. *Genet. Resour. Crop. Evol*. 50:65–74.

Lee, J.K., Nitta, M., Kim, N.S., Park, C.H., Yoon, K.M., Shin, Y.B., et al. 2002. Genetic diversity of perilla and related weedy types in Korea determined by AFLP analyses. *Crop Sci*. 42:2161–2166.

Longvah, T., and Doesthale, Y.G. 1991. Chemical and nutritional studies on hanshi (*Perilla frutescens*), a traditional oilseed from northeast India. *J. Am. Oil Chem. Soc*. 68:781–784.

Longvah, T., and Doesthale, Y.G. 1998. Effect of dehulling, cooking and roasting on the protein quality of *Perilla frutescens* seed. *Food Chem*. 63:519–523.

Longvah, T., Deosthale, Y.G., and Uday-Kumar, P. 2000. Nutritional and short term toxicological evaluation of perilla seed oil. *Food Chem*. 70:13–16.

Misra, L.N., and Husain, A. 1987. The essential oil of *Perilla ocimoides*: a rich source of rosefuran. *Planta Med*. 53:379–380.

Morton, J.F. 1991. Food, medicinal and industrial uses of perilla, and its ornamental and toxic aspects. In *Progress on terrestrial and marine natural products of medicinal and biological interest*. Edited by J.M. Pezzuto et al. Proceedings of a symposium, Department of Pharmacy, Medicinal Chemistry and Pharmacognosy, University of Illinois, Chicago, IL. pp. 34–38.

Nguyễñ-Xuân-Duñg, La-Dinh-Moi, Lu'u-Dam-Cu, and Leclercq, P.A. 1995. Essential oil constituents form the aerial parts of *Perilla frutescens* (L.) Britton. *J. Essent. Oil Res*. 7:429–432.

Nitta, M., Lee, J.K., and Ohnishi, O. 2003. Asian *Perilla* crops and their weedy forms: their cultivation, utilization and genetic relationships. *Econ. Bot*. 57:245–253.

Parke, M. 1984. The pleasures of purple mint. *Horticulture*, 62(8):14–17.

Preston, K.A. 1999. Can plasticity compensate for architectural constraints on reproduction? Patterns of seed production and carbohydrate translocation in *Perilla frutescens*. *J. Ecol*. 87:697–712.

Publications and Information Directorate. 1966. Perilla. In *The wealth of India; a dictionary of Indian raw materials and industrial products, Vol. 7*. Publications and Information Directorate, Council of Scientific & Industrial Research, New Delhi, India. pp. 311–313.

Standal, B.R., Ako, H., and Standal, G.S.S. 1985. Nutrient content of tribal foods from India: *Flemingia vestita* and *Perilla frutescens*. *J. Plant Foods*, 6:147–153.

Tominaga, T., and Nitta, M. 1998. Variation in seed germination of cultivated and weedy perilla (*Perilla frutescens* var. *frutescens*). *Weed Res*. 43:43–48.

Vadivel, B., Govindarasu, P., and Sampath, V. 1981. Effect of seed treatments with gibberellic acid on germination of *Perilla frutescens* Brit. *Indian Perfumer*, 25(2):4–5.

Yu, H.C., Kosuna, K., and Haga, M. (*Eds*). 1997. *Perilla—the genus Perilla. Medicinal and aromatic plants—industrial profiles*. Harwood Academic Publishers, Amsterdam, the Netherlands. 191 pp.

Yuba, A., Honda, G., Koezuka, Y., and Tabata, M. 1995. Genetic analysis of essential oil variants in *Perilla frutescens*. *Biochem. Genet*. 33:341–348.

Specialty Cookbooks

Corson, T. 2008. *The zen of fish. The story of sushi from Samuria to supermarket.* HarperCollins, New York. 372 pp.
Dekura, H., Treloar, B., Yoshii, R. 2004. *The complete book of sushi.* Periplus, Singapore. 240 pp.
Fujii, M. 2005. *The enlightened kitchen: fresh vegetable dishes from the temples of Japan.* Kodansha International, Tokyo, Japan. 107 pp.
Hisamatsu, I. 2005. *Quick & easy tsukemono: Japanese pickling recipes.* JOIE Inc., Tokyo, Japan. 104 pp.
Price, D.C., and Kiwano, M. 2004. *Authentic recipes from Korea.* Periplus, Singapore. 112 pp.

77 Persimmon

Family: Ebenaceae (ebony family)

NAMES

Scientific Names: *Diospyros* Species

- Oriental (Japanese) persimmon—*D. kaki* Thunb.
- American persimmon—*D. virginiana* L.
- Date plum—*D. lotus* L.

- The English word "persimmon" is of American Indian derivation, specifically of Algonquian origin, approximating the Cree *pasimian*, meaning dried fruit. American Indians did in fact dry the fruit of the American persimmon for future consumption.
- The Oriental persimmon is also known in English as Chinese date plum, Japanese persimmon, and khaki. In some countries, including Spain and Italy, the name is *kaki*. Most of the time the word "persimmon" is used without a preceding adjective, it denotes *D. kaki*.
- The American persimmon is also known as possumwood. Obsolete names include Indian date and Virginian date plum.
- The date plum is also known as date plum persimmon. The fruit is the size of a date and has a flavor like the date but is more like a cherry than a typical market plum in size.
- The "black persimmon," better known as the black sapote (*D. digyna* Jacq.; see Chapter 86), is a Mexican fruit tree related to the persimmon. The skin and flesh of black sapote turns brownish-black when the fruit is soft and ripe. The fruit is eaten fresh and used in breads, preserves, and desserts, such as ice cream and mousse.
- Israelis created the name "Sharon fruit" for persimmons that they grew on the plains of Sharon in Israel. When the fruit was exported to England, this name proved unwelcome as it was reminiscent of the 1980s British slang term "Sharon" for "a tarty, lower-class woman."
- The genus name *Diospyros* is derived from the Greek *Dios*, for the god Zeus (Jupiter) + *pyros*, wheat or grain, alluding to the edible fruit, indicating that the persimmon is the "fruit of the gods."
- *Kaki* in the scientific name *D. kaki* is a Japanese name for persimmon.
- *Virginiana* in the scientific name *D. virginiana* means "Virginian" and is a part of many plant names. However, the "Virginia" in question is not the modern states of Virginia or West Virginia, but the territory of Virginia, granted by the English King James I (1566–1625), in three charters, the first of which was issued in 1606. The region included depends on the charter that was issued but in any event included a very large region of the eastern United States.
- *Lotus* in the scientific name *D. lotus* traces to the Latin *lotus*, which in turn comes from the Greek *lotos*, referring to plants of unknown identification. Presumably when the Swedish botanist Linnaeus coined the name for the species, he had in mind the food use associated with some species that had been called lotus. The word lotus discussed here is certainly not a reference to the genus *Lotus*, which includes bird's-foot trefoil, a temperate-region forage plant, nor is it several water lilies of the genera *Nelumbo* and *Nymphaea*, which are referred to as lotus.

FIGURE 77.1 Oriental persimmon (*Diospryos khaki*). Left: (From Nicholson, 1885–1889.) Right: (From Henderson, 1890.)

FIGURE 77.2 American persimmon (*Diospryos virginiana*). (From Michaux, 1850, vol. 2, plate 93.)

PLANT PORTRAIT

Of the approximately 400 species of *Diospryos*, several are cultivated for their edible fruit, called persimmons or date plums. The black heartwood of several species is the "ebony" of commerce. The species described in this chapter produce trees that are either male or female. Of course, females are needed for fruit production, but males are frequently not necessary, especially for the Oriental persimmon.

The Oriental persimmon is the main persimmon of interest. It is a deciduous, long-lived tree, 4.5 to 18 m (15–60 ft.) high, native to Japan, China, Burma, and northern India. Oriental persimmons have been consumed for at least 2000 years and are so popular that they have affectionately been called "the apple of the Orient." The tomato-like or plum-like fruit has thin, smooth, glossy,

FIGURE 77.3 Date plum (*Diospryos lotus*). (From Pallas, 1784–1831.)

yellow, orange, red, or brownish-red skin and yellow, orange, or dark-brown, juicy, gelatinous flesh. The fruit is seedless or contains four to eight flat, brown seeds approximately 2 cm (3/4 in.) long. Generally, the flesh is bitter and astringent until fully ripe, when it becomes soft, sweet, and pleasant, but dark-fleshed types may be nonastringent, crisp, sweet, and edible even before fully ripe. There are thousands of cultivars. The Oriental persimmon is grown predominantly in subtropical and warm temperate regions. It has been described as the most popular fruit in the world in terms of the quantity consumed, with more than 1 million metric tons produced annually. The major producers are China (with more than 70% of world production in recent years), Japan, Brazil, Korea, and Italy. Minor producers include Israel, the United States, New Zealand, Australia, Spain, Georgia (the country established in 2009), Egypt, India, and Chile. In the United States, the tree is best adapted to central and southern California, Arizona, Texas, Louisiana, Mississippi, Georgia, Alabama, southeastern Virginia, and northern Florida. Most commercial production of persimmons in the United States is in California.

The American persimmon is a tree, typically 6 to 20 m (20–65 ft.) tall, found throughout southeastern North America. American Indians ate the fruits fresh or dried them like prunes. Early North American settlers made a tea from the fruits, which was used as a gargle for sore throats and to treat heartburn, diarrhea, and upset stomach. Persimmon beer was popular in nineteenth century America. The plum-like fruit is variable in size, 1.3 to 5 cm (1/2–2 in.), usually orange, ranging to black, and usually with a heavy waxy bloom on the skin. Unripe fruits are astringent and inedible, but they become very sweet and pleasantly flavored when fully ripe in the fall. The better varieties have larger, tastier fruit. The fruits should be allowed to ripen on the trees because they do not ripen well when picked early. The heartwood of the tree is nearly black, and the wood is extraordinarily strong and heavy. The American persimmon is cultivated commercially to a small extent.

The date plum is sometimes cultivated for its edible fruit in Italy, west Asia, China, and Japan. The fruit is the size of a date, yellow or yellow-brown, turning black or blue-black when ripe. It is consumed both fresh and dried.

CULINARY PORTRAIT

Oriental persimmons are mostly eaten fresh but are also canned and made into sauces, jellies, and jams. The texture and the flavor of the ripe fruit have been compared with that of apricots. The flesh can be puréed, with a few drops of lemon juice to prevent discoloration. Persimmons or their purée make a delicious addition to ice cream, custards, sorbets, puddings, cookies, cakes, pies, and the like. Persimmon goes well with rice, cheese, seafood, and poultry. In Japan, sake is sometimes added to the fresh fruit. Another way of making the taste more interesting is to squeeze lemon or lime juice over the fruit. Some cultivars are suitable for making dried fruit, reminiscent of dried figs.

In North American supermarkets, most Oriental persimmons come in two varieties: Fuyu and Hachiya. Ripe Fuyu persimmons look like flattened tomatoes and are crisp, whereas the acorn-shaped Hachiya is very soft and juicy. The best persimmons are round and plump, with deep red undertones and a glossy, smooth skin. Fruits with blemishes, bruises, or cracked skin and those that are missing the green leaves at the top should be avoided. Ripe persimmons should be purchased only if they will be eaten immediately (because the fruit ripens quickly, it is most often sold green to ripen at home). Otherwise, firmer fruits should be obtained and allowed to ripen. Unripe Hachiya persimmons taste very bitter and are so astringent that they have been described as capable of sucking all the moisture from one's mouth. The astringency will go away as the fruits ripen. Persimmons can be ripened at room temperature in a paper bag. Ripe Hachiya fruit are very soft to the touch, feel heavy, and have a juicy soft pulp. They should be stored in a refrigerator when ripe. Fuyu persimmons should be washed, the core and leaves removed, and sliced or eaten whole, like an apple. Hachiya persimmons should be washed and sliced in half. The seeds (which are inedible) should be removed, and the flesh spooned out of the skin.

American persimmons are very rarely found in the marketplace and usually need to be collected from the wild. Their pulp makes delicious nut bread, puddings, and pies and is occasionally made into vinegar, molasses, and beer by home cooks. In southern Indiana, where American persimmon trees are common, restaurants and home cooks may feature such delicacies as persimmon bread and rolls, and persimmon pudding.

"Bezoars" are rounded, layered stones found in internal organs such as the stomach, gall bladder, and kidneys of large herbivorous mammals, primarily sheep and goats. A bezoar is most often formed from an undigested mass of ingested material that remains in the stomach. Such stones can obstruct the passage of food into the intestine, and surgery is usually necessary to remove them. Bezoars retrieved from the stomachs of animals were once believed to be magical antidotes against poison (the word bezoar comes from the Persian for "protection against poison"). One of the crown jewels of England is a bezoar set in gold. Bezoars of vegetable origin are most frequent in areas where persimmons are popular. So common is the persimmon as a cause of the formation of bezoars that the term "diospryobezoar" was once proposed for persimmon bezoars. Persimmons seem extraordinarily able to cause bezoars because their tannins, when liberated in the stomach, can consolidate vegetable material into a mass, especially when the stomach is empty. Fortunately bezoars are relatively rare, although people who have had stomach surgery or suffer from slow stomach action seem prone to developing them. Because of the possibility of developing bezoars, it has been suggested that gorging on persimmons is not a good idea, especially on an empty stomach.

Culinary Vocabulary

- A "saketini" is often defined as a martini made with sake instead of vodka (to the horror of martini purists), but the term is also used simply to indicate a cocktail made with sake and any liquor. Given that sake and the Oriental persimmon are both Japanese specialties, it should come as no surprise that "persimmon saketinis" (flavored with persimmon) are encountered in some Japanese and specialty restaurants.

CURIOSITIES OF SCIENCE AND TECHNOLOGY

- The "Victorian language of flowers" was a secret coded language in Victorian times in England, with flowers and plants symbolic of certain messages, so when the flower or plant was mentioned in a letter, those who knew the code could understand the hidden information. "Persimmon" meant "bury me amid nature's beauties."

- In American folklore, persimmon seeds were cut in half to predict weather in the winter. If the resulting seeds were spoon- or shovel-shaped, it foretold of a hard winter with deep snow.
- The town of Mitchell in southern Indiana is the home of what has been claimed to be the world's only persimmon-pulp canning factory, known as Dymple's Delight (named for Dymple Green).
- Astringency of many fruits is due to "phenolic" chemicals called tannins. Tannins are useful for "tanning" animal hides because these chemicals have the ability to bind protein molecules together. The dry, puckery sensation caused by tannins in the mouth results from the compounds cross-linking the proteins on the surface of the tongue and palate and in the saliva, causing the surfaces to constrict and their lubrication to fail. Persimmons have considerable tannin, but tannins are also present in peaches, bananas, dates, and other fruits when they are immature, presumable discouraging animals from eating them until the seeds are mature, at which point the fruit becomes ripe and loses its astringency.
- The Japanese use the astringent tannins of persimmons to clarify sake (rice wine). This is done by adding the juice of unripe persimmons to the wine, resulting in the precipitation of any suspended protein.
- Shortages during the American civil war resulted in Southerners using the large seeds of American persimmon as buttons.
- An indication that food plants have high status is when colors are named for them. "Persimmon" as a color is said to closely resemble the color of a very ripe persimmon fruit and has been described as a medium orange-red. However, as pointed out earlier, the fruit color is somewhat variable.
- The wood of the American persimmon was once famous for its use in golf clubs. Today, persimmon clubs are still available at premium prices, but synthetic materials are commonly used instead.

KEY INFORMATION SOURCES

Ames, G.K. 1999. *Persimmon production*. Appropriate Technology Transfer for Rural Areas (Organization), Fayetteville, AR. Irregularly paginated.

Bellini, E., and Giordani, E. (Eds.). 2001. *Proceedings First Mediterranean Symposium on Persimmon, Faenza, Italy, November 23–24, 2001*. Centre International de Hautes Études Agronomiques Méditerranéennes, Zaragoza, Spain. 126 pp.

Benharroch, D., Krugliak, P., Porath, A., Zurgil, E., and Niv, Y. 1993. Pathogenetic aspects of persimmon bezoars, a case-control retrospective study. *J. Clin. Gastroenterol.* 17:149–152.

Collins, R. (Ed.). 2003. *Proceedings of the 2nd International Persimmon Symposium, September 10–13, 2000, Queensland, Australia*. International Society for Horticultural Science, Leuven, Belgium. 244 pp.

Condit, I.J. 1919. The kaki or Oriental persimmon. *Calif. Agric. Exp. Stn. Bull.* 316:231–266.

English, J. 1982. A practical review of persimmons. *Annu. Rep. Northern Nut Grow. Assoc.* 73:36–38.

Fletcher, W.F., and Gould, H.P. 1942. *The native persimmon*. Revised edition. U.S. Department of Agriculture, Washington, DC. 22 pp.

Gazit, S., and Levy, Y. 1963. Astringency and its removal in persimmons. *Israel J. Agric. Res.* 13(3):125–132.

George, A.P., and Nissen, R.J. 1985. The persimmon as a subtropical fruit crop. *Qld. Agric. J.* 111(3):133–140.

George, A.P., Mowat, A.D., Collins, R.J., and Morley-Bunker, M. 1997. The pattern and control of reproductive development in non-astringent persimmon (*Diospyros kaki* L.): a review. *Sci. Hortic.* 70:93–122.

Goodell, E. 1982. Two promising fruit plants [*Diospryos virginiana* and *Actinidia arguta*] for northern landscapes. *Arnoldia*, 103–133.

Griffith, E., Griffith, M.E., and McDaniel, J.C. 1982. *Persimmons for everyone*. North American Fruit Growers, Place of publication not stated, distributed by Dorothy Nichols, Arcola, MO. 145 pp.

Ikegami, T. 1967. Morphological studies on the origin of *Diospryos kaki* in Japan. *Mem. Osaka Kyoiku Univ.* 16(2):55–58.
Itoo, S. 1980. Persimmon. In *Tropical and subtropical fruits*. Edited by S. Nagy and P.E. Shaw. AVI Publishing, Westport, CT. pp. 442–468.
Itoo, S. 1986. Persimmon. In *CRC handbook of fruit set and development*. Edited by S.P. Monselise. CRC Press, Boca Raton, FL. pp. 355–370.
Kim, T.C., and Ko, K.C. 1997. Classification of persimmon (*Diospyros kaki* Thunb.) cultivars on the base of horticultural traits. *Acta Hortic*. 436:77–83.
Kitagawa, H., and Glucina, P.G. 1984. *Persimmon culture in New Zealand*. Science Information Publication Centre, Wellington, N.Z. 74 pp.
LaRue, J.H., Opitz, K.W., and Beutel, J.A. 1982. *Growing persimmons*. Cooperative Extension, University of California, Berkeley, CA. 11 pp.
Maisenhelder, L.C. 1971. *Common persimmon* (Diospyros virginiana). U.S. Department of Agriculture, Forest Service, Washington, DC. 6 pp.
Morton, J. 1987. Japanese persimmon. In *Fruits of warm climates*. Creative Resource Systems, Winterville, NC. pp. 411–416.
Ng, F.S.P. 1978. *Diospryos roxburghii* and the origin of *Diospryos kaki*. *Malaysian Forester*. 41:43–50.
Ng, F.S.P. 1991. *Diospryos kaki* L.f. In *Plant resources of South-East Asia. Vol. 2. Edible fruits and nuts*. Edited by E.W.M. Verheij and R.E. Coronel. Pudoc, Leiden, the Netherlands. pp. 154–157.
Spongberg, S.A. 1979. Notes on persimmons, kakis, date plums, and chapotes. *Arnoldia*. 39:290–310.
Subhadrabandhu, S. (Ed.). 1997. *First International Persimmon Symposium, Chiang Mai City, Thailand, July 17–19, 1996*. International Society for Horticultural Science, Leuven, Belgium. 416 pp.
Sweedman, R. 1989. *Persimmon growing*. 2nd ed. N.S.W. Agriculture & Fisheries, Sydney, Australia. 12 pp.
Whitsom, M. 2007. The pleasing persimmon. *The Lady-Slipper (Kentucky Native Plant Society)*, 22(1):2–4.
Yonemori, K., and Kitajima, A. 2008. *Diospyros kaki*, persimmon. In *The encyclopedia of fruit & nuts*. Edited by J. Janick and R.E. Paull. CABI, Wallingford, Oxfordshire, U.K. pp. 326–336.
Yonemori, K., Sugiura, A., and Yamada, M. 2000. Persimmon genetics and breeding. *Plant Breed. Rev.* 19:191–225.

Specialty Cookbooks

Note: Euell Gibbon's classic *Stalking the Wild Asparagus* (1962) is a source of recipes for the American persimmon.

Bear Wallow Books. 1978. *Old fashioned persimmon recipes*. Bear Wallow Books, Nashville, IN. 32 pp.
California Fruit Exchange. 1964. *Fresh persimmon recipes*. Blue Anchor, 40(4):21–22.
Filut, M.A. 1985. *People pleasing persimmon recipes*. Circulation Service, Leawood, KS. 38 pp.
Francis, W.O. 1959. *Persimmon pudding*. Ohio Valley Folk Research Project, Ross County Historical Society, Chillicothe, OH. 2 pp.
Green, D. 1996. *Persimmon goodies*. Dymple's Delight Persimmon Products. Mitchell, IN. 23 pp.
Greene, M. 1983. *Persimmon recipe book*. Ozarks Mountaineer, Branson, MO. 32 pp.
Hazelton, J.W. 1999. *Persimmons (kaki) from seed to supper: for good growing, good eating and good health*. Jack's Bookshelf, Inc., St. Petersburg, FL. 56 pp.
Lum, L.A. 1992. *The French-Icarian persimmon tree cookbook*. Pollard Pr., Jacksonville, FL. 189 pp.

78 Pomegranate

Family: Punicaceae (pomegranate family)

NAMES

Scientific Name: *Punica granatum* L.

- The word "pomegranate" is from the Latin *pomum granatum*, meaning "apple of many seeds," an apt description. The Latin *granatum*, "(many) grained" or "(many) seeds," is also in the scientific name *P. granatum* and additionally turns up in grenadine, the name for the thick, sweet syrup made from pomegranate.
- The genus name *Punica* was the Roman name for the ancient North African city-state Carthage, from where the best pomegranates came to Italy. The Romans called pomegranates "Carthaginian apples." *Punicus* strictly refers to Phoenicia in Asia Minor but in ancient Rome was more frequently used to mean Carthage.

PLANT PORTRAIT

The pomegranate is an attractive shrub or small tree, 6 to 10 m (20–33 ft.) high, much-branched, more or less spiny, and extremely long-lived (specimens grown at Versailles, France have survived for two centuries). The plant has a strong tendency to sucker from the base. The leaves are deciduous, or occasionally evergreen in hot climates. The fruit is about the size of an orange (typically about 7.5 cm or 3 in. in diameter) and has a thin, hard, red (occasionally yellowish) skin. The protruding blossom end of the fruit gives it the appearance of a dullish Christmas tree ornament. The fruit is internally segmented into irregular compartments separated by tough membranes. Inside are hundreds of small seeds, each surrounded by juicy translucent red pulp (sometimes colorless, sometimes purple-red or pink). The seeds and their pulp are the edible portion of the fruit. The seeds represent about half of the weight of the fruit. There are many varieties, of which 'Wonderful', which originated in Florida, is the major cultivar of California and Israel and the kind most likely to be found in North American supermarkets. In Japan, pomegranate trees are commonly used for bonsai because of their attractive flowers and the unusual twisted bark developed by older specimens. Dwarf forms (often called "Nana") are also available for those wishing to grow plants in containers.

The pomegranate tree is native from Iran to the Himalayas in northern India and has been cultivated since ancient times throughout the Mediterranean region of Asia, Africa, and Europe. Pomegranates may have been cultivated in Persia (Iran) for the last 5000 years. The fruit was featured in Egyptian mythology and art, praised in the Old Testament of the Bible and in the Babylonian Talmud, and carried by desert caravans for the sake of its thirst-quenching juice. The most important growing regions are Egypt, China, Afghanistan, Pakistan, Bangladesh, Iran, Iraq, India, Burma, and Saudi Arabia. Pomegranates are also raised in tropical Africa and Central and South America. There is limited cultivation in the dry areas of California and Arizona.

The pomegranate is adapted to mild-temperate and subtropical regions and can be grown outdoors as far north as Utah and Washington, DC, although it does not fruit often in the latter locations. The plants are sometimes grown as hedges. Pomegranates are popular southern garden plants, and dwarf, ornamental varieties, some with golf-ball–sized fruit, are also grown in tubs in northern areas so that they can be brought indoors overwinter. Pomegranate flowers are typically vermillion, but ornamental varieties may have white flowers, and double-flowered forms also have been selected.

FIGURE 78.1 Pomegranate (*Punica granatum*) flowering branch. (From Engler and Prantl, 1889–1915.)

CULINARY PORTRAIT

Eating a pomegranate requires time and patience, often lacking in modern times, explaining why the fruit has lost some of its appeal as a fresh fruit. Pomegranates can be consumed fresh by deeply scoring them several times vertically and breaking them apart to reveal the clusters of juice sacs, which can be lifted out of the rind and eaten. The leathery rind and the internal membranes separating the sections are inedible. Many people consume only the juicy, refreshing, tangy-sweet flesh surrounding the pips (seeds) and discard the latter. Pomegranates have been very important in Iranian cuisine for thousands of years. The fruit can be used to enhance salads, soups, sauces, vegetables, cheeses, poultry, and seafood.

Pomegranate juice is a very popular beverage and is widely made into the syrup known as grenadine for use in alcoholic cocktails as well as nonalcoholic kinds of mixed refreshing beverages. Grenadine is the main ingredient in a Shirley Temple cocktail. The sweet syrup is also used to flavor a wide variety of desserts, including ice cream and sorbets. Store-bought "grenadine" may be merely a confection of red food coloring. For home use, the juice can be extracted from the fruit by reaming the halved fruits on an ordinary orange-juice squeezer (although it might be hard to find one today in the average kitchen). Pomegranate juice can be sipped right out of the fruit. Roll the pomegranate on a flat surface, pressing gently to pop the seeds inside. Poke a hole near the top and insert a straw. Be aware that the rich, scarlet juice will stain. Pomegranate juice is used to flavor and color sauces in the Middle East. In northern India, fresh grenadine pomegranate juice is used to marinate meat, the proteolytic enzymes acting as a meat tenderizer.

Squeezing out pomegranate juice tends to result in tannin exuding from the membranes, resulting in astringency (tannins are mouth-puckering substances that are chiefly responsible for astringency in plant foods). Some of the tannin can be removed by stirring in a little dissolved gelatin, which reacts with the tannin to produce an insoluble compound (a cloud forms that can be filtered away). Allowing the fruit to shrivel before being crushed also reduces tannins in the juice.

In the Middle East, pomegranate seeds are dried and used as a spice, especially for meat dishes, and in Mexico the seeds are also used as a topping or garnish. Seedless varieties are available but are

FIGURE 78.2 Pomegranate (*Punica granatum*). (From Köhler, 1883–1914.) Cross section of fruit at upper right, long section of fruit at lower left.

not popular. Experienced pomegranate fanciers either prefer the crunch of the seeds, which are pleasant in texture, or have learned to suck off the juicy flesh surrounding the seeds and spit them out.

Pomegranates are picked when fully mature, as they will not ripen after harvest. The fruit is equal to the apple in having a long storage life and can be kept in a refrigerator for several weeks.

Nutritionally, the pomegranate has much to offer. A single fruit may contain 40% of an adult's daily requirements for vitamins A, B_3 (niacin), C, and E, and it also supplies folic acid, natural estrogens, and potassium. Pomegranate is considered to be a rich source of antioxidants—chemicals that protect the body against free radicals that result in aging. Because the skin can directly absorb some nutritional chemicals, pomegranate is sometimes included in shampoos and face creams.

Culinary Vocabulary

- A "Bacardi cocktail" is one made with lime juice, sugar, grenadine, and Bacardi light rum. In 1936, the New York Supreme Court ruled that authentic "Bacardi cocktails" had to be made with the rum made by Bacardi Imports, Inc., of Miami, Florida, which held the registered trademark on the brand "Bacardi."
- *Koliva* is a traditional Greek food made from boiled wheat, cinnamon nuts, and pomegranate seeds, pressed into cakelike forms and covered with powdered sugar. This ancient preparation symbolizes the cycle of life, and so is often served at Greek funerals.
- *Dibs rumman* (pronounced deebs room-man) is a Hindi term for pomegranate syrup, used in Indian cuisine.

CURIOSITIES OF SCIENCE AND TECHNOLOGY

- Archaeologist Sir William Cristal discovered the earliest known menu, carved in hieroglyphics on stone tablets, in 1922 while excavating the pyramid that contained the tomb of an Egyptian prince. The menu was for the meal that was presented to celebrate the birth of the Prince's twin sons, one of whom was to later become Ramses III, probably the most powerful and famous of all Egyptian Pharaohs. According to the menu, there were two first courses—garlic in sour cream and barley soup. This was followed by salmon from the Tigris River. The main course consisted of roast pig and goat cheese. Honey cakes, fresh dates, and pomegranates were served for dessert.

- The Hittites (a conquering people of Asia Minor and Syria during the second millennium BC) had a law code, which included fines for damaging pomegranate trees (as well as some other crops, including grape vines).
- In ancient Rome, brides wore headdresses made from twigs of the pomegranate. The juice was considered to promote fertility. In Judaism, the pomegranate is also a symbol of fertility, relating to the first commandment of the Torah, to be fruitful and multiply. An Oriental practice was to burst pomegranates in the bedchamber when the newlyweds entered, signifying that the marriage should be blessed with many children. In Turkey, a bride was expected to throw the fruit on the ground, with the number of seeds being released indicative of the number of children she would have. Arab brides also smashed pomegranates in their tents to promote the birth of many children. Similarly in Chinese and Indian cultures, the pomegranate symbolizes fertility. Some Chinese women offer pomegranates to the Goddess of Mercy in the hope of becoming pregnant, and pomegranates are thrown on the bedroom floors of newlyweds. In Christianity, the pomegranate is a symbol of the Resurrection and also the Virgin Mary (both representing aspects of life-giving forces, and so also associated with fertility).
- An old custom in Morocco was to squeeze the juice of pomegranate on the horns of oxen that plowed fields, or on the blades of the plows, to promote good crops. This is one of many examples of a plant associated with human fertility also being associated with crop fertility.
- The pomegranate is the oldest symbol of Judaism, predating the Star of David by hundreds of years. Pomegranate-shaped pommels (knobs) are used to cover the wooden handles of the Torah (the Jewish scriptures on scrolls that are read in synagogues). Moses was ordered to put embroidered pomegranates at the bottom of the high priest's robes. The golden bells decorating the Holy Temple in Jerusalem were pomegranate-shaped, and pomegranate images also appeared on mosaic floors, in stone friezes, and on ancient coins. The pomegranate has been called the "national fruit" of Israel, but this designation has also been given to the cactus pear (*Opuntia ficus-indica* (L.) Mill.; see Chapter 14).
- Fruitcakes were first made in Roman times, the recipe generally calling for pomegranates as the chief constituent.
- The Moors were mixed Arab and Berber (a North African people) conquerors of Spain in the eighth century. They introduced the pomegranate and a method of tanning leather with pomegranate juice to Spain, resulting in the deep, rich, blood-red of cordovan leather (so-named because it was originally manufactured in the region of Cordoba, Spain).
- Women in some countries stain their teeth red by eating pomegranate flowers. By contrast, some Polynesians dye their teeth black with the rind. In Italy, pomegranate leaves were once made into a mouthwash that was considered good for loose teeth.
- The French used the word *grenade* for an explosive shell that strewed metal particles over a wide area and grenadiers for the special regiments (founded in 1791) of soldiers who launched these new weapons, basing the words on the seed-scattering characteristic of a pomegranate when it is smashed.
- As noted earlier, *granatum* in the scientific name *P. granatum* means "(many) grained," a poetic way of expressing the thought that there are many seeds present, and the old Latin name for pomegranate was *pomum granatum* ("many-seeded apple"). The historical association of the Latin word *granatum* with the pomegranate is thought to have given rise to the word garnet based on the deep red color of pomegranate seeds. Garnets are gemstones

of various colors, including red. They are used as semiprecious gemstones (particularly as the birthstone for January) and as an abrasive, particularly in "garnet paper." Red garnets have even been used as bullets (e.g., in the southwestern United States) because of the curious thought that their red color predisposed them to making wounds bloodier.

- In Iran, cut-open pomegranates are sometimes stomped by a person wearing special shoes in a clay tub and the expressed juice is run through outlets into clay troughs.
- Studies published in 2001 in *Nutrition Science News* suggested that pomegranate juice lowers the risk of heart disease by lowering cholesterol that contributes to the clogging of arteries. The antioxidants in pomegranates—bioflavonoids—were described as having three times the activity of those in red wine and green tea, and so the fruit may also be useful in combating disease and slowing the aging process.
- The pomegranate family, Punicaceae, includes only one genus and two species, including the well-known pomegranate and the little-known *P. protopunica* Balf., which is peculiar to the island of Socotra in the Indian Ocean.

KEY INFORMATION SOURCES

Artes, F., Villaescusa, R., and Tudela, J.A. 2000. Modified atmosphere packaging of pomegranate. *J. Food Sci.* 65:1112–1116.

Ashton, R. 2006. *Incredible pomegranate: plant and fruit.* Third Millennium, Tempe, AZ. 162 pp.

Beach, M. 1980. Dwarf and other pomegranates. *Light Garden*, 17(1):13–15.

Ben-Aire, R., Segal, N., and Guelfat-Reich, S. 1984. The maturation and ripening of the 'Wonderful' pomegranate. *J. Am. Soc. Hort. Sci.* 109:898–902.

Butani, D.K. 1976. Insect pests of fruit crops and their control. 21. Pomegranate. *Pesticides*, 10(6):23–26.

Chace, E.M., Church, C.G., and Poore, H.D. 1930. *The wonderful variety of pomegranate: composition, commercial maturity, and by-products.* U.S. Department of Agriculture, Washington, DC. 16 pp.

Elyatem, S.M., and Kader, A.A. 1984. Post-harvest physiology and storage behaviour of pomegranate fruits. *Sci. Hortic.* 24:287–298.

Goor, A. 1967. The history of the pomegranate in the Holy Land. *Econ. Bot.* 21:215–229.

Hodgson, R.W. 1917. The pomegranate. *Bull. Calif. Agric. Exp. Stn.* 276:163–192.

Jaidka, K., and Mehra, P.N. 1986. Morphogenesis in *Punica granatum* (pomegranate). *Can. J. Bot.* 64:1644–1653.

Levin, G.M., and Ashton, R.W. 2006. *Pomegranate.* Third Millennium, Tempe, AZ. 129 pp.

Mars, M., and Marrakchi, M. 1999. Diversity of pomegranate (*Punica granatum* L.) germplasm in Tunisia. *Genet. Resour. Crop Evol.* 46:461–467.

Melgarejo-Moreno, P., Martinez-Nicolas, P.J., and Tome, J.M. (Eds.). 2000. *Production, processing and marketing of pomegranate in the Mediterranean region: advances in research and technology. Proceedings of the Symposium, Orihuela, Spain, 15–17 October 1998.* Centre International de Hautes Études Agronomiques Méditerranéennes, Paris, France. 253 pp.

Melgarejo, P., Martinez-Valero, R., Guillamon, J.M., Miro, M., and Amoros, A. 1997. Phenological stages of the pomegranate tree (*Punica granatum* L.). *Ann. Appl. Biol.* 130:135–140.

Mitra, S.K. 1999. Pomegranate. In *Tropical horticulture, vol. 1.* Edited by T.K. Bose, S.K. Mitra, A.A. Farooqui, and M.K. Sadhu. Naya Prokash, Calcutta, India. pp. 338–343.

Mohan Kumar, G.N. 1990. Pomegranate. In *Fruits of tropical and subtropical origin: composition, properties and uses.* Edited by S. Nagy. Florida Science Source Inc., Lake Alfred, FL. pp. 328–347.

Moore, S. 2004. The pomegranate. *Herbarist*, 70:25–30.

Morton, J. 1987. Pomegranate. In *Fruits of warm climates.* Creative Resource Systems, Winterville, NC. pp. 352–355.

Nath, N., and Randhawa, G.S. 1969. Classification and description of some varieties of *Punica granatum* L. *Indian J. Hortic.* 16(4):191–201.

Patil, A.V., and Karale, A.R. 1985. Pomegranate. In *Fruits of India, tropical and subtropical.* Edited by T.K. Bose. Naya Prokash, Calcutta, India. pp. 537–548.

Perez-Vicente, A., Serrano, P., Abellan, P., and Garcia-Viguera, C. 2004. Influence of packaging material on pomegranate juice colour and bioactive compounds, during storage. *J. Sci. Food Agric.* 84:639–644.

Roy, S.K., and Waskar, D.P. 1997. Pomegranate. In *Postharvest physiology and storage of tropical and subtropical fruits.* Edited by S K Mitra. CAB International, Wallingford, Oxon, U.K. pp. 365–374.

Seeram, N.P., Schulman, R.N., and Heber, D. (Eds.). 2006. *Pomegranates: ancient roots to modern medicine*. CRC Press, Boca Raton, FL. 264 pp.

Sharaf, A., Fayez, M.B.E., and Negm, A.R. 1967. Pharmacological properties of *Punica granatum* L. *Qual. Plant. Mater. Veg*. 14:331–336.

Shulman, Y., Fainberstein, L., and Lavee, S. 1984. Pomegranate fruit development and maturation. *J. Hortic. Sci*. 59:265–274.

Sood, D.R., Singh-Dhindsa, K., and Wagle, D.S. 1982. Studies on the nutritive value of pomegranate (*Punica granatum*). Haryana *J. Hortic. Sci*. 11:175–179.

Srivastava, H.C., and Bisla, S.S. 1984. Floral biology of *Punica granatum* L. *Indian J. Agric. Res*. 18:102–106.

Stover, E., and Mercure, E.W. 2007. The pomegranate: a new look at the fruit of paradise. *HortScience*, 42:1088–1092.

Sudiarto, and Rifai, M.A. 1991. *Punica granatum* L. In *Plant resources of South-East Asia. Vol. 2. Edible fruits and nuts. Edited by* E.W.M. Verheij and R.E. Coronel. Pudoc, Leiden, the Netherlands. pp. 270–272.

Turner, G.W., and Lersten, N.R. 1983. Apical foliar nectary of pomegranate (*Punica granatum*: Punicaceae). *Am. J. Bot*. 70:475–480.

Vidal, A., Fallarero, A., Pena, B.R., Medina, M.E., Gra, B., Rivera, F., et al. 2003. Studies on the toxicity of *Punica granatum* L. (Punicaceae) whole fruit extracts. *J. Ethnopharmacol*. 89:295–300.

Specialty Cookbooks

Kleinberg, A. 2005. *Pomegranates:70 celebratory recipes*. Ten Speed Press, Berkeley, CA. 112 pp.

79 Quinine

Family: Rubiaceae (madder family; more familiarly, this is occasionally called the "coffee family")

NAMES

Scientific Names: *Cinchona* species

- Quinine—*C. calisaya* Wedd. (*C. ledgeriana* (Howard) Bern. Moens ex Trimen; "*C. officinalis*" of many authors, although the species correctly given this name does not have significant quinine in its bark, and so is of limited interest).
- Redbark—*C. pubescens* Vahl (*C. succirubra* Pav. ex Klotzsch).

- The name "quinine" (used for both the plants and the medicinal chemical extracted from the bark) is based on the Spanish *quina*, cinchona bark, which in turn is based on the Peruvian native (Quechua) word *quina-quina* (literally "bark of bark") for cinchona bark.
- Pronunciation: principally KWY-nine and kwee-NINE; also kwi-NEAN (British) and several other versions.
- Several species unrelated to *Cinchona* have "quinine" in one or more of their names (e.g., jojoba, *Simmondsia chinensis* (Link) C.K. Schneid., is sometimes called quinine plant; quinine bush is *Alstonia constricta* F. Muell. of Australia), either because they were used as antimalarial medicines or merely because some part of the plant is very bitter, like quinine.
- Quinine (*C. calisaya*) is also known as brown Peru bark, calisaya, Chinabark, crown bark, ledgerbark, Peruvian bark, and yellow cinchona. *Calisaya* in the scientific name *C. calisaya* is from the Spanish *calisaya*, which is perhaps from Calisaya, the seventeenth century South American Indian who revealed the properties of cinchona bark to the Spaniards.
- Redbark (*C. pubescens*) is also known as brown quinine, quinine, red cinchona, and red Peruvian bark. "Redbark" is for the reddish bark of the commonly harvested "succiruba strain" of the species, otherwise the bark is brown. *Pubescens* in the scientific name *C. pubescens* is Latin for hairy, referring to the plant hairs that are frequently present on the branches, flowering parts, and lower surfaces of the leaves.
- The genus *Cinchona* commemorates Francisca Henríquez de Ribera (1576–1639), the Countess of Chinchón, wife of the 4th Count of Chinchón and Spanish Viceroy of Peru, who in 1638 fell desperately ill with malaria. According to legend, she was cured using the ancient herbal remedy of cinchona bark. At her instigation, the bark was collected for malaria sufferers and later exported to Spain.

PLANT PORTRAIT

The approximately two or three dozen species of *Cinchona* are evergreen trees, mostly native to the highlands of the Andes Mountains of South America from Bolivia to Colombia. Only the widespread *C. pubescens* is native outside the Andes, occurring also in some mountainous regions of southern Central America and the coastal range of Venezuela. The species noted here are cultivated for their "Peruvian bark," the source of the antimalarial quinine.

Cinchona calisaya is a shrub, to 3 m (10 ft.) high, or a tree, to 15 m (49 ft.) in height. Its flowers are white to pinkish or purplish, approximately 13 mm (1/2 in.) long. The capsule-like fruits are

FIGURE 79.1 Quinine (*Cinchona calisaya*). (From Köhler, 1883–1914.)

FIGURE 79.2 Redbark (*Cinchona pubescens*). (From Köhler, 1883–1914.)

egg-shaped, 5 to 25 cm (1/5–1 in.) long. The species is native to the eastern slopes and foothills of the Andes from central Peru to central Bolivia.

Cinchona pubescens is a tree that grows up to 10 m (33 ft.) high, with a trunk up to 20 cm (about 8 in.) in diameter at breast height. Its flowers are pink or purplish, approximately 13 mm (1/2 in.) long. Its capsular fruit is oblong, 13 to 41 mm (1/2–1½ in.) in length. The species is native from the mountains of central Costa Rica and northeastern Venezuela along the Andes south to central Bolivia.

Quinine is a chemical (an alkaloid) present in the bark of *Cinchona* trees, primarily the species listed earlier. Pure quinine is usually a white powder. Before the 1600s, Peruvian Indians used an infusion of the bark to treat fever, and it was recognized by Europeans that the preparation was also effective at treating and preventing malaria. Not until 1820, 200 years after the bark was introduced into Europe for the treatment of malaria, was quinine isolated and identified by the French chemists J.B. Caventou and P.J. Pelletier. In 1944, Robert Woodward and William Doering synthesized quinine from coal tar. Natural quinine has been replaced by chemically synthesized drugs such as Chloroquine and Mefloquine (Larium), and quinine is now rarely used to treat malaria. However, natural quinine has remained a treatment of choice for a severe form of malaria known as falciparum malaria. In recent years, with the emergence of drug-resistant strains of malaria, quinine is becoming more important in the treatment of malaria. Another alkaloid chemical called quinidine was discovered to be beneficial for heart patients, and quinidine is used today to treat irregular heart beat. The sales demand for this drug still generates the need for harvesting natural quinine bark because scientists have been unsuccessful in synthesizing this chemical without using the natural quinine found in cinchona bark.

Wanton, uncontrolled harvesting of the bark led to severe decimation of the trees in their native habitat. Because the trees were so valuable, Peruvian officials strictly prohibited their export. However, in the 1860s, some British and Dutch adventurers smuggled a few seedlings out of the country, and these were used to set up large plantations in Java and India. Up until the Second World War, these plantations supplied almost 95% of the world's requirements of quinine. Ironically, seeds from these plantations were later returned to Central America to establish plantations. Today, Africa is the leading supplier of quinine bark, with much smaller amounts from Peru, Bolivia, and Ecuador. The soft-drink industry, which requires quinine for tonic water, supports some of the present cultivation.

Malaria is based on the Italian *mala*, bad + *aria*, air. Malaria has been known since the dawn of recorded history, approximately 6000 years ago. Because there are no references to malaria in the medical recordings of the Mayans or Aztecs, it is likely that European settlers and slavery brought malaria to the New World within the last 500 years. Worldwide, approximately 300 million people are infected, and at least 1 million die each year from the disease. Malaria is characterized by attacks of fever (sometimes more than 40°C or 104°F), chills, nausea, severe headache, mild delirium, gastric pains, and vomiting. Malaria is caused principally by the protistan parasite *Plasmodium vivax*, of which there are several strains differing in severity (other species include *P. ovale*, and *P. malariae*). Malaria is only transmitted by the female *Anopheles* mosquito (the male feeds on nectar and plant juices). Quinine (and other drugs such as chloroquine) combats malaria by binding strongly to blood proteins and forming complexes that are toxic to the malarial parasite. In 1957, the countries of the world attempted to eradicate the malarial parasite from the Earth, but by 1974, it was realized that this could not be done.

In the past, *Cinchona* bark was prepared by grinding it down to a fine powder and mixing it with water or wine. Today, quinine is generally taken in tablet form, occasionally intravenously by injection. Quinine is commonly used to treat muscle spasms, particularly to relieve night-time leg cramps.

Excessive medicinal doses of quinine can lead to "cinchonism," characterized by ringing in the ears, temporary deafness, blurred vision, nausea, and abdominal upset. In severe cases, an overdose may result in circulatory collapse, kidney failure, and coma.

CULINARY PORTRAIT

Cinchona bark is used in the manufacture of fortified apéritif wines and other alcoholic drinks, and particularly in carbonated soft drinks called tonics, which are often mixed with alcoholic beverages. Dubonnet is a trademarked French apéritif made from semi-dry white or red fortified wine flavored with herbs and quinine. Quinine from the bark is used as a flavoring, primarily to provide the bitter taste in tonic water. Quinine water or tonic water is carbonated water with a very small amount of quinine, flavored with fruit extracts and sugar. Although "gin and tonic" was originally consumed to prevent malaria, quinine has remained the desired bitter taste for gin and tonic as well as for vodka and tonic. A standard recipe for a gin and tonic cocktail: 45 to 60 mL (1½–2 oz.) of gin, 170 mL (6 oz.) of tonic, ice, and fresh lime juice or a twist of lime or lemon, all in a highball glass.

"Bitters," generally made up of a bitter-tasting herb, water, and alcohol, have long been used medicinally to stimulate appetite. Bitters were widely available during the Prohibition period in the United States, although the alcohol content could be as high as 35%. Several countries are known for their brands of bitters used to flavor alcoholic beverages: Peychaud's and Abbot's (United States), Amer Picon and Secrestat (France), Boonekamp's (Holland), Campari and Fernet Branca (Italy), Law's Peach Bitters (England), Pommeranzen (Germany), and Unicum (Hungary), to name a few. The formula that has been most competitive with quinine water is Angostura Bitters™ (with manufacturing now based in Trinidad), said to be "made from a secret blend of more than 40 tropical herbs, plant extracts and spices." Johann Gottlieb Benjamin Siegert (1796–1830), is known as the inventor. He studied medicine in Berlin and tended to the wounded at the 1815 Battle of Waterloo. An adventurer, Siegert then went to Venezuela, sailing up the Orinoco River to the western Venezuelan port of Angostura (which means "straits" in Spanish; renamed Ciudad Bolívar in 1846). He became surgeon general to Simon Bolivar and his troops during the liberation of South America and also found time to concoct Angostura Bitters. Siegert invented Angostura Bitters in 1824 as a medicine, but it is mainly used today to give a twist of bitterness to cocktails and other mixed drinks. Although the composition of Angostura Bitters has remained a trade secret, "Angostura" has become a generic term for a bitter bark used very much like quinine as a tonic and to reduce fever. Angostura is obtained from angostura, *Angostura trifoliata* (Willd.) T.S. Elias (*Galipea officinalis* Hancock, *Cusparia febrifuga* Humb. ex DC., *C. trifoliata* (Willd.) Engl.). However, labels on Angostura Bitters have stated that it includes no bark of the angostura tree.

FIGURE 79.3 Gathering cinchona bark in a Peruvian forest. (From Figuier, 1867.)

Culinary Vocabulary

- The bark of *Cinchona* trees used as a source of medicinal and culinary quinine is called "Peruvian bark" because much of it came originally from Peru and "Jesuits' bark" or "sacred bark" because the Jesuits played an important part in its dispersal to Spain. The bark of the uprooted tree is beaten loose, peeled by hand, and dried quickly.
- Quinine water is also known as tonic (pronounced TAHN-ik) water, Indian tonic water, or just tonic.
- "Calisay" (pronounced cah-lee-sah-e) is a beverage made from *Cinchona* extract in Spain's Catalonia region (see information in the Names section for *C. calisaya*).
- "Punt e Mes" (pronounced poohnt a mess) is a sweet Italian vermouth flavored with quinine.
- Qunquina (pronounced kin-dee-nah) is a quinine-flavored Spanish aperitif wine.
- VAT is short for "vodka and tonic" (also, "value added tax").

CURIOSITIES OF SCIENCE AND TECHNOLOGY

- In sixteenth-century Europe, malaria indiscriminately killed peasants, kings, princes, and popes. Although today malaria is a common disease of tropical lands, especially Africa, up to the seventeenth century, it affected vast parts of Europe. Until the importation of cinchona bark, patients had one principal cure available—bleeding.
- In the 1600s, for a period after it was realized that *Cinchona* bark prevented and cured malaria, the cost of the bark powder was often matched by its weight in gold.
- The support of the Vatican for the use of cinchona bark worked against its acceptance in some regions, particularly in England. Many Protestants feared it was part of a "Popish plot" against them. In the late seventeenth century, the Englishman Robert Talbor used these fears to establish his reputation as a "feverologist." Agreeing that the "Jesuit's powder" should be avoided, he claimed to have a secret remedy for malaria. He successfully cured King Charles II of England of malaria, and afterward King Louis XIV of France bought the recipe for the cure, under the condition that the constituents not be made public until after Talbor's death. When the recipe was published in 1682, the secret medicine was revealed. It was cinchona bark.
- One of the first military expenditures of the Continental Congress, around 1775, was for $300 to buy quinine to protect General Washington's troops.
- In 1927, J. Wagner von Jauregg (1857–1940), of Austria was awarded the Nobel Prize in Medicine for his work in treating syphilis using malaria. Patients were inoculated with a type of malaria to produce fevers that would destroy the temperature-sensitive syphilis bacteria. After three or four cycles of the fever, the patient was administered quinine to cure the malaria. Today, syphilis is treated with antibiotics.
- After painkillers, the plant that has done most to relieve human suffering is probably quinine. Malaria has had an extraordinarily harmful effect in human history. The falls of Rome and of ancient Greece have both been attributed to malaria. Alexander the Great died from it. Africa repelled European settlers for centuries because of its prevalence.
- A "new" antimalarial drug called qinghaosu was derived from the sweet wormwood (*Artemisia annua* L.) in the 1970s. The plant (now a common weed around the world) has been used in China for more than 2000 years to treat fevers associated with malaria. The drug is effective against most deadly forms of *falciparum* malaria, including strains resistant to chloroquine.

KEY INFORMATION SOURCES

Andersson, L., and Kawasaki, M.L. 1997. *A revision of the genus* Cinchona *(Rubiaceae–Cinchoneae)*. New York Botanical Garden, Bronx, NY. 115 pp.

Beljaars, P.R., Rondags, T.M.M., Moerer, J., and Baosch, N.F. 1979. Examination of the quinine content in tonic water. *De Warenchemicus*, 9:12–17.

Draper, T. 1947. What is the connection of the Countess Anna de Chinchon and cinchona bark? *J. Hist. Med. Allied Sci.* 2:577–578.

Duran-Reynals, M.L. de Ayala. 1946. *The fever bark tree; the pageant of quinine*. Doubleday, Garden City, NY. 275 pp.

Flavor and Extract Manufacturers. 1979. *Scientific literature review of quinine salts in flavor usage*. Flavor and Extract Manufacturers' Association of the United States. 124 pp. ["Prepared pursuant to contract no. 223-74-2224 with the Public Health Service, Food and Drug Administration, Department of Health, Education and Welfare."]

Gonzalez, R., Merchan, R., Crespo, J.F., and Rodriguez, J. 2002. Allergic urticaria from tonic water. *Allergy*, 57:52.

Grammiccia, G. 1987. Notes on the early history of *Cinchona* plantations. *Acta Leiden*. 55:5–13.

Guerra, F. 1977. The introduction of cinchona in the treatment of malaria. *J. Trop. Med. Hyg*. 80:118–122, 135–139.

Harten, A.M. van. 1969. Cinchona (*Cinchona* spp.). In *Outlines of perennial crop breeding in the tropics*. Edited by F.P. Ferwerda and F. Wit. Misc. Paper 4, Agricultural University, Wageningen, the Netherlands. pp. 111–128.

Harten, A.M van. 1995. Quinine: *Cinchona* spp. (Rubiaceae). In *Evolution of crop plants*. 2nd ed. Edited by J. Smartt and N.W. Simmonds. Longman Scientific & Technical, Burnt Mill, Harlow, Essex, U.K. pp. 435–438.

Honigsbaum, M. 2001. *The fever trail: the hunt for the cure for malaria*. Macmillan, London. 333 pp.

Hunter, C.S. 1986. In vitro propagation and germplasm storage of cinchona. In *Plant tissue culture and its agricultural applications*. Edited by L.A. Withers and P.G. Alderson. Butterworths, London. pp. 291–301.

Jaramillo-Arango, J. 1949. A critical review of the basic facts in the history of cinchona. *J. Linn. Soc. Bot*. 53:272–309.

Jarcho, S., and Torti, F. 1993. *Quinine's predecessor: Francesco Torti and the early history of cinchona*. Johns Hopkins University Press, Baltimore, MD. 354 pp.

Kaay, H.J. van der, and Overbosch, D. (Eds.). 1987. *Proceedings of the symposium on the use of quinine and quinidine in malaria, Amsterdam, March 9–11, 1987*. Institute of Tropical Medicine, Leiden, the Netherlands. 221 pp.

Lombard, F.F. 1947. *Review of literature on cinchona diseases, injuries, and fungi*. U.S. Department of Agriculture, Washington, D.C. 70 pp.

McHale, D. 1986. The cinchona tree. *Biologist*, 33:45–53.

Moreau, R.E. 1945. *An annotated bibliography of cinchona-growing from 1883–1943*. East African Agricultural Research Institute, Nairobi, Amani, Tanganyika territory. 41 pp.

Philip, K. 1995. Imperial science rescues a tree: global botanic networks, local knowledge and the transcontinental transplantation of cinchona. *Environ. Hist*. 1:173–200.

Prendergast, H.D.V., and Dolley, D. 2001. Jesuits' bark (cinchona) and other medicines. *Econ. Bot*. 55:3–6.

Richardson, R.W. 1944. *Cinchona in the Americas*. Research Division, Social and Geographic Unit, U.S. Office of Inter-American Affairs, Washington, D.C. 20 pp.

Rocco, F. 2003. *The miraculous fever tree: malaria and the quest for a cure that changed the world*. HarperCollins, New York. 348 pp.

Sharma, A., Tewari, R., Ganniyal, A.K., and Virmani, O.P. 1987. Cinchona: a review. *Curr. Res. Med. Aromatic Plants*, 9:34–56.

Staba, E.J. 1988. Alkaloid production from cinchona cell and organ systems. In *Genetic manipulation of woody plants*. Edited by J.W. Hanover and D.E. Keathley. Plenum, New York. pp. 313–328.

Staritsky, G., Huffnagel, E., Dharmadi, A., and Dalimoenthe, S.L. 2001. *Cinchona* L. In *Plant resources of South-East Asia. Vol. 12(1). Medicinal and poisonous plants 1*. Edited by L.S. de Padua, N. Bunyapraphatsara and R.H.M.J. Lemmens. Backhuys Publishers, Leiden, the Netherlands. pp. 198–205.

Swift, R.M., Griffiths, W., and Cammera, P. 1989. False positive urine drug screens from quinine in tonic water. *Addict. Behav*. 14:213–216.

Taylor, N. 1945. *Cinchona in Java; the story of quinine*. Greenberg, New York. 87 pp.

Uskokovic, M.R., and Grethe, G. 1973. The cinchona alkaloids. *Alkaloids Chem. Physiol*. 14:181–223.

Verpoorte, R., Schripsema, J., and Leer, T. van der. 1988. Cinchona alkaloids. *Alkaloids*, 34:331–398.
Wagner, G.H., Diffey, B.L., and I've, F.A. 1994. 'I'll have mine with a twist of lemon': quinine photosensitivity from excessive intake of tonic water. *Br. J. Dermatol.* 131:734–735.
Winters, H.F. 1950. *Cinchona propagation*. Porto Rico Agricultural Experiment Station Bulletin, vol. 47. Mayagüez, Puerto Rico. 26 pp.
Wolf, L.R., Otten, E.J., and Spadafora, M.P. 1992. Cinchonism: two case reports and review of acute quinine toxicity and treatment. *J. Emerg. Med.* 10:295–301.

Specialty Cookbooks

Note: Virtually all bartending books provide recipes using quinine in the preparation of various beverages. Numerous guides to preparation of alcoholic beverages are cited in Chapter 87.

80 Rooibos Tea

Family: Fabaceae (Leguminosae; pea family)

NAMES

Scientific Name: *Aspalathus linearis* (Burm. f.) R. Dahlgren

- The word rooibos (pronounced "roy-boss") means "red bush" in Afrikaans. Both "rooibos" and "rooibos tea" are commonly used for the plant species and for the tea prepared from it.
- Rooibos tea is also known as red tea, reflecting the principal use of the plant—production of a tea that turns reddish when it is prepared by fermentation (as explained below). The tea itself is also called red tea.
- The genus name *Aspalathus* is from the Greek *aspalathos*, a word that in ancient times referred to a scented, spiny shrub, the bark and roots of which produced a fragrant oil used in the preparation of spiced wine.
- *Linearis* in the scientific name *A. linearis* is Latin for linear, an allusion to the long, narrow shape of the leaves.

PLANT PORTRAIT

Aspalathus linearis is one of the approximately 278 species of the genus *Aspalathus*, which is confined to South Africa. *Aspalathus linearis* is extremely variable, and distinct geographical forms have been recognized, differing in appearance and chemical content. The plant is an erect or straggling shrub, 1.35 to 2 m (4.4–6.6 ft.) tall, with red-brown branches. The needlelike leaves are 1.5 to 6 cm (0.6–2.4 in.) long. The flowers are yellow (rarely reddish, maroon, or mauve), approximately 6.5 mm (0.26 in.) long, and borne in short clusters. The seedpods are approximately 1.5 cm (0.6 in.) long and usually contain one seed (occasionally two) which is propelled away from the plant when the pod splits open. There appears to be a domesticated form, variously referred to as "Nortier" tea, Rocklands type, and red tea type, that is widely grown. It is considered superior to the wild forms, and when wild plants are collected and sold, they are often combined with the cultivated form to achieve an acceptable taste.

Rooibos tea is native to the mountain slopes of western Cape Province, South Africa, particularly in the Cederberg (Cedarberg) Mountains. Rooibos tea is uniquely obtained from the mountain areas within, and closely bordering, the northwestern regions of Western Cape Province. Production is centered in a region 200 to 300 km (124–186 miles) north of Cape Town, within and adjacent to the Cederberg and Olifantsrivier mountain ranges, and on the Bokkeveld plateau. The plant has not been successfully cultivated commercially elsewhere.

Rooibos tea was used medicinally and as a tea for centuries by Indigenous Africans (primarily the mountain-dwelling Khoi tribe and the San people). They are said to have adopted a process used originally by Malay slaves to harvest and prepare the plants, involving chopping branches off with axes, bruising them with hammers, and leaving them to "ferment" (an oxidation, not a genuine fermentation) in heaps before drying them in the sun. Modern technology copies these techniques but is mechanized.

In the twentieth century, rooibos tea acquired popularity as a tea preparation, primarily in South Africa but also in other nations. Apartheid and the associated trade restrictions slowed exports for

FIGURE 80.1 Harvester inspecting rooibos tea plant on a farm, Klipfontein, Graafwater, South Africa. (Photo courtesy of Bernd Hohn, F.L. Michaelmis GmbH, Bremen Germany (with additional thanks to Dr. Frans van der Westhuizen, King's Products, South Africa, for facilitating contact with Mr. Hohn.)

a period. Only in the last decade has the tea attracted widespread commercial interest. Rooibos tea is a particularly popular drink in South Africa, where it is considered to be the unofficial national beverage, and makes up of the order of 20% of the "tea" market of the country. Annual production of rooibos tea in 2003 was reported as 10,600 tonnes, of which 6400 tonnes were exported (production in 1966 was only 524 tonnes). In 2004, it was estimated that the total value of rooibos tea was 180 million Rands (18,600,000 American dollars), and approximately 60% of the national harvest was exported.

Most rooibos tea marketed today is from cultivated plants. Wild plants are still collected to a small extent and are often sold as a specialty product. There are several hundred rooibos tea farmers, with farms ranging in size from a few hectares to more than 5000 hectares (12,400 acres), although there are only a few large farms. A very small number of firms specialize exclusively in the production and marketing of rooibos tea in South Africa. Rooibos tea processing is dominated by eight large companies, with Rooibos Ltd. having 75% of market share. Two community, Indigenous-African-managed cooperatives (Wupperthal and Heiveld) in the neighboring Cederberg and Bokkeveldberg mountains produce organically grown product (both cultivated and wildcrafted) that is sold under international "Fairtrade" labels (see Chapter 1). Together, they supply approximately 100 tonnes of certified organic product, most of this exported. In 2005, the Wupperthal cooperative paid 3 €/kg (dried product; ca. $9.27/lb.) to its farmers; that same year, Rooibos Ltd. paid 1.9 €/kg (ca. $5.87/lb.), with only 15% certified organic. The industry is very labor intensive, employing thousands of people.

In addition to the culinary uses of rooibos tea, some medicinal extracts, to be consumed orally, are claimed to alleviate hay fever, asthma, allergies, insomnia, and other conditions. There are also skin cosmetics and a formulation for thinning hair on the basis of the theory that antioxidants from the plant can penetrate the skin and have a therapeutic effect.

CULINARY PORTRAIT

Green and red kinds of rooibos tea are consumed. The distinctive red color of the red form develops when the harvested leaves and twigs are bruised during the usual preparation procedure. The red color is the result of oxidation of the constituent polyphenols. The harvested material is then spread in a layer to allow "fermentation" (oxidation, as noted earlier) to occur for 8 to 24 hours, during which the material is turned. The unfermented product remains green in color and is referred to as green rooibos. This has a higher content of antioxidants but a poorer taste.

Rooibos tea is used primarily as a tasty tea, its lack of caffeine, low concentration of tannins, high level of ascorbic acid (vitamin C), and considerable content of antioxidants (particularly flavonoids) all touted as contributing to health. The alleged health-giving properties have been a selling point, much as with "green tea" (common tea prepared without fermentation). Rooibos tea is consumed hot, iced, plain, or sweetened, with or without milk or other additives. It is also used to a minor extent to flavor edible products, such as wines, juices, and cordials.

There have been reports of bacterial (coliform and *Salmonella*) contamination in some samples of rooibos tea. Water is added before the fermentation stage, and contaminated water may have been a chief problem in the past. To avoid *Salmonella* contamination (unusual for vegetable products), pasteurization is now carried out. The Perishable Products Export Control Board of South Africa certifies that rooibos tea export products have been pasteurized and meet standards of quality with respect to bacteria and impurities.

Culinary Vocabulary

- Some authorities define "tea" exclusively by reference to the common (Chinese) tea plant, *Camellia sinensis* (L.) Kuntze. So-called "herbal teas" like rooibos tea are defined as "tisanes." However, this is a rather arbitrary distinction.

CURIOSITIES OF SCIENCE AND TECHNOLOGY

- A species of black ant is known to collect the seeds of rooibos tea. Because the capsules explosively eject the seeds, they are difficult to collect. Indigenous people sometimes trailed the ant to its nest, then robbed the ant colony of its seed collection, claimed to number approximately 25,000 in one case.
- Anique Theron of South Africa found that when she gave rooibos tea to her crying baby it relieved the child's colic and crying. This earned her the Gold Medal of Geneva, Switzerland, as "the best woman inventor of 1997."
- *The No. 1 Ladies' Detective Agency* is the basis of a series of best-selling novels (the first in 1998, 12 as of 2011) by Zimbawean-born author Alexander McCall Smith. Because the principal character welcomed clients and visitors with a cup of rooibos tea, it stimulated a flood of interest in this beverage that was obscure until recently.

KEY INFORMATION SOURCES

Amin, R., Blumenthal, M., and Silverman, W. 2005. Rooibos tea trademark dispute settled. *HerbalGram*, 68:60–62.

Binns, T., Bek, D., Nel, E., and Ellison, B. 2007. Sidestepping the mainstream: fair-trade rooibos tea production in Wupperthal, South Africa. In *Alternative food geographies. Representation and practice*. Edited by D. Maye, L. Holloway, and M. Kneafsey. Elsevier, Amsterdam, the Netherlands. pp. 331–349.

Cheney, R.H., and Scholtz, E. 1963. Rooibos tea, a South African contribution to world beverages. *Econ. Bot.* 17:188–196.

Coetzee, C., Jefthas, E., and Reinten, E. 1999. Indigenous plant genetic resources of South Africa. In *Perspectives on new crops and new uses*. Edited by J. Janick. ASHS Press, Alexandria, VA. pp. 160–163.

Dahlgren, R. 1964. The correct name of the "rooibos" tea plant. *Bot. Notiser*, 117:188–196.

Dahlgren, R. 1968. Revision of the genus *Aspalathus*. II. The species with ericoid and pinoid leaflets. 7. subgenus *Nortieria*. With remarks on rooibos tea cultivation. *Bot. Notiser*, 121:165–208.

Dahlgren, R. 1988. Crotalarieae (*Aspalathus*). *Flora of Southern Africa*, 16(36):84–90.

Dos, A., Ayhan, Z., and Sumnu, G. 2005. Effects of different factors on sensory attributes, overall acceptance and preference of rooibos (*Aspalathus linearis*) tea. *J. Sens. Stud.* 20:228–242.

Gadow, A. von, Joubert, E., and Hansmann, C.F. 1997. Comparison of the antioxidant activity of rooibos tea (*Aspalathus linearis*) with green, oolong and black tea. *Food Chem.* 60:73–77.

Gess, S.K. 2000. Rooibos. Refreshment for humans, bees and wasps. *Veld & Flora*, 86:19–21.

Gess, S.K., and Gess, F.W. 1994. Potential pollinators of the Cape Group of Crotalarieae (*sensu* Polhill) (Fabales: Papilionaceae), with implications for seed production in cultivated rooibos tea. *Afr. Entomol.* 2:97–106.

Heerden, F.R. van, Wyk, B.-E. van, Viljoen, A.M., and Steenkamp, P.A. 2003. Phenolic variation in wild populations of *Aspalathus linearis* (rooibos tea). *Biochem. Syst. Ecol.* 31:885–895.

Joubert, E., and Ferreira, D. 1996. Antioxidants of rooibos tea—a possible explanation for its health promoting properties? *S. Afr. J. Food Sci. Nutr.* 8(3):79–83.

Joubert, E., and Schulz, H. 2006. Production and quality aspects of rooibos tea and related products. A review. *J. Appl. Bot. Food Qual.* 80:138–144.

Joubert, E., and Villiers, O.T. de. 1997. Effect of fermentation and drying conditions on the quality of rooibos tea. *Int. J. Food Sci. Technol.* 32:127–134.

Krafczyk, N., and Glomb, M.A. 2008. Characterization of phenolic compounds in rooibos tea. *J. Agric. Food. Chem.* 56:3368–3376.

Marnewick, J.L. 2009. Rooibos and honeybush: recent advances in chemistry, biological activity and pharmacognosy. In *African natural plant products: new discoveries and challenges in chemistry and quality*. Edited by H.R. Juliani, J.E. Simon, and C.-T. Ho. American Chemical Society, Washington, D.C. pp. 277–294.

McKay, D.L., and Blumberg, J.B. 2006. A review of the bioactivity of South African herbal teas: rooibos (*Aspalathus linearis*) and honeybush (*Cyclopia intermedia*). *Phytother. Res.* 21:1–16.

Morton, J.F. 1983. Rooibos tea, *Aspalathus linearis*, a caffeineless, low-tannin beverage. *Econ. Bot.* 37:164–173.

Nel, E., Binns, T., and Bek, D. 2007. 'Alternative foods' and community-based development: rooibos tea production in South Africa's West Coast Mountains. *Appl. Geogr.* 27:112–129.

Ollier, C. 2006. Le rooibos, thé rouge (red bush tea). *Phytothérapie*, 4:188–198 (in French).

Plessis, H.J. du, and Roos, I.M.M. 1986. Recovery of coliforms, *Escherichia coli* type I and *Salmonella* species from rooibos tea (*Aspalathus linearis*) and decontamination by steam. *Phytolphylactica*, 18:177–181.

Small, E., and Catling, P.M. 2009. Blossoming treasures of biodiversity 29. Rooibos tea—a new beverage for the world. *Biodiversity*, 10(2–3):113–119.

Swanepoel, M.L. 1987. *Salmonella* isolated from rooibos tea. *S. Afr. Med. J.* 71:369–370.

Turpie, J.K., Heydenrych, B.J., and Lambeth, S.J. 2003. Economic value of terrestrial and marine biodiversity in the Cape Floristic Region: implications for defining effective and socially optimal conservation strategies. *Biol. Conserv.* 112:233–251.

Wilson, N.L.W. 2005. Cape natural tea products and the U.S. market: rooibos rebels ready to raid. *Rev. Agric. Econ.* 27:139–148.

Specialty Cookbooks

Errey, S. 2005. *Rooibos revolution: recipes for nature's healing tea*. Bellissimo Books, Vancouver, BC. 64 pp.

Rooibos Tea Board. 1986. *Tasty treats with rooibos*. Rooibos Tea Board, Clanwilliam, South Africa. 93 pp.

Rooi Tea Control Board. 1979. *80 rooi tea wonders*. 2nd ed. Clanwilliam, South Africa. 61 pp.

81 Rose Apple

Family: Myrtaceae (myrtle family)

NAMES

Scientific Name: *Syzygium jambos* (L.) Alston (*Eugenia jambos* L.)

- The name "rose apple" was coined, as for many other tropical plants, using familiar English names of well-known plants. The rose apple is not at all related to the common apple or the rose, but the fruit is the size and shape of a small apple and is very distinctly rose scented.
- The rose apple is also known as jambos, Malabar plum, and plum rose.
- The name Malabar plum is also used for the related jambolan, *Syzygium cumini* (L.) Skeels, a cherry-sized, astringent fruit of Asia.
- 'Pacific Rose' is a recently developed New Zealand cultivar of the common apple. It has been called "rose apple," which is likely to result in confusion with the fruit usually designated by this name.
- Several other Asian species of *Syzygium* with edible fruits are also sometimes called rose apple, although generally when used without qualification this name is applied to *S. jambos*. Other "rose apples" include the following:
  - Malay rose apple (Malay apple, mountain apple, Otaheite apple)—*S. malaccense* (L.) Mer. & L.M. Perry
  - Semarang rose apple (Java apple, Semarang apple, jambu)—*S. samarangense* (Blume) Merr. & L.M. Perry
  - Watery rose apple (bellfruit, water apple)—*S. aqueum* (Burm. f.) Alston
- A number of names sometimes applied to the rose apple are ambiguous, often applied locally to other species of *Syzygium*; these names may include plum rose, water apple, and Malay apple.
- "Lily-pillies," a term used in Australia, are succulent-fruited trees belonging to the family Myrtaceae (the family also has plants with other types of fruits), and this includes species of *Syzgium*. The origin of the name lilly pilly is unknown. It may have been coined by English settlers who based the word on an aboriginal name. There are various spellings such as "lilly pilly" and "lilli pilli," with or without hyphens. Although the term originally was based on native Australian plants, it has also been applied to imported plants such as the rose apple.
- See Chapter 27 for the origin of the genus name *Syzygium*.
- *Jambos* in the scientific name *S. jambos* is based on *jambu*, a Sanskrit-derived Hindu name of the rose apple. The word *Jambos* was once recognized as a genus name (the plants designated are now included in *Syzgium*). In Southeast Asia and India, rose apple and related fruits are often called jambos.

PLANT PORTRAIT

The rose apple is an evergreen shrub or, more often, a tree 7.5 to 12 m (25–40 ft.) tall. The leaves are glossy and leathery, often wine-colored when young. The flowers are large and showy, 5 to 10 cm (2–4 in.) wide, creamy-white or greenish-white, and occur in clusters of four or five. Most

FIGURE 81.1 Rose apple (*Syzygium jambos*) in fruit. (From Curtis, 1834, vol. 61, plate 3356.)

of the flower consists of hundreds of conspicuous stamens, up to 4 cm (1½ in.) long. The fruits are 2.5 to 5 cm (1–2 in.) wide, almost round or somewhat longer than wide, with skin that is smooth, thin, and greenish or dull-yellow flushed with pink when ripe. The flesh is yellowish, sweet, rose-scented, and firm, with a crisp, almost crunchy texture when ripe and fresh. There are one to four hard, round seeds, which rattle inside the fruit. The seed are said to be poisonous. The fruits are very light in weight because they are hollow.

The rose apple is native to the East Indies and Malaya, and is cultivated and naturalized in many parts of India, Southeast Asia, and the Pacific Islands. It was introduced into Jamaica in 1762, in Florida before 1877, and later in California. Rose apples are adapted to tropical or near-tropical climates but have survived cold temperatures down to –4°C (25°F) and have been grown as ornamentals as far north in California as San Francisco. In the colder limits of its cultivation range, the plant will usually not flower or bear fruit. Generally, the rose apple is a very minor fruit, and orchards are not profitable. There has been little breeding, and almost no cultivated varieties are available. This tree is mostly grown for fruit purposes in home gardens. The rose apple usually is propagated by seeds, but for those considering growing plants from the seeds that are available in a fruit, it should be appreciated that the seeds remain viable for only a short time after being removed from the fruit, and often the resulting plant will not produce fruit as tasty as the one from which the seeds were obtained. The plant is considered to be too large to grow in containers.

CULINARY PORTRAIT

Rose apple is considered good but not outstanding as a fresh fruit (many judge it insipid), but it is prized for use in jellies, jams, and preserves (often in combination with other fruit). It is also cooked with sugar to make a rose-flavored dessert such as custard or pudding and is stewed or preserved in syrup. The large, hollow seed cavity is sometimes employed to stuff the fruits for baking. Rose apples bruise quite easily and are highly perishable and should be used soon after picking. Because it spoils quickly and is not competitive with other fresh fruit, the rose apple is rarely available in

Western markets. Fruit extract can be used to make a sweet smelling rose water that is considered equal to the best made from rose petals. The flowers can also be candied.

Culinary Vocabulary

- A "pink rose apple" (known as *chompoo* in Thailand) is a pink variety of rose apple with striped skin, consumed raw in salads and desserts.
- "Apple roses" have nothing to do with "rose apples"; they are ornamental pastries, prepared using cored, halved apples, sliced and decorated to look like roses, and often served with a filling and a covering glaze.

CURIOSITIES OF SCIENCE AND TECHNOLOGY

- In India, some tribes traditionally use rose apple water to ritually wash the face of newborn infants.
- Rose apple fruit is regarded as a tonic for the brain and liver in India.
- The flexible branches of rose apple have been used in Puerto Rico to make hoops for large sugar casks as well as for weaving large baskets.
- Seeds of rose apple have several embryos. Although most seeds produce one to three seedlings, as many as eight plants may arise from a single seed. The seedlings are likely to be locked in a competition in which only one survives. Growers of rose apple separate the seedlings so that each one will grow well.
- In Guatemala, rose apple trees have been planted as a hedge around coffee plantations, the plant drastically pruned to promote dense growth.
- The rose apple is considered to be an invasive weed in some countries. For example, in the Galapagos Islands, attempts are being made to eliminate it. In Florida, there is concern that the tree could develop into another of the many weeds in the state, but this has not yet occurred.
- A one-dollar stamp featuring the rose apple was issued by Pitcairn Island and released at the opening of the London (England) Stamp Exhibition in 2000. (Pitcairn Island is a British colony in the South Pacific, where the descendants of the Mutiny on the Bounty found refuge. The main island (there are also three uninhabited islands) has an area of only 450 hectares (approximately 1¾ square miles). It is one of the remotest of the world's inhabited islands, with a population in 2001 of less than 50. One of the major industries is its role as the nominal source for issuing stamps for collectors.)

KEY INFORMATION SOURCES

Bhatnagar, N., Dave, Y., and Rao, T.V.R. 1997. Epicarpic studies in the developing fruit of *Eugenia jambos* L. (Myrtaceae). *J. Phytol. Res.* 10:9–14.

Cai-Qui, L.M., Lan, L., and Mingliang, C. 1996. Analysis of nutritional composition of *Syzygium jambos* in Guizhou Province. *Acta Nutrimenta Sinica.* 18:235–237 (in Chinese).

Chan, H.T., Jr., and Lee, C.W. Q. 1975. Identification and determination of sugars in soursop, rose apple, mountain apple and Surinam cherry. *J. Food Sci.* 40:892–893.

Djipa, C.D., Delmee, M., and Quetin-Leclercq, J. 2000. Antimicrobial activity of bark extracts of *Syzygium jambos* (L.) Alston (Myrtaceae). *J. Ethnopharmacol.* 71:307–313.

El-Siddig, K., Luedders, P., Ebert, G., and Adiku, S.G.K. 1998. Response of rose apple (*Eugenia jambos* L.) to water and nitrogen supply. *J. Appl. Bot.* 72:203–206.

Fisch, B.E. 1976. The roseapple. *California Rare Fruit Growers Yearbook.* 8:100–111.

Francis, J.K. 1990. Syzygium jambos *(L.) Alst., Myrtaceae: rose apple, myrtle family.* U.S. Forest Service, Southern Forest Experiment Station, Institute of Tropical Forestry, Rio Piedras, Puerto Rico. 4 pp.

Guedes, C.M., Pinto, A.B., Moreira, R.F.A., and De Maria, C.A.B. 2004. Study of the aromatic compounds of rose apple (*Syzgium jambos* Alston) fruit from Brazil. *Eur. Food Res. Technol.* 219:460–464.

Lee, P.L., Swords, G., and Hunter, G.L.K. 1975. Volatile components of *Eugenia jambos* L., rose-apple. *J. Food Sci*. 40:421–422.

Lingen, T.G. van. 1991. *Syzygium jambos* (L.) Alston. In *Plant resources of South-East Asia. Vol. 2. Edible fruits and nuts*. Edited by E.W.M. Verheij and R.E. Coronel. Pudoc, Leiden, the Netherlands. pp. 296–298.

Martins, A.B.G., and Antunes, E.C. 2000. The air layering propagation of rose apple (*Syzygium jambos* (Linn.) Alston). *Rev. Bras. Frutic*. 22:205–207 (in Portuguese, English summary).

Maurya, K.R. 1977. Rose apple for home garden. *Inten. Agric*. 15(2):15–16.

Morton, J. 1987. Rose apple. In *Fruits of warm climates*. Creative Resource Systems, Winterville, NC. pp. 383–386.

Rao, B.G.V.N., and Nigam, S.S. 1969. Chemical examination of the essential oil from *Eugenia jambos* Linn. *Perfumery Essent. Oil. Rec*. 60:282–286.

Rattanapanone, N., Chongsawat, C., and Chaiteep, S. 2000. Fresh-cut fruits in Thailand. *HortScience*, 35:543–546.

Saha, A.K. 1969. Some investigations of vegetative propagation of Jaman (*Syzygium jambos* L.). *Allahabad Farmer*, 43:187–189.

Slowing, K., Carretero, E., and Villar, A. 1996. Anti-inflammatory compounds of *Eugenia jambos*. *Phytother. Res*. 10(Suppl.):S126–S127.

Slowing, K., Sollhuber, M., Carretero, E., and Villar, A. 1994. Flavonoid glycosides from *Eugenia jambos*. *Phytochemistry*, 37:255–258.

Vernin, G., Vernin, G., Metzger, J., Roque, C., and Pieribattesti, J.C. 1991. Volatile constituents of the Jamrosa aroma *Syzygium jambos* L. Aston from Reunion Island. *J. Essent. Oil Res*. 3:83–97.

Whistler, W.A. 1988. A revision of *Syzygium* (Myrtaceae) in Samoa. *J. Arnold Arbor*. 69:167–192.

Young, H., and Paterson, V.J. 1990. The flavour of fruits. In *The flavour of exotic fruit. Food flavours, Part C*. Edited by I.D. Morton and A.J. MacLeod. Elsevier, Amsterdam, the Netherlands. pp. 281–286.

Specialty Cookbooks

Bhumichitr, V. 2008. *Gourmet Thai in minutes: over 120 inspirational recipes*. Kyle Books, Lanham, MD. 160 pp.

Ezard, T. 2007. *Lotus: Asian flavors*. Periplus, Singapore. 256 pp.

Sullivan, C. 1995. *Classic Jamaican cooking: traditional recipes and herbal remedies*. 3rd ed. Serif, London. 192 pp.

82 Roselle

Family: Malvaceae (mallow family)

NAMES

Scientific Name: *Hibiscus sabdariffa* L.

- The English name "roselle" appears to be a corruption of the French word *oseille*, sorrel (*Rumex* species), which has a similar taste.
- Roselle is also called African mallow, East Indian sorrel, Florida cranberry, Guinea sorrel, hibiscus, hibiscus tea flower, Indian sorrel, Jamaica dogwood, Jamaica sorrel, Jamaica tea, Jamaica, Java jute, jelly okra, lemon bush, mesta (applied to the fiber type; the seed oil may be called mesta oil), Natal sorrel, Queensland jelly plant, red sorrel, red tea, rosella, rosella hemp (also refers to the fiber from the plant), rouselle, rozelle, rozelle hemp plant, sorrel (better reserved for true sorrel, *Rumex* species), sour-sour, Sudanese tea, thorny mallow, and various-leaved hibiscus.
- "White sorrel" is a variety of roselle (i.e., red sorrel) with green (instead of red) calyx lobes.
- The fiber from the species has been called India rosella hemp, rosella fiber, rosella hemp, rozelle hemp, and Pusa hemp. Occasionally, some of these names are applied to the plant.
- The tropical African red-leaf hibiscus or false roselle (*H. acetosella* Welw. ex Hiern) is grown as an ornamental in Florida and elsewhere and is often confused with *H. sabdariffa*. Unlike the latter, its floral calyx is not fleshy and eaten, although the young leaves are consumed for their acid flavor, usually cooked with rice or vegetables.
- The genus name *Hibiscus* is based on an old Greek and Latin name for some large mallow plant.
- *Sabdariffa* in the scientific name *H. sabdariffa* was the name by which the roselle was first known. The name is of Turkish origin, suggesting that the plant was transported to the West through the early Turkish Empire.

PLANT PORTRAIT

Roselle is an erect shrub, native to the Old World tropics from India to Malaysia, but probably transported in early times to sub-Saharan Africa. Although normally grown as an annual, it is a perennial. There are both edible kinds and types used as sources of fiber. Forms that are grown for their fiber are tall, less branched, and sometimes more than 3.5 m (11 ft.) in height, rarely up to 5 m (16 ft.), whereas culinary varieties are much branched, bushy, and generally 1 to 2 m (3–6½ ft.) tall. The stems are green or red. The young plants have leaves that are unlobed, but as the plant grows, the later-developed leaves are shallowly to deeply, palmately three or five parted (sometimes seven parted). The large flowers have pale yellow petals (sometimes suffused with pink or red) and a dark red eye. The sepals of flowers are just below the petals and are collectively called the calyx. This is the most edible part of roselle. Its sepals vary from dark purple to bright red (sometimes white) at maturity, and they are quite fleshy, especially when the flower turns into a fruit. The calyx increases from 1 to 2 cm (3/8–3/4 in.) before the flower is fertilized to approximately 5.5 cm (2 in.) in length at maturity. At maturity, when the calyces are edible, the succulent calyx encloses a seed pod. At the base of the five-lobed calyx are 8 to 12 "epicalyx" bracts (tiny leaves), which are also edible, although the base is usually discarded. Some forms of roselle contain a red pigment that gives a brilliant red color

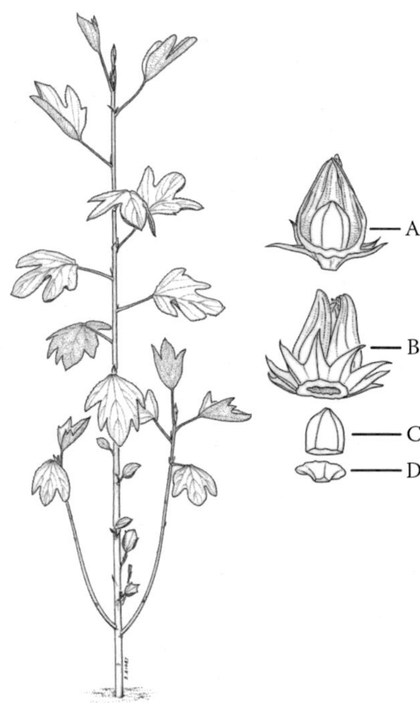

FIGURE 82.1 Roselle (*Hibiscus sabdariffa*), by S. Rigby. Left: vegetative stem. (A) Fruit with front bracts removed to show enclosed seed capsule. (B) Edible portion. (C) Inedible seed capsule removed through bottom. (D) Bottom of calyx.

to culinary products made from the plant, other forms are completely green. Edible types of roselle are usually succulent, have well-developed lateral branches, and lack a hairy covering, whereas types grown for fiber are less succulent, have few lateral branches, and are often hairy or prickly.

Roselle may have been domesticated in western Sudan before 4000 BC. It was not recorded in Europe until 1576. It seems to have been carried from Africa to the New World by slaves for use as a food plant. The species was called Jamaican sorrel in 1707 in Jamaica, where the regular use of the calyces as food seems to have first been practiced. The use of the plant as "greens" was known in Java as early as 1658. Taken to the New World, roselle was cultivated in Mexico, parts of Central America, the West Indies, and in the late nineteenth century in southern Florida, Texas, and California. Today, roselle is grown for culinary purposes in tropical and subtropical areas, particularly in the Sudan, China, Thailand, and Mexico, and is exported from the first three countries. Large amounts were grown for jam in Australia about the end of the nineteenth century. As a food plant, it grows well in Florida and California and has been cultivated for food purposes as a garden curiosity as far north as Michigan.

Roselle is primarily cultivated as a fiber plant of the tropics. The fiber is obtained from the inner bark. The species is grown for fiber in the drier parts of west and central Africa (particularly Nigeria) and in tropical America, but the major areas of production are in India and several other Southeast Asian countries. The fiber strands are up to 1.5 m long and are useful for cordage and in manufacturing burlap. Roselle is sometimes also grown for seed oil, which is edible. Different parts of roselle are used medicinally.

CULINARY PORTRAIT

Roselle is a multipurpose food plant. Its seeds are eaten roasted or ground into meal. Although somewhat bitter, the meal is high in protein. The pleasantly acidic leaves and young shoots are

FIGURE 82.2 Flowering branch of roselle (*Hibiscus sabdariffa*), by M. Jomphe.

consumed raw or cooked, either as a sour-flavored vegetable or as a condiment. Cooked roselle goes well with other vegetables, meat, and fish. The juice (from leaves, stems, or calyxes) is squeezed out and used in fresh or fermented drinks and in flavoring extracts and syrup. Fresh calyces are used for wine, jelly, syrup, gelatin, beverages, pudding, and cakes, and dried calyces are used for tea (especially popular in Egypt), jelly, marmalade, ices, ice cream, sherbets, butter, pies, sauces, tarts, and other desserts. The fleshy calyces are also eaten in chutneys, preserves, fruit salads, and in other confections, much like cranberries. In southern Florida, where roselle is called Florida cranberry, the fleshy calyces are used in preparing sauce or jelly. The calyces contain more than 3% pectin, and it is generally not necessary to add pectin to make a firm jelly. In India, the calyces are used in making soups, puddings, cakes, and for coloring syrups and liqueurs. In the West Indies, roselle is used to color and flavor rum.

In preparing culinary products with roselle, it should be kept in mind that the juice is quite acidic, so that aluminum, copper, and iron containers are best avoided. Acidic materials interact with these metals and tend to produce toxic ions of the metals.

Culinary Vocabulary

- In Switzerland, the edible calyx is called *karkadé*. The calyx is also known as *karkadé* or *carkadé* in North Africa and the Near East as well as in the pharmaceutical and food flavoring trades of Europe. The calyces have also been called karkade flowers in the commercial trade.
- In stores, dried roselle may be labeled "Jamaica flowers," "hibiscus flowers," or "flor de Jamaica."
- In Mexico and elsewhere, *Jamaica* (pronounced hah-my-ka) is a traditional, red, Mexican beverage made primarily from roselle. *Agua fresca de Jamaica* ("roselle flower water") is another roselle beverage popular in much of Latin America.

- "Sudan tea" is much more of a medicinal than a culinary preparation. This is an infusion of the calyx, combined with salt, pepper, asafetida, and molasses in East Africa as a cure for biliousness.

CURIOSITIES OF SCIENCE AND TECHNOLOGY

- Roselle is cultivated throughout West Africa, especially in Sierra Leone. Red varieties of roselle are used by traditional healers as a cure for patients. However, in some regions, the healers are not themselves allowed to eat roselle, as it is understood that if they do they will die.
- In promoting the use of roselle in the early twentieth century, making wine from it was considered one of its good points. With Prohibition of alcoholic beverages in the United States (1919–1933), a different but rather resourceful argument was used: "the prohibition wave has produced a greatly increased demand for soft drinks, and roselle sirup as a flavoring extract might well be considered for introduction to the public by the dispensers of soft drinks."
- An experiment showed that chickens become less inebriated from consuming alcohol after eating roselle extracts. Curiously, in Guatemala, roselle-"ade" is a favorite remedy for hangovers.
- Roselle seeds have been used as an allegedly aphrodisiac coffee substitute.
- Roselle flowers are short-lived and are generally open for just 1 day.
- Approximately 90% of the fresh, edible floral calyx is water.

KEY INFORMATION SOURCES

Abdallah, M.S., El-Kholy, S.A., and M Sa'ad, F. 1976. *Taxonomic studies in the flora of Egypt: culinary cultivars of* Hibiscus sabdariffa *L. in Egypt*. Herbarium, Agricultural Research Centre, Cairo, Egypt. 20 pp.

Ahmad, S., and Vossen, H.A.A. van der. 2003. *Hibiscus sabdariffa* L. In *Plant resources of South-East Asia. Vol. 17. Fibre plants*. Edited by M. Brink and R.P. Escobin. Backhuys Publishers, Leiden, the Netherlands. pp. 162–167.

Babatunde, F.E., and Auwalu, B.M. 2003. Analyses of gowth, yield and agronomic factors of red variant roselle (*Hibiscus sabdariffa* L.). *Adv. Hortic. Sci.* 17:102–104.

Beattie, J.H. 1937. *Production of roselle*. U.S. Department of Agriculture, Washington D.C. 4 pp.

Boonkerd, T., Songkhla, B.N., and Thephuttee, W. 1993. *Hibiscus sabdariffa* L. In *Plant resources of South-East Asia. 8. Vegetables*. Edited by J.E. Siemonsma and K. Piluek. Pudoc Scientific Publishers, Wageningen, the Netherlands. pp. 178–180.

Commonwealth Agricultural Bureaux. 1970–1979. *Roselle* (Hibiscus sabdariffa)*:1956–75*. Commonwealth Agricultural Bureaux Annotated Bibliography Plant series no. G432. C.A.B., Farnham Royal, U.K. 9 pp.

Cooper, B. 1993. The delightful *Hibiscus sabdariffa*. *Tea Coffee Trade J.* 165(1):100–102.

Crane, J.C. 1949. Roselle—a potentially important plant fiber. *Econ. Bot.* 3:89–103.

Edmonds, J.M. 1991. *The distribution of* Hibiscus *L. section* Furcaria *in tropical East Africa*. International Board for Plant Genetic Resources, Rome, Italy. 60 pp.

El-Adawy, T.A., and Khalil, A.H. 1994. Characteristics of roselle seeds as a new source of protein and lipid. *J. Agric. Food Chem.* 42:1896–1900.

Jirovetz, L., Jager, W., Remberg, G., Espinosa-Gonzalez, J., Morales, R., Woidich, A., et al. 1992. Analysis of the volatiles in the seed oil of *Hibiscus sabdariffa* (Malvaceae) by means of GC-MS and GC-FTIR. *J. Agric. Food Chem.* 40:1186–1187.

Kerharo, J. 1971. Senegal bisap (*Hibiscus sabdariffa*) or Guinea sorrel or red sorrel. *Plant. Med. Phytother.* 5:277–281.

Lambourne, J. 1913. The roselle. *Agric. Bull. Federated Malay States*, 2:59–67.

Morton, J.F. 1975. Renewed interest in roselle (*Hibiscus sabdariffa* L.), the long-forgotten "Florida cranberry." *Proc. Fla. State. Hortic. Soc.* 87:415–425.

Morton, J.F. 1987. Roselle. In *Fruits of warm climates*. Edited by C.F. Dowling Jr. Media Inc., Greensboro, NC. pp. 281–286.

Mozaffari-Khosravi, H., Jalali-Khanabadi, B.-A., Afkhami-Ardekani, M., Fatehi, F., and Noori-Shadkam, M. 2008. The effects of sour tea (*Hibiscus sabdariffa*) on hypertension in patients with type II diabetes. *J. Hum. Hypertens*. 23:48–54.

Paull, R.E. 2008. *Hibiscus sabdariffa*, sorrel. In *The encyclopedia of fruit & nuts*. Edited by J. Janick and R.E. Paull. CABI, Wallingford, Oxfordshire, U.K. pp. 466–468.

Riley, J.M. 1973. The roselle. *California Rare Fruit Growers Yearbook*, 5:102–104.

Sarojini, G., Rao, K.C., Tulpule, P.G., and Lakshminarayana, G. 1985. Effects of processing on physico-chemical properties and fatty acid composition of *Hibiscus sabdariffa* seed oil. *J. Am. Oil Chem. Soc.* 62:728–730.

Sermsri, N., Duyapat, C., and Murata, Y. 1987. Studies on roselle (*Hibiscus sabdariffa* var. *altissima* L.) cultivation in Thailand: II. Effect of planting time, harvesting time and climatic factors on fiber yield. *Jpn. J. Crop Sci*. 56:64–69.

Sermsri, N., Tipayarek, S., and Murata, Y. 1987. Studies on roselle (*Hibiscus sabdariffa* var. *altissima* L.) cultivation in Thailand: I. Growth analysis. *Jpn. J. Crop Sci*. 56:59–63.

Sharaf; A. 1962. The pharmacological characteristics of *Hibiscus sabdariffa* L. *Planta Med.* 10:48–52.

Wester, P.J. 1911. Contributions to the history and bibliography of the roselle. *Bull. Torrey Bot. Club*, 38:91–98.

Wester, P.J. 1912. Roselle, its cultivation and uses. *Philipp. Agric. Rev*. 5:123–132.

Wester, P.J. 1920. The cultivation and uses of roselle. *Philipp. Agric. Rev*. 13:89–99.

Whallon, A.P. 1954. Roselle: a forgotten plant. *Org. Gard.* 1(4):64–66.

Wilson, F.D. 1994. The genome biogeography of *Hibiscus* L. section *Furcaria* DC. *Genet. Resourc. Crop Evol.* 41:13–25.

Wilson, F.D. 1999. Revision of *Hibiscus* section *Furcaria* (Malvaceae) in Africa and Asia. *Bull. Nat. Hist. Mus. Bot*. 29(1):47–79.

Wilson, F.D., and Menzel, M.Y. 1964. Kenaf (*Hibiscus cannabinus*), roselle (*Hibiscus sabdariffa*). *Econ. Bot.* 18:80–91.

Wong, P.K., Yusof, S., Ghazali, H.M., and Che Man, Y.B. 2002. Physico-chemical characteristics of roselle (*Hibiscus sabdadriffa* L.). *Nutr. Food Sci*. 32(2/3):68–73.

Specialty Cookbooks

El-Baz, M. 2006. *The essence of herbal and floral teas*. iUniverse, New York. 86 pp.

Harris, J.B. 1998. *The Africa cookbook: tastes of a continent*. Simon & Schuster, New York. 382 pp.

Orosa, M.Y. 1932. *Roselle recipes*. Bureau of Printing, Manila, Philippines. 24 pp.

Wolfe, L. 1985. *Recipes: the cooking of the Caribbean islands*. Macmillan, London. 143 pp.

83 Saffron

Family: Iridaceae (iris family)

NAMES

Scientific Name: *Crocus sativus* L.

- The word "saffron" is derived from the Arabic *za'faran*, meaning yellow. Before the advent of coal-tar dyes, saffron was used to produce a popular yellowish-orange dyestuff.
- Saffron is also known as common saffron, dye-saffron, saffron crocus, and true saffron.
- "Mexican saffron" is a spice preparation of flowers of safflower (*Carthamus tinctorius* L.). "Portuguese saffron" is also safflower.
- Saffron should not be confused with other species called "crocus" because some, such as autumn crocus or meadow saffron (also known as "naked ladies"), *Colchicum autumnale* L., are extremely poisonous.
- The genus name *Crocus* is based on *krokus*, the Greek name for saffron.
- *Sativus* in the scientific name *C. sativus* is Latin meaning sown or cultivated.

PLANT PORTRAIT

Saffron is a perennial herb 15 to 30 cm (6–12 in.) tall, which grows from *corms* (condensed stems) that are produced annually. The corms are produced underground and are somewhat depressed, globular, onion shaped, white, and fleshy. Each corm produces six to nine narrow leaves. In the fall, saffron gives rise to one (sometimes two or even three) lily-shaped flowers that are lavender, violet, red-purple, bluish, or white. Each corm flowers only once (in the year following its development), giving rise to new corms at its base for the next year before withering and dying. Saffron is a sterile plant (it does not produce viable seeds) and is known only in cultivation. *Crocus cartwrightianus* Herb. is a very similar wild species of Greece that is sometimes cultivated. This wild plant, which is probably an ancestor of saffron, was gathered in antiquity and is still harvested for saffron.

The female reproductive part of flowering plants, known as the pistil, is made up of the ovary, style, and stigma. Stigmas are the receptive platforms for pollen and are sticky in many plants. Pollen grains germinate on the stigma and grow through the style to reach and fertilize the ovules (eggs) in the ovary. Saffron styles are extremely long, commencing at about ground level. The ovaries are subterranean in saffron, although in fertile species of *Crocus* seed pods develop so that they are above soil level, allowing seed distribution. The stigmas of saffron are divided into three long lobes, atop the style. The brilliantly colored stigmatic lobes of saffron flowers are the economically important part. The tops of the styles are often collected too, but their presence in high-quality saffron is limited.

Iran is the world's largest supplier of saffron, responsible for the majority of all production. Spain furnishes much of the world's supply of saffron (because of the value of saffron, it is called "red gold" there), and the best is said to come from La Mancha. Other significant saffron producers include Kashmir and Greece, and several other countries commercially cultivate small amounts. Saudi Arabia, Kuwait, and the United States are the leading consumers of saffron. Throughout its recorded history, saffron has been a commodity of great commercial value. Today, it is the world's most expensive spice, averaging approximately $2200.00 per kg ($1000.00 per lb.), occasionally selling for up to five times as much. Symbolically, a small amount colors a large dish a brilliant gold.

FIGURE 83.1 Saffron (*Crocus sativus*). (From Köhler, 1883–1914.)

Saffron has been used as a condiment, medicine, dyestuff, and perfume since antiquity and was known by the classical Egyptians, Greeks, Jews, Romans, Hindus, and Muslims. It or its wild ancestor was recorded in Crete of 1600 BC. Saffron is mentioned in classical Greek and Roman literature and in subsequent writings, including that of Shakespeare. It has been said to be mentioned in the Old Testament, but this may be a misinterpretation. It was introduced to Spain by the Arabs in the tenth century. Saffron reached France and Germany in the eleventh century and England in the fourteenth century. It became very popular for flavoring and coloring food and even for use as a perfume. There were harsh penalties, including death, during the Middle Ages to discourage adulteration of saffron. Today, adulteration continues to be a problem, even urine being added to increase color. Saffron is often mixed with raw materials of similar color, such as the flowers of marigolds (species of *Calendula* and *Tagetes*), safflower, and other plants. Petals and styles of the saffron flower are also used to adulterate saffron. Buying powdered saffron, which is much more difficult to keep from being adulterated, is a bad idea, and it is far preferable to purchase whole threads.

CULINARY PORTRAIT

Saffron's major use is in coloring and flavoring foods, including cheese, butter, cream, sauce, meat, poultry, seafood, soups, stews, preserves, puddings, confectionery, pastry, liquor, and nonalcoholic beverages. Some liqueurs, such as chartreuse, may contain saffron. Saffron is an unusually versatile spice, combining well with both sweet and savory dishes. Saffron seems to go especially well with rice. It is considered an essential ingredient of several well-known dishes (see the Culinary vocabulary section). Spain flavors many of its unique food preparations with this condiment. The Arabs, said to be saffron's greatest consumers, use it with lamb, chicken, and rice dishes. Saffron cakes and loaves are very popular in parts of the United Kingdom. Saffron is also popular in Iranian, Moroccan, central Asian, Indian, and European cuisines.

Commercial saffron is a compressed, very aromatic, matted mass of slim, dark orange to reddish-brown, occasionally yellow strands 2 to 3 cm (0.8–1.2 in.) long. The deeper the color the better the quality. Ground saffron is also available, but as it may be adulterated, the thread form is a more reliable product. Whole material can be used, or individual strands can be added intact or crushed. Because the odor is so sharp and penetrating and in view of the possibility of toxicity, no more than called for in recipes should be used. Saffron threads should not be added directly to foods; they should be placed in a little warm water until it takes up the strong yellow color and aroma, and the liquid then poured into the dish being prepared. Because saffron is so expensive, knowledgeable chefs never use wooden utensils when mixing saffron because wood tends to absorb it. The same is true of rubber and other porous materials. When added to butter or oil, mixtures with saffron should not be browned at high temperatures as this will result in deterioration of the flavor. The odor of saffron is tenaciously and characteristically pungent, sweet, spicy, and flowery; the taste is pleasantly spicy and bitter. People who are not used to it may find it unpleasant or too strong, masking and overpowering the other flavors of the dish and producing a medicinal taste.

Saffron is less widely used today than in the past, when it was often added to soups, sauces, and dishes for Lent. The high cost has encouraged the alternative use of turmeric (see Chapter 9) and the synthetic colorant tartrazine. Turmeric is an inexpensive and effective saffron substitute, but only a little should be used because its acrid flavor can easily overwhelm a dish.

Culinary Vocabulary

- Dishes that characteristically use saffron include French bouillabaisse (shellfish and fish, a famous soup of Marseille, and indeed the most famous of all Mediterranean fish dishes); several Spanish dishes: Arroz con Pollo (chicken and rice), Bacalao a la Vizcaina (codfish), zarzuela (fish and rice), and the Spanish national dish paella (meat, fish, rice, vegetables); the Italian risotto Milan style (rice cooked in meat stock); and the more expensive curries in the East, such as chicken biriana (biriani or biryani is a highly spiced Indian rice dish flavored with saffron and layered with meat such as lamb or chicken, and served on special occasions).
- Chelow kabab, the national dish of Iran, consists of saffroned rice and kebabs (meats, especially lamb and beef, wrapped in bread, accompanied by vegetables and condiments).
- Schwenkfelder is a traditional wedding cake of the Pennsylvania Dutch, which is flavored with saffron.

CURIOSITIES OF SCIENCE AND TECHNOLOGY

- Because of its expense, saffron was used by those wishing to flaunt their wealth. In the first century, the emperor Nero had saffron sprinkled on the streets of Rome for his entry into the city. Heliogabalus, another Roman emperor, in the third century, is said to have bathed in saffron-scented water. In India, saffron was used to show the caste markings of the wealthy.
- Saffron was a sacred plant in Egypt at the time of the Pharaohs. The Egyptians dedicated saffron to Thoth, the god of wisdom and magic, and called saffron "the blood of Thoth." Tiny bits of saffron stigmas have been found in Egyptian mummies. Cleopatra (69–30 BC) used saffron in her cosmetics.
- The Phoenicians thrived between the second century BC and the first century AD. They prepared moon-shaped cakes flavored with saffron in honor of their fertility goddess, Ashtoreth, and thought that the saffron would make them more potent and fertile.
- Metrodora was a Greek female physician who, between 550 and 650, wrote what is believed to be the oldest surviving medical text by a woman. In this she gave a herbal cure for hemorrhoids, which included finely ground pine cones, saffron, and wine-soaked dates.
- Medieval scribes burnished saffron upon foil for use in illuminated manuscripts in the same manner that gold was used for the purpose.

- King Henry I of England (1068–1135), became very fond of saffron, supplied by returning crusaders. When the ladies of his court started using up the entire saffron supply to dye their hair, he forbade them to use his favorite spice. Similarly, ladies risked the wrath of the fathers of the early Christian Church by using saffron to give their hair a yellow tint.
- In China (indeed in much of the world), the golden color of saffron led to its association with royalty and divinity. Statues of the Buddha were frequently colored gold, and so were the robes of Buddhist priests (with turmeric because it is much cheaper than saffron). To indicate their importance, some Chinese rubbed saffron on themselves.
- Swans and peacocks were sometimes served "endored" at lavish Christmas feasts in the Middle Ages. The flesh was painted gold with saffron dissolved in melted butter.
- Saffron came to be viewed as capable of reviving spirits. It was even put in canaries' drinking water so that they would sing more cheerfully.
- Romantic stories have been told of how saffron was brought to England despite the efforts of Arabs to maintain a monopoly in the Near East. By legend, saffron was smuggled into England at the end of the fourteenth century by a pilgrim from the Holy Land who risked his life by hiding a corm in his hollow staff. The same story is told about how silkworm eggs were smuggled into England, so it seems one good legend stimulates another.
- In Ireland, it used to be believed that saffron had a sanitizing effect, and so it was thought unnecessary to frequently launder clothes that were dyed with saffron. King Henry the VIII (1491–1547), banned the Irish from dyeing their linen with saffron to encourage them to wash their linen more frequently.
- In 1444 in Nürnberg, Germany, 1456 people went to the stake for adulterating saffron.
- Approximately 160,000 flowers are needed to produce 1 kg of the highest-quality saffron (approximately 75,000 flowers for 1 lb.), all hand-picked. An experienced harvester can pick 10,000 to 12,000 flowers in a day. The season of available flowers only lasts 7 to 10 days, and during this time a harvester can pick a kilogram (2.2 lb.) of material, at most.
- Saffron was a highly esteemed medicine in the past. In 1671, a book entitled *Crocologia*, with almost 300 pages, was published, providing instructions for the treatment of diseases using saffron. Illnesses that have been treated with saffron include menstrual problems, dysentery, measles, jaundice, fever, scarlet fever, smallpox, colds, asthma, enlargement of liver, enlargement of spleen, infections of urinary bladder and kidney, melancholia, insomnia, cholera, diabetes, cancer, and more. Saffron was also used as an abortifacient, often killing the women to whom it was administered. The species is used to a minor extent today in modern Chinese medicine, and occasionally in India and the Far East as a tonic, but is essentially obsolete in modern Western medicine. Excessive use can cause inflammation of mucous membranes, cramps, and bleeding. Symptoms of poisoning can include narcosis and extravagant gaiety.
- The reddish-yellow pigment of saffron, crocin, is so effective that one part can color up to 150,000 parts of water unmistakably yellow.
- Biotechnology may offer an alternative to cultivating saffron. It has been suggested that any compound as expensive as saffron deserves to be considered for production by plant cell culture technology. Attempts have been made to induce high-value metabolites of saffron in tissue culture, with a view to large-scale production of saffron by artificially grown tissues.

KEY INFORMATION SOURCES

Agayev, Y.M., 2002. New features in karyotype structure and origin of saffron, *Crocus sativus* L. *Cytologia*, 67:245–252.

Alonso, G.L., Salinas, M.R., and Saéz, J.R. 1998. Crocin as coloring in the food industry. *Recent Res. Dev. Agric. Food Chem.* 2:141–153.

Basker, D. 1993. Saffron, the costliest spice: drying and quality, supply and price. *Acta Hortic.* 344:86–97.

Basker, D., and Negbi, M. 1983. Uses of saffron. *Econ. Bot.* 37:228–236.

Caiola, M.G., and Chichiricco, G. 1991. Structural organization of the pistil in saffron *Crocus sativus* L. *Isr. J. Bot. Basic Appl. Plant Sci.* 40:199–207.

Caiola, M.G., Caputo, P., and Zanier, R. 2004. RAPD analysis in *Crocus sativus* L. accessions and related *Crocus* species. *Biol. Plant*, 48:375–380.

Dhar, A.K., and Mir, G.M. 1997. Saffron in Kashmir. VI. A review of distribution and production. *J. Herbs Spices Med. Plants*, 4(4):83–90.

Escribano, J., Alonso, G.L., Coca-Prados, M., and Fernández, J.A. 1996. Crocin, safranal and picrocin from saffron (*Crocus sativus* L.) inhibit the growth of human cancer cells in vitro. *Cancer Lett.* 100:23–30.

Fernández, J.A., and Abdullaev, F.I. (Eds.). 2004. *Proceedings of the 1st International Symposium on Saffron Biology and Biotechnology, Albacete, Spain, October 22–25, 2003.* International Society for Horticultural Science, Wageningen, the Netherlands. 499 pp.

Hassnain, F.M. 1998. *Saffron cultivation in Kashmir.* Rima Pub. House, New Delhi, India. 159 pp.

Homes, J., Legros, M., and Jaziri, M. 1987. *In vitro* multiplication of *Crocus sativus* L. *Acta Hortic.* 212:675–676.

Ingram, J.S. 1969. Saffron (*Crocus sativus* L.). *Trop. Sci.* 11:177–184.

Mathew, B. 1977. *Crocus sativus* and its allies (Iridaceae). *Plant Syst. Evol.* 128:89–103.

Mathew, B. 1982. *The crocus: a revision of the genus* Crocus *(Iridaceae).* Timber Press, Portland, OR. 127 pp. + plates.

McGimpsey, J.A., Douglas, M.H., and Wallace, A.R. 1997. Evaluation of saffron (*Crocus sativus* L.) production in New Zealand. *N. Z. J. Crop Hortic. Sci.* 25:159–168.

Nair, S.C., Kurumboor, S.K., and Hasegawa, J.H. 1995. Saffron chemoprevention in biology and medicine: a review. *Cancer Biother.* 10:257–264.

Negbi, M. 1999. *Saffron:* Crocus sativus *L.* Harwood Academic, Amsterdam, the Netherlands. 154 pp.

Rees, A.R. 1988. Saffron—an expensive plant product. *Plantsman.* 9:210–217.

Rios, J.L., Recio, M.C., Giner, R.M., and Manez, S. 1996. An update review of saffron and its active constituents. *Phytother. Res.* 10:189–193.

Sampathu, S.R., Shivashankar, S., and Lewis, Y.S. 1984. Saffron (*Crocus sativus* Linn.)—cultivation, processing, chemistry and standardization. *CRC Crit. Rev. Food Sci. Nutr.* 20:123–157.

Souret, F.F., and Weathers, P.J. 1999. Cultivation, in vitro culture, secondary metabolite production, and phytopharmacognosy of saffron (*Crocus sativus* L.). *J. Herbs Spices Med. Plants*, 6(4):99–116.

Souret, F.F., and Weathers, P.J. 2000. The growth of saffron (*Crocus sativus* L.) in aeroponics and hydroponics. *J. Herbs Spices Med. Plants*, 7(3):25–35.

Tammaro, F., and Marra, L. (Eds.). 1990. *Proceedings of the International Conference on Saffron* (Crocus sativus *L.*), *L'Aquila, Italy, October 27–29, 1989.* Università Degli Studi, L'Aquila, Italy. 335 pp.

Tarantilis, P.A., and Polissiou, M.G. 1997. Isolation and identification of the aroma components from saffron (*Crocus sativus*). *J. Agric. Food Chem.* 45:459–462.

Tsimidou, M., and Biliaderis, C.G. 1997. Kinetic studies of saffron (*Crocus sativus* L.) quality deterioration. *J. Agric. Food Chem.* 45:2890–2898.

Velasco-Negueruela, A. 2001. Saffron. In *Handbook of herbs and spices.* Edited by K.V. Peter. Woodhead Publishing, Cambridge, U.K. pp. 276–286.

Visvanath, S., Ravishankar, G.A., and Venkataraman, L.V. 1990. Induction of crocin, crocetin, picrocrocin, and safranal sythesis in callus cultures of saffron—*Crocus sativus* L. *Biotechnol. Appl. Biochem.* 12:336–340.

Willard, P. 2001. *Secrets of saffron: the vagabond life of the world's most seductive spice.* Beacon Press, Boston, MA. 225 pp.

Specialty Cookbooks

Humphries, J. 1996. *The essential saffron companion.* Ten Speed Press, Berkeley, CA. 160 pp.

Loukie, W. 1999. *Saffron, garlic & olives.* Fisher Books, Tucson, AZ. 137 pp.

Strong-Church, V.A., and Church, J. 1980. *Mediterranean gold: the historical romance of saffron, the world's most precious spice, and an international guide to culinary alchemy.* Strong-Church Enterprises, Pleasant Hill, CA. 96 pp.

84 Sago Palm

Family: Arecaceae (Palmae; palm family)

NAMES

Scientific Name: *Metroxylon sagu* Rottb. (*M. rumphii* (Willd.) C. Mart.)

- The name "sago" is based on the Malay/Javanese word for the sago palm, *sagu*, a term which is in the scientific name *M. sagu*.
- Pronunciation: SAY-goh.
- The sago palm is also known as the true sago palm (in contrast to other plants used to produce "sago starch").
- "Sago starch" is usually obtained from the sago palm, but also sometimes from other palms (such as species of *Caryota* and *Corypha*), and from cycads. *Cycas revoluta* is often called sago palm or king sago palm (see Chapter 32).
- The genus name *Metroxylon* is based on the Greek *metra*, heart of a tree + *xylon*, wood, referring to the large pith (soft center of the trunk).

PLANT PORTRAIT

"Sago" is an edible starch, extracted from the pith-like center of several East Asian palms, mostly sago palm, but also as noted above from cycads. The sago palm is a tree, 6 to 15 m (20–49 ft.) high with large leaves composed of many leaflets. The stems are often protected by sharp thorns, but there are also thornless varieties. The sago palm is a plant of hot humid tropics, growing wild in freshwater swamps of Southeastern Asia and Oceania, and probably originating from New Guinea and/or the Moluccas area. It flowers and produces fruit just once, about the age of 15 years, and dies (at least the main stem dies; the plant suckers abundantly, producing new plants from the base). Before flowering, the plants build up starch reserves in the center of the trunk, and at this stage they are cut down and the soft material is scooped out. The starch is extracted by repeated washing and straining and is dried to yield sago meal and sago of various textures ranging from granulated to "pearl sago." A single sago palm yields 150 to 300 kg (331–661 lb.), sometimes as much as 400 kg (880 lb.) of starch. Use of the sago palm has a very long history, and indeed it has been described as humankind's oldest food plant. Sago palm is harvested from both wild and cultivated trees in Southeast Asia and Oceania. Malaysia and Indonesia are the chief exporting countries. Sago starch is an important dietary staple in some parts of eastern Asia and is exported for use in foods and for stiffening textiles. As a global food resource, sago is quite minor, the crop producing only approximately 1.5% of the world's production of starch.

Aside from its food uses, in its native distribution area tribal peoples widely use sago palms as construction material for huts, fencing, furniture, tools, and toys. A "porridge" made from sago starch is used as an adhesive, as a sealant for cracked pottery, and as a dressing for wounds.

CULINARY PORTRAIT

Sago starch is very bland, and sweeteners are often used to provide flavor. "Tapioca pudding" can be made with starch from either sago or cassava (*Manihot esculenta* Crantz), and sometimes cassava starch is called sago starch. Sago is best known in Western cooking in the form of sago pudding.

FIGURE 84.1 Sago palm (*Metroxylon sagu*). (From Rumphius, 1741–1750.) Note young plant developing as a sucker near the base of the main plant.

It was also once commonly added to thicken soups and sauces and made into dishes for invalids. However, the popularity of sago in Western nations has declined. The grainy texture of sago pudding has become much less accepted than today's creamy puddings. Sago remains an important staple carbohydrate for many people in Southeast Asia, Oceania, and the Pacific Islands, where the plants grow. Although native inhabitants in these regions use sago to make gruels and other simple dishes, in Southeast Asia it is also used for making noodles, confectionery, and various baked products. A recent novelty is the addition of processed pea size sago pearls to iced tea drinks in which they float and are eaten.

Culinary Vocabulary

- "German sago" is a potato flour that has been colored, either with iron oxide or with sugar burnt to a yellowish white.
- "Sap Sago" is a brand of cheese (sometimes sold as "Swiss Green Cheese"), manufactured in Switzerland, where it is known as schabziger, since the eighth century.

CURIOSITIES OF SCIENCE AND TECHNOLOGY

- The Sawiyano people of East Sepik Province in Papua New Guinea use what are called sago spathe paintings in the initiation of young boys and for special men's ceremonies. A "spathe" is a leafy covering of the young flowering axis of the palm family as well as several other families, and this is simply peeled away from the plant and used as a canvas for painting (the "bark" of the plant is also similarly used; technically, palm plants do not have a true bark). Stylized, often intricately patterned sago spathe paintings, which are often housed in museums today, are part of the Sawiyano story of creation. It is believed that the creator, Awoufaise, first brought such paintings into being then created the Sawiyano (male) ancestors from the paintings.

- The "sape," often called "the boat lute" in the West, is a four-string, short-necked lute with an elongated body, used by the Kayan and Kenyah tribes of East Malaysia. According to legend, it accompanied a warrior headhunting dance called "Ngajat." Originally, sape strings were made from the sago palm, but they are now made with nylon.
- Sago grubs are the larvae of the capricorn beetle (species of the genus *Rhynchophorus*). They are very common in sago palms and frequently are collected as food by aboriginal groups living where the trees are found. The Asmat people of New Guinea rely on the sago palm and its grubs. During special festivals (which has included preparation for headhunting expeditions), an especially beautiful sago palm is selected and is dressed as a woman in a skirt of leaves. The tree is then cut down and attacked as if it were a fallen enemy. Holes are cut in the trunk to allow the sago beetles to lay their eggs, and the grubs are collected 6 weeks later and welcomed triumphantly.
- The fruits of *Metroxylon* species were once commonly used as a source of "vegetable ivory" and are still used for the purpose. The seeds are hard and ivory-like and were once widely carved into buttons in Europe. At present, some seeds are being used by indigenous peoples in Alaska instead of sea-mammal ivory to carve traditional crafts for tourists.
- If sago palms experience prolonged flooding, they often develop pneumatophores—roots that function as respiratory organs on top of the soil. These channel oxygen down to the underground roots, allowing them and the plants to survive.
- SAGO is an acronym for System Administrative and General Offices.
- "Sago spleen" is a diseased condition of the spleen in which the spleen develops tissue with scattered gray translucent bodies looking like grains of sago.

KEY INFORMATION SOURCES

Barreau, J. 1959. The sago palm and other food plants of marsh dwellers on the South Pacific Islands. *Econ. Bot.* 13:151–162.

Cecil, J.E. 1982. *The sago starch industry: a technical profile based on a preliminary study made in Sarawak.* Tropical Products Institute, Overseas Development Administration, London. 30 pp.

Ellen, R. 2004. Processing *Metroxylon sagu* Rottboell (Arecaceae) as a technological complex: a case study from South Central Seram, Indonesia. *Econ. Bot.* 58:601–625.

Ellen, R. 2006. Local knowledge and management of sago palm (*Metroxylon sagu* Rottboell) diversity in South Central Seram, Maluku, eastern Indonesia. *J. Ethnobiol.* 26:258–298.

Flach, M. 1997. *Sago palm,* Metroxylon sagu *Rottb.* International Plant Genetic Resources Institute, Rome, Italy. 76 pp.

Flach, M., and Schuiling, D.L. 1989. Revival of an ancient starch crop: a review of the agronomy of the sago palm. *Agroforest. Syst.* 7:259–281.

Food and Agriculture Organization. 1986. *The development of the sago palm and its products.* Report of the FAO/BPPT consultation, Jakarta, Indonesia, January 16–21, 1984. Food and Agriculture Organization, Rome, Italy. 252 pp.

Hackett, C. 1984. *Tabular descriptions of crops grown in the tropics. 2. Sago palm (*Metroxylon sagu *Rottb.).* CSIRO Institute of Biological Resources, Division of Water and Land Resources, Canberra, Australia. 51 pp.

Kainuma, K., Okazaki, M., Toyoda, Y., and Cecil, J.E. (Eds.). 2002. New frontiers of sago palm studies. In *Proceedings of the International Symposium on Sago (SAGO 2001): A New Bridge Linking South and North, Held on October 15–17, 2001, at the Tsukaba International Congress Center, Epochal Tsukuba, Japan.* Universal Academy Press, Tokyo, Japan. 388 pp.

Karim, A.A., Tie, A.P.L., Manan, D.M.A., and Zaidul, I.S.M. 2008. Starch from the sago (*Metroxylon sagu*) palm tree—properties, prospects, and challenges as a new industrial source for food and other uses. *Compr. Rev. Food Sci. Food Saf.* 7:215–228.

Rauwerdink, J.B. 1986. An essay on *Metroxylon*, the sago palm. *Principes*, 30:165–180.

Ruddle, K. 1978. *Palm sago: a tropical starch from marginal lands.* University Press of Hawaii, Honolulu, HI. 207 pp.

Schuiling, D.L. 1995. The variability of the sago palm and the need and possibilities for its conservation. *Acta Hortic.* 389:41–66.

Schuiling, D.L. 1996. *Metroxylon* Rottboell. In *Plant resources of South-East Asia. Vol. 9. Plants yielding non-seed carbohydrates*. Edited by M. Flach and F. Rumawas. Backhuys Publishers, Leiden, the Netherlands. pp. 116–120.

Schuiling, D.L., and Flach, M. 1985. *Guidelines for the cultivation of the sago palm*. Department of Tropical Crop Science, Wageningen Agricultural University, Wageningen, the Netherlands. 34 pp.

Schuiling, D.L., and Jong, F.S. 1996. *Metroxylon sagu* Rottboell. In *Plant resources of South-East Asia. Vol. 9. Plants yielding non-seed carbohydrates*. Edited by M. Flach and F. Rumawas. Backhuys Publishers, Leiden, the Netherlands. pp. 121–126.

Schuiling, D.L., Jong, F.S., and Flach, M. 1993. *Exploitation and natural variability of the sago palm (*Metroxylon sagu *Rottb.): report of a Sarawak and all-Indonesia study tour, January–February 1992*. Department of Agronomy, Wageningen Agricultural University, the Netherlands. 82 pp.

Stanton, W.R., and Flach, M. (Eds.). 1980. Sago, the equatorial swamp as a natural resource. In *Proceedings of the Second International Sago Symposium, Kuala Lumpur, Malaysia, September 15–17, 1979*. Nijhoff, The Hague, the Netherlands. 244 pp.

Subhadrabandhu, S., and Sdoodee, S. (Eds.). 1995. *Proceedings Fifth International Sago Symposium, Hat Yai, Songkhla, Thailand, 27–29 January 1994*. International Society for Horticultural Science, Leuven, Belgium. 278 pp.

Tan, H.T. 1982. Sago palm. A review. *Abstr. Tropic. Agric*. 8(9):9–23.

Tan, K. (Ed.). 1977. Sago-'76. In *Papers of the First International Sago Symposium: The Equatorial Swamp as a Natural Resource, July 5–7, 1976, Kuching, Malaysia*. Kemajuan Kanji, Kuala Lumpur, Malaysia. 330 pp.

Tarver, H.M., and Austin, A.W. 2000. Sago. In *The Cambridge world history of food*. Edited by K.F. Kiple and K.C. Ornelas. Cambridge University Press, Cambridge, U.K. pp. 201–207.

Tie, P.L.A., Karim, A.A., and Manan, D.M.A. 2008. Physicochemical properties of starch in sago palms (*Metroxylon sagu*) at different growth stages. *Starch/Starke*, 60:408–416.

Tomlinson, P.B. 1971. Flowering in *Metroxylon* (the sago palm). *Principes*, 15:49–62.

Tsiung, N.T., Liong, T.Y., and Siong, K.H. (Eds.). 1991. Towards greater advancement of the sago industry in the '90s. In *Proceedings of the Fourth International Sago Symposium, August 6–9, 1990, Kuching, Sarawak, Malaysia*. Ministry of Agriculture and Community Development, Kuching, Sarawak, Malaysia. 225 pp.

Yamada, N., and Kainuma, K. (Eds.). Sago-'85. In *The Third International Sago Symposium, Tokyo, Japan, May 20–23, 1985*. The Sago Palm Research Fund, Tropical Agricultural Research Center, Tskuba, Japan. 233 pp.

Specialty Cookbooks

General Foods Corporation. 1978. *Minute tapioca favorites*. 10th ed. General Foods Corp., White Plains, NY. 23 pp.

New, R. 1986, 1988. *Sago: the development of sago products in Papua New Guinea*. Liklik Buk Information Centre, Papua New Guinea. 56 pp.

Standard Brands. 1942. *Royal recipe parade: royal gelatins and puddings, 171 recipes, tapioca puddings*. Standard Brands, New York. 48 pp.

85 Sapodilla

Family: Sapotaceae (sapote family)

NAMES

Scientific Name: *Manilkara zapota* (L.) P. Royen (*Achras zapota* L., *Lucuma mammosa* C.F. Gaertn., *M. achras* (Mill.) Fosberg, *M. zapotilla* (Jacq.) Gilly, *Sapota achras* Mill., *S. zapotilla* (Jacq.) Coville)

- The name "sapodilla" is derived from the Spanish word *zapotilla*, meaning "small sapote" (for additional information, see Chapter 86).
- Suggested pronunciation: sap-oh-DEE-yuh.
- The sapodilla is also known as chicle, chicle tree, chico, chicopote, chicosapote, chicozapote, dilly, naseberry, nispero, sapodilla plum, sapota, sapote, and tree potato (the fruit resembles a potato).
- The name "sapote" is more often applied to other species (see Chapter 86).
- The "black sapodilla" is a Central American fruit tree, *Diospryos digyna* Jacq.
- The genus name *Manilkara* is based on the Malay name for these plants.
- *Zapota* in the scientific name *M. zapota* is from the Spanish *zapote*, referring to a persimmon-like fruit, from the Nahuatl (Aztec) *tzapotl*, a general term applied to several soft, sweet fruits. "Sapote" is based on the same root.

PLANT PORTRAIT

The sapodilla is an evergreen tropical tree growing 12 to 40 m (39–131 ft.), with a trunk diameter sometimes reaching 3.5 m (11 ft.). Large trees can yield 2000 to 3000 fruits in a year. The fruits are round, ovoid, ellipsoid, or conical, flattened at the stem end, 5 to 10 cm (2–4 in.) wide. The immature fruit is hard, gummy, and very astringent, and the thin skin is coated with a sandy brown, rough scurf until fully ripe. The flesh is light brown, yellowish brown, or reddish brown, tender, often gritty like a pear, becoming soft and very juicy, with a sweet pear-like flavor. The best varieties have a smooth (nongritty) texture. The fruit is somewhat similar to apple, with a central core of 2 to 12 flat, rectangular seeds (some fruit are seedless). Sapodilla is native to the Yucatan and possibly other parts of southern Mexico as well as Central America, where it has been cultivated since ancient times. It has been introduced throughout tropical America, the West Indies, and parts of tropical Asia. There are numerous cultivated varieties. The fruit is produced commercially in India, Philippines, Sri Lanka, Malaysia, Australia, Mexico, Venezuela, Guatemala, and some other Central American countries. India is the largest grower. The tree is widely planted in southern Florida, and the fruit is marketed locally and shipped to North American markets. In Mexico and Central America, it is considered to be one of the best of tropical fruit. Although uncommonly sold north of Mexico, sapodilla is becoming popular as a specialty fruit in restaurants in North America. Sapodilla fruit does not travel well and is best ripened on the tree and for these reasons is generally marketed near where it is grown. In areas remote from its centers of cultivation, sapodilla is not well known.

The bark contains a white, gummy latex, chicle, which is obtained by tapping. Chicle, which contains 15% rubber and 38% resin, is the traditional base of chewing gum. The dried latex was chewed by the Mayas and was introduced into the United States about 1866 (the first commercial chewing gum in the United States was made from spruce gum). In 1930, at the peak of production,

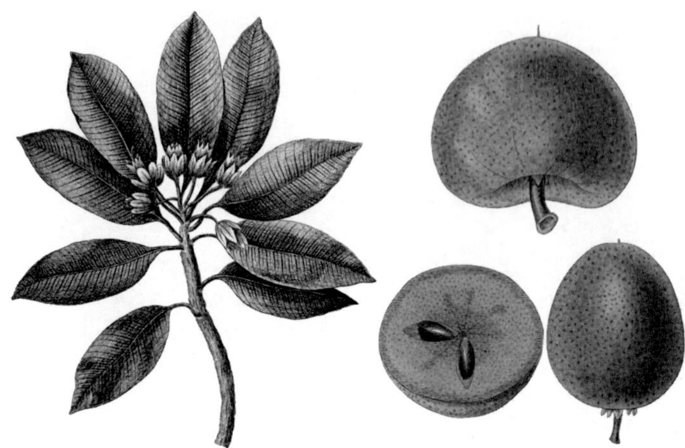

FIGURE 85.1 Sapodilla (*Manilkara zapota*). Left: flowering branch. (From Lamarck and Poiret, 1744–1829, plate 255.) Right: fruit. (From Curtis, 1831, vol. 58, plate 3112.)

approximately 7400 metric tons (approximately 7000 tons) of chicle were imported into the United States. For many years, chicle was the chief ingredient in chewing gum, but it has been replaced by latex from other species and by synthetic gums, although some countries, including Mexico, Venezuela, and Guatemala, still grow sapodilla for chicle.

CULINARY PORTRAIT

Sapodillas are best matured on the tree, as noted above. However, the fruit may be picked when immature for shipment in cold storage. The unripe fruit will ripen, and as unripe sapodillas are inedible before maturity because of a high content of bitter tannins, one should be sure that the fruit is ripe (an unripe fruit appears green when scratched with the thumbnail, whereas in a ripe fruit the scratch shows yellow). When ripe, sapodillas are said to melt in the mouth. Ripe sapodillas are generally cut in half, and the juicy, fragrant flesh eaten with a spoon as a dessert. Chilling improves the flavor, and indeed ripe fruit can be stored in the refrigerator. Gritty-fleshed varieties have been compared with "crunchy brown sugar" (one variety is in fact named 'Brown Sugar'). The sweet pulp has been compared with apricot and honey smelling of jasmine. The skin is not eaten, and the fruit is normally served peeled (the skin peels off easily). The fruit may simply be cut in half and the flesh scooped out with a spoon. The flesh can be added to fruit cups and salads and made into a sauce. In Asia, sapodillas are sometimes fried or stewed, and the flavor of cooked sapodillas has been described as delicious. Cooking with sugar changes the brown color of the flesh to red. Sapodilla fruit is sometimes used to flavor ice cream, sauces, and sorbets and occasionally is cooked to make jam. The fruit is often puréed or made into juice and sometimes canned. It is dried in India. An excellent wine can be prepared from sapodilla.

Care must be taken not to swallow the seeds, which have a protruding hook that might lodge in the throat. The seeds contain chemicals that can cause abdominal pain and vomiting if more than six are eaten. For the sake of safety, an apple corer can be used to remove the center core along with the seeds. The bitter kernels of the seeds have been used to make tea.

Culinary Vocabulary

- The sapodilla fruit is known by many curious names, including beef apple (a phrase also used for an apple cultivar, but mostly for beef dishes), chicoo (chikoo), dilly fruit (notably in Florida), and marmale plum.

CURIOSITIES OF SCIENCE AND TECHNOLOGY

- The latex from the sapodilla tree is used in tropical regions as a crude filling for tooth cavities.
- Sapodilla wood is famous for its durability. Carvings more than 1000 years of age have been found in Mayan ruins, the wood having withstood the tropical climate for this long period.
- More than 75% of the weight of a sapodilla fruit is water.
- General Antonio López de Santa Anna (1794–1876), was a Mexican general and politician. He was president of Mexico (1833–1836) and attempted to crush the Texan Revolution by seizing the Alamo and slaughtering its defenders in 1836, but later that year he was defeated and captured by Sam Houston (1793–1863), American general and president of the Republic of Texas. Santa Anna subsequently experienced periods of control of Mexico and exile from the country. About 1866, while he was at Staten Island, New York awaiting clearance to enter the United States during a period of exile after the Mexican revolution, he brought chicle gum with him, in the hope that this would lead to his fortune for the purpose of making tires. According to legend, Santa Anna gave inventor Thomas Adams Sr. the idea of selling chicle in the United States. In 1871, Adams received the first patent on a gum-making machine and began mass producing a chicle-based gum. His first gum, called "Adams New York No. 1—Snapping and Stretching," was pure chicle with no flavoring. This gum became the first product ever advertised on billboards. In 1884, Adams produced a licorice-flavored gum called "Adams' Black Jack," the first successful flavored gum in North America (an earlier venture with sassafras-flavored gum was a failure). The Adams family went on to acquire the famous brand of gum Chiclets, which in more recent times became the property of, in sequence, American Chicle, Warner-Lambert, and Pfizer Inc. Santa Anna died penniless; Adams died rich.
- A stick of chewing gum typically consists of approximately 60% sugar, 20% corn syrup, and flavorings, delivering approximately 9 calories. Gum chewing is said to be beneficial by stimulating the flow of saliva and thus aiding digestion. Gum chewing is also reputed to calm nervous tension.

KEY INFORMATION SOURCES

Aziz, I.A., Nordin, I.M., Shukor, A.R.A., and Izham, A. 2001. Predicting storage life of sapodilla (*Manilkara zapota* L.) by non-destructive technique. *J. Trop. Agric. Food Sci.* 29:93–97.

Balerdi, C.F., and Shaw, P.E. 1998. Sapodilla, sapote and related fruit. In *Tropical and subtropical fruits*. Edited by P.E. Shaw, H.T. Chan, Jr., and S. Nagy. Agscience, Auburndale, FL. pp. 78–136.

Broughton, W.J. and Wong, H.C. 1979. Storage conditions and ripening of chiku fruits *Achras sapota* L. *Sci. Hortic.* 10:377–385.

Campbell, C.W., and Malo, S.E. 1973. Performance of sapodilla cultivars and seedling selections in Florida. *Proc. Trop. Reg. Am. Soc. Hortic. Sci.* 17:220–226.

Campbell, C.W., Malo, S.E., and Goldweber, S. 1975. *The sapodilla. Fruit crops facts sheet.* University of Florida, Agricultural Extension Service, Gainesville, FL. 2 pp.

Chundawat, B.S. 1998. *Sapota:* Manilkara achras *(Mill.) Fosberg*. Agrotech Pub. Academy, Udaipur, India. 127 pp.

Coronel, R.E. 1992. *Manilkara zapota* (L.) P. van Royen. In *Plant resources in South-East Asia. Vol. 2. Edible fruits and nuts*. Edited by E.W.M. Verheij and R.E. Coronel. Prosea, Bogor, Indonesia. pp. 220–223.

Coronel, R.E. 1999. The sapodilla in Southeast Asia. *WANATCA Yearbook*, 23:3–9.

Egler, F.E. 1947. The role of botanical research in the chicle industry. *Econ. Bot.* 1:188–209

Ganjyal, G.M., Hanna, M.A., and Devadattam, D.S.K. 2003. Processing of sapota (sapodilla): drying. *J. Food Sci.* 68:517–520.

Gilly, C.L. 1943. Studies in the Sapotaceae. II. The sapodilla–nispero complex. *Trop. Woods*, 73:1–22.

Gonzalez, L.G., and Feliciano, P.A., Jr. 1953. The blooming and fruiting habits of the Ponderosa chico. *Philipp. Agric.* 37:384–398.

Gurulingaiah, K. 1976. *Select bibliography on sapota*. University of Agricultural Sciences, University Library, Bangalore, India. 30 pp.

Hegde, V., and Venkatesh, Y.P. 2002. Oral allergy syndrome to sapodilla (*Achras zapota*). *J. Allergy Clin. Immunol.* 110:533–534.

Ingle, G.S., Khedkar, D.M., and Dabhade, R.S. 1981. Ripening studies in sapota fruit (*Achras sapota* Linn.). *Indian Food Packer*, 35(6):42–45.

Kader, A.A. 1999. Sapotes (sapodilla and mamey sapote). Recommendations for maintaining postharvest quality. *Perishables Handling Quart.* 100(Nov.):21–22.

Karling, J.S. 1942. Collecting chicle in the American tropics. *Torreya*, 42:38–50, 104–113.

Lakshminarayana, S. 1980. Sapodilla and prickly pear. In *Tropical and subtropical fruits. Composition, properties, and uses*. Edited by S. Nagy and P.E. Shano. AVI Publishing, Westpoint, CT. pp. 415–441.

Lakshminarayana, S., and Moreno-Rivera, M.A. 1979. Proximate characteristics and composition of sapodilla fruits grown in Mexico. *Proc. Fla. State Hortic. Soc.* 92:303–305.

Lakshminarayana, S., and Subramanyam, H. 1966. Physical, chemical and physiological changes in sapota fruit (*Achras sapota* Linn. (Sapotaceae)) during development and ripening. *J. Food Sci. Technol.* 3:151–154.

Macleod, A.J., and de Troconis, N.G. 1982. Volatile flavor components of sapodilla fruit (*Achras sapota* L.). *J. Agric. Food Chem.* 30:515–517.

Malo, S.E. 1967. A successful method for propagating sapodilla trees. *Proc. Fla. State Hortic. Soc.* 80:373–376.

Matthews, J.P., and Schultz, G.P. 2009. *Chicle: the chewing gum of the Americas. From the ancient Maya to William Wrigley*. University of Arizona Press, Tucson, AZ. 142 pp.

Mickelbart, M.V. 1996. Sapodilla: a potential crop for subtropical climates. In *Progress in new crops*. Edited by J. Janick. ASHS Press, Alexandria, VA. pp. 439–446.

Moore, H.E., and Stearn, W.T. 1967. The identity of *Achras zapota* L. and the names for the sapodilla and the sapote. *Taxon*, 16:382–395.

Morton, J. 1987. Sapotaceae—Sapodilla. In *Fruits of warm climates*. Creative Resource Systems, Winterville, NC. pp. 393–398.

Paull, R.E. 2008. *Manilkara zapota*, chiku. In *The encyclopedia of fruit & nuts*. Edited by J. Janick and R.E. Paull. CABI, Wallingford, Oxfordshire, U.K. pp. 828–831.

Piatos, P., and Knight, R.J., Jr. 1975. Self-incompatibility in the sapodilla. *Proc. Fla. State Hortic. Soc.* 88:464–465.

Pino, J.A., Marbot, R., and Aguero, J. 2003. Volatile components of sapodilla fruit (*Manilkara achras* L.). *J. Essent. Oil Res.* 15:374–375.

Redclift, M. 2004. *Chewing gum: the fortunes of taste*. Routledge, New York. 208 pp.

Rowe-Dutton, P. 1976. *Manilkara achras*—sapodilla. The propagation of tropical fruit trees. *CAB Hortic. Rev.* 4:475–512.

Ruehle, G.D. 1951. *The sapodilla in Florida*. University of Florida, Agricultural Experiment Stations, Gainesville, FL. 14 pp.

Schofield, P.B. 1983. Sapotaceae—sapodilla. In *Tropical tree fruits for Australia*. Information Series Q183018. Edited by P.E. Page. Queensland Dep. Primary Industries, Brisbane, Queensland, Australia. pp. 209–217.

Singh, P.D.A., Simon, O.R., and West, M.E. 1984. Acute toxicity of seeds of the sapodilla (*Achras sapota* L.). *Toxicon*, 22:145–147.

Velez-Colon, R., Beauchamp de Caloni, I., and Martinez-Garrastazu, S. 1989. Sapodilla (*Manilkara sapota* L. V. Rogen, *Achras sapota* Linn.) variety trials at southern Puerto Rico. *J. Agric. Univ. P.R.* 73:255–264.

Specialty Cookbooks

Cuervo-Lorens, M.E. 2004. *Mexican culinary treasures: recipes from Maria Elena's kitchen*. Hippocrene, New York. 266 pp.

Grigson, J. 2007. *Jane Grigson's fruit book*. Bison Books, London. 514 pp.

Rare Fruit Council. 1981. *Tropical fruit recipes: rare and exotic fruits*. Rare Fruit Council International Inc., Miami, FL. 180 pp.

86 Sapote

NAMES

"Sapote" is the name of several Central American trees and their fruits. Those mentioned here belong to four different plant families.

- The name "sapote" is from the Spanish *zapote*, a persimmon-like fruit (such as the black sapote, discussed later), from the Nahuatl (Aztec) *tzapotl*, a general term applied to several soft, sweet fruits. "Sapota" is a variant of sapote. The sapodilla or chicle (*Manilkara zapota*; see Chapter 85) is also called sapota and sometimes sapote.

Pouteria Species

Family: Sapotaceae (sapote family)

- Mamey sapote—*P. sapota* (Jacq.) H.E. Moore & Stearn
- Yellow sapote—*P. campechiana* (Kunth) Baehni
- Green sapote—*P. viridis* (Pittier) Cronquist
- Mamey (mammee) sapote (*P. sapota*) has long been known simply as sapote but is also called marmalade plum, marmalade tree, mamey (mammee) apple, and red sapote. It is discussed in past literature under several scientific synonyms, including *P. mammosa* (L.) Cronquist, *Lucuma mammosa* of some authors, *Achradelpha mammosa* Cook, *Vitellaria mammosa* Radlk., *Calocarpum mammosum* Pierre, *Casimiroa sapota* (Jacq.) Merr., and *Sideroxylon sapota* Jacq. This is the only important species of the genus and the only one discussed in some detail below.
- The mamey sapote should not be confused with the mammee apple (also known as the mammee, mammy, mammy apple, and San Domingo apricot), *Mammea americana* L., which belongs to a different family (the Clusiaceae or Guttiferae, also known as the Garcinia family). The fruit of this native tree of the West Indies is now produced in the tropics and subtropics around the world, including Florida. The fruit is the size of a small grapefruit, tastes like a tart apple while still green, and somewhat like an apricot when ripe and stewed.
- The yellow sapote (*P. campechiana*) is also known as Amarillo, canistel, eggfruit, egg-fruit tree, sapote borracho, and by the scientific synonyms *Lucuma nervosa* A. DC. and *L. salicifolia* Kunth. The tree is 5 to 17 m (16–56 ft.) high. The species probably originated in Central America, possibly in the tropical forest of Mexico. The fruits are orange-yellow, 5 to 7.5 cm (2–3 in.) in diameter, and are sold and consumed locally.
- The green sapote (*P. viridis*, also known as *Calocarpum viride* Pittier) is a tree native from Guatemala to Costa Rica. Its fruits are pale red-brown, sweet, and juicy. They are much esteemed among Indians of Costa Rica and are sold in markets in the country.
- The genus name *Pouteria* is based on an indigenous Indian name, the Guiana (Galibi) *pourama-pouteri*.
- *Sapota* in the scientific name *P. sapota* is based on a native South American name, as discussed above.

FIGURE 86.1 Sapote species. Left: mamey sapote (*Pouteria sapota*). (From Bailey, 1900–1902.) Center: yellow sapote (*P. campechiana*). (From Bailey, 1916.) Right: white sapote (*Casimiroa edulis*). (From Bailey, 1916.)

Casimiroa Species

Family: Rutaceae (rue family)

- White sapote—*C. edulis* La Llave & Lex.
- Woolly-leaf white sapote—*C. tetrameria* Millsp.

- The white sapote (*C. edulis*) is also known as Mexican apple. *Edulis* in the scientific name *C. edulis* means edible.
- The woolly-leaf white sapote (*C. tetrameria*) is also known as yellow sapote and matasano. It is native from Yucatan to Costa Rica but is not widely cultivated.
- *Casimiroa sapota* Oerst. grows wild in southern Mexico and Nicaragua and is commonly cultivated in parts of Mexico. It does not have a well-known English name and is known in Spanish (and sometimes in English) as *matasano* (a name also applied to the previous species). It is of limited significance as a source of fruit.
- The genus name *Casimiroa* commemorates Casimiro Gomez de Ortega (1740–1818), a Spanish physician and botanist, and director of the Madrid Botanic Garden from 1771 to 1801.

Diospyros

Family: Ebenaceae (ebony family)

- Black sapote—*D. digyna* Jacq.

- The black sapote has also been called black persimmon (particularly in Hawaii). The dark ripe fruit with a chocolate-brown pulp (as well as the black bark) is responsible for the "black" in the name. The species is in the same family as the persimmon, and the fruit is reminiscent, hence the name "black persimmon." The name "chocolate fruit" is used in Australia, allegedly because the ripe fruit tastes like chocolate pudding, but more likely to overcome the poor marketing image of the fruit to date.
- The genus name *Diospyros* is based on the Greek *Dios*, of (the Greek God) Zeus or Jove, and *pyros*, wheat or grain, alluding to the edible fruits.
- *Digyna* in the scientific name *D. digyna* is based on the Greek *di*, two + *gyna*, female, a botanical way of saying that there are two ovaries in the flower.

Quararibea

Family: Bombacaceae (bombax family)

- South American sapote—*Q. cordata* (Bonpl.) Vischer

- The name "South American sapote" was contrived because the name "sapote" is applied to many other plants, and there is no generally accepted English vernacular name (other than "sapote").
- The South American sapote is also known as chupa-chupa and matisia.
- The name chupa-chupa is the colloquial name in Colombia and Peru.
- The name matisia reflects the old scientific name *Matisia cordata* Humb. & Bonpl.
- The genus name *Quararibea* was recognized by the botanist Fusée Aublet in 1775 and was based on an indigenous South American word, *guarariba*, for the species *Quararibea guyanensis* Aubl.
- *Cordata* in the scientific name *Q. cordata* is Latin for heart-shaped, a reference to the leaves.

PLANT PORTRAIT

The most important species are detailed in the following.

Mamey Sapote

This is a Mexican and Central American tree, native from southern Mexico to northern Nicaragua. It grows only in tropical and near tropical conditions, often to 18 m (60 ft.) in height, sometimes to 40 m (130 ft.), and sometimes developing a trunk 1 m (3 ft.) thick, which may be buttressed at the base. The leaves are evergreen or deciduous. The fruit is round, ovoid or elliptic, often somewhat pointed at the apex (resulting in an American football shape), 7.5 to 23 cm (3–9 in.) long, 0.2 to 3 kg (1/2–6½ lb.) in weight. The skin is rough, dark-brown, firm, thick, leathery, and semiwoody, whereas the flesh is salmon-pink, orange, red, or reddish-brown, soft, sweet, and pumpkin- or almond-like in flavor. There is one (typically) or up to four large pointed seeds in the fruit. The mamey sapote is frequently cultivated in the New World south of the United States as well as in areas of southern Asia, including the Philippines and southern Vietnam. In southern Florida the cultivation of sapote increased with immigration from Cuba because many of the Cuban immigrants like the fruit.

White Sapote

This is the only important plant of the *Casimiroa* species known as sapote. It is a subtropical, evergreen tree, growing 4.5 to 18 m (5–60 ft.) in height. The tree is native to Mexico and Central America and is cultivated there, occasionally in northern South America, the Mediterranean region, India, the East Indies, New Zealand, and South Africa. There is or has been some limited commercial cultivation in Florida, California, Hawaii, Puerto Rico, and elsewhere. Mature trees can produce 900 kg (1 ton) of fruit annually. The orange-sized fruit is round (spherical), oval or ovoid, more or less five-lobed, 6 to 11 cm (2½–4½ in.) wide and up to 12 cm (4¾ in.) long. The inedible paper-thin skin is green, yellowish, or golden, whereas the flesh is creamy-white or yellow, with many tiny yellow oil glands. The taste is usually sweet but often somewhat bitter (advertising literature has described it as "a mild flavor combination of peaches, lemons, mangos, coconut, caramel, and vanilla, depending on the variety"). The best varieties are said to taste like a combination of peach, banana, and pear. Up to six hard, white seeds, 2.5 to 5 cm (1–2 in.) long, are present. The seed kernels are bitter and narcotic and are said to be fatally toxic if eaten raw by humans or animals.

Black Sapote

The tree is a tropical evergreen, growing as high as 25 m (80 ft.). The fruit is nearly round, 5 to 12.5 cm (2–5 in.) wide. The young fruit is bright-green and shiny, ripening to a muddy-green and indeed nearly black when fully ripe. The interior is a mass of glossy, brown to very dark-brown or almost black, jelly-like, sweet pulp, with a flavor between an apricot and a mango. One to ten flat brown seeds, 2 to 2.5 cm (3/4–1 in.) long, may be present, or the fruit may be seedless. The black sapote is native to both coasts of Mexico and the forested lowlands of Central America, where it is frequently cultivated. It is grown as a curiosity in many tropical regions but is popular only in its native region, especially in Mexico, where the fruits are regularly marketed. The black sapote fruit is very soft when fully ripe. When eaten fresh, it is usually seasoned with lime or orange juice.

South American Sapote

This is a large, semideciduous tree growing as high as 46 m (150 ft.), sometimes developing buttresses at the base of the trunk. It is native to the upper Amazon regions of Brazil, Peru, and Ecuador and is little known outside of South America. The fruit is sold in local markets of the upper Amazon. The fruit is round, ovoid or elliptic, 10 to 14.5 cm (4–5¾ in.) long, and weighing as much as 800 g (28 oz.). The rind is thick, leathery, greenish-brown, and downy. The flesh is orange-yellow, soft, juicy, and sweet. There are two to five seeds, up to 4 cm (1½ in.) long. Fibrous fruit are sometimes encountered. The fruit is highly esteemed in its native areas and is mainly eaten fresh out of hand.

CULINARY PORTRAIT

Sapotes are sweet, pulpy fruits that are usually eaten fresh out of hand and made into purées (like apple sauce), jams, preserves, compôtes, sherbets, and smoothies. Puréed fruit can be used as a sauce, in beverages, drizzled over various foods, or used as filling for cakes and pastries. Most sapotes make good additions to fruit salads. The fruit can be sliced and served with a little lemon or lime juice and topped with whipped cream. In Latin America, the seeds of mamey sapote are sometimes ground and used in confections. Sapotes are occasionally found in specialty supermarkets. Ripe fruit can be refrigerated, but not for more than a week. Most sapotes can be wrapped in plastic and frozen, but some become bitter. After freezing, the fruit may be half-thawed, and the sherbet-like flesh scooped out and consumed directly or used to flavor various dishes. Sapotes are often unsuitable for cooked dishes (some species are firmer and withstand heat better).

Culinary Vocabulary

- "Spanish sherbet" is the name given to sapote pulp imported from Central America, which has been prepared and distributed commercially by a dairy firm in Miami.
- "Chocolate pudding fruit" is the attractive name given to the fruit of the black sapote. Some recipes in fact specify the fruit for making a kind of chocolate pudding.

CURIOSITIES OF SCIENCE AND TECHNOLOGY

- The mamey sapote is credited with sustaining Hernando Cortez (1485–1547) and his Spanish army during their march from Mexico City to Honduras.
- Seeds of the white sapote were burned and powdered, and the preparation used to induce sleep by the ancient Aztecs. The seeds contain chemicals that have been demonstrated to have relaxant properties. In modern Mexico, the preparation continues to be used as a tranquilizer. Fagarine, one of the constituents, slows and regularizes heart rate and is used in modern medicine to control muscle spasms.
- In Costa Rica, the ground seeds of mamey sapote have been used as a linen starch.
- In Guatemala and El Salvador, the oil from the seeds of mamey sapote has been used as a tonic to prevent baldness (a doubtful usage).

- As in grapefruit, the seeds of mamey sapote may crack open and sprout within overmature fruits.
- The crushed bark and leaves of black sapote can blister skin and have been used medicinally in various ways in folk medicine, including the treatment of leprosy and ringworm.

KEY INFORMATION SOURCES

Casimiroa Species

- White sapote—*C. edulis*
- Woolly-leaf white sapote—*C. tetrameria*

Batten, D.J. 1984. White sapote. In *Tropical tree fruits for Australia*. Edited by P. Page. Queensland Dept. of Primary Industries, Brisbane, Australia. pp. 171–174.

Chambers, R.R. 1984. White sapote varieties: progress report. *California Rare Fruit Growers Yearbook*, 16:56–64.

Crane, J.H. 2008. *Casimiroa edulis*, white sapote. In *The encyclopedia of fruit & nuts*. Edited by J. Janick and R.E. Paull. CABI, Wallingford, Oxfordshire, U.K. pp. 770–772.

Englehart, O.J. 1977. Reviewing *Casimiroa edulis*. *California Rare Fruit Growers Yearbook*, 9:35–36.

George, A.P., and Nissen, R.J. 1991. *Casimiroa edulis* Llave & Lex. In *Plant resources of South-East Asia. Vol. 2. Edible fruits and nuts*. Edited by E.W.M. Verheij and R.E. Coronel. Pudoc, Leiden, the Netherlands. pp. 113–114.

George, A.P., Nissen, R.J., and Wallace, D.J. 1988. The casimiroa. *Qld. Agric. J.* 114(1):57–62.

Higham, D.C. 1981. The white sapote and wooly-leaf white sapote. *Rare Fruits Council Newsl.* 10:1–8.

Morton, J.F. 1962. The drug aspects of the white sapotes. *Econ. Bot.* 16:288–294.

Popenoe, F.W. 1911. The white sapote. *Pomona Coll. J. Econ. Bot.* 1(2):83–90.

Schroeder, C.A. 1954. Fruit morphology and anatomy in the white sapote. *Bot. Gaz.* 115:248–254.

Thompson, P. 1972. The white sapote. *California Rare Fruit Growers Yearbook*, 1972:6–20.

Diospyros Species

- Black sapote—*D. digyna*

Fisch, B.E. 1975. Black sapote. *California Rare Fruit Growers Yearbook*, 7:91–97.

Ng, F.S.P. 1991. *Diospryos digyna* Jacq. In *Plant resources of South-East Asia. Vol. 2. Edible fruits and nuts*. Edited by E.W.M. Verheij and R.E. Coronel. Pudoc, Leiden, the Netherlands. pp. 152–154.

Mickelbart, M.V., and Marler, T.E. 1998. Growth, gas exchange, and mineral relations of black sapote (*Diospyros digyna* Jacq.) as influenced by salinity. *Sci. Hortic.* 72:103–110.

Miller, W.R., Sharp, J.L., and Baldwin, E. 1998. Quality of irradiated and nonirradiated black sapote (*Diospyros digyna* jacq.) after storage and ripening. *Proc. Annu. Meet Fla. State Hortic. Soc.* 110:215–218.

Pouteria Species

- Mamey sapote—*P. sapota*
- Yellow sapote—*P. campechiana*
- Green sapote—*P. viridis*

Almeyda, N., and Martin, F.W. 1979. Mamey sapote. Neglected fruit with much promise. *World Farming*, 21:12–15.

Arenas-Ocampo, M.L., Evangelista-Lozano, S., Arana-Errasquin, R., Jimenez-Aparicio, A., and Davila-Ortiz, G. 2003. Softening and biochemical changes of sapote mamey fruit (*Pouteria sapota*) at different development and ripening stages. *J. Food Biochem.* 27:91–107.

Campbell, C.W., and Lara, S.P. 1982. Mamey sapote cultivars in Florida. *Proc. Fla. State. Hortic. Soc.* 95:114–115.

Crane, J.H. 2008. *Pouteria sapota*, mamey sapote. In *The encyclopedia of fruit & nuts*. Edited by J. Janick and R.E. Paull. CABI, Wallingford, Oxfordshire, U.K. pp. 839–842.

Crane, J.H. 2008. *Pouteria viridis*, green sapote. In *The encyclopedia of fruit & nuts*. Edited by J. Janick and R.E. Paull. CABI, Wallingford, Oxfordshire, U.K. pp. 843–844.

Diaz-Perez, J.C., Bautista, S., Villanueva, R., and Lopez-Gomez, R. 2003. Modeling the ripening of sapote mamey (*Pouteria sapota* (Jacq.) H.E. Moore and Stearn) fruit at various temperatures. *Postharvest Biol. Technol.* 28:199–202.

Morera, J.A. 1994. Sapote (*Pouteria sapota*). In *Neglected crops. 1492 from a different perspective*. Edited by J.E. Hernández-Bermejo and J. Léon. Food and Agriculture Organization of the United Nations, Rome, Italy. pp. 103–109.

Morton, J.F. 1983. Why not select and grow superior types of canistel? [*Pouteria campechiana*]. *Proc. Am. Soc. Hortic. Sci.* 27(A):43–52.

Morton, J.F. 1991. *Pouteria campechiana* (Kunth) Baehni. In *Plant resources of South-East Asia. Vol. 2. Edible fruits and nuts*. Edited by E.W.M. Verheij and R.E. Coronel. Pudoc, Leiden, the Netherlands. pp. 258–259.

Oyen, L.P.A. 1991. *Pouteria sapota* (Jacq.) H.E. Moore & Stearn. In *Plant resources of South-East Asia. Vol. 2. Edible fruits and nuts*. Edited by E.W.M. Verheij and R.E. Coronel. Pudoc, Leiden, the Netherlands. pp. 259–262.

Whitman, W.F. 1965. The green sapote, a new fruit of South Florida. *Proc. Fla. State Hortic. Soc.* 78:330–335.

Quararibea Species

- South American sapote—*Q. cordata*

Clement, C.R. 1982. The South American sapota, *Quararibea cordata* (Humb. & Bonpl.). *Proc. Trop. Reg. Am. Soc. Hortic. Sci. Annu. Meet.* 25:427–432 (in Portuguese).

Hodge, W.H. 1960. The South American 'sapote'. *Econ. Bot.* 14:203–206.

Santos, R.P. 2008. *Quararibea cordata*, South American sapote. In *The encyclopedia of fruit & nuts*. Edited by J. Janick and R.E. Paull. CABI, Wallingford, Oxfordshire, U.K. pp. 184–187.

Whitman, W.F. 1977. South American sapote. *Proc. Fla. State Hortic. Soc.* 89:226–227.

Specialty Cookbooks

A World School Publication. 2006. *Organic cooking: eating well: 300 simple organic gourmet recipes for a healthier life*. AuthorHouse, Bloomington, IN. 328 pp.

Gravette, A. 2002. *Classic Cuban cuisine*. Fusion Press, London. 232 pp.

Pitkin, J.M. (*Ed.*). 2000. *Great chefs of the Caribbean: from the television series Great Chefs of the Caribbean*. Cumberland House Publishing, Nashville, TN. 246 pp.

Raichlen, S. 2000. *Steven Raichlen's healthy Latin cooking: 200 sizzling recipes from Mexico, Cuba, Caribbean, Brazil, and beyond*. Rodale, Emmaus, PA. 416 pp.

Scott-Aitken, L. 2005. *Raw food recipes: no meat no heat*. G. St. Jean, Laval, QC. 112 pp.

Van Waerebeek-Gonzalez, R. 1999. *The Chilean kitchen: authentic, homestyle foods, regional wines, and culinary traditions of Chile*. HP Books, New York. 324 pp.

87 Sarsaparilla

Family: Smilacaceae (sarsaparilla or catbrier family; in older and some current literature, the Smilacaceae are placed in the Liliaceae, the lily family)

NAMES

Scientific Names: *Smilax* Species

- Mexican sarsaparilla—*S. aristolochiifolia* Mill. (*S. medica* Schltdl. & Cham.)
- Ecuadorian sarsaparilla—*S. febrifuga* Kunth
- Honduran sarsaparilla—*S. regelii* Killip & C.V. Morton (*S. officinalis* F.J. Hanb. & Flueck., *S. ornata* Hook., *S. utilis* Hemsl., *S. grandifolia* Reg.)

- The English word "sarsaparilla" (commonly misspelled sasparilla and sometimes as sassparilla) is based on the Spanish *zarza*, bramble or bush + *parilla*, little vine, literally "little vine-bush," an appropriate name coined by the Spanish when they arrived in Central America and found the local inhabitants using this climbing plant.
- Suggested pronunciation: saes-[e]-puh-ri'-luh.
- Mexican sarsaparilla is also known as gray sarsaparilla, Tampico sarsaparilla, and Veracruz sarsaparilla.
- Honduran sarsaparilla is also known as brown sarsaparilla and Jamaica sarsaparilla.
- (East) Indian sarsaparilla is *Hemidesmus indicus* (L.) W.T. Aiton of the milkweed family (Asclepiadaceae), a plant of India and Ceylon used for medicinal purposes. As noted in the next section, several other species are also called sarsaparilla and are sometimes used medicinally.
- The Roman naturalist Pliny the Elder (23–79) used the Latin name smilax in reference to a "bindweed" (*Smilax aspera* L.), which appears to have led Linnaeus to use the name *Smilax* when he first recognized the genus. The genus name *Smilax* is also an ancient Greek name of an evergreen oak, a bean, and a yew, none of which is related to the modern genus *Smilax*. Several species of the genus *Asparagus* are called "smilax" by florists, particularly *A. asparagoides* (L.) W. Wight.

PLANT PORTRAIT

The genus *Smilax* includes more than 200 species, distributed in the tropical regions of both hemispheres; many of the species have medicinal properties. Some are also used as wild sources of vegetables, some for leaf greens, and others for starchy tubers. The dried roots of several tropical American species of *Smilax*, particularly the three listed above, are the principal sources of sarsaparilla, used as a flavoring and medicinally. The tropical American sarsaparilla species are perennial, woody, climbing vines with thick rootstalks and thin roots about a meter (or yard) long, which are dried for use as sarsaparilla. Sometimes various other (nontropical American) species are substituted, for example, the North American species known as catbrier, particularly *S. rotundifolia* L., which was used during the American civil war to make sarsaparilla beverages. China sarsaparilla (*S. china* DC.), a species of China and Japan, has been used since ancient times in Asia as a kind of medicinal sarsaparilla. Other plants used as substitutes for sarsaparilla include the Virginia sarsaparilla (*Aralia nudicaulis* L.), bristly sarsaparilla (*A. hispida* Vent.), and the American spikenard

FIGURE 87.1 Mexican sarsaparilla (*Smilax aristolochiifolia*). (From Köhler, 1883–1914.)

(*A. racemosa* L.), which are North American plants of the family Araliaceae (ginseng family). "Sarsaparilla" has often been made without true sarsaparilla as flavoring, for example, with birch oil and sassafras (some mistakenly believe that sarsaparilla normally comes from sassafras). Native Americans used sarsaparilla to treat a wide variety of complaints, including skin diseases, stomach problems, rheumatism, fevers, and venereal diseases, and these uses were taken up by European settlers. Sarsaparilla is widely marketed today as a herbal supplement, touted to alleviate a variety of ailments, particularly impotence.

CULINARY PORTRAIT

Sarsaparilla root has been used commercially as a foaming ingredient and flavoring in root beer and other soft drinks for almost 2 centuries and is a traditional ingredient of "root beer." Root beer is an original American soft drink, first made by indigenous peoples of the New World. Farmers during the eighteenth and nineteenth centuries also made elementary root beers, using available wild herbs and roots and bark of trees. However, Charles E. Hires (1851–1937), is generally credited with creating the root beer beverage familiar today and has been called "the Father of Root Beer." He first experimented with sarsaparilla in his Philadelphia pharmacy in 1869, sold root beer made with sarsaparilla as a cold drink at the Centennial Exposition in 1876 in Philadelphia, and organized his own company in 1890. A list of ingredients from a 1922 pamphlet about Hires Root Beer included birch bark, chirreta, dog grass, ginger, "Hires special plant," hops, juniper berries, licorice, sarsaparilla, sugar, vanilla, wintergreen, and maté (other lists claim as many as 26 ingredients). He intended to name his new concoction "root tea" but used the name "root beer" to appeal to the large market of hard-drinking Pennsylvania miners. (The Hires Root Beer Company was precluded from patenting the name "root beer" in 1879, when Congress passed a law stating that no word in the dictionary could be registered—a law that was repealed in 1920.) Because it was not made with alcohol, the temperance movement strongly supported the drinking of root beer. Hires himself was active in the temperance and Quaker movements. Prohibition (1920–1933) in the United States, sometimes called "The Great Dry Spell," contributed to the popularity of root beer. Hires became the largest

FIGURE 87.2 Honduran sarsaparilla (*Smilax regelii*). (From Harter, 1988.)

manufacturer of the beverage in the world and extremely wealthy. Today, "root beer" sales in the United States are almost 2 billion dollars annually, and root beer has a loyal following. In modern times, so-called sarsaparilla beverages tend to use artificial flavorings.

Culinary Vocabulary

- A "black cow" or "brown cow" is usually a float made with root beer and vanilla ice cream. Sometimes a "chocolate cow" or "brown cow" substitutes chocolate ice cream for vanilla.
- Sarsi is a very popular sarsaparilla-based root beer marketed in Australia, The Philippines, and other Southeast Asia countries.

CURIOSITIES OF SCIENCE AND TECHNOLOGY

- In the nineteenth century, American cowboys often ordered sarsaparilla (root beer) because it was both the most widely used treatment for syphilis and was also considered a male aphrodisiac. Sarsaparilla was generally the "good guy" drink of the early American Western movies (it was like ordering a glass of milk in a saloon while everyone else was drinking whisky), most movie goers unaware that in fact it was usually requested in the hope of curing venereal diseases.
- Dr. Butler's Ale, first made around the beginning of the seventeenth century in London, England, was a famous alcoholic beverage which used sarsaparilla and was one of the earliest "root beers." Its medicinal values were extensively advertised, giving people a good excuse to drink.
- A&W Root Beer is the most popular root beer in North America. It began when Roy W. Allen (died 1968) opened his first root beer and hamburger stand in Lodi, California, in 1919, using a recipe purchased with 10 cents from a pharmacist in Arizona. In 1922, Frank Wright joined Allen in the business, and the brand name was born—A&W Root Beer. By 1923, A&W had established the first car hop restaurant in the United States, reshaping American fast food culture. (For those not old enough to remember, car hops are waiters or waitresses who serve patrons in their cars at a drive-in restaurant.)

FIGURE 87.3 Ecuadorian sarsaparilla (*Smilax febrifuga*). (From Harter, 1988.)

- It has been estimated that as many as 2000 different brands of root beer have entered the market since C.E. Hires marketed his brand. National brands in the United States include A&W, Hires', Barq's, Mug, and Dad's Old Fashioned. Some of the more creative names include "Witches' Brew" (from Salem, Massachusetts, where women identified as witches were once hung) and "Big Al's Sarsaparilla" (from Lansing, Michigan, honoring gangster Al Capone, who used to frequent the local restaurants). "Root 66" Root Beer (from central Virginia) commemorates Route 66, the legendary "Highway of Dreams," or "The Main Street of America," which started in Chicago and ended in Santa Monica, Los Angeles, crossing eight states and three time zones (the highway was officially decommissioned in 1985, but there are still some stretches in existence).

KEY INFORMATION SOURCES

Bernardo, R.R., Pinto, A.V., and Parente, J.P. 1996. Steroidal saponins from *Smilax officinalis*. *Phytochemistry*, 43:465–469.

Devys, M., Alcaide, A., Pinte, F., and Barbier, M. 1969. Pollinastanol in the fern *Polypodium vulgare* L. and the sarsaparilla *Smilax medica* Schlecht and Cham. *Acad. Sci. Compt. Rend. Ser.* D 269:2033–2035 (in French).

Enkema, L.A. 1952. *Root beer; how it got its name; what it is; how it developed from a home-brewed beverage to its present day popularity*. Hurty-Peck & Co., Indianapolis, IN. 10 pp.

Haywood, J.K. 1990s. *The healing powers of sarsaparilla*. J.K. Haywood, Kingston, St. Vincent. 19 pp.

Hood & Co. 1880–1889. *Hood's sarsaparilla laboratory illustrated*. Condensed edition. C.I. Hood & Co., Lowell, MA. 16 pp.

Kudritskaya, S.E., Fishman, G.M., Zagorodskaya, L.M., and Chikovani, D.M. 1988. Carotenoids of sarsaparilla. *Chem. Nat. Compd.* 23:635 (in English and Russian).

Lloyd, J.U. 1930. *Sarsaparilla preparations*. W. Phillips and Co., Cincinnati, IA. 3 pp.

Morrison, T.1992. *Root beer: advertising and collectibles*. Schiffer, West Chester, PA. 128 pp.

Morrison, T. 1997. *More root beer advertising and collectibles*. Schiffer Pub., Atglen, PA. 160 pp.

Quarantiello, L.E. 1997. *The root beer book: a celebration of America's best-loved soft drink*. Limelight Books, Lake Geneva, WI. 96 pp.

Rafatullah, S., Mossa, J.S., Ageel, A.M., Al-Yahya, M.A., and Tariq, M. 1991. Hepatoprotective and safety evaluation studies on sarsaparilla [*Smilax regelii*]. *Int. J. Pharm.* 29:296–301.

Robles, G., and Villabolos, R. 1998. *Plantas medicinales del género Smilax en Centroamérica: actas de la reunión celebrada del 22 al 25 de setiembre de 1997 en Turrialba, Costa Rica.* Centro Agronómico Tropical de Investigación y Enseñanza, Turrialba, Costa Rica. 178 pp (in Spanish).

Yates, D., and Yates, E. 2003. *Ginger beer & root beer: heritage 1790 to 1930.* Donald Yates Publishers, Homerville, OH. 356 pp.

Specialty Cookbooks

Bolton, R. (Ed.). 2008. *Cheerio! A book of punches and cocktails. How to mix them.* Create Space, Scotts Valley, CA. 100 pp. [Reprint of 1928 book.]

Bolton, R. (Ed.). 2008. *Chicago bartenders 1945 bar guide.* Create Space, Scotts Valley, CA. 92 pp. [Reprint of 1945 book.]

Bolton, R. (Ed.). 2008. *The cocktail book. A sideboard manual for gentlemen.* Create Space, Scotts Valley, CA. 98 pp. [Reprint of 1926 book.]

Bolton, R. (Ed.). 2008. *Wehman Bros. bartender's guide.* Create Space, Scotts Valley, CA. 100 pp. [Reprint of 1912 book.]

Bolton, R. (Ed.). 2008. *The old Waldorf Astoria bar book.* Create Space, Scotts Valley, CA. 190 pp. [Reprint of 1935 book.]

Bolton, R. (Ed.). 2008. *Modern American drinks: how to mix and serve all kinds of cups and drinks.* Create Space, Scotts Valley, CA. 134 pp. [Reprint of 1895 book.]

Brown, R. (Ed.). 2008. *Cooling cups and dainty drinks.* Create Space, Scotts Valley, CA. 242 pp. [Reprint of 1869 book.]

Brown, R. (Ed.). 2008. *The gentleman's table guide: wine cups, American drinks, punches, summer & winter beverages.* Create Space, Scotts Valley, CA. 96 pp. [Reprint of 1871 book. by E. Ricket and C. Thomas.]

Bullock, T. 2008. (foreword by R. Brown). *The ideal bartender.* Create Space, Scotts Valley, CA. 64 pp. [Reprint of 1917 book.]

Bullock, T., and Frienz, D.J. 2002. *Classic cocktails: over 170 drinks from yesteryear that you can enjoy today.* Gramercy, New York. 112 pp.

Cresswell, S.E. 1998. *Homemade root beer, soda, & pop.* Storey Books, Pownal, VT. 121 pp.

Frienz, D.J. 2001. *173 pre-Prohibition cocktails: potations so good they scandalized a president.* Howling at the Moon Press, Jenks, OK. 112 pp.

Renfrow, C. 1995. *A sip through time: a collection of old brewing recipes.* Cindy Renfrow, Pottstown, PA. 334 pp.

Spaziani, G., and Halloran, E.J. 2000. *The home winemaker's companion: secrets, know-how, and recipes for making 115 great-tasting wines.* Storey Publishing, Pownall, VT. 265 pp.

Vargas, P., and Gulling, R. 1997. *Cordials from your kitchen: easy elegant liqueurs you can make and give.* Storey Publishing, Pownall, VT. 176 pp.

Vargas, P., and Gulling, R. 1999. *Making wild wines & meads:125 unusual recipes using herbs, fruits, flowers & more.* Revised edition. Storey Publishing, Pownall, VT. 176 pp.

Wright, H.S. 2005. *Old time recipes for home made wines, cordials and liqueurs from fruits, flowers, vegetables and shrubs.* Kessinger Publishing, Whitefish, MT. 152 pp. [Reprint of 1909 book.]

88 Sea Buckthorn

Family: Elaeagnaceae (oleaster family)

NAMES

Scientific Name: *Hippophae rhamnoides* L.

- "Buckthorn" in the name "sea buckthorn" reflects similarity with the true buckthorns, spiny shrubs of the genus *Rhamnus*. The "sea" in the name points out that sea buckthorn can tolerate rather saline soils and so often grows beside the sea. "Seaberry" is an occasionally used name.
- In England, sea buckthorn was sometimes called "sallow thorn." "Sallow" in the name has been explained as based on the light color of the leaves.
- The genus name *Hippophae* has been most widely interpreted as based on the classical Latin *hippo*, horse, + *phaeos*, to shine, literally "shining horse," a name that was coined in ancient times after it was noted that feeding the leaves to horses made their hair shiny. According to Greek legend, the mythical flying horse Pegasus preferred the foliage of sea buckthorn to all other foods.
- In much of the older literature, the genus name is spelled *Hippophaë* (the two dots over the e is a dieresis). Such "diacritical signs" are no longer used in scientific names for plants, with just one permissible exception: when a vowel preceding another vowel is pronounced (e.g., *Isoëtes* is pronounced i-so-ee-tees, not i-soe-tees).
- *Rhamnoides* in the scientific name *H. rhamnoides* is based on the Latin (derived from Greek) words *rhamnus*, buckthorn + *oides*, resembling, that is, resembling the (true) buckthorn.

PLANT PORTRAIT

All of the species of the genus *Hippophae* are called sea buckthorns, but just how many species there are is unsettled. Sea buckthorns are native to much of north-temperate Eurasia, including China, Mongolia, Russia, Great Britain, France, Denmark, the Netherlands, Germany, Poland, and Scandinavia. The most widespread, *H. rhamnoides*, has been variously divided into six or more geographically separated subspecies, but some specialists have decided that some of these deserve to be called separate species. There is general agreement that there is just one species in Europe, *H. rhamnoides*, and that it has several subspecies, but the classification of the remaining plants in the genus, distributed in the mountains of Central Asia, the Hindu-Kush Himalaya Mountains, and adjoining areas in China, is more controversial. The latter plants are variously treated as constituting as many as six additional species.

Sea buckthorn plants typically grows 2 to 4 m (6½–13 ft.) in height, although some in China have reached 18 m (59 ft.), and others grow no higher than 50 cm (20 in.). There are both male and female plants, and the latter develop berries that are round to almost egg-shaped and up to 1 cm (3/8 in.) long. The fruit is usually orange, but sometimes yellow or red. Unlike the majority of fruits that fall away from the maternal plant at maturity because the fruit stalk develops a weak area (an "abscission zone"), the berries remain on the bushes all winter until eaten by birds.

Records of the use of sea buckthorn for medicinal purposes trace back a thousand years or so. The berries have been employed for food for at least centuries in Eurasia. However, selection of improved

FIGURE 88.1 Sea buckthorn (*Hippophae rhamnoides*). (From Curtis, 1905, vol. 131, plate 8016.)

fruit varieties has only been carried out since about 1950. Cultivated varieties have been released from the former USSR, Mongolia, the former East Germany, and Finland, and breeding work is being conducted in Russia, China, Scandinavia, Canada, and other countries. While there has been great interest in sea buckthorn as a fruit crop in Eurasia for more than half a century, it is only in recent years that North Americans have seriously initiated commercial cultivation. Sea buckthorn is mainly known in North America as an attractive ornamental shrub with silvery deciduous leaves and colorful berries.

Sea buckthorn has been used medicinally in China for at least 12 centuries, and sea buckthorn oil is currently employed clinically in hospitals in Russia and China. The pulp and the seeds in the berries contain medicinally valuable essential oil. Medicinal oil has also been obtained from the young branches and leaves and incorporated into an ointment for treating a wide variety of types of skin damage, including burns, bedsores, eczema, and radiation injury. The oil is also taken internally for diseases of the stomach and intestine. In Europe and Asia, there are more than a dozen commercial products available for medicinal use on the skin or internally. Such strictly medicinal usage has not developed significantly yet in North America. However, there is interest in the healthful cosmetic properties of sea buckthorn in North America. Sea buckthorn berries are relatively high in "essential fatty acids," which are important for the maintenance of a healthy skin. This likely accounts for some of the value of the oil as a skin conditioner. Cosmetic preparations are well known in Russia and China. Sea buckthorn oil absorbs ultraviolet light, and because the oil is also known to be useful for promoting skin health, it is particularly suitable for sun care cosmetics. The Body Shop, well known for its ecologically friendly cosmetics, has marketed sea buckthorn oil sunscreen products, both for sun-blocking and tan-enhancement.

China, Russia, and Mongolia are the largest producers of sea buckthorn. In China, fruit is harvested from more than 1 million ha (2.5 million acres) of wild sea buckthorn and almost 300,000 ha (750,000 acres) of cultivated plants. The main product is oil for medical and cosmetic purposes, but 50 different food products are also produced. Hundreds of sea buckthorn products are now made from the berries, oil, leaves, and bark. The range of products now available in Eurasia includes juice, jellies, liquors, candy, vitamin C tables, ice cream, tea, biscuits, food colorants, cosmetics, shampoos, and medicines.

CULINARY PORTRAIT

The berries are too acidic to eat fresh for most palates but make excellent juice, jellies, marmalades, sauces, and liqueurs. Orange sea buckthorn berries are typically used for edible products. The juice may be used by itself or blended with other juices; the flavor of the juice has been described as resembling passion fruit. The fruits have a distinctive, sourish taste and a unique aroma reminiscent of pineapple.

Sea buckthorn fruits are among the most nutritious and vitamin-rich of all berries. The contents of vitamins C and E are very high; both of these are antioxidants. Sea buckthorn berries are also very rich in a variety of other antioxidant chemicals (several carotenoids, including beta-carotene (provitamin A), flavonoids, certain enzymes, and other substances).

Culinary Vocabulary

- In Byelorussia, buckthorn fruit juice is known as Russian pineapple because of its pineapple-like taste.

CURIOSITIES OF SCIENCE AND TECHNOLOGY

- It has been estimated that there is enough vitamin C in the berries of the sea buckthorn plants of the world to meet the dietary requirements for this vitamin of the entire human population.
- One of the uses of sea buckthorn oil has been to treat burns resulting from radiation. In 1986, the World's worst nuclear power accident occurred at Chernobyl in the Ukraine (at that time part of the USSR). Many of the victims were treated with sea buckthorn.
- Birds have been shown to be very effective at distributing the seeds of sea buckthorn. It has been demonstrated that germination of the seeds is six times greater when they have passed through a bird's gut.
- During the Second World War, the Germans valued sea buckthorn (called the *seaberry*) so highly for its vitamin C content that they placed it on a list of plants to be protected as essential to the war effort.
- China designated its sea buckthorn sports drinks "Shawikang" and "Jianibao" as the official beverages for its athletes attending the Seoul Olympic Games in 1988. Juices made with sea buckthorn have a reputation for enhancing stamina and vitality.
- The Russian cosmonauts also were supplied with sea buckthorn beverages, consistent with the belief that this would enhance their health and resistance to stress to meet the substantial physical challenges they faced. It has been claimed that sea buckthorn was the first fruit juice in space.
- The sea buckthorn produces about as many male plants as females (some studies have shown that there are more males). As with most species in the world, the only function of the males is to fertilize the females so that offspring will be produced. It does seem necessary to have many males of the sea buckthorn in nature because the plants of this wind-pollinated plant are often scattered and the males have to produce a great deal of pollen as very little of it manages to reach the females. On the other hand, in cultivated orchards, the males can be planted very near the females, and so much fewer are required. In practice, in sea buckthorn orchards, each male plant is surrounded by a "harem" of approximately seven female plants.
- Technically, sea buckthorn "berries" are not true berries, in fact they are not even true fruits. Botanically, a "fruit" is derived from tissues that make up the ovary, the part of the female flower that contains the ovules or eggs. In the sea buckthorn, the single fertilized ovule produces what is technically a "nut" (a hard-coated fruit with just one seed), and this

is what people would normally call the seed or pit. The fleshy tissue that covers this in the sea buckthorn is actually derived from the petal tissue of the original flower, and so botanically it is not part of the true fruit.

- In Russia, lotions to prevent hair loss, made with sea buckthorn, have been marketed. This is related to the ancient observation that sea buckthorn makes the hair of horses shiny and healthy.
- Sea buckthorn has acquired an enormous reputation for preventing soil erosion and is widely planted for soil conservation and reclamation. China is confronted with grave problems of soil erosion, with 2 million sq. km (770,000 sq. mi.) of eroding land. China has estimated its annual soil losses to be around 5 billion tons and the annual loss of cultivated land approximately 70,000 hectares (173,000 acres). The use of vegetation has been perceived as the most promising tool to control land degradation in China. Sea buckthorn is one of the species successfully used on a large scale, particularly in northern China, to control desertification, to conserve land and water resources, and to integrate economic exploitation with ecological rehabilitation. Approximately a million hectares (2.5 million acres) of sea buckthorn have been planted in China, most of it for soil and water conservation.
- Nitrogen is the most important element for plant nutrition, and for most plants in nature there just is not enough. Although nitrogen is 80% of the atmosphere, this is in an inert form that is unavailable to plants. Certain privileged plants have developed symbiotic (mutually beneficial) associations with bacteria, which "fix" atmospheric nitrogen (convert it to a form that can be used by the plants). In most cases, this occurs in swellings of the roots called nodules, where the bacteria are housed. It is well known that members of the pea family (Leguminosae or Fabaceae) generally are provided with fixed nitrogen through the cooperation of relatively advanced bacteria of the genus *Rhizobium*. Except for the pea family, however, very few other plant species receive such assistance. The sea buckthorn is one of these rare cases. Its partner is a bacterium of the genus *Frankia*, which belongs to a primitive class of bacteria called the Actinomycetes. Because these form long, thread-like branched filaments, they were considered to be fungi until fairly recently. Only approximately two dozen other plant genera have developed nitrogen-fixing relationships with the Actinomycetes, in contrast with the hundreds of genera in the pea family that have symbiotic relationships with the more advanced nitrogen-fixing bacteria.
- A living windbreak is a linear arrangement of plants, primarily trees and shrubs, established for the purposes of reducing harmful effects of local wind flow. Dependent on their purpose, windbreaks may be called shelterbelts, timberbelts, hedgerows, living snow fences, and conservation buffers. All serve to prevent soil erosion caused by wind. Of course, plants that serve as windbreaks must be resistant to the drying effects and physical injuries caused by wind, and sea buckthorn is extremely suited to this task. Millions of sea buckthorn shrubs have been grown on the Canadian prairies since the 1930s for windbreak purposes.

KEY INFORMATION SOURCES

Baker, R.M. 1996. The future of the invasion shrub, sea buckthorn (*Hippophae rhamnoides*), on the west coast of Britain. *Asp. Appl. Biol.* 44:461–468.

Bartish, I.V., Jeppsson, N., and Nybom, H. 1999. Population genetic structure in the dioecious pioneer plant species *Hippophae rhamnoides* investigated by RAPD markers. *Mol. Ecol.* 8:791–802.

Bartish, I.V., Jeppsson, N., Nybom, H., and Swenson, U. 2002. Phylogeny of *Hippophae* (Eleagnaceae) inferred from parsimony analysis of chloroplast DNA and morphology. *Syst. Bot.* 27:41–54.

Bernath, J., and Foldesi, D. 1992. Sea buckthorn (*Hippophae rhamnoides* L.): a promising new medicinal and food crop. *J. Herbs Spices Med. Plants*, 1:27–35.

Beveridge, T., Li, T.S.C., Oomah, B.D., and Smith, A. 1999. Sea buckthorn products: manufacture and composition. *J. Agric. Food Chem.* 47:3480–3488.

Jana, S., Schroeder, W.R., and Barl, B. 2002. *Sea buckthorn cultivar and orchard development in Saskatchewan.* Revised edition. Agriculture Development Fund, University of Saskatchewan, Regina, SK. 67 pp.

Jeppson, N. 2008. *Hippophae rhamnoides*, sea buckthorn. In *The encyclopedia of fruit & nuts*. Edited by J. Janick and R.E. Paull. CABI, Wallingford, Oxfordshire, U.K. pp. 339–343.

Jeppsson, N., Bartish, I.V., and Persson, H.A. 1999. DNA analysis as a tool in sea buckthorn breeding. In *Perspectives on new crops and new uses*. Edited by J. Janick. ASHS Press, Alexandria, VA. pp. 338–341.

Kieper, R., and Gullacher, D. 2002. *Mechanical harvesting of sea buckthorn: an inter-provincial study*. Saskatchewan Agriculture Development Fund and Prairie Agriculture Machinery Institute, SK, Canada. 35 pp.

Li, T.S.C. 1999. Sea buckthorn: new crop opportunity. In *Perspectives on new crops and new uses*. Edited by J. Janick. ASHS Press, Alexandria, VA. pp. 335–337.

Li, T.S.C. 2002. Product development of sea buckthorn. In *Trends in new crops and new uses*. Edited by J. Janick and A. Whipkey. ASHS Press, Alexandria, VA. pp. 393–398.

Li, T.S.C., and Beveridge, T.H.J. 2003. *Sea buckthorn (*Hippophae rhamnoides *L.): production and utilization*. NRC Research Press, Ottawa, ON. 133 pp.

Li, T.S.C., and Beveridge, T.H.J. 2007. *Sea buckthorn: a new medicinal and nutritional botanical*. Agriculture and Agri-Food Canada, Publication 10320E. Ottawa, ON. 89 pp.

Li, T.S.C., and McLoughlin, C. 1997. *Sea buckthorn production guide*. Canada Sea Buckthorn Enterprises Ltd., Peachland, BC. 19 pp.

Li, T.S.C., and Schroeder, W.R. 1996. Sea buckthorn (*Hippophae rhamnoides* L.): a multipurpose plant. *HortTechnology*, 6:370–380.

Li, T.S.C., and Schroeder, W.R. 1999. *A growers guide to sea buckthorn*. PFRA Shelterbelt Centre, Agriculture and Agri-Food Canada, Indian Head, SK. Irregularly paginated.

Olander, S. 1998. *A bibliography of seabuckthorn*. Swedish University of Agricultural Sciences, Uppsala, Sweden. 52 pp.

Pearson, M.C., and Rogers, J.A. 1962. Biological flora of the British Isles. No. 85. *Hippophae rhamnoides* L. *J. Ecol.* 50:501–509.

Persson, H.A., and Nybom, H. 1998. Genetic sex determination and RAPD marker segregation in the dioecious species sea buckthorn (*Hippophae rhamnoides* L.). *Hereditas*, 129:45–51.

Pesonen, H., and Stark, T. 1999. *The seabuckthorn book*. Yellow Heart Products, Lake Worth, FL. 81 pp.

Ranwell, D.S. (Ed.). 1972. *The management of sea buckthorn (*Hippophae rhamnoides *L.) on selected sites in Great Britain*. Report of the Hippophaë Study Group. Natural Environment Research Council, London. 55 pp.

Rongsen, L. 1992. *Seabuckthorn: a multipurpose plant species for fragile mountains*. International Centre for Integrated Mountain Development, Kathmandu, Nepal. 62 pp.

Rousi, A. 1971. The genus *Hippophae* L. A taxonomic study. *Ann. Bot. Fennici*, 8:177–227.

Secretariat, International Symposium on Sea Buckthorn. 1989. *Proceedings of international symposium of sea buckthorn (Hippophae rhamnoides L.), Oct. 19–23, 1989, Xian, China.* Secretariat, International Symposium on Sea Buckthorn, Xian, China. 421 pp.

Small, E., Catling, P.M., and Li, T.S.C. 2002. Blossoming treasures of biodiversity:5. Sea buckthorn (*Hippophae rhamnoides*)—an ancient crop with modern virtues. *Biodiversity*, 3(2):25–27.

Trajkovski, V., and Jeppsson, N. 1999. Domestication of sea buckthorn. *Botanica Lithuanica Suppl.* 2:37–46.

Yao, Y., and Tigerstedt, P.M.A. 1993. Isozyme studies of genetic diversity and evolution in *Hippophae*. *Genet. Resour. Crop Evol.* 40:153–164.

Yao, Y., and Tigerstedt, P.M.A. 1994. Genetic diversity in *Hippophae* L. and its use in plant breeding. *Euphytica*, 77:165–169.

Yao, Y., and Tigerstedt, P.M.A. 1995. Geographical variation of growth rhythm, height, and hardiness and their relations in *Hippophae rhamnoides*. *J. Am. Soc. Hortic. Sci.* 120:691–698.

Yao, Y., Tigerstedt, P.M.A., and Joy, P. 1992. Variation of vitamin C concentration and character correlation between and within natural sea buckthorn (*Hippophae rhamnoides* L.) populations. *Acta Agric. Scand.* 42:12–17.

Specialty Cookbooks

Boutenko, V. 2009. *Green smoothie revolution: the radical leap toward natural health*. North Atlantic Books, Berkeley, CA. 175 pp.

Scherb, M. 2009. *A taste of heaven: a guide to food and drink made by monks and nuns*. Tarcher/Penguin, New York. 218 pp.

Stewart, A. 2006. *The flavours of Canada: a celebration of the finest regional foods*. Raincoast Books, Vancouver, BC. 224 pp.

89 Spinach (Exotic Species)

This chapter features

Malabar spinach (*Basella alba*)
New Zealand spinach (*Tetragonia tetragonioides*)
Water spinach (*Ipomoea aquatica*)

The three species discussed are culinary equivalents to common spinach (*Spinacia oleracea* L.), although they have distinctive tastes and qualities. They can be substituted in recipes for common spinach and used similarly. Although spinach has a reputation for being the archetypical despised vegetable and the most universally disliked vegetable by children in the Western World, it is nevertheless one of the most universally encountered foods. Spinach is eaten by itself as a separate dish, intermixed with lettuce and other vegetables in mixed salads, and incorporated into a very wide variety of culinary preparations, more commonly cooked, but sometimes also raw. As with many greens, when cooking spinach alone, it is advisable to wash just before using (prolonged soaking in water can soften leaves) and to use a small amount of liquid because the leaves generally release considerable fluid naturally. Cooking should be brief, generally only a few minutes. In simmered dishes, spinach is best added at the end of the cooking cycle, although in some Asian dishes spinach is cooked for long periods, for example in curries.

In addition to the species discussed here, there are other exotic spinach-like species. Some examples follow. Egyptian spinach (*Corchorus olitorius* L.), also known as bush okra, is widely grown, from the Middle East to tropical Africa. Tree spinach refers to certain *Cnidoscolus* species, especially *Cnidoscolus chayamansa* McVaughn (also known as chaya), a tropical American species that furnish spinach-like vegetables. "Australian spinach" is (1) *Chenopodium murale* L., a wild plant of Eurasia—its leaves are eaten as a potherb and the seeds are used to prepare sauces (*C. auricomum* Lindl. and *C. erosum* R. Br. are also called Australian spinach, and used occasionally as spinach); (2) *Tetragonia implexicoma* (Miq.) Hook f., native to Australia and New Zealand—its leaves are also cooked and eaten as a spinach substitute.

MALABAR SPINACH

Family: Basellaceae (basella family)

NAMES

Scientific Name: *Basella alba* L. (*B. rubra* L.)

- The "spinach" in "Malabar spinach" indicates the spinach-like character of this vegetable. Malabar is the coast region of southwestern India on the Arabian Sea, part of the area where Malabar spinach may have originated.

- Malabar spinach is also known as basella, Ceylon spinach, climbing spinach, Indian spinach, Surinam spinach, and vine spinach.
- Malabar spinach is sometimes called Malabar nightshade. However, the true nightshades are species of the genus *Solanum*, some of which are poisonous vines. The viny rather than the poisonous nature of the plants led to the name "Malabar nightshade."
- Malabar spinach can usually be purchased in Asian food specialty stores, especially in Chinatown and Vietnamese markets. Some Asian names under which it may be found include *lo kwai, luo kai, saan choy, shan tsoi, shu chieh* (Chinese); *tsuru murasa kai* (Japanese); *mong toi* (Vietnamese); *paag-prung* (Thai); and *gondola, genjerot, jingga* (Indonesian).
- The genus name *Basella* is the Latinized version of the name used in India.
- *Alba* in the scientific name *B. alba* is Latin for white. The flowers are usually white (but also sometimes red, or violet).

Plant Portrait

Malabar spinach is a short-lived, twining perennial vine producing fleshy, edible leaves. The ancestors of the cultivated forms are unknown, and it is not clear if genuinely wild forms are present in nature or if all wild plants represent escaped cultivated plants. Probably Malabar spinach is native to southern Asia and Africa. It is now distributed throughout the tropics, both in cultivation and as an escape. It was introduced to Europe in 1688. By the nineteenth century, it was grown in French gardens, using superior varieties that had been obtained from the Orient. This vegetable is grown to a small extent in North America but is important basically in tropical and subtropical areas, particularly in eastern Asia and India. In tropical areas, it spreads aggressively, commonly exceeding 5 m (16 ft.) in length. Three types are available in the tropics for cultivation, each distinguishable by its leaf shape and color. These include forms with (1) dark-green, oval leaves; (2) red, oval-rounded leaves and red stems (*B. rubra* L. of some authors); and (3) dark-green, heart-shaped leaves (*B. cordifolia* Lam. of some authors).

FIGURE 89.1 Malabar spinach (*Basella alba*). (From Vilmorin-Andrieux, 1885.)

Malabar spinach is much less likely to accumulate dirt or sand on its leaves than most crinkled varieties of common spinach because the wrinkling of the latter tends to trap particles. Also, because Malabar spinach is a vine, unlike common spinach the leaves are well off the ground and so less likely to accumulate particles of soil. Malabar spinach is widely described as "crunchy," but the attractive texture is due to its naturally succulent, thick leaves, whereas the "crunchiness" of common spinach may sometimes be due to particles of dirt that did not get washed off.

Culinary Portrait

Malabar spinach leaves and young shoots are cut, cooked, and eaten like spinach or mixed into soups and meat stews. In Southeast Asia and southern China, the vegetable is widely used in stir fries. In Vietnam, it is often incorporated into soups. In tropical Africa, it is used as a leafy vegetable. In Bengal (India), the leaves are cooked with chopped onions and hot chilies, then fried in a little mustard oil. Although not a true spinach, the large, succulent leaves are mild and delicate in taste and remarkably spinach-like in flavor. This vegetable is rather glutinous or mucilaginous (or "slimy") in texture when cooked, but this is generally not considered objectionable. Indeed, in India, Malabar spinach is used like the mucilaginous okra to thicken prepared dishes. This thickening feature is also desirable in sauces. Some cooks add a little vinegar to reduce mucilage. Overcooking increases the mucilaginous feature and should be avoided. Younger leaves and the top 15 cm (6 in.) of branch tips can be eaten fresh or cooked, whereas older material is best cooked. Malabar spinach may be refrigerated for a few days but does not keep well.

Culinary Vocabulary

- The expression "slippery soup" is often used to indicate a slippery floor or ground that makes walking treacherous or a tricky situation that makes maneuvering difficult. It is also used for soups with a slippery or slimy character. The Chinese in particular have a "slippery soup" that highlights the slimy character of Malabar spinach, combining this with, for example, hard-boiled eggs, bean curd, and ginger.

Curiosities of Science and Technology

- The dark red juice of Malabar spinach was used as a dye and ink in ancient China. Women used the juice as a rouge. The dye was also used as ink for making seal impressions, but only by the highest officials.
- There are two widespread classes of plants that differ in the way they carry out photosynthesis (a third class is known in plants of dry environments, such as cacti). Most familiar temperate zone plants are called C_3 plants because the first stable compound formed when carbon dioxide is processed is a three carbon compound, that is, C_3. C_4 plants are so named because the first organic compound incorporating CO_2 is a four carbon compound. Malabar spinach is one of many tropical plants having the C_4 type of photosynthesis, which allows for increased growth if temperature and light intensity are sufficiently high.
- The famous student of primates Dian Fossey (1932–1985), noted that Malabar spinach is one of the foods consumed by gorillas.

Key Information Sources

Anonymous. 2004. How to grow alugbati [*Basella alba* Linn.]. *Agriculture*, 8(7):16–17.

Chattopadhyay, T.K. 1977. Response of *Basella alba* to sources and doses of nitrogen nutrition. *Prog. Hortic.* 9(2):71–75.

Cyunel, E. 1989. *Basella alba* L.: in vitro culture and the production of betalains. Biotechnol. *Agric. For. Berlin*, 7:47–68.

Demidov, A.S., Khrzhanovskii, Y.A.V., Shaidorov, Y.I., and Geodakyan, R.O. 1991. Cultivating *Basella rubra* L. as salad greens. *Rastitel'nye Resursy*, 27(3):124–129 (in Russian).

Devi, H.M., and Pullaiah, T. 1975. Life history of *Basella rubra* and taxonomic status of the family Basellaceae. *J. Indian Bot. Soc*. 54:154–166.

Enriquez, F.G., Kawada, K., and Matsui, T. 2000. Effects of storage temperature on the keeping quality of Malabar spinach (*Basella alba* L.). *Food Preserv. Sci*. 26:211–217.

Glassgen, W.E., Metzger, J.W., Heuer, S., and Strack, D. 1993. Betacyanins from fruits of *Basella rubra*. *Phytochemistry*, 33:1525–1527.

Kameoka, H., Kubo, K., and Miyazawa, M. 1991. Volatile flavor components of Malabar nightshade *Basella rubra* L. *J. Food Comp. Anal*. 4:315–321.

Lacroix, C., and Sattler, R. 1988. Phyllotaxis theories and tepal-stamen superposition in *Basella rubra*. *Am. J. Bot*. 75:906–917.

Militiu, A. 1980. *Basella rubra*, a little-known leafy vegetable. Cultivation, protein and water content. *Prod. Veg. Hortic. Bucuresti*, 29(7):8–9 (in Romanian).

Miura, H., Yamazaki, H., and Nishijima, T. 1997. Post-sown priming with a potting mixture to improve emergence of Malabar spinach, *Basella alba* L. *J. Jpn. Soc. Hortic. Sci*. 66:513–517.

Murakami, T., Hirano, K., and Yoshikawa, M. 2001. Medicinal foodstuffs. XXIII. Structures of new oleanane-type triterpene oligoglycosides, basellasaponins A, B, C, and D, from the fresh aerial parts of *Basella rubra* L. *Chem. Pharm. Bull*. (Tokyo), 49:776–779.

Paiva, W.O.D., and Menezes, J.M.T. 1989. Agronomic evaluation of the performance of Indian spinach *Basella alba* L. synonym *Basella rubra* in Ouro Preto d'Oeste Rondonia Brazil. *Acta Amazonica,* 19:3–8, (in Portuguese).

Perera, K.D.A., and Pinto, M.E.R. 1984. Agronomic studies in Ceylon spinach. *Trop. Agric*. 140:61–68.

Rahmansyah, M. 1993. *Basella alba* L. In *Plant resources of South-East Asia. Vol. 8. Vegetables*. Edited by J.E. Siemonsma and K. Piluek. Pudoc Scientific Publishers, Wageningen, the Netherlands. pp. 93–95.

Sastry, T.C.S., Agrawal, S., and Kavethaker, K.Y. 1982. Note on *Basella alba* Linn. var. *rubra* (Linn.) Stewart (Basellaceae). *Ind. J. Forest. Dehra Dun*. 5:152.

Sinnadurai, S. 1970. The effect of nitrogen fertilizers on Indian spinach (*Basella alba* L.). *Ghana J. Agric. Sci*. 3(1):51–52.

Steenis, C.G.G.J. van. 1957. *Basella*. In *Flora Malesiana, Series 1, Vol. 5*. Edited by C.G.G.J. Steenis et al. Noordhoff-Kolff, Djakarta, Indonesia. pp. 300–302.

Tiwari, S.C., Igbokwe, P.E., and Collins, J.B. 1988. *Malabar spinach (*Basella alba *L.)*. Information Sheet No. 1321. Mississippi Agricultural and Forestry Experiment Station, MS. 2 pp.

Winters, H.F. 1963. Ceylon spinach (*Basella rubra*). *Econ. Bot*. 7:195–199.

Xu, H.Q. 1986. Studies on fertilization and embryogenesis of red vinespinach *Basella rubra*. *Acta Bot. Sin*. 28:361–367 (in Chinese).

Specialty Cookbooks

Note: Malabar spinach is a particularly popular vegetable in South Asian cooking.

Lau, A. L.-Y. 2001. *Asian greens: a full-color guide featuring 75 recipes*. St. Martin's Griffin, New York. 112 pp.

NEW ZEALAND SPINACH

Family: Aizoaceae (ice-plant family, carpet-weed family, fig-marigold family)

Names

Scientific Name: *Tetragonia tetragonioides* (Pall.) Kuntze (*T. expansa* Murray)

- New Zealand spinach is not at all related to common spinach. It owes its name simply to the fact that its European discoverer, Captain Cook (see next section), reported it from New Zealand as a "spinach-like" vegetable that was eaten by the early Australian explorers, partly to avoid scurvy.

- In Australia, New Zealand spinach is known as "Warrigal greens," or less frequently "Warrigal cabbage," "Botany Bay greens," and "Botany Bay spinach," reflecting the geographical preference of Australians for names based on places in their own country.
- In the markets of London, England, New Zealand spinach has been called "patent spinach."
- New Zealand spinach has also been called "summer spinach," presumably because it grows well during (hot) summer weather.
- Other names for New Zealand spinach: Cook's cabbage, sea spinach, and tetragon.
- The genus name *Tetragonia* is derived from the Greek *tetra*, four and *gonia*, angle, referring to the four-angled fruits.
- *Tetragonioides* in the scientific name means "similar to the genus *Tetragonia*" so that the name *T. tetragonioides* translates as "the *Tetragonia* that is similar to *Tetragonia*." The description "similar to the genus *Tetragonia*" seemed desirable when the species was originally placed in another genus, *Demidovia*.

Plant Portrait

New Zealand spinach is a short-lived perennial herb, cultivated as an annual. It is native to New Zealand, Australia, some Pacific islands, Japan, and southern South America. The plant was found growing wild in New Zealand in 1770 by the famous British naturalist Sir Joseph Banks (1743–1820), while he was accompanying Captain James Cook on his first voyage around the world. Banks subsequently observed it along the seacoast of southern and western Australia and Tasmania. New Zealand spinach was brought to Europe in 1770. Selections have been made resulting in plants with larger leaves than found in wild forms. Although the cultivated forms are used similarly to common spinach, New Zealand spinach has the added benefit of continuing to grow well during hot weather. Unlike true spinach, which develops a compact rosette of leaves, New Zealand spinach produces a leafy, branching stem, growing 30 to 60 cm (1–2 ft.) tall, but the shoot tips and young leaves are generally cut when 10 to 20 cm (4–8 in.) tall. The plants often cover an area more than 1 m (approximately 1 yd.) in diameter. The leaves are thick, triangle-shaped, dark green on top, and glistening underneath. New Zealand spinach is cultivated in most parts of the tropics and temperate regions as a minor crop.

FIGURE 89.2 New Zealand spinach (*Tetragonia tetragonioides*). (From Vilmorin-Andrieux, 1885.)

Culinary Portrait

New Zealand spinach leaves are used like spinach. The leaves and young shoots are steamed with a little water until tender and consumed as a hot vegetable, or added to soups, stews, pastas, or meat dishes. The raw leaves can be used in salads. The flavor is similar to but milder than that of common spinach. New Zealand spinach has been said to taste like a cross between spinach and sorrel, with an overtone of green apples.

New Zealand spinach contains soluble oxalates, which have poisoned livestock in parts of Australia. Sheep have died after consuming large quantities from lush stands. Ingestion has also led to urinary calculi (kidney or urinary tract stones) in sheep. There is no danger of poisoning in humans provided that the plant is eaten in moderate amounts. It is recommended that New Zealand spinach should be blanched for 3 minutes to remove soluble oxalates and the water discarded. (Rhubarb is another example of a plant that contains considerable oxalate.) When cooked, New Zealand spinach has a more creamy consistency than spinach.

Culinary Vocabulary

- "Bushfood" or "bush tucker" refers to food collected from the wild in Australia. New Zealand spinach has been called a bush food because it has been collected from the wild, but today better-tasting cultivated varieties are available.

Curiosities of Science and Technology

- It has been claimed that New Zealand spinach is the only potherb that Europeans have derived from Australasia (this was true until recently).
- New Zealand spinach was first grown in Europe at the world's premier botanical garden, Kew Gardens in London, in 1772. Ironically, it has never become popular in British gardens.
- As noted earlier, New Zealand spinach was first brought to Europe as a result of Captain James Cook's first voyage in 1770. However, Cook and his crew did not appreciate at the time that the vegetable could have alleviated the scurvy that the crew suffered. On Cook's second voyage, a botanist named Foster again collected New Zealand spinach, fed it to the crew, and discovered that it indeed was a treatment for scurvy.
- New Zealand spinach, like other members of the ice-plant family, possesses minute dots on the leaves that reflect the sun, making it appear that plant has a coating of ice (this is why many species in the family are called ice plants). The tiny dots are in fact huge cells that protrude from the epidermis, causing light dispersion such that the plants seem to be covered by glittering ice crystals. In some species, it has been demonstrated that these "epidermal bladder cells" are storage sites where the plants gets rid of excessive salt.

Key Information Sources

Ahmed, A.K., and Johnson, K.A. 2000. The effect of the ammonium: nitrate nitrogen ratio, total nitrogen, salinity (NaCl) and calcium on the oxalate levels of *Tetragonia tetragonioides* Pallas. Kunz. *J. Hortic. Sci. Biotech.* 75:533–538.

Banadyga, A.A. 1977. Greens or "potherbs"—chard, collards, kale, mustard, spinach, New Zealand spinach. Agric. *Info. Bull. U.S. Dept. Agric.* 409:163–170.

Brooker, S.G. 1986. Food and beverages from NZ native plants. *Food Technol. N. Z.* 21(7):30, 32–33, 37, 39, 41.

Colbert, T. 1978. New Zealand spinach. *Pacific Hortic.* 39(3):13–14.

Gorini, F. 1982. New Zealand spinach. Leaf vegetables. *Informatore di Ortoflorofrutticoltura*, 23(1):3–5 (in Italian).

Hudezek, H., and Kraxner, U. 1982. Vegetable species for culture in the open, pt. 5: [*Tetragonia expansa, Arracacia esculenta, Lactuca indica, Oxalis tuberosa*]. *Gemuese*, 18:288–290 (in German).

Jadczak, D., and Orlowski, M. 2000. Influence on planting density and the length of harvested shoots on the yield of New Zealand spinach. *Annales Universitatis Mariae Curie Sklodowska. Sectio EEE Horticultura*, 8(Suppl.):197–203 (in Polish).

Jankowiak, J. 1976. New Zealand spinach: a rewarding hot-weather greens crop. *Org. Gard. Farm.* 23(5):76–77.

Jaworska, G., and Kmiecik, W. 2000. Comparison of the nutritive value of frozen spinach and New Zealand spinach. *Polish J. Food Nutr. Sci.* 9(4):79–84.

Jaworska, G., and Slupski, J. 2001. The value of New Zealand spinach for freezing. *Zywnosc*, 8(2):92–102 (in Polish).

Kays, S.J. 1975. Production of New Zealand spinach (*Tetragonia expansa* Murr.) at high plant densities. *J. Hortic. Sci.* 50:135–141.

Kays, S.J., and Austin, M.E. 1975. Use of growth regulators for increased quality of New Zealand spinach. *HortScience*, 10:416–417.

Kemp, M.S., Burden, R.S., and Brown, C. 1979. A new naturally occurring flavanone from *Tetragonia expansa*. *Phytochemistry*. 18:1765–1766.

Melin, J. 1978. New Zealand spinach. *Jardins France*, 2:50 (in French).

Myers, C. 1991. New Zealand spinach. Crop Sheet SMC-025. In *Specialty and minor crops handbook*. Edited by C. Myers. The Small Farm Center, Division of Agriculture and Natural Resources, University of California, Oakland, CA. 2 pp.

Priszter, S. 1978. *The New Zealand spinach* (Tetragonia tetragonioides *(Pall.) O. Ktze.)*. Akademiai Kiado, Budapest, Hungar. 56 pp (in Hungarian).

Siemonsma, J.S. 1993. *Tetragonia tetragonioides* (Pallas) O. Kuntze. In *Plant resources of South-East Asia. Vol. 8. Vegetables*. Edited by J.E. Siemonsma and K. Piluek. Pudoc Scientific Publishers, Wageningen, the Netherlands. pp. 269–271.

Tashiro, T., Kobayashi, T., and Hino, K. 1978. Hydrocarbons of spinach and New Zealand spinach. *J. Jpn. Soc. Food Sci. Technol.* 25:153–157 (in Japanese, English summary).

Tolken, H.R. 1981. Tetragoniaceae *Tetragonia eremaea*, *Tetragonia tetragonioides*. In *Flora of Central Australia*. Edited by J. Jessop. Reed, Sydney, N.S.W., Australia. pp. 38–39.

Wacker, H.D. 1977. New Zealand spinach all summer long. *Gartenpraxis*, 3:134–135 (in German).

Wilson, C., Lesch, S.M., and Grieve, C.M. 2000. Growth stage modulates salinity tolerance of New Zealand spinach (*Tetragonia tetragonioides*, Pall.) and red orach (*Atriplex hortensis* L.). *Ann. Bot.* 85:501–509.

Specialty Cookbooks

Albi, J. 1996. *Greens glorious greens: more than 140 ways to prepare all those great-tasting, super-healthy, beautiful leafy greens*. St. Martin's Griffin, New York. 288 pp.

Ballister, B. 2002. *The fruit and vegetable stand: the complete guide to the selection, preparation and nutrition of fresh produce*. Revised edition. Overlook TP, Woodstock. 455 pp.

Conran, T., and Clevely, A.M. 2000. *The chef's garden*. Soma Books, San Francisco, CA. 144 pp.

Murrills, A. 2000. *Food city: Vancouver*. Polestar, Victoria, BC. 272 pp.

WATER SPINACH

Family: Convolvulaceae (morning-glory family)

Names

Scientific Name: *Ipomoea aquatica* Forsk. (*I. reptans* Poir.)

- The name "water spinach" is an appropriate description for this species: it is sometimes cultivated in water and is spinach like, although it is not a true spinach.
- Water spinach is also called Chinese convolvulus (the genus *Convolvulus* includes morning glory), swamp cabbage, swamp morning glory, and water convolvulus. It is also known

in the marketplace by the Asian names green engtsai, kancon, and kang kong (kangkong, kankong, kangcong).

- According to an old Taiwanese story, the ancestors of the people of Taiwan wanted to leave China to escape the reign of a tyrant. They delayed their voyage to find a vegetable that would remain fresh during the long voyage. Finally, they discovered the water spinach, which thrived in their earthenware water containers, and survived the journey to Taiwan. To remind their children of the voyage, they named the vegetable *ong tsoi*; *ong* means a large earthenware vessel for holding water, whereas *tsoi* means a vegetable. This is currently the main name of water spinach throughout China. (A less romantic interpretation is based on a second meaning of *ong*, "to bury plant propagules under the soil," the way that the plant is reproduced.)
- The genus name *Ipomoea* is formed from the Greek *ips*, worm, vine-like plant + *homoios*, similar to. Many of the species are twining vines, hence the phrasing "like a vine" or "like a worm."
- *Aquatica* in the scientific name *I. aquatica* is Latin for growing in water, descriptive of the natural habitat of the species.

Plant Portrait

Water spinach is an important, green, leafy, tropical, and subtropical Asian vegetable. Domestication of water spinach took place in tropical Asia, possibly India. The species is known to have been cultivated in China by AD 300. It is grown widely in Southeast Asia, Taiwan, Malaysia, Australia, and some parts of Africa. The plant is considered to be Vietnam's national vegetable. Water spinach has also been introduced to Hawaii, Brazil, Central America, and several Caribbean Islands. Two forms are cultivated, an aquatic form (sometimes called variety *aquatica*) and an upland form (sometimes called variety *reptans*). The aquatic form is a short-lived, herbaceous perennial that grows in or near water as a semiaquatic plant. The leaves are smooth edged, with long stalks, and the stems are hollow. The plant lies prostrate or floats. It is cultivated in the southern part of India and Southeast Asia. The upland form is an annual herb cultivated as a vegetable on dry or marshy land. It has narrower

FIGURE 89.3 Water spinach (*Ipomoea aquatica*). Lower left: vegetative branch by B. Flahey. Upper right: flowering branch by M. Jomphe.

leaves than the aquatic variety and the stems are generally upright. It is an important market vegetable in Malaysia, Indonesia, and other Southeast Asian countries. Where grown locally, the plant is marketed in bundles, resembling common watercress, with small leaves attached to the main stem. Canned water spinach is often seen on store shelves. Water spinach is also a valuable livestock fodder, fed to cattle and swine in the Philippines, Malaysia, and Fiji. It is also used as fish food.

Ipomoea aquatica is one of 16 aquatic plants that are listed on the *U.S. Federal Noxious Weed List.* The U.S. Congress passed the *Plant Protection Act*, which provides for penalties for knowing violations, such as moving a federal noxious weed without a permit. Civil penalties range from \$1000 to \$250,000 per violation. Many independent dealers, pet shops, and even some mail order firms that sell aquatic plants for use in aquaria and ornamental ponds are unaware that the plant is on the list (note additional information below).

Culinary Portrait

Water spinach leaves are used in the tropics as a cooked or fresh vegetable. Almost all parts of the young plants are eaten. Because older stems become fibrous, young succulent tips and tender stems are preferred. These are eaten fresh in salads or cooked like spinach. Cooking in oil is common. Spices should be added to enhance the bland flavor. The vegetable is mildly laxative. Harvested material deteriorates quickly and is best used shortly after picking.

Culinary Vocabulary

- *Pahk boong fai daeng*, a Thai specialty dish made with water spinach, translates as "water spinach on fire."

Curiosities of Science and Technology

- In China, elderly people who have rheumatism are often not allowed to consume water spinach, in fear that this will worsen their condition. However, this appears to be a traditional belief without any scientific basis. Compare another traditional Chinese food prohibition: those with wounds must abstain from eating seafood, especially prawns and crabs, which are thought to slow the healing of wounds, possibly making them worse.
- Water spinach has several desirable medicinal and nutritional properties. It has been documented to be protective against high blood pressure and nosebleeds. This vegetable is also a rich source of carotenoids that have been implicated in prevention of skin tumors. Owing to its high iron content, it is often fed to patients who are suffering from certain types of anemia.
- Water spinach has been listed by the U.S. Department of Agriculture as a potentially noxious weed in the United States, although it is easily obtained through American seed catalogs. It has become established as an aquatic weed in several areas in Florida, where its production is discouraged. Fear of water spinach as a potential weed in Florida is partly due to the remarkable growth rate it has demonstrated in the state. Plants have been observed growing 10 cm (4 in.) daily and climbing to a height of more than 21 m (69 ft.).
- Water spinach takes up some organic and inorganic heavy metals from waste water and so can be used in waste water treatment in the tropics.

Key Information Sources

Bruemmer, J.H., and Roe, B. 1980. Protein extraction from water spinach (*Ipomoea aquatica*). *Proc. Annu. Meet. Fla. State Hortic. Soc.* 92:140–143.

Chen, B.H., and Han, L.H. 1990. Effects of different cooking methods on the yield of carotenoids in water convolvulus (*Ipomoea aquatica*). *J. Food Prot.* 53:1076–1078.

Cornelis, J., Nugteren, J.A., and Westphal, E. 1985. Kangkong (*Ipomoea aquatica* Forssk.): an important leaf vegetable in South-East Asia. Review article. *Abstr. Trop. Agric.* 10(4):9–21.

Datta, S.C., and Biswas, K.K. 1970. Germination-regulating mechanisms in aquatic angiosperms. 1, *Ipomoea aquatica* Forsk. *Broteria*, 39:175–185.

Datta, P.C., and Saha, N. 1976. Development of leaf forms in *Ipomoea reptans* (Linn.) Poir. *Ann. Bot.* 40:837–843.

Devi, A.J.U., Parabia, M.H., and Reddy, M.N. 1990. Morphological and anatomical studies of the seedlings of *Ipomoea aquatica* Forsk. *Feddes Rep.* 101:391–394.

Edie, H.H., and Ho, B.W.C. 1969. *Ipomoea aquatica* as a vegetable crop in Hong Kong. *Econ. Bot.* 23:32–36.

Gothberg, A., Greger, M., Holm, K., and Bengtsson, B.E. 2004. Influence of nutrient levels on uptake and effects of mercury, cadmium, and lead in water spinach. *J. Environ. Qual.* 33:1247–1255.

Jain, S.K., Gujral, G.S., and Vasudevan, P. 1987. Potential utilization of water spinach (*Ipomoea aquatica*). *J. Sci. Ind. Res.* 46:77–78.

Kameoka, H., Kubo, K., and Miyazawa, M.1992. Essential oil components of water convolvulus (*Ipomoea aquatica* Forsk.). *J. Essent. Oil Res.* 4:219–222.

Karnchanawong, S., and Sanjitt, J. 1995. Comparative study of domestic wastewater treatment efficiencies between facultative pond and water spinach pond. *Water Sci. Technol.* 32:263–270.

Linnemann, A.R., Louwen, J.M., Straver, G.H.M.B., and Westphal, E. 1986. Influence of nitrogen on sown and ratooned upland kangkong (*Ipomoea aquatica* Forssk.) at two planting densities. *Netherlands J. Agric. Sci.* 34:15–23.

Oliver, J.D. 1992. *A review of the literature:* Ipomoea aquatica *and* Ipomoea fistulosa. Florida Department of Natural Resources, Tallahassee, FL. 14 pp.

Ose, K., Chachin, K., and Imahori, Y. 1995. Browning mechanism of water convolvulus (*Ipomoea aquatica* Forsk.) stored at low temperature. *ACS Symp. Ser. Am. Chem. Soc.* 600:178–187.

Ose, K., Chachin, K., and Ueda, Y. 1999. Relationship between the occurrence of chilling injury and the environmental gas concentration during storage of water convolvulus (*Ipomoea aquatica* Forsk.). *Acta Hortic.* 483:303–310.

Oyer, E.B. 1978. Lesser known vegetables with apparent ability to withstand stress conditions. *A.S.A. Spec. Publ. Am. Soc. Agron.* 32:155–160.

Park, K.W., Han, K.S., and Won, J.H. 1993. Effect of propagation method, planting density and fertilizer level on the growth of water spinach (*Ipomoea aquatica*). *J. Korean Soc. Hortic. Sci.* 34:241–247 (in Korean).

Rai, U.N., and Sinha, S. 2001. Distribution of metals in aquatic edible plants: *Trapa natans* (Roxb.) Makino and *Ipomoea aquatica* Forsk. *Environ. Monit. Assess.* 70:241–252.

Snyder, G.H., Morton, J.F., and Genung, W.G. 1981. Trials of *Ipomoea aquatica*, nutritious vegetable with high protein- and nitrate-extraction potential. *Proc. Annu. Meet. Fla. State Hortic. Soc.* 94:230–235.

Stephens, J.M. 1994. *Kangkong—*Ipomoea aquatica *Forsk., also* Ipomoea reptans *Poir.* University of Florida Cooperative Extension Service, Institute of Food and Agricultural Sciences, Gainesville, FL. 1 p.

Van, T.K., and Madeira, P.T. 1998. Random amplified polymorphic DNA analysis of water spinach (*Ipomoea aquatica*) in Florida. *J. Aquat. Plant Manag.* 36:107–111.

Vora, A.B., Patel, J.A., and Ravkishore, C.V.N. 1988. Ecophysiological studies in *Ipomoea aquatica* Forsk. *J. Environ. Biol.* 9(1 Suppl.):119–122.

Westphal, E. 1993. *Ipomoea aquatica* Forsskal. In *Plant resources of South-East Asia. Vol. 8. Vegetables.* Edited by J.E. Siemonsma and K. Piluek. Pudoc Scientific Publishers, Wageningen, the Netherlands. pp. 181–184.

Specialty Cookbooks

Note: Asian cookery is more likely to use water spinach in recipes than any other kind of spinach.

Daks, N. 2007. *Wok cooking made easy. Delicious cooking in minutes.* Periplus, Singapore. 128 pp.

McDermott, N. 2004. *Quick & easy Thai:70 everyday recipes.* Chronicle Books, San Francisco, CA. 168 pp.

Moskowitz, I.C., and Romero, T.H. 2007. *Veganomicon: the ultimate vegan cookbook.* Da Capo Press, New York. 336 pp.

Olizon-Chikiamco, N. 2003. *Filipino homestyle dishes: delicious meals in minutes.* Periplus, Singapore. 96 pp.

Trang, C. 2006. *The Asian grill: great recipes, bold flavors.* Chronicle Books, San Francisco, CA. 168 pp.

90 Stevia

Family: Asteraceae (Compositae; sunflower family)

NAMES

Scientific Name: *Stevia rebaudiana* (Bertoni) Bertoni

- The English name stevia is simply the genus name *Stevia*. The genus was named for P.J. Esteve, a sixteenth-century Spanish botanist who died about 1566. Esteve was a professor of botany in Valencia Spain.
- Stevia is also known as honey grass, honey-yerba, Rebaudi's stevia, sweet hemp, sweet herb (of Paraguay), sweet plant (of Paraguay), and sugar leaf. Most of these alternative names for stevia are based on translations of (South American) Indian names that have appeared in journal articles.
- The florist's "stevia" is another species of the sunflower family, *Piqueria trinervia* Cav. ("*Stevia serrrata*" of some authors).
- *Rebaudiana* in the scientific name *S. rebaudiana* commemorates a Paraguayan chemist, Ovidio Rebaudi, the first to study the sweet chemicals of stevia, in the 1880s.

PLANT PORTRAIT

Stevia is a herbaceous or semi-woody shrub-like plant growing up to 1 m (approximately 1 yd.) tall. It produces small white flowers with a pale purple throat and seeds like those of dandelions. The shoots generally die after they set seed, but this perennial plant reproduces well in the spring from new shoots that arise at the base. Stevia is native to northeastern Paraguay (claims that it occurs as a native plant in nearby portions of Brazil and Argentina have not been verified), mainly in the Ypane River watershed, where it occurs on the edges of marshes and in grasslands. Plants are often available from nurseries specializing in herbs (seeds are infrequently available). In north temperate areas at some distance from the equator, the plants will not flower because they are adapted to a day length schedule close to the equator, their native location. Remaining vegetative will actually increase the content of sweet chemical in the leaves.

Stevia is most notable for its production of sweet chemicals, some more than 300 times as sweet as sugar, which add no calories to one's diet. The leaves are very sweet, approximately 30 times as sweet as sugar. The Guarani Indians of the Paraguayan highlands have used stevia for centuries for medicine (notably to treat diabetes and mild skin abrasions) and flavor (especially to sweeten bitter drinks such as maté, a tea-like drink made from the leaves of *Ilex paraguariensis* A. St-Hil.). Although the Conquistadors brought it to the attention of Europeans, stevia remained obscure to Western civilization until the late nineteenth century. Stevia was experimentally cultivated in southern England during the Second World War in an attempt to reduce dependence on imported sugar. In the 1970s, Japan became highly proficient in growing the plant and extracting the sweet chemicals from the leaves, and stevia acquired the status of a major food sweetener, which now holds about one-half of that country's sweetener market (including sugar), with sales worth more than $200 million. Stevia is cultivated commercially, especially in China, also in Taiwan, Laos, Thailand, Korea, Japan, Malaysia, Indonesia, Brazil, occasionally in Paraguay, and to a minor extent in the southeastern United States and southern Canada. Stevia sweetening agents are consumed in Brazil, China, Israel, Japan, Korea, and Paraguay. Stevia is most popular in Japan, where extracts are

FIGURE 90.1 Stevia (*Stevia rebaudiana*), a greenhouse-grown plant. (Photograph by E. Johnson.)

used to sweeten diet soft drinks, sugarless gum, soy sauce, pickled vegetables, and confectionary products. In Brazil some yogurts are sweetened with stevia. Interest in developing the species is growing rapidly in other countries.

In Paraguay, *S. rebaudiana* is used as a tea for the treatment of diabetes, apparently following reports in the literature that the plant is hypoglycemic (reduces blood sugar levels). As noted earlier, this is an ancient use of the plant by the Native Americans of Paraguay. Stevia is now often recommended (especially by those marketing it, but also by some reputable health authorities) for use by diabetics and hypoglycemics. Stevia has been demonstrated not to increase blood sugar levels as does sugar, but research is not conclusive that stevia is completely harmless for people with diabetes, who are advised to seek medical guidance before consuming it.

Diet-conscious consumers are attracted to stevia sweeteners because they have no calories. Many people like the idea that stevia is derived from a natural source, in contrast to synthetic sweetening agents, about which there are doubts about how safe they are, despite government approval. Because of lingering suspicion that stevia has toxic properties, stevia extracts could not be used legally in commercial manufacturing in North America until December of 2008, when a ruling by the United States Drug Administration reversed previous policies (but specifically only authorized a purified stevia extract, rebaudioside A). Usage remains banned (except in some cases as a "dietary supplement") in some countries, but there is expectation that they too may discontinue prohibitions of stevia. An early scientific report found that an extract of the leaves had an antifertility effect in female rats, and this seemed consistent with another report that Paraguayan Matto Grosso Indian tribes have used stevia as an oral contraceptive. The women were said to drink a daily decoction in water of powdered leaves and stems to achieve this purpose. This alleged contraceptive activity of the plant has been a controversial issue, but concern has diminished in recent times.

Stevia has been marketed for food use in jurisdictions where it has been illegal to incorporate it into commercial foods and despite the medical questions that have been raised in the past. This has been done by the widespread subterfuge of selling it as a herbal health supplement and as a skin conditioner, usage that often managed to bypass the prohibitions against stevia as food. Several books dedicated to stevia were published in the 1990s, and in the late 1990s, the U.S. Food and Drug

Administration launched legal actions against some of these books, including cookbooks, which were alleged to be in violation of laws intended to keep stevia from being used in food.

CULINARY PORTRAIT

Stevia rebaudina extracts containing stevioside are approved as food additives in Japan, South Korea, Brazil, Argentina, and Paraguay, and as of 2009, extracts have also been used commercially in the United States. As noted earlier, herbal stevia (i.e., raw leaves, usually ground to a powder) is widely available in much of Europe and elsewhere where extracts cannot be used in commercial food.

In South America, whole or ground dried leaves are used directly as a sweetener, and the liquid obtained after soaking the leaves in water is used to make preserves. The leaves are sometimes chewed. Manufacturers have also incorporated powdered stevia into herbal tea preparations. Two or three leaves are considered sufficient to sweeten a cup of coffee or tea, and it is recommended that once sweetened, the leaves be removed.

Stevia sweeteners tend to have a bitter metallic aftertaste, limiting their use. New technologies are under development to produce a more acceptable sweetener for use in beverages and baked goods. To some, stevia has a slight licorice flavor, which often requires getting used to. Stevia may be purchased as cut or powdered dried leaf, as a crude green powder, or as a brownish liquid extract with an unpleasant, licorice-like taste. Using the powder directly is inadvisable because the sweetening power is so high people put in too much and find that it is just too sweet. The white powder may be made into a liquid concentrate by dissolving it in water (approximately a 10% solution by volume—1 teaspoon powder: 3 tablespoons water), which is best stored in the refrigerator and dispensed from a dropper bottle. One teaspoon of this solution is approximately equal to a cup of sugar. Different brands of stevia may have different concentrations of stevioside, and it is advisable to check the manufacturer's advice on relative sweetening power.

Stevia does not have some of the properties of sugar, which make the latter desirable for some culinary applications, particularly with respect to texture. Sugar promotes texture of cooked fruit, it enhances elasticity of gluten in cakes and egg preparations, it combines with pectin in fruit to form jelly, makes meat more tender, and adds bulk and texture to ice cream. Many cookbooks dedicated to stevia recommend adding ingredients (e.g., apple sauce, egg whites, fruit juice, yogurt, at a ratio of 1/4 to 1/2 cup for each cup of replaced sugar) to substitute for the bulk otherwise contributed by sugar. Also unlike sugar, stevia does not caramelize with heat (caramelization is the oxidation of sugar, achieved by heating until it melts and becomes thick and dark). Accordingly, stevia-baked goods do not brown very much and may be checked for doneness by touching and not by color. Sugar is a preservative agent in many foods, for example, in jellies and jams sugar binds water, thereby inhibiting the growth of microorganisms. Stevia lacks this ability, and some foods prepared with it should not be stored for lengthy periods. Whereas sugar seems to be a useful taste enhancer for almost all kinds of foods, strong-flavored foods seem to benefit particularly from stevia. For example, stevia blends well with citrus fruit flavors such as lemon and with cranberry.

Culinary Vocabulary

- Several major soft drink companies have trade names for sweeteners based at least in part on stevia. Truvia is the consumer brand of a sweetener made partly with rebaudioside A from stevia and developed in association with Coca-Cola. PureVia is PepsiCo's brand of another preparation of rebaudioside A in combination with other ingredients.
- The phrase "artificial sweeteners" is somewhat ambiguous (see discussion below regarding "synthetic sweeteners" and the fact that "artificial," i.e., chemically synthesized compounds may also exist naturally in some species). Stevia preparations are "natural"

sweeteners because rebaudioside A (and other steviosides) are natural compounds in stevia plants. Nevertheless, in food science, there is at present no universally accepted definition of "natural."

CURIOSITIES OF SCIENCE AND TECHNOLOGY

- There is intense competition for the sweetener market between natural and synthetic sweetening agents. The possible economic advantages of chemically synthesized sweeteners are indicated by cyclamates which, before they were banned in the United States in October of 1970, reached a production figure of 247,000 metric tons/year (272,000 tons/year). Ten cents worth of cyclamate had the sweetening power of a dollar's worth of sucrose. Today, saccharin (Sweet-n-Low) and aspartame (Equal, NutraSweet, Spoonful, Candarel, and Equal-Measure) are common synthetic sweeteners in North America. Aspartame is "nutritive" (has calories) but is 180 times sweeter than sucrose, so very little needs to be used. Saccharin is noncaloric and is passed through the body unchanged. Aspartame is considered to be the closest in taste to sugar, whereas saccharin has a bitter aftertaste. Diet drinks lack the sugar that contributes to tooth decay but can still be harmful to the teeth in the long run because the acidic quality of diet drinks tends to erode tooth enamel. "Bulk sweeteners," such as xylitol, sorbitol, and mannitol, are artificial sweeteners that have about the same caloric value as sugar and replace it in many commercial foods. (Just because something is "artificial," i.e., chemically synthesized, does not mean it is also not available in natural form; sorbitol occurs naturally in cherries and plums.) Xylitol and sorbitol are used in "sugar-free" gums that have been proven to help prevent tooth decay. A study published in the *Journal of the American Paediatrics* reported in 2000 that chewing xylitol-sweetened gum could reduce the risk of ear infection in children by up to 40%.
- Aztec sweet herb (*Phyla scaberrima* (Juss. ex Pers.) Moldenke (*Lippia dulcis* Trévir.)) is the source of a noncarbohydrate sweetener, hernandulcin, which is approximately 1000 times sweeter than sucrose. Unfortunately, it is definitely poisonous. Like stevia, it is in the sunflower family.
- A wild bramble of China, "sweet tea" or *Rubus suavissimus* S.K. Lee (*R. chingii* Hu), produces *rubusoside*, an intensely sweet compound structurally related to some of the sweetening chemicals of stevia.
- Miracle fruit (see Chapter 64) is an interesting plant that is not sweet but changes the flavor of acid foods such as lemon and rhubarb into a delicious sweetness. The chapter on miracle fruit also presents additional sweet-tasting plants.

KEY INFORMATION SOURCES

Akashy, H., and Yokoyama, Y. 1975. Dried leaf extracts of *Stevia*. Toxicological tests. *Shokuhin Kogyo*, 18:34–43 (in Japanese, partly translated into English).

Blumenthal, M. 1992. AHPA petitions FDA for approval of *Stevia* leaf sweetener. *HerbalGram*, 26:22, 55.

Bonvie, L., Bonvie, B., and Gates, D. 1997. *A tale of incredible sweetness & intrigue*. B.E.D. Publications Co., Atlanta, GA. 79 pp.

Brandle, J. 1999. Genetic control of rebaudioside A and C concentration in leaves of the sweet herb, *Stevia rebaudiana*. *Can. J. Plant Sci*. 79:85–92.

Brandle, J.E., and Rosa, N. 1992. Heritability for yield, leaf: stem ratio and stevioside content estimated from a landrace cultivar of *Stevia rebaudiana*. *Can. J. Plant Sci*. 72:1263–1266.

Brandle, J.E., Starratt, A.N., and Gijzen, M. 1998. *Stevia rebaudiana*: its agricultural, biological, and chemical properties. *Can. J. Plant Sci*. 78:527–536.

Cardello, H.M.A.B., Da Silva, M.A.P.A., and Damasio, M.H. 1999. Measurement of the relative sweetness of stevia extract, aspartame and cyclamate/saccharin blend as compared to sucrose at different concentrations. *Plant Foods Hum. Nutr*. 54:119–130.

Carneiro, J.W.P., Muniz, A.S., and Guedes, T.A. 1997. Greenhouse bedding plant production of *Stevia rebaudiana* (Bert.) Bertoni. *Can. J. Plant Sci.* 77:473–474.

Chalapathi, M.V., Thimmegowda, S., Rao, G.G.E., Devakumar, N., and Chandraprakash, J. 1999. Influence of fertilizer levels on growth, yield and nutrient uptake of ratoon crop of stevia (*Stevia rebaudiana*). *J. Med. Arom. Plant Sci.* 21:947–949.

Crammer, B., and Ikan, R. 1986. Sweet glycosides from the stevia plant. *Chem. Br.* 22:915–916, 918.

Goettemoeller, J., and Ching, A. 1999. Seed germination in *Stevia rebaudiana.* In *Perspectives on new crops and new uses.* Edited by J. Janick. ASHS Press, Alexandria, VA. pp. 510–511.

Haebisch, E.M.A.B. 1992. Pharmacological trial of a concentrated crude extract of *Stevia rebaudiana* (Bert.) Bertoni in healthy volunteers. *Arq. Biol. Tecnol.* 35:299–314.

Handro, W., and Ferriera, C.M. 1989. *Stevia rebaudiana* (Bert.) Bertoni: production of natural sweeteners. In *Biotechnology in agriculture and forestry. Vol. 7. Medicinal and aromatic plants—II.* Edited by Y.P.S. Bajaj. Springer-Verlag, Berlin. pp. 468–487.

Healy, W., and Graper, D. 1989. Flowering of stevia. *Acta Hortic.* 252:137–142.

Hemsley, W.B. 1909. *Stevia rebaudiana.* In *Icones plantarum. Fourth series, Vol. 9.* Edited by W.J. Hooker. Dulau & Co., London. Tabula 2816.

Kinghorn, A.D. (Ed.). 2002. Stevia. *The genus* Stevia. Taylor & Francis, London. 211 pp.

Lewis, W.H. 1992. Early uses of *Stevia rebaudiana* (Asteraceae) leaves as a sweetener in Paraguay. *Econ. Bot.* 46:336–337.

Martelli, A., Frattini, C., and Chialva, F. 1985. Unusual essential oils with aromatic properties—I. Volatile components of *Stevia rebaudiana* Bertoni. *Flavour Fragrance J.* 1:3–7.

Metivier, J., and Viana, A.M. 1979. The effect of long and short day length upon the growth of whole plants and the level of soluble proteins, sugars and stevioside in leaves of *Stevia rebaudiana. Bert. J. Exp. Bot.* 30:1211–1222.

Mohede, J., and Son, R.T.M. van 1999. *Stevia rebaudiana* (Bertoni) Bertoni. In *Plant resources of South-East Asia. Vol. 13. Spices.* Edited by C.C. de Guzman and J.S. Siemonsma. Backhuys, Leiden, the Netherlands. pp. 207–211.

Norina, L.V., Bailey, W.C., and Timcke, K. 2003. Stevia (*Stevia rebaudiana* Bertoni): its potential for New Zealand. *Food New Zealand,* 3(4):29–35.

O'Brien Nabors, L., and Inglett, G.E. 1986. A review of various other alternative sweeteners. In *Alternative sweeteners.* Edited by L. O'Brien Nabors and R.C. Gelardi. Marcel Dekker, New York. pp. 309–323.

Phillips, K.C. 1987. Stevia: steps in developing a new sweetener. In *Developments in sweeteners. 3.* Edited by T.H. Grenby. Elsevier Applied Science, London. pp. 1–43.

Planas, G.M., and Ku , J. 1968. Contraceptive properties of *Stevia rebaudiana. Science,* 162:1007.

Robinson, B.L. 1930. Observation on the genus *Stevia. Contrib. Gray Herb. Harv. Univ.* 90:36–58.

Robinson, B.L. 1930. The stevias of Paraguay. *Contrib. Gray Herb. Harv. Univ.* 90:79–90.

Schardt, D. 2000. Stevia: a bittersweet tale. *Nutr. Action Health Lett.* 27(3):3.

Shock, C.C. 1982. *Experimental cultivation of Rebaudi's* Stevia *in California.* University of California, Davis Agronomy Program Report No. 122. 9 pp.

Shock, C.C. 1982. Rebaudi's stevia: natural noncaloric sweeteners. *Calif. Agric.* 1982(Sept–Oct):4–5.

Small, E., and Catling, P.M. 2001. Blossoming treasures of biodiversity:1. Stevia (*Stevia rebaudiana* (Bertoni) Bertoni)—how sweet it is! *Biodiversity,* 2(2):22.

Soejarto, D.D., Compadre, C.M., Medon, P.J., Kamath, S.K., and Kinghorn, A.D. 1983. Potential sweetening agents of plant origin. II. Field search for sweet-tasting *Stevia* species. *Econ. Bot.* 37:71–79.

Soejarto, D.D., Kinghorn, A.D., and Farnsworth, N.R. 1982. Potential sweetening agents of plant origin. Organoleptic evaluation of *Stevia* leaf herbarium samples for sweetness. *J. Nat. Prod.* 45:590–599.

Strauss, S. 1995. The perfect sweetener? *Technol. Rev.* 98:18–20.

Tamura, Y., Nakamura, S., Fukui, H., and Tabata, M. 1984. Comparison of *Stevia* plants grown from seeds, cuttings and stem tip cultures for growth and sweet diterpene glycosides. *Plant Cell Rep.* 3:180–182.

Tateo, F., Fugazza, M., Faustle, S., Bianchi, A., Tateo, S., Berte, F., and Bianchi, L. 1990. Technical and toxicological problems connected with the formulation of low-energy foods. 2. Mutagenic and fertility-modifying activity of extracts and constituents of *Stevia rebaudiana* Bertoni. *Rev. Soc. Ital. Sci. Aliment.* 19:1–2, 13–22 (in Italian).

Toyoda, K., Matsui, H., Shoda, T., Uneyama, C., Takada, K., and Takahashi, M. 1997. Assessment of the carcinogenicity of stevioside in F344 rats. *Food Chem. Toxicol.* 34:597–603.

Van Hooren, D.L., and Lester, H.R. 1992. Stevia drying in small scale bulk tobacco kilns. In *Methods to utilize tobacco kilns for curing, drying and storage of alternate crops, final report*. Edited by D.L. Van Hooren and H.R. Lester. Ontario Ministry of Agriculture and Food, Delhi, ON. pp. 159–164.

Yao, Y., Ban, M., and Brandle, J. 1999. A genetic linkage map for *Stevia rebaudiana*. *Genome*, 42:657–661.

Specialty Cookbooks

DePuydt, R. 1998. *Baking with stevia II: more recipes for the sweet leaf*. Sun Coast Enterprises, Oak View, CA. 106 pp.

DePuydt, R.E. 2002. *Stevia: naturally sweet recipes for desserts, drinks, and more*. The Book Publishing Company, Summertown, TN. 224 pp.

Goettemoeller, J. and Cavaciuti, S. 1998. *Stevia sweet recipes: sugar-free-naturally*. Vital Health Publishing, Ridgefield, CT. 196 pp.

Jobs, L. 2005. *Sensational stevia desserts*. Healthy Lifestyle Publishing LLC, Valley Forge, PA. 120 pp.

Kirkland, J. 2000. *Low-carb cooking with stevia, the naturally sweet & calorie-free herb*. Crystal Health Publishing, Arlington, TX. 256 pp.

Kirkland, J., and Kirkland, T. 2000. *Sugar-free cooking with stevia: the naturally sweet & calorie-free herb*. 3rd ed. Crystal Health Publishing, Arlington, TX. 288 pp.

Richard, D. 1996, 1999. *Stevia rebaudiana: nature's sweet secret*. 3rd ed. Vital Health Pub., Bloomington, IL. 76 pp.

Sahelian, R., and Gates, D. 1999. *The stevia cookbook: cooking with nature's calorie-free sweetener*. Avery Publishing Group, Garden City Park, NY. 171 pp.

91 Sweetsop and Soursop

Family: Annonaceae (custard apple family)

These two related fruits are examined together because of their similarities.

This chapter features

Sweetsop (*Annona squamosa*)
Soursop (*Annona muricata*)

SWEETSOP

Names

Scientific Name: *Annona squamosa* L.

- The name "sweetsop" (or sweet sop) contrasts with the related soursop (or soursop), treated below. "Sop" indicates the loose texture of the fruit (the word has the same root as soup and was once used to indicate food soaked in liquid). Specifically, the white flesh of both the sweetsop and the soursop seems to resemble milk-soaked bread, known in the past as "sops." The name sweetsop is sometimes (erroneously) applied to *A. reticulata* L., a lesser-known fruit species of *Annona*.
- The sweetsop is also known as anona, applebush, bullock's head, cinnamon apple, custard apple, scaly custard apple, and sugar apple.
- The use of "apple" in some of the names indicates the size of the fruit.
- "Custard" in the name custard apple is based on the creamy texture and appearance of the ripe flesh. The name custard apple arose in the seventeenth century from the resemblance to custard and apples. *Annona reticulata* is also called custard apple.
- "Scaly" in the name scaly custard apple reflects the scaly appearance of the fruit. *Squamosa* in the scientific name *A. squamosa* is Latin for scaly.
- The name custard apple is applied to other species of *Annona*, most notably cherimoya (see Chapter 24).
- For information on the genus name *Annona*, see Chapter 24.

Plant Portrait

The sweetsop is a deciduous tree 3 to 6 m (10–20 ft.) in height. The fruit is nearly round, egg-shaped, or conical, 6 to 10 cm (2⅓–4 in.) long, with a thick, knobby, yellow-green, greenish, or pinkish rind. Most types of sweetsop have green skin but dark red varieties are available. The flesh is creamy-white, juicy, and sweet, in segments like an orange, with one or more seeds in each segment. The seeds are black or dark brown, approximately 1.3 cm (1/2 in.) long, and inedible. Some seedless fruit is available, but seeded varieties are generally tastier. The sweetsop probably originated in the West Indies or Central America and requires tropical or almost tropical conditions. It is widely

FIGURE 91.1 Sweetsop (*Annona squamosa*). (From Lamarck and Poiret, 1744–1829, plate 494.)

cultivated in tropical South America, occasionally northward in the New World and in Australia. The Spanish and Portuguese brought sweetsop from the New World to Asia, and today cultivation is most extensive in India. The sweetsop is extremely popular in the tropics, especially where the cherimoya cannot be grown. It is not cultivated commercially in the United States but is widely grown in southern Florida by homeowners for fruit. There is a large literature concerned with the uses of a variety of extracts from the sweetsop, especially the seeds, for potential medicinal applications and as protectants for crop plants against insects.

Culinary Portrait

Sweetsop is a delicious tropical fruit, usually eaten fresh out of hand when ripe, discarding the poisonous seeds. The taste is very attractive, one reviewer calling it "liquid sugar," another a cross between a pear and a coconut. The fruit is used to make sherbets and flavor ice cream. Sweetsop fruit is not suitable for many cookery applications; the pulp should not be cooked when used to make ice cream but does make good pies and preserves. The strained pulp makes an excellent drink mixed with milk and can be processed to produce wine and juice. The fruit may be stored in a refrigerator for up to a week, or it may be frozen. The potential uses are similar to those of the cherimoya (see Chapter 24). The many seeds present in most fruits are annoying to remove because the rind adheres strongly to the flesh (which should be scooped out), and some desirable flesh adheres to the seeds (many people suck the flesh off the seeds).

Culinary Vocabulary

- Sweetsop fruit is sometimes sold as "anon" (based on the genus name *Annona*).

Curiosities of Science and Technology

- The poisonous seeds of sweetsop are powdered and have been used for a variety of purposes in India: fish poison, insecticide, as a paste applied to the head to kill lice (taking care to avoid the eyes because the paste can cause blindness), and occasionally applied to the uterus to cause abortion.
- In Mexico, sweetsop leaves are rubbed on floors and placed in hen's nests to repel lice.

Key Information Sources

Andrade, E.H.A., Zoghbi, M. das G.B., Maia, J.G.S., Fabricius, H., and Marx, F. 2001. Chemical characterization of the fruit of *Annona squamosa* L. occurring in the Amazon. *J. Food Comp. Anal.* 14:227–232.

Anonymous. 1995. *Production and marketing of sweetsop*. Special Publication No. 1. Taidong Agricultural Improvement Station, Taiwan. 48 pp (in Chinese).

Beerh, O.P., Giridhar, N., and Raghuramaiah, B. 1983. Custard apple (*Annona squamosa*). I. Physico-morphological characters and chemical composition. *Indian Food Packer*, 37(3):77–81.

Bhakuni, D.S., Tewari, S., and Dhar, M.M. 1972. Aporphine alkaloids of *Annona squamosa. Phytochemistry*, 11:1819–1822.

Broughton, N.J., and Guat, T. 1979. Storage conditions and ripening of the custard apple *Annona squamosa* L. *Sci. Hortic.* 10:73–82.

Cogez, X., and Lyannaz, J.P. 1993. Manual pollination of the sugar apple (*Annona squamosa*). *Trop. Fruits Newsl.* 7:5–6.

Joy, B., and Rao, J.M. 1997. Essential oils of the leaves of *Annona squamosa* L. *J. Essent. Oil Res.* 9:349–350.

Kumar, R., Hoda, M.N., and Singh, D.K. 1977. Studies on the floral biology of custard apple (*Annona squamosa* Linn.). *Indian J. Hortic.* 34:252–256.

Leal, F. 1990. Sugar apple. In *Fruits of tropical and subtropical origin: composition, properties and uses*. Edited by S. Nagy, P.E. Shaw, and W.F. Wardowski. Florida Science Source Inc., Lake Alfred, FL. pp. 149–158.

Maahdeem, H. 1994. Custard apples. (*Annona* spp.). In *Neglected crops. 1492 from a different perspective*. Edited by J.E. Hernández-Bermejo and J. Léon. Food and Agriculture Organization of the United Nations, Rome, Italy. pp. 85–92.

Morton, J. 1987. Sugar apple. In *Fruits of warm climates*. Creative Resource Systems, Winterville, NC. pp. 67–92.

Mukerjea, T.D., and Govind, R. 1958. Studies on indigenous insecticidal plants. Pt. II. *Annona squamosa. J. Sci. Ind. Res.* 17C(1):9–15.

Pal, D.K., and Sampath-Kumar, P. 1995. Changes in the physico-chemical and biochemical compositions of custard apple (*Annona squamosa* L.) fruits during growth, development and ripening. *J. Hortic. Sci.* 70:569–572.

Pelissier, Y., Marion, C., Dezeuze, A., and Bessiere, J.M. 1993. Volatile components of *Annona squamosa* L. *J. Essent. Oil Res.* 5:557–560.

Prakash, G.S., Reddy, B.M.C., and Dass, H.C. 1982. Performance of different cultivars of *Annona squamosa* and some hybrids. *Lal Baugh*, 27:21–25.

Rajput, C.B.S. 1985. Custard apple. In *Fruits of India, tropical and subtropical*. Edited by T.K. Bose. Naya Prokash, Calcutta, India. pp. 479–486.

Rathore, D.S. 1999. Custard apples. In *Tropical horticulture*, vol. 1. Edited by T.K. Bose, S.K. Mitra, A.A. Farooqui, and M.K. Sadhu. Naya Prokash, Calcutta, India. pp. 351–358.

Sadhu, M.K., and Ghosh, S.K. 1976. Effects of different levels of nitrogen, phosphorus and potassium on growth, flowering, fruiting and tissue composition of custard apple. *Indian Agric.* 20:297–301.

Vishnu-Prasanna, K.N., Sudhakar-Rao, D.V., and Krishnamurthy, S. 2000. Effect of storage temperature on ripening and quality of custard apple (*Annona squamosa* L.) fruits. *J. Hortic. Sci. Biotech.* 75:546–550.

Visneswariah, K., Jayaram, M., Krishnaprasad, N.K., and Majumder, S.K. 1971. Toxicological studies of the seeds of *Annona squamosa. Indian J. Exp. Biol.* 9:519–521.

Vithanage, H.I.M.V. 1984. Pollen-stigma interactions: development and cytochemistry of stigma papillae and their secretions in *Annona squamosa* L. (Annonaceae). *Ann. Bot.* 54:153–167.

Wu, M.C., and Tsay, L.M. 1998. Activity of softening enzymes during storage of sugar apple (*Annona squamosa* L.) at different temperatures. *Food Preserv. Sci.* 24:319–323.

Wu, M.C., Chen, C.H., and Chen, C.S. 1999. Effects of different storage temperatures on change of fruit composition of sugar apple (*Annona squamosa* L.). *Food Preserv. Sci.* 25:149–154.

Specialty Cookbooks

Brennan, J. 1984. *The original Thai cookbook*. Perigree Trade, New York. 384 pp.

DeMers, J., and Fuss, E. 2005. *Authentic recipes from Jamaica*. Periplus, Singapore. 112 pp.

SOURSOP

Names

Scientific Name: *Annona muricata* L.

- The name "soursop" (or sour sop) contrasts with the related sweetsop, described above, which is comparatively sweet. However, the soursop is rarely "sour," albeit the taste is somewhat tart.
- The soursop is generally known in most Spanish-speaking countries as guanabana (*guanábana*, pronounced gwahn-AH-bahn-ah), a name that is also used in English, and is certainly more melodious and attractive than the name soursop, which has been a concern of companies trying to market the fruit in northern countries.
- The name guanabana should not be confused with cassabanana (see Chapter 21).
- The soursop is also known as graviola and prickly custard apple.
- "Wild soursop" is the name of an African species, *A. senegalensis* Pers., used for various edible purposes.
- *Muricata* in the scientific name *A. muricata* is Latin for roughened with short hard points, descriptive of the fruit surface.

Plant Portrait

The soursop is a tropical evergreen tree growing to a height of 4 to 9 m (13–30 ft.). The fruits are oval, heart shaped, or rather irregular, 10 to 30 cm (4–12 in.) long, up to 15 cm (6 in.) wide, and as heavy as 6.8 kg (15 lb.). The skin is dark green in the immature fruit, becoming slightly yellowish-green with maturity. On the surface of the fruit are stubby, soft, pliable, fleshy spines, the tips of which break off easily when the fruit is ripe. The interior flesh is white or creamy, granular, with fibrous, juicy segments. Some of these segments contain a single oval, smooth, hard, black seed, 1.25 to 2 cm (1/2–3/4 in.) long, although most of the segments are seedless. The soursop is probably native to the West Indies. It has spread throughout the humid tropics and is cultivated commercially in numerous tropical countries around the world, including Mexico, India, Southeast Asia, and Polynesia. It is grown commercially in southern Florida to a limited extent. There are many named varieties.

FIGURE 91.2 Soursop (*Annona muricata*). Left and center: branch and fruit. (From Jacquin, 1764–1771, plate 5.) Right: long section of fruit. (From Engler and Prantl, 1885–1915.)

Culinary Portrait

The soursop is a well-known fruit throughout much of the tropical world, but is rarely found fresh in northern countries. The best-tasting, least-fibrous soursops are cut in sections and the flesh eaten raw with a spoon. Soursop skin is bitter and inedible while the pulp has a pineapple-like aroma, with a unique, musky, acidic flavor, which has been described as a combination of pineapple and strawberry with a hint of cinnamon, or reminiscent of the black currant. The pulp can be shredded and added to fruit cups or salads or chilled and served as dessert with sugar and a little milk or cream. The seeds are slightly toxic and must be sieved out (see the information for sweetsop). Commercially produced soursop is usually processed into ice creams, sherbets, and drinks. Soursop beverages are widespread throughout the tropics, and soursop custard, sherbet, ice cream, and milk shakes are also made. Soursop ice cream, marketed under its Spanish name guanabana, is sold in some gourmet supermarkets. The pulp is also used in salad dressings and sauces. Preserved, canned soursop in syrup can be found in many ethnic markets in the larger cities of North America and is sometimes served in Mexican restaurants. The canned pulp can be puréed or blended in the home and is easily transformed into a delicious desert, although fresh pulp is more desirable. The pulp freezes well and retains flavor. Immature soursops are cooked as vegetables, used in soup, or roasted or fried in various countries.

The fruit, picked when full grown but slightly yellow-green, is easily bruised and punctured and must be handled with care. Air shipment is necessary to reach northern countries in a marketable condition. When ripe, the fruit is soft enough to yield to slight pressure and at this stage can be stored 2 or 3 days longer in a refrigerator, sealed in plastic wrap. The skin will blacken and become unsightly while the flesh is still unspoiled and usable. The flavor of overripe fruit has deteriorated.

Culinary Vocabulary

- Popular soursop beverages are called *champola* in several countries, including Brazil, Cuba, and the Dutch East Indies, and *carato* in Puerto Rico.

Curiosities of Science and Technology

- In British Guiana, an old cure for drunkenness required applying a solution of soursop leaves and lime juice to the head.
- The leaves of soursop have a reputation for promoting sleep. For this purpose in the Netherlands Antilles, they are placed in pillowslips, scattered on the bed, or used to make a boiled tea.
- In the Virgin Islands, soursop fruit is used as a bait in fish traps.
- Pulverized soursop seeds and boiled leaf teas are used against head lice in tropical regions.
- Chemicals in soursop seeds called annonaceous acetogenins have potential value as pesticides and antitumor agents, and research for such products is in progress.

Key Information Sources

Awan, J.A., Kar, A., and Udoudoh, P.J. 1980. Preliminary studies on the seeds of *Annona muricata* Linn. *Plant Foods Hum. Nutr.* 30:163–168.

Aziz, P.A., and Yusof, S. 1994. Physico-chemical characteristics of soursop fruit (*Anona muricata*) during growth and development. *ASEAN Food J.* 9:147–150.

Benero, J.R., Collazo de Rivera, A.L., and de George, L.M.I. 1974. Studies on the preparation and shelf-life of soursop, tamarind, and blended soursop-tamarind soft drinks. *J. Agric. Univ. Puerto Rico*, 58:99–104.

Iwaoka, W.T., Zhang, X., Hamilton, R.A., Chia, C.L., and Tang, C.S. 1993. Identifying volatiles in soursop and comparing their changing profiles during ripening. *HortScience*, 28:817–819.

Jirovetz, L., Buchbauer, G., and Ngassoum, M.B. 1998. Essential oil compounds of the *Annona muricata* fresh fruit pulp from Cameroon. *J. Agric. Food Chem.* 46:3719–3720.

Juliano, J.B. 1935. Morphological contribution on the genus *Anona* Linnaeus. *Philipp. Agric.* 24:528–541.

Koesriharti. 1991. *Anona muricata* L. In *Plant resources of South-East Asia. Vol. 2. Edible fruits and nuts.* Edited by E.W.M. Verheij and R.E. Coronel. Pudoc, Leiden, the Netherlands. pp. 75–78.

Little, R.J., Elbert, L., and Wadsworth, F. 1989. Notes on insect pests of soursop (guanabana), *Annona muricata* L., and their natural enemies in Puerto Rico. *J. Agric. Univ. Puerto Rico,* 73:383–389.

MacLeod, A.J., and Pieris, N.M. 1981. Volatile flavor components of soursop (*Annona muricata*). *J. Agric. Food Chem.* 29:488–490.

McComie, L.D. 1987. The soursop (*Annona muricata* L.) in Trinidad: its importance, pests and problems associated with pest control. *J. Agric. Soc. Trinidad Tobago*, 87:42–55.

Morton, J.F. 1966. The soursop, or guanabana (*Annona muricata* Linn). *Proc. Fla. State Hortic. Soc.* 79:355–366.

Morton, J. 1987. Soursop. In *Fruits of warm climates*. Creative Resource Systems, Winterville, NC. pp. 75–80.

Mosca, J.L., Alves, R.E., and Filgueiras, H.A.C. 1999. Harvest and postharvest handling of sugar-apple and soursop: current research status in Brazil and review of recommended techniques. *Acta Hortic.* 485:273–280.

Onimawo, I.A. 2002. Proximate composition and selected physicochemical properties of the seed, pulp and oil of sour sop (*Annona muricata*). *Plant Foods Hum. Nutr.* 57:165–171.

Paull, R.E. 1982. Postharvest variation in composition of soursop (*Annona muricata* L.) fruit in relation to respiration and ethylene production. *J. Am. Soc. Hortic. Sci.* 107:582–585.

Paull, R.E. 1990. Soursop fruit ripening—starch breakdown. *Acta Hortic.* 269:277–281.

Paull, R.E. 1998. Soursop. In *Tropical and subtropical fruits*. Edited by P.E. Shaw, H.T. Chan, Jr., and S. Nagy. Agscience, Auburndale, FL. pp. 386–400.

Paull, R.E., Deputy, J., and Chen, N.J. 1983. Changes in organic acids, sugars, and headspace volatiles during fruit ripening of soursop (*Annona muricata* L.). *J. Am. Soc. Hortic. Sci.* 108:931–934.

Peters, M., Badrie, N., and Commissiong, E. 2001. Processing and quality evaluation of soursop (*Annona muricata* L) nectar. *J. Food Qual.* 24:361–374.

Sánchez-Nieva, F., Igaravídez, L., and López-Ramos, B. 1953. *The preparation of soursop nectar*. Agricultural Experiment Station, Rio Piedras, Puerto Rico. 19 pp.

Sánchez-Nieva, F., Hernandez, I., and Iguina de George, L.M. 1970. Frozen soursop purée. *J. Agric. Univ. Puerto Rico*, 54:220–236.

Umme, A., Asbi, B.A., Salmah, Y., Junainah, A.H., and Jamilah, B. 1997. Characteristics of soursop natural purée and determination of optimum conditions for pasteurization. *Food Chem.* 58:119–124.

Worrell, D.B., Carrington, C.M.S., and Juber, D.J. 1994. Growth, maturation and ripening of soursop (*Annona muricata* L.) fruit. *Sci. Hortic.* 57:7–15.

Specialty Cookbooks

Burke, V. 2001. *Walkerswood Caribbean kitchen*. Simon & Schuster, London. 80 pp.

Macbean, V. 2001. *Coconut cookery*. Frog Books, Berkeley, CA. 200 pp.

Shields, D. 1998. *Caribbean light: all the flavors of the island without all the fat*. Doubleday, New York. 324 pp.

Snook, K., and Penniman, E. 2007. *A taste of Mustique*. Macmillan Caribbean Publishing, Oxford. 194 pp.

92 Tamarind

Family: Fabaceae (Leguminosae; pea family)

NAMES

Scientific Name: *Tamarindus indica* L.

- The English name "tamarind" and the genus name *Tamarindus* are based on the Arabic *tamr hindī* (*tamar-u'lHind*), "date of India," because of the resemblance of the dried fruit to dried dates. (The same root led to the feminine name Tamara.)
- Pronunciation: TAM-uh-rihnd.
- The tamarind is also known as Indian date, Indian tamarind, kilytree, and Madeira mahogany.
- Some tropical lumber trees of the genus *Dialium* (of the pea family), including *D. cochinchinese* Pierre, *D. guineense* Willd., and *D. indicum* L., are known as "velvet tamarind" because of the tamarind taste of the fruit pulp. The fruit pods are more or less edible, and are sometimes consumed.
- "Malabar tamarind" is *Garcinia cambogia* Desr., also known as cambodge. The fruits of this tropical tree of India and Malaysia are used as a spice and medicine.
- The "Manila tamarind" is *Pithecellobium dulce* (Roxb.) Benth., also known as guayamochil and sweet inga. This tropical tree produces pods that are used for various edible products.
- "Tamarins" are species of small South American marmosets with silky fur and long tails. Golden lion tamarins (frequently misspelled "tamarinds") are very popular "monkeys" in zoos but are endangered in their native home of Brazil, where deforestation has eliminated 98% of their forest habitat.
- *Indica* in the scientific name *T. indica* means "Indian," reflecting the mistaken belief that the tree originally came from India, although in fact its native area is Africa. The tree was brought to India by Arab traders long ago, and became well established there.

PLANT PORTRAIT

The tamarind is a slow-growing, massive tree, sometimes reaching a height of 30 m (100 ft.) and a trunk circumference of 2.5 m (8 ft.). The tree is long-lived, sometimes remaining productive after 200 years of age. The fruits are flattish, cinnamon-brown or grayish-brown beanlike pods 5 to 20 cm (2–8 in.) long. When young, the pods have tender skin, green, highly acid flesh, and soft, whitish seeds. The mature pods have a brittle, easily removed shell and brown or reddish-brown pulp, usually with 3 to 12 hard, glossy-brown seeds. With further aging, the pulp dehydrates to a sticky paste. A mature tree may produce as much as 225 kg (500 lb.) of fruit, of which the pulp constitutes one-third to one-half. The stalks that hold the fruit to the tree are so tough that they cannot be broken off by hand without damaging the fruit and need to be clipped off. The tamarind is native to tropical Africa, thought to have originated in Madagascar. It has been said to be the only important spice of African origin. It has been grown since antiquity in India and has been spread to other tropical and subtropical regions. India remains the principal country of production and consumption of tamarind. Thailand is the second largest producer, with significant amounts also in Mexico and Costa Rica.

FIGURE 92.1 Tamarind (*Tamarindus indica*). (From Köhler, 1883–1914.)

CULINARY PORTRAIT

The bittersweet, very acidic pulp is the most important food item produced by the tamarind. Most tamarind trees produce sour fruit, but some sweet-fruited types are known. In north-central Thailand, a cultivar with relatively sweet pulp is grown that is eaten fresh. Dehydrated pulp is used in chutneys, curries, sour soups, pickles, relishes, and sauces. Tamarind is an essential ingredient of Worcestershire sauce, which is the principal way that most people in Western nations have encountered the fruit. Tamarind also occurs in HP sauce. Tamarind pulp concentrate is popular as a flavoring in East Indian and Middle Eastern cuisines, in the same way that lemon juice is widely used in Western culture. Sugared, de-seeded tamarind pulp is made into sweet confections and beverages in many tropical cultures. Tamarind pulp is also used in jellies, jam, sherbet, ice cream, fruit preserves, and wine. The strong, acidic flavor complements the taste of fruits. Tamarind pulp also is added to soups, sauces, stews, marinades, cakes, candies, meat, game, and fish. East Indian and some other Asian markets sell tamarind in various forms, such as jars of concentrated pulp with seeds, canned paste, and whole pods dried into bricks or ground into powder. The pulp needs to be strained through a sieve (to remove bits of seed and fibers) before it is combined with other ingredients. The pulp and paste can be stored in a refrigerator almost indefinitely when kept in plastic wrap. In recipes calling for tamarind, lemon juice can usually be substituted (the juice of one lemon equals 1 tablespoon of tamarind paste in 1/5 cup of water), although the taste is not the same. Tamarind is laxative and is in fact used for this purpose.

In addition to the pulp, the young, sour pods are cooked as seasoning with rice, fish, and meats in India. In some areas, the unripe fruits are roasted and eaten, and the fully ripe fresh fruit is consumed directly. The seeds are sometimes eaten after roasting or boiling and are occasionally made into flour or meal and baked as cakes. The flowers and leaves are also sometimes cooked as vegetables in countries that produce tamarind.

Culinary Vocabulary

- "Tamarind liquid" is an Asian and South African flavoring prepared by soaking tamarind pulp in water.

- *Dakhar* is a Senegalese beverage made with tamarind juice and sugar.
- *Vindaloo* (pronounced VEN-deh-loo) is a spicy Indian meat or chicken dish, flavored with tamarind juice and other spices, and served over rice.
- *Jugo* or *Fresco de Tamarindo* is a favorite tamarind drink in many Latin America countries.
- "Tamarind balls" are a popular, sweet-and-sour, Caribbean coastal delicacy, made by kneading tamarind pulp (with the seeds) and sugar into balls, and allowing them to air dry. They are consumed as sweet confections.

CURIOSITIES OF SCIENCE AND TECHNOLOGY

- Tamarind is so sour that Italian traveler Marco Polo (1254–1324), claimed the Malabar pirates made their victims swallow a mixture of it and sea water so that they would vomit any pearls they may have swallowed in an attempt to conceal them.
- Most plants have difficulty surviving in the shade of a tamarind tree, and falling tamarind leaves have a corrosive effect on fabrics in damp weather. These observations gave rise to the belief that it is harmful to sleep or to tie a horse beneath a tamarind. In India, some people fear the tamarind and think that evil spirits such as Yamadutaka, the messenger of the death god Yama, live inside the tree, which is why nothing can grow under it.
- In Asian temples and shrines, tamarind pulp is often used as a mild abrasive and cleaner on brass furnishings.
- The bark of the tamarind tree is sometimes used as chewing gum.
- Tamarind lumber has been sold in North America as "Madeira Mahogany."
- The capital city of Senegal, Dakar, is based on *dakar*, the local name of the tamarind.
- Tamarind pulp is said to have more sugar and fruit acid on a volume basis than any other fruit.
- Tamarind leaflets fold up at night (the leaves usually have 10 to 40 leaflets, arranged on opposite sides of the midrib).
- Much like hickory branches have traditionally been used to enforce discipline in North America, in Asia a tamarind switch has been used for corporal punishment.

KEY INFORMATION SOURCES

Bhat, S.G. 1966. Tamarind seed oil: its properties and composition. *Indian Oil Soap J.* 32:53–57.

Bhattacharya, S., Bal, S., Mukherjee, R K., and Bhattacharya, S. 1994. Functional and nutritional properties of tamarind (*Tamarindus indica*) kernel protein. *Food Chem.* 49:1–9.

Bueso, C.E. 1980. Soursop, tamarind, and chironja. In *Tropical and subtropical fruits: composition, properties and uses*. Edited by S. Nagy and R.E. Shaw. AVI Publishing Company, Westport, CT. pp. 375–406.

Caluwé, E. De, Halamová, K., and Van Damme, P. 2009. Tamarind *(Tamarindus indica* L.): a review of traditional uses, phytochemistry and pharmacology. In *African natural plant products: new discoveries and challenges in chemistry and quality*. Edited by H.R. Juliani, J.E. Simon, and C.-T. Ho. American Chemical Society, Washington, DC. pp. 85–110.

Coronel, R.E. 1991. Tamarindus indica L. In *Plant resources of South-East Asia. Vol. 2. Edible fruits and nuts*. Edited by E.W.M. Verheij and R.E. Coronel. Pudoc, Leiden, the Netherlands. pp. 298–301.

Dalla Rosa, K.R. 1993. Tamarindus indica: *a widely adapted, multipurpose fruit tree*. Agroforestry Information Service of the Nitrogen Fixing Tree Association, Paia, HI. 2 pp.

Fisch, B.E. 1974. The tamarind. *California Rare Fruit Growers Yearbook*, 6:221–250.

Gunasena, H.P.M., Hughes, A., Haq, N., and Smith, R.W. 2000. *Tamarind:* Tamarindus indica *L.* International Centre for Underutilised Crops, University of Southampton, Southampton, U.K. 170 pp.

Hasan, S.K., and Ijaz, S. 1972. Tamarind review. *Sci. Ind.* 9:131–137.

Hernandez-Unzon, H.Y., and Lakshminarayana, S. 1982. Developmental physiology of tamarind fruit (*Tamarindus indica*). *HortScience*, 17:938–940.

Hughes, A. 2008. *Tamarindus indica*, tamarind. In *The encyclopedia of fruit & nuts*. Edited by J. Janick and R.E. Paull. CABI, Wallingford, Oxfordshire, U.K. pp. 400–405.

Indian Council of Forestry Research and Education. 1992–1995. *Tamarind (*Tamarindus indica*)*. Indian Council of Forestry Research and Education, Forest Research Institute, Dehra Dun, India. 16 pp.

Khurana, A.L., and Ho, C.T. 1989. HPLC analysis of nonvolatile flavor components in tamarind, *Tamarindus indica* L. *J. Liq. Chromatogr.* 12:419–430.

Kiambi, K. 1992. *Tamarind,* Tamarindus indica. Kenya Energy and Environment Organizations, Nairobi, Kenya. 16 pp.

Lanhers, M.C., Fleurentin, J., and Guillemni, F. 1996 *Tamarindus indica* L. *Ethnopharmacologia*, 18:42–57.

Lee, P.K., Swords, G., and Hunter, G.L.K. 1975. Volatile constituents of tamarind (*Tamarindus indica* L.). *J. Agric. Food Chem.* 23:1195–1199.

Lefevre, J.-C. 1971. Review of the literature on the tamarind. *Fruits*, 26:687–695 (in French).

Marangoni, A., Alli, I., and Kermasha, S. 1988. Composition and properties of seeds of the tree legume *Tamarindus indica. J. Food Sci.* 53:1452–1455.

Mishra, R.N. 1997. *Tamarindus indica* L.: an overview of tree improvement. In *Proceedings of the National Symposium on* Tamarindus indica *L., June 27–28, 1997*. Edited by A.P. Tirupathi. Forest Dept. A.P., India. pp. 51–58.

Morton, J.F. 1958. The tamarind, its food, medicinal and industrial uses. *Proc. Fla. State Hortic. Soc.* 71:288–294.

Morton, J.F. 1987. Tamarind. In *Fruits of warm climates*. Creative Resource Systems, Winterville, NC. pp. 115–121.

Nagarajan, B. 1999. *Breeding systems and hybridisation techniques in tamarind*. Institute of Forest Genetics and Tree Breeding, Coimbatore, India. 8 pp.

Parrotta, J.A. 1990. Tamarindus indica *L., Leguminosae (Caesalpinioideae): tamarind, legume family*. U.S. Forest Service, Southern Forest Experiment Station, Institute of Tropical Forestry, Rio Piedras, Puerto Rico. 5 pp.

Rao, P.S. 1959. Tamarind. In *Industrial gums*. Edited by R.L. Whistler. Academic Press, New York. pp. 461–504.

Rao, Y.S., Mathew, M., and Potty, S.N. 1999. Tamarind (*Tamarindus indica* L.) research—a review. *Ind. J. Arecanut Spices Med. Plants*, 1(4):127–145.

Saideswara, R.Y., and Mathew, M. 2001. Tamarind. In *Handbook of herbs and spices*. Edited by K.V. Peter. Woodhead Publishing, Cambridge, U.K. pp. 287–296.

Shankaracharya, N.B. 1998. Tamarind—chemistry, technology and uses—a critical appraisal. *J. Food Sci. Technol.* 35:193–208.

Wong, K.C., Tan, C.P., Chow, C.H., and Chee, S.G. 1998. Volatile constituents of the fruit of *Tamarindus indica* L. *J. Essent. Oil Res.* 10:219–221.

Specialty Cookbooks

Note: More than a thousand cookbooks marketed today have recipes using tamarind. The following is a small selection.

Raichlen, S. 2000. *Barbecue! Bible: sauces, rubs, and marinades, bastes, butters, and glazes*. Workman, New York. 304 pp.

Roden, C. 1999. *Tamarind & saffron: favourite recipes from the Middle East*. Viking, London. 209 pp.

93 Tobacco

Family: Solanaceae (potato family)

NAMES

Scientific Name: *Nicotiana tabacum* L.

- The word "tobacco" is from *tabaco*, one of the words encountered in Haiti by Christopher Columbus and transferred to Spanish. *Tabacum* in the scientific name *N. tabacum* is based on the Spanish word. The original North American Indian (Taino) term on which the Spanish word was based is uncertain and perhaps referred to the plant, a pipe used to smoke tobacco, or a roll of tobacco wrapped in a corn (maize) leaf. The island of Tobago has been said to have been named by the Indians who named the plant.
- *Nicotiana tabacum* is known both as tobacco and common tobacco. The name for tobacco is very similar in virtually all languages, the result of the plant being spread throughout the world in relatively recent times.
- The genus name *Nicotiana* commemorates Jean Nicot (1530–1600), French ambassador to Portugal, who obtained tobacco from sailors returning to Lisbon from America and introduced it into France in 1560. Nicot popularized tobacco in Europe for medicinal purposes and had in mind using it as a snuff rather than as a fumitory (smoking substance). Nicot's contemporary André Thevet (1502–1590), a monk who brought tobacco from Brazil to France, popularized its use in France for smoking and may actually have been more deserving of recognition than Nicot.

PLANT PORTRAIT

What is "food?" In a biological sense, "food" provides needed chemicals that the body can use to produce energy, build or repair tissues, and conduct the complex biochemical transformations that sustain life. However, "food plants" certainly include spices and herbs, most of which are not really significant nutritionally because they are basically present in quite small amounts to provide flavor and stimulate appetite. Some plants traditionally considered to be food plants merely deliver high amounts of stimulant chemicals (although almost inevitably plants also provide at least some desirable chemicals such as antioxidants). These include beverages with the chief purpose of providing high amounts of caffeine (such as coffee, tea, maté, cola drinks, and guarana) and masticatories with the same stimulant function (such as betel nut, coca, and khat). Peruvian Indians widely used tobacco as a chew in pre-Columbian times in the Andes, specifically to alleviate hunger, just like coca was used. In addition to delivering stimulation, many of these beverages and masticatories also have a taste that is desired, albeit acquired. As a stimulant masticatory, chewing tobacco is at least debatably a food.

Tobacco is the world's most harmful plant. Nevertheless, it qualifies as a "food plant" because it is used as a spice, albeit rarely, as pointed out in this chapter. In early times in South America, some Indians actually classified tobacco as a food—occasionally using it as an additive (e.g., to *chicha*, the maize beer consumed by Peruvian Indians). At the end of the twentieth century and beginning of the twenty-first century, a culinary tobacco fad developed in some of the most exclusive restaurants of the world. For example, in the fashionable Michel Rostang restaurant in Paris, the dessert menu has featured "Havana cigar with Cognac mousse," a cigar-shaped creation actually made with

FIGURE 93.1 Tobacco (*Nicotiana tabacum*, left) and wild tobacco (*N. rustica*, right). (From Köhler, 1883–1914.)

crumbled Cuban cigars. In New York, the Italian restaurant Serafina Sandro offered a "Tobacco Special Menu," with such delicacies as filet mignon in a tobacco-wine sauce, gnocchi garnished with tobacco, tobacco-flavored panna cotta (literally cooked cream) for dessert, and a beverage of tobacco-infused grappa. The Parisian chocolate firm Richart has produced some chocolates with tobacco content. At the Trump World Tower in Manhattan, a tobacco-flavored Manhattan cocktail has been served. Other tobacco-laced cocktails in the United States have included the "Nicotini" (tobacco-flavored martini) and the "Black Lung" (tobacco-flavored Kahlua). A single shot of a tobacco-flavored cocktail has been dubbed "a quick puff." In France, tobacco-flavored alcoholic beverages have been called "light cigarettes."

Tobacco is a herbaceous annual (or less commonly a biennial), growing 1 to 3 m (3–10 ft.) in height. The stem tends to become woody at its base. The flowers are white, yellow, pink, purple, or red, 3 to 6 cm (1.2–2.4 in.) long. Leaves vary in size depending on variety, and the lower ones are often longer than 50 cm (20 in.). The leaves of cigar-wrapper varieties are thin, fine textured, and small veined, whereas those of plug and pipe tobacco varieties are usually coarser, tougher, and thicker. The plant is modified by removing the top, usually when 8 to 12 leaves have developed, as well as branches that are consequently stimulated to grow. This causes the remaining leaves to increase in size by as much as 50%, increases nicotine content, and encourages even ripening of the leaves. *Nicotiana tabacum* is thought to be a hybrid, with its ancestors distributed in the eastern Andes Mountains of South America. Tobacco from this species is produced in most temperate and tropical regions of the world, and is a major crop of many nations.

In addition to *N. tabacum*, *N. rustica* L., known as wild tobacco and Aztec tobacco, was domesticated for tobacco production, although it is hardly grown commercially in North America. Like the main tobacco species (*N. tabacum*), it is thought to have arisen as a hybrid, probably from ancestral species in Peru, and it does not have a natural native distribution. The name "wild tobacco" is a misnomer because the species arose in cultivation and has no natural wild distribution area. When Europeans came to North America in the fifteenth and sixteenth centuries, both *N. tabacum* and *N. rustica* were widely cultivated by the Indians. *Nicotiana rustica* was the first species to be bought back to the Old Word, but *N. tabacum* was imported soon afterward and quickly became much more popular. *Nicotiana rustica* is a course South American annual, hardier than *N. tabacum*, and is "naturalized" (a word referring to plants that grow "wild" outside of cultivation and outside of their native range) in eastern North America, occurring as far north as southern Ontario. *Nicotiana rustica* was widely grown by North America Indians and was cultivated for tobacco by them as recently as the middle of the twentieth century. Tobacco was first cultivated by European colonists in Virginia in 1612, and the species first grown was *N. rustica*. It is now very rarely cultivated in North America, although it is grown in Europe and Asia for smoking tobacco and as a source of the insecticide nicotine. The dried leaves of *N. rustica* contain up to 10% nicotine, whereas those of regular tobacco usually have 1.5% to 4% (although up to 8% is possible).

Smoking tobacco is not entirely an American (i.e., New World) invention. Australian aborigines also used several Australian species of *Nicotiana* for the purpose, and it appears that people native to the southwestern Pacific may have also used local species of the genus. Perhaps 10 species of *Nicotiana* containing the alkaloid nicotine were used by native peoples for religious and medicinal purposes. The distribution ranges of these species were extended by such usage before European colonization. The earliest record of the use of tobacco is in a fifth-century Mayan temple in Mexico, where a bas-relief (sculpture with very little depth on a wall) shows a priest smoking.

Tobacco has extremely serious health hazards and has killed more humans in history than any other plant substance. A recent analysis suggested that approximately 500 million people alive today will eventually be killed by tobacco use. Smoking tobacco is a cause of emphysema and lung cancer, heart disease, and numerous other disorders. It has been said that smoking is the largest cause of preventable death. It has been estimated that each cigarette smoked takes 5 minutes off one's life, a figure that indicates that smoking a pack of 20 cigarettes a day shortens an expected year of life by 1 month. Smoking is one of the most common causes of "accidental" fires. It is also a cause of social unrest, with the addictive qualities of tobacco causing otherwise thoughtful people to subject innocent passers-by to the deleterious effects of the disgusting habit.

Tobacco is used as a fumitory (i.e., for smoking) in pipes, cigars, and cigarettes, inhaled as a powder (snuff is simply finely ground tobacco), chewed as plugs, and savored by covering the gums with a fine powder. "Smokeless" tobacco includes snuff that is inhaled and all material that is taken orally—so called "spit tobacco." Chewing tobacco is in several forms. Finely shredded tobacco is dipping tobacco (really not much different from snuff), pressed bricks and cakes are called plugs, and ropelike strands are called twists.

Tobacco was chewed by New World Indians. In the United States, chewing tobacco was so prevalent in the late 1800s that more than 90% of tobacco factories only made chewing tobacco, and spittoons were far more common than ash trays. Chewing tobacco was so popular during the latter nineteenth and early twentieth centuries that 12,000 brands were registered in the United States. Today, the practice of chewing tobacco has remained most popular in the United States, where it has been associated with cowboys and sports, notably baseball, where the stimulant value of nicotine may have been the motivation. At the end of the twentieth century, smokeless tobacco was used by approximately 15 million Americans, and in the United States approximately 6% of men 18 years and older (chewing tobacco is primarily a male practice) were using some form of spit tobacco.

Chewing tobacco is addictive and is not, contrary to the belief of many, safer than smoking tobacco, although the set of health hazards are somewhat different than for smoking tobacco. The nicotine in chewing tobacco has similar adverse affects as that in cigarettes. Tobacco placed in the mouth is associated particularly with cancers and other pathological conditions of the mouth, tongue, lips, and throat. Studies have found that one-half to three-quarters of spit tobacco users have oral lesions. Chewing tobacco users are four times more likely than nonusers to develop tooth decay.

FIGURE 93.2 Left: planting tobacco in colonial America. (From Billings, 1875.) Right: picking tobacco. (From Jumelle, 1901.) As discussed in the text, the workers are removing the tops of the plants to increase leaf size and quality.

CULINARY PORTRAIT

In the United States, the Food and Drug Administration does not permit tobacco as a food ingredient because it not safe; hence, it cannot be used by commercial manufacturers to produce and sell edible products. Restaurant food, however, is not regulated by the agency. The use of pipe and cigar tobacco by some chefs to impart spiciness to food is limited and likely to remain so, given the toxicity and negative image of tobacco. Nicotine gum is a prescription product which, like the nicotine patch, is intended to assist tobacco users to overcome their addiction gradually.

Chewing tobacco users place a wad or plug in the cheek or between the lower lip and gums. A typical user holds a wad in his cheek for 30 minutes at a time and uses the product in this manner throughout the day. The nicotine is absorbed through the oral mucous membranes.

Candy cigarettes—white candy sticks or chocolate and gum versions wrapped in white paper with a brown end looking like a filter, all packaged in realistic if small cigarette-like packages—are still marketed in North America, despite the suspicion that they prime children to become smokers when they grow up, or even sooner. Until subjected to public shaming and legislation preventing such obvious encouragement of youngsters to take up smoking, cigarette manufacturers allowed the makers of candy cigarettes to mimic their packaging and use their brand names, so packs of the candy versions were nearly indistinguishable from real ones. A very recent development is the manufacture of so-called "edible tobacco" products by tobacco companies; these are flavored preparations laced with nicotine, to be used much like chewing tobacco. Some of these products have been dubbed "tobacco candy," and concern has been raised about their potential appeal to young people.

The World Health Organization's Farm and Agriculture Organization in 1981 raised the possibility of extracting protein from tobacco plants as a means of alleviating world hunger. The project has not subsequently received significant funding, in considerable part because tobacco has become a pariah crop in the eyes of many people. The project has been the subject of research at North Carolina State University. The possibility remains of producing nicotine-free "tobacco-burgers," either as simulated meat, or actual meat of animals raised and fed on tobacco.

Tobacco is the subject of considerable genetic engineering for use as a platform for the production of pharmaceutical proteins, which have immense medicinal value as antibodies, vaccines, growth regulators, and for other therapeutic purposes. There has been research on development of low-nicotine forms, which could be useful for the production of edible vaccines from the plant (e.g., Kim et al., 2005).

Culinary Vocabulary

- A "Kentucky breakfast" is an obsolete colloquial slur meaning three cocktails and a chew of tobacco. (Kentucky used to be called the "Tobacco State." At one time, two-thirds of the U.S. crop was produced in Kentucky.)
- A "Mexican breakfast" is a Southern United States colloquial slur for a cigarette and a glass of water, usually prescribed for those with a hangover.
- The expression "spice tobacco" is used in the tobacco trade to refer to smoking tobacco prepared with blends of different kinds of tobacco and sometimes also other plants to achieve a novel aroma. This is not a culinary phrase.

CURIOSITIES OF SCIENCE AND TECHNOLOGY

- In 1492, Christopher Columbus was given a gift from the native Arawaks of what he recorded as "certain dried leaves" when he and his men set foot in the New World for the first time on the beach of San Salvador Island, or Samana Cay in the Bahamas, or Gran Turk Island. Columbus politely had the tobacco brought back to his ship, but afterward, not realizing what he had been given, ordered it thrown away.
- In 1492 in Cuba, Columbus's crewmen Rodrigo de Jerez and Luis de Torres were the first Europeans to observe smoking, including the use of smoldering rolls of leaves inserted into the nostrils. Jerez became the first European to smoke tobacco and indeed became addicted. When he returned to Spain and smoke was observed billowing from his mouth and nose, the Holy Inquisition imprisoned him for 7 years.
- The use of tobacco after its introduction into the Old World was severely condemned by many authorities. In 1604, British King James I (1566–1625), wrote an unsigned pamphlet (*A Counterblaste to Tobacco*) attacking tobacco and associating its use with syphilis. He described smoking tobacco as "a custome lathsome to the eye, hateful to the nose, harmefull to the braine, dangerous to the lungs, and the blacke stinking fume thereof, neerest resembling the horrible Stigian smoke of the pit that is bottomlesse."
- From the 1500s to the 1700s, tobacco was sometimes prescribed by doctors to treat ailments such as headaches, toothaches, arthritis, and bad breath. The idea that tobacco smoke is beneficial persisted down to modern times among New World peoples. It is now known that secondhand smoke can be deadly; nevertheless, New World medicine men have blown the smoke of cigars or pipes over the skin of the sick to cure them. In the Antilles (West Indies excluding the Bahamas), medicine men placed the lighted end of a cigar that was 60 cm (2 ft.) long in their mouth, and forcefully blew the smoke out the other end of the cigar, over the patient (a practice that has been observed in recent times in South America).
- In the time of William Shakespeare (1564–1616), tobacco cost its weight in silver shillings in London, England. It was an extravagance of dandies who held smoke ring-blowing contests.
- Although the claim that he introduced tobacco to England is untrue, Sir Walter Raleigh (1554–1618), certainly popularized smoking. Before being executed in 1618, he is said to have called for his pipe. Raleigh was buried with a tin of tobacco and his favorite pipe.
- Tobacco was acceptable legal tender in several Southern colonies in early America. In Virginia, taxes were paid in tobacco, making it literally a "cash crop" (an expression that really means a crop sold for money, rather than one consumed on the farm).
- Jamestown, Virginia, was founded in 1607, the first permanent English settlement in the New World. In 1619, the first shipment of mail-order brides arrived. The men had to pay

54 kg (120 lb.) of tobacco for the passage of their new wives. By 1621 newly arrived brides were selling for 68 kg (150 lb.) of tobacco.

- Catherine de Medici (1519–1589), Queen of France, was the first woman in Europe to use tobacco, which she took in a mixture of snuff.
- In 1632, Massachusetts outlawed public smoking.
- During 1633, Turkish Sultan Murad IV executed as many as 18 tobacco users daily.
- In seventeenth-century China, importers of tobacco were beheaded.
- In 1634, Czar Alexis of Russia enacted new penalties for smoking: whipping, a slit nose, and transportation to Siberia for the first offence and execution for the second. The ban was lifted in 1676.
- In 1647, Connecticut banned public smoking but allowed citizens to smoke once a day, provided that they did so alone. In 1650, Connecticut only allowed smoking with a physician's prescription
- In 1657, Switzerland prohibited smoking.
- In 1724, Pope Benedict XIII took up smoking and repealed the regulations against smoking by the clergy.
- In 1759, George Washington (1732–1799), first president of the United States, harvested his first tobacco crop. It was poor, and by 1761 he was deeply in debt.
- The American revolutionary war of 1776 was often called "The Tobacco War" because American were deeply in debt to the British tobacco merchants they dealt with. American tobacco helped finance the American Revolution by serving as collateral for loans from France.
- In 1830, Prussia required cigars smoked in public to be contained in a wire mesh cage to prevent sparks from setting fire to ladies' dresses.
- The superstition in the British army never to light three cigarettes off the same match started during the Boer War (1899–1902) when the flame gave an enemy sniper time to sight his rifle, and the soldier lighting the third cigarette was often shot.
- In the nineteenth century, a man tried to smuggle tobacco leaves wrapped around his body under his clothes into France. As he perspired, nicotine was absorbed into his body, making him seriously ill and eventually killing him (Holmes, C. 2006. Plants to poison mind & body. *Herbarist*, 72: 38–43).
- The tobacco industry of the world produces about 4 trillion (4,000,000,000,000) cigarettes annually, for which consumers spend about $100 billion ($100,000,000,000). Although smoking is declining in much of the Western World, it is increasing elsewhere. One-third of the world's male population currently smokes.
- Nicotine, first chemically isolated in 1807, is extremely toxic. One or two drops (60–120 mg) of pure nicotine placed on the skin can kill an adult human. "Green tobacco sickness" has been recorded among farmworkers who become poisoned by nicotine absorbed through their skin when they handled the plants; the conditions sometimes requires hospitalization. A typical cigar contains enough nicotine to kill two people if injected into their bodies. Species of *Nicotiana* have been used as dart-poison ingredients in South America.
- Tomato and tobacco plants are members of the same family, the Solanaceae, which explains why the two species can be grafted on to each other. When tomato tops are grafted on to the roots of tobacco plants, nicotine (albeit in small amounts) is transferred to and accumulates in the tomatoes. When tobacco tops are grafted on to tomato roots, nicotine is not transferred to the roots. These experiments (and others) indicate that nicotine precursors in tobacco plants are normally synthesized in the roots and transferred to the leaves, and not vice versa. In a 1999 episode of the television series *The Simpsons*, Homer grows and tries to profit from a hybrid crop of tobacco and tomato called "tomacco," which produces highly addictive tomatoes.

- Nicotine is present in trace (insignificant) amounts in tomatoes, potatoes, and green peppers, all in the Solanaceae, the same family of plants in which tobacco is placed.
- Nicotine has been used as a pesticide, indeed one that is so toxic that is rarely used for this purpose today. It may seem surprising, therefore, that some animals are specialist feeders on tobacco. Particularly notable is the Carolina sphinx moth or hawkmoth (*Manduca sexta*), which has a larval stage commonly referred to as a tobacco hornworm, although it is not a "worm" (it is a larva).
- In African recipes, "Player's #3 cigarette tin" is very widely cited as a unit of measurement (approximately equal to 1 cup).
- The word "cigar" is based on the cicada, the Spanish name for the cigar-shaped insect. Inhabitants of Cuba were smoking tobacco in this form when Europeans first arrived.

KEY INFORMATION SOURCES

Akehurst, B.C. 1981. *Tobacco*. 2nd ed. Longman, New York. 764 pp.

Baker, J.B., and Berry, W. 2004. *Tobacco harvest: an elegy*. University Press of Kentucky, Lexington, KY. 88 pp.

Chaplin, J.F. 1976. *Tobacco production. Revised edition*. U.S. Department of Agriculture, Agricultural Research Service, Washington, DC. 77 pp.

Chojar, A.K. 2002. *Tobacco cultivation and marketing*. Deep & Deep Publications, Rajouri Carden, New Delhi, India. 406 pp.

Clark, M. 2001. Hmm, hot and spicy. It's what? Tobacco?! *New York Times* (Internet), Jan. 31.

Clayton, E.E. 1958. The genetics and breeding progress in tobacco during the last 50 years. *Agron. J.* 50:352–356.

Cordry, H.V. 2001. *Tobacco: a reference handbook*. ABC-CLIO, Santa Barbara, CA. 419 pp. [Anti-tobacco]

Gadani, F., Ayers, D., and Hempfling, W. 1995. Tobacco: a tool for plant genetic engineering research and molecular farming. Part II. *Agro. Food Ind. Hi. Tech.* 6(2):3–6.

Gage, C.E. 1942. *American tobacco types, uses, and markets*. U.S. Department of Agriculture, Washington, DC. 129 pp.

Garner, W.W. 1946. *The production of tobacco*. Blakiston, Philadelphia, PA. 516 pp.

Gately, I. 2002. *Tobacco: a cultural history of how an exotic plant seduced civilization*. Grove Press, New York. 403 pp.

Gerstel, D.U., and Sisson, V.A. 1995. Tobacco. In *Evolution of crop plants*. 2nd ed. Edited by J. Smartt and N.W. Simmonds. Longman Scientific & Technical, Burnt Mill, Harlow, Essex, U.K. pp. 458–463.

Gilman, S.L., and Zhou, X. (Eds.). 2004. *Smoke: a global history of smoking*. Reaktion Books, London. 408 pp.

Goodman, J. 2004. *Tobacco in history and culture: an encyclopedia*. Charles Scribner's Sons, Detroit, MI. 2 vols.

Goodspeed, T.H. 1954. *The genus* Nicotiana; *origins, relationships, and evolution of its species in the light of their distribution, morphology, and cytogenetics*. Chronica Botanica, Waltham, MA. 536 pp.

Harana, I., and Vermeulen, H. 2000. *Nicotiana tabacum* L. In *Plant resources of South-East Asia. Vol. 16. Stimulants*. Edited by H.A.M. van der Vossen and M. Wessel. Backhuys, Leiden, the Netherlands. pp. 93–99.

Haustein, K.O. 2003. *Tobacco or health? Physiological and social damages caused by tobacco smoking*. Springer, New York. 446 pp.

Hawks, S.N., and Collins, W.K. 1983. *Principles of flue-cured tobacco production*. Hawks-Collins Book, Raleigh, NC. 358 pp.

Hull, J.W. 2002. *Tobacco in transition*. Southern Office, Council of State Governments, Atlanta, GA. 80 pp.

Johnson, J. 2002. *Growing and processing tobacco at home: a guide for gardeners*. J. Johnson, Gauter, MS. 210 pp.

Kim, Y.S., Kim, M.Y., Kang, T.J., Kwon, T.H., Jang, Y.S., and Yang, M.S. 2005. Expression of the green fluorescent protein (GFP) in tobacco containing low nicotine for the development of edible vaccine. *J. Plant Biotech.* 7:97–103.

Liemt, G. van. 2001. *The world tobacco industry: trends and prospects*. International Labour Organisation, Geneva, Switzerland. 52 pp.

Mackay, J., and Eriksen, M.P. 2002. *The tobacco atlas*. World Health Organization, Geneva, Switzerland. 128 pp.

McMurtrey, J.E., Bacon, C.W., and Ready, D. 1942. *Growing tobacco as a source of nicotine*. U.S. Department of Agriculture, Washington, DC. 39 pp.

Moyer, D.B. 2005. *The tobacco book: a reference guide of facts, figures, and quotations about tobacco*. Sunstone Press, Santa Fe, NM. 494 pp.

Papenfus, H.D., and Quin, F.M 1984. Tobacco. In *The physiology of tropical field crops*. Edited by P.R. Goldsworthy and N.M. Fisher. John Wiley & Sons, Chichester, U.K. pp. 607–636.

Small, E., and Catling, P.M. 2006. Blossoming treasures of biodiversity:22. Tobacco: another old star with a new act. *Biodiversity*, 7(3–4):47–54.

U.S. Department of Health and Human Services. 2000. Reducing tobacco use: a report of the Surgeon General. Department of Health and Human Services, U.S. Public Health Service, Washington, DC. 462 pp.

Utoma, B.I., and Rahayuu, E. 2000. *Nicotiana rustica* L. In *Plant resources of South-East Asia. Vol. 16. Stimulants*. Edited by H.A.M. van der Vossen and M. Wessel. Backhuys, Leiden, the Netherlands. pp. 91–93.

Uy, D., and Sison, V.A. 1995. Tobacco. In *Evolution of crop plants*. Edited by J. Smartt and N.W. Simmonds. Longman Scientific & Technical, Burnt Mill, Harlow, Essex, U.K. pp. 458–463.

Wernsman, E.A., and Rufty, R.C. 1988. Tobacco. In *Principles of cultivar development*, vol. 2. Edited by W.R. Fehr, E.L. Fehr, and H.J. Jessen. Macmillan, New York. pp. 669–698.

World Health Organization. 2003. *WHO Framework Convention on Tobacco Control*. WHO, Geneva, Switzerland. 36 pp.

Wyman, S. (Knight Ridder Tribune News Service). 2003. Enter the nicotini: smoking ban leads to nicotine-infused drink. *Brandenton Herald*, Sept. 1.

Specialty Cookbooks

Note: Cookbooks and recipes for tobacco are not available.

94 Tree Tomato (Tamarillo)

Family: Solanaceae (potato family)

NAMES

Scientific Name: *Solanum betaceum* Cav. (*Cyphomandra betacea* (Cav.) Sendtn., *C. crassifolia* (Ortega) Kuntze; "*C. crassicaulis* (Ortega) Kuntze" is a mistaken version; L. Bohs, personal communication)

Cyphomandra betacea is overwhelmingly the most frequent scientific name encountered in the literature. The publications of L. Bohs (cited in the Key information sources section) argue that the old name *Solanum betaceum* should be reinstated because phylogenetic studies indicate that *Cyphomandra* is part of *Solanum*; some modern authorities have followed this viewpoint, and I have accepted this treatment. Alternatively, *Solanum* may deserve to be split into several groupings, one of which is *Cyphomandra*, in which case the name *C. betacea* should be retained. The new delimitation of the very large genus *Solanum* seems to require adjusting to different scientific names than are now almost universally used. For example, the name of the tomato (traditionally *Lycopersicon esculentum* Mill.) is now *Solanum lycopersicum* L.

- The name "tree tomato" is based on the traditional Andean name, *tomate de árbol*, tree tomato.
- The name "tamarillo" was invented in 1967 in New Zealand for promoting the tree tomato. The name has been interpreted as having been inspired by the Spanish word *amarillo* (meaning yellow, a frequent color of the fruits) and a variation of the Maori word *tama* (meaning leadership). Tamarillo has become the standard commercial name for the fruit. The tamarillo should not be confused with the tomatillo (see Chapter 16). Suggested pronunciations: tam-uh-RIHL-oh, tam-uh-REE-oh.
- For information on the genus name *Solanum*, see Chapter 75.
- The genus name *Cyphomandra* is based on Greek *kyphoma*, hunch back + *andros*, male, referring to the hump-shaped anthers of the flowers.
- *Betaceum* in the scientific name *S. betaceum* is Latin for "like a beet," a reference to the fruit color.

PLANT PORTRAIT

The tree tomato is a perennial, semi-woody, evergreen or partly deciduous bush or tree, growing 1.8 to 5.5 m (6–18 ft.), or rarely up to 7.6 m (25 ft.) in height. The trunk of the tree is slim, rarely larger than 10 cm (4 in.) near the ground. The plant bears fruit 18 months to 2 years after seeding and may continue to produce for 5 to 6 years. The fruits dangle on long stalks, singly or in clusters of 3 to 12. The egg-shaped fruits are pointed at both ends, approximately 5 to 10 cm (2–4 in.) long and 3.8 to 5 cm (1½–2 in.) wide. Skin color is solid deep purple, blood red, orange, or yellow, and the surface may have faint, dark, longitudinal stripes. Flesh color may be orange-red, orange, yellow, or cream-yellow. The pulp is black in dark purple and red fruits and yellow in yellow and orange fruits. Large red- and golden-fruited strains were developed in New Zealand and are notably superior to the wild forms of the species. The blackish seeds occur in two lengthwise compartments buried in the pulp and are bitter, thin, nearly flat, circular, and larger and harder than those of the true tomato. The skin

FIGURE 94.1 Tree tomato (*Solanum betaceum*). (From Turner, 1894.)

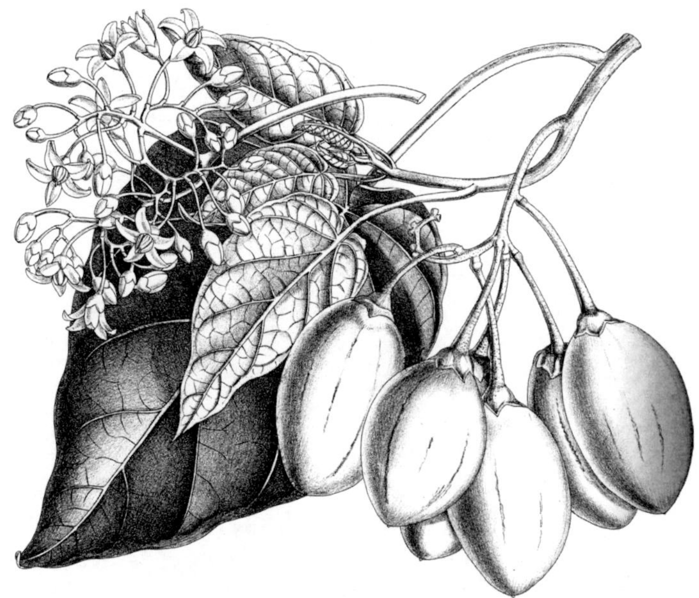

FIGURE 94.2 Tree tomato (*Solanum betaceum*). (From Curtis, 1899, vol. 125, plate 7682.)

is leathery, pliable, and unpleasant in flavor. The outer layer of the flesh is somewhat firm, succulent, and bland, whereas the pulp is soft, juicy, sweet, and tart.

The species is native to the Andes of Peru and probably also Chile, Ecuador, and Bolivia. The tree tomato was domesticated before the discovery of America, and it is uncertain whether wild populations still exist. The plant is grown as a fruit crop in semitropical areas throughout the world, most notably in New Zealand, also in South and Central America, Jamaica, Africa (particularly

Kenya), southern and Southeastern Asia, and Australia. The plants are frost sensitive but are grown to a limited extent in California and occasionally in Florida. The branches of the tree tomato are very brittle and may snap off under slight pressure, so protection from wind is necessary.

CULINARY PORTRAIT

Tree tomatoes are low in acid and the flavor is generally moderately agreeable, bland or semisweet, suggestive of a mild or underripe tomato. The taste has been described as a cross between a passion fruit and a tomato. Ripe fruit can be eaten raw, perhaps sprinkled with a little salt or sugar, or lime or lemon juice. The ripe tree tomatoes can simply be halved lengthwise, and the flesh and pulp scooped out and consumed with a spoon. The bitter, inedible skin can be easily removed by pouring boiling water over the fruit and letting it stand for 4 minutes before peeling. The pulp near the skin also tends to be bitter and is less suitable for eating fresh than the bulk of the pulp. After the tough skin and hard seeds are removed, the fruits are used in salads, sandwiches, stews, soups, stuffings, pies, chutneys, preserves, sauces, relishes, blended juices, and pickles. Unripe fruit is often cooked and used as a vegetable. In most recipes calling for tomatoes, tree tomatoes can be substituted, but the flavor can overpower dishes so they should be used in moderation. The flavor complements meat, poultry, and fish. The peeled fruits can be puréed in a blender or by cooking and strained to remove the seeds. Puréed fruit can be used to flavor cocktails, ice cream, yogurt, and sorbet. The fruit is high in pectin and is easily made into jam or jelly but oxidizes and discolors without special treatment during processing. Lipsham Liqueurs Ltd., a relatively small, family-based company in New Zealand, offers a tamarillo-based liqueur. The fruit should not be cut on a wooden or other permeable surface as the juice will make an indelible stain.

When buying tree tomatoes, the outer skin should be free of blemishes and defects and the fruit should have smooth, shiny, well-colored skin. Only those that are firm to the touch should be purchased. Hard fruits need to ripen at room temperature until they soften somewhat (unripe tree tomatoes tend to be bitter), and they can then be consumed or refrigerated. The fruit can be stored in a refrigerator in a perforated plastic bag for up to 2 weeks, but temperatures below 3°C (38°F) can cause the skin to discolor.

Culinary Vocabulary

- In Indonesia, the tree tomato is known as *teron Beland*, "Dutch eggplant."

CURIOSITIES OF SCIENCE AND TECHNOLOGY

- For years in the middle part of the twentieth century, advertisements touted a remarkable "tree tomato" claiming to produce huge quantities of tomatoes from seeds that could be purchased. The suggestion was that the fruit produced were regular tomatoes, but the plant in question was the tree tomato, which has fruit that is in fact disappointing when conventional tomatoes are expected. However, it is true that tree tomatoes are large producers: mature commercial trees can average yields up to 20 kg (44 lb.) of fruit per season.
- In 1999 in New Zealand, a protest took place against a trial of transgenic tree tomatoes (called tamarillos) conducted by HortResearch at the Kerikeri Research Centre. The research was intended to engineer resistance against the aphid-transmitted mosaic virus of the tree tomato, first discovered in New Zealand, and constituting a major problem for the industry. Some agricultural species in the potato family are susceptible to the virus, so countries are prone to banning the fruit for fear that the virus will be transmitted to their crops.
- Botanically, tree tomatoes are fruits, but when it comes to how they are eaten, they are both fruits and vegetables. Red or purplish fruits have a more tart taste than their yellow

or orange counterparts, and because of this they are more frequently used as a vegetable than as a fruit. The sweeter, orange and yellow tree tomatoes are more suitable for eating as dessert fruit.

- In New Zealand, the leading country of tree tomato production, children have learned to squeeze the fruit until the contents have been well-softened, bite off the stem end, and squeeze the contents directly into their mouths.

KEY INFORMATION SOURCES

Anonymous. 1984. What U.S. and Japanese customers think of red tamarillos. *N. Z. J. Agric.* 1984(Aug.):34–36.

Barghchi, M. 1998. In vitro regeneration, plant improvement and virus elimination of tamarillo (*Cyphomandra betacea* (Cav.) Sendt.). In *Tree biotechnology: towards the millennium.* Edited by M.R. Davey, P.G. Alderson, K.C. Lowe, and J.B. Power. Nottingham University Press, Nottingham, U.K. pp. 173–185.

Bohs, L. 1989. Ethnobotany of the genus *Cyphomandra* (Solanaceae). *Econ. Bot.* 4:143–163.

Bohs, L.A. 1994. Cyphomandra (*Solanaceae*). Flora Neotropical Monograph 63. New York Botanical Garden, Brooklyn, NY. 175 pp.

Bohs, L. 1995. Transfer of *Cyphomandra* (Solanaceae) and its species to *Solanum. Taxon*, 44:583–587.

Bohs, L. 2007. Phylogeny of the Cyphomandra clade of the genus *Solanum* (Solanaceae) based on ITS sequence data. *Taxon*, 56:1012–1026.

Boyes, S., and Struebi, P. 1997. Organic acid and sugar composition of three New Zealand grown tamarillo varieties (*Solanum betaceum* (Cav.)). *N. Z. J. Crop Hortic. Sci.* 25:79–83.

Clark, C.J., and Richardson, A.C. 2002. Biomass and mineral nutrient partitioning in a developing tamarillo (*Cyphomandra betacea*) crop. *Sci. Hortic.* 94:41–51.

Clark, C.J., Smith, G.S., and Gravett, I.M. 1989. Seasonal accumulation of mineral nutrients by tamarillo. 1. Leaves. *Sci. Hortic.* 40:119–131.

Clark, C.J., Smith, G.S., and Gravett, I.M. 1989. Seasonal accumulation of mineral nutrients by tamarillo. 2. Fruit. *Sci. Hortic.* 40:203–213.

Dadlani, S.A., and Chandal, K.P.S. 1970. The little-grown tree tomato. *Indian Hortic.* 14(2):13–14.

Dawes, S.N., and Callaghan, M.E. 1970. Composition of New Zealand fruit. 1. Tamarillo (*Cyphomandra betacea* (Cav.) Sendt.). *N. Z. J. Sci.* 13:447–451.

Fisch, M.B. 1974. The tree tomato. *California Rare Fruit Growers Yearbook*, 6:268–290.

Fletcher, W.A. 1979. *Growing tamarillos.* Bull. 307. New Zealand Ministry of Agriculture, Wellington, New Zealand. 27 pp.

Heatherbell, D.A., Reid, M.S., and Wrolstad, R.E. 1982. The tamarillo: chemical composition during growth and maturation. *N. Z. J. Sci.* 25:239–243.

Hume, E. P., and Winters, H.F. 1948. Tomatoes from a tree. *Foreign Agric.* 12(6):121–122.

Hume, E.P., and Winters, H.F. 1949. The palo de tomate or tree tomato. *Econ. Bot.* 3:140–142.

Lewis, D.H., and Considine, J.A. 1999. Pollination and fruit set in the tamarillo (*Cyphomandra betacea* (Cav.) Sendt.). 1. Floral biology. *N. Z. J. Crop Hortic. Sci.* 27:101–112.

Lewis, D.H., and Considine, J.A. 1999. Pollination and fruit set in the tamarillo (*Cyphomandra betacea* (Cav.) Sendt.). 2. Patterns of flowering and fruit set. *N. Z. J. Crop Hortic. Sci.* 27:113–123.

Martindale, W.L. 1974. Tomatoes from a tree. *J. Agric. Victoria*, 72(10):347–349.

McLennan, M. 1972. Tamarillos truly are versatile. *N. Z. . J. Agric.* 124(4):1, 53–55.

Morton, J.F. 1982. The tree tomato, or "tamarillo," a fast-growing, early-fruiting small tree for subtropical climates. *Proc. Annu. Meet. Fla State Hortic. Soc.* 95:81–85.

Morton, J. 1987. Tree tomato. In *Fruits of warm climates.* Creative Resource Systems, Winterville, NC. pp. 437–440.

Muthukrishnan, C.R., and Palanisamy, K.P. 1969. Utilization of tree tomato for the preparation of some edible products. *Farm Fact.* 3(10):11–13.

Prasad, T., Prasad, K., and Singh, D. 1990. Embryology and seed development in *Cyphomandra betacea* Sendt. Solanaceae. *J. Phytol. Res.* 3:79–84.

Pratt, H.K., and Reid, M.S. 1976. The tamarillo: fruit growth and maturation, ripening, respiration, and the role of ethylene. *J. Sci. Food Agric.* 27:399–404.

Pringle, G.J., and Murray, B.G. 1991. Interspecific hybridisation involving the tamarillo *Cyphomandra betacea* (Cav.) Sendt. (Solanaceae). *N. Z. J. Crop Hortic. Sci.* 19:103–111.

Pringle, G.J., and Murray, B.G. 1991. Reproductive biology of the tamarillo *Cyphomandra betacea* Cav. Sendt. Solanaceae and some wild relatives. *N. Z. J. Crop Hortic. Sci.* 19:263–274.
Prohens, J., and Nuez, F. 2000. The tamarillo (*Cyphomandra betacea*): a review of a promising small fruit crop. *Small Fruit Rev.* 1(2):43–68.
Prohens, J., Ruiz, J.J., and Nuez, F. 1996. Advancing the tamarillo harvest by induced postharvest ripening. *HortScience*, 31:109–111.
Rodriguez-Amaya, D.D., Bobbio, P.A., and Bobbio, F.O. 1983. Carotenoid composition and vitamin A value of the Brazilian fruit *Cyphomandra betacea. Food Chem.* 12:61–66.
Sydenham, F. 1943. Tree-tomato culture. *N. Z. J. Agric.* 15:93–94.
Torrado, A., Suarez, M., Duque, C., Krajewski, D., Neugebauer, W., and Schreier, P. 1995. Volatile constituents from tamarillo (*Cyphomandra betacea* Sendtn.) fruit. *Flavour Fragrance J.* 10:349–354.
Verhoeven, G. 1991. *Cyphomandra betacea* (Cav.) Sendtner. In *Plant resources of South-East Asia. Vol. 2. Edible fruits and nuts.* Edited by E.W.M. Verheij and R.E. Coronel. Pudoc, Leiden, the Netherlands. pp. 144–146.
Wong, K.C., and Wong, S.N. 1997. Volatile constituents of *Cyphomandra betacea* Sendtn. fruit. *J. Essent. Oil Res.* 9:358–359.

Specialty Cookbooks

Bilton, J. 2009. *Jan Bilton's tamarillo cookbook.* Revised edition. Irvine Holt, Blenheim, New Zealand. 143 pp.
Fisch, M.B. 1974. Tree tomato recipes. *California Rare Fruit Growers Yearbook*, 6:302–309.
Lazy Eight Ranch. 1980–1989. *Tamarillo cookery.* Lazy Eight Ranch, Takaka, New Zealand. 8 pp.

95 Turmeric

Family: Zingiberaceae (ginger family)

NAMES

Scientific Name: *Curcuma longa* L. (*C. domestica* Valeton)

- The English name "turmeric" is derived from the obsolete French *terre-mérite*, which is based on medieval Latin *terra merita*, meritorious earth, possibly so-named because ground turmeric resembles (ocher) mineral pigments. An obsolete spelling for turmeric is tumeric.
- Turmeric is also known as Indian saffron and has been sold fraudulently on occasion as saffron (see Chapter 83). Curcumin gives turmeric its color and is very much like the chemical (glycoside) in saffron that gives the latter spice its yellow color.
- Occasionally, turmeric has been called "yellow ginger."
- The so-called "white turmeric" is a closely related plant, zedoary (see Chapter 102).
- The genus name *Curcuma* is a Latinization of the Arabic *kurkum*, meaning "yellow color," probably referring to the color of turmeric, although when Linnaeus coined the scientific name, he could have also had the yellow color of the flowers in mind.
- *Longa* in the scientific name *C. longa* is Latin for long, an allusion to the (comparatively) long rhizomes.

PLANT PORTRAIT

Turmeric, which is related to ginger, is a large-leaved, perennial herb growing 45 to 120 cm (1½–4 ft.) in height. The plants have five to eight large leaves, which are 30 to 70 cm (12–28 in.) long. The species probably originated in southeastern or southern Asia and is cultivated in tropical countries, especially in India, China, and the East Indies. In India, approximately 60 cultivars are recognized. The thick, rounded, underground stems (rhizomes) have a tough brown skin and bright orange to reddish brown flesh and are 2 to 6 cm (0.8–2.4 in.) long and 1 to 1.7 cm (0.4–0.7 in.) thick. India produces almost the entire world's crop and uses 80% of it. Erode in southern India is the world's largest trading center of turmeric, leading to its alternative names, Yellow City and Turmeric City. Turmeric spice usage dates back nearly 4000 years, to the Vedic culture in India. Turmeric has also been used in Indian (Ayurvedic) and Chinese medicine for thousands of years. The spice has been employed in Europe since the Middle Ages. It contains the yellow pigment curcumin, which on exposure to air oxidizes into vanillin, the active principle in vanilla. The pulverized rhizome constitutes the spice turmeric, generally prepared by cooking, dehydrating, and grinding to a powder.

Turmeric has been employed as a dye for millennia. The strong dye stains most everything it touches, including skin, clothes, furniture, and even some cooking utensils, which is probably one reason the spice is not popular in Western cuisine. The color fades notably on exposure to direct sunlight.

CULINARY PORTRAIT

Turmeric is similar to ginger root in size but is sweeter, more delicate, and more fragrant, with a pleasantly bitter taste. Unlike ginger, turmeric is rarely sold fresh and is mostly available dried—either whole or ground. It is used to color and season sauces (notably, Worcester[shire] sauce), salads,

FIGURE 95.1 Turmeric (*Curcuma longa*). (From Köhler, 1883–1914.)

fruit syrups, liqueurs, cheese, butter, margarine, lentils, and rice. It is also added to white mustard seeds to given a bright yellow color to mustard. The spice is also used to flavor curries, chutneys, poultry, eggs, fish and other seafood, and cooked vegetable dishes, and is an important ingredient in mulligatawny soup. Western cuisine generally does not use turmeric directly, but as part of several spice mixtures and sauces. It is often used in place of the much more expensive saffron, for its color more than for its flavor, which is musky, warm, and peppery. Saffron gives a more orange color and is less bitter. Cooking increases the bitterness and darkness of turmeric. Because of the strong taste, turmeric is best used sparingly to avoid overwhelming dishes. Turmeric powder should be kept in an airtight container and stored in a cool dry place. The aroma and color degenerate if ground turmeric is stored too long. Prepared turmeric is orange-yellow in color and has a very distinct, aromatic, spicy aroma. However, the color varies somewhat according to variety and preparation and is not a necessary indicator of quality.

Turmeric is most often used in curries, for which it is the key spice. Curry can be made up of hundreds of ingredients, but in Western culture curry powder is typically a pungent seasoning blend of cumin, coriander, turmeric, cayenne, and cardamom. This composition reflects a formula concocted by the British to replicate the taste of Indian cooking.

Fresh turmeric leaves are used in some regions of Indonesia as a flavoring.

Culinary Vocabulary

- Turmeric prepared from the central tubers is known as round turmeric, whereas that made from the more cylindrical lateral tubes is called finger turmeric.
- The traditional South Asian dish *momos* (based on the Nepali word for meat dumplings) is spiced with turmeric.
- *Masala* means spice or spices in Indian cooking. It usually includes several spices and is a staple of traditional Indian and Middle Eastern cuisines. There are innumerable kinds of masala, the most general being *garam masala*, and this will usually contain turmeric.

CURIOSITIES OF SCIENCE AND TECHNOLOGY

- In Thailand, turmeric was used to dye the robes of the Buddhist monks their rich yellow color.
- According to the old "Doctrine of Signatures," plants were supposed to indicate their ability to affect cures by some evident characteristic. Because turmeric is yellow, it was presumed to cure yellow jaundice.
- In the 1990s, several companies attempted to patent turmeric's medicinal properties in the United States. The Indian Council of Scientific and Industrial Research filed claim in the United States that turmeric is an Indian discovery and cannot be patented. In 1995, the U.S. Patent and Trademark Office upheld India's claim.
- Egan et al. (2004) reported that, at least in mice, turmeric could alleviate cystic fibrosis, a life-threatening disease characterized by thick, sticky mucus that clogs the lungs and pancreas, generally resulting in death of humans in their early 30s. Earlier studies have suggested that turmeric might also be of value in treating inflammatory bowel disease, cancer, alcohol-related liver disease, and Alzheimer's disease.
- "Turmeric paper" is a kind of litmus paper. It is paper saturated with turmeric, used to detect the presence of alkalis (basic compounds), which turn the paper brown, or boric acid, which turns it red-brown.
- Turmeric is widely viewed as a deterrent to ants, useful for protecting plants in gardens. This view has been described as an old wives' tale and as based on anecdotal evidence only, but the following publication provided experimental verification of ant deterrence: Mekuriaw, Y., Moudgal, R.P., and Seyoum, T. 2006. Modification and comparative evaluation of the existing and modified hay-box brooder in Mecha woreda of Amhara region, Ethiopia. *Livestock Research for Rural Development*, 18(5), Article #63. http://www.lrrd.org/lrrd18/5/meku18063.htm (URL current November 2010).

KEY INFORMATION SOURCES

Ammon, H.P.T., and Wahl, M.A. 1991. Pharmacology of *Curcuma longa*. *Planta Med.* 57:1–7.

Bambirra, M.L.A., Junqueira, R.G., and Gloria, M.B.A. 2002. Influence of post harvest processing conditions on yield and quality of ground turmeric (*Curcuma longa* L.). *Brazil. Arch. Biol. Technol.* 45:423–429.

Burtt, B.L. 1977. The nomenclature of turmeric and other Ceylon Zingiberaceae. *Notes R. Bot. Gard. Edinburgh*, 35:209–215.

Dahal, K.R., and Idris, S. 1999. *Curcuma longa* L. In *Plant resources of South-East Asia. Vol. 13. Spices*. Edited by C.C. de Guzman and J.S. Siemonsma. Backhuys, Leiden, the Netherlands. pp. 111–116.

Egan, M.E., Pearson, M., Weiner, S.A., Rajendran, V., Rubin, D., Glockner-Pagel, J., et al. 2004. Curcumin, a major constituent of turmeric, corrects cystic fibrosis defects. *Science*, 304:600–602.

Govindarajan, V.S. 1980. Turmeric—chemistry, technology and quality. *CRC Crit. Rev. Food Sci. Nutr.* 12:199–301.

Iqbal, S.H., and Nasim, G. 1986. Vesicular-arbuscular mycorrhiza in roots and other underground portions of *Curcuma longa*. *Biologia*, 32:223–227.

Johnston, B.A., and Webb, G. 1997. Turmeric patent overturned in legal victory. *HerbalGram*, 41:11–12.

Ketkar, H.D., and Tulpule, S.H. 1981. An instance of rare seed setting [in] *Curcuma longa* L. (tumeric), a vegetatively propagated spice crop, India. *Curr. Sci.* 50:862–863.

Krishnamurthy, K.H. 1992. *Ginger and turmeric*. Books for All, Delhi, India. 118 pp.

Krishnamurthy, M.N., Padmabai, R., Natarajan, C.P., and Kuppuswamy, S. 1975. Colour content of turmeric varieties and studies on its processing. *J. Food Sci. Technol.* 12:12–14.

Krishnamurthy, N., Mathew, A.G., Nambudiri, E.S., Shivashankar, S., Lewis, Y.S., and Natarajan, C.P. 1976. Oil and oleoresin of turmeric. *Trop. Sci.* 18:37–45.

Majeed, M., Badmaev, V., and Murray, F. 1996. *Turmeric and the healing curcuminoids: their amazing antioxidant properties and protective powers*. Keats Pub., New Canaan, CT. 47 pp.

Mathai, C.K. 1976. Variability in turmeric (*Curcuma* species) germplasm for essential oil and curcumin. *Qualitas Plantarum*, 25:227–230.

Nair, M.K. (Ed.). 1982. Proceedings of the National Seminar on Ginger and Turmeric, Calicut, India, April 8–9, 1980. Central Plantation Crops Research Institute, Kasaragod, Kerala, India. 250 pp.

Nethsingha, C., and Paskaranathan, U. 1976. *Turmeric—a literature survey*. Ceylon Institute of Scientific and Industrial Research, Colombo, Sri Lanka. 37 pp.

Peter, K.V., and Kandiannan, K. 1999. Turmeric. In *Tropical horticulture*, vol. 1. Edited by T.K. Bose, S.K. Mitra, A.A. Farooqui, and M.K. Sadhu. Naya Prokash, Calcutta, India. pp. 683–691.

Pfeiffer, E., Hoehle, S., Solyom, A.M., and Metzler, M. 2003. Studies on the stability of turmeric constituents. *J. Food Eng.* 56:257–259.

Philip, J. 1983. Studies on growth, yield and quality components in different turmeric types. *Indian Cocoa Arecanut Spices J.* 6:93–97.

Philip, J., and Nair, P.C.S. 1983. Morphological and yield characters of turmeric types. *Indian Spices*, 20(2):13–21.

Randhawa, G.S., and Mahey, R.K. 1988. Advances in the agronomy and production of turmeric in India. *J. Herbs Spices Med. Plants*, 3:71–101.

Ratnambal, M.J. 1986. Evaluation of turmeric accessions for quality. *Qualitas Plantarum*, 36:243–252.

Rethinam, P., Sivaraman, K., and sushama, P.K. 1994. Nutrition of turmeric. In *Advances in horticulture. Plantation and spice crops*. Part 1, Vol. 9. Edited by K.L. Chadha and P. Rethinam. Malhotra Publishing House, New Delhi, India. pp. 477–490.

Sasikumar, B. 2001. Turmeric. In *Handbook of herbs and spices*. Edited by K.V. Peter. Woodhead Publishing, Cambridge, U.K. pp. 297–310.

Shah, J.J., and Raju, E.C. 1975. General morphology, growth and branching behaviour of the rhizomes of ginger, turmeric and mango ginger. *New Botanist*, 2:59–69.

Shamina, A., Zachariah, T.J., Sasikumar, B., and George, J.K. 1998. Biochemical variation in turmeric (*Curcuma longa* Linn.) accessions based on isozyme polymorphism. *J. Hortic. Sci. Biotech.* 73:479–483.

Sherlija, K.K., Unnikrishnan, K., and Ravindran, P.N. 2001. Anatomy of rhizome enlargement in turmeric (*Curcuma longa* L.). *J. Econ. Taxon. Bot.* 19:229–235.

Sopher, D.E. 1964. Indigenous use of turmeric in Asia and Oceania. *Anthropos*, 59:93–127.

Vijayalakshmi, K., Subhashini, B., and Koul, S.V. 1997. *Plants in pest control: turmeric and ginger*. Centre for Indian Knowledge Systems, Chennai, India. 34 pp.

Wardini, T.H., and Prakoso, B. 2001. *Curcuma* L. In *Plant resources of South-East Asia. Vol. 12(1). Medicinal and poisonous plants 1*. Edited by L.S. de Padua, N. Bunyapraphatsara and R.H.M.J. Lemmens. Backhuys Publishers, Leiden, the Netherlands. pp. 210–219.

Zakaullah, and Parveen, S. 1980. A review of the diseases of turmeric *Curcuma longa*. *Pak. J. For.* 30:191–195.

Specialty Cookbooks

Note: Turmeric is mentioned in thousands of cookbooks marketed today. Recipes for the foods of India particularly specify this spice.

Baljekar, M. 2002. *Curry: fire and spice. Over 150 great curries from India and Asia*. Lorenz, London. 256 pp.

Dhillon, K. 2008. *The curry secret: how to cook real Indian restaurant meals at home*. Right Way, London. 128 pp.

Iyer, R. 2002. *The turmeric trail: recipes and memories from an Indian childhood*. St. Martin's Press, New York. 252 pp.

McDermott, N. 1999. *The curry book: memorable flavors and irresistible recipes from around the world*. Houghton, Mifflin, Harcourt, Boston, MA. 272 pp.

Panjabi, C. 2006. *50 great curries of India*. Revised edition. Kyle Cathie, London. 192 pp.

Trang, C., Thompson, D., Owen, S., and Singh, V. 2006. *Curry cuisine. Fragrant dishes from India, Thailand, Vietnam and Indonesia*. DK Publishing, New York. 352 pp.

96 Vietnamese Herbs

This chapter features

Rao om (*Limnophila aromatica*)
Rau ram (*Polygonum odoratum*)
Vap ca (*Houttuynia cordata*)
Vietnamese balm (*Elsholtzia ciliata*)

Although the Vietnamese use herbs in cooking to finish off dishes, for the most part the cuisine is characterized by the use of raw herbs as accompaniments to foods. Especially in southern Vietnam, fresh herbs are added to various dishes. At a meal, diners help themselves to their favorite herbs from bowls, often pinching off leaves from stems that are left behind and adding their choice of herbs to flavor their individual dishes (or sometimes preparing a hand roll of herbs to eat). Collectively, culinary herbs are known in Vietnamese as *rau thom*, literally "fragrant leaves." *Rau* refers to leafy vegetables and *thom* means fragrant. Many Vietnamese herbs begin with the word rau, as exemplified by rau om, rau ram, and *rau kinh gió'i* (Vietnamese balm), discussed later in this chapter. Vap ca is another flavoring herb widely used in Vietnamese food. Individually, these four herbs have limited importance outside of Southeast Asia, and they are dealt with collectively here because together they represent a significant class of food plant in the world of exotic cuisines.

Pho (pronounced "fuh," like the word foot without the "t") is a traditional Vietnamese soup that has been described as "one of the great and economical treats in the world of food … a proverbial meal in a bowl." It has also been described as "Vietnam's answer to fast food." The dish consists of long, chewy rice noodles and various cuts of beef, all immersed in a savory broth, usually with slices of onion, chopped green onion, and fresh coriander. Pho is generally available in Vietnamese restaurants throughout the English-speaking world. When served, a plate of garnishes will be provided on the side that one can add to the pho to change the taste of any given spoonful. The side plate typically contains cut lime, crunchy bean sprouts, Thai basil, and other herbs such as described in the following sections. There will probably also be several condiments on the table that one can use to personalize the pho and possibly also incendiary green peppers that many Vietnamese like to eat with the dish. One food reviewer, enchanted with "Pho Houses" advised that to have "phun with pho … pinch a single leaf of Vietnamese balm and add it to a spoonful of noodle, beef and broth and slurp it into your mouth. Pick up a green pepper, bite off a little portion and chew it along with the soup. It won't kill you."

RAU OM

Family: Scrophulariaceae (figwort or snapdragon family; some authorities have recommended transferring the genus *Limnophila* to the Plantaginaceae, plantain family)

Names

Scientific Name: *Limnophila aromatica* (Lam.) Merr. (*L. chinensis* (Osbeck) Merr. subsp. *aromatica* (Lam.) T. Yamaz., *L. aromaticoides* Yang & Yen, *L. gratissima* Blume, *L. punctata* Blume, *L. punctata* var. *subracemosa* Benth., *Ambulia aromatica* Lam.)

- The English name "rau om" is the name of the plant in southern Vietnam.
- Rau om is also known as Asian limno, finger leaf, finger grass, rice paddy herb, and swamp leaf.
- In Vietnam, where rau om is most popular, it is generally cultivated in flooded rice fields, hence the name "rice paddy herb."
- The name "swamp leaf" also is indicative of the watery habitat.
- The names "finger leaf" and "finger grass" are allusions to the small (finger-sized) leaves.
- Asian names include: *ngo om*, *rau ngo* (encountered in Hawaii), *rau om* (Vietnamese); *phak khayaeng* (Thai); *keukeuhan* (Sundra); *beremi* and *shiso-kusa* (Malay, Japanese); *shui fu rong* (Mandarin); and *seui fa* (Cantonese).
- The genus name *Limnophila* is derived from the Greek *limne*, marsh and *philos*, loving, a reference to the wet habitat.
- In addition to the plant genus *Limnophila*, there is a genus of stone flies also called *Limnophila*. Plant names are independent of animal names, and there are other cases of the same name independently used for a plant and for an animal.
- *Aromatica* in the scientific name *L. aromatica* is Latin for aromatic.

Plant Portrait

Rau om is a tropical and subtropical herb, occurring in Japan, China, Taiwan, Malaysia, Indonesia, India, East Pakistan, Micronesia, and Australia. The species is a small, tender, perennial (sometimes annual), trailing, semiaquatic herb 20 to 100 cm (8–39 in.) tall (usually less than 50 cm or 20 in.). The flowering stalks are up to 15 cm (6 in.) long and have many blue flowers in the axils of upper leaves, which mature into small capsules (flowers and fruit are rarely produced in cultivation). The odor has been described as similar to turpentine, or more attractively, floral, and like a mixture of cinnamon and cloves. Rau om is a minor culinary herb in parts of Asia and is sometimes encountered in North American ethnic markets and restaurants. In Oriental groceries it is easily recognized by having whorls of three leaves along the stems.

In addition to its culinary value, the plant is a suitable candidate for ornamental use in a hanging basket. The species is also cultivated as an aquatic plant in fish tanks. In Asia, rau om is used to treat many ailments. In China, it is used for the treatment of intoxication and pain; in Indochina, to treat wounds; in Malaysia, chiefly as a poultice on sore legs, but also to promote appetite, as an expectorant to clear mucus from the respiratory tract, and to treat fever; and in Indonesia, as an antiseptic or cleanser for worms.

Culinary Portrait

The whole plant is eaten raw as greens or steamed in Asia, and the dried herb is also occasionally used. The leafy stems are consumed as a side dish with rice and used fresh to garnish and flavor

FIGURE 96.1 Rau om (*Limnophila aromatica*), by S. Rigby.

soups, curries, dips, and sweet and sour preparations. In southern Vietnam, the herb is traditionally served raw as a garnish accompanying spicy fish soups. Ra om also goes well with meats. In Malaysia, the plant has been eaten like spinach, either raw or cooked. The flavor is reminiscent of cinnamon and cumin, with citrus overtones, especially of lemon. The distinct aroma and flavor are considered very desirable in several sweet and sour Vietnamese dishes and in Thai cooking. The unique flavor of rau om has led to its being labeled as "the secret ingredient in famous Vietnamese hot and sour soups," typically made with tamarind and cantaloupe (see *canh chua ca* in the Culinary vocabulary section of Rau Om). Although the herb is unfamiliar to Western tastes (some have described the flavor as "soapy"), it makes an interesting, tasty, aromatic garnish and conversation piece.

Culinary Vocabulary

- *Canh chua ca* (pronounced gahn choo-ah gah) is a traditional South Vietnamese sour (or sweet and sour) fish soup—the emblematic dish of the city of Saigon. Rau om is considered to be an indispensable ingredient.

Curiosities of Science and Technology

- The durian (see Chapter 33) is a large tropical tree producing heavy fruits with a taste that has been described as a mixture of old cheese and onions, flavored with turpentine. In the Dutch East Indies, rau om, itself having a distinctly turpentine taste, is considered a good accompaniment, preventing the malodorous belching that often results when durian is eaten alone.
- Rau om is grown as an aquatic plant in fish aquaria by hobbyists, but purchasing the plant for this purpose requires knowledge of the kinds that are available and supply sources that reliably specify what kinds they can furnish. The relatively short-leaved, common form of the species, with opposite leaves that develop similarly above and below water, is considered less suitable for aquaria, often not growing particularly well. By contrast, a form with narrower, longer leaves that become reddish on the undersides thrives in aquaria and looks more attractive. The former is often termed var. *aromatica*, the latter var. *hippuroides* or var. *aromaticoides*.

Key Information Sources

(Also see references in Kuebel and Tucker (1988) in the Rau Ram section.)

Adams, D. 1996. New and unusual herbs for the garden. *Herbarist*, 62:51–52.

Bui, M.-L., Grayer, R.J., Veitch, N.C., Kite, G.C., Tran, H., and Nguyen, Q.-C.K. 2004. Uncommon 8-oxygenated flavonoids from *Limnophila aromatica* (Scrophulariaceae). *Biochem. Syst. Ecol*. 32:943–947.

Chandran, R., and Bhavanandan, K.V. 1986. Cytological investigation of the family Scrophulariaceae I. *Limnophila. Cytologia*, 51:261–270.

Dutta, N.M. 1975. A revision of the genus *Limnophila* of eastern India. *Bull. Bot. Soc. Bengal*, 29(1):1–7.

Kukongviriyapan, U., Luangaram, S., Leekhaosoong, K., Kukongviryapan, V., and Preeprame, S. 2007. Antioxidant and vascular protective activities of *Cratoxylum formosum*, *Syzygium gratum* and *Limnophila aromatica*. *Biol. Pharm. Bull*. 30:661–666.

Ochse, J.J., and Bakhuizen van den Brink, R.C. 1980. *Vegetables of the Dutch East Indies*. A. Asher & Co., Amsterdam, the Netherlands. 1016 pp.

Philcox, D. 1970. A taxonomic revision of the genus *Limnophila* R. Br. (Scrophulariaceae). *Kew Bull*. 24:101–170.

Schmelzer, G.H. 2001. *Limnophila* R.Br. In *Plant resources of South-East Asia. Vol. 12(2). Medicinal and poisonous plants 2*. Edited by J.L.C.H van Valkenburg and N. Bunyapraphatsara. Backhuys Publishers, Leiden, the Netherlands. pp. 341–344.

Sribusarakum, A., Bunyapraphatsara, N., Vajragupta, O., and Watanabe, H. 2004. Antioxidant activity of *Limnophila aromatica* Merr. *Thai J. Phytopharmacy*, 11(2):11–17.

Tucker, A.O., Maciarello, M.J., Hendi, M., and Wheeler, K.A. 2002. Volatile leaf and stem oil of commercial *Limnophila chinensis* (Osb.) Merrill ssp. *aromatica* (Lam.) Yamazaki (Scrophulariaceae). *J. Essent. Oil Res*. 14:228–229.

Wannan, B.S., and Waterhouse, J.T. 1985. A taxonomic revision of the Australian species of *Limnophila* Scrophulariaceae. *Aust. J. Bot*. 33:367–380.

Yamazaki, T. 1978. On *Limnophila aromatica* (Lam.) Merr. and *L. chinensis* (Osb.) Merr. *J. Jpn. Bot*. 53:312–313 (in Japanese, English summary).

Yang, Y.P., and Yen, S.H. 1997. Notes on *Limnophila* (Scrophulariaceae) of Taiwan. *Bot. Bull. Acad. Sinica*, 38:285–295.

Specialty Cookbooks

(Also see cookbooks listed for Lychee, Longan, and Rambutan; Rau Ram; and Lemon Grass.)

Note: Rau om recipes appear primarily in Vietnamese cookbooks, occasionally in Cambodian and Thai recipes. Because the herb is rarely available in Western stores, Asian cookbooks written specifically for English-speaking audiences rarely include the herb, often recommending that cumin seeds or cilantro be substituted, along with a little lemon. However, many experts state that there is no good substitute for rau om.

Hoyer, D. 2009. *Culinary Vietnam*. Gibbs Smith, Layton, UT. 224 pp.

My Tran, D. 2003. *The Vietnamese cookbook*. Capital Books, Sterling, VA. 120 pp.

Riviere, M., de Bourqknecht, D., Lallemand, D., and Smend, M. 2008. *Cambodian cooking: a humanitarian in collaboration with Act of Cambodia*. Periplus Editions, Singapore. 96 pp.

RAU RAM (VIETNAMESE CORIANDER)

Family: Polygonaceae (buckwheat family; also known as smartweed family and knotweed family)

NAMES

Scientific Name: *Polygonum odoratum* Lour. (*Persicaria odorata* (Lour.) Soják)

(The name *Persicaria odorata* is advocated in Wilson, K.L. 1988. *Polygonum* sensu lato (Polygonaceae) in Australia. *Telopea*, 3: 177–182.)

- The name "rau ram" has been transferred to English from the original Vietnamese, *rau râm*. The word is often pronounced "zow-zam" or "zow-ram" by the Vietnamese.
- Additional English names for rau ram are Cambodian mint, fragrant knotweed, hot mint, laksa plant, polygonum, Vietnamese cilantro, Vietnamese coriander, Vietnamese mint, and Vietnamese parsley.
- The name Vietnamese parsley is also occasionally applied to water dropwort (see Chapter 48).
- Additional Asian names include *yuht naahm heung choi, laak saa yihp* (Cantonese); *luam lows* (Hmong); *chi krassang tomhom, xang-hum* (Khmer); *phak pheo, phak phew* (Laotian); daun kesum, *daun laksa, daun kesom* (Malay); *yue nan xiang cai, la sha ye* (Mandarin); and *pak pai, pa pao, phak phai, chan chom, hom chan* (Thai).
- The name laksa plant is based on the use of rau ram in Singapore for the noodle dish "laksa" (the Sanskrit *laksha* means "hundred thousand" and indicates that the dish has many ingredients).
- The genus name *Polygonum* is based on the Greek *polys*, many + *gonu*, joint, an allusion to the conspicuously swollen nodes of the stem.
- *Odoratum* in the scientific name *P. odoratum* means producing a smell.
- *Polygonum odoratum* is sometimes confused with a similar aquatic plant called water pepper, *Persicaria hydropiper* L. (*Persicaria hydropiper* (L.) Spach). The name *P. odoratum* should not be confused with *Polygonatum odoratum* (Mill.) Druce, angular Solomon's seal, a Eurasian herb of the lily family that is sometimes cultivated as an ornamental.

PLANT PORTRAIT

Rau ram is a fragrant, tender, sprawling, Indochinese herb with red stems that grow to approximately 30 cm (1 ft.) in height and turn up at the tips. The green leaves are marked with red. The plant spreads rapidly on moist ground by creeping rootstocks. Indeed, in subtropical areas as found in its native Vietnam, rau ram can become quite weedy. Roots develop from the leaf nodes. The lower stems become woody. The flowers are small, white, red, or sometimes purplish-pink depending on growing conditions. In Vietnam and in the Philippines, flowers are developed profusely in the first year, but when cultivated in Western countries, flowers may not be commonly produced and develop only after vernalization (exposure to a cold period). Rau ram is a tender, short-lived perennial but can be grown as an annual. It is an important condiment crop in Southeast Asia and has been spread since the 1960s by Southeast Asian immigrants, especially the Vietnamese, to the United States, Australia, Philippines, France, and elsewhere. The plant is also used in Latin America, both for condiment and medicine. Limited culture has been initiated in the southern United States in response to demand for it by immigrants. In greenhouse culture in the United States, leaves can be harvested year-round. Although familiar in Indochina, rau ram is generally not known to most people in Western countries.

"Kesom oil," rich in aliphatic acids, is a volatile oil extracted from *Polygonum minus* Huds. The similar essential oil from rau ram has been termed "Australian kesom oil" and is used to a minor extent as a flavoring.

FIGURE 96.2 Rau ram (*Polygonum odoratum*), by S. Rigby.

Culinary Portrait

The foliage of rau ram is used especially to flavor Vietnamese dishes. The odor has been described as like fresh coriander foliage (hence the name Vietnamese coriander) with a hint of lemon or like cilantro (cilantro is coriander foliage). The penetrating odor permeates an area when used in cooking, and because it is not universally admired, it is not often used in Vietnamese or other Southeast Asian restaurants catering to Western tastes. The flavor is milder than that of cilantro, hot and peppery but refreshing. The taste has been characterized as "extremely soapy flavored, with fruity overtones" or, more attractively, "like lemon and coriander leaves with a slight radish-like pungent aftertaste." Although it often seems that a taste for rau ram needs to be acquired, many who dislike cilantro can tolerate rau ram. This herb is not easily acquired in Western countries but may be available in Southeast Asian grocery stores.

Rau ram can be eaten fresh as a component of salads, or used as a garnish. In Asia, rau ram is combined with water dropwort (see Chapter 48) to flavor a sauerkraut-like cabbage preparation in brine. Small amounts go well with chicken salad, potato salad, and devilled eggs. Only the young, green leaves are used, as the older, redder leaves are too hot in flavor. Rau ram is employed as a condiment to add taste to cooked dishes, especially fish, fowl, and meat dishes. Fresh or cooked leaves are used with fish, mussels, clams, and oysters, as well as turtle and frog dishes. Rau ram is also consumed with rice or vegetable dishes. Vietnamese traditionally eat rau ram with embryonic duck eggs (see *hot vit lon* in the Culinary Vocabulary section of Vap Ca). It has been suggested that only a small amount be used for flavoring purposes, as a little goes a long way. The flavor is easily destroyed with cooking and overchopping.

In Vietnam, pregnant women avoid consuming rau ram because the fresh herb is suspected of having abortifacient properties.

Culinary Vocabulary

- One of the dishes well known for using rau ram is *Đu'a Cân*, a pickled preparation resembling sauerkraut.

Curiosities of Science and Technology

- Because of its alleged antiaphrodisiac properties, rau ram is widely consumed by Buddhist priests to help retain their celibacy.
- In Vietnam, rau ram is used to treat wounds and snake bites.

Key Information Sources

Bond, R. E. 1989. Rau ram. *Herbarist*, 55:34–37.

Corlett, J.L., Clegg, M.S., Keen, C.L., and Grivetti, L.E. 2002. Mineral content of culinary and medicinal plants cultivated by Hmong refugees living in Sacramento, California. *Int. J. Food Sci. Nutr.* 53:117–128.

Do, N.T. 1993. *Polygonum odoratum* Lour.—rau ram. *J. Biol.* 16(4):44–45.

Dung, N.X., Hac, L.V., and Leclercq, P.A. 1993. Volatile constituents of the aerial parts of Vietnamese *Polygonum odoratum* L. *J. Essent. Oil Res.* 7:339–340.

Hunter, M. 1996. Australian kesom oil—a new essential oil for the flavour and fragrance industry. *Agro. Food Ind. Hi Tech.* 7(5):26–28.

Hunter, M.V., Brophy, J.J., Ralph, B.J., and Bienvenu, F.E. 1997. Composition of *Polygonum odoratum* Lour. from southern Australia. *J. Essent. Oil Res.* 9:603–604.

Kuebel, K. R., and Tucker, A. O. 1988. Vietnamese culinary herbs in the United States. *Econ. Bot.* 42:413–419.

Nanasonbat, S., and Techchuen, N. 2009. Antimicrobial, antioxidant, and anticancer activities of Thai local vegetables. *J. Med. Plants Res.* 3:443–449.

Oyen, L.P.A., and Huyen, D.D. 1999. *Persicaria odorata* (Lour.) Soják. In *Plant resources of South-East Asia. Vol. 13. Spices*. Edited by C.C. de Guzman and J.S. Siemonsma. Backhuys, Leiden, the Netherlands. pp. 170–172.

Potter, T.L., Fagerson, I.S., and Craker, L.E. 1993. Composition of Vietnamese coriander leaf oil. *Acta Hortic.* 344:305–311.

Starkenmann, C., Ludmila, L., Niclass, Y., and Rouguet, D. 2006. Comparison of volatile constituents of *Persicaria odorata* (Lour.) Soják (*Polygonum odoratum* Lour.) and *Persicaria hydropiper* L. Spach (*Polygonum hydropiper* L.). *J. Agric. Food. Chem.* 54:3067–3071.

Vastano, B.C., Rafi, M.M., DiPaolo, R.S., Zhu, N., Ho, C.T., Rella, A.T., et al. 2002. Bioactive homoisoflavones from Vietnamese coriander or pak pai (*Polygonatum* [sic] *odoratum*). In *Quality management of nutraceuticals. Proceedings of a symposium, Washington, DC, August 2000.* Edited by C.T. Ho and Q.Y. Zheng. American Chemical Society, Washington, DC. pp. 269–280.

Yaacob, K.B. 1987. Kesom oil—a natural source of aliphatic aldehydes. *Perfumer Flavorist*, 12(5):27–30. [Obtained from *Polygonum minus* Huds.]

Specialty Cookbooks

Note: Rau ram is a Vietnamese specialty but also occurs in other parts of Asia, especially in Malaysia and Singapore. There are a hundred or so books that deal with Vietnamese cooking, of which the following are examples. Also, see recipes listed in the Rau om section.

Ngo, B., and Zimmerman, G. 1986. *The classic cuisine of Vietnam.* New American Library, New York. 250 pp.

Rauthier, N. 1989. *The foods of Vietnam.* Stewart, Tabori and Chang, New York. 239 pp.

Trang, T., Hirsheimer, C., and Yan, M. 1999. *Authentic Vietnamese cooking food from a family table.* Simon & Schuster, New York. 255 pp.

VAP CA (CHAMELEON PLANT)

Family: Saururaceae (Lizard's-tail family)

NAMES

Scientific Name: *Houttuynia cordata* Thunb.

- The English name "vap ca" is one of the Vietnamese names for the plant, more often called *giâp cá* in Vietnam. Giâp probably originates from the Cambodian word for this plant and means a fishy smell. Cá means fish. Other Vietnamese names include *diep ca* and *rau giap ca*. The name vap ca has been used since the early 1990s by the Richters Herb Company (Goodwood, Ontario, Canada), which played an important role in popularizing the culinary use of the plant in North America.

- Vap ca is also known in English as chameleon plant, Chinese lizard tail, coriander-scented houttuynia, dokukami (see Japanese names), fishwort, heartleaf, and houttuynia. The Asian names given below are likely to be useful for interpreting restaurant meals in Asian countries and sometimes in ethnic restaurants in Western countries.
- The name "chameleon plant" is based on the cultivated tricolored variety 'Chameleon' (apparently the same plant as those called 'Tricolor', 'Variegata', and 'Court Jester'), which has attractive red, cream, and green markings on the foliage. "Chameleon plant" is a much more widely known name than "vap ca" in English-speaking countries.
- The names fishwort (literally "fish plant"), fish plant, fish odor plant, fish mint, and fish scale mint (fishscale mint) reflect the fishy odor of the herb.
- The name "Chinese lizard tail" is a result of the species belonging to the lizard's tail family, in which leaves tend to be somewhat reminiscent of the shape of a lizard's tail.
- The name "heartleaf" is descriptive of the shape of the leaves.
- In Assam (India), the name is *masundari*.
- The official name in Japan is *gyoseiso*, but it is also called *dokudami* (meaning "detoxicant"), *dokudami zoku, shihyao, chung-yao*, and *jyu-yaku*.
- Another Japanese name is *kaeruppa*, meaning "frog leaf," reflecting a belief that a half-dead frog can be resuscitated by eating a leaf.
- In Korea, vap ca is called *osaengch'o*.
- In Thailand, it is called *pluu-kao* (*phluu kae*, or *puu khao*) and *phak khao thong*.
- The Laotian name is *phak khao thong*.
- There are several interesting Chinese names for vap ca, including *yu xing cao* (fishy smelling herb), *zhu bi kong* (pig's nostrils, used in Sezchuan), *chou giao mai* (stinky wheat, used in Zhejiang), *gou re cai* (dog ear vegetable or dog's ear vegetable, used in southern China), and *gou xing cao* (doggy smelling herb or herb that smells like a dog, used in Gansu). Another Chinese name is *chu tsai*.
- The genus name *Houttuynia* commemorates Martin Houttuyn (1720–1794), a Dutch naturalist.
- In addition to the plant genus *Houttuynia*, there is an animal genus *Houttuynia* (plant names are independent of animal names). The latter is a genus of tapeworms and includes a species found in ostriches.
- *Cordata* in the scientific name *H. cordata* means heart-like, that is, having two lobes like a heart, which is descriptive of the shape of the leaves.

Plant Portrait

Vap ca is a common wild plant in temperate and tropical regions of eastern Asia, from the Himalayas through Nepal, China, and Java to Japan, often growing in damp shady places in mountainous districts. It is a semi-aquatic perennial herb, 15 to 60 cm (6–24 in.) in height, occasionally as tall as 1 m (approximately 1 yd.). The leaves are heart shaped and dark green with some dark red-violet coloring. The stems are similarly colored and may also be purplish-red. All parts of the plant are strongly aromatic, and in flower vap ca has a fetid smell. The species has a clump-forming rootstock and may spread widely through its buried or partially buried rhizomes (underground stems). The plant is showiest in early summer, when it develops its flowers. The flowering branch bears four to six snowy white or green-white, cross-shaped floral bracts at the base of a stout spike of small, naked, crowded flowers. These have been said to superficially resemble white buttercups (genus *Ranunculus*) or the flowers of dogwood (genus *Cornus*). A Japanese and a Chinese chemical race,

FIGURE 96.3 Vap ca (*Houttuynia cordata*). (From Curtis, 1827, vol. 54, plate 2731.)

each differently scented, have been described. The Japanese type smells like oranges or ginger as well as cilantro (edible-leaved coriander). The Chinese type smells like raw meat, fish, and cilantro. Both types are cultivated in Vietnam and by transplanted Vietnamese in Texas and elsewhere. Vap ca is available as a curiosity through some garden supply outlets. It is little known in Western nations as a culinary herb.

Vap ca has a very long history in Asia as a medicinal plant. It has been used in China for medicinal purposes for 2000 years and continues to be employed today in Asian medicine.

Vap ca is often grown as an ornamental, frequently as a ground cover. However, the plant is so invasive that if it is planted by the edge of a pool it should be contained. Large pots will serve this purpose. In northern areas, the plant is insufficiently hardy to survive so that its potential weediness is not a problem.

Culinary Portrait

The leaves or shoots are eaten fresh or cooked (often steamed), in salads, soups, and fish stews. Vap ca is popular in Vietnamese stews and sweet and sour dishes. Roots and fruits are also considered edible. The roots are cleaned and used as a spicy substitute for bean sprouts, especially in southwestern China. The taste of the herb is pungent, generally reminiscent of coriander or cilantro. However, the odor has been described as "like a decayed fishy smell" and "like sour citrus." Some Vietnamese judge the aroma to be too strong and avoid the use of the herb. The taste may be more acceptable in young material, produced in the spring and summer. The plant tends to have become bitter by the autumn.

Culinary Vocabulary

- One of the Vietnamese dishes well known for using rau ram as a garnish is *hot vit lon*—boiled fertilized duck eggs cooked 3 days before they hatch (the same use is noted earlier for rau ram).

Curiosities of Science and Technology

- Vap ca develops seeds without fertilization. The pollen grains are nonfunctional, so sexual reproduction does not occur. Very few flowering plants have completely or almost completely given up producing seeds by the normal sexual procedure. The most familiar example is dandelion.

Key Information Sources

Chen, Y., Wu, W., and Liu, S. 2001. Effects of organic fertilizer on output and quality and nutrient uptake by *Houttuynia cordata* Thunb. *J. Sichuan Agric. Univ*. 19:245–248 (in Chinese).

Garcia, A.A. 1986. A cultivated Saururaceae (*Houttuynia cordata*). *Darwiniana*, 27:569–570 (in Spanish).

Hayashi, K., Kamiya, M., and Hayashi, T. 1995. Virucidal effects of the steam distillate from *Houttuynia cordata* and its components on HSV-1, influenza virus, and HIV. *Planta Med*. 61:237–241.

Jong, H.S., Min, J.K., Hyuk D.K., and In, H.P. 2003. Antimicrobial activity of fractional extracts from *Houttuynia cordata* root. *J. Korean Soc. Food Sci. Nutr*. 32:1053–1058 (in Korean, English summary).

Jong, T.-T., and Jean, M.-Y. 1993. Alkaloids from *Houttuynia cordata*. *J. Chin. Chem. Soc*. 40:301–303.

Jong, T.-T., and Jean, M.-Y. 1993. Constituents of *Houttuynia cordata* and the crystal structure of vomifoliol. *J. Chin. Chem. Soc*. 40:399–402.

Kim, K.Y., Chung, H.J., and Chung, D.O. 1997. Chemical composition and antimicrobial activities of *Houttuynia cordata* Thunb. *Korean J. Food Sci. Technol*. 29:400–406 (in Korean, English summary).

Liu, Y., and Deng, Z. 1979. Investigation of the chemical constituents of the essential oil of *Houttuynia cordata* Thunb. *Acta Bot. Sin*. 21:244–249.

Ohga, Y., Shigyou, A., Ohmori, K., and Nakamura, S. 1995. Cultivation methods of *Houttuynia cordata* Thunb. and *Artemisia dubia* Wall and processing method of their tea-like products. *Bull. Fukuoka Agric. Res Center*. 14(March):50–53 (in Japanese, English summary).

Pröbstle, A. 1994. Phytochemische und pharmakologische Untersuchungen von *Houttuynia cordata* Thunb. [Phytochemical and pharmacological analyses of *Houttuynia cordata* Thunb.] Doctoral dissertation. Faculty of Chemistry and Pharmacy, Ludwig Maximilians University, Munich, Germany. 160 pp (in German).

Pröbstle, A., and Bauer, R. 1992. Aristolactams and a 4,5-dioxoaporphine derivative from *Houttuynia cordata*. *Planta Med*. 58:568–569.

Pröbstle, A., Neszmelyi, A., Jerkovich, G., Wagner, H., and Bauer, R. 1994. Novel pyridine and 1,4-dihydropyridine alkaloids from *Houttuynia cordata*. *Nat. Prod. Lett*. 4:235–240.

Rohweder, O., and Treu-Koene, E. 1971. Structure and morphological significance of the inflorescence of *Houttuynia cordata* Thunb. (Saururaceae). *Zurich Naturforsch Ges Vierteljahrsschr*, 116:195–212 (in German, English summary).

Takahashi, M. 1986. Microsporogenesis in a parthenogenetic species, *Houttuynia cordata* Thunb. (Saururaceae). *Bot. Gaz*. 147:47–54.

Taylor, R.L. 1976. *Houttuynia cordata*, dokudami. *Davidsonia*, 7(4):63.

Tucker, S.C. 1981. Inflorescence and floral development in *Houttuynia cordata* (Saururaceae). *Am. Fern J*. 68:1017–1032.

Tutupalli, L.V., and Chaubal, M.G. 1975. Saururaceae. V. Composition of essential oil from foliage of *Houttuynia cordata* and chemosystematics of Saururaceae. *Phytochemistry*, 38:92–96.

Wei W., Youliang, Z., Li, C., Yuming, W., Zehong, Y., and Ruiwu, Y. 2005. PCR-RFLP analysis of cpDNA and mtDNA in the genus *Houttuynia* in some areas of China. *Hereditas*, 142:24–32.

Wu, W., Zheng, Y.L., Ma, Y., Yang, R., Zhong, J., Ren, F., and Liu, T. 2003. Analysis on yield and quality of different *Houttuynia cordata*. *Zhongguo Zhongyao Zazhi*, 28:718–720, 771 (in Chinese).

Yamazaki, T. 1978. Structure of the flower and inflorescence of *Houttuynia cordata* Thunb. *Bot. Mag*. 91:69–82.

Zeng, H., Jiang, L., and Zhang, Y. 2003. Chemical constituents of volatile oil from H*outtuynia cordata* Thunb. *J. Plant Res. Environ*. 12(3):50–52 (in Chinese).

Specialty Cookbooks

Brissenden, R. 2007. *Southeast Asian food: classic and modern dishes from Indonesia, Malaysia, Singapore, Thailand, Laos, Cambodia, and Vietnam*. Periplus, Singapore. 571 pp.

Wright, C.A. 2005. *Some like it hot: spicy favorites from the world's hot zones*. Harvard Commons Press, Boston, MA. 453 pp.

VIETNAMESE BALM

Family: Labiatae (Lamiaceae; mint family)

NAMES

Scientific Name: *Elsholtzia ciliata* (Thunb.) Hyl. (*E. cristata* Willd., *E. patrinii* (Lepech.) Garcke)

- The name "Vietnamese balm" is based on the distinct resemblance of the flavor to lemon balm (*Melissa officinalis* L.) and the fact that the plant came to the attention of people in the West from Vietnamese sources.
- Vietnamese balm has also been called crested late-summer mint, elsholtzia, late summer mint, lepechin, mint bush, mint shrub, and Vietnamese mint.
- The Vietnamese name for Vietnamese balm is *rau kinh gió'i* (or *xiang ru, rau kinh gio*), a name that may be useful when attempting to obtain this herb in Vietnamese grocery outlets.
- The genus name *Elsholtzia* commemorates a German physician and botanist with the imposing name of Johann Siegesmund Elsholtz (1623–1688).
- The genus *Elsholtzia* should not be confused with *Eschscholzia*, a small genus of the poppy family, particularly known for the ornamental California poppy (*E. californica* Cham.).
- *Ciliata* in the scientific name *E. ciliata* is Latin for ciliate, that is, furnished with fine hairs.

PLANT PORTRAIT

Vietnamese balm is an annual (perennial in some climates), lemon-scented, herbaceous, bushy plant growing 30 to 90 cm (1–3 ft.) tall. It produces attractive spikes of light-purple flowers in late summer or fall, if not killed by frost. This native of temperate central and eastern Asia has escaped from cultivation to become established in parts of Europe and North America. The species is naturalized in the eastern United States, where it has been collected in Maine, Vermont, Massachusetts, and Connecticut. In Canada, it is occasionally found along roadsides and in old fields in Manitoba, Ontario, Quebec, and New Brunswick. Vietnamese balm has been used for many years in East Asia as a medicinal and culinary plant. It was also formerly grown in eastern Europe, with some cultivation still occurring in Austria. In areas of Texas where Vietnamese emigrants have settled, it is occasionally sold in Vietnamese stores. Vietnamese balm has negligible commercial importance in North America. Living plants are available from some nurseries, and sometimes it is possible to establish plants from fresh sprigs purchased from Asian produce markets.

In addition to its culinary uses, Vietnamese balm is cultivated as an ornamental, considered attractive for its very aromatic foliage and interesting, upright habit. It has insecticidal properties and is said to be a good plant for attracting bees from bee hives.

CULINARY PORTRAIT

Young plants of Vietnamese balm are finely cut and eaten raw in salads, boiled as a potherb, or used as an aromatic condiment in vegetable and meat dishes and soups. The taste is pungent, whereas the aroma has been described as "a vibrant lemon scent with sweet floral undertones." This herb can also be added to seafood dishes or brewed for tea. In India, the seeds are sometimes eaten raw, and oil from the seeds is used for cooking purposes.

FIGURE 96.4 Vietnamese balm (*Elsholtzia ciliata*). Left: (From Hallier, 1880–1888, vol. 18, plate 1773.) Right: (From Bailey, 1916.)

Culinary Vocabulary

- Pho, described above, is the most famous Vietnamese noodle dish. However, *bún riêu cua*, a tangy crab- and tomato-based noodle soup, is also considered to be exceptionally representative of the best of Vietnamese cuisine. As with most Vietnamese dishes, it is accompanied by fresh herbs, including Vietnamese balm leaves (*kinh gioi*).

Curiosities of Science and Technology

- Vietnamese balm has been used medicinally in Asia. In Japan, the oil was used to promote urine flow and to reduce fever. In China, the plant was employed to treat stomach problems, to expel gas from the alimentary canal, and to promote urine flow. It has also been used in China to treat nosebleed and burning of the feet. In contemporary Western medicine, such uses are considered obsolete.

Key Information Sources

Bakova, N.I., Dmitriev, L.B., Mashanov, V.I., Dimitrieva, V.L., and Grandberg, I.I. 1988. Characteristics of the essential oils of various *Elsholtzia ciliata* Thunb. biotypes. *Izv. Timiryazev. Skh. Akad.* 10(2):162–169 (in Russian).

Barannikova, T.A. 1979. *Elsholtzia patrinii*, an essential oil plant. *Dokl. T. S. Kh. A. Timiriazevsk*, 251:68–73 (in Russian).

Bisht, J.C., Pant, A.K., Mathela, C.S., Kubold, U., and Vostrowsky, O. 1985. Constituents of essential oil of *Elsholtzia strobilifera*. *Plant Med.* 51:412–414.

Dmitriev, L.B., Klyuev, N.A., Mumladze, M.G., Zamureenko, V.A., Esvandzhiya, G.A., and Grandberg, I.I. 1984. Essential oil from *Elsholtzia patrinii*. *Izvestiya Timiryazevskoi Sel'skokhozyaistvennoi Akademii*, 3:171–175 (in Russian).

Kharina, T.G., Kalinkina, G.I., Dembitskii, A.D., and Maksimenko, N.B. 1995. Morphobiological peculiarities and qualitative composition of essential oil of *Elsholtzia ciliata* (Thunb.) Hyl. (South of Tomsk district). *Rastitel'nye Resursy*, 31(3):58–64 (in Russian).

Kubold, U., Vostrowsky, O., Bestmann, H.J., Bisht, J.C., Pant, A.K, Melkani, A.B., et al. 1987. Terpenoids from *Elsholtzia* species. II. Constituents of essential oil from a new chemotype of *Elsholtzia cristata*. *Plant Med.* 53:268–271.

Lee, Y.H., Lee, I.R., Won, W.S., and Park, C.H. 1988. Flavonoids of *Elsholtzia cristata*. *Archives Pharm. Res.* 11:247–249.

Miske, D. 1994. A dozen lemon-fresh herbs. *Cornell Plant.* 49(1):7–13.

Nguyen, X.D., Len, V.H., Le, H.H., and Leclercq, P.A. 1996. Composition of the essential oils from the aerial parts of *Elsholtzia ciliata* (Thunb.) Hyland. from Vietnam. *J. Essent. Oil Res.* 8:107–109.

Ren, Z., and Feng, Z. 1994. Beekeeping value of *Elsholtzia* spp. honey plants in China. *J. Bee.* 10:27–28 (in Chinese).

Sohn, K., Song, J.S., Chae, Y.A., and Kim, K.S.1998. The growth and essential oil of *Elsholtzia ciliata* (Thunb.). Hylander. *J. Kor. Soc. Hort. Sci.* 39:809–813 (in Korean, English summary).

Sohn, K., Song, J.S., and Kim, K.S.1998. Morphological observation of glandular trichomes of *Elsholtzia ciliata* (Thunb.) Hylander by scanning electron microscope. *J. Kor. Soc. Hort. Sci.* 39:814–818 (in Korean, English summary).

Watanabe, Y., and Hirokawa, H. 1968. Auto-ecological studies on the annual weeds in Tokachi Japan. III. The effect of time of emergence on the seed production *Elsholtzia ciliata*, *Polygonum* spp., *Rorippa islandica*, *Setaria faberi*, *Commelina communis*, *Echinochloa crus-galli*, *Digitaria violascens*, *Chenopodium album*, *Stellaria media*. *Res. Bull. Hokkaido Nat. Agric. Exp. Stn.* 93:16–22 (in Japanese).

Zheng, S.Z., Shen, X.W., and Lu, R.H. 1990. The chemical constituents of *Elsholtzia ciliata* (Thunb.) Hyland. *Acta Bot. Sin.* 32:215–219 (in Chinese).

Specialty Cookbooks

Creasy, R. 1999. *Edible herb garden*. Periplus, North Clarendon, VT. 105 pp.

Nguyen, A., Beisch, L., and Cost, B. 2006. *Into the Vietnamese kitchen: treasured foodways, modern flavours*. Ten Speed Press, Berkeley, CA. 344 pp.

97 Wasabi

Family: Brassicaceae (Cruciferae; mustard family)

NAMES

Scientific Name: *Wasabia japonica* (Miq.) Matsum. (*Eutrema wasabi* Maxim.)

- The Japanese name *wasabi* means "mountain hollyhock." The English name wasabi was simply adopted from the Japanese, and the Japanese name also served as the root for the genus name *Wasabia*.
- Pronunciation: In Japan, WAH-sah-bee is correct, but some American dictionaries permit the accent on the second syllable.
- Wasabi is also known as Japanese horseradish. In parallel, horseradish is sometimes known as Western wasabi in Japan.
- In Japan, wasabi is known as *seiyo*. When conventional horseradish was first introduced, it became known as *seiyo wasabi* because the pungency was reminiscent of wasabi. Products manufactured from Western horseradish, not genuine wasabi, are sometimes exported as "wasabi," which is misleading.
- The Japanese nickname of wasabi, *namida*, means tears, reflecting the strong pungent nature of the herb.
- *Japonica* in the scientific name *W. japonica* is Latin for Japanese.

PLANT PORTRAIT

Wasabi is a semi-aquatic perennial herb. Its native range is limited to regions of the Russian island of Sakhalin, north of Hokkaido, and to the major Japanese islands of Honshu, Shikoku, and Kyushu. The species grows wild in wet, cool mountain river valleys along stream beds and on river sand bars. A related, similar but smaller species, *W. tenuis* (Miq.) Matsum. (*Eutrema tenue* (Miq.) Makino), grows wild in Japan. Wasabi is typically not taller than 50 cm (20 in.). The leaves and leaf stalks are quite brittle and break easily. As the older leaves fall off, prominent scars are produced on the rhizome (underground or underwater stem, commonly called a "root"), giving a characteristic appearance. The rhizomes can be 2 to 5 cm (3/4–2 in.) thick and grow to a length of 5 to 40 cm (2–16 in.). Genuine "roots" that come off the rhizome are generally not sold commercially but are harvested where the plants are grown and used as a condiment. Wasabi can be cultivated on land and in water, but the latter is considered to produce a much higher quality product. Several cultivars are grown in Japan. The crop has been raised in Japan for more than a thousand years in cool streams or artificial water beds, much like watercress. The conditions for growing wasabi differ dramatically from the requirements of most crops and are quite demanding. Wasabi seeds and rhizomes that can be used to establish plants can be obtained from several commercial firms, but few people can provide the environment necessary for the plants to grow well. Requirements include considerable shade and cold well-oxygenated water. Commercial cultivation occurs outside of Japan in Taiwan, North Korea, New Zealand, the Pacific Northwest of the United States and southwestern British Columbia. Fresh wasabi fetches approximately $220/kg in Japan ($100 a pound) and approximately $100 kg ($45 a pound) in North America.

FIGURE 97.1 Wasabi (*Wasabia japonica*), by B. Flahey.

CULINARY PORTRAIT

In Japan, wasabi has been called "the king of the edible wild plants." It is one of the three most important condiments in Japan (the others are grated horseradish and mustard) and is a staple of the country's cuisine. Wasabi is primarily a Japanese condiment, mostly used to flavor sashimi (raw fish dishes), sushi fish dishes (sushi is cooked, seasoned rice, along with other ingredients), and soba (noodles). As sushi becomes more popular in Western countries, wasabi's popularity is increasing. In Japanese restaurants, wasabi is typically served ground and placed in the corner of the plate in a tiny cone. In addition to traditional Japanese dishes, wasabi can be used as a condiment for grilled and roasted meats and vegetables and can be added to salad dressings, marinades, and dips. A variety of wasabi food products are marketed. Wasabi wines and liqueurs are sold in some Japanese specialty stores as novelties.

In Japan, wasabi is often used fresh, but in other countries, it is generally available dried, as a pale green powder, or in the form of a green paste with a very strong smell and taste. Wasabi has a distinct flavor that many, including the Japanese, consider superior to common horseradish. The same chemical that produces the pungency of horseradish, sinigrin, is responsible for the pungency of wasabi.

Wasabi is such an expensive condiment that the genuine article is rarely available. True wasabi is almost never offered in restaurants outside of Japan, and even in that country it has been found that only about 5% of restaurants provide the real thing. Indeed, powdered horseradish is used to adulterate wasabi. Such fraudulent substitution is inevitable because of the limited supply and great cost. Generally, the dab of pale-green paste served with sushi is imitation wasabi, usually a combination of horseradish, mustard, cornstarch, and artificial food coloring. An easy way to tell if genuine wasabi is being served is based on the fact that the isothiocyanate, which produces the sensation of heat, dissipates rapidly, and a fresh lump on the plate will lose most of its heat within

15 minutes. In a high-end restaurant that serves genuine wasabi, a waiter may be expected to come to the table occasionally to freshen up the condiment. Only enough material that can be consumed within approximately 15 minutes should be grated.

Wasabi powder is sold in shops specializing in Oriental foods, but the fresh rhizome is superior, although the flavor deteriorates rapidly after cutting. Occasionally, fresh wasabi is available but is quite expensive. Fresh, unshriveled rhizomes should be chosen, and these can be stored in damp towels in the refrigerator for up to 30 days. They may be rinsed in cold water every few days, and spoiled areas trimmed away when necessary. To use, first cut the rhizome just below the leaf bases, trim away bumps or rough areas, scrub with a stiff brush, peel with a knife, and grate. Wasabi is prepared by grating the fresh rhizome against a rough surface in much the same way that horseradish is prepared. Some Japanese sushi chefs will only use a sharkskin grater, which produces ground wasabi with a smooth, soft, and aromatic finish. (After use, a sharkskin grater should be rinsed under cold running water and left to air dry.) The rhizome is best grated with a circular motion. Holding the rhizome at a 90° angle to the grater is thought to produce an ideal size of particle and to minimize the root surface that is exposed to the air (if ground too finely, the flavor dissipates too quickly). Although a sharkskin grater is ideal, a ceramic grater with fine nubs is considered a good alternative. A stainless steel grater may also be used, using a side where the spikes are small. The grated wasabi is piled into a ball and allowed to stand at room temperature for 5 to 10 minutes to allow flavor and heat to develop. (The flavor will dissipate notably after approximately 15 minutes and is of limited desirability after approximately 4 hours.)

Wasabi powder normally contains ground yellow mustard to improve the pungency. To reconstitute wasabi powder, combine one part powder and three-fourth part water and use soon. The pungent taste does not develop until wasabi powder has been combined with water for several minutes. Dry wasabi powder tastes bitter. "Wasabi paste" in tubes is popular in Japan and typically actually contains some genuine wasabi, but mostly horseradish.

In Japan, wasabi leaves (especially the leaf stalks) are sometimes pickled fresh in sake brine or soy sauce, and dried leaves are used to flavor foods such as salad dressing, soups, cheese, and crackers. The leaves may become available in regions of the world that take up wasabi cultivation. Where they can be obtained, the same guidelines used for selecting salad greens can be used: choose those that look fresh, with no sogginess or wilting and with a uniform color. Wasabi leaves can be stored in a refrigerator but should be used as soon as possible.

There have been a few, isolated reports in the United States of serious adverse reactions to wasabi, involving paleness, confusion, profuse sweating, and collapse after eating a large serving; these reports suggested the response may be serious for those with weakened blood vessels in the heart or brain. However, it is unclear that genuine wasabi was involved, and the use of this condiment is not considered to be a threat to the health of most people.

Culinary Vocabulary

- An "Angry Red Planet" is a Bloody Mary cocktail with wasabi added.
- Consumers should not be misled by the Japanese words *seiyo-wasabi* ("western horseradish"), *kona wasabi* ("horseradish powder"), and *wasabi daikon* ("radish wasabi"), which are preparations made from common horseradish, not genuine wasabi.

CURIOSITIES OF SCIENCE AND TECHNOLOGY

- Medicinal values of chemicals extracted from wasabi were first documented in a tenth century Japanese medical encyclopedia. Natural chemicals (isothiocyanates) in wasabi may kill microbes responsible for food poisoning, a factor that might have led to the widespread use of the condiment with raw fish dishes in Japan.

- The isothiocyanates in wasabi may also have other medicinal properties. Research has suggested they may help treat or prevent blood clotting, asthma, and tooth decay. Isothiocyanates are also present in common horseradish, which can be expected to have the same effects.
- Roy Carver is the entrepreneur who first brought wasabi commercial cultivation and production to the United States. In 1991, Carver and a team of engineers and scientists set out to recreate in Florence, Oregon, the growing conditions demanded by wasabi. The Japanese producers refused to share critical crop information with Carver, but he nevertheless was able to acquire the necessary expertise. Carver's team created a sophisticated irrigation system that pumps 114,000 L (3000 gallons) of water a minute through beds of round river rock in shaded greenhouses in which the wasabi plants now grow. The farm established in Oregon appears to be the largest of its type in the world.
- One of the reasons that has been advanced for the evolution of sex in living things is that it provides a mechanism for escaping diseases. In nature, it is very difficult to become free of a disease that has become established in the body of a plant or animal, but reproductive cells often are not infected, and when a disease-free sperm unites with a similarly uninfected egg, the new organism starts off healthy. Wasabi is easily grown from cuttings but is known to rapidly accumulate diseases when propagated vegetatively, and the diseases cannot be eliminated from the cuttings. Several viruses in particular have proven difficult to control. In Japan, crops established from cuttings are rotated with crops established from disease-free seed.
- Many nerve cells (neurons) in the human body are specialized to react only to certain chemicals, producing unique sensations, such as pungency and burning. For example, the distinctive taste stimulations produced by wasabi and chili pepper are sensed by different populations of neurons. Curiously, tetrahydrocannabinol, the psychoactive component of marijuana, activates both the wasabi and chili pepper neurons.
- Wasabi is the basis of an experimental smoke alarm for the deaf. Smoke triggers the device to spray wasabi vapor, and when placed near a sleeping subject the acrid sensation in the nasal passages has proven capable of quickly waking up people.
- In early March 2007, it was reported that a spill of wasabi had occurred on the International Space Station from a tube of the condiment being used by astronaut Sunita Williams. Space station crew members had been given special packs of their favorite foods to help endure their months in space. Unfortunately, under the weightless conditions of space, cleanup of spilled food is challenging, and it took a week to remove the wasabi.

KEY INFORMATION SOURCES

Chadwick, I., Lumpkin, T.A., and Elberson. L.R. 1993. The botany, uses and production of *Wasabia japonica* (Miq.) (Cruciferae). Matsum. *Econ. Bot.* 47:113–135.

Depree, J.A., and Savage, G.P. 1996. Storage properties of a wasabi-flavoured mayonnaise. *Proc. Nutr. Soc. N. Z.* 21:142–150.

Depree, J.A., Howard, T.M., and Savage, G.P. 1997. Wasabi—Japanese horseradish. *ASEAN Food J.* 12(1):33–42.

Depree, J.A., Howard, T.M., and Savage, G.P. 1999. Flavour and pharmaceutical properties of the volatile sulphur compounds of wasabi (*Wasabia japonica*). *Food Res. Int.* 31:329–337.

Etoh, H., Nishimura, A., Takasawa, R., Yagi, A., Saito, K., Sakata, K., et al. 1990. Omega-methylsulfinylalkyl isothiocyanates in wasabi, *Wasabia japonica* Matsum. *Agric. Biol. Chem.* 54:1587–1589.

Follett, J.M. 1987. Production of *Wasabia japonica* in Japan. *Proceedings of the International Plant Propagator's Society*, 36:443–447.

Hara, M., Mochizuki, K., Kaneko, S., Iiyama, T., Ina, T., Etoh, H., and Kuboi, T. 2003. Changes in pungent components of two *Wasabia japonica* Matsum. cultivars during the cultivation period. *Food Sci. Technol. Res.* 9:288–291.

Hassan, M., Iyanaga, K., Fujime, Y., and Okuda, N. 1999. Sand-gravel hydroponic beds for wasabi. *Thai J. Agric. Sci.* 32:253–261.

Hodge, W.H. 1974. Wasabi—native condiment plant of Japan. *Econ. Bot.* 28:118–129.

Hosokawa, K., Oikawa, Y., and Yamamura, S. 1999. Clonal propagation of *Wasabia japonica* by shoot tip culture. *Planta Med.* 65:676.

Kinae, N., Masuda, H., Shin, I.S., Furugori, M., and Shimoi, K. 2000. Functional properties of wasabi and horseradish. *Biofactors*, 13:265–269.

Kumagai, H., Kashima, N., Seki, T., Sakurai, H., Ishii, K., and Ariga, T. 1994. Analysis of volatile components in essential oil of upland wasabi and their inhibitory effects on platelet aggregation. *Biosci. Biotech. Biochem.* 58:2131–2135.

Martin, R.J., and Deo, B. 2000. Preliminary assessment of the performance of soil-grown wasabi (*Wasabia japonica* (Miq.) Matsum.) in New Zealand conditions. *N. Z. J. Crop Hortic. Sci.* 28:45–51.

Masuda, H., Inoue, T., and Kobayashi, Y. 2003. Anticaries effect of wasabi components. In *Oriental foods and herbs: chemistry and health effects. Proceedings of a symposium, Orlando, April 2002*. Edited by C.T. Ho, J.K. Lin, and Q.Y. Zheng. American Chemical Society, Washington, DC. pp. 142–153.

Masuda, H., Naohide, K., Woo, G.J., and Shin, I.S. 2004. Inhibitory effects of gochoonangi (*Wasabia japonica*) against *Helicobacter pylori* and its urease activity. *Food Sci. Biotech.* 13:191–196.

Masuda, H., Harada., Y., Inoue, T., Kishimoto, N., and Tano, T. 1999.Wasabi, Japanese horseradish, and horseradish: relationship between stability and antimicrobial properties of their isothiocyanates. In *Flavor chemistry of ethnic foods. Proceedings of a meeting held during the Fifth Chemical Congress of North America, Cancun, November 1997*. Edited by F. Shahidi and C.T. Ho. Kluwer Academic Publishers, New York. pp. 85–96.

Masuda, H., Harada, Y., Tanaka, K., Nakajima, M., and Tabeta, H. 1996. Characteristic odorants of wasabi (*Wasabia japonica* matum), Japanese horseradish, in comparison with those of horseradish (*Armoracia rusticana*). In *Biotechnology for improved foods and flavors*. Edited by G.R. Takeoka, R. Teranishi, P.J. Williams, and A. Kobayashi. American Chemical Society, Washington, DC. pp. 67–78.

Morimitsu, Y., Nakamura, Y., Osawa, T., and Uchida, K. 2002. Wasabi: a traditional Japanese food that contains an exceedingly potent glutathione S-transferase inducer for RL34 cells. In *Free radicals in food: chemistry, nutrition, and health effects*. Edited by M.J. Morello, F. Shahidi, and C.T. Ho. American Chemical Society, Washington, DC. pp. 301–309.

Morimitsu, Y., Hayashi, K., Nakagawa, Y., Horio, F., Uchida, K., and Osawa, T. 2000. Antiplatelet and anticancer isothiocyanates in Japanese domestic horseradish, wasabi. *Biofactors*, 13:271–276.

Ohi, M., Isoda, H., and Ohsawa, K. 1994. Production of intergeneric hybrids of *Eutrema wasabi* Maxim. and *Armoracia rusticana* ph. Gaerth., B. Mey. et Scherb. by ovule culture. *J. Jpn. Soc. Hortic. Sci.* 63:603–610.

Palmer, J. 1990. Germination and growth of wasabi *Wasabia japonica* Miq. Matsumara. *N. Z. J. Crop Hortic. Sci.* 18:161–164.

Potts, S.E., and Lumpkin, T.A. 1997. Cryopreservation of *Wasabia* spp. seeds. *Cryo. Lett.* 18:185–190.

Shin, I.S., Masuda, H., and Naohide, K. 2004. Bactericidal activity of wasabi (*Wasabia japonica*) against *Helicobacter pylori*. *Int. J. Food Microbiol.* 94:255–261.

Small, E., and Catling, P.M. 2004. Blossoming treasures of biodiversity:11. Wasabi (*Wasabia japonica*)—a hot crop from cold mountains. *Biodiversity*, 5(2):29–32.

Sultana, T., McNeil, D.L., Porter, N.G., and Savage, G.P. 2003. Investigation of isothiocyanate yield from flowering and non-flowering tissues of wasabi grown in a flooded system. *J. Food Comp. Anal.* 16:637–646.

Sultana, T., Porter, N.G., Savage, G.P., and McNeil, D.L. 2003. Comparison of isothiocyanate yield from wasabi rhizome tissues grown in soil or water. *J. Agric. Food Chem.* 51:3586–3591.

Sultana, T., Savage, G.P., McNeil, D.L., Porter, N.G., Martin, R.J., and Deo, B. 2002. Effects of fertilisation on the allyl isothiocyanate profile of above-ground tissues of New Zealand-grown wasabi. *J. Sci. Food Agric.* 82:1477–1482.

Suzuki, T., and Yamaguchi, M. 2003. Stimulatory effect of wasabi leafstalk extract (*Wasabia japonica* Matsum.) on bone calcification: interaction with bone anabolic factors in mouse calvaria tissue in vitro. *Food Sci. Technol. Res.* 9:87–90.

Taniguchi, M., Nomura, R., Kamihira, M., Kijima, I., and Kobayashi, T. 1988. Effective utilization of horseradish and wasabi by treatment with supercritical carbon dioxide. *J. Ferment. Bioeng.* 66:347–353.

Yamaguchi, M., Ma, Z.J., and Suzuki, T. 2003. Anabolic effect of wasabi leafstalk (*Wasabia japonica* Matsum.) extract on bone components in the femoral-diaphyseal and -metaphyseal tissues of aged female rats in vitro and in vivo. *J. Health Sci.* 49:123–128.

Yu, E.Y., Pickering, I.J., George, G.N., and Prince, R.C. 2001. In situ observation of the generation of isothiocyanates from sinigrin in horseradish and wasabi. *Biochim. Biophys. Acta.* 1527(3):156–160.

Specialty Cookbooks

Greene, M. 1969. *Recipes: the cooking of Japan*. Time-Life Books, New York. 120 pp.

Kobayashi, K. 2000. *The quick and easy Japanese cookbook: great recipes from Japan's favorite TV cooking show host*. Kodansha, Tokyo, Japan. 104 pp.

Kazuko, E., and Fukuoka, Y. 2007. *Sushi and traditional Japanese cooking*. Lorenz, London. 224 pp.

Kosaki, T., Wagner, W., and Hutton, W. 1995. *The food of Japan: authentic recipes from the land of the rising sun*. Periplus Editions, Singapore. 132 pp.

98 Water Chestnut

This chapter features

Chinese water chestnut (*Eleocharis dulcis*)
European water chestnut (*Trapa natans*)

As presented in the following, two unrelated plants used as food are called water chestnut. A quite different plant from the species treated in this chapter is also called "water chestnut." This is *Pachira aquatica* Aubl. of the Bombacaeae (bombax family), a native tree of Mexico and northern South America. It is also called Malabar chestnut, Guiana chestnut, and provision. This water-loving tree produces large fruits that are roasted and taste somewhat like chestnuts.

CHINESE WATER CHESTNUT

Family: Cyperaceae (sedge family)

NAMES

Scientific Name: *Eleocharis dulcis* (Burm. f.) Trin. ex Hensch. (*E. tuberosa* (Roxb.) Schult., *Scirpus tuberosus* Roxb.)

- The chestnut brown skin color together with the "chestnutty" flavor and texture of the white flesh gave rise to the name "water chestnut." "Water nut" is a similar name.
- The Cantonese name *matai* for the Chinese water chestnut means "horse's hoof," which is much like the shape of the dark brown, flat-bottomed corms (a term explained below). Occasionally, the name horse's hoof has been used in English. Other foreign language names include *hon matai*, *kweilin matai*, *pi chi*, *pi tsi*, *sui matai*, and *kuro-kuwai*.
- The genus name *Eleocharis* is based on the Greek *elos*, a marsh, and *charis*, grace, indicating that the plants are graceful and live in marshes.
- *Dulcis* in the scientific name *E. dulcis* is Latin for sweet, pleasant.

PLANT PORTRAIT

The Chinese water chestnut, which is the most important water chestnut from a culinary viewpoint, is a rush-like plant native to Southeast Asia. It grows to a height of 1 to 1.5 m (3–5 ft.). The plant spreads by creeping rhizomes (underground stems), which by suckering produce additional plants through the summer months. Possibly the Chinese water chestnut has been cultivated for as long as 10,000 years based on carbon dating of seeds that were found by archaeologists in Thailand. The plants are cultivated for their round (but vertically compressed) turnip-shaped tubers that look like the corms of gladioli and are used as a vegetable. The vegetable is not a nut (which is a type of hard fruit); it is a corm (a swollen portion of stem) which grows underground in water. The crop is grown in ponds, much like rice, and the corms (often called tubers) are harvested by scooping them off the bottom with forks. The water chestnut's brownish-black skin resembles that of a true chestnut (*Castanea* species), but its flesh is white and juicy. Chinese water chestnut is grown extensively in China, and most of the supply consumed in Western nations is imported from China. Commercial crops of water chestnut are also grown in Japan, Taiwan, Thailand, and Australia. Very limited cultivation has been carried out in Florida, California, and Hawaii.

Culinary Portrait

Chinese water chestnuts are eaten raw or cooked. Raw water chestnuts are often served as an appetizer or consumed out of hand as a snack. The flavor is bland with a hint of sweetness, and the whitish flesh is firm, crisp, sweet, juicy, and fragrant. The nutty flavor has been compared with coconut, apple, and macadamia nuts. Cooked water chestnut has been likened in taste to cooked corn. The vegetable retains its crunchiness and crispness even after prolonged cooking. Water chestnuts are very popular in Asian cooking, especially in stir-fried dishes where their crunchy texture is valued. They can be combined with vegetables such as bamboo shoots and snow peas as well as soy sauce and other seasonings to make a stir-fry; added in chopped form to soups, salads, rice, and stuffings; and wrapped whole with bacon and baked and served as an appetizer or as a side dish. The fresh corms can be peeled with the fingers and eaten like a fresh fruit. Chinese water chestnuts are available fresh in most Chinese markets. Canned water chestnuts are available, either whole or sliced, from many supermarkets, but the fresh are far superior.

When purchasing fresh Chinese water chestnuts, it is best to choose those that are firm, with no sign of shriveling. If not used immediately, they should be refrigerated, tightly wrapped in a plastic bag, and stored for no longer than a week. Some authorities recommend keeping fresh Chinese

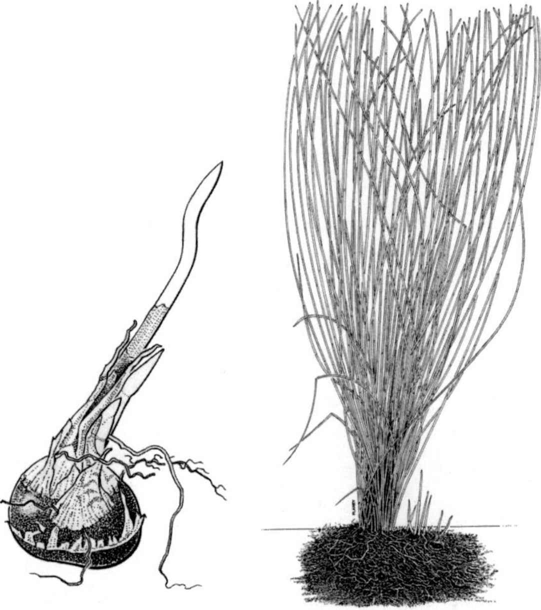

FIGURE 98.1 Chinese water chestnut (*Eleocharis dulcis*), by B. Flahey. An edible corm ("tuber"), which has developed a young stem and roots, is shown at the left. A mature clump of shoots is at the right.

water chestnuts in a bowl of water, which should be changed every day to keep the product in good condition while being stored. The water chestnuts should be thoroughly washed, then peeled before use by first cutting off the top and bottom and then removing the remaining skin. To prevent peeled water chestnuts from discoloring, they can be immersed in acidulated water, and a little lemon juice can be added to the cooking water.

Water chestnut powder, also called water chestnut flour, is a ground form of dried water chestnuts, like powdered starch. This is used as a thickener in Asian cooking, for example, to thicken sauces and also to provide a shiny glaze and a crispy coating to deep-fried foods. Like cornstarch, water chestnut powder is mixed with a small amount of water before being added to a hot mixture to be thickened. Water chestnut powder is available in Asian markets and in some health food stores but is more expensive than most other starches.

Culinary Vocabulary

- *Sub gum* is a base for many Chinese dishes, consisting of water chestnuts, bamboo shoots, and fresh mushrooms.
- *Mah tai goh* (pronounce mah tie go), a Chinese snack or dessert, is a pan-fried chunk of sweet water chestnut pudding cake.

Curiosities of Science and Technology

- Chinese herbal medicine is based on the concept of *yin* and *yang* forces of Daoist herbal theory. Yang represents masculinity, strength, and heat, and yin by contrast is feminine, mild, and cold. The Chinese water chestnut is considered yin, or cooling, for medical applications.
- In China, a paste made of Chinese water chestnuts is reputed to help children pass accidentally swallowed coins.

Key Information Sources

Aqua-nut (Firm). 1996. *Growing Chinese water chestnuts*. 3rd ed. Aqua-nut, Lauriston, Vic., Australia. 48 pp.

Brecht, J.K., Bergsma, K.A., Sanchez, C.A., and Snyder, G.H. 1992. Harvest maturity and storage temperature effects on quality of Chinese water chestnuts (*Eleocharis dulcis*). In *Second International Symposium on Specialty and Exotic Vegetable Crops*. Edited by D.N. Maynard. International Society for Horticultural Science, Wageningen, the Netherlands. pp. 313–319.

Diver, S. 2000. *Chinese water chestnut*. Appropriate Technology Transfer for Rural Areas (Organization), Fayetteville, AR.

Hodge, W.H. 1956. Chinese water chestnut or matai—a paddy crop of China. *Econ. Bot*. 10:49–65.

Kanes, C.A., and Vines, H.M. 1977. Storage conditions for Chinese water chestnuts, *Eleocharis dulcis* (Burm f.) Trin. *Acta Hortic*. 62:151–160.

Kays, S.J., and Sanchez, M.G.C. 1985. Storage of Chinese water chestnut (*Eleocharis dulcis* (Burm. f.) Trin.) corms. In *Postharvest handling of vegetables*. Edited by C.T. Phan. International Society for Horticultural Science, Wageningen, the Netherlands. pp. 149–159.

Leeper, G.F., and Williams, A.K. 1976. Peeling of Chinese waterchestnuts. *J. Food Sci*. 41:86–88.

Li, M., and Midmore, D.J. 1999. Estimating the genetic relationships of Chinese water chestnut (*Eleocharis dulcis* (Burm. f.) Hensch) cultivated in Australia, using random amplified polymorphic DNAs (RAPDs). *J. Hortic. Sci. Biotech*. 74:224–231.

Li, M., Kleinhenz, V., Lyall, T., and Midmore, D.J. 2000. Response of Chinese water chestnut (*Eleocharis dulcis* (Burm. f.) Hensch) to photoperiod. *J. Hortic. Sci. Biotech*. 75:72–78.

Mazumdar, B.C. 1985. Water chestnut—the aquatic fruit. *World Crops*, 37:42–44.

McGeachin, R.B., and Stickney, R.R. 1979. *Culture of Chinese waterchestnuts in the southeastern United States*. Proceedings of the 33rd Annual Conference Southeastern Association of Fish and Wildlife Agencies, Nashville, TN. pp. 606–610.

Mercado, B.T., Osotsapar, Y., Malabayabas, C.A., and Dilig, A.D. 1976. *The botany and culture of* Eleocharis dulcis *Trin*. Terminal report. Laguna College, Laguna, Philippines. 38 pp.

Midmore, D.J., and Cahill, G. 1998. *The Chinese waterchestnut industry: a situation analysis and industry strategy: a report for the Rural Industries Research & Development Corporation*. Rural Industries Research and Development Corp., Barton, A.C.T., Australia. 34 pp.

Osotsapar, Y., and Mercado, B.T. 1976. Morphology and anatomy of *Eleocharis dulcis* (Burm. f.) Trin. *Kalikasan*, 5:332–340.

Paisooksantivatana, Y. 1996. *Eleocharis dulcis* (Burm. f.) Trinius ex Henschel. In *Plant resources of South-East Asia. Vol. 9. Plants yielding non-seed carbohydrates*. Edited by M. Flach and F. Rumawas. Backhuys Publishers, Leiden, the Netherlands. pp. 97–100.

Pandey, V.N., and Srivastava, A.K. 1991. Yield and nutritional quality of leaf protein concentrate from *Eleocharis dulcis* Burm. f. Hensch. *Aquat. Bot.* 41:369–374.

Parker, M.L., and Waldron, K.W. 1995. Texture of Chinese water chestnut: involvement of cell wall phenolics. *J. Sci. Food Agric.* 68:337–346.

Parr, A.J., Waldron, K.W., Ng, A., and Parker, M.L. 1996. The wall-bound phenolics of Chinese water chestnut (*Eleocharis dulcis*). *J. Sci. Food Agric.* 71:501–507.

Peng, L., and Jiang, Y. 2004. Effects of heat treatment on the quality of fresh-cut Chinese water chestnut. *Int. J. Food Sci. Technol.* 39:143–148.

Preacher, J. 1983. Backyard water chestnuts. *Plants & Gardens*, 39(2):39–41.

Shiam, R., and Pratap, K. 1986. Ecological studies of Chinese water chestnut *Eleocharis dulcis*. *Acta Bot. Indica*, 14:77–82.

Shiam, R., Pratap, K., and Chand, G. 1986. A contribution to the ecological studies of *Eleocharis dulcis* Trin. *Ind. J. Forest. Dehra Dun*, 9:16–22.

Twigg, B.A., Stark, F.C., and Kramer, A. 1957. Cultural studies with matai. *Am. Soc. Hort. Sci. Proc.* 70:266–272.

Specialty Cookbooks

Note: Chinese water chestnuts are standard fare in Chinese food, but thanks to their crunchy texture, excellent cooking qualities, neutral taste, and universal availability (at least in cans), they have become regular components of numerous recipes in thousands of cookbooks representing a very wide variety of ethnic and Western cuisines. However, cookbooks dedicated to Chinese water chestnuts do not seem to be available.

EUROPEAN WATER CHESTNUT

Family: Trapaceae (water chestnut family)

Names

Scientific Name: *Trapa natans L.*

- Picturesque local names like "devil nut" and "death flower" reflect the general dislike for the European water chestnut. The plant has also been called "water caltrop" after the metal burs designed to puncture horses' feet in warfare (see below). "Water nut" is another occasional name.
- In the Old World, the European water chestnut is often known under the name "Jesuit's nut" because the seeds were often used in making rosaries.
- The European chestnut is native to Europe, Asia, and Africa and indeed is cultivated mainly in Asia; hence, the adjective "European" is only partly accurate. By contrast, the Chinese chestnut is an Asian native and is mainly a product of China; hence, its name is more appropriate.
- The genus name *Trapa* is an abridgement of *calcitrapa*, Latin for a foot trap or caltrop, so-named because of the spines on the fruit. A caltrop is a device with four metal points so

arranged that when any three are on the ground, the fourth projects upward as a hazard to the hoofs of horses or to pneumatic tires.

- *Natans* in the scientific name *T. natans* is Latin for "floating."

Plant Portrait

The European water chestnut is native to Europe, Asia, and Africa. It is an annual aquatic plant with long, cord-like, rarely branching stems that can attain lengths of up to 5 m (16 ft.). Its woody fruits (the "chestnuts") are nut-like, 5 to 10 cm (2–4 in.) wide, and are armed with four hard, barbed spines approximately 1.3 cm (1/2 in. long) that are sharp enough to penetrate shoe leather, and these can ruin a beach for recreation. (Some of the Asian edible species of *Trapa* have two rather than four spines on the fruits.) The European water chestnut is a pest in North America, like many introduced species that have spread wildly at the expense of more valued plants. It can wipe out the natural aquatic plants in an area, prevent nearly all recreational water use where it occurs, and create breeding grounds for mosquitoes. European water chestnut was first recorded in North America near Concord, Massachusetts, in 1859. Wild populations have since become established in many locations in the northeastern United States, and more recently in southeastern Canada. To help control its distribution, the sale of all species of *Trapa* is banned from most of the southern United States. Ironically, European water chestnut is an important food in the Old World. The fruits are rich in protein and were once commonly eaten. They were a staple in northern Europe 5000 to 10,000 years ago. Some types of water chestnuts of the genus *Trapa* are cultivated for food in northern India and Southeast Asia and are also used medicinally.

Culinary Portrait

The fruits of the European water chestnut can be eaten fresh, boiled, or roasted. The meat of the "nut" inside is floury, with a pleasant, slightly sweet taste. In Indochina, flour is made from the seeds and combined with sugar and honey to make a pastry. Because of possible toxicity, it is recommended that the fruits be boiled for an hour before consumption.

FIGURE 98.2 Harvesting European water chestnut (*Trapa natans*) in China during the nineteenth century, from Flores des serres (1845–1880, vol. 8).

FIGURE 98.3 European water chestnut (*Trapa natans*). (From Engler and Prantl, 1889–1915.) Note the difference in underwater and floating leaves.

FIGURE 98.4 European water chestnut (*Trapa natans*). (From Nicholson, 1885–1889.)

Culinary Vocabulary

- Although European water chestnut is not an important food source anywhere today, it is relatively significant in India. In eastern India, where it is widely cultivated in fresh water lakes, it is called *singhara* and *paniphal*. Dried, ground flour from the fruits is called *singhare ka atta* and is often consumed during religious rituals.

Curiosities of Science and Technology

- European water chestnut has two kinds of leaves: finely divided, featherlike submerged leaves, and undivided floating leaves at the surface of the water. Such "dimorphism" of foliage is fairly common in rooted aquatic plants and is adaptive. The divided underwater leaves allow water currents to flow easily through the foliage without providing resistance that could result in the plants being uprooted. The floating leaves give buoyancy to the plant and because they are closest to the sun and have a greater area exposed to the sun they provide considerable photosynthate for the plant.
- European water chestnut provides an interesting example of how a plant that is so damaging in some parts of the world that legislation has been enacted against it is considered so rare in other parts of the world that it is protected. *Trapa natans* has been introduced to North America and Australia, where it has become so invasive it has been listed as a noxious weed, requiring control measures. However, the plant is rare in most of Africa and regions of Europe (e.g., Belgium, France, Germany, Holland, Sweden), and in France and parts of Turkey, the species is protected by legislation.

Key Information Sources

Cozza, R., Galanti, G., Bitonti, M.B., and Innocenti, A.M. 1994. Effect of storage at low temperature on the germination of the waterchestnut (*Trapa natans* L.). *Phyton*, 34:315–320.

Daniel, P., Vajravelu, E., and Thiyagaraj, J.G. 1983. Considerations on *Trapa natans* from peninsular India. *J. Econ. Taxon. Bot.* 4:595–602.

Groth, A.T., Lovett-Doust, L., and Lovett-Doust, J. 1996. Population density and module demography in *Trapa natans* (Trapaceae), an annual, clonal aquatic macrophyte. *Am. J. Bot.* 83:1406–1415.

Hizukuri, S., Takeda, Y., Shitaozono, T., Abe, J., Ohtakara, A., Takeda, C., and Suzuki, A. 1988. Structure and properties of water chestnut (*Trapa natans* L. var. *bispinosa* Makino) starch. *Starch*, 40:165–171.

Kadono, Y., and Schneider, E.L. 1986. Floral biology of *Trapa natans* var. *japonica*. *Bot. Mag.* 99:435–439.

Karg, S. 2006. The water chestnut (*Trapa natans* L.) as a food resource during the 4th to 1st millennia BC at Lake Federsee, Bad Buchau (southern Germany). *Environ. Archaeol.* 11:125–130.

Kurihara, M., and Ikusima, I. 1991. The ecology of the seed in *Trapa natans* var. *japonica* in a eutrophic lake. *Vegetatio*, 97:117–124.

Menegus, F., Cattaruzza, L., Scaglioni, L., and Ragg, E. 1992. Effects of oxygen level on metabolism and development of seedlings of *Trapa natans* and two ecologically related species. *Physiol. Plant.* 86:168–172.

Paull, R.E. 2008. *Trapa natans*, water chestnut. In *The encyclopedia of fruit & nuts*. Edited by J. Janick and R.E. Paull. CABI, Wallingford, Oxfordshire, U.K. pp. 901–904.

Parekh, J., and Chanda, S. 2007. In vitro antimicrobial activity of *Trapa natans* L. fruit rind extracted in different solvents. *Afr. J. Biotech.* 6:766–770.

Shilov, M.P., and Mikhailova, T.N. 1970. Ecological and phytocenotic characteristics of the water chestnut (*Trapa natans* L.) in river water bodies of the Vladimir Province. *Ecology*, 5:409–413.

Taha, R.M. 2003. *Trapa natans* L. In *Plant resources of South-East Asia. Vol. 12(3). Medicinal and poisonous plants 3*. Edited by R.H.M.J. Lemmens and N. Bunyapraphatsara. Backhuys Publishers, Leiden, the Netherlands. pp. 405–406.

Tsuchiya, T., and Iwakuma, T. 1993. Growth and leaf life-span of a floating-leaved plant, *Trapa natans* L., as influenced by nitrogen flux. *Aquat. Bot.* 46:317–324.

Vuorela, I., and Aalto, M. 1982. Paleobotanical investigations at a neolithic dwelling site in southern Finland with special reference to *Trapa natans*. *Ann. Bot. Fennici*, 19:81–92.

Specialty Cookbooks

Note: European water chestnuts are used substantially in the same way as Chinese water chestnuts. Up until the late nineteenth century, European water chestnuts were fairly commonly used as food in Europe, but today use as food is largely confined to Asia. Cookbooks dedicated to European water chestnut are not available, and recipes are very rare in English but can be found in books dealing with wild plants harvested as food.

99 Wax Gourd

Family: Cucurbitaceae (gourd family)

NAMES

Scientific Name: *Benincasa hispida* (Thunb.) Cogn. (*B. cerifera* Savi)

- Wax gourd is also called ash gourd, ash pumpkin, Chinese fuzzy gourd, Chinese preserving melon, Chinese squash, Chinese vegetable marrow, Chinese watermelon, Chinese wax gourd, Chinese winter melon, Christmas melon, fuzzy melon, hairy cucumber, hairy gourd, hairy melon, jointed gourd, moqua wax melon, tallow gourd, ton-kwa, tunka, white gourd, white pumpkin, winter gourd, winter melon, and zit-kwa.
- The name cassabanana is sometimes mistakenly applied to the wax gourd (see Chapter 21).
- The name "winter melon" is also applied to certain "true melons" (*Cucumis melo* L.), including honeydew melons.
- The genus name *Benincasa* commemorates an Italian nobleman and patron of botany, Count Giuseppe Benincasa (died 1596).
- *Hispida* in the scientific name *B. hispida* means "rough-hairy," which is descriptive of the plant in general and the fruit in particular.

PLANT PORTRAIT

Wax gourd is a pumpkin-like, tendril-bearing, annual, herbaceous vine with large, soft, hairy leaves. The vines may grow to more than 6 m (20 ft.) in length. The plants are usually allowed to spread over the ground like pumpkin or winter squash but can also be grown upright on a trellis. The fruits are very distinctive: hairy and covered with a thick layer of white wax at maturity. There is amazing variation in size, shape, hairiness, and taste of fruits, especially in southern China. Fruits may be round, egg-shaped, or dumbbell-shaped (constricted in the center), as long as 1.2 m (4 ft.), and weighing from less than 2 kg (4½ lb.) to more than 35 kg (77 lb.). The immature fruits usually have greenish skin but are sometime purplish-blue. The flesh is solid white and contains cucumber-like seeds. The younger the fruit, the firmer the flesh. Generally, different varieties are harvested for immature fruits and for mature fruits. Small, solid, green fruit are considered particularly desirable.

The cultivated forms may have originated in eastern Asia or Southeast Asia. Wax gourd has been grown as a vegetable in China since the year 500. The plant is now cultivated throughout the Old World tropics and recently has been introduced in warmer parts of the New World. The wax gourd is grown commercially to some extent on Oriental vegetable farms in South Florida and elsewhere in North America. Although the wax gourd is esteemed in Asia, particularly China, India, The Philippines, Thailand, and Vietnam, it is scarcely known in North America and most of Europe. Nevertheless, it is often available in large Asian markets in Western World cities, such as in San Francisco.

CULINARY PORTRAIT

The culinary uses of wax gourd are varied. Both immature and mature fruits are picked and used raw like cucumber or more commonly cooked as a vegetable—boiled, braised, or stir-fried. The immature fruits are much tastier. The flesh has been characterized as bland in taste. It is a good addition to

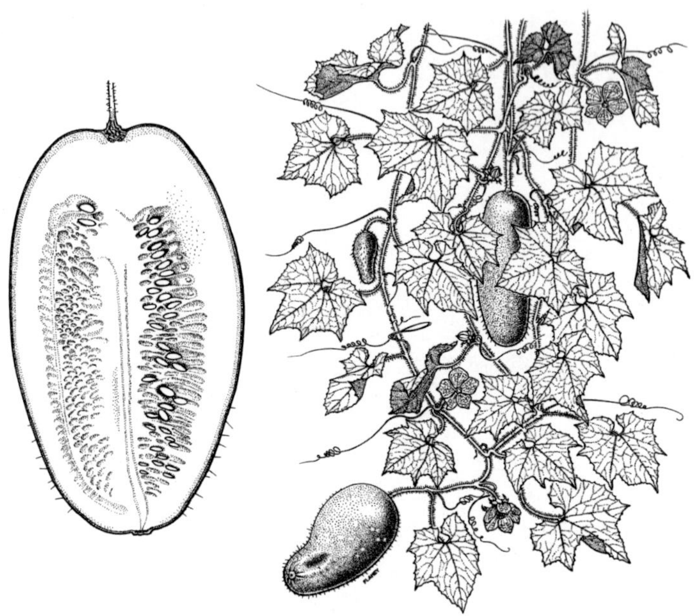

FIGURE 99.1 Wax gourd (*Benincasa hispida*), by B. Flahey.

soups and stews. The immature fruits of many varieties can be sliced and eaten raw, and indeed in Asian markets, slices of fresh wax gourd are often sold as a treat. The flesh of some varieties is used in curries or is sugar-coated and eaten as a sweet in India, Malaysia, and Cuba. In southern and eastern Asia, ripe fruits are used in making preserves and sweet pickles. In Asia, the seeds are sometimes fried or roasted and eaten like pumpkin seeds, and the young leaves and flower buds are occasionally cooked. Wax gourd will keep for more than 6 months if stored at 13 to 15°C (55–59°F) and 70% to 75% relative humidity. The flavor can change during storage, becoming acidic and less pleasant. While being cooked, wax gourd gives off an odor that has been likened to flatulence.

Culinary Vocabulary

- "Sweetheart cake" or "wife cake" is a traditional Chinese pastry made with wax gourd. According to legend, the cake symbolizes the successful search and reunification of a man for his wife, who had been sold into slavery.
- *Petha* is a soft, translucent candy popular in North India and Pakistan, made with wax gourd.
- "Winter melon pond soup" is a Chinese banquet specialty, prepared by using the gourd as a vessel in which a soup is prepared by several hours of steaming, and diners consume both the many ingredients in the soup along with the winter melon flesh.

CURIOSITIES OF SCIENCE AND TECHNOLOGY

- Wax gourd vines grow rapidly in warm weather. Some vines have been measured as averaging an extension of 60 cm (5 ft.) in a day.
- It was once believed that a fresh slice of wax gourd rubbed on the skin would instantly relieve prickly heat.
- Wax gourd fruit is covered by a white, chalky wax, and this is believed to deter decay microorganisms, promoting long shelf life of the fruit, which is sometimes stored for as long as a year without refrigeration.

- The whitish coating of wax from wax gourds has been used to make candles.
- In China, wax gourd is sometimes grown on the banks of village ponds. A bamboo framework is erected over the water, and the plants are encouraged to grow upon the frame so that the fruits hang over the water. This method ensures an abundance of water for the plants and saves land space.
- In China, globose or ovoid wax gourd fruits are hollowed out from one end and used as soup pots; they are filled with soup ingredients, capped with the cut lid of the gourd, and steamed, sometimes for up to 6 hours (see "Winter melon pond soup" in the Culinary vocabulary section).
- Some varieties of wax gourd fruits are more than 96% water.
- Wax gourd roots are very resistant to various soil-borne diseases, and less-resistant crops, such as melons, are sometimes grafted onto the wax gourd roots.
- In Thailand, male elephants experiencing must (musth, a periodic state of frenzy of bull elephants usually connected with the rutting season) are almost always dealt with by feeding them wax gourds.
- Rennet is a complex of enzymes in the stomach of mammals that digests milk. Rennet is used to coagulate milk in cheese production. Rennets are obtained from slaughtered livestock animal stomachs, but because of insufficient supply, a number of plants, including the wax gourd, have been found to provide rennet substitutes for commercial production of cheese.

KEY INFORMATION SOURCES

Bates, D.M., Merrick, L.C., and Robinson, R.W. 1995. Minor cucurbits. In *Evolution of crop plants*. 2nd ed. Edited by J. Smartt and N.W. Simmonds. Longman Scientific & Technical, Burnt Mill, Harlow, Essex, U.K. pp. 105–111.

Dase, X. 2001. Production actuality and breeding trend of wax gourd. *Acta Agric.* 13(2):60–63 (in Chinese, English summary).

Eskin, N.A.M., and Landman, A.D. 1975. Study of milk clotting by an enzyme from ash gourd (*Benincasa cerifera*). *J. Food Sci.* 40:413–414.

Gorini, F. 1977. White gourd or wax gourd. Fruit vegetables. *Informatore di Ortoflorofrutticoltura*, 18(2):6–8 (in Italian).

Hamid, M.M., Sana, M.C., and Begum, R.A. 1989. Physio-morphology and yield of different ash gourd (*Benincasa hispida* Cong.) lines. *Bangladesh J. Agric.* 14:51–55.

Kamalanathan, S. 1972. Studies on sex expression and sex ratio in ash-gourd (*Benincasa hispida*). *Madras Agric. J.* 59:486–495.

Marr, K.L., and Mei, X.Y. 2001. *Benincasa hispida* (Cucurbitaceae) the "pumpkin" of Asian creation stories? *Econ. Bot.* 55:575–577.

Maynard, D.N., and Paris, H.S. 2000. *Benincasa hispida*, wax gourd. In *The encyclopedia of fruit & nuts*. Edited by J. Janick and R.E. Paull. CABI, Wallingford, Oxfordshire, U.K. pp. 276–278.

Mini, C., Joseph, P.M., and Rajan, S. 2000. Effect of fruit size on seed quality of ash gourd (*Benincasa hispida*). *Seed Res.* 28:215–216.

Morton, J.F. 1971. The wax gourd, a year-round Florida vegetable with unusual keeping quality. *Proc. Fla. State Hortic. Soc.* 84:104–109.

Myers, C. 1991. Wax gourd, ash gourd, winter melon, Christmas melon. Crop Sheet SMC-036. In *Specialty and minor crops handbook*. Edited by C. Myers. The Small Farm Center, Division of Agriculture and Natural Resources, University of California, Oakland, CA. 2 pp.

Pandey, S., Kumar, S., Rai, M., Mishra, U., and Singh, M. 2008. Assessment of genetic diversity in Indian ash gourd (*Benincasa hispida*) accessions using RAPD markers. In *Cucurbitaceae 2008, Proceedings of the IXth Eucarpia Meeting on Genetics and Breeding of Cucurbitaceae, INRA, Avigon, France, May 21–24, 2008*. Edited by M. Pitrat. Institut Nationale de la Recherche Agronomique, Avignon, France. pp. 59–70.

Peicong, G. 1984. Growth and fruiting in the wax gourd (*Benincasa hispida* Cogn.). III. The relation of the rate of fruiting, the size of fruit and the yield per unit area to the planting densities of the wax gourd. *J. South China Agric. Coll. Kuangchou*, 5(3):48–53 (in Chinese, English summary).

Raj, N.M., Prasanna, K.P., and Peter, K.V. 1993. Ash gourd, *Benincasa hispida* (Thunb.) Cogn. In *Genetic improvement of vegetable crops*. Edited by G. Kalloo and B.O. Bergh. Pergamon Press, New York. pp. 235–238.
Ramesh, M., Gayathri, V., Rao, A.V.N.A., Prabhakar, M.C., and Rao. C.S. 1989. Pharmacological actions of fruit juice of *Benincasa hispida*. *Fitoterapia*, 60:241–248.
Randhawa, K.S., Singh, M., and Arora, S.K. 1982. Floral features of *Benincasa hispida* Thunb. and Cogn. with special reference to anthesis, dehiscence, stigma receptivity and pollen germination under different media. *Haryana J. Hortic. Sci.* 11:266–269.
Randhawa, K.S., Singh, M., Arora, S.K., and Singh, P. 1983. Varietal variation in physical characters and chemical constituents of ash-gourd fruits (*Benincasa hispida* Thunb. and Cogn.). *J. Res. Punjab Agric. Univ*. 20:251–254.
Ranote, P.S., Singh, M., Randhawa, K.S., Sekhon, K.S., and Arora, S.K. 1983. Suitability of different varieties of ash gourd for candy making. *J. Res. Punjab Agric. Univ. Ludhiana*, 20:132–134.
Rifai, M.A., and Reyes, M.E.C. 1993. *Benincasa hispida* (Thunberg ex Murray) Cogniaux. In *Plant resources of South-East Asia. Vol. 8. Vegetables*. Edited by J.E. Siemonsma and K. Piluek. Pudoc Scientific Publishers, Wageningen, the Netherlands. pp. 95–97.
Siddique, M.I., Ahmed, M., Awan, J.A., Ur-Rehman, S., and Ahmed, A. 1990. Production of wax gourd candy by using high fructose syrup. *J. Food Sci. Technol*. 27:205–208.
Vogel, G. 1993. Biographies of vegetable species. 12: *Benincasa hispida*. *Gartenbau Magazin*, 2(7):46–47 (in German).
Walters, T.W., and Decker-Walters, D.S. 1989. Systematic re-evaluation of *Benincasa hispida* (Cucurbitaceae). *Econ. Bot*. 41:274–278.
Wu, C.M., Liou, S.E., Chang, Y.H., and Chiang, W. 1987. Volatile compounds of the wax gourd *Benincasa hispida* Cogn. and a wax gourd beverage. *J. Food Sci*. 52:132–134.

Specialty Cookbooks

Creasy, R. 2000. *The edible Asian garden*. Periplus, Singapore. 108 pp.
Srisawat, P., and Kongpon, S. 1998. *The elegant taste of Thailand: Cha Am cuisine*. 2nd ed. SLG Books, Berkeley, CA. 224 pp.
Suzuki, T. 2000. *Japanese homestyle cooking*. Japan Publications Trading, Tokyo, Japan. 160 pp.
Xayavong, D., and Bear, J. 2000. *Taste of Laos: Lao/Thai recipes from Dara restaurant*. SLG Books, Berkeley, CA. 128 pp.

100 Wonderberry and Garden Huckleberry

Family: Solanaceae (potato family)

NAMES

Scientific Names: *Solanum* Species

- Wonderberry—*S. retroflexum* Dunal (*S. ×burbankii* Bitter).
- Garden huckleberry—*S. scabrum* Mill. (*S. melanocerasum* All., *S. intrusum* Soria).

- Wonderberry is also known as sunberry. The origin of these names is discussed in the next section. There are many African names for wonderberry, including gsoba and msoba.
- There is a variety called Wonderberry of the ornamental shrub *Pyracantha koidzumii* (Hayata) Rehder (*Pyracantha* species are known as firethorn). There is also a selection of illicit marijuana plant known as wonderberry. The acai berry (see Chapter 3) has also, in recent times, been referred to as "wonder berry."
- The garden huckleberry should not be confused with the plants usually known as huckleberries (species of *Vaccinium* and *Gaylussacia*). The expression "garden huckleberry" more often than not is applied to huckleberries, not to *S. scabrum*.
- Confusingly, some nurseries distribute the garden huckleberry under the name wonderberry.
- For information on the genus name *Solanum*, see Chapter 75.
- *Retroflexum* in the scientific name *S. retroflexum* is based on the Latin *retro*, backwards + *flexum*, bent; the sepals on the fruits are often reflexed, and the stalk of the fruit is often recurved.
- *Scabrum* in the scientific name *S. scabrum* is Latin for rough or gritty to the touch, referring to the hairiness of the leaves (actually rather limited compared with related species).

PLANT PORTRAIT

The wonderberry is thought to be native to Africa, occurring throughout the continent. It has been introduced as a weed to some parts of Australia and North America. The species is an annual or short-lived perennial. The plant is a hairy herb growing as tall as 80 cm (31 in.). The petals of the small flowers are white, with a yellow to green base, and usually there is a distinctive purple central strip on each petal. The berries are purple (black fruited-forms are sometimes included in the species), 7 to 9 mm (approximately 1/3 in.) in diameter. In Africa, the fruits are eaten both raw and cooked, and the leaves are cooked. The American plant breeder and horticulturist Luther Burbank (1849–1926), introduced the species into the United States as the "sunberry" at the beginning of the twentieth century, and for some time it received acclaim for the preparation of pies, jams, and sauces. Burbank sold the sunberry to John Lewis Childs, who changed the name to wonderberry. A controversy raged for years in the American press regarding just what it was that Burbank had introduced, and his veracity was often questioned. Although Burbank claimed that his wonderberry was a hybrid between the garden huckleberry and *Solanum villosum* Mill., a weed introduced to North America from southern Europe, it has been demonstrated that this is incorrect.

FIGURE 100.1 Wonderberry (*Solanum retroflexum*). (From Symon, 1981, drawn by Maria Szent-Ivany. Reproduced with the permission of the Board of the Botanic Gardens & State Herbarium (Adelaide, South Australia).

FIGURE 100.2 Garden huckleberry (*Solanum scabrum*). (From Stevels, 1990.) Reproduced with the permission of Prof. Dr. L.J.G. van der Maesen, Wageningen University, National Herbarium of the Netherlands.

It was probably an accidental introduction to his garden. The intriguing story of the wonderberry is detailed in Charles Heiser's book (see Key Information Sources section).

The garden huckleberry is an annual, growing as tall as 1 m (approximately 1 yd.). Its flowers are similar to those of the wonderberry, and it has deep-purple berries 15 to 17 mm (approximately 1/2 in.) broad. The species is of obscure origin but is considered to be a native of Africa by many and is widely cultivated throughout that continent. It has been introduced as a cultivated garden plant to Europe, North America, Australia, and New Zealand and sometimes escapes from cultivation.

Closely related to the above species is a group of species called "black nightshades", (which includes *Solanum nigrum* L., known as the black, garden, or common nightshade, and several dozen other related species). These occur throughout the world, often as weeds. Most of these are frost-susceptible and die with the onset of freezing temperatures in temperate regions, but many will grow as perennials in warm climates. The species are frequently very difficult to distinguish and require study. Black nightshade species have a considerable reputation as being poisonous, and indeed the alkaloid solanine is often present in toxic concentrations, especially in the immature fruits. Boiling may destroy the toxin in the leaves. It may be that the development of poisonous levels of toxins in some of the species of black nightshades depends on climate, season, and soil type. In any event, it appears that at least some varieties of several of the species are regularly consumed as food without harmful effects and are sometimes cultivated in home gardens. They are not encountered in Western food markets nor are they cultivated commercially. Because of the possibility that they could pose weed problems (which they do in many regions) and because the fruits are inferior to the common commercial berries, there seems little likelihood that the black nightshade species will be developed as commercial fruits. However, the use of the leaves as a green vegetable is important in the diet of many developing nations, particularly in Africa and Southeast Asia. Black nightshades are also widely used as herbal medicines in developing nations and deserve study for the possibility of applications in Western medicine. Seeds can be obtained from garden catalogs advertised on the Web.

CULINARY PORTRAIT

In Africa and some other regions of the world, the berries of wonderberry and garden huckleberry are often eaten raw. In North America, their fruits are rarely encountered, and when consumed they are cooked in pies, jams, and preserves, and sometimes substituted for raisins in plum puddings. Garden huckleberry fruits are said to freeze well for use in pies made in winter. The fruits of the garden huckleberry have been used as a commercial source of food coloring for fruit juices and apple sauce. In Africa, the foliage is commonly consumed, and indeed the garden huckleberry is one of the most important leafy vegetables in West and especially Central Africa.

CURIOSITIES OF SCIENCE AND TECHNOLOGY

- Luther Burbank once offered a $10,000.00 prize to anyone who could disprove his account of the origin of the wonderberry (which in fact, as noted above, was erroneous). No one ever collected.
- In Zambia, the garden huckleberry is used by aboriginal groups as a source of ink.
- In Kenya, boiled leaves of the black nightshade species are sometimes consumed by pregnant women in the expectation that this will result in the birth of children with attractive dark eyes and smooth skin.
- In Bohemia (an ancient kingdom, now in the Czech Republic), black nightshade leaves were placed in the cradles of infants in the belief that this promoted sleep.
- Although many of the black nightshade species are self-pollinating, some are cross-pollinated by insects. Some of the species are adapted to pollination by "buzz" or vibratile

pollinators (certain species of bees and Syrphid flies). These insects use a curious trick—they alight on the flowers and shiver their flight muscles (producing an audible buzz), causing the pollen to squirt out of the flowers (the insects consume some of the pollen, but enough is left to be transferred to other flowers for pollination).

- Black nightshade species grow well in fertile soils, especially those high in nitrogen. The Saxon name *mixplenton* (meaning plant of dung heap) was applied to the common black nightshade in Britain because it grows well on soils where animal manures have been deposited. This ability to grow well on manured soil was very important in the development of some agricultural crops historically. Such plants would have been noticed on the manured soils around early campsites where they would have thrived, people would have used these conveniently available plants for various purposes, and therefore they were likely among the first crops to have been domesticated.

KEY INFORMATION SOURCES

General

Bhiravamurty, P.V., and Rethy, P. 1883. Taximetric studies of the *Solanum nigrum* complex. *Proc. Indian Nat. Sci. Acad. B. Biol. Sci.* 49:661–666.

Edmonds, J.M. 1972. A synopsis of the taxonomy of *Solanum* sect. *Solanum* (*Maurella*) in South America. *Kew Bull.* 27:95–114.

Edmonds, J.M. 1977. Taxonomic studies on *Solanum* section *Solanum* (*Maurella*). *Bot. J. Linn. Soc.* 75:141–178.

Edmonds, J.M. 1979. Biosystematics of *Solanum* sect. *Solanum* (*Maurella*). In *The biology and taxonomy of the Solanaceae*. Edited by J.G. Hawkes, R.N. Lester, and A.D. Skelding. Academic Press, London. pp. 529–548.

Edmonds, J.M., and Chweya, J.A. 1997. *Black nightshades:* Solanum nigrum *L. and related species: promoting the conservation and use of underutilized and neglected crops*. International Plant Genetic Resources Institute, Rome, Italy. 113 pp.

Ganapathi, A. 1987. Phyletic relationship of some tetraploid taxa of the *Solanum* L. section *Solanum Maurella*. *Isr. J. Bot. Basic Appl. Plant Sci.* 36:31–39.

Heiser, C.B. 1969. *Nightshades the paradoxical plants*. W. H. Freeman, San Francisco, CA. 200 pp. (Republished as: Heiser, C.B. 1987. *The fascinating world of the nightshades: tobacco, mandrake, potato, tomato, pepper, eggplant, etc.* Dover Publications, New York. 200 pp.)

Jacoby, A., and Labuschagne, M.T. 2006. Hybridization studies of five species of the *Solanum nigrum* complex found in South Africa and two cocktail tomato cultivars. *Euphytica*, 149:303–307.

Rao, G.R., and Kumar, A. 1984. Meiotic studies in species-hybrids of the *Solanum nigrum* complex. *Cytologia*, 49:33–38.

Schilling, E.E., and Andersen, R.N. 1990. The black nightshades (*Solanum* section *Solanum*) of the Indian subcontinent. *Bot. J. Linn. Soc.* 102:253–259.

Siemonsma, J.S., and Jansen, P.C.M. 1993. *Solanum americanum* Miller. In *Plant resources of South-East Asia. Vol. 8. Vegetables.* Edited by J.E. Siemonsma and K. Piluek. Pudoc Scientific Publishers, Wageningen, the Netherlands. pp. 252–255.

Symon, D.E. 1981. A revision of the genus *Solanum* in Australia. *J. Adelaide Bot. Gard.* 4:1–367. [Pages 37–56 are especially relevant.]

Wonderberry (*Solanum retroflexum*)

Ganapathi, A., and Rao, G.R. 1980. Genetic system and interrelationship between *Solanum retroflexum* and *Solanum nodiflorum* of *Solanum nigrum* complex. *Curr. Sci.* 49:598–599.

Ganapathi, A., and Rao, G.R. 1986. The crossability and genetic relationship between *Solanum retroflexum* Dun. and *S. nigrum* L. *Cytologia*, 51:757–762.

Jacoby, A., Labuschagne, M.T., and Viljoen, C.D. 2003. Genetic relationships between Southern African *Solanum retroflexum* Dun. and other related species measured by morphological and DNA markers. *Euphytica*, 132:109–113.

Rao, G.R., and Kumar, A. 1981. Cytology of *Solanum nigrum* L., *Solanum retroflexum* Dunn. and their hybrids. *Proc. Plant Sci. India Acad. Sci. Bangalore*, 90:227–230.

Van Schalkwyk, A., and Berger, D.K. 2008. Genetic relationships between South African *Solanum retroflexum* and other related species using partial 18S sequencing. *S. Afr. J. Bot.* 74:391.

Garden Huckleberry (*Solanum scabrum*)

Adesina, S.K., and Gbile, Z.O. 1984. Steroidal constituents of *Solanum scabrum* subsp. *nigericum*. *Fitoterapia*, 6:362–363.

Apio Olet, E., Heun, M., and Lye, K.A. 2006. A new subspecies of *Solanum scabrum* Miller found in Uganda. *Novon*, 16:508–511.

Ganapathi, A., and Rao, G.R. 1987. Phylogenetic relationships in the evolution of *Solanum scabrum*. *Genome*, 29:639–642.

Ganapathi, A., and Rao, G.R. 1987. Cytogenetic relationship between *Solanum scabrum* Mill. and *S. americanum* Mill. *Cytologia*, 52:91–96.

Gbile, Z.O., and Adesina, S.K. 1985. Taxonomy and chemistry of Nigerian *Solanum scabrum*. *Fitoterapia*, 1:11–16.

Lehmann, C., Biela, C., Töpfl, S., Jansen, G., and Vögel, R. 2007. *Solanum scabrum*—a potential source of a coloring plant extract. *Euphytica*, 158:189–199.

Mikolas, V. 1991. *Solanum scabrum* Miller, new record, a new species in flora of Slovakia, Czechoslovakia. *Biologia*, 46:31–36 (in Czech).

Price, C., and Wrolstad, R.E. 1995. Anthocyanin pigments of Royal Okanogan huckleberry juice. *J. Food Sci.* 60:369–374. ['Royal Okanogan' was claimed to be a variety of *Vaccinium* huckleberry, but this paper proved that the juice supplied came from the garden huckleberry.]

Stevels, J.M.C. 1990. *Légumes traditionnels du Cameroun, un étude agro-botanique*. Wageningen Agricultural University Papers 90-1. Wageningen Agricultural University, Wageningen, the Netherlands. 262 pp (in French).

Specialty Cookbooks

Abukutsa-Onyango, M. O. 2009. African leafy Solanaceae vegetable (*Solanum scabrum*) in cream. *Sol Newsletter*, 22:7. [A recipe that uses leaves, not fruit.]

101 Yard-Long Bean

Family: Fabaceae (Leguminosae; pea family)

NAMES

Scientific Name: *Vigna unguiculata* (L.) Walp. subsp. *sesquipedalis* (L.) Verdc. (*V. sinensis* (L.) Savi ex Hassk. subsp. *sesquipedalis* (L.) Van Eselt.)

- The names yard-long bean, spaghetti pole, snake bean, garter bean (presumably for the garter snake), long-podded cowpea, and Chinese long bean (*dow guak* in China) obviously were coined to indicate the remarkable length of these beans.
- Yard-long bean is also known as asparagus bean, asparagus pea, pea bean, and Chinese pea, names that reflect the opinion that the taste is reminiscent of asparagus or pea. Most people do not think the taste is like asparagus.
- The name asparagus pea is also used for the winged bean, *Psophocarpus tetragonolobus* (L.) DC., an important bean plant of tropical regions; and also for the winged pea, *Lotus tetragonolobus* L. (*Tetragonolobus purpureus* Moench), a bean plant that is cultivated particularly in Europe.
- Still additional names for yard-long bean are Bodi bean, Chinese bean, and Peru bean.
- The name "Bodi bean" presumably reflects the Indian or African home of the species (perhaps based on one of Bodinayakkanur India, Bodi in Benin, or the Bodi people of Ethiopia).
- The genus name *Vigna* commemorates the Italian Dominico Vigna (?–1647), professor of botany and director of the botanical garden in Pisa.
- *Unguiculata* in the scientific name *V. unguiculata* is Latin for "with a small nail" or "with a small claw," a technical way of indicating that the petals of the flowers have small stalks.
- *Sesquipedalis* in the scientific name *V. unguiculata* subsp. *sesquipedalis* is from the Latin *sesqui*, one and a half, and *pes*, foot, that is, 1½ ft. long, a more modest reference to the length of the pods than "yard-long bean."

PLANT PORTRAIT

Yard-long bean originated either in India, or more probably Africa. It is usually a strong, climbing, annual vine growing up to 4 m (13 ft.) in length. Cultivated varieties that grow as bushes rather than vines are also available. The plants produce large yellow to violet-blue flowers, and pencil-thin pods 45 to 90 cm (1½–3 ft.) long, the pods usually hanging down in pairs. Each pod typically has 15 to 20 seeds, which are 8 to 12 mm (0.3–0.5 in.) long, black in African forms, brown in U.S. forms, and wrinkled. This plant of the warm tropics is cultivated mainly in the Far East, mostly in Bangladesh, India, Indonesia, Pakistan, and Philippines. It is also grown in the Caribbean, Africa, and parts of Europe. More recently, it has found its way to North America where it is now available through garden catalogs. Yard-long bean is an important vegetable in Southeast Asia, particularly with Chinese market gardeners. North American supermarkets frequently stock yard-long beans grown in California and Mexico. There are numerous varieties, including green- and white-podded forms. There is also a red-podded form, unusual in that the pods retain their red color when cooked. Although the yard-long bean can develop yard-long pods, the pods are usually picked when no more than half this length, when they taste best. However, in parts of China, the pod is allowed to mature until pea-sized seeds are produced.

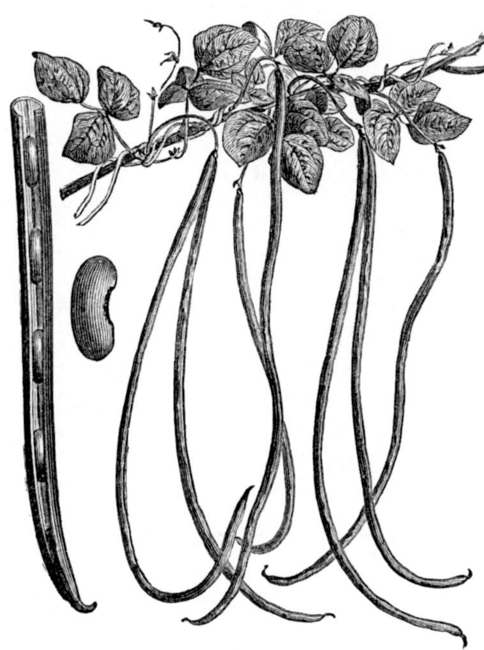

FIGURE 101.1 Yard-long bean (*Vigna unguiculata* subsp. *sesquipedalis*). (From Vilmorin-Andrieux, 1885.)

CULINARY PORTRAIT

Yard-long beans have a flavor similar to but not as sweet as that of a green bean and reminiscent of its cousin, the black-eyed pea (*Vigna unguiculata* (L.) Walp. subsp. *unguiculata*). Some people find the taste similar to asparagus, and the seeds have been said to taste like a cross of navy beans and asparagus. The texture of the pod is more pliable and not as crisp or sweet as that of a green bean. Nevertheless, yard-long beans are usually eaten fresh, like green beans. The young leaves and stem are also used as a vegetable, and the plant is employed as animal feed. Yard-long beans can be found in most Asian markets and some supermarkets with specialty produce sections.

Tips: Select those that are young (slender and very flexible), crisp, and blemish-free; the seeds should not have matured. Refrigerate in a plastic bag for up to 5 days. Wash just before using. Trim off and discard each end. Yard-long beans are most often cut into 5-cm (2-in.) lengths and sautéed or stir-fried, but the following methods are available: (1) Boil in 2.5 cm (1 in.) of water until tender-crisp (4 to 7 minutes for cut beans, 5 to 10 minutes for whole beans). (2) Steam for 10 to 15 minutes (whole beans) or 8 to 12 minutes for cut beans. (3) To microwave, cook 0.5 kg (1 lb.) of cut beans with 60 mL (1/4 cup) water, on high power for 7 to 12 minutes. (4) To stir-fry, add a tablespoon of salad oil to 2.5 cm (1 in.) pieces, let sit for 1 minute, then add 2 to 4 tablespoons liquid; cover and cook for 4 to 7 minutes until tender-crisp. Overcooking makes yard-long beans mushy. Generally, yard-long beans require more cooking than conventional string (green, French, or snap) beans.

Culinary Vocabulary

- The word "bean" is applied both to seeds and pods (which are fruits). The English word bean has been traced with certainty only within the confines of the English language but probably came from German. The word was spelled the same and had the same meaning a thousand years ago in English. A variety of seed-like products of similar size and shape are also termed beans, such as coffee beans and jelly beans. Seed pods are also often called beans. Many edible bean pods have been called "string beans" because stringy strands are present or were once present but have been removed by breeding, but the yard-long bean is naturally stringless. Edible bean pods have also been called "snap beans" because, at least when they are young and tender, the pods can be easily snapped or broken into pieces before cooking. Some yellow kinds are called "wax beans" based on the waxy appearance of the pods or "butter beans" based on the buttery color. Additional names applied to edible bean pods are French bean, green bean, and salad bean.

CURIOSITIES OF SCIENCE AND TECHNOLOGY

- In Asia, yard-long bean vines are often grown using corn (maize) plants as support. The practice of using corn to support bean vines of the genus *Phaseolus* was carried out for thousands of years by the Indians of North America, and it is intriguing that when corn was brought back to the Old World after Columbus's discovery of the New World, inhabitants of the Old World quickly perceived that they could also grow their own native bean vines on corn.
- In Asian folk medicine, the leaves of yard-long bean are sometimes boiled with rice and applied to the ear to treat earache.
- The yard-long bean twines counterclockwise, viewed growing toward the observer. Some species of vines climb by twisting counterclockwise, others clockwise, but the direction is usually consistent for each species.

KEY INFORMATION SOURCES

Allam, E.K., and Elshamy, M.R. 1973. Effect of virus infection on the quality of cowpea and asparagus bean seeds. *Seed Sci. Technol.* 1:825–832.

Carada, V.B., and Cueva, M.L. de la. 1994. Evaluation of the elite cultivars of yardlong bean in the Philippines. *Plant Industry Bull.* 9(1–4):43–58.

El-Shal, M.A., Mustafa, A.M., Abdelkader, M.M., and Said, M.S. 1977. Studies on identification and production of asparagus beans (*Vigna sesquipedalis* Fruw). *Mansoura J. Agric. Sci.* 2(1):218–225.

Grubben, G.J.H. 1993. *Vigna unguiculata* (L.) Walp. Cv. Group Sesquipedalis. In *Plant resources of South-East Asia. Vol. 8. Vegetables*. Edited by J.E. Siemonsma and K. Piluek. Pudoc Scientific Publishers, Wageningen, the Netherlands. pp. 274–278.

Ha, T.J., Lee, M.-H., Park, C.-H., Pae, S.-B., Shim, K.-B., Ko, J.-M., et al. 2010. Identification and characterization of anthocyanins in yard-long beans (*Vigna unguiculata* ssp. *sesquipedalis* L.) by high-performance liquid chromatography with diode array detection and electrospray ionization/mass spectrometry (HPLC–DAD–ESI/MS) analysis. *J. Agric. Food Chem.* 58:2571–2576.

Hegde, V.S., and Mishra, S.K. 2009. Landraces of cowpea, *Vigna unguiculata* (L.) Walp., as potential sources of genes for unique characters in breeding. *Genet. Res. Crop Evol.* 56:615–627.

Heij, G. 1989. Research experiences with yard-long bean, *Vigna sinensis* ssp. *sesquipedalis*, in glasshouses in the Netherlands. In *First international symposium on diversification of vegetable crops (Angers, France)*. Edited by J.Y. Peron and G.W.H. Welles. International Society for Horticultural Science, Wageningen, the Netherlands. pp. 305–311. [Also *Acta Hortic.* 242:305–312.]

Mak, C. 1987. Comparative performance of vegetable cowpea and long bean. *Malaysian Appl. Biol.* 216:323–325.

Mak, C., and Yap, T.C. 1977. Heterosis and combining ability of seed protein, yield, and yield components in [yard-]long bean. *Crop Sci.* 17:339–341.

Nangju, D. 1979. Effect of trellis planting and harvest methods on vegetable cowpea *Vigna unguiculata* ssp. *sesquipedalis*. *J. Am. Soc. Hortic. Sci.* 104:294–297.

Pandey, R.K., and Westphal, E. 1989. *Vigna unguiculata* (L.) Walp. In *Plant resources of South-East Asia. Vol. 1. Pulses*. Edited by L.J.G. van der Maesen and S. Somaatmadja. Pudoc, Leiden, the Netherlands. pp. 77–81.

Philippines University Los Banos, College, Laguna. Institute of Plant Breeding. 2000. Bush sitao [*Vigna unguiculata* subsp. *sesquipedalis*] production guide. *Agriculture*, 4(7).

Sinaga, R.M., and Marpaung, L. 1992. The effect of packaging system and storage temperature on quality and keepability of yardlong bean. *Buletin Penelitian Hortikultura*, 24(1):107–115 (in Indonesian).

Soedomo, R.P. 1992. Seed harvest technology of yardlong beans (*Vigna sesquipedalis* (L.) Fruhw). *Buletin Penelitian Hortikultura*, 24(1):16–28 (in Indonesian).

Thiraporn, P., Takano, T., and Choomsai, A. 1987. High temperature resistance of yard long bean (*Vigna unguiculata* L. Walp. sub. sp. *sesquipedalis* L. Verdc.) seeds. *Acta Hortic*. 215:95–104.

Van Horn, M., and Myers, C. 1991. Chinese long bean, yard-long bean, asparagus bean. Crop Sheet SMC-0010. In *Specialty and minor crops handbook*. Edited by C. Myers. The Small Farm Center, Division of Agriculture and Natural Resources, University of California, Oakland, CA. 4 pp.

Wang, S., Li, P., and Yao, Y. 199. Collection and utilization of snapbean and yardlong bean germplasm resources. *Crop Genet. Resour*. 1:12–14 (in Chinese).

Wei, S., and Xu, X. 1993. Yardlong bean. In *Proceedings of the FAO/UNDP project RAS/89/040 Workshop on underexploited and potential food legumes in Asia, Chiang Mai, Thailand, Oct. 31–Nov. 3, 1990*. Edited by C. Narong, C.L.L. Gowda, and L. Paisan. Food & Agriculture Organization, Regional Office for Asia and the Pacific, Bangkok (Thailand). [RAPA Publication (FAO). no. 1993/7; FAO Accession No: XF95:342245 (available on microfiche)]. pp. 116–119.

Zhang W., Wang, Y., and Lin, M, 1994. Genetic distance estimation and cluster analysis of some yardlong bean resources. *Acta Hortic. Sin*. 21(2):180–184 (in Chinese, English summary).

Specialty Cookbooks

Deseran, S. 2001. *Asian vegetables: from long beans to lemongrass, a simple guide to Asian produce, plus 66 delicious, easy recipes*. Chronicle Books, San Francisco, CA. 156 pp.

Green, A. 2000. *The bean bible: a legumaniacs guide to lentils, peas, and every edible bean on the planet!* Running Press, Philadelphia, PA. 320 pp.

102 Zedoary

Family: Zingiberaceae (ginger family)

NAMES

Scientific Name: *Curcuma zedoaria* (Christm.) Roscoe (*C. heyneana* Val., *C. mangga* Val. & Van Zijp., *C. pallida* Lour., *C. xanthorrhiza* Roxb., *C. zerumbet* Christm., *Amomum zedoaria* Christm.)

- The word "zedoary" is based on the Arabic *zadwaar*, from the Persian (Farsi) *zedwaar*, names for the spice.
- Zedoary is also called "white turmeric" and is in fact closely related to turmeric (see Chapter 95).
- Zedoary is sometimes called Indian arrowroot (for other plants called Indian arrowroot, see Chapter 7).
- It has been claimed that zedoary is the spice which has the most variant spellings of its name. The name has been traced from the original Arabic through such variations as zedoar, zeduale, citoval, setwall, cetewale, citouart, and dozens more. Because of all these variations, it has been said to be the likeliest candidate for "mystery spices" of old recipes.
- One of the most unpronounceable foreign names for a spice is the Tamil word for zedoary: *karppurakkiccilikkilangku*.
- An old German name for zedoary is *Christus-wurzel*, Christ's root.
- Because of the mango-like fragrance, zedoary is called *amb halad* in many Indian languages (*amb* means mango).
- "Yellow zedoary" is the occasional name of *C. aromatica* Salisb., a wild species of India, sometimes cultivated there for its rhizome, and used as a flavoring agent and condiment and the source of a yellow dye. This species is also sometimes encountered in Western countries as an ornamental.
- For information on the genus name *Curcuma*, see Chapter 95.

PLANT PORTRAIT

Zedoary is a herbaceous perennial herb growing as tall as 1.5 m (5 ft.). The flowering stalk emerges in the spring, directly from the rhizome (underground stem), before the leaves develop. The flowers of varieties used for the spice are white, pale yellow, or pinkish, but other colors have been developed in ornamental forms. The flowering stalk has colorful, red and green bracts. The leaf blades are up to 80 cm (31 in.) long. The rhizome (underground stem) is large and tuberous with many branches, and a thin brown skin, with hard, white, yellow, or orange flesh. The rhizome is the source of the spice zedoary, which has a smell similar to the odor of turmeric and mango. Zedoary is sold as a powder (*kentjur* in Chinese shops), or dried and sliced, showing up with a gray surface and a yellow to gray-white interior. There are two types of rhizomes sold in Indian markets—"round," which is small and stout like ginger, and "long," which is long and slender like turmeric.

Zedoary is thought to be native to northeastern India, and has been widely grown since antiquity. The dried rhizomes were an important item of trade in Asia as early as the sixth century. Zedoary

FIGURE 102.1 Zedoary (*Curcuma zedoaria*). (From Köhler, 1883–1914.)

was thought to be an effective general antidote by Arab physicians since this early time. During the Middle Ages, it was considered to be a digestive aid in Europe (indeed, modern research has found that it promotes digestive functions). Familiar in medieval Europe, the spice has become almost unknown in Western countries. It continues to be used as a medicine in China and Japan. Zedoary is extensively cultivated in India, Southeast Asia, and China, primarily for its medicinal properties. The distillation of essential oil from the rhizomes for use in perfumery was started in Europe in the Middle Ages. Essential oil extracted from the plant is still used to a minor extent in manufacturing perfume and cosmetics as well as for flavoring liquor.

CULINARY PORTRAIT

The spice zedoary, made from the dried, pulverized rhizomes, is used as a condiment, and to flavor bitters and liqueurs. Zedoary has a fragrant, camphor-like smell and a warm, bitter, aromatic taste which is somewhat reminiscent of ginger. The flavor has been said to be a cross between musky ginger and tangy green mango. Zedoary is extremely rare in the West today, having been displaced by the superior ginger. Because of the strong taste, it is not usually used as a spice alone but rather tends to be used in spice mixtures. Combined with turmeric and ginger, it makes a tasty spice paste for lamb and chicken curries. India and Indonesia are the primary centers of culinary use of zedoary. In India, zedoary is usually used fresh or pickled. In Indonesia, it is more usually employed as a dried spice, often in curry powder, especially for seafood dishes. In Thailand the young, starchy rhizome is eaten as a vegetable. Although not particularly pleasant in the mouth, the rhizomes are occasionally chewed in India and Java. The young shoots (vegetative and flowering) are sometimes eaten as a cooked vegetable or used in salads in Indonesia. The aromatic leaves have a lemony flavor and are employed in cookery, especially with fish and other foods in Java.

Culinary Vocabulary

- *Shoti* is a starch preparation based on extracts from the rhizomes, prepared and marketed in India. This resembles arrowroot and is used for thickening culinary dishes. Shoti is easily digested and nutritious, so it is widely used as part of an Eastern regimen for the sick or for the very young.

CURIOSITIES OF SCIENCE AND TECHNOLOGY

- In Chinese herbal medicine, yang represents masculinity, strength, and heat, and yin by contrast is feminine, mild, and cold. Food and spice species possess some degree of yin or yang, and there is a frequent concern to balance "cooling" and "heating" foods to maintain good health. Zedoary is considered to be mildly warm.

KEY INFORMATION SOURCES

Burtt, B.L. 1977. *Curcuma zedoaria. Gardens' Bull.* 30:59–62.

Fatima-Navarro, D. de, Souza, M.M. de, Neto, R.A., Golin, V., Niero, R., Yunes, R.A., et al. 2002. Phytochemical analysis and analgesic properties of *Curcuma zedoaria* grown in Brazil. *Phytomedicine*, 9:427–432.

Garg, S.N., Bansal, R.P., Gupta, M.M., and Kumar, S. 1999. Variation in the rhizome essential oil and curcumin contents and oil quality in the land races of turmeric *Curcuma longa* of North Indian plains. *Flavour Fragrance J.* 14:315–318.

Hong, C.H., Kim, Y., and Lee, S.K. 2001. Sesquiterpenoids from the rhizome of *Curcuma zedoaria. Arch. Pharm. Res.* 24:424–426.

Ibrahim, H., and Jansen, P.C.M. 1996. *Curcuma* Roxburgh. In *Plant resources of South-East Asia. Vol. 9. Plants yielding non-seed carbohydrates.* Edited by M. Flach and F. Rumawas. Backhuys Publishers, Leiden, the Netherlands. pp. 72–74.

Ibrahim, H., and Jansen, P.C.M. 1996. *Curcuma zedoaria* (Christmann) Roscoe. In *Plant resources of South-East Asia. Vol. 9. Plants yielding non-seed carbohydrates.* Edited by M. Flach and F. Rumawas. Backhuys Publishers, Leiden, the Netherlands. pp. 76–78.

Joshi, S., Singh, A.K., and Dhar, D.N. 1989. Isolation and structure elucidation of potential active principle of *Curcuma zedoaria* rhizomes. *Herba Hungarica*, 28:95–98.

Lai, E.Y.C., Chyau, C.C., Mau, J.L., Chen, C.C., Lai, Y., Shih, C.F., and Lin, L.L. 2004. Antimicrobial activity and cytotoxicity of the essential oil of *Curcuma zedoaria. Am. J. Clin. Med.* 32:281–290.

Leonel, M., Sarmento, S.B.S., and Cereda, M.P. 2003. New starches for the food industry: *Curcuma longa* and *Curcuma zedoaria. Carb. Polymers*, 54:385–388.

Leong-Škorničková, J., Šída, O., Sabu, M., and Marhold, K. 2008. Taxonomic and nomenclatural puzzles in Indian *Curcuma*: the identity and nomenclatural history of *C. zedoaria* (Christm.) Roscoe and *C. zerumbet* Roxb. (Zingiberaceae). *Taxon*, 57:949–962.

Mau, J.L., Lai, E.Y.C., Wang, N.P., Chen, C.C., Chang, C.H., and Chyau, C.C. 2003. Composition and antioxidant activity of the essential oil from *Curcuma zedoaria. Food Chem.* 82:583–591.

McClatchey, W. 1993. Traditional uses of *Curcuma longa* (Zingiberaceae) in Rotuma. *Econ. Bot.* 47:291–296.

Medical Economics Company. 2000. *Herbal monographs (valerian–zedoary). PDR for herbal medicines.* Medical Economics Company, Montvale, NJ. pp. 783–847.

Montaldo, A. 1986. *Ginger and zedoary: important aspects of their production.* IICA, Quito, Ecuador. 10 pp (in Spanish).

Nasim, G., and Zahoor, R. 1995. Vesicular arbuscular mycorrhizae in plants of medicinal importance: 1. Zedoary and 2. turmeric. *Scientific Khyber*, 8(2):55–64.

Sakurane, J., Komamura, H., Isonokami, M., Tadokoro, T., Nakamura, T., and Yoshikawa, K. 1999. A case of contact dermatitis due to *Curcuma longa* (turmeric). *Environ. Dermatol.* 6:237–242.

Sasaki, Y., Fushimi, H., and Komatsu, K. 2004. Application of single-nucleotide polymorphism analysis of the trnK gene to the identification of *Curcuma* plants. *Biol. Pharm. Bull.* 27:144–146.

Sasaki, Y., Fushimi, H., Cao, H., Cai, S.Q., and Komatsu, K. 2002. Sequence analysis of Chinese and Japanese *Curcuma* drugs on the 18S rRNA gene and trnK gene and the application of amplification-refractory mutation system analysis for their authentication. *Biol. Pharm. Bull.* 25:1593–1599.

Srimal, R.C. 1997. Turmeric: a brief review of medicinal properties. *Fitoterapia*, 68:483–493.

Syu, W.J., Shen, C.C., Don, M.J., Ou, J.C., Lee, G.H., and Sun, C.M. 1998. Cytotoxicity of curcuminoids and some novel compounds from *Curcuma zedoaria*. *J. Nat. Prod.* 61:1531–1534.

Tyas, K.N. 1999. *Curcuma zedoaria* Christmann Roscoe. *Prosea*, 2(7):43–45 (in Indonesian).

Uchimura, C., Higashi, T., and Nakazono, A. 1983. Studies on the cultivation of zedoary (*Curcuma zedoaria* Roscoe). 1. Growth habits and absorptions of mineral of nutrients. *Kyushu Agric. Res.* 45:41 (in Japanese).

Uchimura, C., Makazono, A., and Tsuyushige, M. 1983. Studies on the cultivation of zedoary (*Curcuma zedoaria* Roscoe). 2. The response of zedoary to planting times, planting densities and the amounts applied fertilizers. *Kyushu Agric. Res.* 45:42 (in Japanese).

Uchimura, C., Nakazono, A., Tsuyushige, M., Higashi, T., Machida, M., Yasuniwa, M., and Misono, A. 1989. Plant characteristics and cultivation systems of zedoary (*Curcuma zedoaria* Rosc.) a medicinal plant. *Bull. Kagoshima Agric. Exp. Stn.* 17:9–23 (in Japanese).

Yusuf, M., Rahman, M.A., Chowdhury, J.U., and Begum, J. 2002. Indigenous knowledge about the use of zingibers in Bangladesh. *J. Econ. Taxon. Bot.* 26:566–570.

Specialty Cookbooks

Bharadwaj, M. 2000 *The Indian spice kitchen*. Hippocrene, New York. 240 pp.

Harbury, K.E. 2004. *Colonial Virginia's cooking dynasty*. University of South Carolina Press, Columbia, SC. 479 pp.

Hess, K. 1996. *Martha Washington's booke of cookery and booke of sweetmeats*. Columbia University Press, New York. 518 pp.

Mansfeld, C., Trotter, C., and Barber, A. 2007. *Spices: recipes to delight the senses*. Periplus, Singapore. 288 pp.

Spiller, E. 2008. *Seventeenth century English recipe books*. Aldershot, Ashgate, U.K. 616 pp.

Appendix 1

Sources of Illustrations Presented in This Book

Note: In addition to drawings from the following sources, this book presents 27 figures drawn by artists associated with the Ottawa Center of Agriculture & Agri-Food Canada, as indicated in the Acknowledgments. Also, sources of several more drawings and photographs are given in their captions. Although the originals of most of the figures corresponding with the illustrations presented in this book are copyright-free, the present versions have been electronically enhanced and/or altered in significant respects and so constitute new works. In many cases, the original works list the illustrations under obsolete plant names; in all cases, the correct names by today's standards are given.

Aitchison, J.E.T. 1888. The botany of the Afghan Delimitation Commission. *Trans. Linn. Soc. London*, 3(1). Printed for the Linnean Society by Taylor and Francis, London. 139 pp. + 48 plates.

André, E. 1890. Le genere *Sicana*. *Rev. Hortic.* 62:515–517.

Bailey, L.H. (Ed.). 1900–1902. *Cyclopedia of American horticulture*. Virtue & Company, Toronto, ON. 4 vols.

Bailey, L.H. (Ed.). 1916. *The standard cyclopedia of horticulture*. Macmillan, Toronto, ON. 3 vols.

Baillon, M.H. 1876–1892. *Dictionnaire de Botanique*. Librarie Hachette, Paris. 4 vols.

Beardsley, A.V. 1896. *Under the hill, Chapter 3*. The Savoy, No. 1 (Jan.). Leonard Smithers, London. p. 167. [Periodical.]

Billings, E.R. 1875. *Tobacco: its history, varieties, culture, manufacture and commerce*. American Publishing Company, Hartford, CN. 486 pp.

Chambers, W., and Chambers, R. 1875. *Chambers's encyclopedia. A dictionary of universal knowledge for the people*. J.B. Lippincott Company, Philadelphia, PA. Vol. 3, p. 103.

Curtis, W. (Ed.). 1787–present. *The Botanical Magazine*. London [Illustrations reproduced from volumes published from 1790 to 1913.]

Delessert, B. 1820–1846. *Icones selectae plantarum quas in systemate universali*. Paris. 5 vols.

Dillenius, J.C. 1774. *Hortus elthamensis*. Cornelius Haak, Leiden, the Netherlands. 2 vols.

Duhamel du Monceau, H.L. 1800–1819. *Traité des arbres et arbustes*. Paris. 7 vols.

Dujardin-Beaumetz and Egasse, E. 1889. *Les plantes médicinales indigènes et exotiques*. Octave Doin, Paris. 845 pp.

Duthie, J.F. 1893. *Field and garden crops of the north-western provinces and Oudh, with illustrations. Part 3*. Department of Land Records and Agriculture, N.W. Provinces and Oudh, India. 65 pp.

Edwards, S.T., and Lindley, J. 1815–1847. *Edward's botanical register*. James Ridgway, London. 33 vols.

Engler, H.G.A., and Prantl, K.A.E. (Eds.). 1889–1915. *Die natürlichen Pflanzenfamilien*, ed. 1. Wilhelm Engelmann, Leipzig, Germany.

Fairlie, J. 1893. Life in the Malay Peninsula. *The Century* (New York), 45(4):577–588.

Figuier, L. 1867. *The vegetable world: being a history of plants, with their botanical descriptions and peculiar properties*. Chapman, London. 576 pp.

Flora and Sylva.Vol. 2, 1904. London. [Monthly review, published as annual volumes.]

Flore des serres et des jardins de l'Europe: annales générales d'horticulture. 1845–1880. L. van Houtte, Gand, Belgium. 23 vols.

Gardener's Chronicle (The). 1874–1956. Serial journal published in London. [Illustrations reproduced from volumes published in 1903 and 1904.]

Gartenflora: Zeitschrift für Garten-und Blumenkunde.1852–1938. Serial journal published in Berlin, Germany. [Illustrations reproduced from volumes published in 1874, 1880, 1881, 1882, 1890, 1894, and 1899.]

Gay, C. 1854. *Atlas de la historia fisica y politica de Chile*. Tomo primero.E. Thunot Y Ca., Paris.

Hallier, E.H. 1880–1888. *Flora von Deutschland*, edition 5 of D.F.L. von Schlechtendal et al. F.E. Köhler, Gera-Untermhaus, Germany. 39 vols.

Harter, J. (Ed.). 1988. *The plant kingdom compendium. A definitive volume of more than 2400 copyright-free engravings*. Bonanza Books, New York. 374 pp.

Hartwich, C. 1911. *Die menschlichen genussmittel*. Tauchnitz, Leipzig, Germany. 878 pp.

Henderson, P. 1890. *Henderson's handbook of plants and general horticulture*. Peter Henderson & Company, Jersey City, NJ. 528 pp.

Hooker, W.J. 1836–1900. *Icones plantarum, or figures with descriptive characters and remarks of new and rare plants selected from the Kew Herbarium*. Dulau, London. 35 vols.

Jackson, J.R. 1890. *Commercial botany of the nineteenth century*. Cassell & Company, London, U.K. 168 pp.

Jacquin, J.F. von. 1819. *Ueber den Ginkgo*. Carl Gerold, Wien, Germany. 8 pp. + plate.

Jacquin, N.J. 1764–1771. *Observationum botanicarum*. Vindobonae, Vienna, Austria. 4 parts.

Jumelle, H. 1901. *Les cultures coloniales, plates industrielles & médicinales*. J.B. Baillière et Fils, Paris. 360 pp.

Lamarck, J.B.A.P.M. de, and Poiret, J.L.M. 1744–1829. *Encyclopédie méthodique*. Botanique. Chez Panckoucke, Paris. 22 vols.

Köhler, H.A. 1883–1914. *Köhler's Medizinal Pflanzen*, 4 vols. Verlag von F. E. Köhler, Germany. [Illustrations reproduced are from vols. 1 to 3, published 1883–1898.]

Linnaeus, C. 1737. *Hortus Cliffortianus*. Amsterdam, the Netherlands. 501 + unnumbered pages.

Loddiges, C. & Sons. 1817–1833. *The botanical cabinet*. J. & A. Arch, Cornhill, U.K. 20 vols.

Loudon, J.C. 1844. *The trees and shrubs of Britain*. Longman, Brown, Green, and Longmans, London. 8 vols.

Marilaun, A.K. von. 1895. *The natural history of plants*. (Translated from German by F.W. Oliver). Henry Holt, New York. 2 vols. (each with 2 parts).

Michaux, F.A. 1850. *The North America sylva; or a description of the forest trees of the United States, Canada, and Nova Scotia*. G.P. Putnam, New York. 3 vols. (Translated from French.)

Mora-Urpí, J. 1994. Pejibaye palm. (*Bactis gasipaes*). In *Neglected crops. 1492 from a different perspective*. Edited by J.E. Hernández-Bermejo and J. Léon. Food and Agriculture Organization of the United Nations, Rome. pp. 211–221.

Morren, C. (Ed.). 1851–1885. *La Belgique Horticole, Journal des jardins, des serres, et des vergers*. Liège, Belgium. 35 vols.

Nicholson, G. (Ed.). 1885–1889. *The illustrated dictionary of gardening, a practical and scientific encyclopedia of horticulture for gardeners and botanists*. L. Upcott Gill, London. 4 vols.

Oeder, G.C., Müller, O.F., and Vahl, M. (Eds.). 1761–1883. *Flora Danica*. Copenhagen, Denmark. 17 vols.

Paillieux, A., and Bois, D. 1892. *Le potager d'un curieux: histoire, culture & usages de 200 plantes comestibles peu connues ou inconnues*. 2e édition. Librairie agricole de la maison rustique, Paris. 588 pp.

Pallas, P.S. 1784–1831. *Flora Rossica*. Imperiali J.J. Weitbrecht, Petropoli (St. Petersburg), Russia. 2 vols.

Paxton, J. (Ed.). 1834–1849. *Paxton's magazine of botany, and register of flowering plants*. Orr and Smith, London. 16 vols.

Rhind, W. 1855. *A history of the vegetable kingdom*. Blackie and Son, London. 720 pp.

Richard, A. 1847 (1851?). *Tentamen florae abyssinicae seu enumeratio plantarum hucusque in plerisque Abyssiniae*. Atlas. Paris.

Ruiz, L.H., and Pavon, J.A. 1802. *Flora Peruviana, et Chilensis, sive descriptiones, et icones, plantarum Peruvianarum et Chilensium*. Vol. 3. Garbrielis de Sancha, Spain.

Rumphius, G.E. 1741–1750. *Herbarium Amboinense*. Joannes Burmannus, Amsterdam. 6 vols.

Schimper, A.F.W. 1903. *Plant geography upon a physiological basis*. Clarendon Press, Oxford, U.K. 839 pp.

Siebold, P.F. de, and Zuccarini, J.G. 1835–1870. *Flora Japonica*. Lugduni Bataborum, Apud auctorem. 893 pp. (2 vols. bound as one.)

Stevels, J.M.C. 1990. *Légumes traditionnels du Cameroun, un étude agro-botanique*. Wageningen Agricultural University Papers 90-1. 262 pp.

Symon, D.E. 1981. A revision of the genus *Solanum* in Australia. *J. Adelaide Bot. Gard.* 4:1–367.

Thomé, O.W. 1885. *Flora von Deutschland Österreich und der Schweiz*. H.V. Verlag, Berlin-Lichterfelde, Germany. Vol. 4.

Thomé, O.W. 1903–1905. *Prof. Dr. Thomé's Flora von Deutschland, Österreich und der Schweiz*. H.V. Verlag, Berlin-Lichterfelde, Germany. [First 4 vols, reissued with amended text.]

Turner, F. 1892. New commercial crops for New South Wales. The cultivation and uses of the carob bean (*Ceratonia siliqua* Linn.). *Agricultural Gazette of New South Wales*, 2:235–240 + plate.

Turner, F. 1893a. New commercial crops for New South Wales. The cultivation of the Australian nut (*Macadamia ternifolia* F.v.M.). *Agricultural Gazette of New South Wales*, 4:3–5 + plate.

Turner, F. 1893b. New commercial crops for New South Wales. The cultivation and uses of the "caper bush" (*Capparis spinosa* Linn.). *Agricultural Gazette of New South Wales*, 4:525–529.

Turner, F. 1894. The tree tomato. *Cyphomandra betacea* Sendtn. *Agricultural Gazette of New South Wales*, 4:525–529.

Vilmorin-Andrieux et Cie. 1884. *Supplément aux fleurs de pleine terre*. Vilmorin-Andrieux et Cie, Paris. 203 pp.

Vilmorin-Andrieux. M.M. 1885. *The vegetable garden*. John Murray, London. 620 pp.

Appendix 2

Selected Literature on Exotic Food Crops

GENERAL

Adams, W.D., and Leroy, T.R. 1992. *Growing fruits and nuts in the South: the definitive guide*. Taylor Pub., Dallas, TX. 202 pp.

Asiedu, J.J. 2010. *Processing tropical crops: a technological approach*. Mission Press, Ndola, Zambia. 266 pp.

Avila, C., and Root, M. 1997. *Sabor! A guide to tropical fruits and vegetables and Central American foods*. Litografía e Imprenta LIL, San José, Costa Rica. 218 pp.

Bacon, J. 1990. *Complete guide to exotic fruits and vegetables*. Xamadu, London. 242 pp.

Ballister, B. 2002. *The fruit and vegetable stand: the complete guide to the selection, preparation and nutrition of fresh produce*. Revised edition. Overlook Press, Woodstock, NY. 455 pp.

Bettencourt, E., Hazekamp, T., and Perry, M.C. 1992. *Tropical and subtropical fruits and tree nuts: annona, avocado, banana and plantain, breadfruit, cashew, citrus, date, fig, guava, mango, passionfruit, papaya, pineapple and others*. International Board Plant Genetic Resoources, Rome, Italy. 337 pp.

Bittenbender, H.C. 1984. *Handbook of tropical fruits and spices*. Michigan State University Department of Horticulture, East Lansing, MI. 127 pp.

Bois, D. 1927. *Les plantes alimentaires chez tous les peuples et à travers les âges*. Paul Lechavalier, Paris. 3 vols (in French).

Bourne, M.J., Lennox, G.W., and Seddon, S.A. 1988. *Fruits and vegetables of the Caribbean*. Macmillan Education, London. 58 pp.

Brooks, R.M., and Olmo, H.P. 1997. *The Brooks and Olmo register of fruit & nut varieties*. 3rd ed. ASH Press (American Society for Horticultural Science), Alexandria, VA. 743 pp.

Brouk, B. 1975. *Plants consumed by man*. Academic Press, New York. 479 pp.

Brücher, H. 1989. *Useful plants of neotropical origin and their wild relatives*. Springer-Verlag, New York. 296 pp.

Chan, H.T., Jr. (Ed.). 1983. *Handbook of tropical foods*. M. Dekker, New York. 639 pp.

Chantiles, V.L. 1984. *The New York ethnic food market guide & cookbook*. Dodd, Mead & Co., New York. 370 pp.

Clarke, B. 2002. *Exotic food ingredients*. Department of Primary Industries, Brisbane, Qsld., Australia. 97 pp.

Cobley, L.S. 1976. *An introduction to the botany of tropical crops*. 2nd ed. Longman, New York. 371 pp.

Cwiertka, K, and Walraven, B. (Eds.). 2001. *Asian food. The global and the local*. University of Hawai'i Press, Honolulu, HI. 190 pp.

Darley, J.J. 1993. *Know and enjoy tropical fruits: tropical fruit and nuts: a cornucopia*. P. & S. Pub., Thuringowa, Qld., Australia. 186 pp.

Duke, J.A. 1981. *Handbook of legumes of world economic importance*. Plenum Press, New York. 345 pp.

Eneji, A.E. 2009. *Agronomy of tropical crops*. Studium Press, Houston, TX. 354 pp.

Facciola, S. 1998. *Cornucopia II. A source book of edible plants*. Kampong Publications, Vista, CA. 713 pp.

Food and Agriculture Organization of the United Nations. 1990. *Utilization of tropical foods: fruits and leaves*. FAO, Rome, Italy. 70 pp.

Gibbon, D., and Pain, A. 1985. *Crops of the drier regions of the tropics*. Longman, New York. 157 pp.

Gulick, P., and Sloten, D.H. van. 1984. *Tropical and subtropical fruits and tree nuts*. International Board Plant Genetic Resources, Rome, Italy. 191 pp.

Hackett, C., and Carolane, J. 1982. *Edible horticultural crops. A compendium of information on fruit, vegetable, spice and nut species*. Academic Press, New York. 673 pp.

Hanelt, P. (Ed.). 2001. *Mansfeld's encyclopedia of agricultural and horticultural crops*. Springer-Verlag, Heidelberg, Germany. 6 vols.

Hedrick, U. P. (Ed.). 1972. *Sturtevant's edible plants of the world*. Dover Publications, New York. 686 pp. [Reprint of original edition published in 1919.]

Heiser, C.B. 1979. *The gourd book*. University of Oklahoma Press, Norman, OK. 248 pp.

Hernández Bermejo, J.E., and Léon, J. (Eds.). 1994. *Neglected crops: 1492 from a different perspective*. FAO Plant Production and Protection Series No. 26. Food and Agriculture Organization of the United Nations, Rome, Italy. 341 pp.

International Board for Plant Genetic Resources. 1986. *Genetic resources of tropical and subtropical fruits and nuts (excluding* Musa*)*. IBGR Secretariat, Rome, Italy. 162 pp.

Janick, J. 1999. New crops and the search for new food resources. In *Perspectives on new crops and new uses*. Edited by J. Janick. ASHS Press, Alexandria, VA. pp. 104–110.

Janick, J., and Paull, R.E. (Eds.). 2008. *The encyclopedia of fruit and nuts*. CABI, Cambridge, MA. 954 pp.

Kennard, W.C., and Winters, H.F. 1960. *Some fruits and nuts for the tropics*. United States Department of Agriculture, Washington, DC. 135 pp.

Kunkel, G. 1984. *Plants for human consumption: an annotated checklist of the edible phanerogams and ferns*. Koeltz Scientific Books, Koenigstein, Germany. 393 pp.

Kurup, G.T., Palaniswami, M.S., Potty, V.P., Padmaia, G., and Kabeerathumma (no initial), and Pillai, S.V. (Eds.). 1996. *Tropical tuber crops: problems, prospects and future strategies*. Oxford & IBH Pub. Co., New Delhi, India. 597 pp.

León, J. (Ed.). 1974. *Handbook of plant introduction in tropical crops*. Food and Agriculture Organization of the United Nations, Rome, Italy. 140 pp.

Li, T.S.C. 2008. *Vegetables and fruits: nutritional and therapeutic values*. CRC, Boca Raton, FL. 304 pp.

Lyle, S. 2006. *Fruit and nuts: a comprehensive guide to the cultivation, uses and health benefits of over 300 food-processing plants*. Timber Press, Portland, OR. 480 pp.

Martin, F.W. (Ed.). 1984. *CRC handbook of tropical food crops*. CRC Press, Boca Raton, FL. 296 pp.

Maye, D., Holloway, L., and Kneafsey, M. 2007. *Alternative food geographies: representation and practice*. Elsevier, Amsterdam, the Netherlands. 358 pp.

McIlroy, R.J. 1963. *An introduction to tropical cash crops*. Ibadan University Press, Iadan, Nigeria. 157 pp.

Mirza, J.I., and Bokhari, M.H. 1996. *Fruits and vegetables of Pakistan*. Ferozsons, Rawalpindi, Pakistan. 66 pp.

Morgan, L. 1997. *The ethnic market food guide: an ingredient encyclopedia for cooks, travelers, and lovers of exotic food*. Berkely Books, New York. 246 pp.

Morris, L.C. 1984. *Tropical fruits, vegetables, root crops and spices: a select bibliography*. Grace, Kennedy, & Company, Kingston, Jamaica. 235 pp.

Morton, J.F. 1971. *Exotic plants*. Golden Press, New York. 160 pp.

National Academy of Sciences. 1975. *Underexploited tropical plants with promising economic value*. Committee on Internal Relations, Washington, DC. 188 pp.

Noel, D. 1985. *Tropical nuts and fruits for Western Australia*. Cornucopia, Subiaco, WA, Australia. 84 pp.

Norman, M.J.T., Pearson, C.J., and Searle, P.J.E. 1984. *The ecology of tropical food crops*. Cambridge University Press, Cambridge, U.K. 369 pp.

Norrington, N.L., and Campbell, C. 2001. *Tropical food gardens: a guide to growing fruit, herbs and vegetables in tropical and sub-tropical climates*. Bloomings, Hawthorn, Vict., Australia. 160 pp.

Nyambo, A. Nyomoro, A., Ruffo, C.K., and Tengnäs. 2005. *Fruits and nuts: species with potential for Tanzania*. Regional Land Management Unit (RELA), World Agroforestry Centre, Nairobi, Kenya. 160 pp.

Opeke, L.K. 1982. *Tropical tree crops*. Wiley, New York. 312 pp.

Purseglove, J.W. 1968. *Tropical crops: dicotyledons*. Wiley, New York. 332 pp.

Purseglove, J.W. 1972. *Tropical crops: monocotyledons*. Wiley, New York. 607 pp.

Rice, R.P., Rice, L.W., and Tindall, M.D. 1990. *Fruit and vegetable production in warm climates*. Macmillan, London. 486 pp.

Rice, R.P., Tindall, M.D., and Rice, L.W. 1987. *Fruit and vegetable production in Africa*. Macmillan, London. 371 pp.

Richardson, J. 1990. *Worldwide selection of exotic fruits and vegetables*. Les Éditions Héritage Inc., Saint-Lambert, QC. 256 pp.

Roecklein, J.C., and Leung, P. 1987. *A profile of economic plants*. Transaction Books, New Brunswick, NJ. 623 pp.

Sanwal, S.K. 2008. *Underutilized vegetable and spice crops*. Agrobios (India), Jodhpur, India. 308 pp.

Schery, R.W. 1972. *Plants for man*. 2nd ed. Prentice-Hall Inc., Englewood Cliffs, NJ. 657 pp.

Schneider, E. 1986. *Uncommon fruits and vegetables. A commonsense guide*. Harper & Row Pub., New York. 547 pp.

Seelig, R.A., and Bing, M.C. 1990. *Encyclopedia of produce*. United Fresh Fruit and Vegetable Association, Alexandria, VA. Irregularly paginated.

Siani, G. 1998. L'exotisme alimentaire un choix de marketing pour les enseignes [Marketing choices by chain stores in the light of increasingly exotic food tastes]. *Direction et Gestion, La Revue des Sciences de Gestion*, 33(174):39–52 (in French).

Simmonds, N.W. (Ed.). 1976. *Evolution of crop plants*. Longman, London. 339 pp. [Includes a number of minor crops not in the 2nd edition, mentioned below.]

Smartt, J., and Simmonds, N.W. (Eds.). 1995. *Evolution of crops plants*. 2nd ed. Longman Scientific & Technical, Burnt Mill, Harlow, Essex, U.K. 531 pp.

Small, E. 2009. *Top 100 food plants: the world's most important culinary crops*. NRC Research Press, Ottawa, ON. 36 pp.

Tanaka, T. 1976. *Tanaka's cyclopedia of edible plants of the world*. Keigaku Publishing Co., Tokyo, Japan. 924 pp.

Thattoppilly, G., Monti, L.M., Mohan Raj, D.R., and Moore, A.W. (Eds.). 1992. *Biotechnology: enhancing research on tropical crops in Africa*. Technical Centre for Agricultural and Rural Cooperation, Wageningen, the Netherlands. 364 pp.

Tindall, H.D., and Sai, F.A. 1976. *Fruits and vegetables in West Africa*. Food and Agriculture Organization of the United Nations, Rome, Italy. 259 pp.

Van Atta, M. 1991. *Growing & using exotic food*. Pineapple Press, Sarasota, FL. 180 pp.

Van der Plas, M. and Verhaegh, I.A.P. 1986. *Marketing of exotic and out-of-season fresh fruit and vegetables in the European Common Market*. Landbouw-Economisch Intituut, LS Den Haag, the Netherlands. 192 pp.

Vaughan, J.G., Geissler, C.A., and Nicholson, B.E. 1997. *The new Oxford book of food plants*. Oxford University Press, Oxford, U.K. 239 pp.

Weir, R.G., and Cresswell, G.C. 1993. *Tropical and subtropical fruit and nut crops*. Inkata Press, Melbourne, Australia. 105 pp.

Welanetz, D. von, and Welanetz, P. von. 1982. *The von Welanetz guide to ethnic ingredients*. J.P. Tarcher, Inc., Los Angeles, CA. 731 pp.

Westphal, E. (General Editor), et al. 1989–2001. *Plant resources of South-East Asia*. Backhuys, Leiden, the Netherlands. 16 "volumes" (some published as more than one book).

Whealy, K., and Thuente, J. 2001. *Fruit, berry, and nut inventory: an inventory of nursery catalogs listing all fruit, berry and nut varieties available by mail order in the United States*. 3rd ed. Seed Savers Exchange, Decorah, IA. 560 pp.

Wiersema, J.H. and León, B. 1999. *World economic plants. A standard reference*. CRC Press, New York. 749 pp.

Williams, C.N. 1975. *The agronomy of the major tropical crops*. Oxford University Press, New York. 228 pp.

Williams, C.N., Chew, W.Y., and Rajaratnam, J.H. 1980. *Tree and field crops of the wetter regions of the tropics*. Longman, London. 262 pp.

Zeven, A.C., and de Wet, J.M.J. 1982. *Dictionary of cultivated plants and their regions of diversity*. Pudoc, Wageningen, the Netherlands. 263 pp.

VEGETABLES

Acrivos, N. 1988. *A guide to tropical and sub-tropical vegetables*. Brevard Rare Fruit Council, Melbourne, FL. 47 pp.

Bacon, J. 1989. *Exotic vegetables A–Z*. Salem House Pub., Topsfield, MA. 128 pp.

Bonar, A. 1986. *Vegetables: a complete guide to the cultivation, uses, and nutritional value of common and exotic vegetables*. Hamlyn, Twickenham, U.K. 160 pp.

Buishand, T., Houwing, H.P., and Jansen, K. 1986. *The complete book of vegetables*. Gallery Books, W.H. Smith Publishers, New York. 180 pp.

Dahlen, J., Phillipps, K. 1983. *A popular guide to Chinese vegetables*. Crown Publishers Inc., New York. 113 pp.

Greenshill, T.M. 1968. *Growing better vegetables, a guide for tropical gardeners*. Evans Brothers, London. 159 pp.

Grubben, G.J.H., Tindall, H.D., and Williams, J.T. 1977. *Tropical vegetables and their genetic resources*. International Board for Genetic Resources, Rome, Italy. 197 pp.

Halpin, A.M. (Ed.). 1978. *Unusual vegetables*. Rodale Press, Emmaus, PA. 443 pp.

Harrington, G. 1978. *Grow your own Chinese vegetables*. MacMillan Pub. Co., New York. 268 pp.

Herklots, G.A. 1972. *Vegetables in south-east Asia*. London George Allen & Unwin Ltd., London. 525 pp.

Holland, B., Welch, A.A., Buss, D.H., and McChance, R.A. 1992. *Vegetable dishes. The second supplement to McCance & Widowson's The composition of foods*. 5th ed. Ministry of Agriculture, Fisheries & Food & Royal Society of Chemistry, Cambridge, U.K. 242 pp.

Horticultural Crops Group of the FAO Plant Production and Protection Division. 1988. *Vegetable production under arid and semi-arid conditions in tropical Africa: a manual*. Food and Agriculture Organization of the United Nations, Rome, Italy. 434 pp.

Hutton, W. 1997. *Tropical vegetables*. Periplus, Singapore. 63 pp.

Kay, D.E. 1973. *Root crops*. Tropical Products Institute, London. 245 pp.

Kays, S.J., and Dias, J.C.S. 1996. *Cultivated vegetables of the world*. Exon Press, Athens, GA. 170 pp.

Lamberts, M. 1990. Latin America vegetables. In *Advances in new crops*. Edited by J. Janick and J.E. Simon. Timber Press, Portland, OR. pp. 378–385.

Larkom, J. 2008. *Oriental vegetables. The complete guide for the gardening cook*. Revised edition. Kodansha America, New York. 232 pp.

Laws, B., and Green, H. 2004. *Spade, skirret and parsnip: the curious history of vegetables*. Sutton Pub., Stroud, U.K. 216 pp.

Lebot, V. 2009. *Tropical root and tuber crops: cassava, sweet potato, yams and aroids*. CABI, Wallingford, U.K. 413 pp.

Maynard, D.N., and Hochmuch, G.J. 1997. *G.J. Knott's handbook for vegetable growers*. 4th ed. Wiley, Hoboken, NJ. 621 pp.

Messiaen, C.-M., MacCrimmon, P.R., and Tindall, H.D. 1992. *Tropical vegetable garden: principles for improved and increased production with application to the main vegetable types*. Macmillan, London. 514 pp.

Munro, D.B., and Small, E. 1997. *Vegetables of Canada*. NRC Press, Ottawa, ON. 417 pp.

National Research Council. 2008. *Lost crops of Africa. Vol. 2. Vegetables*. National Academic Press, Washington, DC. 380 pp.

Nonnecke, I. L. 1989. *Vegetable production*. Van Nostrand Reinhold, New York. 657 pp.

Ochse, J.J. 1977. *Vegetables of the Dutch East Indies (edible tubers, bulbs, rhizomes, and spices included): survey of the indigenous and foreign plants serving as pot-herbs and side-dishes*. Australian National University Press, Canberra, Australia. 1005 pp. [A reprint of the original 1931 edition.]

Onwueme, I.C., and Charles, W.B. 1994. *Tropical root and tuber crops: production, perspectives, and future prospects*. FAO, Rome, Italy. 228 pp.

Oomen, H.A.P.C., and Grubben, G.J.H. 1977. *Tropical leaf vegetables in human nutrition*. Koninklijk Instituut voor de Tropen, Amsterdam, the Netherlands. 133 pp.

Organ, J. 1960. *Rare vegetables for garden and table*. Faber and Faber Limited, London. 184 pp.

Phillips, R., and Rix, M. 1993. *Vegetables*. Random House, New York. 270 pp.

Plucknett, D.L. 1979. *Small-scale processing and storage of tropical root crops*. Westview Press, Boulder, CO. 461 pp.

Rubatzky, V.E., and Yamaguchi, M. 1996. *World vegetables. Principles, production and nutritive values*. 2nd ed. Chapman and Hall, New York. 843 pp.

Rupp, R. 1987. *Blue corn & square tomatoes: unusual facts about common vegetables*. Storey Communications, Pownal, VT. 222 pp.

Sabota, C., and Sharma, G. 1995. Production potential of exotic vegetables in the Southeastern United States. *J. Sustainable Agric*. 7 (2/3): 25–39.

Schippers, R.R. 2000. *African indigenous vegetables: an overview of the cultivated species*. Natural Resources Institute, Chatham, U.K. 222 pp. [Revised version published as CD in 20002.]

Schneider, E. 2001. *Vegetables from amaranth to zucchini: the essential reference*. William Morrow/ HarperCollins, New York. 777 pp.

Stephens, J.M. 1982. *Know your minor vegetables*. Revised edition. Florida Cooperative Extension Service, Institute of Food and Agricultural Sciences, University of Florida, Gainesville, FL. 93 pp.

Tindall, H.D. 1983. *Vegetables in the tropics*. AVI Publishing Co., Westport, CT. 533 pp.

United States Agricultural Research Service. 1977–1981. *Vegetables for the hot, humid tropics*. United States Agricultural Research Service, New Orleans, LA and Mayagüez, Puerto Rico. 8 vols.

FRUIT

Akinnifesi, F.K., Leakey, R.R.B., Ajayi, O., Sileshi, G., Tchoundjeu, Z., Matakala, P., and Kewsiga, F.R. (Eds.). 2008. *Indigenous fruit trees in the tropics: domestication, utilization and commercialization*. CABI, Wallingford, U.K. 438 pp.

Allen, B.M. 1965. *Malayan fruits: an introduction to the cultivated species*. B. Moore for Eastern Universities Press, Singapore. 245 pp.

Allen, B.M. 1975. *Common Malaysian fruits, with Tamil and Thai names*. Longman, Singapore. 64 pp.

Arun, A. 2007. *Tropical fruits: diseases and pests*. 2nd ed. Kalyani, Ludhiana, India. 216 pp.

Bay-Petersen, J. (Ed.). 1987. *Postharvest handling of tropical and subtropical fruit crops*. Proceedings of a seminar held in Taichung (Taiwan) in November 1987: Food and Fertilizer Technology Center for the Asian and Pacific Region, Taipei, Taiwan. 135 pp.

Benson, A.H. 1914. *Fruits of Queensland*. A.J. Cumming, Government Printer, Brisbane, Australia. 102 pp. [2008 reprint by Dodo Press.]

Bircher, A.G., and Bircher, W.H. 2001. *Encyclopedia of fruit trees and edible flowering plants in Egypt and the Subtropics*. American University in Cairo Press, Cairo, Egypt. 568 pp.

Boning, C.R. 2006. *Florida's best fruiting plants*. Pineapple Press, Sarasota, FL. 232 pp.

Bose, T.K. (Ed.). 1985. *Fruits of India: tropical and subtropical*. Naya Prokash, Calcutta, India. 637 pp.

Cassidy, M., Rice, L.W., Nzima, M.D.S., and Fenner, R. 1982–1988. *Tropical and subtropical fruit production in Zimbabwe*. Horticultural Promotion Council, Harare, Zimbabwe. 33 pp.

Cheema, G., Bhat, S.S., and Naik, K.C. 1954. *Commercial fruits of India with special reference to western India*. Macmillan, Calcutta, India. 422 pp.

Chundawat, B.S. 1990. *Arid fruit culture*. Oxford & IBH Publishing, New Delhi, India. 208 pp.

Decker, K.J. 2005. The appeal of exotic fruits. *Food Products Design*, 15(8):102–122.

Eiseman, F.B. Jr., and Eiseman, M. 1994. *Fruits of Bali*. Periplus, Berkeley, CA. 64 pp.

Feree, D.C., and Chandler, L.E. (Eds.). 1998. *A history of fruit varieties: The American Pomological Society: One hundred and fifty years, 1848–1998*. Good Fruit Growers, Yakima, WA. 196 pp.

Harris, K.S., and Henry, L.M. 2008. *LMH official dictionary of Caribbean exotic fruits*. LMH Publishing Ltd., Kingston, Jamaica. 39 pp.

Heaton, D.D. 2006. *A consumer's guide on world fruit*. BookSurge Publishing, Charleston, SC. 251 pp.

Hessayon, D.G. 1990. *The fruit expert*. Waltham Cross, Herts, U.K. 128 pp.

Hewitt, S. 2005. *Fruit trees of the Caribbean*. Macmillan Caribbean, Oxford, U.K. 146 pp.

Hutton, W. 2004. *Handy pocket guide to tropical fruits*. Periplus, Singapore. 62 pp.

Jackson, D., and Looney, N.E. (Eds.). 1999. *Temperate and subtropical fruit production*. 2nd ed. CABI, Wallingford, Oxon, U.K. 332 pp.

Janick, J., and Moore, J.N. (Eds.). 1997. *Fruit breeding*. Wiley, New York. 3 vols.

Jarimopas, B. 2008. *Tropical fruits in developing countries: quality of production*. Nova Science Publishers, New York. 96 pp.

Jeans, H. 1972. *About tropical fruits*. Thorsons, London. 64 pp.

Jensen, M. 2001. *Trees and fruits of Southeast Asia: an illustrated field guide*. Orchid Press, Bangkok, Thailand. 234 pp.

Martin, F.W., and Cooper, W.C. 1977. *Cultivation of neglected tropical fruits with promise*. U.S. Department of Agriculture, Agricultural Research Service, Southern Region, New Orleans, LA. 17pp.

Martin, F.M., Campbell, C.W., and Ruberté, R.M. 1987. *Perennial edible fruits of the tropics: an inventory*. U.S. Government Printing Office, Washington, DC. 247 pp.

Mazumdar, B.C. 2004. *Minor fruit crops of India: tropical and subtropical*. Daya Publishing House, New Delhi, India. 145 pp.

Mohlenbrock, R.H. 1980. *You can grow tropical fruit trees*. Great Outdoord Pub. Co., St.Petersberg, FL. 77 pp.

Morton, J.F. 1987. *Fruits of warm climates*. Creative Resource Systems, Winterville, NC. 506 pp.

Nagy, S., Shaw, P.E., and Wardowski, W.F. (Eds.). 1990. *Fruits of tropical and subtropical origin. Composition, properties and uses*. Florida Science Source, Inc., Lake Alfred, FL. 391 pp.

Nakasone, H.Y., and Paul, R.E. 1998. *Tropical fruits*. CABI International, Wallingford, Oxon, U.K. 445 pp.

Oshiro, K. 2000. *Growing fruits in Hawaii. Also herbs nuts and seeds. A how-to guide for the gardener*. Bess Press, Honolulu, HI. 76 pp.

Piper, J. 1899. *Fruits of South-East Asia: facts and folklore*. Oxford University Press, New York. 194 pp. + plates.

Popenoe, W. 1920. *Manual of tropical and subtropical fruits: excluding the banana, coconut, pineapple, citrus fruits, olive, and fig*. Macmillan, New York. 474 pp. (Reprinted by Kessinger Publishing LLC, 2008.)

Popenoe, W. 1924. *Economic fruit-bearing plants of Ecuador*. Government Printing Office, Washington, DC. Various paginations.

Ray, P.K. 2002. *Breeding tropical and subtropical fruits*. Springer-Verlag, New York. 338 pp.

Reich, L. 2008. *Uncommon fruits for every garden*. Timber Press, Portland, OR. 288 pp.

Roach, F.A. 1985. *Cultivated fruits of Britain: their origin and history*. Blackwell, New York. 349 pp.

Richardson, M.W. 1976. *Tropical fruit recipes: rare and exotic fruits of Florida*. Rare Fruit Council International, Miami, FL. 160 pp.

Ruck, H.C. 1975. *Deciduous fruit tree cultivars for tropical and sub-tropical regions*. Commonwealth Agricultural Bureaux, Oxford, U.K. 91 pp.

Sampson, J.A. 1986. *Tropical fruits*. 2nd ed. Longman Scientific & Technical, Burnt Mill, Harlow, Essex, U.K. 336 pp.

Schrock, D. (Ed.). 2008. *All about citrus & subtropical fruits*. 2nd ed. Meredith Books, Des Moines, IA. 127 pp.

Shaw, P.E., Chan, H.T., Jr., and Nagy, S. (Eds.). 1998. *Tropical and subtropical fruits*. Agscience, Auburndale, FL. 569 pp.

Simmons, A.F. 1972. *Growing unusual fruit*. Walker, New York. 354 pp.

Tankard, G.J. 1990. *Tropical fruit: an Australian guide to growing and using exotic fruits*. Viking O'Neal, Ringwood, Vict., Australia. 152 pp.

Tate, D. 2007. *Tropical fruit*. Archipelago Press, Singapore. 96 pp.

Van Aken, N. 1995. *The great exotic fruit book. A handbook of tropical and subtropical fruits, with recipes*. Ten Speed Press, Berkeley, CA. 149 pp.

Wardowski, F., Nagy, S., and Shaw, P.E. (Eds.). 1990. *Fruits of tropical and subtropical origin: composition, properties, uses*. Florida Science Source, Lake Alfred, FL. 391 pp.

Yaacob, O., and Subhadrabandhu, S. 1995. *The production of economic fruits in South-east Asia*. Oxford University Press, New York. 419 pp.

HERBS AND SPICES

Charalambous, G. 1994. (Ed.). *Spices, herbs and edible fungi*. Elsevier, New York. 764 pp.

Crockett, J.U., and Tanner, O. 1977. *Herbs*. Time-Life Books, Alexandria, VA. 160 pp.

Day, A., and Stuckey, L. 1964. *The spice cookbook*. David White Co., New York. 623 pp.

Duke, J.A., Bogenschutz-Godwin, M.J., duCellier, J., and Duke, P.-A.K. 2002. *CRC handbook of medicinal herbs*. 2nd ed. CRC Press, Boca Raton, FL. 870 pp.

Duke, J.A. 2002. *CRC handbook of medicinal spices*. CRC Press, Boca Raton, FL. 348 pp.

Farrell, K. T. 1990. *Spices, condiments, and seasonings*. 2nd ed. AVI/Van Nostrand Reinhold, New York. 414 pp.

Green, A. 2006. *Field guide to herbs & spices: how to identify, select and use virtually every seasoning at the market*. Quirk Books, Philadelphia, PA. 313 pp.

Grieve, M. 1931. *A modern herbal*. Penguin Books, New York. 912 pp. [Reprinted 1978.]

Grieve, M. 1934. *Culinary herbs and condiments*. Harcourt Brace, New York. 209 pp. [Reprinted by Dover Publications, 1954.]

Hill, T. 2004. *The contemporary encyclopedia of herbs and spices: seasonings for the global kitchen*. Wiley, Hoboken, NJ. 432 pp.

Hutton, W. 2003. *Handy pocket guide to tropical herbs & spices*. Periplus, Hong Kong. 64 pp.

McCormick [& Company, Inc.]. 1979. *Spices of the world cookbook*. McGraw-Hill, New York. 431 pp.

Morton, J.F. 1977. *Herbs and spices*. Golden Press, New York. 160 pp.

Mulherin, J. 1988. *The Macmillan treasury of spices and natural flavorings: a complete guide to the identification and uses of common spices and natural flavorings*. Macmillan, New York. 144 pp.

Ochatt, S., and Jain, S.M. (Eds.). 2007. *Breeding of neglected and under-utilized crops: spices and herbs*. Science Publishers, Enfield, NH. 447 pp.

Parry, J.W. 1945. *The spice handbook*. Chemical Publishing, Brooklyn, NY. 254 pp.

Peter, K.V. (Ed.). 2001. *Handbook of herbs and spices*. Woodhead Publishing, Cambridge, U.K. 319 pp.

Peter, K.V. (Ed.). 2004. *Handbook of herbs and spices*. Vol. 2. Woodhead Publishing, Cambridge, U.K. 360 pp.

Purseglove, J. W., Brown, E. G., Green, C. L., Robbins, S. R. 1981. *Spices*. Longman Inc., New York. 2 vols.

Raghaven, S. 2006. *Handbook of spices, seasonings, and flavorings*. 2nd ed. CRC, Boca Raton, FL. 330 pp.

Rosengarten, F., Jr. 1969. *The book of spices*. Livingston Pub., Philadelphia, PA. 489 pp.

Rosengarten, F., Jr. 1973. *The book of spices*. Jove Publications Inc., New York. 475 pp. [Revised, abridged edition of Rosengarten 1969.]

Saville, C., and Creasy, R. 1997. *Exotic herbs: a compendium of exceptional culinary herbs*. Henry Holt, New York. 308 pp.

Seidemann, J. 2005. *World spice plants: economic usage, botany, taxonomy*. Springer-Verlag, Berlin. 592 pp.

Simon, J.E., Chadwick, A.F., and Craker, L.E. 1984. *Herbs: an annotated bibliography, 1971–1980*. The Shoe String Press, Inc., Hamden, CT. 770 pp.

Small, E. 2006. *Culinary herbs*. 2nd ed. NRC Research Press, Ottawa, ON. 1036 pp.

Staples, G.W. and Kristiansen, M.S. 1999. *Ethnic culinary herbs. A guide to identification and cultivation in Hawai'i*. University of Hawai'i Press, Honolulu. 123 pp.

Stobart, T. 1977. *Herbs, spices and flavourings*. Penguin Books, New York. 320 pp.

Tan, H.T.W. 2005. *Herbs and spices of Thailand*. Time Editions Marshall Cavendish, Singapore. 127 pp.

Teuscher, E. 2006. *Medicinal spices: a handbook of culinary herbs, spices, spice mixtures and their essential oils*. Medpharm, Stuttgart, Germany. 459 pp.

Weiss, E.A. 2002. *Spice crops*. CABI, Wallingford, Oxon, U.K. 411 pp.

NUTS

Dewan, M.L. 1992. *Nut fruits for the Himalayas*. Concept Pub. Co., New Delhi, India. 184 pp.

Duke, J.A. 1988. *CRC Handbook of nuts*. CRC, Boca Raton, FL. 343 pp.

International Trade Centre. 1994. *Tropical nuts: a study of market opportunities in the United Kingdom*. International Trade Centre, UNCTAD/GATT, Geneva. 43 pp.

Janick, J., and Moore, J. N. (Eds.). 1997. *Fruit breeding: nuts*. Wiley, New York. 278 pp. [This is vol. 3 of Janick and Moore (1997), listed in Fruits, above.]

Malhotra, S.P. 2008. *World edible nuts economy*. Concept Pub. Co., New Delhi, India. 538 pp.

Menninger, E.A. 1977. *Edible nuts of the world*. Horticultural Books, Stuart, FL. 175 pp.

Rosengarten, F., Jr. 1984. *The book of edible nuts*. Walker and Company, New York. 384 pp.

Wickens, G.E. 1995. *Edible nuts*. Food and Agriculture Organization of the United Nations (FAO), Rome, Italy. 198 pp.

Wilkinson, J. 2005. *Nut grower's guide: the complete handbook for producers and hobbyists*. Landlinks Press, Collingwood, Vic., Australia. 228 pp.

Index of Common Names

C

D

E

I

J

K

Q

R

S

T

X

Y

Z

Index of Scientific Names

L

M

N

O

P

Index of Culinary Terms and Dishes